BIOCHEMICAL MECHANISMS IN HEART FUNCTION

Developments in Molecular and Cellular Biochemistry

Series Editor: Naranjan S. Dhalla, Ph.D., FACC

1. V.A. Najjar (ed.): *Biological Effects of Glutamic Acid and Its Derivatives.* 1981 ISBN 90-6193-841-4
2. V.A. Najjar (ed.): *Immunologically Active Peptides.* 1981 ISBN 90-6193-842-2
3. V.A. Najjar (ed.): *Enzyme Induction and Modulation.* 1983 ISBN 0-89838-583-0
4. V.A. Najjar and L. Lorand (eds.): *Transglutaminase.* 1984 ISBN 0-89838-593-8
5. G.J. van der Vusse (ed.): *Lipid Metabolism in Normoxic and Ischemic Heart.* 1989 ISBN 0-7923-0479-9
6. J.F.C. Glatz and G.J. van der Vusse (eds.): *Cellular Fatty Acid-Binding Proteins.* 1990 ISBN 0-7923-0896-4
7. H.E. Morgan (ed.): *Molecular Mechanisms of Cellular Growth.* 1991 ISBN 0-7923-1183-3
8. G.J. van der Vusse and H. Stam (eds.): *Lipid Metabolism in the Healthy and Diseased Heart.* 1992 ISBN 0-7923-1850-1
9. Y. Yazaki and S. Mochizuki (eds.): *Cellular Function and Metabolism.* 1993 ISBN 0-7923-2158-8
10. J.F.C. Glatz and G.J. van der Vusse (eds.): *Cellular Fatty-Acid-Binding Proteins, II.* 1993 ISBN 0-7923-2395-5
11. R.L. Khandelwal and J.H. Wang (eds.): *Reversible Protein Phosphorylation in Cell Regulation.* 1993 ISBN 0-7923-2637-7
12. J. Moss and P. Zahradka (eds.): *ADP-Ribosylation: Metabolic Effects and Regulatory Functions.* 1994 ISBN 0-7923-2951-1
13. V.A. Saks and R. Ventura-Clapier (eds.): *Cellular Bioenergetics: Role of Coupled Creatine Kinases.* 1994 ISBN 0-7923-2952-X
14. J. Slezák and A. Ziegelhöffer (eds.): *Cellular Interactions in Cardiac Pathophysiology.* 1995 ISBN 0-7923-3573-2
15. J.A. Barnes, H.G. Coore, A.H. Mohammed and R.K. Sharma (eds.): *Signal Transduction Mechanisms.* 1995 ISBN 0-7923-3663-1
16. A.K. Srivastava and J.-L. Chiasson (eds.): *Vanadium Compounds: Biochemical and Therapeutic Applications.* 1995 ISBN 0-7923-3763-8
17. J.M.J. Lamers and P.D. Verdouw (eds.): *Biochemistry of Signal Transduction in Myocardium.* 1996 ISBN 0-7923-4067-1
18. E.-G. Krause and R. Vetter (eds.): *Biochemical Mechanisms in Heart Function.* 1996 ISBN 0-7923-4118-X

KLUWER ACADEMIC PUBLISHERS – DORDRECHT / BOSTON / LONDON

Biochemical Mechanisms in Heart Function

Edited by

ERNST GEORG KRAUSE

Max Delbruck Center for Molecular Medicine
Robert-Rossle-Straße 10
13122 Berlin-Buch
Germany

and

ROLAND VETTER

Max Delbruck Center for Molecular Medicine
Robert-Rossle-Straße 10
13122 Berlin-Buch
Germany

Kluwer Academic Publishers
Dordrecht / Boston / London

Library of Congress Cataloging-in-Publication Data

Biochemical mechanisms in heart function / edited by Ernst-Georg
Krause, Roland Vetter.
p. cm. -- (Developments in molecular and cellular
biochemistry ; v. 18)
Includes index.
ISBN 0-7923-4118-X
1. Heart--Molecular aspects. 2. Heart--Physiology. 3. Heart-
-Pathophysiology. 4. Biochemistry. I. Krause, Ernst-Georg.
II. Vetter, Roland, Dr. III. Series.
QP114.M65B54 1996
612.1'73--dc20 96-23973
CIP

ISBN 0-7923-4118-X

Published by Kluwer Academic Publishers,
P.O. Box 17, 3300 AA Dordrecht, The Netherlands.

Kluwer Academic Publishers incorporates
the publishing programmes of
D. Reidel, Martinus Nijhoff, Dr W. Junk and MTP Press.

Sold and distributed in the U.S.A. and Canada
by Kluwer Academic Publishers,
101 Philip Drive, Norwell, MA 02061, U.S.A.

In all other countries, sold and distributed
by Kluwer Academic Publishers Group,
P.O. Box 322, 3300 AH Dordrecht, The Netherlands.

Printed on acid-free paper

Printed in the Netherlands

Molecular and Cellular Biochemistry:

An International Journal for Chemical Biology in Health and Disease

CONTENTS VOLUMES 160 & 161, July/August 1996

BIOCHEMICAL MECHANISMS IN HEART FUNCTION
E.G. Krause and R. Vetter, guest editors

Molecular and Cellular Biochemistry **160/161:** 1, 1996.

Preface

This special issue of Molecular and Cellular Biochemistry contains original research papers as well as invited reviews dedicated, on the occasion of the 40th anniversary of the inauguration of the Heart Research Group in Berlin-Buch that today forms a part there of the Max Delbrück Center for Molecular Medicine, to Professor Albert Wollenberger, founder of the Heart Research Group and for 21 years its head.

The papers in this issue are written by researchers working in the field of cardiovascular research who together with Albert Wollenberger share the belief that an integrative application of advances in molecular and cellular biology will lead to new concepts for treatment and prevention of cardiovascular diseases. We hope that this special will serve as a good source of information in this regard.

We wish to thank all of the contributors for their help and cooperation. We also wish to thank Mrs. Verona Kuhle for her secretarial help. We are grateful to Dr. Naranjan S. Dhalla, Editor-in-Chief of Molecular and Cellular Biochemistry for his interest and encouragement, and for agreeing to publish this issue in honor of Albert Wollenberger.

ERNST-GEORG KRAUSE and ROLAND VETTER
Max Delbrück Center for Molecular Medicine
Robert-Rössle-Straße 10
13122 Berlin-Buch
Germany

ALBERT WOLLENBERGER, Professor, Ph.D. (Harvard), Dr. Sc. Med. (Berlin)

The dedication is accorded to Prof. Wollenberger with high esteem for his pioneering studies on molecular and cellular cardiology and with gratitude by his former associates for his leadership and guidance. In particular, Prof. Ernst-Georg Krause and his colleagues Liane Will-Shahab and Wolfgang Schulze recall with great pleasure the exciting times of their collaboration with Prof. Wollenberger.

PART I

CALCIUM MOVEMENTS AND CELLULAR METABOLISM

Molecular and Cellular Biochemistry **160/161**: 5–12, 1996.

Control of cardiac performance by Ca-turnover

Jiří Šimurda, Milena Šimurdová, Pavel Bravený and Georges Christé[1]

Department of Physiology, Masaryk University, 662 43 Brno, Czech Republic and [1]INSERM U 121, 18 Ave Doyen Lépine, F-69500 Bron, France

Abstract

A quantitative model of Ca-turnover in cardiac cells that incorporates negative feedback modulation of sarcolemmal calcium transport (via Ca channels and Na/Ca exchange) has been designed. The Na/Ca exchange current was expressed as $I_{NaCa} = I_{NaCar} + \Delta I_{NaCa}$. The component I_{NaCar} reflects slow changes of Ca^{2+} and Na^+ concentrations and depends on the Na/K pump. ΔI_{NaCa} is the fast component related to the Ca^{2+} transient. The single input to the model is an arbitrary sequence of intervals between excitations; outputs are sequences of calcium amounts transferred among the compartments during individual intervals. The model operates with a combination of discrete variables (amounts of Ca transferred during contraction, relaxation and rest) and continuous variables – slow changes in ionic concentrations. Since the model is not formalistic but respects the nature of the underlying elements of the system, it enables us to simulate the known effects of cardiotropic drugs or to predict their unknown mechanisms by visualizing the changes in individual Ca compartments. By altering the parameters, the model also simulates the known species and tissue differences in rate-dependent phenomena. (Mol Cell Biochem **160/161**: 5–12, 1996)

Key words: excitation-contraction coupling, sodium-calcium exchange, calcium current, mammalian heart muscle, rate dependency, quantitative model

Introduction

At maximum work load, the normal heart is able to shorten the cycle length about three times, from the average 0.85 sec at rest to some 0.3 sec. This interval corresponds to the resting duration of the electrical and mechanical systole alone. Thus, the prerequisite of undiminished stroke volume at high rates is a considerable abbreviation of electrical and consequently of mechanical systole. It allows, first, for adequate diastolic filling, and second, for the recovery of contractility (mechanical restitution) which is significantly accelerated at the same time (Fig. 1). Both features, shortening and enhancement of the heart contraction upon increased heart rate, are primarily due to a beat-to-beat autoregulation of the cellular calcium turnover. It is more complex and more intriguing than in any other tissue. In essence, it is a combination of two forms of Ca cycling, one between the extracellular space and the cytoplasm – similar to that in smooth muscle cells – and the other between the cytoplasm and the intracellular (reticular) stores – similar to that in skeletal muscle cells.

Each of the two cycles consist of several, usually doubled steps. Ca enters cardiac cells during depolarization through specific L-type channels and by means of reverse Na-Ca exchange (e.g. Levi *et al.* [1]). Ca efflux is carried out predominantly by Na-Ca exchange as a rule during most of the cycle. The slow sarcolemmal Ca pump plays a negligible role in the heart [2]. Powerful active Ca uptake into the sarcoplasmic reticulum (SR) maintains the resting concentration below 10^{-7} mol/l, and at the same time, builds up a store of Ca ready for rapid release upon excitation. This process, opening of the Ca release channels of SR, is generally believed to be triggered by Ca influx (Ca-Ca release mechanism). However, the very depolarization may have a role, very likely more than a rudimentary one [3, 4]. These individual compartments and steps in-between have been discovered gradually during the last three decades and have been well documented by a variety of methods: Voltage clamp, patch clamp, intracellular Ca monitoring, ion sensitive microelectrodes, mechanography, specific blocking agents, laser light scattering, X-ray microanalysis, etc. Most of them have been fittingly described by partial quantitative models [5–11]. A model of the functioning total is, however, a problem of its own.

The present models are largely focused on revealing the interplay of known and putative elements of the autoregula-

Address for offprints: P. Bravený, Department of Physiology, Masaryk University, 662 43 Brno, Czech Republic

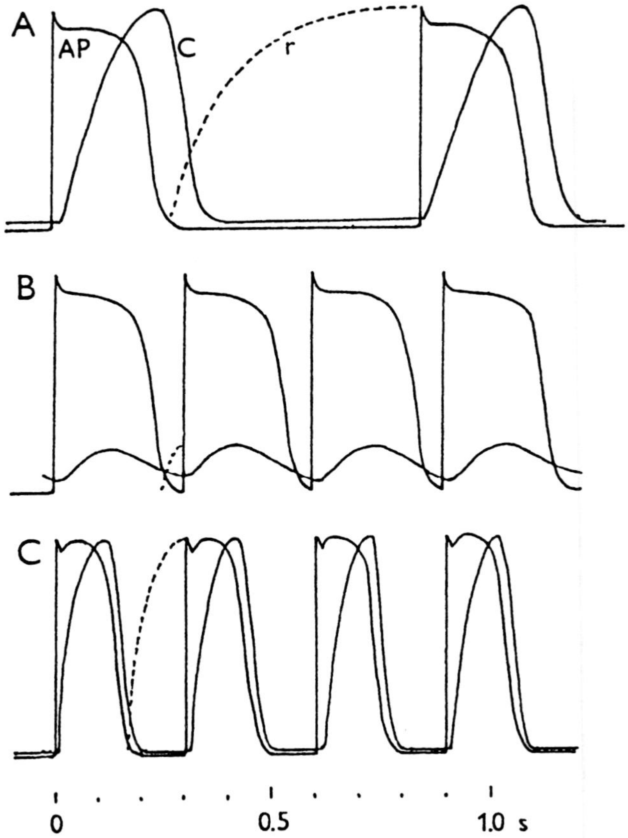

Fig. 1. Schematic representation of ventricular action potentials (*AP*) and contraction (*C*) together with reconstructed restitution of contractility (*r*). A) basal pattern at heart rate 70 p.min. B) at 200 p.min: the unchanged AP and r would critically limit force development and filling time. C – actual pattern at 200 p.min.: abbreviation of *AP* and increased rate of *r*.

tory system. The most adequate approach, the continuous description of the whole system requires a considerable number of nonlinear differential equations [12, 13]. It is quite complicated, time consuming and thus, impractical for on-line verification of experimental data.

We made an attempt to design a convenient hybrid model comprising of a set of discrete variables for fast events and of a few continuous variables for slow events. This model appears to offer a simple yet still realistic insight into the mechanism of physiologically important autoregulation of cardiac contraction. In this article we describe the principles of our approach. It is based on a quantitative formulation of keeping the intracellular Ca concentration constant, and of reacting to a deviation from the steady state.

The model

Control of membrane calcium transport

To formulate the first problem quantitatively, it is useful to express the amount of calcium M_{CaT} transported across the cell membrane during the cycle T by membrane ionic currents. Taking into account only the two main transport systems, Ca channels and Na/Ca exchanger with 3:1 stoichiometry one obtains:

$$M_{CaT} = \int_0^T J^{SL} dt = -\frac{1}{2F}\int_0^T I_{Ca}\, dt + \frac{1}{F}\int_0^T I_{NaCa}\, dt, \quad (1)$$

where J^{SL} is total sarcolemmal (SL) calcium flux, I_{Ca} and I_{NaCa} are electrical currents mediated by Ca channels and Na/Ca exchanger, F is the Faraday charge. If $M_{CaT} > 0$, calcium enters the cell and vice versa.

At any frequency, in steady state:

$$M_{CaT} = 0 \text{ and } \frac{1}{2F}\int_0^T I_{Ca}\, dt = \frac{1}{F}\int_0^T I_{NaCa}\, dt. \quad (2)$$

The principle of the control of intracellular calcium concentration at a constant level corresponding to a given frequency is based on the following condition which should be implicitly comprised by any model. If the amount of the released Ca^{2+} varies randomly both transport mechanisms are modified, thus interfering with condition (2). The direction of the deviation must ensure that $M_{CaT} \neq 0$ tends to resume equilibrium. This is the essence of negative feedback which regulates the constant level of intracellular calcium and thus, both resting tension and contractility.

Provided the dependence of the integrals in equation (1) on $[Ca^{2+}]_i$ is described quantitatively, condition (2) becomes an equation which describes the course of the rate-dependence of steady state contractions. At any frequency the amount of released Ca^{2+} – and hence, of contraction – automatically settles at a level where both integrals are identical. The currents also depend on other parameters (e.g. on membrane voltage and on K^+ and Na^+ concentrations, or more precisely, on their activities). If the model involves the variability of these parameters, it is necessary to respect additional conditions of equilibria (e.g. $M_{NaT} = 0$, $M_{KT} = 0$ and $Q_T = 0$, where M_{NaT} and M_{KT} are the transported amounts of Na^+ and K^+, and Q_T is the passing electrical charge).

The experimental evidence for this feedback mechanism is given by the correlations between the currents associated with the Ca-transporting systems (I_{Ca}, I_{NaCa}) and $[Ca^{2+}]_i$-transient or contraction as a rough indicator of released Ca^{2+} [5, 14, 15].

Modulation of I_{Ca}

In voltage clamp experiments on ventricular trabeculae, a conditioning series of small depolarizations (20 mV from resting potential) strikingly reduces contraction. The following test series of 7 depolarizing pulses to a given positive voltage produces a prominent positive 'staircase'. At any

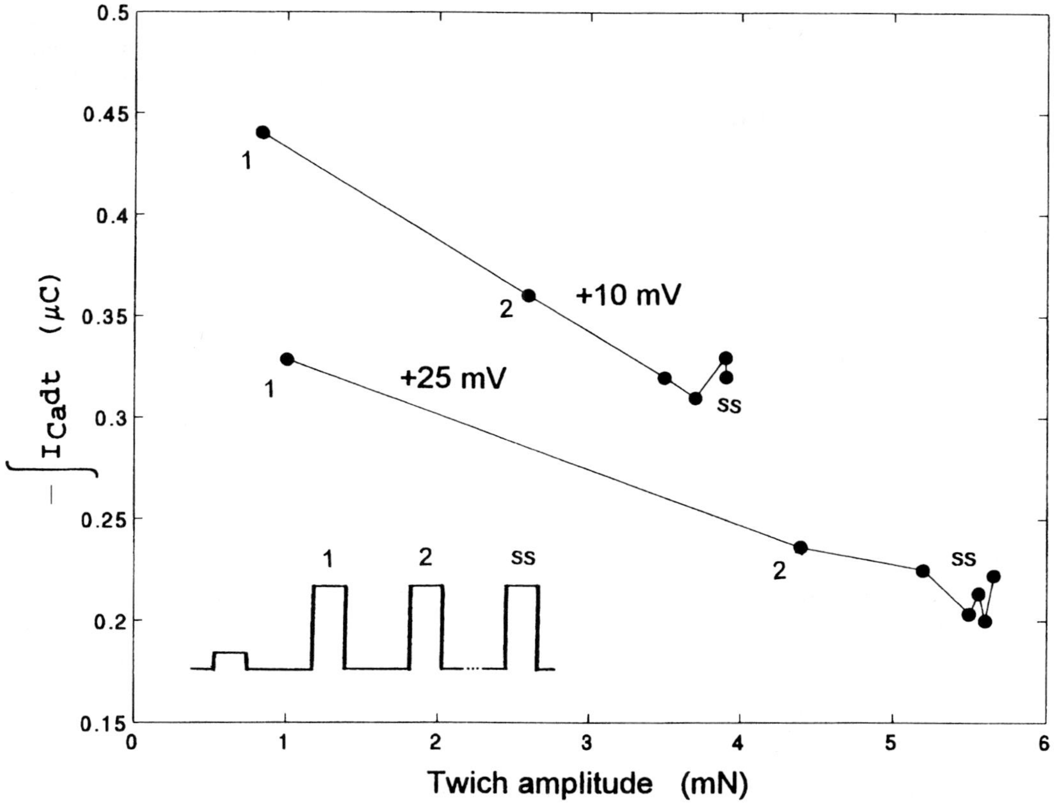

Fig. 2. The relationship between the time integral of calcium current (ordinate) and contraction (abscissa) during ascending staircase at two levels of depolarization. Inset: experimental protocol (ss = steady state). Cat papillary muscle, 31°C.

level of test clamp depolarizations, the time integral of I_{Ca} (carrying Ca into the cell) decreases with increasing force of contraction (Fig. 2). Thus, the modulation of I_{Ca} by transiently increased intracellular free calcium concentration meets the condition of the negative feedback.

Under conditions of free running action potentials, this negative feedback is further amplified by $[Ca^{2+}]_i$ dependent modulation of potassium currents. An enhanced Ca^{2+} release increases potassium currents which, in turn, abbreviates the action potential duration and, thereby, curtails Ca^{2+} entry.

Modulation of I_{NaCa}

With respect to the role of the Na/Ca exchanger in the feedback regulation of $[Ca^{2+}]_i$ it is useful to express $[Ca^{2+}]_i$ as the sum of its resting, 'diastolic' value $[Ca^{2+}]_{ir}$ and of the 'systolic' increment $\Delta[Ca^{2-}]_i$. Accordingly, the current I_{NaCa} may be dissociated into the respective components I_{NaCar} and ΔI_{NaCa}. The latter is due to the transient $[Ca^{2+}]_i$ increase and may be recorded directly in response to depolarizations even below the I_{Ca} threshold [16]. Figure 3 shows the correlation between the time integral of ΔI_{NaCa} and contraction under three extracellular calcium concentrations. The nonlinear relations may be approximated by quadratic polynomials and, as expected, also meet the condition of a negative feed-back.

The results fit well with the expression for ΔI_{NaCa} [16] derived from a modified kinetic diagram of the Na/Ca exchanger as outlined by Mullins [17].

Description of the model

Our aim is to present the basic principle of the model. The details will be published elsewhere.

The initial relations follow from the law of mass conservation written in the integral form and applied to the movements of Ca^{2+} among compartments. Particularly, if MCa^{2+} means the total amount of myoplasmic free calcium ions, then

$$MCa^{2+}(t) = MCa^{2+}(0) + \int_0^t J^{SL}dt + \int_0^t J^{SR}dt + \int_0^t J^{b}dt \quad (3)$$

where J^{SL} , J^{SR} and J^b are the total fluxes of Ca^{2+} into the myoplasm across SL, SR membrane and from the passive Ca^{2+} buffer bindings (namely troponin C). The influx is considered positive, the efflux negative. If $t = 0$ is regarded the beginning of the n-th cycle then $MCa^{2+}(0)$ corresponds to the diastolic level of Ca^{2+} and $MCa^{2+}(t)$ corresponds to the Ca^{2+} transient. The fluxes J^{SL}, J^{SR} and J^b may be further dissociated into their components:

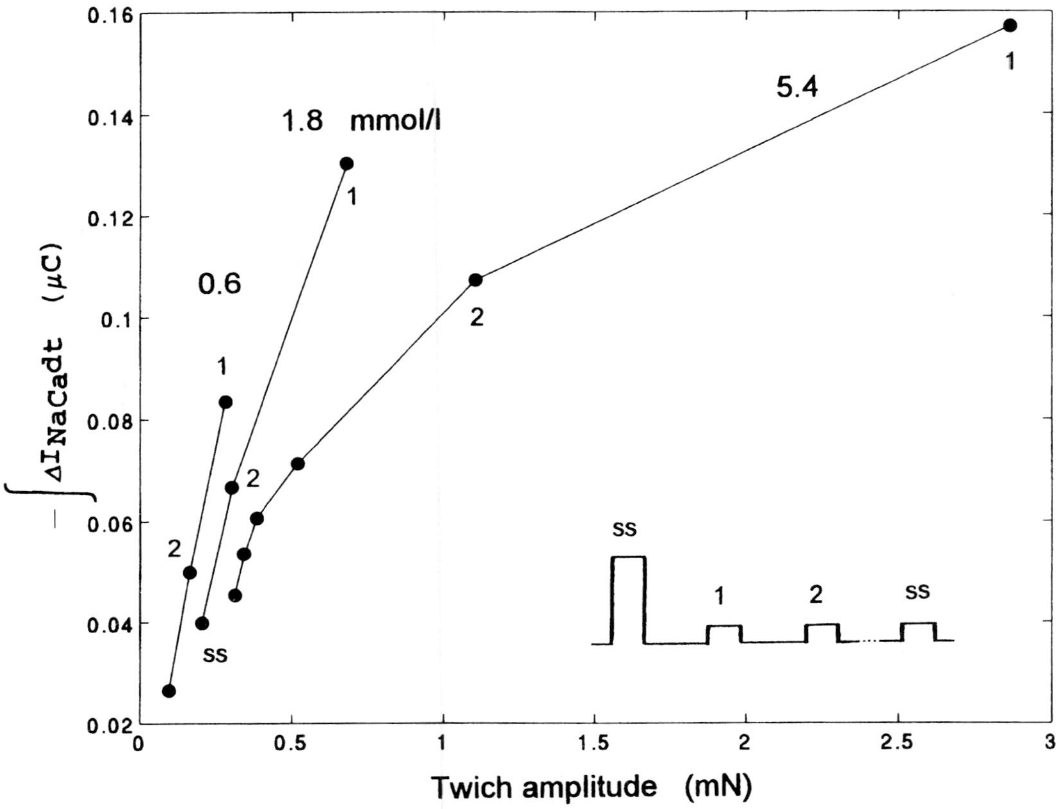

Fig. 3. The relationship between the time integral of the fast component of Na/Ca exchange current (ordinate) and contraction (abscissa) during descending staircase at three different extracellular Ca^{2+} concentrations (0.6, 1.8, 5.4 mmol/l). Inset: experimental protocol (ss = steady state). Dog right ventricular trabecula, 31°C.

$$J^{SL} = \frac{-I_{Ca}}{2F} + \frac{I_{NaCar}}{F} + \frac{\Delta I_{NaCa}}{F},$$

$$J^{SR} = J^{SR}_{up} + J^{SR}_{rel} + J^{SR}_{leak}, \tag{4}$$

$$J^{b} = J^{b}_{up} + J^{b}_{rel}$$

(*up* stands for uptake and *rel* for release). We found it useful to separate every cycle into three imaginary partial intervals $(T = T_{s1} + T_{s2} + T_{d})$. The systolic events (contraction and relaxation) are regarded to occur during the intervals T_{s1} and T_{s2}, respectively, the diastolic ones during T_d. In fact, these processes partly overlap. In agreement with the experimental data [18], it is possible to consider the amount of free Ca ions $MCa^{2+}(t)$ at the onset and at the end of each partial interval to be negligible with respect to the transported amounts (the integrals in equation (3)). With this presumption and with regard to (4) the equation (3) yields

$$-\frac{1}{2F}\int_0^{T_{s1}} I_{Ca}\,dt + \frac{1}{F}\int_0^{T_{s1}} \Delta I_{NaCa}\,dt + \int_0^{T_{s1}} J^{SR}_{rel}\,dt + \int_0^{T_{s1}} J^{SR}_{up}\,dt + \int_0^{T_{s1}} J^{b}_{up}\,dt = 0,$$

$$\frac{1}{F}\int_{T_{s1}}^{T_{s2}} \Delta I_{NaCa}\,dt + \int_{T_{s1}}^{T_{s2}} J^{b}_{rel}\,dt + \int_{T_{s1}}^{T_{s2}} J^{SR}_{up}\,dt + \int_{T_{s1}}^{T_{s2}} J^{b}_{up}\,dt = 0, \tag{5}$$

$$\frac{1}{F}\int_{T_{s2}}^{T_{d}} I_{NaCar}\,dt + \int_{T_{s2}}^{T_{d}} J^{b}_{leak}\,dt + \int_{T_{s2}}^{T_{d}} J^{SR}_{up}\,dt + \int_{T_{s2}}^{T_{d}} J^{b}_{up}\,dt = 0.$$

These expressions involve the components of fluxes which differ from zero within the given interval. If the time course of any process (inactivation and recovery from inactivation of I_{Ca}, decay of ΔI_{NaCa}, electrical and mechanical restitution) is approximated by exponential functions, most of the time integrals may be expressed analytically. Thus, the model becomes rather simple but still maintains the physiological

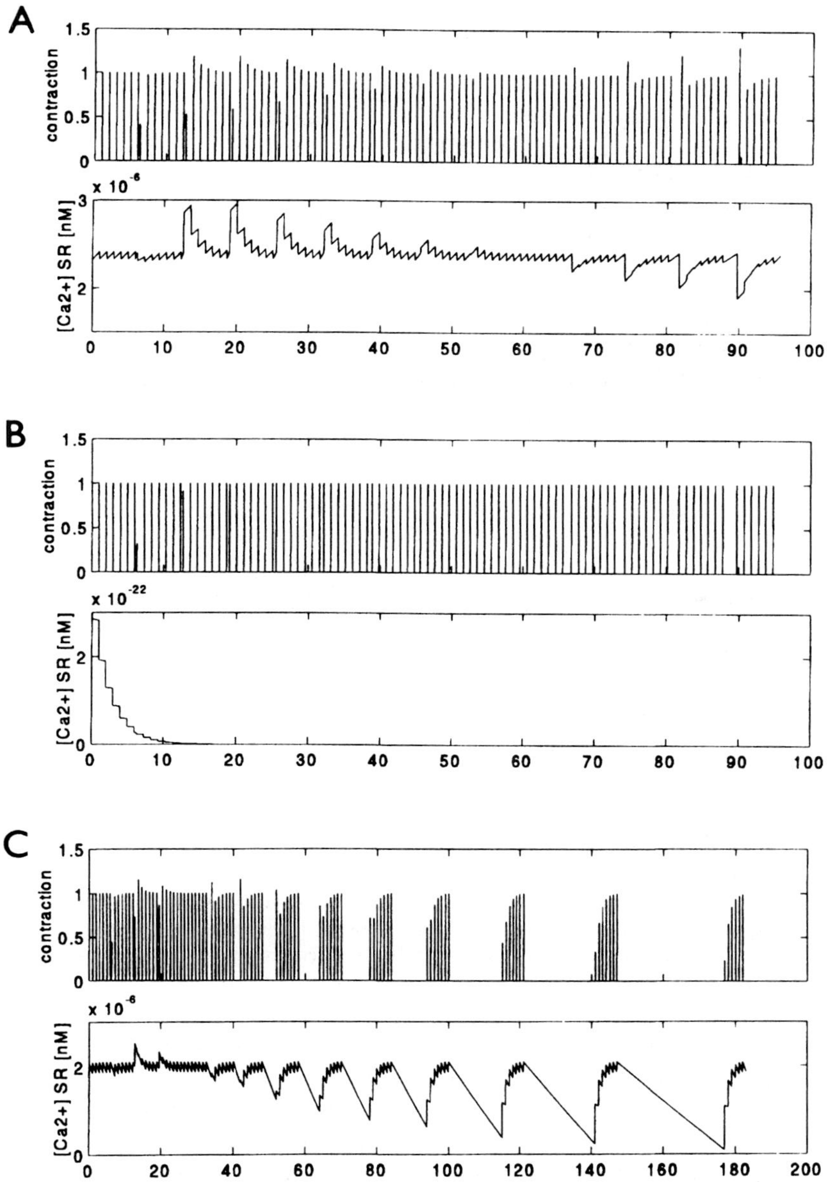

Fig. 4. The simulated response to a change of a single interval. At basal rate 1 Hz every seventh interval is changed. The outputs are: force of contraction (upper panels) and reticular Ca content (lower panels). A. control conditions, B. effect of thapsigargin (block of reticular Ca uptake), C. effect of ryanodine (leak of Ca from SR).

meaning of all parameters involved. The model may be considered a hybrid one since it contains both discrete quantities (Ca^{2+} transferred during partial intervals among the compartments) and continuous quantities. The calculation of their respective integrals during each cycle is based on numeric solution of differential equations. When the equations involving the Na/K and Na/Ca exchangers are introduced, the time courses of $[Na^+]_i$ and $[Ca^{2+}]_{ir}$ are obtained.

In a similar way, the transition to discrete quantities was applied as far as other compartments (sarcoplasmic reticulum, passive buffers) were concerned. The contents of the bound calcium in these compartments varies in three steps during each cycle. The formalism routinely employed in the theory of regulatory systems made it possible to emphasize the involved feedback mechanisms accordingly.

Discussion

The reported model operates with a single input – an arbitrary sequence of intervals between excitations. The results of simulations are sequences of Ca^{2+} quanta transferred among the compartments during individual intervals. The linkage between the discrete and continuous part of the model

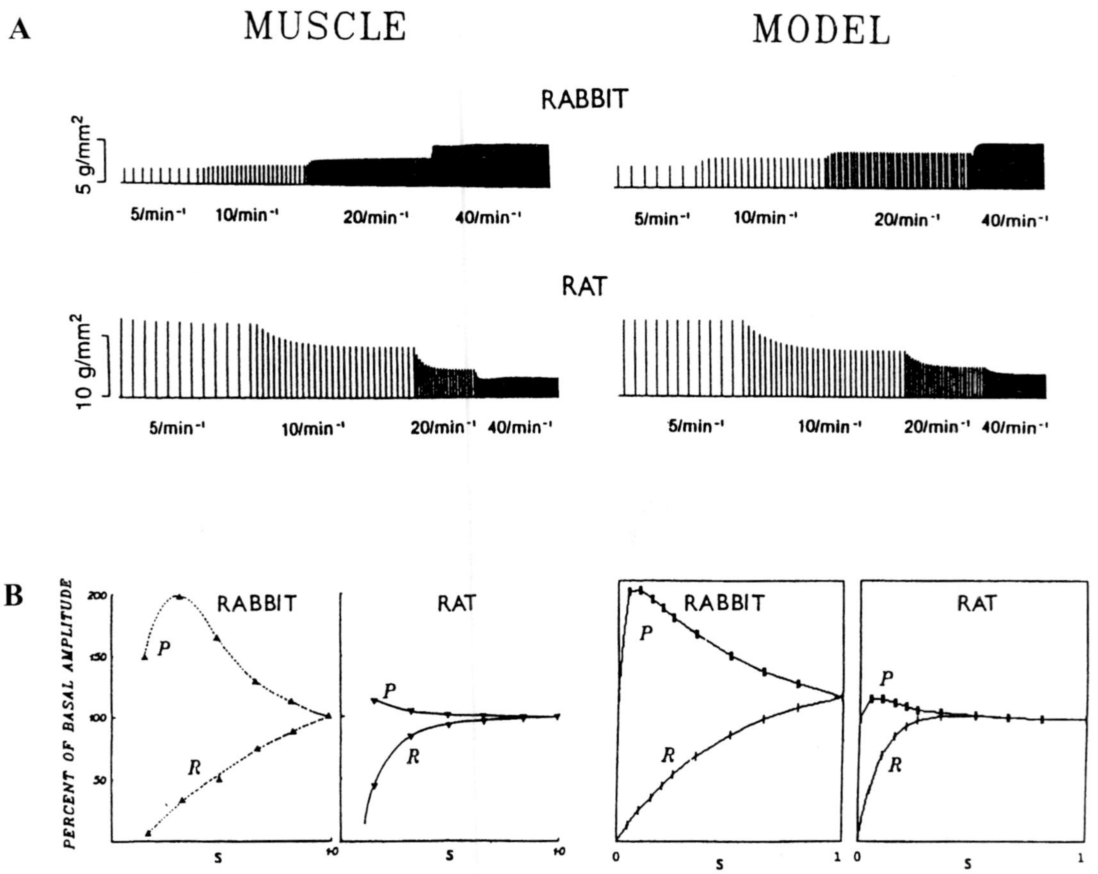

Fig. 5. The comparison of experimental results (muscle) and simulation (model) in the rabbit and rat heart. A. frequency-force relation (experimental data from [19]). B. interval-strength relations; *R* – restitution, *P* – potentiation (data from [20]). The species difference at low rates (A) is mainly due to different time course of the slow component of Na/Ca exchange – I_{NaCar}. At faster rates (B) the differences are related mainly to the rapid component ΔI_{NaCa} and to time constant of mechanical restitution.

is a crucial and sensitive step. In our model the discrete quantities (Ca^{2+} and Na^{+} transported across the surface membrane during the cycle) determine the initial conditions for differential equations that describe the diastolic changes of $[Na^{+}]_i$ and $[Ca^{2+}]_{ir}$. The structure of the model may be easily changed in order to verify the alternative interpretations of individual processes. The main parameters of the model are: (1) the time constants relating to the time course of the current components and of the electrical and mechanical restitution; (2) the constants which determine the magnitude of the current components and the compartmentalization of cytoplasmic calcium; (3) the constants requisite for the description of concentration-dependence of the currents generated by the Na/K and Na/Ca exchangers.

The primary objective of the model was to get a better insight into the frequency-dependence of electrical and mechanical behaviour of the heart. Since it is not formalistic but respects the nature of the underlying elements of the system, the model makes it possible to simulate the known effects of cardiotropic drugs and agents (Fig. 4) or to predict their unknown mechanisms by visualizing the changes in individual Ca compartments. By altering the individual parameters the model simulates also the well known prominent species differences in the frequency effect (Fig. 5) and tissue differences (atria vs. ventricles).

Finally, we would like to illustrate in Fig. 6 the simulation of the significant adaptation of the heart cell functions shown in Fig. 1. Both parallel phenomena, the increased rate of mechanical restitution (A) and diminished action potential duration (B) are accomplished by a single change at the input: By an increase of heart rate.

Conclusion

The present model of complex ion transports in the heart muscle cell combines the elements of a discrete model (for fast, systolic events) and of a continuous one (for slow, diastolic events). In spite of considerable simplification, the model reflects the real processes within the excitation-contraction coupling. It helps to verify their conjectural mechanisms on one hand, and to predict the mechanism of various

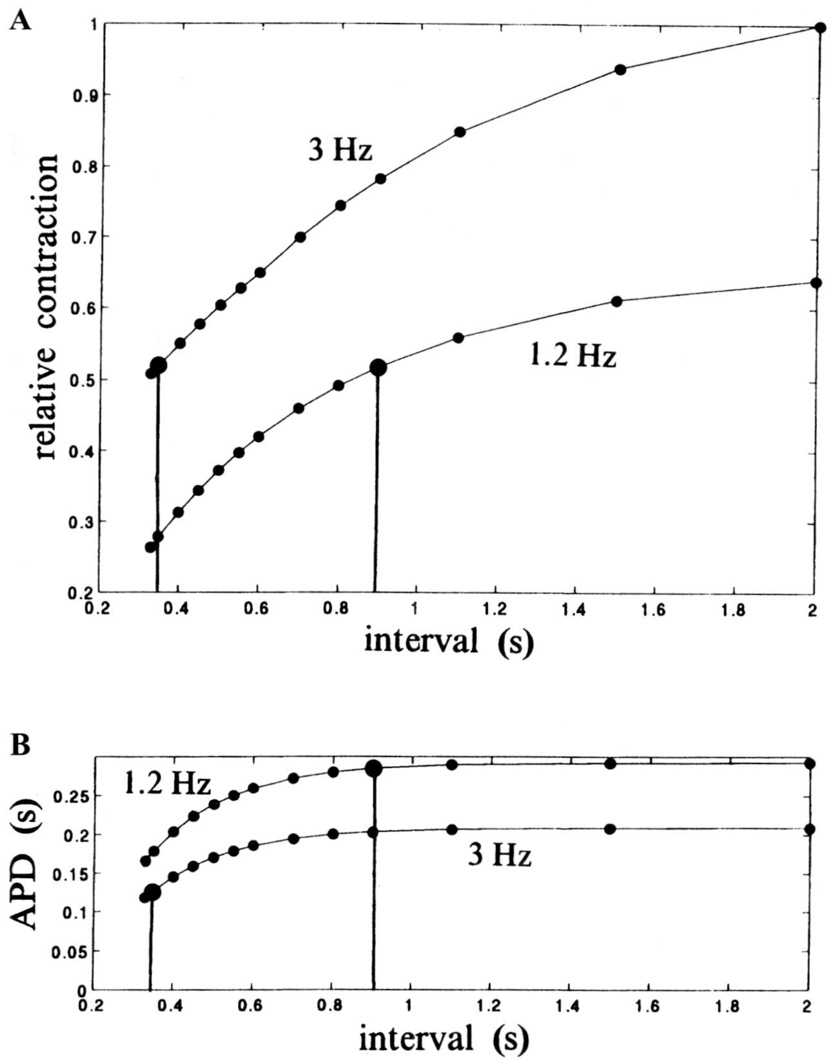

Fig. 6. The simulated force (A) and action potential duration (B) restitution at two basal rates. Vertical bars denote the steady state values. The simulations correspond to the pattern shown in Fig. 1A and C.

cardiotropic effects on the other. Due to the relative simplicity, the model can operate rapidly, virtually on-line.

Acknowledgements

This work was supported by the Granting Agency of the Czech Republic (Project 305/93/0169).

References

1. Levi AJ, Brooksby P, Hancox JC: One hump or two? The triggering of calcium release from the sarcoplasmic reticulum and the voltage dependence of contraction in mammalian cardiac muscle. Cardiovasc Res 27: 1743–1757, 1993
2. Bers DM, Bassani JWM, Bassani RA: Competition and redistribution among calcium transport systems in rabbit cardiac myocyes. Cardiovasc Res 27: 1772–1777, 1993
3. Lederer WJ, Cannel MB, Cohen NM, Berlin JR: Excitation-contraction coupling in heart muscle. Mol Cell Biochem 89: 115–119, 1989
4. Arlock P, Noble MIM, Wohlfart B: Cardiac cell membrane repolarization is required for onset of mechanical restitution in papillary muscle. Acta Physiol Scand 142: 113–118, 1991
5. Wier WG, Yue DT: Intracellular calcium transients underlying the short-term force-interval relationship in ferret ventricular myocardium. J Physiol 376: 507–530, 1986
6. Langer GA: The role of calcium at the sarcolemma in the control of myocardial contractility. Can J Physiol Pharmacol 65: 627–631, 1987
7. Apell HJ: Electrogenic properties of the Na,K pump. J Membrane Biol 110: 103–114, 1989
8. Sipido KR, Wier WG: Flux of Ca^{2+} across the sarcoplasmic reticulum of guinea-pig cardiac cell during excitation-contraction coupling. J Physiol 435: 605+630, 1991
9. Isenberg G, Han S, Schiefer A, Wendt-Gallitelli MF: Changes in mito-

chondrial calcium concentration during the cardiac contraction cycle. Cardiovasc Res 27: 1800–1809, 1993

10. Bassani JWM, Bassani RA, Bers DM: Relaxation in rabbit and rat cardiac cells: Species-dependent differences in cellular mechanisms. J Physiol 476.2: 279–293, 1994
11. Bassani RA, Bers DM: Rate of diastolic Ca release from the sarcoplasmic reticulum of intact rabbit and rat ventricular myocytes. Biophys J 68: 2015–2022, 1995
12. DiFrancesco D, Noble D: A model of cardiac electrical activity incorporating ionic pumps and concentration changes. Phil Trans R Soc Lond B 307: 353–698, 1985
13. Hilgemann DW, Noble D: Excitation-contraction coupling and extracellular calcium transients in rabbit atrium: Reconstruction of basic cellular mechanisms. Proc R Soc Lond B 270: 163–205, 1987
14. Fozzard HA: Heart:Excitation-contraction coupling. Ann Rev Physiol 39: 201–220, 1977
15. Yue DT: Intracellular [Ca^{2+}] related to rate of force development in twitch contraction of heart. Am J Physiol 252: H760–H770, 1987
16. Šimurda J, Šimurdová M, Bravený P, Šumbera J: A contraction-related component of slow inward current in dog ventricular muscle and its relation to Na^+-Ca^{2+} exchange. J Physiol 456: 49–70, 1992
17. Mullins LJ: A mechanism of Na/Ca transport. J Gen Phsiol 70: 681–695, 1977
18. Isenberg G, Wendt-Gallitelli MF: Binding of calcium to myoplasmic buffers contributes to the frequency-dependent inotropy in heart ventricular cells. Basic Res Cardiol 87: 411–417, 1992
19. Kort AA, Lakatta EG: Spontaneous sarcoplasmic reticulum calcium release in rat and rabbit cardiac muscle: Relation to transient and rested state. Circ Res 63: 969–979, 1988
20. Kruta V, Bravený P: Potentiation of contractility in the heart muscle of the rat and some other mammals. Nature 187: 327–328, 1960

Molecular and Cellular Biochemistry **160/161:** 13–25, 1996.

Deformation of the Bowditch staircase in Ca^{2+}-overloaded mammalian cardiac tissue – a calcium phenomenon?

Matthias Löhn, Gerhard Szymanski and Fritz Markwardt
Julius-Bernstein-Institut für Physiologie, Martin-Luther-Universität Halle-Wittenberg, Halle, Germany

Abstract

Concentrations of 1–4 μmol l^{-1} isoproterenol cause both in right ventricular papillary muscles and in enzymatically isolated myocytes of the guinea-pig a Ca^{2+} overload-induced state which is functionally characterized by biphasic (multiphasic) twitches and biphasic (multiphasic) intracellular calcium transients, respectively, during excitation-contraction coupling. This state was stabilized in the *in vitro* experiments for some hours by a co-ordination of the interstimulus interval, the temperature of the superfusion fluid and the addition of calcium agonists. The functional stability is the precondition for the reproducibility of the experimental results particularly after the application of long-lasting stimulation programmes. Changes in the shape of biphasic contractions were compared with changes in the time course of biphasic intracellular calcium transients using three manipulations of a different kind: (1) the interruption of the steady pacing rhythm, (2) the variation of the interstimulus interval, (3) the addition of ryanodine. It was shown that: (1) The BOWDITCH staircase in calcium overloaded multicellular preparations is changed in that each individual component of the twitch passes through its own staircase. A homologous behaviour can be observed in the configuration of the phasic and tonic component of biphasic intracellular calcium transients. (2) At different driving frequencies the relative proportion of the two components of a biphasic twitch corresponds to the time integrals of the two components of biphasic intracellular calcium transients. (3) Ryanodine suppresses both the first component of the biphasic twitch and the phasic component of the biphasic intracellular calcium transient.

The SR Ca^{2+}-ATPase inhibitor thapsigargin increases the second component of the biphasic calcium transient. This supports the hypothesis that the size of the tonic component is in part determined by intracellular calcium reuptake. The results of both kinds of experiments would be compatible with the assumption that in calcium overloaded mammalian cardiac cells calcium reaches the contractile system directly as well as via two intracellular stores ('extended two-Ca-store concept'). (Mol Cell Biochem **160/161:** 13–25, 1996)

Key words: e-c coupling, Ca-overloaded myocytes, biphasic Ca-transients, ryanodine, thapsigargin, isoproterenol

Introduction

Contrary to the frog heart, in the mammalian myocardium the calcium flux from the extracellular to the intracellular space is not immediately available as a whole for binding to troponin C. There is abundant evidence, both from indirect observations and from direct measurements that in mammalian heart muscle the excitation-contraction coupling involves a complex interplay of transmembrane and intracellular calcium movements [1–5]. The calcium entering the mammalian cardiac cell during the plateau phase of the action potential is first retained in intracellular storage structures. One theory summarizing the bulk of experimental evidence which has attracted much favourable attention is the 'two-Ca-store concept' [6] and its mathematical generalization as 'n-store-m-phase model' [7] allowing satisfactorily to simulate the following inotropic phenomena: The post-rest potentiation, the postextrasystolic potentiation, staircase phenomena of all kind, inotropic heart nerve effects provided that decrease or increase, respectively, in the force of contraction is caused by the electrotropic effects of the heart nerves, the transient changes of the developed tension after steps of the

Address for offprints: M. Löhn, Julius-Bernstein-Institut für Physiologie, Magdeburgerstr. 6, 06097 Halle/Saale, Germany interstimulus interval, and the steady-state interval-strength-relationship.

The simple basic idea of the 'two-Ca-store concept' is that during the slow inward current calcium is taken up by the store S_1, and simultaneously released from store S_2. During the diastole it moves between the two stores which are leaking continuously. Via increasing the intracellular cAMP concentration, β-adrenergic agonists are capable to increase the magnitude of the voltage dependent calcium inward current and, consequently, to increase the 'triggered' calcium release from sarcoplasmic reticulum. One of the most interesting features occurring as a concomitant phenomenon of prolonged membrane depolarization is the gradual conversion of the normal monophasic contraction shape to a biphasic one. It is brought about, for instance, by the application of 'mild' concentrations of the β-adrenergic receptor agonist isoproterenol.

In the presence of 0.1–0.4 μmol l^{-1} isoproterenol, the time course of the recovery after rest is not the BOWDITCH-staircase in the common sense (see Figs 9 and 10 [8]) but a step-by-step metamorphosis of the twitch from the monophasic 'late' type (late component predominant) to the biphasic 'early' type (early component predominant) [9]. This experimental finding can be simulated by computation based on the 'two-calcium-store concept' with the only supplementary assumption that one part of the calcium influx, which manifests as the slow inward current, directly, i.e. under circumvention of the stores S_1 and S_2, respectively, reaches the myofilaments (see 'path b' in Fig. 8 and 'state 1' in Fig. 9 in [10]). A critical verification of the predictions made by this "extended two-calcium-store concept' is lacking.

In most of the experimental studies concerning multiphasic heart muscle contractions the attention was focussed on changes of the relative proportion of the different phases [9, 11–22]. One of the disregarded investigative problems concerned with biphasic (multiphasic) calcium transients is the increase of the fluorescence time integral of the tonic (late) component not only relatively but also absolutely with increasing interstimulus interval.

The simplest hypothesis in terms of the 'extended 2-Ca-store concept' would be that the time constant of the process characterizing the calcium elimination from the contractile element increases. As was shown by analog computation in that case not only the tonic component becomes predominant as compared to the phasic (early) one but also the area beneath the tonic component absolutely increases (see Fig. 13 in [10]).

The present paper extends our earlier investigations on multiphasic heart muscle contractions with the aim (1) of comparing changes in the shape of biphasic contractions with changes in the time course of biphasic intracellular calcium transients, and (2) to discover whether the assumption of a 'use dependence' of the SR-calcium pump could be supported by a well-aimed intervention into the mechanism which removes calcium from troponin C.

Materials and methods

Cell isolation

Isolated guinea-pig myocardial cells were prepared by a collagenase digestion as described by Kao *et al.* [23]. Guinea-pigs of both sex (body weight between 180–420 g) were killed by cervical dislocation. The heart was removed and perfused for 5 min at a constant perfusion pressure (40 cm H_2O) by a LANGENDORFF perfusion system with nominal Ca^{2+}-free solution containing (mmol l^{-1} NaCI 135, KCI 5.4, $MgCI_2$ 4, NaH_2PO_4 0.33, HEPES 10, glucose 15; pH 7.2 (NaOH). Thereafter the perfusion solution was changed to an enzyme solution (nominal Ca^{2+}-free solution plus 0.5 mg ml^{-1} fatty acid free bovine albumin (SIGMA), plus 0.3–0.5 mg ml^{-1} collagenase (type Worthington CLS II, Biochrom KG, Berlin, Germany), pH 7.2 (NaOH). After 6–15 min, the perfusion flow increased due to digestion of the vessel structure. The ventricles were cut off, minced into small pieces and placed into an incubation solution (mmol l^{-1} K-glutamate 140, HEPES 10, EGTA 5, glucose 2; pH 7.2 (KOH). The preparation was mechanically agitated by a PASTEUR-pipette which promotes the detachment of single cells from the tissue. The incubation solution was replaced with medium solution containing: MEM 10.73 g l^{-1}, fetal calf serum 0.8 g l^{-1}, Penicillin (Jenapharm, Jena, Germany) 2000 IU l^{-1}, Streptomycin (Jenapharm, Jena, Germany) 4 mg l^{-1}; pH 7.4 (NaOH). All chemicals were purchased from SERVA, if not otherwise mentioned. The cells were stored for maximal 8 h within the medium at 22°C without oxygen supply. All experiments were done at 22°C.

Measurement of free intracellular Ca^{2+} concentration

For registration of the kinetics of the free cytosolic Ca^{2+}-concentration ($[Ca^{2+}]_i$) in myocardial cells, the following optical system was chosen: An inverted microscope, a xenon-arc-light source and a beamsplitter (NIKON, Düsseldorf, Germany) as well as a photomultiplier (SMT, Seefeld, Germany). Changes of the $[Ca^{2+}]_i$ were measured with the calcium sensitive dye Fluo-3. For experiments directed to qualitative observations the cells were loaded for 30 min with 3 μmol l^{-1} of the membrane permeable ester Fluo-3/AM (Molecular Probes, Eugene, USA) at 22°C. Pluronic F-127 (Molecular Probes) was added, acting as dye solubilizator. The excitation of Fluo-3 was performed at 470–490 nm, and the emission of fluorescence light was measured at 520–560 nm. For quantitative comparisons the contamination of the fluorescence signals by contraction was avoided by ratiometric fluorescence measurement using the two fluorochromes Fluo-3/AM (1 µmol l^{-1}) and Fura-red/AM (9 µmol l^{-1}). The cells were loaded 10 min and 1 h, respectively, at 22°C. For excitation a wavelength of 460–490 nm was used. Fluorescence signals were collected at wavelengths between 520–540 nm (Fluo-3) and > 610 nm (Fura-red) by means of two photomultipliers.

Data acquisition

The data acquisition (fluorescence-photocurrents) was performed via an AD/DA converter (Labmaster, Axon Instruments, Foster City, USA) by an IBM-compatible computer with the commercial computer programme 'PCLAMP' (Axon Instruments). Curve fittings were done with commercial software: 'SIGMAPLOT' (Jandel, Madera, USA) and 'LOTUS' (Lotus Development, München, Germany). Statistical data were presented as mean ± S.D.

Right ventricular guinea-pig papillary muscles

Preparation, electrical stimulation, mechanical recording, in vitro milieu conditions

The experiments were performed on right ventricular papillary muscles of guinea-pigs (body weight about 250 g). The animals were killed by exsanguination after they had been stunned by a blow on the head. Suitable muscles ranging in weight between 2–5 mg were dissected under stereomicroscopic control in a dissecting disc filled with O_2-saturated physiological solution at 37°C. The contractions were nearly isometric, all experiments were performed with the initial length set to l_{max}.

The muscles were allowed to adapt to the *in vitro* conditions for 2 h (1 h at 37°C, 1 h at 22–24°C). As a rule rectangular electrical pulses of 2 ms duration about 20% above threshold were used as stimuli. The tension developed was recorded either horizontally by means of a strain gauge transducer (construction Dr. G. Küchler, Berlin HU) or vertically with the force transducer F 30/372 (Hugo Sachs Elektronik, Germany). The organ bath was perfused with a modified Tyrode solution containing (mmol/l): NaCI 150, KCI 5.4, $CaCI_2$ 2.5, $MgCI_2$ 0.5, glucose 11, HEPES 5 (ACRIS FERAK), pyruvic acid sodium salt 5 (MERCK); pH 7.4 continuously measured in the inflow; medium gassed by vigorous bubbling out of the chamber with carbogen; double distilled water; Isoproterenol (SIGMA); BAY-K 8644 (BAYERZAG); CGP 28-392 (CIBA GEIGY). The stabilization of the biphasic (multiphasic) contractions, a basic requirement for the reproducibility of long-time experiments, was achieved by a precision adjustment of the temperature, the interstimulus interval, the isoproterenol concentration and the concentration of calcium-agonists. This empirical procedure lasted as a rule between 1–2 h. The stabilization of biphasic contractions and calcium transients, respectively, was successful at isoproterenol concentrations between 0.1–0.2 µmol l^{-1}, at temperatures of the superfusion fluid between 20–24°C and concentrations of Ca-agonists of 0.1 µmol l^{-1}.

Measurement of the twitch tension and of time integral of the calcium transient

The developed tension of the quasiisometrically mounted papillary muscle was measured in monophasic contractions from the level of the diastolic tonus, in multiphasic contractions from the level of relaxation residue of the preceding component up to the peak tension.

As to the measuring of approximative time integral of the intracellular calcium concentration the biphasic transient was divided by a vertical subsidiary line, as is shown in the Fig. 1.

Results

Preliminary remark

Figure 2 summarizes records of monophasic, biphasic and threephasic Ca^{2+}-transients (A–E) obtained from isolated myocytes, and isometric contractions (F–K) obtained from right ventricular papillary muscles, respectively. As described in the methods section, it was possible to stabilize the shape both of the biphasic/multiphasic Ca^{2+}-transient and of the biphasic/multiphasic twitch.

Post-rest recovery

In Fig. 3 the relation K_2/K_1 (developed tension of the second component divided by the developed tension of the first component of biphasic twitches) of a right ventricular guinea-pig papillary muscle is plotted as a function of the time after 5 min rest. The line is a curve fitted by eye. On the left is shown the steadyness of twitches before 5 min rest, on the right the typical post-rest staircase of calcium-overloaded mammalian cardiac tissue.

In Fig. 4 the K_2-mean starting value before the rest of seven biphasic calcium transients was adjusted to about 140% of K_1 (left part of the diagram). After 5 min rest the fluorescence time integral of the second component amounting approximately the fivefold of the first component declines to the steady-state value before rest (right part of the diagram). It should be noted that a similarity exists between the post-rest

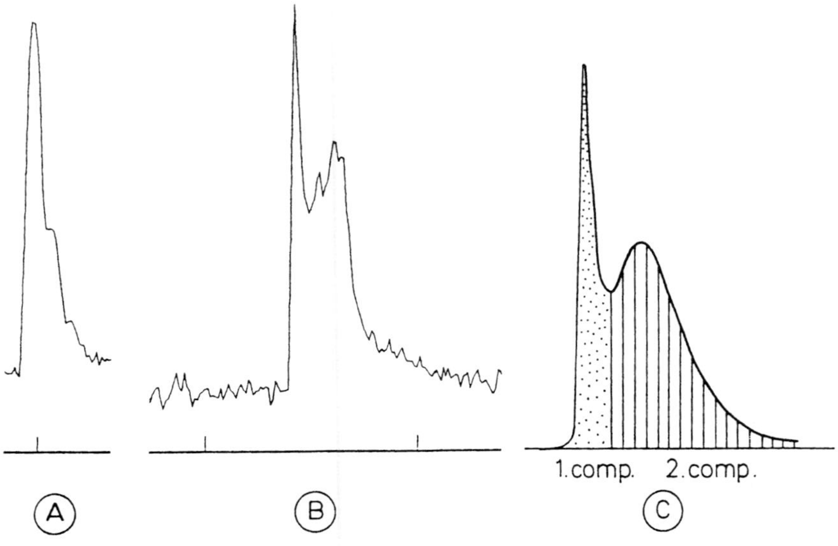

Fig. 1. Graphic representation explaining the method to determinate the approximate value of the time integral for the components of the calcium transient: The division of the idealized signal (C) is defined by a subsidiary line orthogonal to the abscissa crossing the minimum of fluorescence between the assumed phasic (dotted) and tonic (hatched) component. In A, B original records of quasimonophasic (A) and clear biphasic (B) calcium transients are shown. The two components of the fluorescence signal (C) are labelled '1. comp' and '2. comp.', respectively. Ordinate: intensity of fluorescence; abscissa: time.

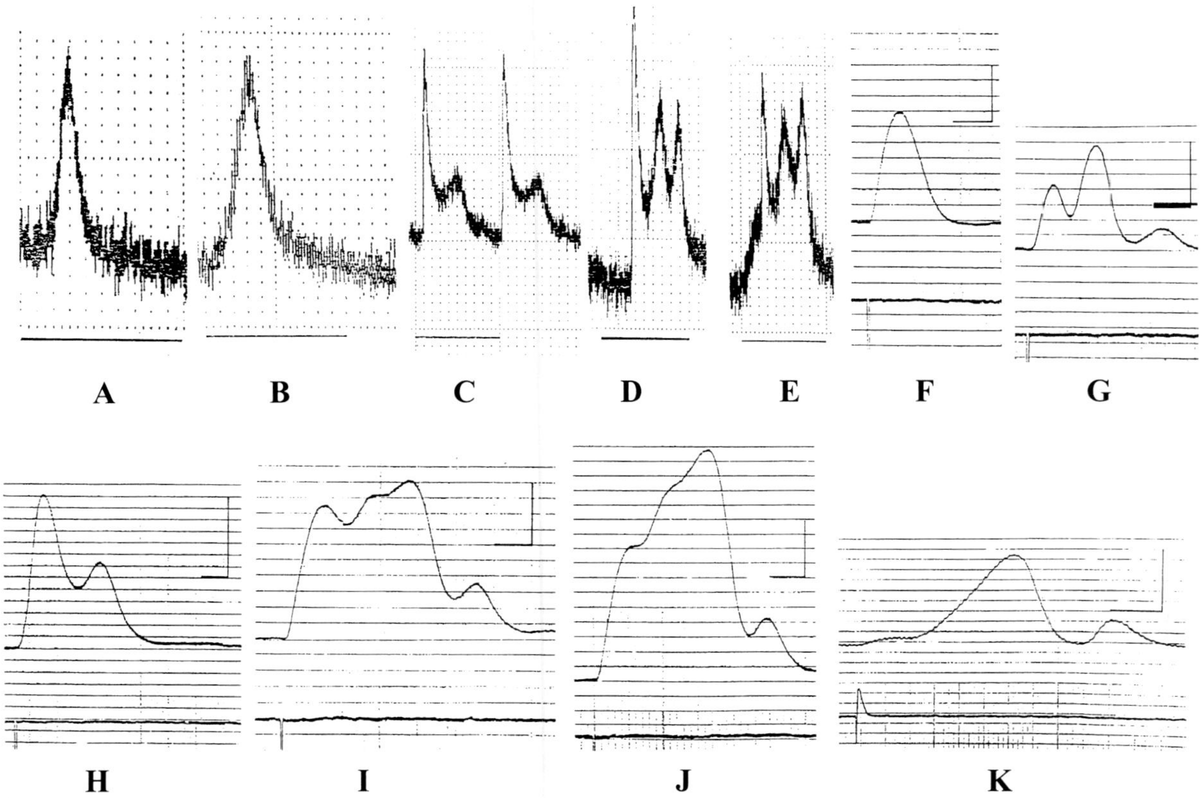

Fig. 2. Phenomenologic survey showing stabilized monophasic (A, B) biphasic (C), threephasic (D, E), calcium transients of isolated myocytes and stabilized monophasic (F), biphasic (G = second component prevailing, H = first component prevailing) and threephasic (I, J) twitches recorded from rightventricular papillary muscles. The last record (K) was taken from a papillary muscle 45 min after superfusion with (1 μmol l^{-1}) ryanodine-containing solution; please note the rudiment of the phasic component preceding the twitch and the first diastolic oscillation. The horizontal bars beneath the records indicate the time calibration: 0.5 sec (A, B), 1 sec (C, D, E), 0.1 sec (F–K); the vertical lines mark 10^{-2} N, bottom trace in F–K: stimulus artifact. Ordinates: intensity of fluorescence (A–E), developed tension (F–K); abscissae: time.

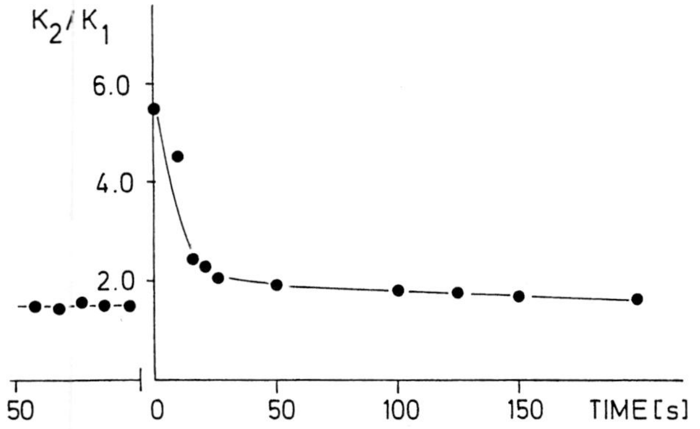

Fig. 3. Typical Bowditch staircase after 5 min rest observable in calcium-overloaded rightventricular guinea-pig papillary muscles. The developed tension of the two components (K_1, K_2) was stabilized at a ratio of about 1.5 (left part of the diagram). After the pause, the increase of the K_2/K_1 ratio amounts 5.5 being thus approximately thrice as great as that observed before the pause. Within about 10 min, the K_2/K_1 ratio decreases to the control value. Isoproterenol 0.15 μmol l^{-1}, BAY-K 8644 0.1 μmol l^{-1}.

recovery of biphasic calcium transients and the post-rest recovery of isoproterenol-induced biphasic contractions of multicellular heart muscle preparations.

Prediction of the 'extended two-Ca-store model'

The simulation of the post-rest staircase using a hardware analogue shows a predominance of the second component accompanied by lacking potentiation of the first transient after rest and a gradual development of the early component which becomes dominant during the staircase, indicating the step by step filling of the store S_2. Both components reach their maximum at steady-state (see Fig. 11D in [10]).

The influence of the interstimulus interval

In multicellular preparations the most striking feature of biphasic (and threephasic, respectively) twitch under steady-state conditions is its configuration to changes in temperature and interstimulus interval.

The results of a typical experiment in guinea-pig right ventricular papillary muscle are illustrated in the two diagrams of the Fig. 5. Each column of the left diagram represents the peak developed tension F of the two components of biphasic contractions at selected interstimulus intervals (10 sec (control 1), 8 , 5, 3 and 10 sec (control 2)), each column of the right diagram represents the corresponding time to peak of the two components. During the experiment, care was taken to ensure steady-state conditions.

Under otherwise constant milieu conditions the steady-state interval-strength relationship of biphasic twitches differs inasmuch from this kind of function observed in mono-

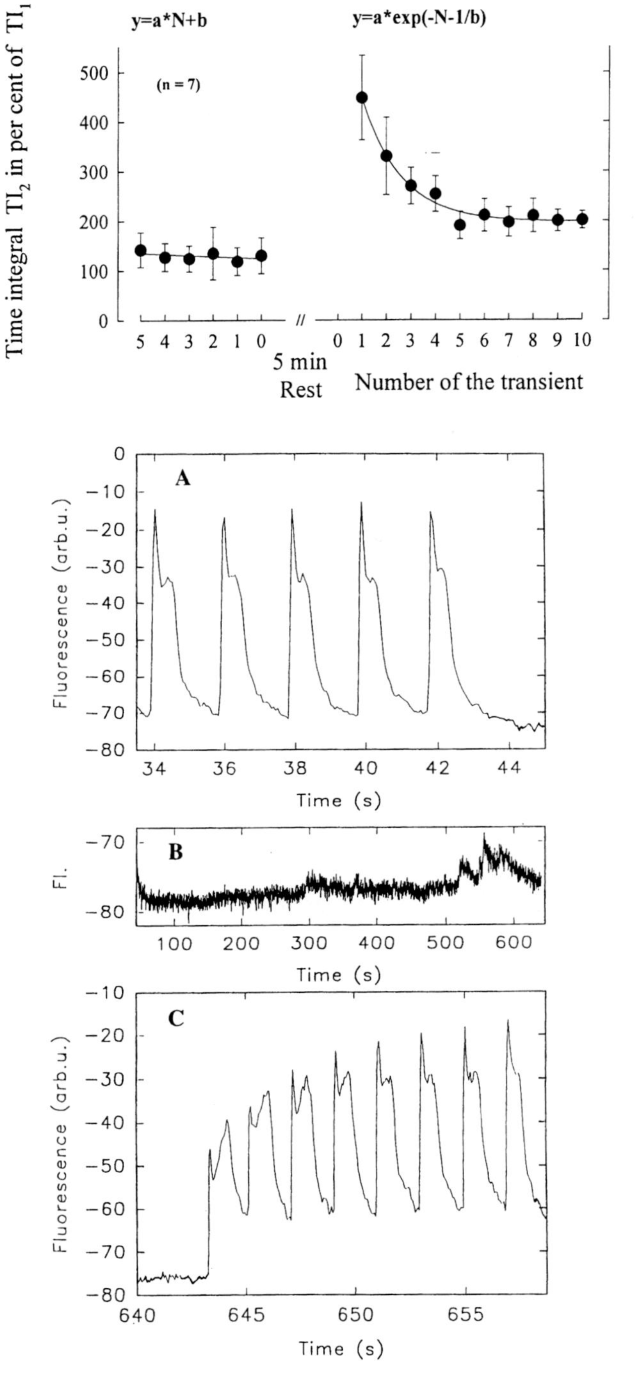

Fig. 4. The typical time course of the Bowditch staircase of stabilized biphasic calcium transients following a 5 min period of quiescence. In the steady-state phase before the rest (left part of the diagram) the starting point of the ratio of size was fixed to a relation of the time integrals Tl_2/Tl_1, approximately equal to 1.5. After 5 min rest the time integral Tl_2 amounting about triple the area beneath T_1, needs a considerable time for equilibration (right part of the diagram). Interstimulus interval 2 sec. The principle of an experiment designed to record the postrest recovery of intracellular calcium transients is summarized in A–C: Steady-state fluorescence signals of stabilized biphasic transients with an interruption of the field stimulation at the 44 sec-moment (A), fluorescence signal during 10 min rest (B), and in typical change of shape during the post-rest staircase (C).

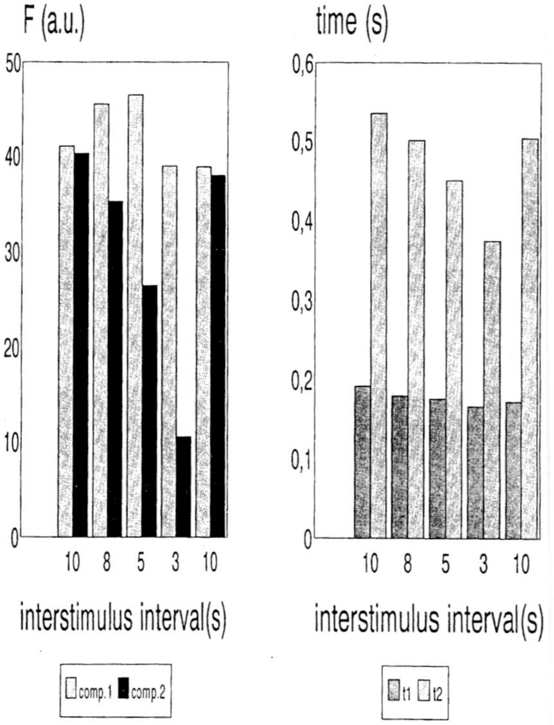

Fig. 5. Relation between the interstimulus interval (abscissa) and the developed tension (F; left diagram) and time to peak ('time'; right diagram) on the ordinate for the two components of stabilized biphasic contractions under steady-state conditions of pacing.

phasic (normal) contractions as an increase of the developed tension with decreasing interstimulus interval refers only to the first component relative to the second one.

This type of experiment was repeated using isolated myocytes. The four records in Fig. 6 show time courses of calcium transients under steady-state conditions of field stimulation at intervals of 1, 3, 10 and 30 sec, respectively. If the observations in multicellular preparations are accepted as representing the interference of two portions of the activator calcium, the present results may plausibly be interpreted along similar lines: The calcium depletion of the sarcoplasmic reticulum during a long interstimulus interval brought about by the leakages could have the consequence that some fraction of the transmembrane calcium influx serves to fill up the storage structures, and for this reason the second component of the calcium transient predominates. This point will be discussed further in a later section. In Fig. 7 the ordinate scale represents the steady-state time integral of fluorescence of the second component in per cent of the first component. With increasing interstimulus interval (abscissa) the second component predominates the first one. At low frequency stimulation the noise level of the recording system was usually about 10 % of the signal which made the distinction between the initial and the late component difficult.

Prediction of the 'extended two-Ca-store model'

Provided that the parameters of the analogue remained unchanged and that the steady-state filling up of the store S_2 is brought to a close, at short intervals the early component representing the contribution of S_2 is predominant. This component decreases with increasing duration of the interstimulus interval (see Fig. 10 in [10]). After all at long intervals only the second component, the analogon to that portion of the slow inward current which reaches the contractile proteins directly, remains visible.

The effect of ryanodine

The results of experiments using right ventricular papillary muscles can be summarized briefly: Ryanodine inhibits the

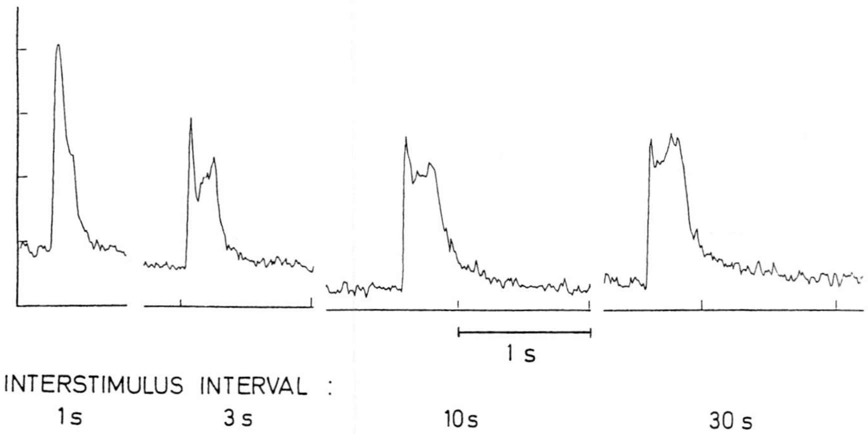

Fig. 6. Principle of an experiment designed to illustrate the metamorphosis of stabilized isoproterenol-induced biphasic calcium transients under steady-state conditions of pacing with different interstimulus intervals. Please note the prevailing phasic component at short intervals and the prevailing tonic component at long intervals. Ordinate: intensity of fluorescence; abscissa: time.

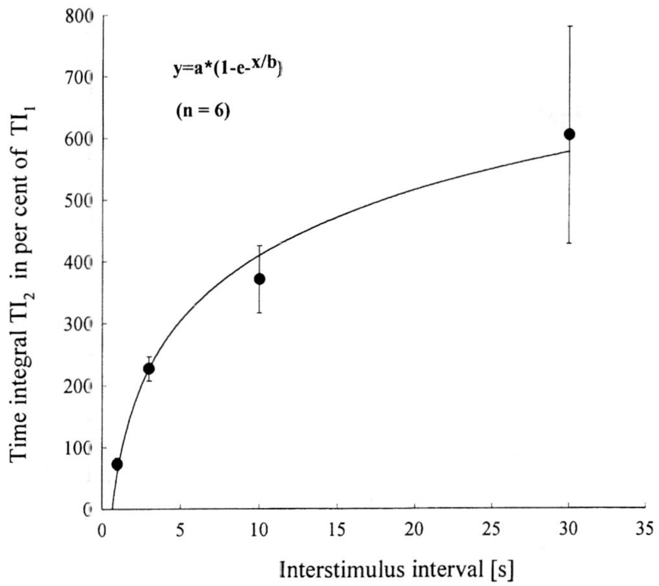

Fig. 7. Diagram showing the influence of the interstimulus interval (abscissa) on the time integral of the second (tonic) component in per cent of the time integral of the first (phasic) component of a stabilized biphasic calcium transient. The bars represent the S.D. of the mean. Number (n) of cells are in parentheses.

first component of biphasic contractions and the two early components of threephasic contractions. The survey of records in Fig. 8 illustrates the effect of replacing the standard medium by a solution containing 1 μmol l^{-1} ryanodine. The uppermost trace shows the time in sec, the middle trace is the isometric twitch. Deflections of the lowermost trace indicate stimulus artifacts. The numbers at the side of each record indicate the time after the superfusion of the papillary muscle with ryanodine-containing solution. This series of eight records shows some transitional stages of the change in shape from isoproterenol-induced biphasic to a monophasic twitch representing only the second (late) component of the contraction (compare this figure with Fig. 1 in [22] and with Fig. 1D in [24]). A similar pattern of behaviour can be obtained from isoproterenol-induced biphasic calcium transients of isolated guinea-pig myocytes after ryanodine (1 μmol l^{-1}) application. As illustrated in Fig. 9, under steady-state conditions of driving the individual configuration of the biphasic transient was adjusted by a suitable co-ordination of temperature and driving frequency so that the time integral of the fluorescence of the first component amounts between 50–80% of the second component. After application of ryanodine there was usually an initial increase of the two components of the calcium transient to levels significantly above the initial square measure followed by a gradual decrease. Although there was a considerable noise in the records and a large scatter in the results the shift in favour of the second component seems to be unequivocal (see Fig. 9). The quotient K_1/K_2 gradually becomes smaller after the application of ryanodine, and after about 10 min the measuring method of the area beneath K_1 makes no sense.

Prediction of the 'extended two-Ca-store model'

The degree of coupling 'q' between S_1 and S_2, defined as $q = \tau_L/(\tau_K + \tau_L))$ with the coupling time constant τ_K, and the leakage time constant τ_L, varies between 0 (uncoupled stores) and 1 (stores maximally coupled). The calcium concentration within the contractile compartment decreases monotonously with increasing coupling time constant (see Fig. 12 in [10]) and in case that the stores are uncoupled (equivalent to $\tau_K = \infty$) only the direct flowing calcium becomes effective, and the biphasic transient degenerates to the late component.

There exists a wealth of evidence that depolarization of the mammalian cardiac cell into the range of the plateau of the action potential under physiological conditions causes a large, rapidly rising plasmic calcium transient which could be shown to be reduced by the plant alkaloid ryanodine by up to 90%. From comparative studies in single cardiac SR release channels incorporated in lipid bilayers [25] and in rat ventricular muscle cells [26] it seems safe to conclude that ryanodine does not block the slow inward current but at concentrations above 100 μmol l^{-1} ryanodine leads to a transition from normal gating to a stable subconductance level of the Ca-release channel [4, 27–33].

Thus it is to be expected that ryanodine in appropriate concentrations not only decreases the intracellular calcium transient but also causes a conversion of the biphasic to a monophasic calcium transient in cardiac myocytes and, in corresponding way, the conversion of the biphasic to a monophasic twitch of multicellular preparations. For high concentrations of ryanodine and sufficient time of action the model predicts that the late component of the calcium transient and the late component of biphasic tension development remains.

The effect of thapsigargin on the second component of biphasic intracellular calcium transients

The observations to be described here provide information about the nature of the mechanism causing an increase of the fluorescence time integral of the tonic component of intracellular calcium transients not only in comparison with the phasic component but also absolutely with increasing interstimulus interval. Experimental results from 8 animals are presented in Fig. 10. This figure displays a comparison of time integrals of the tonic component of biphasic calcium transients at interstimulus intervals of 1 and 300 sec. The 300 sec value is significantly different from the 1 sec controls recorded before and after the 300 sec interstimulus interval

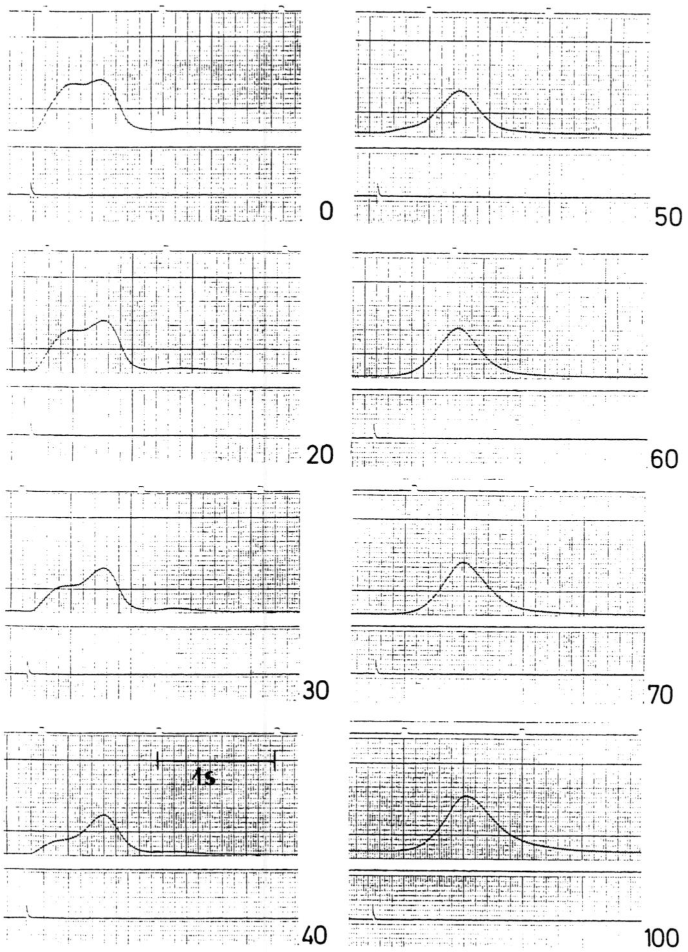

Fig. 8. Records showing the step by step transition from a stabilized biphasic contraction of a papillary muscle to a monophasic twitch after ryanodine (1 μmol l^{-1}) application. The numbers at the right of each record indicate the time in minutes after the beginning of the superfusion with ryanodine-containing solution. Ordinates: developed tension; abscissa: time. Interstimulus interval 5 sec.

series (p = 1%; Student's test).

A possible interpretation of this result would be the 'use dependence' of the calcium removal from the cytosol. The following observations on the effect of thapsigargin support this possibility to some degree.

In the preceding sections we have concerned largely with measuring the relative proportion of the phasic and tonic component of intracellular calcium transients. Using the absolute value of the time integral of the tonic component as a standard of assessment for the functioning of the SR-calcium pump, it appeared to be necessary to take the *in vitro* cell aging into account for each individual population of isolated myocytes which were transferred from the storage fluid to the chamber. Figure 11 relates the time integral of the tonic component in per cent of the initial value to the time in seconds after the cell transfer. From such 'calibration curves' the multiplication factors for the correction of the fluorescence time integrals for any particular cell suspension can be determined. Figure 12 shows how the isolated tonic component of the intracellular calcium transient behaves after the application of thapsigargin. The plot in Fig. 13 illustrates the effect of thapsigargin upon the tonic component of calcium transients recorded from myocytes of twelve animals. The ordinate represents the time integral of the tonic component in per cent of the control, the abscissa indicates the time after thapsigargin application. Although there is a fair degree of scatter of the results, the interesting point that emerges from this graph is that the half-life period to the new steady state of calcium overload amounts about two minutes.

Prediction of the 'extended two-Ca-store model'

The reuptake of Ca^{2+} into the sarcoplasmic reticulum mediated by an ATP-dependent calcium pump (SR-calcium ATPase) which seems to be concentrated in the longitudinal component of the SR [34], can be characterized in the 'extended two-Ca-store model' by a time constant τ_p [10]. From

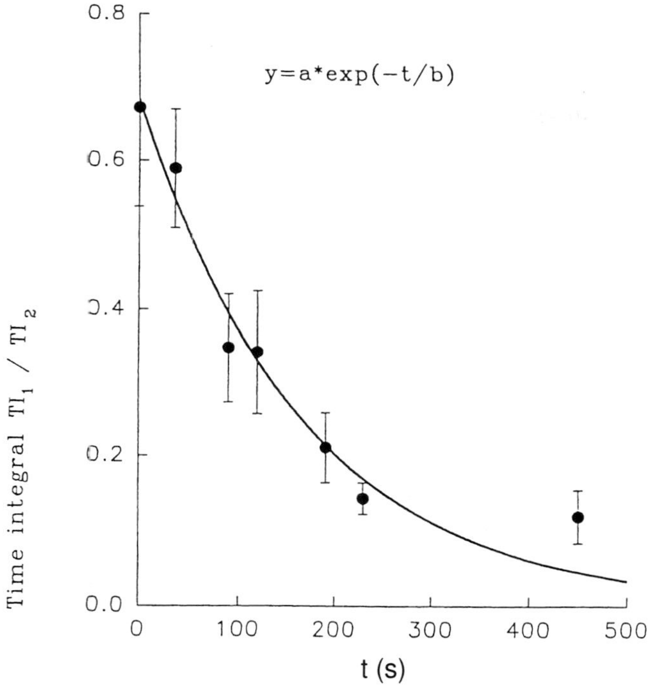

Fig. 9. The quotient: time integral of the phasic component divided by the time integral of the tonic component of stabilized biphasic calcium transients as function of the time in s after ryanodine superfusion of the myocyte. The starting value of the time integral of the first component was stabilized between 50–80% of the time integral of the second component. Interstimulus interval 2 sec; n = 8.

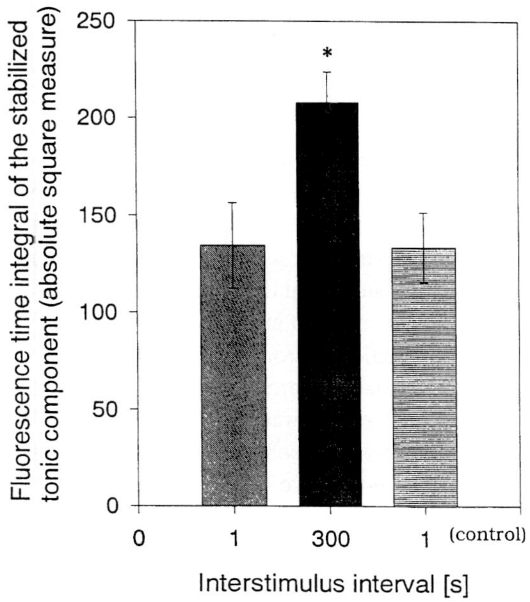

Fig. 10. Time integral of isolated, stabilized tonic component of intracellular calcium transients measured at two steady-state interstimulus intervals (1 and 300 sec). Ordinate scale in absolute square measure; eight myocytes. Value given are mean ± standard deviation. Statistical significance: p = 1% (Student's test).

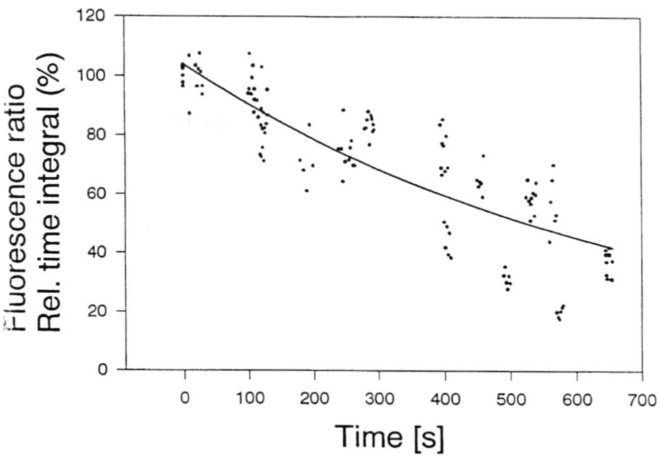

Fig. 11. An example of an *in vitro* cell aging 'calibration curve' gained from a particular population of enzymatically isolated myocytes. The ordinate scale means ratiometrically (Fluo-3/Fura-red) recorded fluorescence time integrals of the tonic components of biphasic intracellular calcium transients expressed in per cent of the initial value. The abscissa shows the time after the cell transfer from the storage solution to the chamber. Nonlinear regression ($y = ae^{-bx}$), a = 104.0396, b = 0.00139. Interstimulus interval 2 sec.

analog simulation of the rate of calcium removal from the contractile element, summarizing the phosphorylation of phospholamban, the binding to calsequestrin and other possible events [35–37] the conclusion to be drawn is (1) that the appearance of the phasic component is most probable at a small τ_p corresponding to a high activity of the SR-calcium ATPase, (2) both components become equal at a distinct time constant, (3) with increasing time constant the time integral of the tonic component increases not only compared to the phasic one but also absolutely, and (4) that a further increase of τ_p results in an incomplete calcium removal (see Fig. 13A–F in [10]). All in all, thapsigargin leads to a relative as well as absolute augmentation of the tonic component of biphasic intracellular calcium transients. This is a puzzling observation for which there is no adequate explanation at this time. There is some evidence, however, to support conclusion that it may be related to an increase of τ_p.

Discussion

There is a considerable literature on the isoproterenol-induced cell necrotization manifesting itself as swelling and clearing of the subsarcolemmal region, as dilatation of the sarcoplasmic reticulum, homogenization of the myofilaments, as herniation of the adjacent cells through the intercalated discs, displacement of the mitochondrial cristae etc. [38–41]. There is, however, little information as to functional peculiarities of the calcium-overload stage. The present investigation has been directed at elucidating the functional alterations observ-

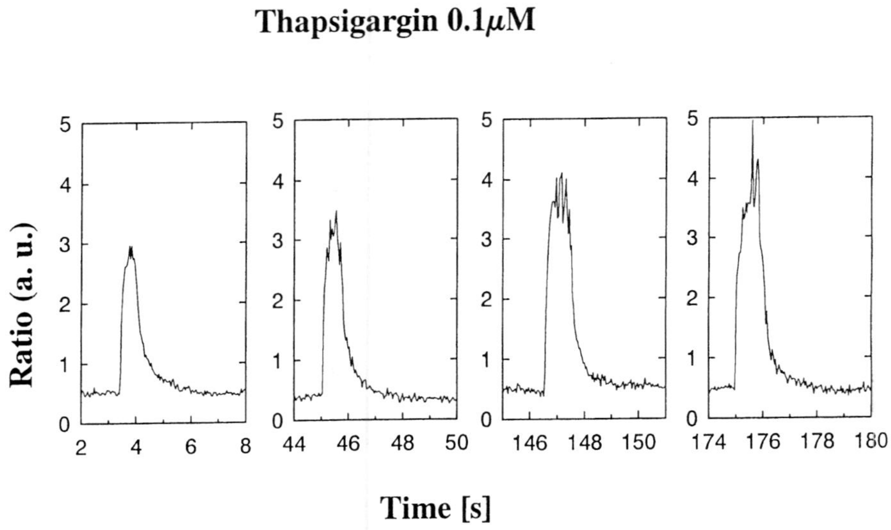

Fig. 12. Changes in size of an isolated tonic component of the calcium transient under the influence of 0.1 μmol l^{-1} thapsigargin. Electrically stimulated ventricular myocyte of a guinea-pig. Ordinate: Fluorescence ratio (Fluo-3/Fura-red). Thapsigargin added at t = 0 (sec). Interstimulus interval 5 sec.

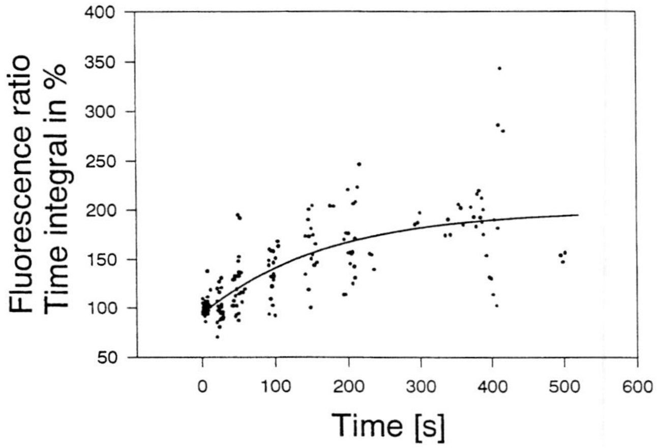

Fig. 13. Thapsigargin effect on the tonic component of intracellular calcium transients; myocytes from 12 animals. Relationship between the time integral of the tonic component of the transient measured ratiometrically (Fluo-3/Fura-red) expressed in per cent of the steady state value in control solution and the time in seconds after thapsigargin application. Nonlinear regression ($y = ae^{-\frac{x}{b}} + c$), a = 11.4139, b = 114.9541, c = 101.6500. Interstimulus intervals: 2 sec, 5 sec.

able in calcium-overloaded cardiac cells and isolated calcium-overloaded multicellular cardiac preparations.

The gradual damage of the cardiac cell is accompanied by a splitting of the excitation-contraction coupling process into two or three components becoming apparent as biphasic or threephasic contractions [9, 15, 19, 21, 22, 42]. The precondition for an analysis of isoproterenol-induced alterations of the contractility is the possibility to applicate long-lasting stimulation patterns as well as well-aimed changes of the milieu conditions, thus, the stabilization of the particular functional state.

Thinking along these lines, we first of all concentrated our attention on the stabilization of the shape of multiphasic twitches. The experimental trick was a fine co-ordination of (1) the isoproterenol concentration (between 1–4 μmol l^{-1}), (2) the temperature of the bathing medium (using PELTIER-batteries), (3) the interstimulus interval and (4) the addition of calcium agonists.

The staircase phenomenon, described by H. P. Bowditch (1871) is changed in characteristic manner in case cardiac tissue comes from 'normal' *in vitro* conditions into a calcium overload state. Normally, if a steady-state series of driving pulses with constant interstimulus intervals is interrupted by a rest longer than the steady-state interstimulus interval, the first contraction of a mammalian papillary muscle is generally increased (post-rest potentiation). During the following series of twitches the developed tension increases in a stepwise manner to a certain maximum while maintaining the monophasic shape of the twitch. There are several pieces of evidence which might lead one to assume that the positive staircase following the potentiated post-rest beat is caused by increased filling of (two) intracellular calcium stores [6, 43, 44].

In calcium-overloaded mammalian cardiac tissue the typical post-rest staircase is characterized by a change of shape of the multiphasic twitch: Each individual component passes through its particular positive staircase reaching a maximum at steady-state conditions. Remarkably, after an interruption of the pacing rhythm the stepwise increase of the first component exceeds the increase of the second (third) component, and consequently the curve, relating the quotient F_{K2}/F_{K1} (force of the late component divided by the force of the early component) to the time following the rest, in a convex manner referring to the abscissa, gradually approaches the steady-state value of the stabilized contraction before the rest.

The original idea of the two-Ca-store concept, assuming

for simplicity two states of functioning, gives neither a satisfactory explanation for the postextrasystolic potentiation under normal conditions nor for the staircase behaviour of multiphasic contractions. The hypothesis that during the plateau phase of the action potential calcium reaches the contractile system directly as well as indirectly (via S_1, S_2) appears to be supported by the results of the present work and by observations occasionally made by other authors.

It seems us to be useful, for a better understanding of the contraction behaviour of calcium overloaded cardiac tissue, to applicate the same stimulation programmes in multicellular preparations and in isolated myocytes and to compare both the contractile answer and the fluorescence signals with predictions of the 'extended two-Ca-store model'.

The first argument supporting the soundness of the functional model mentioned above is the accordance of the analogue simulations of the calcium concentrations within the contractile element (CE) (see Fig. 11D in [10]) with the time course of the post-rest staircase both from biphasic fluorescence signals of myocytes and from biphasic twitches of right ventricular papillary muscles, presuming that the time integral of each component of the calcium transient can serve as a homologon of the developed tension. The second argument supporting the assumption of a direct calcium flux from sarcolemma to the contractile proteins in addition to the roundabout way via the storage structures results from a comparison both of the biphasic fluorescence signals and of biphasic contractions with analogously simulated CE-calcium concentrations at different interstimulus intervals and unchanged parameters of the hardware analogue. The good correlation between the behaviour of biphasic fluorescence signals of isolated myocytes and biphasic twitches of papillary muscles at different driving intervals appear to be a possible indication of a causal relationship between the phases of the myoplasmic calcium concentration and the components of the force development.

One point, however, seems to us to be sufficiently important to merit a particular mention. If the records of the calcium transients are compared with the analogue simulations of the CE-calcium concentration at different steady-state interstimulus intervals the question arises: Why the fluorescence time integral of the second component increases not only relatively but also absolutely with increasing interstimulus interval? A possible explanation in terms of the 'extended-two-Ca-store' concept would be that with increasing time constant τ_P characterizing the calcium elimination from the CE not only the late component becomes predominant as compared to the early one but also the area beneath the late component increases absolutely (see Fig. 13 in [10]). It was the idea, to test this point by a application of thapsigargin which is considered to be a potent specific inhibitor of the SR Ca^{2+}-ATPase in cardiac cells [45–47]. The greater values of the fluorescence time integrals after thapsigargin application may be explained by the simple assumption that the time constant of the calcium reuptake from the contractile element into the sarcoplasmic reticulum increases under the influence of this inhibitor of the SR Ca^{2+} pump.

The third support for a plausible interpretation of the Bowditch staircase observable in calcium overloaded cells in terms of the 'extended two-Ca-store concept' is provided by the observation that the plant alkaloid ryanodine, which is believed to modulate the calcium release from both skeletal and cardiac muscle cells, reduces only the first component of the biphasic calcium transient as well as of the biphasic contraction.

There exists a wealth of pharmacological and physiological evidence that ryanodine at µmolar concentrations binds to low affinity sites in the SR-calcium release channel, thereby causing a stable subconductance state and a decrease of the open state probability of the channel [4, 25, 32, 48].

Such a mechanism influencing the SR-calcium release would explain the negative ryanodine effect exclusively on the first component.

It is important to keep in mind that just as the original idea of the two-Ca-store hypothesis – the 'extended concept' represents a functional, not a formal model. There is however, one essential point which must be considered regarding conclusions drawn from an apparently plausible model concept: The mathematical abstraction is based on strong simplification. In particular this applies to the assumption of a constant rectangular shape of the action potential. Actually the heart muscle cell 'solves' all the time a set of additional differential equations at the cell membrane and, as has been proved, a change of the shape of the action potential can have sensible effects on various partial events of the excitation-contraction coupling process. For instance, the L-type calcium current I_{Ca} is modulated by Ca^{2+} in a complex way [49, 50], both its inactivation and potentiation depend on $(Ca^{2+})_i$ [51, 52], its peak amplitude is potentiated and its activation is retarded by repetitive I_{Ca} activation [53, 54, 55], the L-type channels seem to be in close functional association with ryanodine receptors [56] etc.

As we take it, the results of the experiments described in this paper, despite the strong simplification mentioned above, have accumulated some evidence in favour of a better understanding of the changes in the relative proportion of the individual components in biphasic (multiphasic) twitches and in biphasic (multiphasic) calcium transients under various milieu conditions and pacing programmes.

Complementary experiments using microelectrodes in the case of the papillary muscles and the whole cell patch clamp technique in the case of the myocytes are in preparation.

Acknowledgement

The authors wish to thank Mrs. Monika Schmidt for excellent technical assistance. Financial support for this work (from the Deutsche Forschungsgemeinschaft and from the BMFT) is acknowledged with graditude.

References

1. Stern MD, Lakatta EG: Excitation-contraction coupling in the heart: the state of the question FASEB J 6: 3092–3100, 1992
2. Callewaert B: Excitation-contraction coupling in mammalian cardiac cells Cardiovasc Res 26: 923–932, 1992
3. Barry WH, Bridge JHB: Intracellular calcium homeostasis in cardiac myocytes Circulation 87: 1806–1815, 1993
4. Ogawa Y: Role of ryanodine receptors Crit Rev Biochem Molec Biol 29: 229–274, 1994
5. Nigglie E, Lipp P: Subcellular features of calcium signalling in heart muscle; what do we learn? Cardiovasc Res 29: 441–448, 1995
6. Wussling M, Szymanski G: Ein Zwei-Ca-Speicher-Modell zur qualitativen Beschreibung von Potentiationserscheinungen am Kaninchenpapillarmuskel. Nova Acta Leopoldina 211/38: 141–173, 1973
7. Koester G: Anwendung des n-Speicher-m-Phasen-Modells auf intrazelluläre Ca-Flüsse im Herzmuskel. Habilitationschrift Martin Luther Universität Halle-Wittenberg Fak für Naturwissenschaften, 1979
8. Bowditch HP: Über die Eigenthümlichkeiten der Reizbarkeit, welche die Muskelfasern des Herzens zeigen. Berichte über die Verhandlungen der königlich sächsischen Gesellschaft der Wissenschaften zu Leipzig. Mathemat.-Phys. Classe 23: 652–689, 1871
9. Szymanski G, Zett L: Factors influencing the shape and the potentiation behaviour of monophasic, biphasic and multiphasic contractions of the guinea-pig ventricular myocardium. Zool Jb Physiol 95: 109–133, 1991
10. Wussling M, Szymanski G: Simulation by two calcium store models of myocardial dynamic properties: potentiation, staircase, and biphasic tension development. Gen Physiol Biophys 5: 135–152, 1986
11. Bravený P, Sumbera J: Biphasic activation of the myocardial contraction Physiol Bohemoslov 21: 73–74, 1972
12. Coraboeuf E: Membrane electrical activity and double component contraction in cardiac tissue J Mol Cell Cardiol 6: 215–225, 1974
13. Allen DG, Jewell BR, Wood EH: Studies of the contractility of mamalian myocardium at low rates of stimulation J Physiol 254: 1–17, 1976
14. Beresewicz A, Reuter H: The effects of adrenaline and theophilline on action potential and contraction of mammalian ventricular muscle under 'rested-state' and 'steady-state' stimulation. Arch Pharmacol 301: 99–107, 1977
15. Seibel K, Karema E, Takeya K, Reiter M: Effect of noradrenaline on a early and late component of myocardial contraction. Naunyn Schmied Arch Pharmacol 305: 65–74, 1978
16. Bogdanow KY, Zakharow Sl, Rosenshtraukh LV: The origin of two components in contraction of guinea-pig papillary muscle in the presence of noradrenaline Can J Physiol Pharmacol 57: 866–872, 1979
17. Endoh M, Motomura S: Inhibition by theophylline of the early component of canine ventricular contraction Am J Physiol 11: H349–H358, 1982
18. King BW, Bose D: Mechanism of biphasic contractions in strontium-treated ventricular muscle Circ Res 52: 65–75, 1983
19. Reiter M, Vierling W, Seibel K: Exitation-contraction coupling in rested-state contractions of guinea-pig ventricular myocardium. Naunyn Schmied Arch Pharmacol 325: 159–169, 1984
20. Honorè E, Challice CE, Guilbault P, Dupuis B: Two components of contraction in guinea-pig papillary muscle Can J Physiol Pharmacol 64: 1153–1159, 1986
21. Malécot CO, Bers DM, Katzung BG: Biphasic contractions induced by milrinone at low temperature in ferret ventricular muscle: Role of the sarcoplasmic reticulum and transmembrane calcium influx Circ Res 59: 151–162, 1986
22. Schüttler K, Szymanski G: Threephasic contractions in mammalian myocardium. Isolation and identification of a late component. Implications for rested state contractions. Biomed Biochim Acta 48: 558–564, 1989
23. Kao RL, Christman EW, Chu SL, Krauhs JM, Tyers GFO, Williams EH: The effects of insulin and anoxia on the metabolism of isolated, mature rat cardiac myocytes. Arch Biochem Biophys 203: 587–599, 1980
24. Kondo N: Voltage – dependence of ryanodine-sensitive component of contraction in A23187- and isoprenalin-treated cardiac muscles. Br J Pharmacol 98: 51–60, 1989
25. Rousseau E, LA Dine J, Liu QY, Meissner G: Activation of the calcium release channel of skeletal muscle sarcoplasmic reticulum, by caffeine and related compounds. Arch Biochem Biophys 267: 75–86, 1988
26. Callewaert B, Cleemann L, Morad M: Epinephrine enhances Ca^{2+} current-regulated Ca^{2+} release and Ca^{2+} reuptake in rat ventricular myocytes. Proc Natl Acad Sci USA 85: 2009–2013, 1988
27. Nayler WG, Daile P, Chipperfield D, Gan K: Effect of ryanodine on calcium in cardiac muscle. Am J Physiol 219: 1620–1626, 1970
28. Fairhurst AS, Hasselbach W: Calcium efflux from a heavy sarcotubular fraction. Effects of ryanodine, caffeine and magnesium. Eur J Biochem 13: 504–509, 1970
29. Jones LR, Besch HR, Sutko JL, Willerson JT: Ryanodine- induced stimulation of net Ca^{2+} uptake by cardial sarcoplasmic reticulum vesicles. J Pharmacol Exp Ther 209: 48–55, 1979
30. Fabiato A: Effects of ryanodine in skinned cardiac cells. Fed Proc 44: 2970–2976, 1985
31. Smith JS, Coronado R, Meissner G: Single channel measurements of the calcium release channel from skeletal muscle sarcoplasmic reticulum. J Gen Physiol 88: 573–588, 1986
32. Meissner G: Ryanodine activation and inhibition of the Ca^{2+} release channel of sarcoplasmic reticulum. J Biol Chem 261: 6300–6306, 1986
33. Smith JS, Coronado R, Meissner G: Sarcoplasmic reticulum contains adenine nucleotide-activated calcium channels. Nature 316: 446–449, 1985
34. Jorgensen AO, Shen ACY, Daly P, MacLennan DH: Localisation of Ca^{2+} + Mg^{2+}-ATPase of the sarcoplasmic reticulum in adult rat papillary muscle J Cell Biol 93: 883–892, 1982
35. Simmerman HKB, Theibert JL, Wegener AD, Jana KR: Sequence analysis of phospholamban: Identification of phosphorylation sites and two major structural domains J Biol Chem 261: 1333–1341, 1986
36. Cala SE, Jones LR: Phosphorylation of cardiac and skeletal muscle calsequestrin isoforms by casein kinase II J Biol Chem 266: 391–398, 1991
37. Steenaart NAE, Ganim JR, DiSalvo J, Kranias EG: The phospholamban phosphatase associated with cardiac sarcoplasmic reticulum is a type 1 enzyme Arch Biochem Biophys 293: 17–24,1992
38. Fleckenstein A, Janke J, Döring HJ, Leder O: Die intrazelluläre Überladung mit Kalzium als entscheidender Kausalfaktor bei der Entstehung nicht – coronarogener Myocard – Nekrosen. Verh dtsch Ges Kreisl Forsch 37: 345–353, 1971
39. Fleckenstein A, Janke J, Döring HJ, Leder O: Myocardial fiber necro-

sis due to intracellular Ca overload – a new principle in cardiac pathophysiology. In: Recent Advances in Studies on Cardiac Structure and Metabolism. Myocardial Biology 4: 563–580, University Park Press, Baltimore, 1974
40. Fleckenstein A: Myocardschutz durch Calziumantagonisten In: H. Just, Tschirkov A und V. Schlosser (Hrsg.). Kalziumantagonisten zur Kardioplegie und Myokardprotektion in der offenen Herzchirurgie. Georg Thieme Verlag Stuttgart-New York, 1982
41. Fleckenstein A: Calcium antagonism: History and prospects for a multifaceted pharmacodynamic principle In: L.H. Opie (ed.). Calcium Antagonists and Cardiovascular Disease. Raven Press, New York: 1984
42. Kondo N: Electromechanical effects of A23187 on guinea-pig ventricular muscle: a dual action of A231 87. J Mol Cell Cardiol 18: 907–915, 1986
43. Kruta V, Bravený P: Possible mechanisms involved in potentation phenomena. Paired pulse stimulation of the heart. P.F. Cranfield, B.F. Hoffman (eds). Plenum Press New York and London, 1968, pp 53–64
44. Wohlfart B: Interval-strength relations of mammalian myocardium interpreted as altered kinetics of activator calcium during the cardiac cycle. Thesis University of Lund, Sweden. 1982
45. Kirby MS, Sagara Y, Gaa S, Inesi G, Lederer WJ, Rogers TB: Thapsigargin inhibits contraction and Ca^{2+}-transient in cardiac cells by specific inhibition of the sarcoplasmic reticulum Ca^{2+} pump J Biol Chem 267: 1245–1251, 1992
46. Wrzosek A, Schneider S, Grueninger S, Chiesi M: Effects of thapsigargin on cardiac muscle cells Cell calcium 13: 281–292, 1992
47. Janczewski AM, Lakatta EG: Thapsigargin inhibits Ca^{2+} uptake, and Ca^{2+} deplets sarcoplasmatic reticulum in intact cardiac myocytes Am J Physiol 265: H517–H522, 1993
48. Nagaski K, Fleischer S: Ryanodine sensitivity of the calcium release channel of sarcoplasmic reticulum. Cell calcium 9: 1–7, 1988
49. Eckert R, Chad J: Inactivation of Ca channels. Progress in Biophysics and Molecular Biology 44: 215–267, 1984
50. Lee KS, Marban E, Tsien RW: Inactivation of calcium channels in mammalian heart cells: joint dependence on membrane potential and intracellular calcium. J Physiol 364: 395–411, 1985
51. Isenberg G: The slow inward current component under the influence of modified $[Ca^{2+}]_i$. Pflügers Archiv 371: 61–69, 1977
52. Marban E, Tsien RW: Enhancement of calcium current during digitalis inotropy in mammalian heart: positive feed-back regulation by intracellular calcium? J Physiol 329: 589–614, 1982
53. Lee KS: Potentiation of the calcium-channel currents of internally perfused mammalian heart cells by repetitive depolarization. Proc Natl Acad Sci USA 84: 3941–3945, 1987
54. Fedida D, Noble D, Spindler AJ: Use-dependent reduction and facilitation of Ca^{2+}-current in guinea-pig myocytes. J Physiol 405: 439–460, 1988
55. Zygmunt AC, Maylie J: Stimulation-dependent facilitation of the high threshold calcium current in guinea-pig ventricular myocytes. J Physiol 428: 653–671, 1990
56. Sham JSK, Cleemann L, Morad M: Functional coupling of Ca^{2+} channels and ryanodine receptors in cardiac myocytes. Proc Natl Acad Sci USA 92: 121–125, 1995

Molecular and Cellular Biochemistry **160/161**: 27–32, 1996.

Cooperative interaction between Ca^{2+} binding sites in the hydrophilic loop of the Na^{+}-Ca^{2+} exchanger

Dmitri O. Levitsky,[1,2] Bodvaël Fraysse,[1] Claude Leoty,[1] Debora A. Nicoll[2] and Kenneth D. Philipson[2]

[1]*Laboratory of General Physiology, URA CNRS 1340, University of Nantes, Nantes, France;* [2]*Departments of Physiology and Medicine and the Cardiovascular Research Laboratory, UCLA School of Medicine, Los Angeles, CA 90095-1760, USA*

Abstract

A high affinity Ca^{2+}-binding domain which is located in a middle portion of the large intracellular loop of the Na^{+}-Ca^{2+} exchanger contains two highly acidic sequences, each characterized by three consecutive aspartic acid residues (Levitsky DO, Nicoll DA, and Philipson KD (1994) J Biol Chem 269: 22847–22852). This portion of the protein provides secondary Ca^{2+} regulation of the exchanger activity. To determine number of Ca^{2+} binding sites participating in formation of the high affinity domain, we isolated polypeptides of different lengths encompassing the domain and measured $^{45}Ca^{2+}$ binding. The fusion proteins containing the high affinity domain were obtained in a Ca^{2+}-bound form and as evidenced by shifts in there mobility in SDS-polyacrylamide gels after EGTA treatment. The Ca^{2+} binding curves obtained after equilibrium dialysis reached saturation at 1 μM free Ca^{2+}, Kd value being approx. 0.4 μM. The Ca^{2+} binding occured in a highly cooperative manner. Upon saturation, the amount of Ca^{2+} ion bound varied from 1.3–2.1 mol per mol protein. Proteins with an aspartate in each acidic sequence mutated lacked the positive cooperativity, had lower Ca^{2+} affinity and bound two to three times less Ca^{2+}. Na^{+}-Ca^{2+} exchangers of tissues other than heart though different from the cardiac exchanger by molecular weight most likely possess a similar Ca^{2+} binding site. It is concluded that, by analogy with Ca^{2+} binding proteins of EF-type, the high Ca^{2+}-affinity domain of the Na^{+}-Ca^{2+} exchanger is comprised of at least two binding sites interacting cooperatively. (Mol Cell Biochem **160/161**: 27–32, 1996)

Key words: Na^{+}-Ca^{2+} exchanger in different tissues, high affinity Ca^{2+} binding

Introduction

The Na^{+}-Ca^{2+} exchanger plays a major role in regulation of Ca^{2+} concentration in cardiac myocytes. This 120 kDa sarcolemmal protein provides exchange of three Na^{+} ions for one Ca^{2+} in the direction dependent on the contractile state of the heart (for recent reviews, see [1–3]). It is coded by two genes NCX1 [4] and NCX2 [5] and present, though in smaller amounts than in heart, in other tissues [6]. The exchanger protein from dog heart contains a high affinity Ca^{2+} binding region which is distinct from the transport site and is located in a large intracellular loop [7]. This region participates in providing a 'secondary Ca^{2+} regulation' of the exchanger activity [8]. An EGTA (ethylene glycol bis(β-aminoethyl ether)-N,N,N',N'-tetraacetic acid)-induced shift in electrophoretic mobility of fragments of the exchanger from dog heart indicated that a change in the protein conformation occured upon binding of Ca^{2+} ions [7]. Single site mutation and deletion analyses of the portions of the exchanger loop evidenced location of the high affinity Ca^{2+} binding site inside a segment of 154 amino acids. This stretch is characterized by two acidic sequences. Neither of them fit a consensus structure of the EF-hand type domains of Ca^{2+} binding proteins [9] which bind cooperatively two to four Ca^{2+} ions.

Address for offprints: DO Levitsky, Laboratoire de Physiologie Générale, URA CNRS 1340, Université de Nantes, 2 rue de la Houssinière, F-44087 Nantes Cedex 03, France

Kofuji *et al.* [6] demonstrated that alternative splicing of a single gene NCX1 could generate a maximum of 32 exchanger isoforms. The variable part of the NCX1 was found downstream the high affinity Ca^{2+}-binding site. If a motif with three consecutive aspartates (Fig. 1) is conserved in all isoforms of the exchanger protein as has been shown for NCX1 from brain [10] and kidney [11] and NCX2 from rat brain and skeletal muscle [5] the regulatory Ca^{2+} binding domain may be a characteristic feature of a typical Na^{+}-Ca^{2+} exchanger.

The aim of the study was to quantify Ca^{2+} binding to the high affinity site and to verify the presence of the binding site in tissues other than heart. The results presented below indicate a cooperative binding of at least two Ca^{2+} ions upon saturation of the high affinity domain of purified fragments of the cardiac exchanger. The purified exchanger fragments as well as full-length exchangers from crude fractions of rat tissues reveal a considerable shift in mobility in SDS (sodium dodecyl sulfate)-gels upon prior Ca^{2+} ion chelation with EGTA.

Materials and methods

Plasmids

The following exchanger DNA constructs in pBluescript SK+ [7] were used for further subcloning (the numbers correspond to position of amino acid residues): wild type fragments 371–525 and 240–532, the 240–532 with aspartate-447 mutated to valine and aspartate-498 mutated to isoleucine. The plasmids were digested with *Bam*HI and *Kpn*I, the DNA pieces were extracted from agarose gel (Geneclean) and ligated into *Bam*HI-*Kpn*I sites of the vector PQE-31 (Qiagen). Ninety percent of the sequence in the constructs contained the exchanger DNA. The new plasmids were used to transform strain Ml5[pREP4].

Expression and purification of fusion proteins

A QIAexpress system (Qiagen) was used for large scale production of fusion proteins with 6 × Histidine tags. 500 ml of LB-broth containing 100 μg/ml ampicillin and 25 μg/ml kanamycin was inoculated with 10 ml of overnight cultures containing the expression plasmids. The new cultures were grown at 37°C until the A_{600} reached 0.7, then 2 mM IPTG (isopropyl-β-D-thiogalactopyranoside) was added to induce expression of fusion proteins. In 6 h the cells were harvested by centrifugation, frozen and stored at –70°C. The cells were solubilized in 5 ml of the solution containing 6 M guanidine hydrochloride, 0.1 M sodium phosphate, 10 mM Tris-HCl (pH 8.0). The lysate was centrifuged at 10, 000 × g for 15 min and the supernatant was mixed with 8 ml of 50% Ni-NTA resin which had been preequilibrated with the above buffer. The mixture was loaded onto a 4 ml Qiagen column. The column was washed with buffers B and C containing 8 M urea, 0.5 mM $CaCl_2$, 0.1 M Na-phosphate, 10 mM Tris-HCl, pH 8.0 and 6.3, respectively. The washing solution was supplemented with 0.5 mM $CaCl_2$. The proteins were eluted in 3–5 ml of buffer C containing 0.25 mM imidazole and further concentrated to the final volume 50–200 μl using Centricon-SR30 and stored at 4°C.

Electrophoresis and $^{45}Ca^{2+}$ overlay

Laemmli SDS-PAGE (sodium dodecylsulfate-polyacrylamide gel electrophoresis) [12] in 12% gels and electrotransfer were performed as described before [7]. The nitrocellulose blots were washed in the solution containing 60 mM KCl, 10 mM imidazole (pH 7.2), 5 mM $MgCl_2$ and incubated for 10 min in the same medium supplemented with 2 μCi/ml of $^{45}Ca^{2+}$ [13]. The nitrocellulose membranes were further rinsed in 50% ethanol, air-dried and exposed to X-ray film overnight. The blots were further stained with amido black.

371
V S M H E V N T E V A E N D P V S K I F F E Q G T Y Q C L E N C G T V A L T I I R R G G D

L T N T V F V D F R T E D G T A N A G S D Y E F T E G T V V F K P G E T Q K E I R V G I I

447
D D D I F E E D E N F L V H L S N V K V S S E A S E D G I L E A N H V S A L A C L G S P S T
V

498 525 532
A T V T I F D D D H A G I F T F E E P V T H V S E S I G I M I V K V L R T S G A R G N V I V P
I

Fig. 1. High-affinity Ca^{2+}-binding site of Na^{+}-Ca^{2+} exchanger from canine dog heart. Substitutions of aspartates in two acidic sequences (underlined) of the construct 240–532 yielded mutants D447V and D498I.

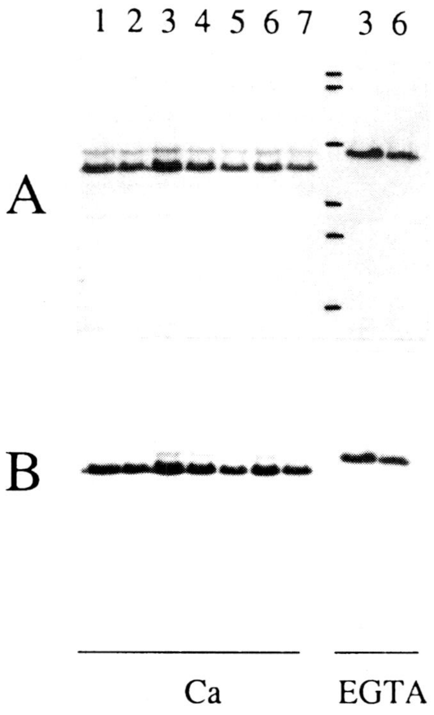

Fig 2. EGTA-induced shift in electrophoretic mobility of purified fusion protein 240–532. The fusion protein containing a 6 histidine tag was isolated and purified as described in Materials and methods. Fractions of 0.6 ml were eluted from a Ni-resin and applied to the 12% SDS-polyacrylamide gel (1–10 µl per well, fractions 1–7, respectively) The electrophoresis sample buffer was supplemented either with 1 mM $CaCl_2$ or 1 mM EGTA. (A) Amido black staining of the nitrocellulose membrane. Positions of mol. wt. markers (106, 80, 49, 33, 28, and 17 kDa) are indicated. (B) $^{45}Ca^{2+}$ overlay.

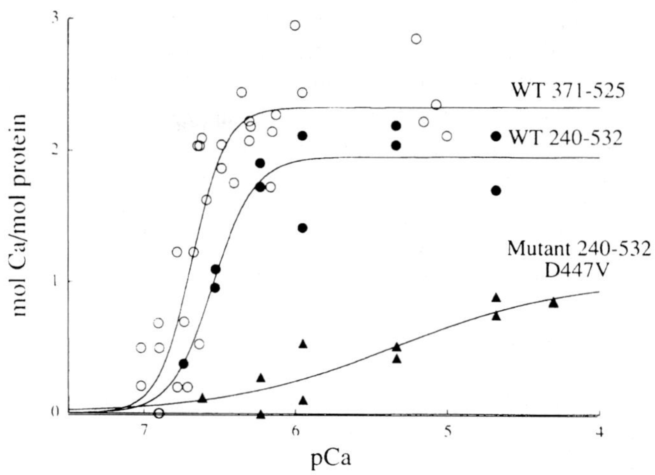

Fig. 3. Free Ca^{2+} dependence of $^{45}Ca^{2+}$ binding to wild type and mutated fragments of the canine heart Na^+-Ca^{2+} exchanger. Equlibrium dialysis of the polypeptides is described in Materials and methods. Curves were drawn after simulation using the Hill equation.

Table 1. Parameters of Ca^{2+} binding to purified fragments of wild type and mutated Na^+-Ca^{2+} exchanger

Construct	mol Ca^{2+}/mol protein	$p^{Ca}50$	^nH
WT 371–525	2.3	6.7	3.8
WT 240–532	2.0	6.5	3.1
Mutant 240–532 D447V	1.0	5.3	0.75
Mutant 240–532 D498I	0.73	6.1	0.83

Equilibrium dialysis

One to two mg of purified protein in 25–250 µl were added to a medium containing 60 mM KCl, 20 mM imidazole (pH 7.2), 5 mM $MgCl_2$, 110 µM $^{45}Ca^{2+}$ (vol. final 4 ml). Aliquots of 480 µl were added to test tubes containing different concentrations of EGTA. The protein solution was added in duplicates to one side of an Equilibrium Microvolume Dialyser module (Hoefer Scientific Instruments, CA). The samples were dialysed overnight at room temperature against the EGTA-containing solutions. Aliquots were taken from both sides for scintillation counting and protein determination. Concentrations of free Ca^{2+} were calculated according to the program of Fabiato and Fabiato [14]. Protein concentration was determined using Lowry assay [15], biuret technique [16], Bio-Rad Protein Assay based on the method of Bradford [17]. Bovine serum albumin and gamma globulin were used as standards. To determine protein concentration according to the amino acid composition, the samples were dialysed against 10 mM NH_4HCO_3 (pH 8.0). Average values obtained by the above methods were taken for calculating moles Ca^{2+} bound per mole protein.

Preparation of homogenates and Western blots

Hundred to three hundred mg of fresh rat tissues were homogenized by a Wheaton glass homogenizer at 0°C in 0.5–1 ml of the medium containing 20 mM HEPES (pH 7.0), 1 mM sodium azide, 1 mM PMSF (phenylmethylsulfonyl fluoride), 1 µg aprotinin and 1 mM DTT (dithiothreitol). The homogenate was centrifuged for 30 min at 2000 g. Supernatant was used immediately for separation in 9% SDS-polyacrylamide gel or frozen and stored at -70°C. Electrotransfer of the proteins to nitrocellulose membrane (Bio-Rad) was done according to Towbin *et al.* [18]. The blot was preincubated with a monoclonal antibody R3F1 against dog heart Na^+-Ca^{2+} exchanger [19] and treated with peroxidase linked anti-mouse Ig (Amersham). The reaction was revealed by ECL Western blotting technique (Amersham).

Results

Quantification of Ca^{2+} binding in purified fragments of dog heart Na^+-Ca^{2+} exchanger

Fragments of wild type Na^+-Ca^{2+} exchanger and mutants of the construct 240–532 (Fig. 1), were isolated as fusion proteins from inclusion bodies in the presence of Ca^{2+} ions. As shown before [7], EGTA added to the sample buffer before SDS-PAGE affects electrophoretic mobility of high affinity Ca^{2+}-binding fragments of the exchanger. This property of the samples allowed us to assess influence of high concentrations of guanidine and urea on functional integrity of the binding domain. Figure 2 shows amido black staining (A) and $^{45}Ca^{2+}$ overlay (B) of a nitrocellulose blot after separation by SDS-PAGE and electrotransfer of fractions of the fusion protein 240–532 eluted from the column in the presence of 0.5 mM $CaCl_2$. Each of the samples gave a major lower band representing the Ca^{2+}-bound form of the protein [4] and a weak band of lower mobility. When the electrophoresis sample buffer was supplemented with 1 mM EGTA, the protein migrated with lower mobility.

The purified fusion proteins were dialysed under equilibrium conditions against media containing $^{45}Ca^{2+}$-EGTA buffers at different free Ca^{2+} concentration. The curves of Ca^{2+} dependence of the Ca^{2+} binding to the wild type proteins 371–525 and 240–532 reached saturation at 1 μM free Ca^{2+} with Kd values 0.2–0.3 μM and two moles Ca^{2+} bound per mol protein (Fig. 3). A fusion protein 240–532 with Asp-447 (the first acidic portion) mutated to Val revealed a rightward shift in the affinity and less Ca^{2+} binding upon saturation of the binding site (Fig. 3). The data on equilibrium dialysis of above fusion proteins as well as of another 240–532 construct in which aspartate of the second acidic portion was substituted by isoleucine are summarized in Table 1. These results indicate the presence of 2 Ca^{2+}-binding sites of high affinity interacting cooperatively in a stretch of 154 amino acids of the intracellular loop of the exchanger. The mutation of one aspartate in any of two acidic portions of the loop drastically reduces the affinity, this being followed by disappearance of one binding site and by the loss of positive cooperativity.

Ca^{2+}-dependent changes in electrophoretic mobility of Na^+-Ca^{2+} exchangers from different tissues

A high affinity monoclonal antibody against dog heart Na^+-Ca^{2+} exchanger [19] and an ECL Western blotting technique allowed us to reveal the presence of homologous proteins in nuclei-free homogenates of different rat tissues. Samples from rat heart and brain gave major bands of apparent mol. wt. of ~120 kDa (Fig. 4A). When 1 mM EGTA was added to the electrophoresis sample buffer, the mobility of the bands increased; an appearance of a new faint band of ~160 kDa was quite reproducible. Homogenates from rat spleen and lung yielded antigens of ~100 kDa in the presence of Ca^{2+} ions. Chelation of Ca^{2+} produced, in contrast, a decrease in the mobility of the major band in the gel which became the same as that in EGTA-treated samples from heart and brain.

The mobility of the Na^+-Ca^{2+} exchangers in the gel can be also modified by a short heating of the samples prior to electrophoresis [20]. When the samples were shortly pretreated at 90°C in the presence of Ca^{2+} ions the mobility of the major bands in heart samples increased while that of a lung sample decreased (Fig. 4B). Preheating made the immunoreactive proteins from both sources migrate with almost the same rate. Judging from the electrophoretic patterns presented in Fig. 4A, we could distribute the Na^+-Ca^{2+} exchangers from different rat tissues into two major groups: of cardiac type (brain) and lung type (spleen, diaphragm, skeletal muscle,

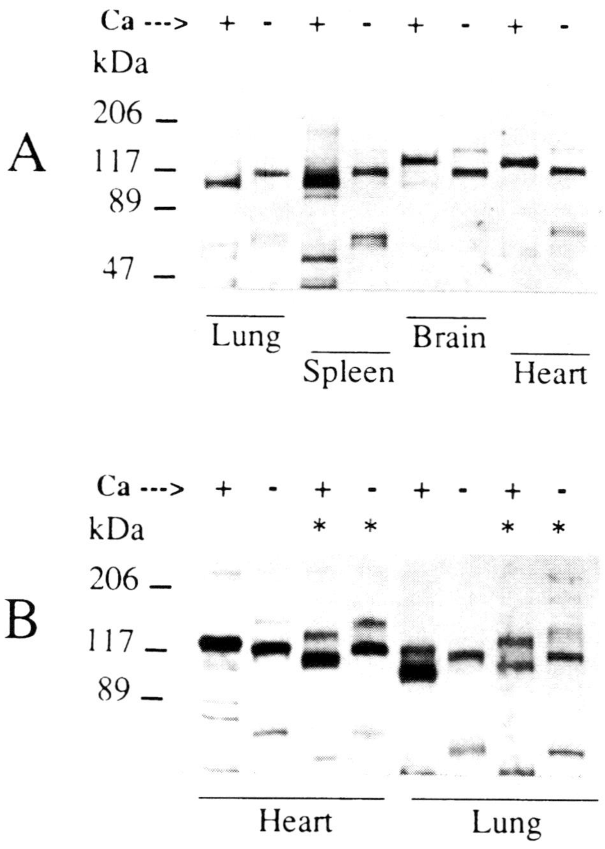

Fig. 4. Western blots of nuclei-free homogenates isolated from rat tissues. (A) The crude tissue fractions from adult rat lung (65 μg protein), spleen, brain and heart (30 μg protein supplemented either with 1 mM Ca^{2+} (+) or 1 mM EGTA (-) and immediately loaded into the gel. It was noted that polypeptides of low molecular weight likely representing proteolytic products of the exchanger also had different mobilities in the presence and absence of Ca^{2+} ions. (B) Samples from lung of 7-day old rat and heart of adult rat (60 and 30 μg protein per well, respectively). Samples pretreated in the sample buffer at 90°C (3 min) are marked by asterisks. The lung exchanger moves slightly faster than the cardiac exchanger.

thoracic aorta and liver). Upon addition of EGTA the mobilities of the samples from the two groups shifted in the opposite direction.

Discussion

The cardiac Na^+-Ca^{2+} exchanger is regulated by cytoplasmic Ca^{2+} [21]. The region responsible for the Ca^{2+} regulation is identical to a high-affinity Ca^{2+} binding domain which has been localized in a stretch of the exchanger intracellular loop comprised of 154 amino acids [7, 8]. This portion of the exchanger contains two acidic sequences with three consecutive aspartate residues [7]. Mutations of the aspartates led to a decrease in the $^{45}Ca^{2+}$ affinity of the cardiac exchanger fragments and a rightward shift in the Ca^{2+} dependence of the outward current in inside-out giant patches of *Xenopus* oocytes injected with the exchanger RNA [7, 8]. $^{45}Ca^{2+}$-overlay data of homogenates from *E. coli* expressing fragments of the exchanger showed also a decrease in the amount of Ca^{2+} bound to the mutated fusion proteins [7] which could reflect existence of more than one Ca^{2+} binding site.

The quantification of the Ca^{2+} binding was performed on pure polypeptides of the exchanger. Accurate calculations were hindered by uncertain values of protein concentration which were different upon using three standard techniques and two standards for calibration (bovine serum albumin and IgG). Comparing with Lowry method, biuret and Bradford techniques gave, respectively, 20% higher and 15% lower values. For calculating molar ratios Ca^{2+} /protein, we used average values of the protein concentration which were determined in the above assays and obtained from the amino acid composition. The data on equilibrium dialysis (Table 1) allowed us to suggest that two binding sites participate in the formation of the high affinity Ca^{2+} binding domain of the cardiac exchanger. Substitution of an aspartate in either of the acidic sequences of the exchanger loop led to disappearance of one of the binding sites.

Though the two acidic sequences of the canine cardiac exchanger (Fig. 1) differ in structure from a Ca^{2+} binding unit of a large superfamily of Ca^{2+} binding proteins of the troponin C type [9] their saturation with Ca^{2+} occurs as well in a cooperative manner (Table 1) which could be followed by a conformational change of the molecule. Two conformational states corresponding to a Ca^{2+}-saturated and a Ca^{2+}-free forms has been demonstrated in SDS-polyacrylamide gels for E-F-hand type Ca^{2+}-binding proteins [13, 22, 23] and for fragments of the Na^+-Ca^{2+} exchanger [7]. The fact that Ca^{2+}-free forms of the proteins moved in the gel either faster or slower than Ca^{2+}-bound forms excluded a pure electrostatic influence of the Ca^{2+} chelation on the electrophoretic mobility.

Western blot analysis of rat tissue homogenates allows us to postulate that Ca^{2+}-induced conformational changes also occur in full-length exchangers. Due to differences in electrophoretic mobilities of the Na^+-Ca^{2+} exchangers in a Ca^{2+}-bound form two groups could be evisioned, exchangers of cardiac type and exchangers of lung type. Ca^{2+}-free forms of the proteins, which were either preheated or not, had similar mobility in the gel. The presence of the third electrophoretic pattern of the heart and lung samples after their preheating indicates that the Na^+-Ca^{2+} exchangers tend to possess multiple conformational states which remain 'frozen' in the gel.

In conclusion, our Western blot data indicate abundance of the Na^+-Ca^{2+} exchanger in rat tissues other than heart. All the immunologically related proteins upon EGTA pretreatment changed their mobility in the SDS-gel which may indicate existence of a similar high-affinity Ca^{2+}-binding domain which is characteristic for canine heart exchanger. It needs to be verified if a phenomenon of the secondary Ca^{2+} regulation reported for the exchangers from heart [21] and brain [5] is fully realized in other tissues.

Acknowledgements

We thank Audrey Fowler (UCLA) for the amino acid analyses. This work was supported by National Institutes of Health Grant HL49101 (to K.D.P.) and by the Foundation Langlois.

References

1. Bers DM: Exitation-contraction coupling and cardiac contractile force. Boston: Kluwer, 1991
2. Philipson KD: The Na^+-Ca^{2+} exchanger: molecular aspects. In: M. Morad (ed.) Molecular Physiology and Pharmacology of Cardiac Ion Channels and Transporters. Boston: Kluwer, 1996
3. Reeves JP: Cardiac sodium-calcium exchange system. In: N. Sperelakis, (ed.) Physiology and Pathophysiology of the Heart, Third Edition. Boston: Kluwer, 1995, pp 309–318
4. Nicoll DA, Longoni S, Philipson KD: Molecular cloning and functional expression of the cardiac sarcolemmal Na^+/Ca^{2+} exchanger. Science 250: 562–565, 1990
5. Li Z, Matsuoka S, Hryshko LV, Nicoll DA, Bersohn MM, Burke EP, Lifton RP, Philipson KD: Cloning of the NCX2 isoform of the plasma membrane Na^+-Ca^{2+} exchanger. J Biol Chem 269: 17434–17439, 1994
6. Kofuji P, Lederer WJ, Schulze DH: Mutually exclusive and cassette exons underlie alternatively spliced isoforms of the Na/Ca exchanger. J Biol Chem 269: 5145–5149, 1994
7. Levitsky DO, Nicoll DA, Philipson KD: Identification of the high affinity Ca^{2+}-binding domain of the cardiac Na^+-Ca^{2+} exchanger. J Biol Chem 269: 22847–22852, 1994
8. Matsuoka S, Nicoll DA, Hryshko LV, Levitsky DO, Weiss JN, Philipson KD: Regulation of the cardiac Na^+-Ca^{2+} exchanger by Ca^{2+}. Mutational analysis of the Ca^{2+}-binding domain. J Gen Physiol 105: 403–420, 1995
9. Marsden BJ, Shaw GS, Sykes BD: Calcium binding proteins. Elucidating the contributions to calcium affinity from an analysis of species variants and peptide fragments. Biochem Cell Biol 68: 587–601, 1989
10. Furman I, Cook O, Kasir J, Rahamimoff H: Cloning of two isoforms of the rat brain Na^+-Ca^{2+} exchanger gene and their functional expres-

sion in HeLa cells. FEBS Letters 319: 105–109, 1993
11. Reilly RF, Shugrue CA: cDNA cloning of a renal Na^{+}-Ca^{2+} exchanger. Am J Physiol 262 (Renal Fluid Electrolyte Physiol 31): F1105–F1109, 1992
12. Laemmli UK: Cleavage of structural proteins during assembly of the head of bacteriophage T4. Nature (London) 227: 680–685, 1970
13. Maruyama K, Mikawa T, Ebashi S: Detection of calcium binding proteins by ^{45}Ca autoradiography on nitrocellulose membrane after sodium dodecyl sulfate gel electrophoresis. J Biochem (Tokyo) 95: 511–519, 1984
14. Fabiato A, Fabiato F: Calculator programs for computing the composition of the solutions containing multiple metals and ligands used for experiments in skinned muscle cells. J Physiol (Paris) 75: 463–505, 1979
15. Lowry OH, Rosebrough NJ, Farr AL, Randall RJ: Protein measurement with the Folin phenol reagent. J Biol Chem 193: 265–275, 1951
16. Itzhaki RF, Gill DM: A micro-biuret method for estimating protein. Anal Biochem 9: 401–410, 1964
17. Bradford M: A rapid and sensitive method for the quantification of microgram quantities of protein utilizing the principle of protein-dye binding. Anal Biochem 72: 248–252, 1976
18. Towbin H, Staehelin T, Gordon J: Electrophoretic transfer of proteins from polyacrylamide gels to nitrocellulose sheets: procedure and some applications. Proc Natl Acad Sci USA 26: 4350–4354, 1979
19. Porzig H, Li Z, Nicoll DA, Philipson KD: Mapping of the cardiac sodium-calcium exchanger with monoclonal antibodies. Am J Physiol 265 (Cell Physiol 34): C748–C756, 1993
20. Durkin JT, Ahrens DC, Pan Y-CC, Reeves JP: Purification and amino-terminal sequence of the bovine cardiac sodium-calcium exchanger: Evidence for the presence of a single sequence. Arch Biochem Biophys 290: 369–375, 1991
21. Kimura J, Noma A, Irisawa H: Na-Ca exchange current in mammalian heart cells. Nature 319: 596–597, 1986
22. Burgess WH, Jemido DK, Kretsinger RH: Interaction of calcium and calmodulin in the presence of sodium dodecyl sulfate. Biochim Biophys Acta 623: 257–270, 1980
23. Brar SK, Siu C-H: Characterization of the cell adhesion molecule gp24 in *Dictylostelium discoideum*. Mediation of cell-cell adhesion via a Ca^{2+}-dependent mechanism. J Biol Chem 268: 24902–24909, 1993

Molecular and Cellular Biochemistry **160/161**: 33–39, 1996.

E-1020, a water soluble imidazopyridine, has direct effects on Ca^{2+}-dependent force and ATP hydrolysis of canine and bovine cardiac myofilaments

Frances M. Powers, Kimberly A. Palmiter and R. John Solaro

Department of Physiology and Biophysics, University of Illinois at Chicago, College of Medicine, Chicago, Illinois 60680, USA

Abstract

E-1020 is a cardiotonic agent that acts as a cyclic-AMP phosphodiesterase inhibitor but also may have actions which alter myofilament response to Ca^{2+}. To identify direct actions of E-1020 on cardiac contractile proteins, effects of E-1020 on myofibrillar Ca^{2+} dependent MgATPase and force generation in chemically skinned fiber bundles were measured. In bovine cardiac myofibrils, E-1020 (100 μM) significantly increased myofilament Ca^{2+} sensitivity and Ca^{2+}-dependent ATPase activity at submaximal pCa values. At pCa 6.75, E-1020 significantly increased ATPase activity in bovine (10–100 μM) and canine (1–100 μM) cardiac myofibrils but had no effect on rat cardiac myofibrils. Moreover, in one population of canine ventricular fiber bundles, E-1020 (0.01–10 μM) significantly increased isometric tension at pCa 6.5 and 6.0, whereas in another population of bundles E-1020 had no effect on tension. In no case was resting (pCa 8.0) or maximal tension (pCa 4.5) increased by E-1020. Measurements of Ca^{2+} binding to canine ventricular skinned fiber preparations demonstrated that E-1020 does not alter the affinity of myofilament troponin C for Ca^{2+}. We conclude that part of the mechanism by which E-1020 acts as an inotropic agent may involve alterations in the responsiveness of contractile proteins to Ca^{2+}. The lack of effect of E-1020 on some preparations may be dependent on isoform populations of myofilament proteins. (Mol Cell Biochem **160/161**: 33–39, 1996)

Key words: E-1020, calcium sensitivity, myofibrillar ATPase activity, skinned fiber bundles, isometric tension

Introduction

Cardiotonic drugs which can substitute for digitalis and catecholamines are being developed for use in the treatment of congestive heart failure [1–3]. Some of these drugs inhibit cyclic AMP (cAMP) phosphodiesterase (PDE) [4–7]; others alter the responsiveness of cardiac myofilaments to Ca^{2+} [8, 9]. Interestingly, many of the PDE inhibitors also affect Ca^{2+} sensitivity of the contractile proteins [8–10]. For example, sulmazole, pimobendan, MCI-154 and EMD 53998, known PDE inhibitors, increase myofibrillar Ca^{2+}-dependent MgATPase activity and augment isometric tension developed in skinned cardiac fibers [11–13].

In work described here, we tested whether the cAMP-specific PDE inhibitor E-1020 (1,2-dihydro-6-methyl-2-oxo-5-(imidazo[1,2-a]pyridin-6-yl)-3-pyridine carbonitrile hydrochloride monohydrate) [14, 15] has direct effects on cardiac myofilaments. Ogawa *et al.* [14] reported that the positive inotropic action of E-1020 was not completely reversed by carbachol. Carbachol, a muscarinic receptor agonist, has been shown to inhibit the cAMP mediated positive inotropic actions [16, 17]. The inability of carbachol to completely inhibit the inotropic effect of E-1020 suggests that E-1020 may have actions as a Ca^{2+} sensitizer.

While the mechanism by which PDE inhibition increases cardiac function is fairly well established [3], the Ca^{2+} sensitizing mechanisms of these drugs have not been fully elucidated. In the case of pimobendan and sulmazole, at least part of the Ca^{2+} sensitization appears to be associated with an increased affinity of the regulatory sites on myofilament troponin C (TnC) for Ca^{2+} [9, 11]. Unlike other PDE inhibitors which are hydrophobic, E-1020 is water-soluble so the

Address for offprints: R.J. Solaro, Department of Physiology and Biophysics, University of Illinois at Chicago, 835 S. Wolcott Avenue, (M/C 901) Rm E202, Chicago, IL 60612, USA mechanism of action may be different from that of Ca^{2+} sensitizing drugs with PDE inhibitory actions.

The purpose of our study was to determine whether E-1020 alters myofilament response to Ca^{2+}. The Ca^{2+} sensitivity of cardiac myofibrillar preparations and skinned bundles of cardiac myocytes was evaluated at various E-1020 concentrations. Calcium binding experiments were also done to elucidate the mechanism by which E-1020 influences myofilament function. We found that E-1020 can increase the MgATPase activity and isometric tension of cardiac myofilaments at submaximal Ca^{2+} concentrations. These effects of E-1020 were independent of changes in Ca^{2+} binding to the regulatory sites on troponin C (TnC). We also speculate that differences in actions of E-1020 among various preparations may be due to differences in isoform populations of myofilament proteins.

Materials and methods

Myofibrils

Cardiac myofibrils were isolated from dog, bovine, euthyroid rat and hypothyroid rat ventricles using the method of Pagani and Solaro [18]. The thyroid state of one group of adult rats were altered to induce the expression of the β-isoform of myosin heavy chain (MHC) within the myocardium and completely suppress the expression of α-MHC [19]. To induce a hypothyroid state, three rats were given 0.025% methimazole in their drinking water daily for four weeks [20]. The ventricular myocardium was minced in 25 mM imidazole (pH 7.0), homogenized and centrifuged. The pellet was homogenized three times in a standard buffer solution (60 mM KCl, 30 mM imidazole and 2.0 mM $MgCl_2$; pH 7.0) and centrifuged. On the third resuspension, 1 mM EGTA was added to the buffer. The pellet was treated with 1 % Triton X-100 (v/v) in standard buffer to remove contaminating membranes. Myofibrils were washed three times by resuspension in standard buffer solution and centrifugation. All solutions were kept on ice (≤ 4°C) and contained leupeptin (0.5 μg/ml), pepstatin A (0.5 μg/ml) and phenylmethylsulfonyl fluoride (0.2 mM) to inhibit proteolytic enzymes. Myofibrillar protein concentration was determined immediately after isolation using the Lowry protein assay as modified by Pagani and Solaro [18]. Myofibrillar ATPase activity was measured within 24 h.

Skinned fiber bundles

Detergent skinned fiber bundles were prepared from dog and beef hearts as described by Pan and Solaro [21]. Muscle fiber bundles (0.5–1.0 mm) were dissected from the trabeculae cordis of the left ventricle then extracted overnight in relaxing solution (10 mM EGTA, 5.5 mM $MgATP^{2-}$, 12 mM creatine phosphate, 2.7 mM Mg^{2+}, 30 mM KCl, 60 mM imidazole (pH 7.0) and 50 μM phenylmethylsulfonyl fluoride) which contained 1% Triton X-100. Fiber bundles were stored at –20°C in equal volumes of relaxing solution and glycerol. At the time of the experiment, fiber bundles were dissected to final diameter of 100 – 200 μm and used in force measurements and Ca^{2+} binding experiments.

Actomyosin ATPase activity

Ca^{2+}-dependent Mg^{2+}-ATPase activity of isolated myofibrils was measured in a reaction mixture containing 0.2–0.5 mg/ml myofibrillar protein, 60 mM KCl, 7 mM $MgCl_2$, 60 mM imidazole (pH 7.0) and 1 mM EGTA. Calcium concentration was varied from pCa 8 to 4.875 (pCa = -log [free Ca^{2+}]). After E-1020 (0–100 μM) was added, the reaction mixture was preincubated at 30°C for 3 or 5 min. The 10 min reaction was initiated with the addition of Na_2ATP (final concentration 5.5 mM) and stopped with ice cold 10% trichloroacetic acid. The reaction mixture was centrifuged at 4°C, 3,000 g for 10 min to pellet the protein. Inorganic phosphate produced by ATP hydrolysis was measured in the supernatant fraction using the methods described by Carter and Karl [22].

Force measurements in canine skinned fiber bundles

Relative force was measured at room temperature (22°C) using the methods described in detail by Pan and Solaro [21]. A bundle of dog skinned cardiac muscle fibers with an average diameter of 200 μm (length = 5–8 mm) was mounted between a force transducer [23] and thin rod with fast setting glue. After equilibration in a high-EGTA relaxing (HR; see below) solution, mean sarcomere length of the fiber bundle was adjusted to 2.02 μm using a laser diffraction pattern [24]. Before Ca^{2+} was added to the bathing solution, the fiber bundle was exposed to a low EGTA relaxing solution (LR; see below) to allow for faster tension development. Maximal isometric tension was measured at the beginning and end of each experiment. If maximal contraction at the end of an experiment was less than 91% of maximal contraction recorded at the beginning of the experiment, data from that fiber bundle were not used. Isometric tension was recorded at a given pCa then the fiber bundle was exposed to cumulative concentrations of E-1020 (10 nM–10 μM). The fiber bundle was placed in HR solution following exposure to Ca^{2+} or Ca^{2+} plus E-1020.

All solutions combined 60 mM imidazole (pH 7.0), 2 mM Mg^{2+}, 5.5 mM $MgATP^{2-}$, 15 mM creatine phosphate, and 10 U/ml creatine kinase. The HR and activating solutions also contained 10 mM EGTA; LR contained 0.1 mM EGTA. A CaEGTA buffer system was used in activating solutions by adjusting free Ca^{2+} to values between pCa 8 and 4. The Ca^{2+} concentration needed for each pCa value was calculated by the BATHE software program (kindly provided by Dr. Robert Godt) using the equations of Godt and Lindley [25]. The final ionic strength of all solutions was adjusted with KCl to 0.15 M.

Ca^{2+}-binding to bovine skinned muscle fiber bundles

A double-isotope technique described by Pan and Solaro [21] was used to measure Ca^{2+}-binding to unrestrained bovine skinned fiber bundles at various Ca^{2+} concentrations in the presence of 0, 3 or 100 μM E-1020. After determining the amount of Ca^{2+} bound to each myocardial fiber bundle, the bundles and bovine serum albumin standards were digested in 0.5 N NaOH at 96°C for 30 min. Protein content was determined using the modified Lowry protein assay [18]. Ca^{2+}-binding measurements were expressed as nmol Ca^{2+} bound per mg protein.

Drugs

E-1020 (1,2-dihydro4-methyl-2-oxo-5-(imidazo[1,2-a]-pyridin-6-yl)-3-pyridine carbonitrile hydrochloride monohydrate) was provided by Eisai Co. Ltd., Tsukuba Research Laboratories, Japan. E-1020 was dissolved in distilled, deionized water immediately before use.

Statistics

Results shown in all figures are means ± S.E. Myofibrillar pCa-ATPase activity measurements were fit to a nonlinear regression analysis of the Hill equation, $Y = [Ca^{2+}]^n/(k^n + [Ca^{2+}]n)$, to derive k, the Ca^{2+} concentration at half maximal activity or force (pCa_{50}), and Hill coefficient $_n$, which characterizes the steepness of the relation. Y is the ATPase activity or force normalized to maximal values. Student's *t* tests were used to compare myofibrillar ATPase activity (paired) and Ca^{2+}-binding (unpaired) at each pCa value and the pCa_{50} of the pCa-ATPase relationship (paired) in the absence and presence of E-1020. An analysis of variance (ANOVA) with repeated measures followed by a Newman-Kuels *post hoc* test was used to identify changes in myofibrillar ATPase activity at pCa 6.75 or tension development of skinned fibers with increasing concentrations of E1020. A value of $p < 0.05$ was considered significant for all comparisons.

Results

Myofibrillar ATPase activity

The effects of E-1020 on Ca^{2+}-stimulated Mg^{2+}-ATPase activity was assessed in beef cardiac myofibrils (Fig. 1). Maximal ATPase activity was not altered by E-1020, but at submaximal Ca^{2+} concentrations ($pCa \geq 5.75$) 100 μM E-1020 increased ATPase activity significantly. The pCa_{50}, a measure of myofibrillar Ca^{2+} sensitivity, was increased significantly in the presence of 100 μM E-1020 (6.16 ± 0.05 under control conditions vs 6.28 ± 0.02 in the presence of E-1020). The Hill *n*, a measure of cooperative interactions among thin filament proteins, was increased from 1.52 ± 0.12 to 1.76 ± 0.10 (P = 0.077) in the presence of E-1020 (100 μM).

Results depicted in Fig. 1 indicated that the effects of E-1020 on myofilament response to Ca^{2+} are greatest at relatively low free Ca^{2+}. Therefore, the concentration-dependent effects of E-1020 on ATPase activity at pCa 6.75 was assessed (Fig. 2). E-1020 significantly increased the myofibrillar ATPase activity in dog (1–100 μM) and bovine (10–100 μM) cardiac preparations. Interestingly, E-1020 (1–120 μM) had no significant effect on myofibrillar response to Ca^{2+} in ventricular myofibrils from euthyroid (data not shown) or hypothyroid rats (Fig. 2).

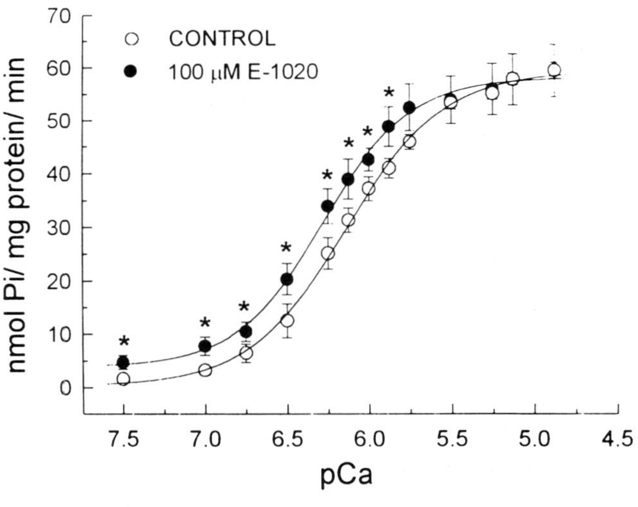

Fig. 1. Effect of 100 μM E-1020 on Ca^{2+}-dependant MgATPase activity of beef cardiac myofibrils. Myofibrils were incubated at 30°C in 5.5 mM MgATP, 2 mM Mg^{2+}, 60 mM imidazole (pH 7.0), 1 mM EGTA and varying pCa values. Ionic strength was adjusted with KCl to 0.125 M. E-1020 was added to the reaction mixture prior to the pre-incubation period. The pCa_{50} and Hill coefficient were 6.16 ± 0.05 and 1.52 ± 0.12 for control and 6.28 ± 0.02 and 1.76 ± 0.10 for 100 μM E-1020. ○, control, N = 9; ●, 100 μM E-1020, N = 9. Values shown are means ± S.E. *p < 0.05 compared to control.

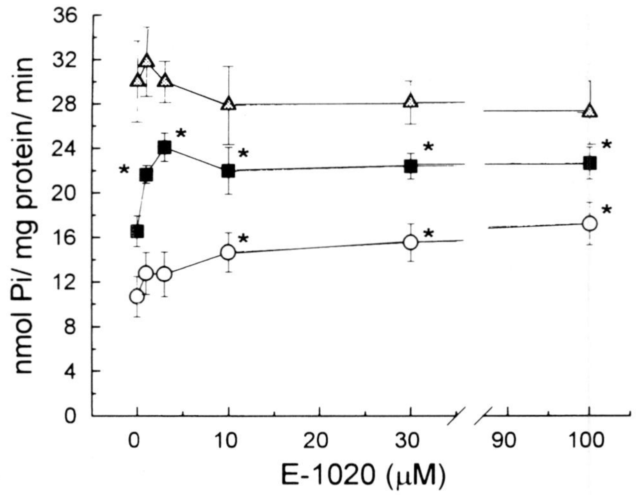

Fig. 2. Effect of E-1020 (1–100 μM) on MgATPase activity of beef (O; N = 6), canine (■; N = 3), and rat (Δ; N = 4) ventricular myofibrillar preparations at pCa 6.75. Incubation conditions for myofilaments were identical to those described in the legend of Fig. 1. (*) Denotes values which are significantly different ($p < 0.05$) from 0 added E-1020. Values shown are means ± S.E.

Isometric force measurements

Isometric tensions developed by skinned cardiac fiber bundles under control conditions are listed in Table 1. The effect of E-1020 on isometric force was measured in eleven dog cardiac skinned fiber bundles at pCa 6.0 and 6 5 and 3–5 fiber bundles at lower pCa values. At low calcium concentrations, muscle fiber bundles appeared to either increase in force in response to E-1020 or E-1020 had no affect on force development. In six fiber bundles, 0.01–10 μM E-1020 increased developed force at pCa 6.5 an average of 132–160% of the force generated at pCa 6.5 when no E-1020 was added (Fig. 3A). Similarly, at pCa 6.0, E-1020 (0.01–10 μM E-1020) significantly increased isometric force in six fiber bundles an average of 125–131% of the force developed at pCa 6.0 with no added E-1020. In five fiber bundles E-1020 did not significantly affect isometric force at pCa 6.5 or 6.0. There were no differences in the relative levels of myofilament activation at pCa 6.5 (2% and 8% maximal tension) and pCa 6.0 (19% and 20% maximal tension) in fiber bundles which increased isometric tension in the presence of E-1020 and those bundles in which E-1020 had no effect. At Ca^{2+} concentrations greater than pCa 6.0 and over the E-1020 concentrations range studied, E-1020 consistently had no effect on force development (Fig. 3B). Importantly, E-1020 did not increase force at pCa 8.0 (n = 8). The effect of E-1020 was reversible upon washout with HR solution at all pCa values studied (not shown). The pCa_{50} of the pCa-force relationship was 5.82 under control conditions and 5.80 in the presence of 100 nM E-1020. The Hill *n* decreased from 2.31–1.94 in the presence of 100 nM E-1020.

Table 1. Tension development in cardiac skinned fiber bundles

pCa	n	Tension (mN/mm^2)	% Maximal Tension
8.00	8	0.49 ± 0.13	0.98 ± 0.28
6.50	11	2.10 ± 7.85	5.69 ± 3.12
6.00	11	9.64 ± 1.98	19.82 ± 2.00
5.75	4	29.75 ± 2.74	53.86 ± 3.46
5.50	5	33.90 ± 4.19	77.99 ± 3.23
4.50	42	50.46 ± 2.30	100

Data a presented as means ± S.E.M. pCa: -log [Ca^{2+}]; n: number of fiber buncles

Ca^{2+}-binding to Bovine Cardiac Myofibrils

To elucidate a possible mechanism of action of E-1020, Ca^{2+} binding to bovine skinned ventricular fibers was studied as a function of pCa. At pCa values higher than 5, essentially all Ca^{2+} binding to skinned fibers is due to the titration of TnC [21]. As reported in Fig. 4, 100 μM E-1020 did not affect Ca^{2+}-binding to myofilament TnC. Similarly, a lower concentration of E-1020 (3 μM; data not shown) did not significantly affect myofilament Ca^{2+} binding at any pCa value. These data indicate that the effects of E-1020 on cardiac myofibrils and skinned fibers were not due to an increased binding of Ca^{2+} to TnC.

Discussion

Although the primary mechanism by which E-1020 elicits positive inotropic effects appears to lay in the inhibition of cAMP-specific PDE [14, 15], our results indicate that the inotropic actions of E-1020 may also involve changes in the response of cardiac myofilaments to Ca^{2+}. Earlier work indirectly suggested that E-1020 affected cardiac contractile elements. In an original paper on E-1020, Ogawa *et al.* [14] studied the inhibitory effect of carbachol on the positive inotropic effect of E-1020 in isolated guinea pig papillary muscles. Carbachol was used to specifically inhibit cAMP mediated events, particularly cAMP-dependent increases in contractile force. At 3×10^{-6} M, carbachol inhibited the E-1020 (10^{-4} M) stimulated developed tension by only 59%. Satoh and Endoh [15] reported a biphasic relationship between E-1020 concentration and contractile force of dog trabeculae, with force increasing more steeply at E-1020 concentrations $\geq 5 \times 10^{-5}$ M. The addition of carbachol ($\leq 3 \times 10^{-6}$ M), inhibited the E-1020 (10^{-7} to 10^{-5} M) stimulated increase in force. At high concentrations of E-1020 ($> 5 \times 10^{-5}$ M), a small component of the E-1020 stimulated increase in contractile force was not inhibited by carbachol. Both studies indicate that E-1020 may have actions which are independent of cAMP-specific PDE inhibition.

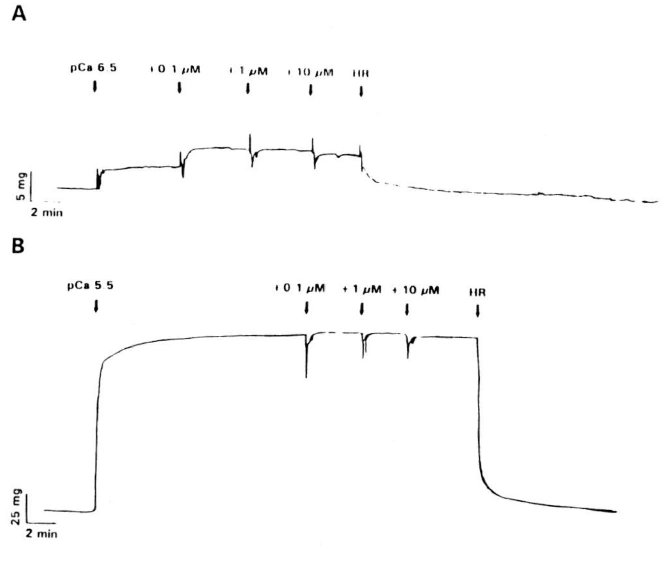

Fig. 3. Effect of E-1020 on isometric force developed at 22°C in skinned dog ventricular fiber bundles at pCa 6.5 (panel A) and at pCa 5.5 (panel B). All solutions contained 60 mM imidazole (pH 7.0), 5.5 mM MgATP, 2 mM Mg2+, 15 mM creatine phosphate, and 10 U/ml creatine kinase. Ionic strength was adjusted with KCl to 0.15 M. HR; High EGTA relaxing solution (see Materials and methods for details).

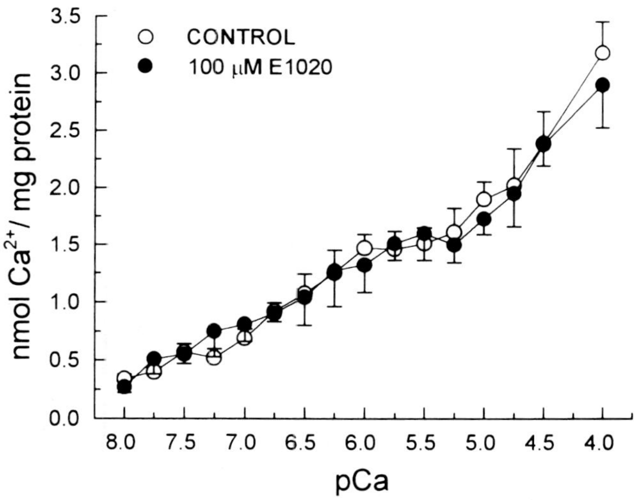

Fig. 4. Ca^{2+}-binding to skinned beef ventricular muscle fibers in the presence of 100 μM E-1020. Solutions (22°C) were the same as those described for Fig. 3 except $^{45}Ca^{2+}$ and ^{3}H-glucose were added. O, control, N = 7; ●, 100 μM E-1020, N = 6. Values shown are means ± S.E. There was no significant difference in Ca^{2+} binding at any pCa value.

The action of E-1020 on myofibrillar ATPase activity is comparable to that of sulmazole, a PDE inhibitor which also modifies myofilament response to Ca^{2+}. At submaximal Ca^{2+} concentrations, 100 μM sulmazole (AR-L 115BS) [11] increased myofibrillar MgATPase activity approximately 30%. E-1020 (1–100 μM) increased canine myofibrillar ATPase 21–71% at pCa 6.75. Neither sulmazole nor E-1020 had an effect on maximal MgATPase activity. The concentration at half the maximal increase in ATPase activity (EC_{50}) was approximately 1 μM and 10 μM E-1020 for canine and bovine myofibrillar preparations respectively. Solaro and Rüegg [11] reported a similar EC_{50} for sulmazole (10 μM) in dog ventricular myofibrils.

In myofibrils isolated from adult rat ventricles, we found no effect of E-1020 on the Ca^{2+}-dependent MgATPase activity. The α-isoform is the predominant ventricular myosin heavy chain (MHC) in the rat, whereas dog and bovine ventricles predominantly express β-MHC. To test whether the effect of E-1020 was dependent on the presence of β-MHC, three rats were made hypothyroid. This treatment resulted in expression of only β-MHC in the rat ventricular myocardium [19]. We observed no effect of E-1020 on Ca2+ dependent MgATPase activity of rat hearts expressing primarily either α- or β-MHC isoforms. Thus, E-1020 does not affect the activation of cardiac contractile proteins of rats in a manner dependent on MHC isoform. Although the reason for the lack of effect of E1020 on some cardiac preparations is not clear, it is possible that actions of E-1020 are dependent on the isoform population of myofilament proteins other than the MHC.

Our results suggest that the mechanism of action of E-1020 may be different than that of PDE inhibiting, Ca^{2+} sensitizing drugs such as pimobendan and sulmazole. In chemically skinned cardiac fiber bundles, Ca^{2+} sensitizers sulmazole and pimobendan consistently increased developed force at submaximal Ca^{2+} concentrations and shifted the force-pCa relation to the left [8, 9]. In addition, the ability of cardiac preparations to respond to sulmazole or pimobendan was not dependent on the species of experimental animal. Although we do report a shift in the pCa-ATPase and pCa-force relations in the presence of E-1020, E-1020 increased force in one population of bundles only at pCa values ≥ 6.0 and the actions of E-1020 on myofibrillar ATPase activity was species dependent. Moreover, there was no effect of E-1020 on TnC Ca^{2+}-binding while sulmazole and pimobendan increase the affinity of Ca^{2+} binding to TnC [9, 11].

Two important questions arise from our findings: (1) How does E-1020 increase force and ATPase activity over a range of relatively high pCa values without affecting TnC-Ca^{2+} binding? and (2) Why is it that E-1020 consistently increased ATPase rate of canine cardiac myofibrils while its effect on submaximal force was not consistent among skinned fiber bundles isolated from dog ventricular trabeculae? A possible unifying interpretation of our data is that E-1020 acts to enhance a process in myofilament activation that is important at high pCa values and dependent on myofilament isoform populations. At high pCa values when the binding of Ca^{2+} to TnC is relatively low, it is presently thought that activation is dependent on strongly bound cross-bridges [26]. These cross-bridges appear to disinhibit the thin filament by moving tropomyosin from its blocking position within neighboring functional units (7 actin: 1 tropomyosin: 1 troponin complex). It is possible that E-1020 may act to increase the interaction energy between functional groups [27, 28] by promoting the actin-myosin reaction at low Ca^{2+} concentrations. Evidence that E-1020 can enhance the actomyosin interaction comes from our studies showing a consistent increase in canine cardiac myofibrillar ATPase activity with the addition of E-1020. In skinned fiber preparations in which E-1020 had a positive effect on tension development, we propose that E-1020 promoted actomyosin interactions resulting in increased submaximal force. In cases of cardiac preparations not showing an effect, we can only speculate that cross-bridge dependent activation and interaction energy between functional units was closer to a maximum in these fiber bundles as possibly determined by isoform populations of or state of myofilament proteins.

Several investigators have shown E-1020 to be an effective cardiotonic agent with vasodilatory properties when used in the treatment of experimental heart failure [29, 30]. In these studies, E-1020 improved cardiac diastolic and systolic function without significantly increasing myocardial oxygen consumption [29]. In addition to its actions as a PDE inhibitor, a direct action of E-1020 on myofilament proteins, which results in an increased contractile force, may contribute to the improved cardiac systolic function observed with E-1020.

Acknowledgements

This study was supported by National Institute of Health grants RO1 HL-22231 (R.J.S.), F32 HL-08656 (F.M.P.) and T32 HL-07692 (F.M.P.; K.A.P).

References

1. Alousi M, Canter JM, Montenaro MJ, Fort DJ, Ferrari RA: Cardiotonic activity of milrinone, a new and potent cardiac bipyridine, on the normal and failing heart of experimental animals. J Cardiovasc Pharmacol 5: 792–803, 1983
2. Farah AE, Alousi M, Schwarz J: Positive inotropic agents. Ann Rev Pharmacol Toxicol 24: 275–328, 1984
3. Colucci WS, Wright RF, Braunwald E: New positive inotropic agents in the treatment of congestive heart failure. N Engl J Med 314: 290–299, 1986
4. Endoh M, Yamashita S, Taira N: Positive inotropic effect of amrinone in relation to cyclic nucleotide metabolism in the canine ventricular muscle. J Pharmacol Exp Ther 221: 775–783, 1982
5. Weishaar RE, Quade M, Schenden JA, Boyd DK, Evans DB: Studies aimed at elucidating the mechanism of action of CI-914, a new cardiotonic agent. Eur J Pharmacol 119: 205–215, 1985
6. Colucci WS, Wright RF, Braunwald E: New positive inotropic agents in the treatment of congestive heart failure. N Engl J Med 314: 349–358, 1986
7. Brunkhorst D, der Leyed H, Meyer W, Nigbur R, Schmidt-Schumacher C, Scholz H: Relation of positive inotropic and chronotropic effects of pimobendan, UD-CG 212 Cl, milirone and other phosphodiesterase inhibitors to phosphodiesterase III inhibition in guinea-pig heart. Naunyn-Schmiedeberg's Arch Pharmacol 339: 575–583, 1989
8. Ruegg JC: Effects of new inotropic agents on Ca^{2+} sensitivity of contractile proteins. Circulation 73 (Suppl. III): 78–84, 1986
9. Fujino K, Sperelakis N, Solaro RJ: Sensitization of dog and guinea pig heart myofilaments to Ca^{2+} activation and the inotropic effect of pimobendan: Comparison with milrinone. Circ Res 63: 911–922, 1988
10. Bohm M, Morano I, Pieske B, Ruegg JC, Wankerl M, Zimmermann R, Erdmann E: Contribution of cAMP-phosphodiesterase inhibition and sensitization of the contractile proteins for calcium to the inotropic effect of pimobendan in the failing heart. Circ Res 68: 689–701, 1991
11. Solaro RJ, Ruegg JC: Stimulation of Ca^{2+} binding and ATPase activity of dog cardiac myofibrils by AR-L 115BS, a novel cardiotonic agent. Circ Res 51: 290–294, 1982
12. Kitada Y, Narimatsu A, Matsumura N, Endo M: Contractile proteins: possible targets for the cardiotonic action of MCI-154, a novel cardiotonic agent? Eur J Pharmacol 134: 229–231, 1987
13. Beier N, Harting J, Jonas R, Klockow M, Lues I, Haeusler G: The novel cardiotonic agent EMD 53 998 is a potent 'calcium sensitizer.' J Cardiovasc Pharmacol 18: 17–27, 1991
14. Ogawa T, Ohhara H, Tsunoda H, Kuroki J, Shoji T: Cardiovascular effects of the new 1,2-dihydro-6-methyl-2-oxo-5-(imidazo[1,2-a}pyridin-6-yl)-3-pyridine carbonitrile hydrochloride monohydrate. Arzneimit -Forsch/Drug Res 39: 33–37, 1989
15. Satoh H, Endoh M: Effects of a new cardiotonic agent 1,2-dihydro-6-methyl-2-oxo-5-[imidazo(1,2-a)pyridin-6-yl]-3-pyridine carbonitrile hydrochloride monohydrate (E-1020) on contractile force and cyclic AMP metabolism in canine ventricular myocardium. Jap J Pharmacol 52: 215–224,1991
16. Endoh M: The time course of changes in cyclic nucleotide levels during cholinergic inhibition of positive inotropic actions of isoprenaline and theophylline in the isolated canine ventricular myocardium. Naunyn-Schmiedeberg's Arch Pharmacol 312: 175–182, 1980
17. Endoh M: Correlation of cyclic AMP and cyclic GMP levels with changes in contractile force of dog ventricular myocardium during cholinergic antagonism of positive inotropic actions of histamine, glucagon, theophylline and papaverine. Jap J Pharmacol 29: 855–864,1993
18. Pagani ED, Solaro RJ: Methods for measuring functional properties of sarcoplasmic reticulum and myofibrils in small samples of myocardium. In: A. Schwartz (ed.) Methods in Pharmacology, Vol. 5. Plenum Publishing Co. 1984, pp 49–61
19. Gustafson TA, Markham BE, Morkin E: Effects of thyroid hormone on a-actin and myosin heavy chain gene expression in cardiac and skeletal muscles of the rat: measurement of mRNA content using synthetic oligonucleotide probes. Circ Res 59: 194–201, 1986
20. Jump DB: Rapid induction of rat liver S14 gene transcription by thyroid hormone. J Biol Chem 264: 4698–4703,1989
21. Pan B-S, Solaro RJ: Calcium binding properties of troponin C in detergent-skinned heart muscle fibers. J Biol Chem 262: 7839–7849, 1987
22. Carter SG, Karl DW: Inorganic phosphate assay with malachite green: An improvement and evaluation. J Biochem Biophys Meth 7: 7–13, 1982
23. Chiu Y-L, Asayama J, Ford LE: A sensitive photoelectric force transducer with a resonant frequency of 6 kHz. Am J Physiol 243: C299–C302, 1982
24. Hibbard MG, Jewell BR: Calcium- and length-dependent force production in rat ventricular muscle. J Physiol 329: 527–540, 1982

25. Godt RE, Lindley BD: Influence of temperature upon contractile activation and isometric force production in mechanically skinned muscle fibers of the frog. J Gen Physiol 80: 279–297, 1982
26. Moss RL: C^{2+} regulation of mechanical properties of striated muscle: Mechanistic studies using extraction and replacement of regulatory proteins. Circ Res 70: 865–884, 1992
27. Hill TL, Eisenberg E, Greene L: Theoretical model for the cooperative equilibrium binding of myosin subfragment 1 to the actin-troponin-tropomyosin complex. Proc Natl Acad Sci USA 77: 3186–3190, 1980
28. Shiner JS, Solaro RJ: The effects of modifiers on enzyme catalysis: A non-classical nearest neighbor approach. Biophys Chem 13: 291–306, 1981
29. Ohhara H, Ogawa T, Takeda M, Katoh H, Daiku Y, Igarashi T: Cardiovascular effects of the new inotropic agent 1,2-dihydro-6-methyl-2-oxo-5-(imidazo[1,2-a]pyridin-6-yl)-3-pyridine carbonitrile hydrochloride monohydrate. Arzneimit -Forsch/Drug Res 39: 38–45, 1989
30. Tanio H, Kumada T, Hayashi M, Himura Y, Nakamura Y, 7Kawai C: Hemodynamic efficacy of E-1020 in comparison with dopamine on acute mitral regurgitation in anesthetized dogs. Jap Circ J 55: 1068–1076, 1991

Molecular and Cellular Biochemistry **160/161:** 41–46, 1996.

Primary cultures of cardiac muscle cells as models for investigation of protein glycosylation

Ursula Henning, Wolf-Peter Wolf [1] and Martin Holtzhauer

Institute of Biochemistry and Molecular Physiology, University of Potsdam, c/o Max Delbrück Center; [1]Department of Molecular Cardiology, Max Delbrück Center, Robert-Rössle-Str. 10, D-13122 Berlin, Germany

Abstract

Primary cardiac cell cultures of newborn rats containing approximately 50% (by cell number) spontaneously contracting cardiomyocytes were used to study the role of protein N-glycosylation for the binding of dihydropyridine (DHP) to the voltage-dependent L-type calcium channel. This binding is not influenced by the accompanying non-muscle cells.

Exposure of the cells up to 6 µg/ml of the N-glycosylation inhibitor tunicamycin for a 44 h period resulted in a decrease of the specific DHP binding sites (B_{max}) to 46.0 ± 17.2% of the untreated control. Similar effects were observed after enzymatic deglycosylation using N-glycosidase F (PNGase F). The results suggest that a posttranslational modification of parts of the cardiac L-type Ca^{++} channel by N-glycosylation is an important determinant for the binding of Ca^{++} antagonists of the DHP-type to the α_1 subunit which itself is not glycosylated. The results suggest a participation of N glycosylation in the assembling of the subunits to the functional channel and/or its turnover. However, a possible effect of tunicamycin on the expression of the Ca channel as an alternative mechanism cannot be excluded. (Mol Cell Biochem **160/161:** 41–46, 1996)

Key words: calcium channel, glycosylation, glycoprotein, cardiac muscle cells, primary culture

Introduction

Protein N-glycosylation plays a key role in protein transport within cells, in directed movement of proteins to cellular compartments, and in cell-cell-interactions. In particular, the localization and orientation of membrane proteins or protein complexes seems to be directed by specific glycosylation of distinct loci within the polypeptide chain. This type of posttranslational modifications is species- and tissue-specific and depends on cellular differentiation, age, and/or metabolic situation [1–3]. Thus, in addition to phosphorylation or acylation, protein glycosylation is a further way to regulate or organize distinct cellular functions both for isolated cells and whole organisms.

Excitable cells such as muscle cells express many different membrane-bound receptors, which are involved in regulation of specific physiologic functions. Some of these receptor molecules, such as monomeric adrenoceptors, are glycoproteins while other are protein complexes with glycosylated subunits, e.g. the voltage-dependent calcium channel (only the subunits α_2 and δ are glycosylated). Especially in these cases, in which the glycoproteins are not involved in cell-cell interactions, the biological significance of protein glycosylation is not very well understood up to now, but there are some hints for participation in directed receptor protein transport [4].

For exploring the role of protein glycosylation of regulatory membrane proteins it should be necessary to combine physiologic and biochemical investigations in an intact, well-defined biosystem. Therefore we used the neonatal rat heart muscle cell, which allows the observation of spontaneous beating frequency as a physiologic parameter for cellular regulation as well as for survival of cells [5–7]. Furthermore, these cells contain the pharmacological well-defined voltage-dependent L-type calcium channel (dihydropyridine receptor), which we used in our assay for measurement of biochemical data [8, 9].

Because protein glycosylation is tissue-specific, in heterologous gene expression systems the extent and the kind of linked oligosaccharides varies in different cell types [10]. Therefore we used as a homologous model primary cultures of cardiac muscle cells for investigating the influence of protein glycosylation on the cardiac calcium channel. The neonatal

Address for offprints: M. Holtzhauer, c/o Max-Delbrück-Centrum, Robert-Rössle-Str. 10, D-13122 Berlin, Germany myocytes offer the opportunity of an easy monitoring of normal physiological functions, because they form during cultivation spontaneously beating cell aggregates. This characteristic behavior is the opposite of that of cultivated adult myocytes, which contract only when they are externally stimulated or when their surface membrane is damaged.

In neonatal rat hearts the content of muscle cells is described to be only about 30% (by cell number), but a 2.5–3-fold enrichment of muscle cells and strong decrease of contaminating fibroblasts, endothelial and mast cells can be achieved by sedimentation procedures [7].

By standardization of the cultivation procedure we demonstrate that primary cultures of neonatal rat heart cells can be used for the investigation of the influence of modulators of intracellular posttranslational protein modifications on the dihydropyridine receptor.

Materials

DMEM tissue culture medium was from Gibco BRL, medium SM20-1 was prepared according to [11], fetal calf serum (FCS) was from Flow Laboratories. All other p.a. chemicals were from local distributors. The plastic tissue culture flasks were from Costar (Germany).

The glycosylation inhibitor tunicamycin (Boehringer Mannheim, Germany) was dissolved in serum-free culture medium and stored in frozen aliquots until use. N-Glycosidase F (PNGase F) was from Boehringer Mannheim, Germany, and was used according to supplier's data sheet.

Antibodies for immunofluroescence: anti-α-actinin, sarcomeric, (monoclonal, Sigma), anti-phospholamban (monoclonal IgG1, BIOMOL, Hamburg, Germany), anti-mouse-IgG, DTAF-labeled (goat, Dianova, Hamburg, Germany). Cell nuclei were visualized by staining with the DNA-specific dye 4′,6′-diamidine-2′-phenylindole dihydrochloride (DAPI) (Boehringer Mannheim, Germany). Dihydropyridine (DHP) ligands were (+)-[Methyl-^{3}H]isopropyl-4-(2,1,3-benzoxadiazol-4-yl)-1,4-dihydro-2,6-dimethyl-5-methoxycarbonyl-pyridine-3-carboxylate [(+)-[methyl-^{3}H]PN 200-110] (specific activity 2.56 TBq/mmol) and nitrendipine (Amersham Buchler (Germany) and Bayer AG, respectively).

Methods

Primary rat heart cell culture

Cardiac cells were prepared from chopped ventricles of 1–2-day-old Sprague-Dawley rats. Enzymatic dissociation by a 0.2% solution of trypsin, seeding, cultivation and counting of spontaneous beating frequency was done as described in [6]. Cells were kept for at least four days in the presence of FCS and of the DNA replication inhibitor 5-fluoro-2′-deoxyuridine [FUdR] [11].

To elucidate the contribution of non-muscle cells, these were partially separated from muscle cells by their different rate of sedimentation and attachment to the surface of the cultivation flask as outlined in [7, 12].

The conditions for testing of the influence of glycosylation inhibitors is described by Henning *et al.* [13].

Cell number, beating rate and DHP binding

Cell number was estimated by microscopic counting under phase contrast as well as by fluorescence microscopy of the labeled nuclei. Spontaneous beating rates were determined at 37°C in a phase contrast microscope and are given as mean from at least five independent loci within the cultivation flask.

For DHP binding, the attached cells were scratched off, homogenized with a Potter homogenisator and diluted in assay buffer normally to a concentration corresponding to 10^6 cells/ml (for experimental details see [13]). The binding experiments were performed in assay buffer (50 mM Tris-HCl, pH 7.4, 1 mM $CaCl_2$; 0.1 mM PMSF) using the DHP (+)-[methyl-^{3}H]PN 200-110 as label in the presence or absence of unlabeled nitrendipine to determine non-specific binding. In the standard assay, for each DHP concentration triplica with about $5 \cdot 10^6$ cells were taken.

Immunofluorescence

Aliquots of the cell suspension (about $3 \cdot 10^4$ cells/20 μl in SM 20 medium) were placed into 12 well multitest slides, dried on air, and fixed with 15% picric acid, 2% formaldehyde. Then the cells were washed sequentially with deionized water, phosphate-buffered saline (PBS), and PGS (PBS, containing 0.5% bovine serum albumin, 0.2% gelatin). Incubation with the antibodies was performed overnight at 4°. After washing, the incubation with the fluorescence-labeled secondary antibody was done for 2 h at room temperature. The fluorescence labeling was examined with appropriate filters using a Zeiss Axioscop (Oberkochen, Germany).

Statistics

The statistic parameters of each experiment were expressed as per cent of the respective control (cultivation of a cell aliquot in the absence of inhibitor), S.D. was calculated using the InStat program (GraphPad), the maximal numbers of drug binding sites were calculated from non-linear regression fits (Prism, GraphPad) to the binding isotherms.

Results

The cellular model

The highest yield of living, spontaneous beating cardiac cells were obtained from 1–2 days old rats using 0.2% trypsin for enzymatic dissection and was in the range of $3.6 \cdot 10^6 \pm 0.6 \cdot 10^6$ cells/heart (n = 21). Taking 3–5-day old rats, the amount of surviving cells drops down to about 80% in comparison to the younger hearts. Also the number of binding sites per cell is reduced.

As determined by immunofluorescence, within the cultivation flask we found $50.2 \pm 3.7\%$ (n = 4) living muscle cells. Thereby we compared the number of cells, labeled with a monoclonal antibody to the (heart)muscle-specific protein phospholamban, with the total number of nuclei. The latter was identical with that obtained by phase contrast microscopy, but in the immunofluorescence experiments the DAPI staining was easier to monitor.

After partial separation of muscular from non-muscular cells by differential sedimentation [7], the nonmuscular fraction contained less than 5% muscle cells, and the myocyte fraction was enriched up to 80%, as checked also by immunofluorescence.

To discriminate the influence of non-muscular cells onto DHP binding data, we determined the drug binding to non-muscular cells as well as the amount of cells, required for a significant signal-to-noise ratio. As shown in Fig. 1, the contribution of non-muscular cells to the DHP binding can be neglected. It can be assumed, that the very small amount of binding as seen in the saturation curve results from the contaminating muscle cells. Furthermore, combination of the binding data from both cellular populations yields that value, obtained for the whole cell assay.

To ascertain the minimum amount of cellular material necessary for binding experiments we determined the specific DHP binding at different cell content in the assay (Fig. 2). The results suggest that at least $0.25 \cdot 10^6$ should be used for each measuring point, but we took about $0.5 \cdot 10^6$ to increase the signal/noise ratio. So in all binding experiments we diluted the cell homogenate to a concentration corresponding to $1 \cdot 10^6$ cells/ml. Whereas the standard deviations of beating rate (f, $\leq \pm 5\%$), maximal DHP binding (B_{max}, $\leq \pm 8\%$), and dissociation constant (K_D, $< \pm 25\%$), respectively, were low within an experiment, the S.D. of these parameters were high between different experiments ($\pm 31\%$, $\pm 36\%$, $\pm 58\%$, respectively). To compare different experiments, therefore, we expressed the data as per cent of the control, i.e. an aliquot of the cells of the same primary culture without any further manipulation.

Tunicamycin prevents protein N-glycosylation in cell cultures [14]. Thereby the susceptibility of cells types to this drug is very different [15]. In a screening assay in 96 well microtitre plates ($3 \cdot 10^4$ cells/well, $9.4 \cdot 10^4$ cells/cm^2) we found that the myocytes tolerate up to 6 µg/ml tunicamycin. When the cells were cultivated for 44 h at this dose, the cell number

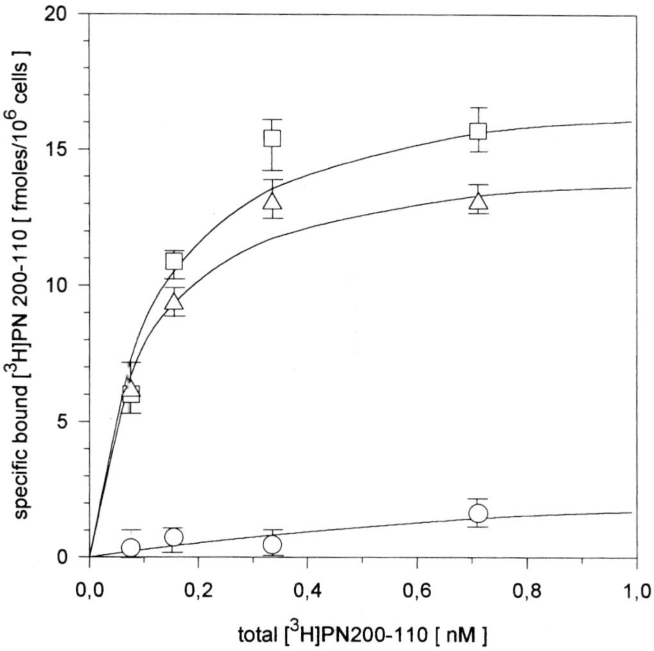

Fig. 1. Binding isotherms of (+)-[methyl-^{3}H]PN 200-110 to fractions of neonatai cardiac cells. (□) enriched muscle cells, (O) non-muscular cells, (Δ) non-fractionated cells of the primary culture.

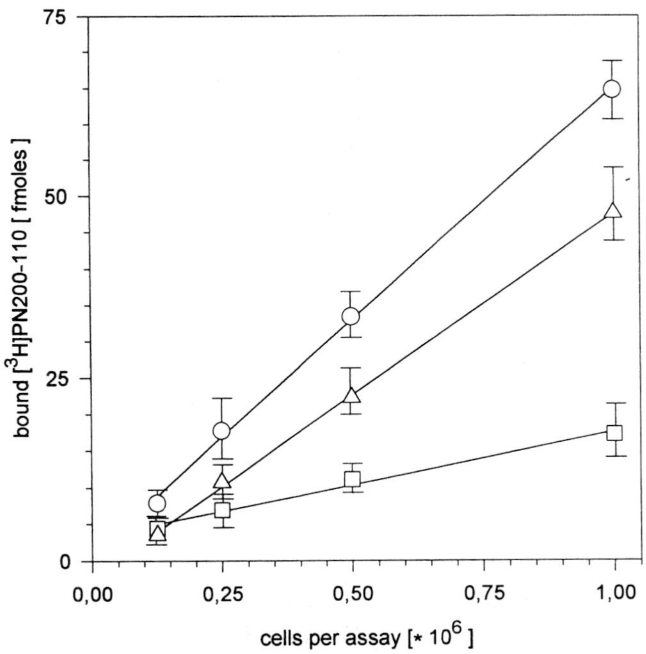

Fig. 2. Correlation of total (O), specific (Δ), and non-specified (□) binding, respectively, of DHP to homogenates of cultivated neonatal heart cells as function of employed material per assay.

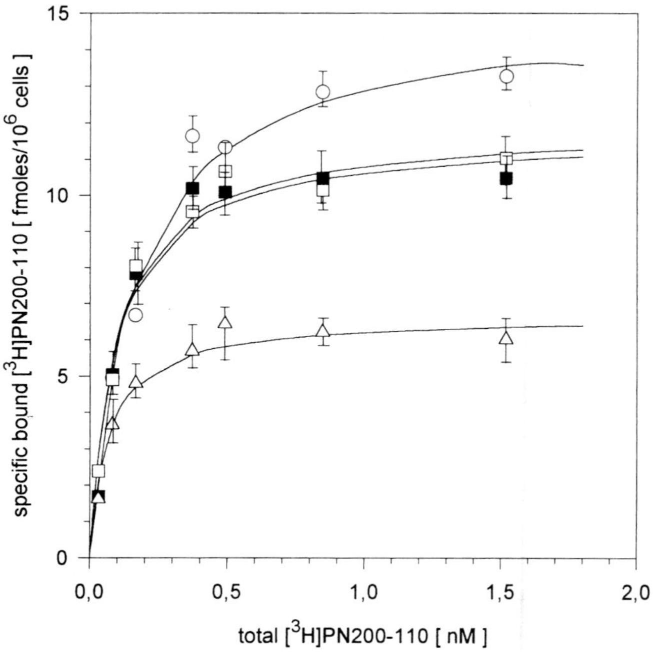

Fig. 3. DHP binding to cardiac cells in the presence of tunicamycin. (○) control (no tunicamycin), (□) 2 μg/ml tunicamycin for 68 hr, starting with seeding of cells, (■) 2 μg/ml tunicamycin for 44 h after a 24 h precultivation without drug, (Δ) 6 μg/ml tunicamycin for 44 h after a 24 h precultivation without drug.

decreased and the myocytes became arrhythmic, formed vacuoles, and showed an abnormal shape; but most of the myocytes survived. On further cultivation for 80 h in fresh medium without inhibitor the cells recovered, i.e. the tunicamycin effect is reversible. To investigate the time needed to establish the action of the inhibitor, we added tunicamycin either at the beginning of the cultivation procedure or after a 24 h period without inhibitor. Both the DHP binding (Fig. 3) and pulsation frequency gave the same results, so we applied a 24 h precultivation before adding the drug to allow the cells to overcome possible membrane alterations caused by enzymatic dissociation.

Deglycosylation experiments

Tunicamycin inhibits formation of N-glycosylation at a very early step of the N-glycane pathway. To look for the influence of the mature, asparagine-bound N-glycanes at the cellular surface we digested them by protein N-glycanase F (PNGase F), which also acts on the L-type calcium channel subunits [16, 17]. After an overnight incubation period with 1 mU/ml, during which the enzyme acts on the whole cell population, we measured the DHP binding and observed a decrease in B_{max} to 59.3 ± 8.2% (n = 3) with respect to the untreated cells, whereas B_{max} is reduced to only 82% at 0.1 mU/ml (Fig. 4). The values obtained after the action of the higher concentration of the enzyme are in the same order of magnitude as B_{max} at 6 μg/ml tunicamycin and somewhat less than that described for the skeletal muscle [17].

Discussion

Primary cultures of neonatal rat heart cells show, as opposited to cells from adult animals, spontaneous rhythmic pulsation. When the cells are transferred in cultivation flasks with fresh medium after enzymatic dissociation, they soon form aggregates of muscle cells ('mini-hearts'), which pulsate under constant conditions with a uniform frequency [71. This feature is an easily observable physiologic parameter while studying the action of drugs.

In our assay we measured the binding of the L-type calcium channel-specific ligand (+)-[methyl-^{3}HIPN 200-110 to cultivated rat heart cells. In heart muscle, this ion channel consists of four subunits (α_1, α_2, β, δ). Only the α_2 and the δ subunit, which are derived from the same mRNA, are glycosylated. The physiologic role of these glycoproteins is not known up to now.

At nanomolar concentrations of the ligand, all specific binding is due to the muscle cells, even though the cell cultures contain about 50% non-muscle cells (cell types and numbers were determined by immunofluorescence microscopy). To get a high signal-to-noise ratio in ligand saturation experiments, we found that about 2.5–5·10^5 cells/assay are sufficient.

To study the influence of the N-glycanes on the ligand binding, we suppressed the intracellular glycosylation by tunicamycin at concentrations which were non-lethal. Even at 6 μg/ml tunicamycin the B_{max} for (+)-[methyl-^{3}H]PN 200-110 dropped down to only about 45% of the control. In comparison, the β-adrenoceptor (monomeric glycoprotein) disappeared completely already at lower concentrations [13]. If the asparagine-bound glycanes are enzymatically removed from intact cells by PNGase F, the number of binding sites for (+)-[methyl-^{3}H]PN 200-110 is also reduced to about 50%. So it can be concluded that the tunicamycin effect on radioligand binding is caused by reduced protein N-glycosylation of cell surface proteins and not by a tunicamycin-dependent suppression of protein synthesis.

As shown in gene expression experiments [18, 19], a magnifying effect on voltage-dependent calcium current and drug action exists if all subunits are co-expressed. Furthermore, in adult myocytes the L type calcium channel is localized within the T-tubular part of the surface membrane in close proximity to the calcium release system of the terminal cisternae of the sarcoplasmic reticulum [20]. We assume that the reduced but not totally depressed binding of the calcium channel ligand is caused by a disturbed interaction between

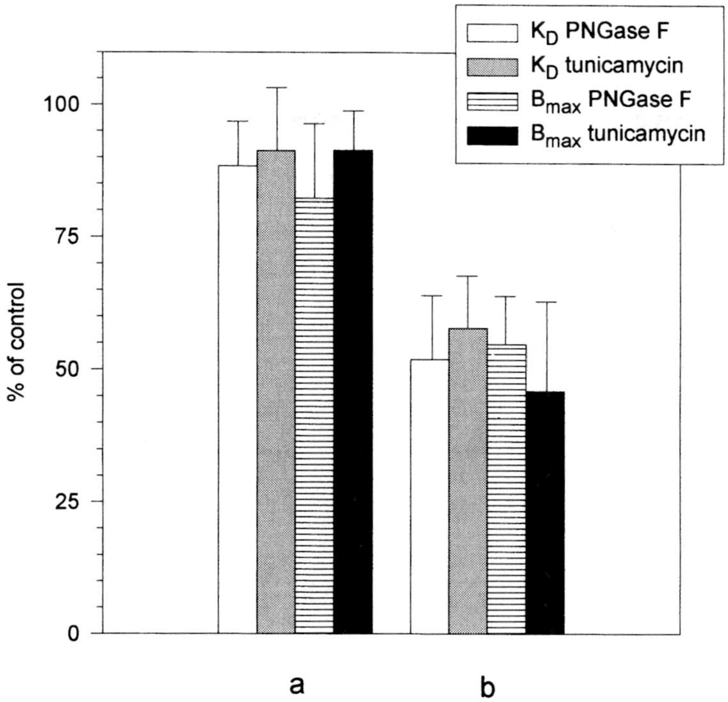

Fig. 4. Effect of enzymatic deglycosylation and tunicamycin on (+)-[methyl-^{3}H]PN 200-110 binding. (a) low doses of enzyme (0.1 mU/ml) and tunicamycin (2 μg/ml), respectively, (b) high doses of enzyme (1 mU/ml) and tunicamycin (6 μg/ml), respectively.

the subunits of the channel and/or its incorrect localization within the surface membrane. The channel subunits, deglycosylated either enzymatically or by blockade of the glycane biosynthesis, could also have a modified fate during the intracellular turnover of these proteins.

Primary cultures of neonatal cardiac myocytes are well defined models for investigation of posttranslational protein modifications, such as e.g. N-glycosylation in a homologous system as indicated by the presented results. The biochemical data will be supplemented by future electrophysiologic and ultrastructural experiments to obtain a deeper knowledge of the role of protein glycosylation in the function of oligomeric membrane proteins.

Acknowledgements

This work was supported by grants of Wissenschaftler-Integrations-Programm (WIP) (M.H.) and Deutsche Forschungsgemeinschaft (DFG) (U.H., M.H.).

References

1. Lis H, Sharon N: Protein glycosylation. Structural and functional aspects. Eur J Biochem 218: 1–27, 1993
2. Kobata A: Structures and functions of the sugar chains of glycoproteins. Eur J Biochem 209: 483–501, 1992
3. Rademacher TW, Parekh RB, Dwek RA: Glycobiology. Annu Rev Biochem 57: 785–838, 1988
4. Rands E, Candelore MR, Cheung AH, Hili WS, Strader CD, Dixon RA: Mutational analysis of beta-adrenergic receptorglycosylation. J Biol Chem 265: 10759–10764, 1990
5. Maltsev VA, Rohwedel J, Hescheler J, Wobus AM: Embryonic stem cells differentiate *in vitro* into cardiomyocytes representing sinusnodal, atrial and ventricular cell types. Mech Dev 44: 41–50, 1993
6. Wallukat G, Wollenberger A: Supersensitivity to b-adrenoceptor stimulation evoked in cultured neonatal rat heart myocytes by L(+)-lactate and pyruvate. J Auton Pharmacol 13: 1–14, 1993
7. Canale ED, Campbell GR, Smolich JJ, Campbell JH: Cardiac Muscle Cells in Culture. In: E.D. Canale, G.R. Campbell, J.J. Smolich, J.H. Campbell (eds). Cardiac Muscle. Springer-Verlag, Berlin, 1986, pp 195–225
8. Lawrence JH, Tomaselli GF, Marban E: Ion channels: Structure and function. Heart Dis Stroke 2: 75–80, 1993
9. Catterall WA, Striessnig J: Receptor sites for Ca^{2+} channel antagonists. Trends Pharmacol Sci 13: 256–262, 1992
10. SantaCruz-Toloza L, Huang Y, John SA, Papazian DM: Glycosylation of shaker potassium channel protein in insect cell culture and in Xenopus oocytes. Biochemistry 33: 5607–5613, 1994
11. Halle W, Wollenberger A: Differentiation and behaviour of isolated embryonic and neonatal heart cells in a chemically definied medium. Am J Cardiol 25: 292–299, 1970
12. Morwinski R, Karsten U, Wallukat G: Mast cells detected in cultures of neonatal rat heart cells. Agents Action 41: C39–C40, 1994
13. Henning U, Wallukat G, Holtzhauer M: Influence of glycosylation inhibitors on dihydropyridine binding to cardiac cells. Mol Cell Biochem 160/161: 83–88, 1996
14. Elbein AD: Glycosidase inhibitors: Inhibitors of N-linked oligosaccharide processing. FASEB J 5: 3055–3063, 1991
15. Larsson O, Carlberg M, Zetterberg A: Selective Killing Induced by an

Inhibitor of N-Linked Glycosylation. J Cell Sci 106: 299–307, 1993
16. Jay SD, Sharp AH, Kahl SD, Vedvick TS, Harpold MM, Campbell KP: Structural characterization of the dihydrophyridine-sensitive calcium channel α_2-subunit and the associated δ peptides. J Biol Chem 266: 3287–3293, 1991
17. Burgess AJ, Norman Rl: The large glycoprotein subunit of the skeletal muscle voltage-sensitive calcium channel. Deglycosylation and development. Eur J Biochem 178: 527–533, 1988
18. Welling A, Kwan YW, Bosse E, Flockerzi V, Hofmann F, Kass RS: Subunit-dependent modulation of recombinant L-type calcium channels. Molecular basis for dihydropyridine tissue selectivity. Circ Res 73: 974–980, 1993
19. Singer D, Biel M, Lotan I, Flockerzi V, Hofmann F, Dascal N: The roles of the subunits in the function of the calcium channel. Science 253: 1553–1557, 1991
20. Jorgensen AO, Shen AC, Arnold W, Leung AT, Campbell KP: Subcellular distribution of the 1,4-dihydropyridine receptor in rabbit skeletal muscle *in situ*: An immunofluorescence and immunocolloidal gold-labeling study. J Cell Biol 109: 135–147, 1989

Molecular and Cellular Biochemistry **160/161**: 47–52, 1996.

Influence of glycosylation inhibitors on dihydropyridine binding to cardiac cells

Ursula Henning, Gerd Wallukat[1] and Martin Holtzhauer

Institute of Biochemistry and Molecular Physiology, University of Potsdam, and [1] Department of Molecular Cardiology, Max Delbrück Center for Molecular Medicine, R.-Rössle-Str. 10, D-13122 Berlin, Germany

Abstract

In primary cultures of neonatal rat heart cells we found a linear correlation between the number of L-type calcium channel-specific dihydropyridine (DHP) binding sites and spontaneous beating frequency (ν).

Formation of glycoproteins in tissue culture was suppressed by different inhibitors of N-glycosylation. This inhibition alters to a different extent the binding of the DHP ligand (+)-[methyl-^{3}H]PN 200-110 and ν. The most severe but reversible effect occurs at 5 μg/ml tunicamycin ($B_{max} \approx 45\%$ and $\nu \approx 6\%$, resp., of control), a slight increase in B_{max} at 0.1–0.5 mM castanospermine and 0.05–2.5 mM deoxymannojirimycin. The other inhibitors gave no significant alterations of B_{max}. (Mol Cell Biochem **160/161**: 47–52, 1996)

Key words: calcium channel, glycosylation inhibitor, cultured cardiac cells, dihydropyridine receptors, glycoproteins

Introduction

L-type voltage-sensitive calcium channels play a central role in the regulation of contractility of cardiac and other excitable cells, mainly by triggering the calcium-induced calcium release from intracellular stores [1], but also by influencing other processes acting at surface and intracellular membranes and modulating contraction frequency as well as force [2].

In myocardial cells, the channel consists of four subunits, α_1, α_2, β and δ and is located at the surface membrane (sarcolemma) [3], especially in the transverse tubules [4]. The calcium ion permeation pathway, the voltage sensor, and the drug binding site(s) are localised in the α_1 subunit [5]. Although the physiological role of the other subunits is not clear as yet, it could be shown in heterologeous gene expression experiments that the assembly of all subunits magnifies in a non-additive way the channel properties such as calcium conductance, channel kinetics, and drug effects [6–10].

In cardiac muscle cells, only the α_2/δ subunits, which are derived from a single messenger RNA [8, 11], are heavily glycosylated. About 25% of the relative molecular mass of the α_2 subunit is due to carbohydrate side chains as determined by SDS PAGE of skeletal muscle preparations [12]. The DNA-derived amino acid sequence of the α_2 subunit contains 12 putative N-glycosylation sites, 5 in the N-terminal probably extracellular part and 7 between the two membrane spanning sequences in the C-terminal half of the molecule, as analysed by the PCGENE computer program. This α_2/δ protein seems to be transcribed from an extremely conserved DNA sequence. As the rabbit skeletal and brain and the human brain α_2/δ subunits show more than 95% amino acid homology [11], it can be concluded that in the heart a corresponding protein exists with a very similar amino acid sequence [3].

Glycosylation of a membrane protein is necessary for transport to the respective membrane, as well as for orientation to extracellular or intracellular space and/or for interactions with other cellular systems [13]. Thus, a number of functionally important proteins of the surface membrane, e.g. adrenergic, cholinergic, and histaminic receptors [14–16], are glycosylated.

N-glycosylation of membrane proteins is a very complex process, which involves a variety of glycosyltransferases and glycosidases in different intracellular organelles [17]. To elucidate its role in subunit function we tried to modulate the oligosaccharide maturation of the α_2/δ glycoprotein using inhibitors of distinct steps of the glycosylation pathway.

In this paper we describe the influence of inhibitors of the

Address for offprints: M. Holtzhauer, c/o Max Delbrück-Centrum, Robert-Rössle-Str. 10, D-13122 Berlin, Germany

N-glycosylation pathway on the binding of a dihydropyridine-type calcium channel blocker and on spontaneous beating frequency as very selective biochemical and general physiological parameters, respectively, using a primary culture of neonatal rat heart cells as a model.

Materials and methods

Materials

The glycosylation inhibitors tunicamycin, castanospermine, swainsonine, deoxymannojirimycin, deoxynojirimycin (Boehringer Mannheim, Germany) were dissolved in serum-free culture medium as twofold to hundredfold concentrated stock solutions and frozen in aliquots until use. (+)-[Methyl-^{3}H]isopropyl-4-(2,1,3-benzoxadiazol-4-yl)-1,4-dihydro-2,6-dimethyl-5-methoxycarbonyl-pyridine-3-carboxylate [(+)-[methyl-^{3}H]PN 200-110] (specific activity 2.56 TBq/mmol) and (-)-4-(3-tert-butylamino-2-hydroxypropoxy)-[5,7-^{3}H] benzimidazol-2-one [(-)-[^{3}H]CGP-12177] (specific activity 1.98 TBq/mmol) were from Amersham Buchler (Germany), nitrendipine from Bayer AG, timolol maleate from Merck, Sharp and Dohme. DMEM tissue culture medium was from Gibco BRL, medium SM20-I was prepared according to [18], foetal calf serum (FCS) was from Flow Laboratories. All other p.a. chemicals were from local distributors. The plastic tissue culture flasks were from Costar (Germany).

Primary rat heart cell culture in the presence or absence of glycosylation inhibitors

Cardiac cells were prepared from chopped ventricles of 1–2-day-old Sprague-Dawley rats. Enzymatic dissociation by a 0.2% solution of trypsin, seeding, cultivation, and counting of spontaneous beating frequency was done as described in [19] for at least four days in the presence of FCS and of the DNA replication inhibitor 5-fluoro-2 -deoxyuridine [FUdR] [18].

In control experiments intended to elucidate the contribution of non-muscle cells, these were partially separated from muscle cells by their different rate of sedimentation and attachment to the surface of the cultivation flask as outlined by [20].

Twenty-four hours after seeding the glycosylation inhibitors were added at the appropriate concentrations together with fresh culture medium with the exception of castanospermine and swainsonine, which were replaced every twelve and twenty-four hours, respectively, due to their limited hydrolytic stability.

Sixty-eight to seventy-two hours after seeding, the spontaneous beating frequency of the formed cell aggregates was counted at 37 °C and the amount of beating cells was estimated. Then the cell layer of up to $25 \cdot 10^6$ cells per 75 cm^2-flask was washed several times with phosphate (PBS) and Tris buffered saline (TBS), the cells were scratched off, homogenised with a Teflon homogeniser in a final volume of 1 ml TBS, and diluted in the assay buffer to a concentration corresponding to about $1.2 \cdot 10^6$ cells/ml.

Binding assays

Aliquots of the homogenate of about $1 \cdot 10^6$ cells in 1 ml (final volume) assay buffer (50 mM Tris·HCl, pH 7.4, 1 mM $CaCl_2$, 0.1 mM PMSF) were incubated with 1,000–240,000 dpm [^{3}H]PN200-110 in the absence or presence of $1 \cdot 10^{-6}$ moles/l nitrendipine (blank) for 1 h at 37°C under yellow light as described in [21]. The binding process was stopped by adding 2 ml ice-cold 10% PEG 6000 in 10 mM Tris-HCl, pH 7.4, also containing 10 mM $MgCl_2$, and filtered through Whatman GF/C filters.

In control binding experiments, the glycosylation inhibitors were added to the binding assay in the same concentrations present during cell cultivation.

The β-adrenoceptor assay was performed parallel to the DHP binding as described by [22] using the radioligand [^{3}H]CGP-12177 in the presence or absence of 1 μM timolol to define blank (nonspecific binding).

The number of binding sites (B_{max}) and dissociation constants (K_D), respectively, were calculated using a non-linear regression fitting option to a rectangular hyperbola of InPlot computer program (GraphPad) from data of at least three independent experiments.

Statistics

To compare the action of different inhibitors, B_{max} and K_D of each individual experiment were expressed as per cent of the respective control (cultivation of a cell aliquote in the absence of inhibitor), S.D. was calculated using the InStat program (GraphPad).

Results

Starting form new-born rat hearts, a primary myocyte culture was established. The yield was about $3 \cdot 10^6$ cells per heart and the cells were seeded to a density of about $4 \cdot 10^5$ cells/cm^2 into tissue culture flasks. Soon after seeding the myocytes, these form dendrite-like connections to other muscle cells and establish so a synchronously beating 'mini-heart.' After a 24 h formation period of a monolayer of about 3000 cells/mm², during which the first interactive structures have been built, the inhibitors were added at non-lethal concentrations. The raster electron micrograph (Fig. 1) gives an impression of a typical neonatal myocardial cell aggregate.

Specific binding of dihydropyridines such as (+)PN 200-110 or nitrendipine in nanomolar concentrations is a very sensitive and selective indicator for voltage-sensitive L-type calcium channels [5]. Pharmacological properties of the channel-drug interactions are characterised by the dissociation constant K_D and the maximal number of binding sites B_{max}.

Spontaneous beating is a significant indicator for heart myocytes from new-born animals. Thus the pulsation was used to identify this cell type in culture. As it is very difficult to standardise primary tissue cultures [23], a relative broad range of frequency of the muscle cell aggregates formed after adhesion onto the cultivation flask surface could be observed.

We found a linear correlation (r = 0.945, n = 9) between the number of DHP binding sites (B_{max}) and pulsation frequency with mean values of 12480 sites/cell and 189 min^{-1}, respectively (Fig. 2). This is a further hint to the essential role of the voltage-sensitive calcium channels in establishing the contractility of cultured cardiac muscle cells. The large differences in pulsation frequency and B_{max} resp., observed only by comparison of the values of different experiments, could be caused by the different developmental stage of the animals used, because during the first days of postnatal development a rapid change in calcium channel properties occurs, as monitored by patch-clamp experiments [24].

The binding data were calculated using a non-linear fit to a binding isotherm (rectangular hyperbola). Scatchard plot analysis of the data gave no different results for B_{max} and K_D, but the fit to the linearized data had a lower correlation coefficient than in the hyperbolic model. A sigmoidal fit or a fit for high and low affinity sites did not show better correlations when compared by a function test (InPlot). A set of typical DHP binding curves as used for calculation of B_{max} and K_D is given in Fig. 3.

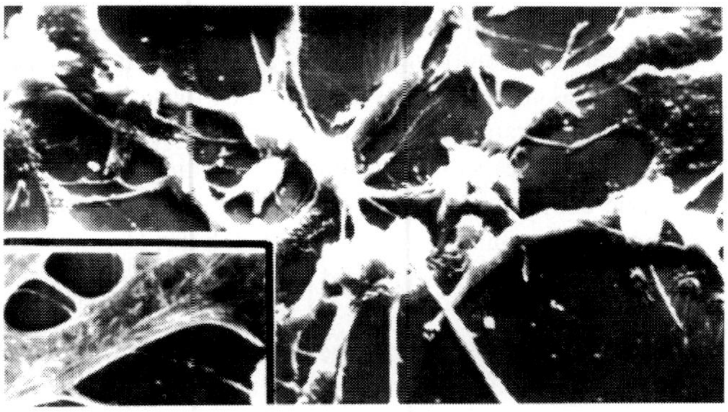

Fig. 1. Scanning electron micrograph of cultured cardiac myocytes. 'Mini-hearts' formed by myocytes from new-born rats about 70 h after seeding (magnification 810-fold). Insert: Connecting strand between two muscle cells containing myofibrils (magniflcation 3,800-fold).

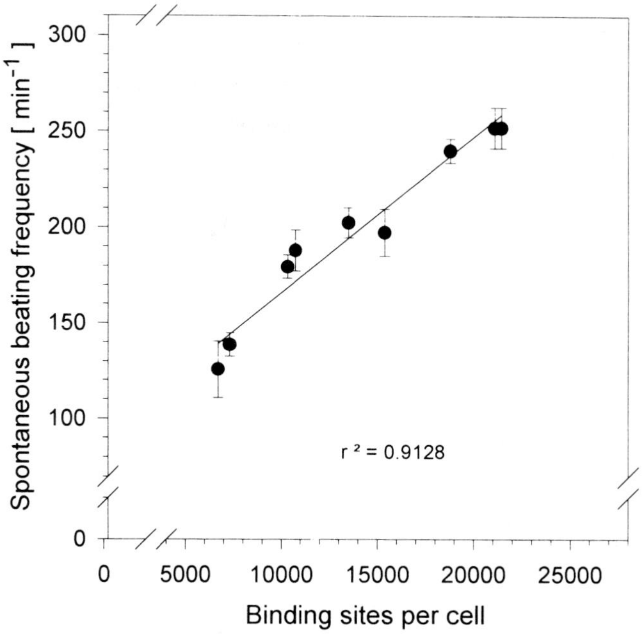

Fig. 2. Correlation of spontaneous beating frequency to the number of DHP binding sites. Values were taken from control experiments in the absence of glycosylation inhibitors.

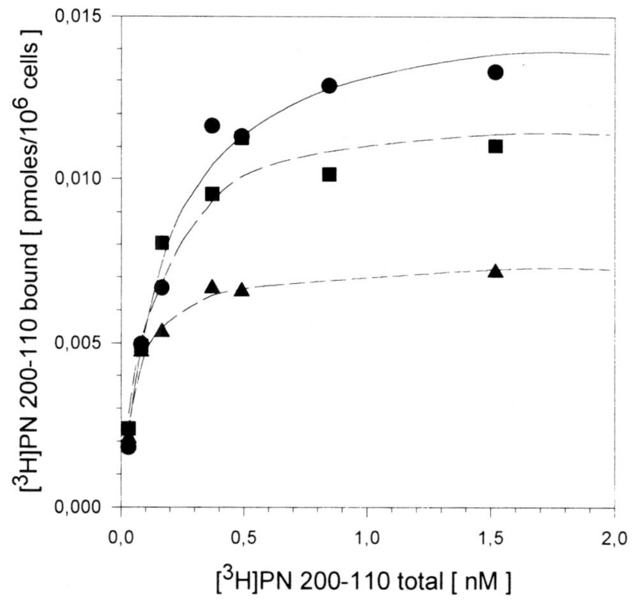

Fig. 3. Binding curves of dihydropyridine radioligand to myocardial cell homogenates. Cells were collected 48 h after cultivation in the presence and absence, respectively, of tunicamycin. ●, control; ■, 2 µg tunicamycin/ml; ▲, 6 µg tunicamycin/ml. Abscissa: total ligand concentration, ordinate: specific bound ligand per 10^6 cells.

The dissociation constant K_D for (+)PN 200-110 in the absence of glycosylation inhibitors was 0.177 ± 0.076 nM (mean ± S.D), independent of the number of binding sites and of frequency. Accompanying non-muscle cells, e.g. mast cells, endothelial cells, or fibroblasts, did not bind the

Table 1. Effects of glycosylation inhibitors on DHP binding

Inhibitor	Concentration in the cultivation medium	B_{max}	K_D % of control (mean ± S.D.)	n [a]
Tunicamycin	6 μg/ml	46.0 ± 17.2[b]	57.8 ± 23.6[c]	7
	0.2 μg/ml	99.7 ± 12.0[d]	87.3 ± 19.2[d]	7
Castanospermine	500 μM	111.7 ± 43.0[d]	92.9 ± 55.6[d]	3
	10 μM	134.5 ± 29.7[d]	123.3 ± 38.3[d]	3
Deoxynojirimycin	2.5 mM	121.3 ± 10.0[d]	83.3 ± 11.2[d]	3
	1 mM	96.9 ± 15.2[d]	79.5 ± 19.1[d]	3
Deoxmannojirimycin	2.5 mM	116.6 ± 26.1[d]	302.1 ± 55.6[b]	3
	0.5 mM	115.4 ± 23.7[d]	291.3 ± 52.8[b]	3
Swainsonine	10 μM	117.5 ± 15.4[d]	107.5 ± 27.0[d]	3
	0.1 μM	107.4 ± 18.9[d]	100.2 ± 13.9[d]	3

[a] no. of independent experiments; [b] $p < 0.01$; [c] $p < 0.05$; [d] not significant

radioligand as proven in control assays (manuscript in press). As determined in a further control experiment, the variation coefficients between different cultivation flasks for frequency, cell number, B_{max}, and K_D were ≤5%, ≤5%, ≤8% and, ≤25%, respectively.

Inhibition of glycosylation with tunicamycin even at high concentrations (6 μg/ml) reduced B_{max} of DHP binding only to 45% of the control (Table 1), whereas the dissociation constant was not significantly altered. In contrast, the binding of the β-adrenergic ligand to the respective receptor already disappeared at 2 μg/ml (Fig. 4) At this concentration, spontaneous beating rate was slightly increased, but dropped down to about 10% of control at 6 μg/ml tunicamycin.

At high doses of tunicamycin (3–6 μg/ml) the cells showed severe morphological alterations and the pulsation became arrhythmic, but further cultivation in the presence of fresh medium without drug for four days restored DHP binding completely in the surviving cells, which then had also a normal phase contrast microscopic appearance (data not shown).

Administration of glycosylation inhibitors during the entire cultivation time in all cultivation media, i.e. starting with seeding of the cells, gave no other results with respect to DHP binding than when the drug was added 24 h after seeding. Consequently, it can be assumed that addition of the inhibitor two days before the binding experiments is sufficient for suppression of glycoprotein *de-novo* synthesis and DHP receptor turnover in the cell type used.

The other inhibitors of the N-glycosylation pathway at concentrations described as effective in cellular assays [25] had no notable effects on the parameters examined (Table 1). Castanospermine, deoxymannojirimycin, deoxynojirimycin, and swainsonine increased slightly, but not significantly B_{max} relative to the respective controls. The dissociation constant was also not influenced by these inhibitors with the exception of deoxymannojirimycin, which at the highest applied concentration (2.5 mM) caused about a threefold increase in K_D (Table 1). Since none of these glycosylation inhibitors had

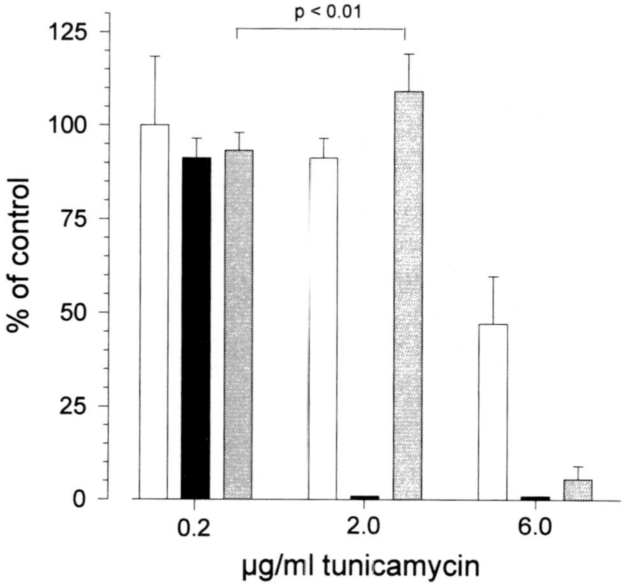

Fig. 4. Effect of tunicamycin on neonatal rat myocytes. Data were normalized to the parallel cultivation in the absence of the inhibitor (control). DHP binding (calcium channel, open bars), CGP binding (β-adrenoceptor, filled bars), beating frequency (hatched bars) after a 2-day-cultivation period in the presence of the drug.

a significant effect on B_{max} of the DHP binding, the observed changes in K_D could be a consequence of a disturbed glycosylation pattern of other proteins of the glycocalyx involved in regulation of the voltage-sensitive calcium channel and of contraction frequency.

Discussion

In eucaryotic cells many membrane proteins are glycosylated. Formation of oligosaccharide side chains in glycoproteins is a very complex process, that is specific for tissue, species, and developmental stage and is important for organisation of proteins in intra- and intercellular interactions [26].

Protein glycosylation is a complex, multistep event of posttranslational modification of the processed protein realised in several compartments of the cell by the action of different glycosyltransferases and glycosidases [13, 25–27]. Glycosidases act especially during the trimming of carbohydrate side chains yielding the main types of glycoprotein-specific oligosaccharides (high mannose, hybrid, complex). These enzymes can be inhibited by specific agents of the glycosylation pathway.

The drugs used in the experiments presented in this paper are inhibitors of protein N-glycosylation, a possible O-glycosylation is not influenced. Tunicamycin blocks N-acetylglucosamine transferase 1, which catalyses the transfer of N-actetylglucosamine to the respective dolicholdiphosphate, a prerequisite for the oligosaccharide transfer to asparagine at the beginning of the protein glycosylation pathway. Castanospermine and 1-deoxynojirimycin inhibit intracellular glucosidases, therefore the trimming of oligosaccharide side chains neither to the high-mannose type nor to the following types will occur, 1-deoxymannojirimycin blocks the development of hybrid or complex type by inhibition of α-mannosidase I and swainsonine, an inhibitor of Golgi mannosidase II, interrupts the transformation to complex type oligosaccharide side chains [25].

The role of the subunits of the L-type calcium channel was investigated by heterologeous gene expression of the respective mRNA's [6, 28]. As the gene products were expressed in vertebrate cells in these experiments, it can be assumed that glycosylation occurred; but it is unknown whether the probably tissue-specific posttranslational modifications (splicing of α_2/δ into the disulfide bridge-linked polypeptides or glycosylation) were done in a muscle-specific manner. So we tried to obtain information on the role of the α_2/δ subunits and their glycosylation by inhibiting different steps of the glycosylation pathway.

During postnatal development of the heart the T-tubular system of the sarcolemma is forming. Parallel to the formation of the T-tubular membrane system the L-type calcium channels are concentrated in this distinct part of the cardiac muscle cell surface membrane [29] as well as in skeletal muscle cells [30]. The observation that the L-type calcium channel is localised near the ryanodine receptor of the sarcoplasmic reticulum [29] could be a consequence of a lectin-like interaction between the calcium channel glycoprotein(s) and the receptor and could represent a biological function of the α_2/δ subunits.

Since tunicamycin is an inhibitor of very early glycosylation events, its attack should result in a severe disorder of glycoprotein formation. For instance, the ability of the cardiac cells to bind a drug to β-adrenoceptors, which are glycoprotein themselves, disappeared at 2 μg tunicamycin per ml cultivation medium, whereas the binding of the specific DHP ligand (−)PN 200-110 to its receptor (α_1 subunit of the calcium channel, not glycosylated) is reduced to only about 45% at a threefold higher concentration. Since the α_1 subunit can bind DHP ligands and voltage-dependent conduct calcium ions also in the absence of the other subunits of the voltage-sensitive calcium channel, as demonstrated by heterologeous gene expression experiments [6], a disturbed interaction between the α_1 and α_2 subunits would not be expected to show such a severe response due to glycosylation inhibition as does the β-adrenoceptor. This could also explain, why K_D, calculated from the binding isotherms, is not really significantly altered during the experiments. Thus, a reduced number of DHP binding sites could be a result of disturbed transport processes from the Golgi apparatus to the outer membrane and/or of an incomplete assembling of the whole calcium channel complex.

The slight increase in B_{max} produced by the action of the other inhibitors could be a consequence of an altered interaction between the α_2/δ and α_1 subunits. As, however, estimation of B_{max} is based on the cell number counted by hemocytometry, this increase could also be due to a decrease of the amount of non-muscle cells (especially mast cells), which are also present in the primary culture of neonatal rat heart cells [31] and which seem to be more susceptible to the inhibitors on account of their higher metabolic activity.

Spontaneous contraction frequency of neonatal myocardial cells, taken as a characteristic physiological parameter of neonatal myocytes, gave another dose dependence than DHP or β-adrenoceptor ligand binding. It showed a slight increase at intermediate concentrations of tunicamycin, at which the β-adrenoceptor binding disappeared, and decreased more strongly than DHP binding at higher inhibitor concentrations. In explanation of this observation it should be considered that beating frequency is not only dependent on the voltage-sensitive calcium channel.

The cell membrane of neonatal rat heart cells is not fully differentiated, e.g. the T-tubular system is not established [32, 33]. Therefore, the role of the α_2/δ subunit as a mediator of the localisation of the voltage-sensitive calcium channel in this cell surface membrane compartment cannot be evaluated in full with the whole-cell model used. Together with electrophysiological parameters of calcium transport this will be investigated in future experiments.

Acknowledgements

This work was supported by grants of Wissenschaftler-Integrations-Programm (WIP) (M.H.) and Deutsche Forschungsgemeinschaft (DFG) (U.H., M.H:). The authors wish to thank Dr. M. Rudolph (University of Potsdam) for raster electron microscopy.

References

1. Glossmann H, Striessnig J: Molecular properties of calcium channels. Rev Physiol Biochem Pharmacol 114: 1–105, 1990
2. Stem MD, Lakatta EG: Excitation-contraction coupling in the heart – The state of the question. FASEB J 6: 3092–3100, 1992
3. Hofmann F, Biel M, Flockerzi V: Molecular basis for Ca^{2+} channel diversity. Annu Rev Neurosci 17: 399–418, 1994
4. Jorgensen AO, Shen AC, Arnold W, Leung AT, Campbell KP: Subcellular distribution of the 1,4-dihydropyridine receptor in rabbit skeletal muscle in situ: An immunofluorescence and immunocolloidal gold-labeling study. J Cell Biol 109: 135–147, 1989
5. Striessnig J, Berger W, Glossmann H: Molecular properties of voltage-dependent Ca^{2+} channels in excitable tissues. Cell Physiol Biochem 3: 295–317, 1993
6. Singer D, Biel M, Lotan I, Flockerzi V, Hofmann F, Dascal N: The roles

of the subunits in the function of the calcium channel. Science 253: 1553–1557, 1991

7. Hullin R, Biel M, Flockerzi V, Hofmann F: Tissue-specific expression of calcium channels. Trends Cardiovasc Med 3: 48–53, 1993
8. Williams ME, Feldman DH, Mccue AF, Brenner R, Velicelebi G, Ellis SB, Harpold MM: Structure and functional expression of alpha1, alpha2 and beta subunits of a novel human neuronal calcium channel subtype. Neuron 8: 71–84, 1992
9. Nargeot J, Dascal N, Lester HA: Heterologous expression of calcium channels. J Membr Biol 126: 97–108, 1992
10. Welling A, Kwan YW, Bosse E, Flockerzi V, Hofmann F, Kass RS: Subunit-dependent modulation of recombinant L-type calcium channels. Molecular basis for dihydropyridine tissue selectivity. Circ Res 73: 974–980, 1993
11. Kim H-L, Kim H, Lee P, King RG, Chin H: Rat brain expresses an alternative spliced form of the dihydropyridine-sensitive L-type calcium channel α_2 subunit. Proc Natl Acad Sci USA 89: 3251–3255, 1992
12. Burgess AJ, Norman Rl: The large glycoprotein subunit of the skeletal muscle voltage-sensitive calcium channel. Deglycosylation and development. Eur J Biochem 178: 527–533, 1988
13. Lis H, Sharon N: Protein glycosylation. Structural and functional aspects. Eur J Biochem 218: 1–27, 1993
14. Stiles GL: Deglycosylated mammalian β_2-adrenergic receptors: Effect on radioligand binding and peptide mapping. Arch Biochem Biophys 237: 65–71, 1985
15. Mitsuhashi M, Payan DG: Receptor glycosylation regulates the affinity of histamine H1 receptors during smooth muscle cell differentiation. Mol Pharmacol 35: 311–318, 1989
16. Habecker BA, Tietje KM, van Koppen CJ, Creason SA, Goldman PS, Migeon JC, Parenteau LA, Nathanson NM: Regulation of expression and function of muscarinic receptors. Life Sci 52: 429–432, 1993
17. Rademacher TW, Parekh RB, Dwek RA: Glycobiology. Annu Rev Biochem 57: 785–838, 1988
18. Halle W, Wollenberger A: Differentiation and behaviour of isolated embryonic and neonatal heart cells in a chemically defined medium. Am J Cardiol 25: 292–299, 1970
19. Wallukat G, Wollenberger A: Supersensitivity to β-adrenoceptor stimulation evoked in cultured neonatal rat heart myocytes by L(+)-lactate and pyruvate. J Auton Pharmacol 13: 1–14, 1993
20. Canale ED, Campbell GR, Smolich JJ, Campbell JH: Cardiac muscle. Springer-Verlag, Berlin, pp 195–225, 1986
21. Haase H, Striessnig J, Holtzhauer M, Vetter R, Glossmann H: A rapid procedure for the purification of cardiac 1,4-dihydropyridine receptors from porcine heart. Eur J Pharmacol 207: 51–59, 1991
22. Haase H, Wallukat G, Vetter R, Will H: Characterization of calcium antagonist receptors in highly purified porcine cardiac sarcolemma. Biomed Biochim Acta 46: S363–9, 1987
23. Henning U: Plaquebildung bei Mumps-Viren – Ein Beitrag zur biologischen Charakterisierung von attenuierten und Wild-Stämmen. [PhD Thesis]. Humboldt University Berlin, 1986
24. Gomez JP, Potreau D, Branka JE, Raymond G: Developmental changes in Ca^{2+} currents from newborn rat cardiomyocytes in primary culture. Pflugers Arch-Eur J Physiol 428: 241–249, 1994
25. Elbein AD: Glycosidase inhibitors: inhibitors of N-linked oligosaccharide processing. FASEB J 5: 3055–3063, 1991
26. Kobata A: Structures and functions of the sugar chains of glycoproteins. Eur J Biochem 209: 483–501, 1992
27. van den Eijnden DH, Joziasse DH: Enzymes associated with glycosylation. Curr Opin Struct Biol 3: 711–721, 1993
28. Welling A, Bosse E, Cavalie A, Bottlender R, Ludwig A, Nastainczyk W, Flockerzi V, Hofmann F: Stable co-expression of calcium channel alpha(1), beta and alpha(2)/delta subunits in a somatic cell line. J Physiol-London 471: 749–765, 1993
29. Wibo M, Bravo G, Godfraind T: Postnatal maturation of excitation-contraction coupling in rat ventricle in relation to the subcellular localization and surface density of 1,4-dihydropyridine and ryanodine receptors. Circ Res 68: 662–673, 1991
30. Jorgensen AO, Shen ACY, Arnold W, Mcpherson PS, Campbell KP: The Ca^{2+}-release channel/ryanodine receptor is localized in junctional and corbular sarcoplasmic reticulum in cardiac muscle. J Cell Biol 120: 969–980, 1993
31. Morwinski R, Karsten U, Wallukat G: Mast cells detected in cultures of neonatai rat heart cells. Agents Action 41: C39–C40, 1994
32. Page E, Earley J, Power B: Normal growth of ultrastructure in rat left ventricular myocardial cells. Circ Res 35: II-12–II-16, 1974
33. Hirakow R, Gotoh T, Watanabe T: Quantitative studies on the ultrastructural differentiation and growth of mammalian cardiac muscle cells. 1. The atria and ventricles of the rat. Acta anat 108: 144–152, 1980

Molecular and Cellular Biochemistry **160/161**: 53–59, 1996.

Comparative study of cardiac electrophysiological effects of atrial natriuretic peptide

Valeria Kecskemeti,[1] Pal Pacher,[1] Csaba Pankucsi[2] and Peter Nanasi[2]

[1]*Department of Pharmacology, Semmelweis University of Medicine, Budapest;* [2]*Department of Physiology, Medical University of Debrecen, Debrecen, Hungary*

Abstract

The effects of atrial natriuretic peptide (ANP) on action potential characteristics were studied in various (human, rabbit, guinea-pig) atrial and guinea-pig right ventricular papillary muscles. ANP (1–100 nM) did not modify the resting membrane potential nor the maximum rate of depolarization phase (V_{max}). Up to 10 nM, ANP dose-dependently decreased the action potential amplitude both in guinea-pig atrial and ventricular muscles, but it did not affect this parameter in the other atrial preparations. ANP caused a dose-dependent, marked decrease of action potential duration (APD) in practically every cardiac preparation studied (exception of guinea-pig left atrium). The strongest effect on APD can be observed in human atrial and guinea-pig ventricular fibers. The K^+ channel blocker 4-aminopyridine (1 mM) and the ATP-dependent K^+ channel inhibitor glibenclamide (10 μM) prevented the effect of ANP on APD in both ventricular atrial preparations. ANP prevented the appearance of isoprenaline (0.5 μM) induced slow AP in K^+ depolarized myocardium. The present data suggest that ANP may inhibit the slow inward Ca^{2+} channel activity and facilitate the K^+ channel activity. (Mol Cell Biochem **160/161**: 53–59, 1996)

Key words: atrial natriuretic peptide, human atrial muscle, cardiac action potential, glibenclamide

Introduction

Atrial natriuretic peptide (ANP) is a cardiac hormone predominantly synthesized and secreted by the cardiac atria [1, 2]. ANP is secreted and stored within specific granules of myocytes as a part of a large prohormone (126 amino acids) and released in the circulation in a form of 28 amino acids [3, 4]. Secretion of ANP is suggested to be due to atrial stretch or volume loading but the exact mechanism of its release remains unclear [1, 2, 5]. Although ANP, as a potent vasodilator participates in the regulation of blood pressure, salt and water balance, very little is known about its direct cardiac actions. Some reports suggest that ANP directly influences cardiac function *in vivo* [6] and it decreases the contractility on isolated heart preparations [7, 8], while others deny such direct cardiac effects of ANP [9, 10].

The electrophysiological effects of ANP on cardiac cells are rather complex and also controversial. ANP was reported to inhibit the basal L-type Ca^{2+} channel current in several species (human atrial cells, rat, chick embryo, rabbit ventricular cells) [11–14] but in frog ventricular myocytes this inhibitory effect occurred only on the isoprenaline-elevated $I_{Ca(}$ [15,16]. ANP has also been reported to block the Na^+ current in rat and guinea-pig ventricular myocytes and in canine Purkinje cells [12, 17]. ANP was found to transform sodium channels into calcium-conducting channels in the former preparation [12]. However, it failed to act on the later one [17]. As far as the effect on the potassium current is concerned, ANP was found to increase the K^+ current in fetal ventricular cells [13] and decrease this current in human atrial cells [11].

In the present study, effects of ANP on action potential parameters of various cardiac preparations from human, guinea-pig and rabbit were examined using the conventional microelectrode technique.

Address for offprints: V. Kecskemeti, Department of Pharmacology, Semmelweis University of Medicine, Budapest, Nagyvarad ter 4, PO Box 370, 1445 Hungary

Materials and methods

Preparations

Guinea-pigs (300–400 g) and New Zealand White rabbits (1.5–2.0 kg) were killed by a blow on the head. The hearts were rapidly removed and the left auricles and right ventricular papillary muscles were isolated. The preparations were placed in a tissue chamber and perfused with oxygenated (95% O_2 - 5% CO_2) Tyrode solution (34 C, pH 7.4). The organs were stimulated by rectangular pulses (1–1.3 Hz, 1 msec and voltage twice the stimulus threshold).

The human right atrial appendages were obtained from the hearts of 9 consecutive patients undergoing corrective cardiac surgery (Thanks Dr Peter Homlay 2nd Department of Surgery of Medical University of Debrecen). Prior to surgery, informed consent and ethical permission was obtained.

Transmembrane potential recording

Transmembrane action potentials (APs) were recorded by glass capillary microelectrodes (5–25 megohms) filled with 3 M KCl with a set of a microelectrode amplifier and a dual-beam storage oscilloscope (Tektronix 2230). APs were stored and calculated by IBM computer. The maximum rate of depolarization (V_{max}) was recorded on the second beam of the oscilloscope as the electronically obtained differential of the rising phase (dV/dt). Details of the technique were described previously [18, 19]. Recently, we have used Intrasys system (Experimetria, Budapest).

For analysis of the effects of drugs on slow APs, the preparations were perfused with 25 mM K^+ Tyrode solution (containing 0.06 mM ascorbic acid). Despite the intense electrical stimulation (10 times threshold strength, the preparations lost their excitability and the fibers depolarized to about -44 mV. Slow APs were induced under these conditions by adding isoprenaline (0.5 μM) to the perfusate. The frequency of stimulation was reduced to 0.3 Hz.

The drugs used were atriopeptin III (Sigma), human ANP (Sigma), isoprenaline (Aludrin, Boehringer Ingelheim), glibenclamide, 4-aminopyridine (Sigma). The composition of the Tyrode solution was (mmoles/liter): 137 NaCl, 4.0 KCl, 1.8 $CaCl_2$, 1.5 $MgCl_2$, 11.9 $NaHCO_3$, 0.42 NaH_2PO_4, 5.5 glucose.

Statistics

All values are presented as mean ± standard error. Statistical analysis were performed with Student's paired *t* test. A p value less than 0.05 was considered significant.

Results

Guinea-pig left atrial and right ventricular papillary muscles

The effects of atriopeptin III(ANP) on different parameters of atrial and ventricular action potentials (APs) at concentrations from 1–100 nM are shown in Fig. 1. ANP caused a slight but not significant increase of the resting membrane potential (RP). Up to 10 nM concentration ANP dose-dependently decreased action potential amplitude (APA). ANP had no significant changes in the maximum rate of rise of depolarization phase (V_{max}) of ventricular AP but at 10 and 100 nM concentrations it attenuated V_{max} in atrial muscles. The effects of ANP on the repolarization phase in the two types of muscles is also different. While the duration of atrial AP remained unchanged the duration of AP (APD) at 20% (APD_{20}), 50% (APD_{50}) and 90% (APD_{90}) in ventricular papillary muscle dose-dependently decreased.

The effects of ANP (100 nM) were studied in papillary muscles pretreated with the K+ channel blocker 4-aminopyridine. Table 1 illustrates that 4-aminopyridine significantly prolonged the APD and prevented the effect of ANP on the repolarization phase.

Rabbit left atrial muscles

Figure 2 shows the effects of atriopeptin III on the left atrial muscle. While 1 nM had no significant changes in AP parameters, 10 and 100 nM caused a slight but statistically not significant decrease of APA. Either RP or V_{max} was not changed by ANP. 10 nM ANP did not modify APD whereas 100 nM significantly shortened the repolarization phase (APD_{20}, APD_{50}, APD_{90} were decreased about in an equal proportion, by 9–10%).Glibenclamide (10 μM), the ATP dependent K^+ channel blocker significantly prolonged the APD_{20}, APD_{50} and APD_{90} by 47, 33 and 29%, respectively and attenuated the V_{max}.

Human right atrial muscles

Figure 3 and Table II show the concentration-dependent effects of human ANP (hANP) on APs of right human atrial muscles. RP, APA, V_{max} were not changed by hANP at all the concentrations tested. hANP caused a significant, dose-dependent shortening of APD_{50}, APD_{75} and APD_{90} but it did not modify the early repolarization phase. As the normalized data show (Fig. 4) these effects of hANP on the half and late repolarization phase are nearly equal.

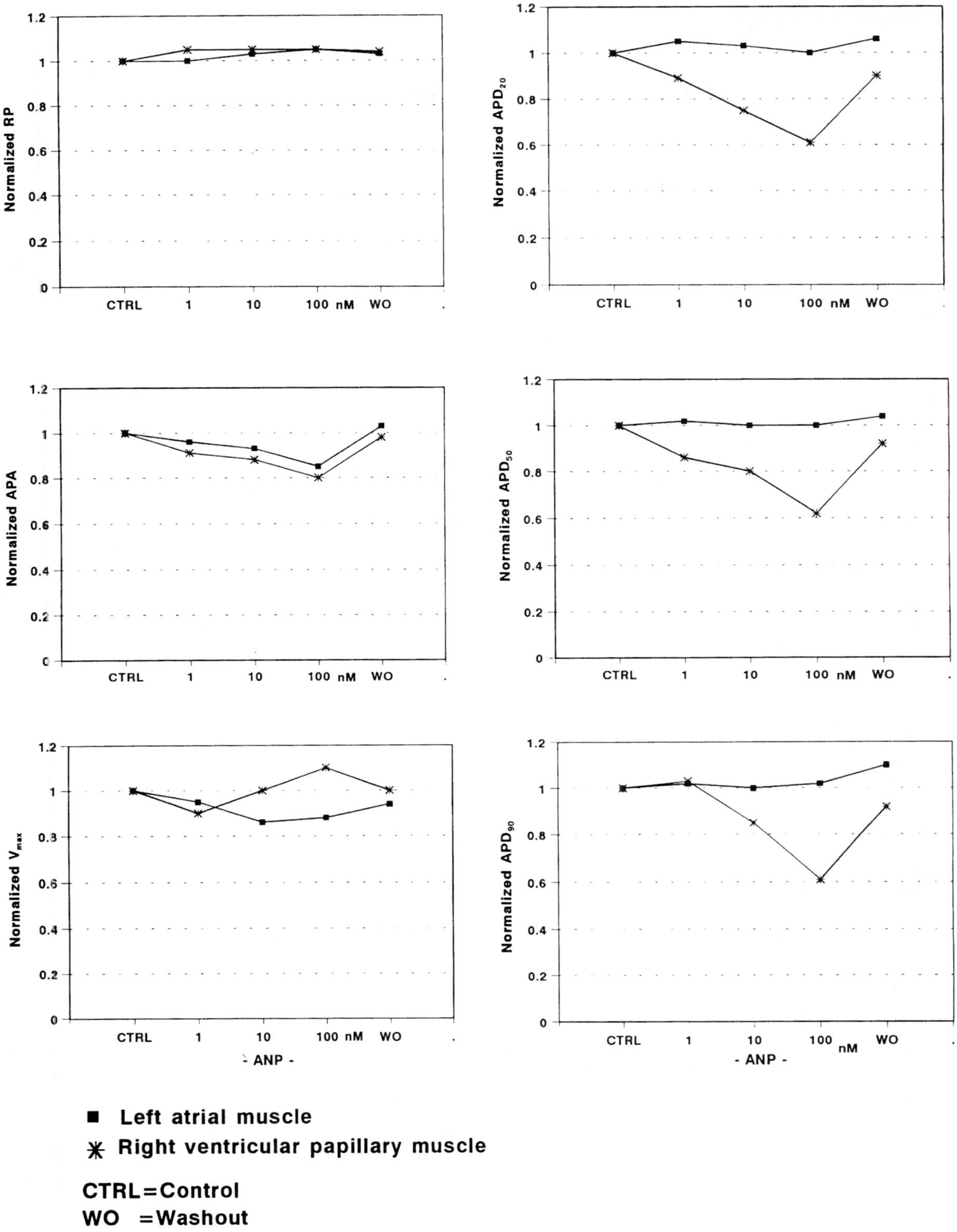

Fig. 1. Concentration-dependent effects of Atriopeptin III on the action potential parameters of guinea-pig atrial and ventricular preparations Normalized data. *Right ventricular papillary muscle; ♦ Left atrial muscle; RP = resting membrane potential; APA= amplitude of action potential; V_{max} = maximum rate of rise of 0 depolarization phase; APD_{20}, APD_{50} and APD_{90} =. action potential duration measured at 20, 50 and 90% repolarization; CTRL = control; WO = washout.

Table 1. Effect of Atriopeptin III. on the parameters of action potential in isolated right ventricular papillary muscle of guinea-pig in the presence of 4-aminopyridine

Treatment	N	RP (mV)	APA (mV)	V_{max} (V/s)	APD at repolarization 20%	50%	90% (ms)
Control	55	–80.2 ± 0.8	104.4 ± 0.9	176.5 ± 4.0	81.1 ± 1.8	135.6 ± 2.5	175.0 ± 3.6
1 mM 4-aminopyridine	44	–81.1 ± 0.9	103.8 ± 1.1	168.0 ± 3.8	96.7* ± 1.6	150.1* ± 2.6	198.6* ± 4.0
+ 100 nM ANP	48	–81.6 ± 1.0	98.6 ± 1.2	170.6 ± 4.2	86.3 ± 1.5	140.7 ± 2.2	180.9 ± 3.8

Data are expressed as mean ± S.E.M. of 5 experiments; *statistically significant difference from control ($p < 0.05$)

Effects on slow action potentials

The effect of ANP on the slow Ca^{2+} current was studied on the slow APs elicited by epinephrine in guinea-pig ventricular papillary muscles depolarized by 25 mM K+ Tyrode solution. In the presence of high external K^+ concentration, the fast Na^+ channels were inactivated, the preparations became inexcitable and the recorded RP was shifted from -82.5 ± 1.1 mV to -44.6 ± 0.9 mV. Table 3 shows that under this condition, isoprenaline (0. 5 μM) induced slow conducting APs but lost the restoring activity in the presence of 100 nMANP.

Discussion

In the present work we have studied the effects of ANP on action potentials of atrial muscles of various mammals (human, rabbit, guinea-pig) and compared them with data obtained in guinea-pig ventricular papillary muscle. The results show that ANP modified differently the parameters of action potentials of various atrial preparations. The amplitude of AP decreased in both atrial and ventricular fibers of guinea-pig, but remained unchanged in rabbit and human atrial muscles. The plateau phase of AP in rabbit atrial and guinea-pig ventricular papillary muscles was shortened, however, it did not change in guinea-pig and human atria. As APA and one part of the plateau phase of cardiac AP can generally be considered as indirect indicators of the activity of slow L-type Ca^{2+} channels, these results suggest the possibility that ANP reduced Ca^{2+} influx via the slow inward Ca^{2+} current. This suggestion was supported by results (Table 3) obtained in K^+-depolarized guinea-pig myocardium and in our preliminary ionic currents study [20]. In high external K^+ concentration, when the fast Na^+ channels were inactivated ANP prevented the appearance of the isoprenaline-induced Ca^{2+} dependent slow AP. Furthermore in preliminary ionic current measurements in frog atrial cells we have found that ANP reduced the epinephrine-elevated Ca^{2+} current [20]. Our results obtained in guinea-pig and rabbit are in agreement with the data of others [11, 13, 14]. ANP was found to decrease the L-type Ca^{2+} current in ventricular cardiocytes of frog [15, 16], rat [12], rabbit [14], chick embryo [13] and in human atrial cells [11]. These data suggested that ANP inhibited the cardiac L-type Ca^{2+} channel activity through the elevation of GMP and then activation of GMP-depedent protein kinase [14, 15, 21–23].

In our study ANP was ineffective on the V_{max} in both atrial and ventricular preparations. V_{max} can be regarded as indirect indicator of the fast Na^+ channel activity. As far as the effect of ANP on fast Na^+ channels is concerned, some papers reported that ANP suppressed Na^+ current in rat and guinea-pig ventricular and canine Purkinje fibers [12, 17]; however, the most of the authors failed to confirm these data [13, 14, 16].

Our results show that the predominant and most consequent electrophysiological effect of ANP on ventricular and atrial muscles is the decrease of the duration of AP. The strongest effect on APD is observed in human atrial and guinea-pig ventricular muscles. The duration of cardiac AP is determined by a fine balance of inward (slow inward current, I_{si}, steady-state or 'window' current) and outward (inward rectifying current,IK,transient outward current, Ito, delayed rectifying current, I_{K1} ATP-dependent current K_{ATP}) currents [24]. The contribution of these currents to the plateau phase of cardiac AP is also different depending on the cardiac tissue and species. The fact that both 4-aminopyridine, a blocker of voltage-dependent K^+ channels and glibenclamide, a blocker of ATP-dependent K^+ channels prevented the ANP-induced APD shortening (Table 1, Fig. 2) suggested that ANP might increase K^+ current. Our findings are supported by the data obtained in chick embryo [13] showing the enhancing effect of ANP on delayed outward K^+ current but are inconsistent with those of Le Grand *et al.* [11] reporting an inhibitory effect of ANP on transient outward K^+ current in human atrial myocytes. It is possible that the ANP- induced inhibition of Ca^{2+} channel activity and stimulation of K^+ channel function contribute to the negative inotropic action of ANP [7, 8] as well as, to the regulation of its own synthesis [25] and then the regulation of cardiac function.

A

B

C

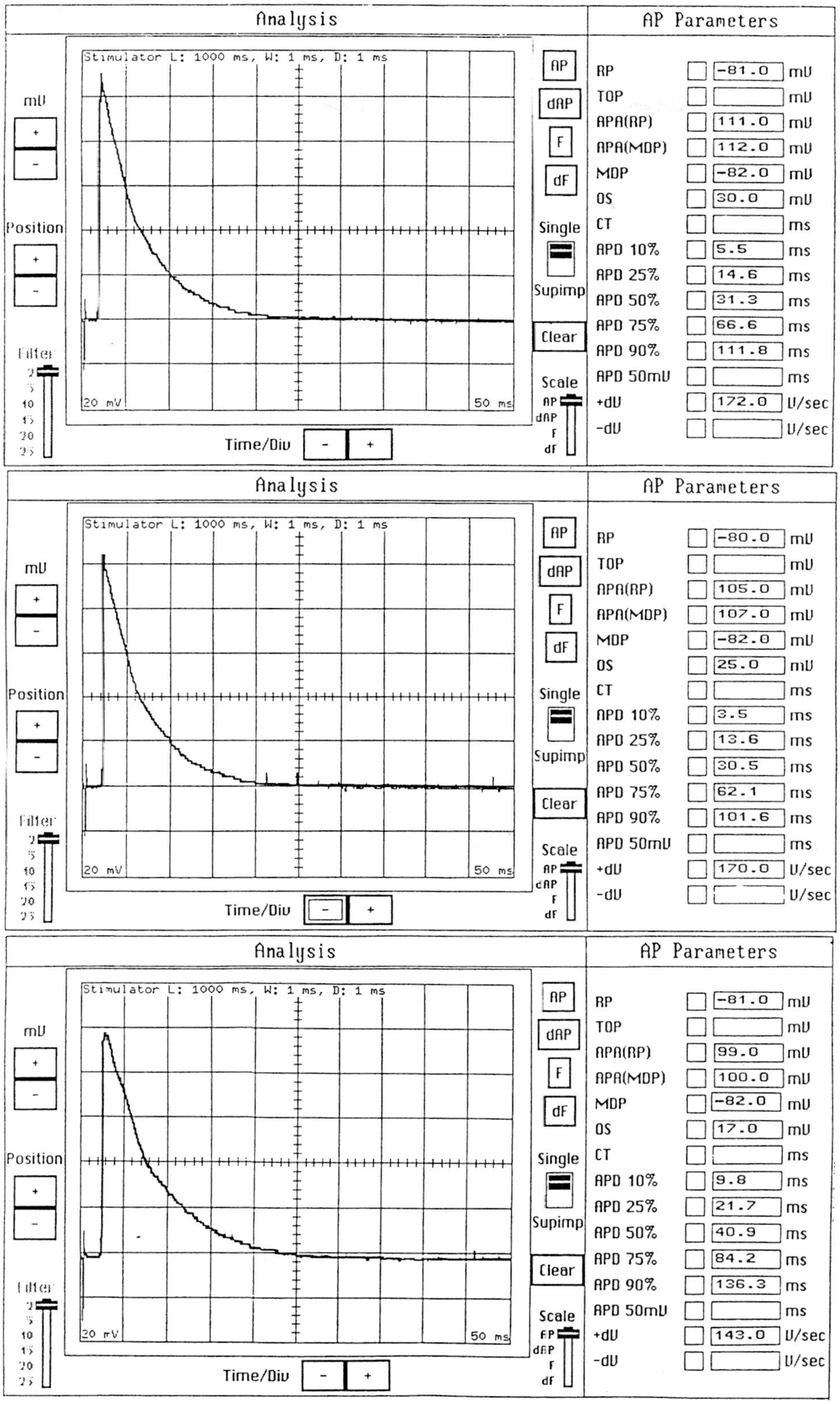

Fig. 2. Effects of Atriopeptin III, and Glibenclamide on action potential characteristics of rabbit left atrial muscle, A panel shows control data, B represents the effect of 100 nM Atriopeptin III and B demonstrates the effect of 100 nM Atriopeptin in the presence of 10 μM glibenclamide.

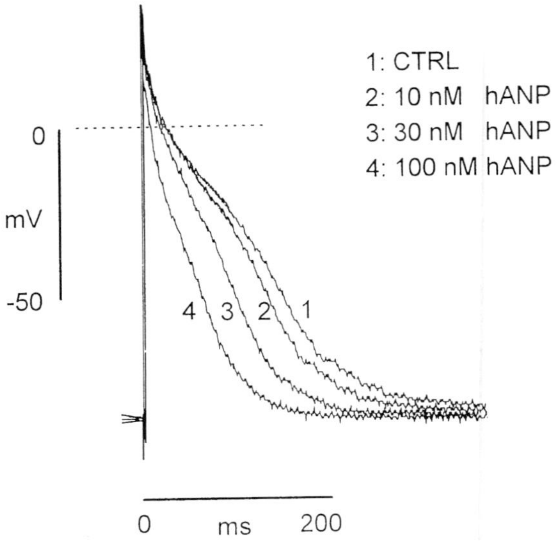

Fig. 3. Effects of human ANP on action potential characteristics of human right atrial fibers Cycle length: 750 ms.

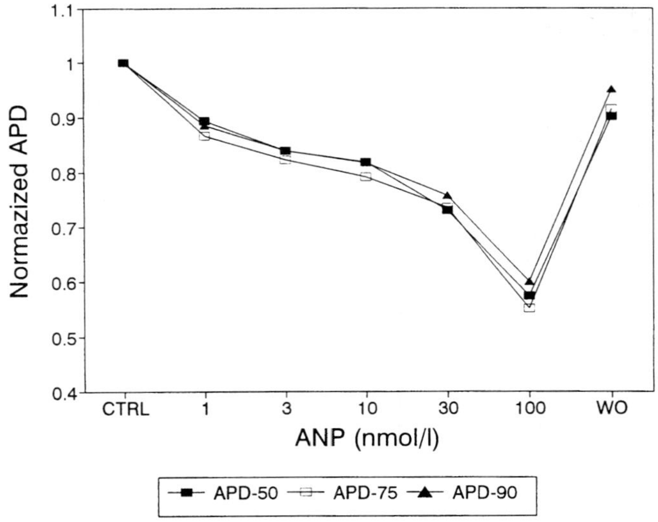

Fig. 4. Normalized data of human ANP effects on action potential duration in human right atrial muscles, APD_{50}, APD_{75} and APD_{90} = action potential duration measured at 50, 75 and 90% repolarization.

Table 2.

Treatment	RP (mv)	APA (mv)	APD50(ms)	APD75(ms)	APD90(ms)
Control	–84.2 ± 1.5	101.7 ± 3.8	59.7 ± 16.1	142.50 ± 18.7	208.7 ± 21.9
10nM hANP	–83.5 ± 1.4	102.3 ± 3.9	45.5* ± 10.9	111.5* ± 14.6	167.9* ± 18.4
30nM hANP	–83.8 ± 1.6	101.6 ± 4.1	42.3* ± 9.8	103.9* ± 13.3	157.3* ± 15.5
100nM ANP	–83.2 ± 1.5	102.3 ± 3.6	29.9* ± 7.7	73.8* ± 9.7	120.5* ± 12.2
Washout	-83.9 ± 1.7	101.9 ± 3.7	53.7 ± 2.4	130.4 ± 6.7	198.5 ± 6.5

Data represent means ± S.E.M. of 7 experiments; RP = resting potential; APA = action potential amplitude; APD = action potential duration measured at during 50%, 75% and 90% repolarization; *Changes in APD are significant ($p < 0.05$)

Table 3. Effect of atriopeptin III. in potassium-depolarized guinea-pig ventricular papillary muscles

	First group			Second group		
	RP (mV)	APA (mV)	Vmax (V/s)	RP (mV)	APA (mV)	Vmax (V/s)
Control	–82.5 ± 1.1	102.0 ± 1.5	185 ± 5.8	–78.5	100.8 ± 1.0	179.3 ± 6.2
25 mM K^+ Tyrode	–44.6 ± 0.9	–	–	–43.8 ± 1.1	–	–
25 mM K^+ Tyrode + 0.5 μM Isoprenaline	–44.2 ± 1.0	60.2 ± 1.2	16.8 ± 2.2	–	–	–
25mM K^+ Tyrode + 100 nM ANP				–43.4 ± 1.0	–	–
25mM K^+ Tyrode + 100 nM ANP + 0.5 μM Isoprenaline				–44.0 ± 0.9	–	–

Results are mean values S.E.M. of 5 experiments: RP = resting potential; APA = action potential amplitude; Vmax = maximum rate of rise of phase 0 depolarization

Acknowledgements

The authors wish to thank Dr. Peter Homlay (2nd Department of Surgery of Medical University of Debrecen) for the atrial preparations and Mrs K. Barna and Mr J. Balogh for the excellent secretarial assistance. This study was supported by Grants ETT T-06018/1993 and ETT T-03 286/1993 of Hungarian Ministry of Health.

References

1. Cantin M, Genest J: The heart and the atrial natriuretic factor. Endocr Rev 6: 107–127, 1985
2. De Bold AJ@ Atrial natriuretic factor: A hormone produced by the heart. Science 230: 767–770, 1985
3. Seidman CE, Duby AD, Choi E, Graham RM, Haber E, Homey C, Smith JA, Seidman JG: The structure of rat preproatrial natriuretic factor as defined by a complementary DNA clone. Science 225: 324–326, 1984
4. Thibault G, Garcia R, Gutkowska J, Bilodeau J, Lazure C, Seidah NG, Chretien M, Genest J, Cantin M: The propeptide Asn1-Tyr 126 is the storage form of rat atrial natriuretic factor. Biochem J 241: 265–271, 1987
5. Needleman P, Blaine EH, Greenwald JE, Mickener ML, Saper CB, Stockman PT, Tolynay HE: The biochemical pharmacology of atrial peptides. Ann Rev Pharmacol Tox 29: 123–154, 1989
6. Seymour AA, Sweet CS, Stabilito LI, Emmert SE: Cardiac and hemodynamic responses to synthetic atrial natriuretic factor in rats. Life Sci 40: 511–519, 1987
7. Neyses L, Vetter H: Action of atrial natriuretic peptide and angiotensin II on the myocardium: studies in isolated rat ventricular cardiomyocytes. Biochem Biophys Res Commun 163: 1435–1443, 1989
8. Vaxelaire JF, Laurent S, Lacolley P, Briand V, Schmidt H, Mishel JB: Atrial natriuretic peptide decreases contractility of cultured chick ventricular cells. Life Sci 45: 41–48, 1989
9. Bergey JL, Kotler D: Effect of atriopeptin I, II and III on atrial contractility, sinus nodal rate (guinea-pig) and agonist- induced, tension in rabbit aortic strips. Eur J Pharmacol 110: 277–281, 1985
10. Criscione L, Burdet R, Hanni , Kamber B, Truog A, Hofbauer KG: Systemic and regional hemodynamic effects of atripeptin II in anesthetized rats. J Cardiovasc Pharmacol 9: 135–141, 1987
11. LeGrand B, Deroubaix E, Coutetil JP, Coraboeuf E: Effects of atrionatriuretic factor on Ca^{2+} current and Ca_i-independent transient outward K^+ current in human atrial cells. Pflügers Arch 421: 486–491, 1992
12. Sorbera LA, Morad M: Atrionatriuretic peptide transforms cardiac sodium channels into calcium-conducting channels. Science 247: 969–973, 1990
13. Bkaily G, Perron N, Wang S, Sculptoreanu A, Jacques D, Ménard D: Atrial natriuretic factor blocks the high-threshold Ca^{2+} current and increases K^+ current in fetal single ventricular cells. J Mol Cell Cardiol 25: 1305–1316, 1993
14. Tohse N, Nakaya H, Takeda Y, Kanno M: Cyclic GMP mediated inhibition of L-type Ca^{2+} channel activity by human natriuretic peptide in rabbit heart cells. Brit J Pharmacol 114: 1076–1082, 1995
15. Fischmeister R, Hartzell HC: Cyclic guanosine 3', 5' monophosphate regulates the calcium current in single cells from frog ventricle. J Physiol 387: 453–472, 1987
16. Gisbert MP, Fischmeister R: Atrial natriuretic factor regulates the calcium current in frog isolated cardiac cells. Circ Res 62: 660–667, 1988
17. Sheets MF, Hanck DA: Atrionatriuretic peptide and calcium-conducting sodium channels. Science 252: 449–450, 1991
18. Kecskemeti V, Kelemen K, Knoll J: Effect of prostaglanding E1 on cardiac transmembrane potential. Eur J Pharmacol 24: 289–296, 1973
19. Kecskemeti V, Braquet P: Cellular electrophysiological effects of platelet-activating factor (PAF) and its antagonist BN 52021 in cardiac preparations. Drug Exptl Clin Res 18: 23–27, 1992
20. Kecskemeti V, Marko R, Kelemen K: Electrophysiological effects of atrial natriuretic peptide (ANP). J Mol Cell Cardiol 27: A138, 1995
21. Levi RC, Alloatti G, Fischmeister R: Cyclic GMP regulates the Ca-channel current in guinea-pig ventricular myocytes. Pflügers Arch 413: 685–687, 1991
22. Méry PF, Lohmann SM, Walter U, Fischmeister R: Ca^{2+} current is regulated by cyclic GMP-dependent protein kinase in mammalian cardiac myocytes. Proc Natl Acad Sci USA 88: 1197–1201, 1991
23. Ono K, Trautwein W: Potentiation by cyclic GMP of β–adrenergic effect on Ca^{2+} current in guinea-pig ventricular cells. J Physiol 443: 387–404, 1991
24. Reuter H: Ionic Channels in cardiac cell membranes. Ann Rev Physiol 46: 473–495, 1984
25. Saito Y, Nakao K, Morii N, Sugawara A, Shiono S, Yamada T, Toh H, Sakamoto M, Kurahashi K, Fujiwara M, Imura H: Bay K 8644, a voltage-sensitive calcium channel agonist facilitates secretion of atrial natriuretic polypeptide from isolated perfused rat hearts. Biochem Biophys Res Commun 138: 1170–1176, 1986

Molecular and Cellular Biochemistry **160/161**: 61–66, 1996.

Effect of TaiCatoxin (TCX) on the electrophysiological, mechanical and biochemical characteristics of spontaneously beating ventricular cardiomyocytes

Elisabeth Fantini, Pierre Athias, Régine Tirosh[1] and Arié Pinson

Laboratoire de Physiologie Faculté de Médecine 21033 Dijon, France; [1]Laboratory for Myocardial Research, The Hebrew University-Hadassah Medical School, POB 12272, Jerusalem 91120, Israel

Abstract

TaiCatoxin (TCX), a complex toxin isolated from Taipan snake venom, is believed to have a specific blocking activity on voltage-dependent cardiac calcium channels. The aim of this study was to investigate the effects of TCX on a broad range of heart muscle cell functions, i.e. electrophysiology, contractility, automaticity and the related biochemical modifications. Myocyte-enriched cultures were prepared from newborn rat heart ventricles. The transmembrane potentials were recorded with glass microelectrodes. The contractions were monitored photometrically. TCX decreased the action potential amplitudes, mainly by lowering the plateau. The action potential duration and the contraction parameters were decreased. Although TCX has a minor overall negative chronotropic effect, it evoked transient but severe arrhythmias and prolonged changes in the intercellular electrical coupling. Moreover, the action of TCX appeared to be dose-dependent. These effects are consistent with a specific blockade of the L-type, voltage-dependent calcium channels, but effects of other components of the toxin complex cannot be excluded. TCX also exhibits phospholipase A_2 activity leading to the release of lysophospholipids and FFA (acyl CoA and acyl carnitine), which have detrimental effects on cellular integrity and function. (Mol Cell Biochem **160/161**: 61–66, 1996)

Key words: cultured cardiomyocytes, ion channel toxin, TaiCatoxin, cardiac action potential, cardiac automacity, arrhythmias, phospholipase A_2

Introduction

Natural toxins are an important tool for the study of ion channels, such as sodium channels [1] . Calcium channels, which are of key importance in cardiac electrophysiology and excitation-contraction coupling, have mainly been studied to date by the use of synthetic drugs, such as dihydropyridines, phenylalkamines or benzothiazepines [2]. However, the use of these compounds is hampered by their limited selectivity and/or membrane solubility. This has led to an intensive search for natural toxins that affect cardiac calcium channels.

Several recent reports suggest that a newly discovered complex toxin purified from Australian Taipan snake venom may selectively block L-type, voltage-dependent calcium channels [3]. The physiological activity of this toxin (named TaiCatoxin, TCX) has only been evaluated in the membrane conductances of single, isolated cardiac myocytes By contrast, the effects of the TCX on the action potential, automaticity and contractility of myocardial cells has not yet been studied.

The aim of this study was to investigate the effects of TCX on the electrophysiology and contractility of rat ventricular myocytes maintained in cell culture and to evaluate the phospholipase activity of TCX.

Our results show that TCX induced aberrations in the electrical and rhythmic activity, which may be partly due to calcium ion fluxes and to the release of FFA and lysophospholipids mediated via its phospholipase A_2 activity.

Materials and methods

Cell culture

Enriched cultures of ventricular cardiomyocytes were obtained from the hearts of neonatal Wistar rats. Hearts were aseptically removed from decapitated rats. The ventricles were minced and a seven-step digestion protocol then carried out with 0.1% trypsin (DIFCO). The cells were collected and suspended in standard culture medium. The cell suspension was enriched in muscular cells by two successive incubation periods for 30 and 90 min [4]. The cardiomyocytes in the suspension were counted and then plated at a density of 2×10^6 cells/dish in 60 mm Petri dishes (Falcon Primaria 3802). The cells were maintained at 37°C and 5% CO_2 in a humidified incubator (Forma Scientific). The culture medium used was Ham's F10 medium (Seromed) containing 10% fetal calf serum (Seromed), 10% human serum and antibiotics (0.1 mg/ml streptomycin, 100 U/ml penicillin). The free Ca^{2+} concentration was adjusted to 1.2 mM in all the media immediately before use by the addition of the appropriate amount of sterile $CaCl_2$ [5].

Physiological parameters

Immediately before experiments, the incubation medium was replaced with Puck's F balanced salt solution, which was covered with a layer of paraffin oil and maintained at 36°C in static bath conditions. The dish was placed in a small experimental chamber mounted on the moving stage of an inverted phase-contrast microscope (Leitz Daivert). Intracellular voltage measurements were carried out using conventional KCl-filled glass-fiber micropipettes (Clark Electromedical) and a high-impedance microelectrode amplifier (Biologic VF 180). Single cell contractions were monitored using a video-camera and a custom-made motion detector, as described previously [6].

The voltage and motion signals were directly digitized. Data analysis was carried out on a PC/AT personal computer using an ASYST-based processing program (ASYST, Kiethley Instr.). Off-line signals were transcribed continuously onto paper using a Siemens ink-jet recorder and stored on magnetic tape by an FM-recorder (Schlumberger MP 5521).

Pharmacology

Concentrated stock solutions of TCX (a gift from Almone Labs., Jerusalem) were prepared in Puck's F saline and kept on ice in the dark until use. Final concentrations were obtained by the addition of 5 µl stock solutions with rapid, gentle mixing. The measurements were carried out before and 25 and 50 min after the addition of toxin. The action potential and contraction parameters were the means of five recordings. Analysis of variance (ANOVA) was carried out on the data, the results of which were expressed as means ± S.E.M.

Phospholipase activity

On the 5th day in culture, cardiomyocytes were incubated overnight with 1 ml medium containing ^{14}C-oleate (0.2 mM, 0.1 µCi/per Petri dish). The cells were then incubated with complete culture medium (pulse chase), followed by washing and incubation for 1 h with Ham F10, either with or without (controls) 10^{-7} M TCX. Lipids were extracted from the cells and the media by the method of Bligh and Dyer [7]. The various PL and NL classes were separated by TLC according to the method of Mayorek *et al.* [8]. The radioactivity was determined by phosphostimulated luminescence autoradiography using a BAS 1000 (Fujix) bio-imaging analyzer.

Results

The effects of 10^{-6} M and 10^{-7} M TCX on the spontaneous action potentials and contractions of rat myocardial cells are shown in Fig. 1 (upper plot). In both series of experiments, the control electrical and contractile activities were similar to those found by our group in previous studies [5, 6]. Action potentials featured high-amplitude and fast sodium-dependent initial upstrokes. Action potentials, after a constant delay, were followed by cell shortening at a stable amplitude. Moreover, the spontaneous beating rates were also regular.

After 10 min in the presence of 10^{-7} M TCX, the action potential amplitude and the upstroke velocity were decreased, the diastolic periods were irregular and early postdepolarization was observed (Fig. 1). After 25 min of this treatment, the action potential amplitude and the upstroke velocity were restored, whereas the postdepolarizations triggered abnormal action potentials leading to extra-contraction. In addition, the slow depolarization due to the electronic conductance among adjacent cells, was slowed down and therefore became detectable (Fig. 1, arrows) just before the fast upstroke of the action potential. This suggests less efficient intercellular coupling. After 50 min, the rhythmic aberrations were reduced, while the slowing down of intercellular coupling persisted.

Increasing the TCX concentration to 10^{-6} M enhanced arrhythmias, action potential was further reduced and upstroke velocity between 10 and 25 min. In particular, abnormal triggered action potentials became more frequent and these arrhythmic activities were accompanied by weakening,

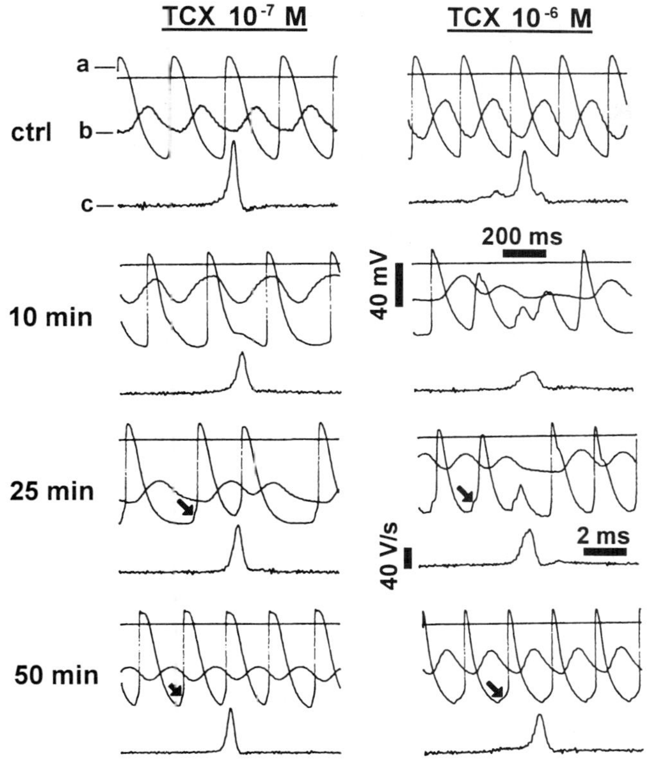

Fig. 1. Simultaneous recording of action potential (a), contraction (b) and the first derivative of the action potential upstroke (c) from rat ventricular cells in culture before (ctrl) and after 10–50 min after the addition of 10^{-7} M TCX (left column) and 10^{-6} M (right column). The traces were recorded at slow (a and b) and fast (c) sweep speeds. The contractions are shown on arbitrary scale. The horizontal baseline indicates zero potential levels in all the panels. Arrows show the initial pre-threshold slow depolarization resulting from intercellular electrical coupling.

or even complete disappearance of contractions. After 50 min with 10^{-6} M TCX, fairly regular activity was resumed, whereas depressed plateau and intercellular coupling persisted, and the action potential was decreased. Finally, Fig. 1 also clearly shows decreased contraction amplitudes at both concentrations tested.

The action potentials after 25 and 50 min of treatment are summarized in Table 1. With 10^{-7}M TCX, the changes in the maximum diastolic potential (MDP) and the overshoot (OS) were not significant. The velocity of the initial rising phase of the action potential (Vmax) remained unchanged, and the moderate decrease in action potential rate (APR) was not significant. TCX at the same concentration did not affect the time parameters of cell contraction (Table 2).

In the presence of 10^{-6} M TCX, Table 1 shows that the resting diastolic potential (MDP) decreased at 25 min, but remained close to control values at the end of the treatment. On the other hand, the overshoot potential was significantly depressed, an adverse effect which persisted until the end of the experiment. As in the presence of 10^{-7} M TCX, the action potential was shorter at 40 and 80% repolarization, an effect which was more pronounced in the presence of 10^{-6} M TCX. The upstroke velocity (Vmax) was significantly slower at 25 min after TCX addition, a decrease which persisted after 50 min treatment. However, at both concentrations, TCX does not seem to have a chronotropic effect on isolated ventricular cardiomyocytes Therefore, the main effect of TCX on the spontaneous contraction rate was to induce significant, albeit transitory, rhythm aberrations (Fig. 1). Finally, in addition to the decrease in contraction amplitude shown in Fig. 1, the contraction duration (CD80) and the shortening time (+Cmax) were significantly reduce at 10^{-6} M TCX.

After 1 h incubation with TCX, lysophospholipid and FFA levels, measured by a bio-imaging analyzer, were increased by 20- and 6-fold, respectively, as compared to controls (Fig. 2). However, after 2 h, the corresponding increases were only 6- and 3-fold, respectively, as compared to controls (not shown)

Although, PLA_2 activity was found with 10^{-7} M TCX, increased PG production and cellular LDH release could not be detected.

Discussion

TCX is a natural peptide toxin, which is assumed to have specific blocking activity on the voltage-dependent calcium channels. To date, the biological actions of TCX have mainly been evaluated via its effects on single-channel conductance [3] or the contractility of heart or blood vessel preparations [9, 10]. This is the first study in which the action of this toxin on mammalian myocardial cells has been investigated with respect to electrophysiology, and spontaneous automaticity, and phospholipase A_2 activity.

Our results show that TCX had a complex range of effects on action potential, contraction amplitude and rate and other parameters. Some of these, such as arrhythmias, were transitory, while others, including decreases in plateau, action potential duration, contraction and intercellular coupling, were sustained. It is common knowledge [11,12] that the cardiac action potential plateau is mainly governed by the L-type, voltage-dependent calcium channels, and is the main feature underlying the long-duration action potential characterizing heart muscle cells. Therefore, it can be reasonably assumed from our results that both the lowering of the plateau and the shortening of the action potential are consistent with the calcium-blocking activity previously attributed to this toxin.

Furthermore, TCX decreased cell contraction. This observation is consistent with the data of De Weille *et al.* [9]. There is now a broad consensus of support for the view that the excitation-contraction coupling in the heart is largely due to the entry of calcium ions during the plateau [13]. We have previously shown that the inhibition of calcium conductances

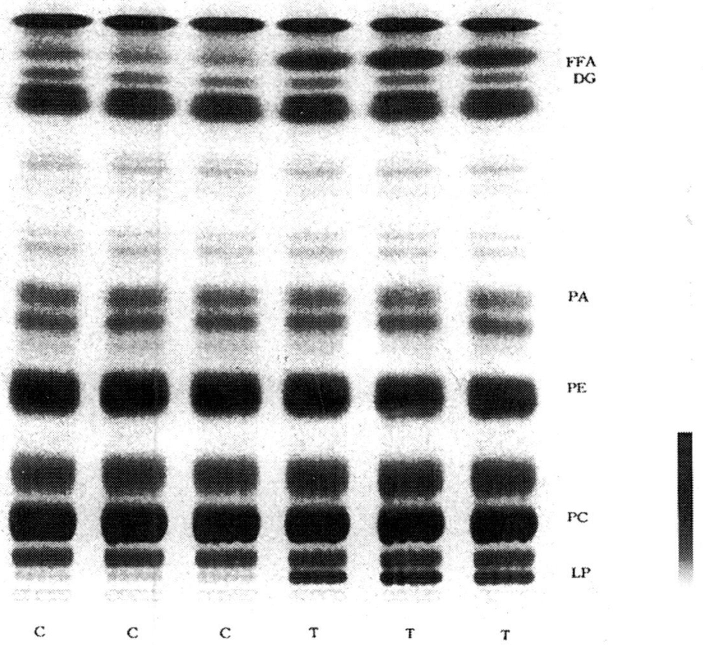

Fig. 2. The effect of 1 h incubation with 10^{-7} M TCX on cellular phospholipid and neutral lipid profiles. C - controls; T - TCX; FFA - free fatty acids; DG - diglycerides; PA phosphatidic acid; PE - phosphatidyl ethanolamine; PC phosphatidylcholine; LP - lysophospholipids. The experiments were carried out in triplicate.

Table 1. Effects of TaiCatoxin (TCX) at 10^{-7} M and 10^{-6} M on the amplitude and time parameters of action potentials recorded from rat ventricular myocytes in cultures

TCX concentration		MDP (–mV)	OS (mV)	ADP80 (ms)	ADP40 (ms)	APR (min^{-1})	Vmax (V/s)
	ctrl	76.7 ± 2.78	18.8 ± 2.68	136.4 ± 7.37^{a}	83.8 ± 3.89^{a}	258.8 ± 21.25	131.8 ± 10.26
1^{0}-7 M	25 min	77.8 ± 1.43	18.4 ± 3.26	117.0 ± 2.48^{b}	61.4 ± 4.03^{b}	252.0 ± 13.05	136.1 ± 12.04
	50 min	79.4 ± 1.67	19.3 ± 2.73	117.2 ± 3.59^{b}	63.8 ± 2.06^{b}	250.8 ± 12.85	142.3 ± 15.10
	p	NS	NS	$p < 0.05$	$p < 0.05$	NS	NS
	ctrl	79.0 ± 3.60^{a}	16.2 ± 3.56^{a}	145.2 ± 7.73^{a}	76.2 ± 5.21^{a}	332.6 ± 19.45	120.8 ± 13.38^{a}
10^{-6} M	25 min	69.8 ± 1.48^{b}	6.4 ± 2.07^{b}	95.0 ± 13.7^{b}	41.4 ± 8.79^{b}	307.6 ± 8.41	38.0 ± 14.40^{b}
	50 min	79.6 ± 2.88^{a}	6.2 ± 3.74^{b}	95.6 ± 9.44^{b}	39.4 ± 4.67^{b}	302.4 ± 7.50	98.6 ± 12.99^{a}
	p	$p < 0.05$	$p < 0.05$	$p < 0.05$	$p < 0.05$	NS	$p < 0.05$

Data are means ± S.E.M. (n = 5). p indicates the significance of the ANOVA; significant differences between control and treatment, individual means are indicated by different letter. See Materials and Methods for statistical analysis. Abbreviations: MDP: maximal diastolic potential; OS: overshoot; APD80, APD40: action potential duration at 80 and 40% repolarization, respectively; APR: action potential rate; Vmax: upstroke velocity.

by various classes of calcium antagonists affects the motion of cultured rat ventricular cells [14]. The negative ionotropic effect of TCX may thus be viewed as an additional factor affecting the membrane calcium channels during the plateau.

The effects of TCX on the action potential profile also suggest cell-to-cell uncoupling (see Fig. 1). The electrical coupling in myocardial tissue is based on gap junctions [15], the permeability of which is known to be modulated by intracellular calcium. However, the junctional conductance is reduced as the free calcium ion concentration in the cytoplasm rises [16]. Several other factors also contribute to the regulation of the gap Junctional conductance, including intracellular pH and cyclic nucleotides [17]. In addition, the gap junction channels are sensitive to the membrane micro-environment including liphophilic molecules and phospholipases in which they are embedded [18]. In this regard, TCX is an ion-channel blocker as well as a neurotoxic phospholipase and a serine protease inhibitor [13]. It is therefore tempting to ascribe the effects of TCX on cell communication to molecular entities other than the calcium-blocking ones. Such action of TCX on membrane structure might also explain why the initial fast upstroke of the action potential, which is gov-

Table 2. Effects of TaiCatoxin (TCX) at 10^{-7} M and at 10^{-6} M on the time parameters of contractions recorded from rat ventricular myocytes in cultures

TCX concentration		CD80 (ms)	tC20 (ms)	$+C_{max}$ (ms)	$-C_{max}$ (ms)
	ctrl	224.2 ± 19.45	56.8 ± 6.88	48.6 ± 4.66	46.6 ± 5.32
10^{-7}	25 min	153.8 ± 15.30	62.0 ± 7.31	64.2 ± 4.33	48.7 ± 2.16
	50 min	155.2 ± 13.77	68.7 ± 2.16	55.8 ± 5.41	46.2 ± 4.47
	p	NS	NS	NS	NS
	ctrl	211.2 ± 15.9[a]	71.4 ± 5.60	54.4 ± 3.83[a]	63.0 ± 7.44[a]
10^{-6} M	25 min	144.2 ± 11.0[b]	65.4 ± 4.56	35.8 ± 3.70[b]	45.8 ± 3.67[b]
	50 min	132 ± 10.95[b]	62.4 ± 4.28	34.8 ± 2.86[b]	46.2 ± 5.80[b]
	p	$p < 0.05$	NS	$p < 0.05$	$p < 0.05$

Data are means ± S.E.M. (n = 5). p indicates the significance of the ANOVA; significant differences between control and treatment, individual means are indicated by different letter. See Materials and Methods for statistical analysis. Abbreviations: CD80: contraction duration at 80% relaxation; tC20: excitation-contraction delay (time from action potential upstroke to 20–80% contraction; $-C_{max}$: time from 20–80% relaxation.

erned by sodium channels, slowed down transitorily at 10^{-7} M and became sustained and significant at 10^{-6} M (see second and third rows in Fig. 1 and Table 1). This slowing down of the fast depolarization phase of the action potential at 10^{-6} M TCX cannot be linked to the voltage dependence of the fast sodium channels since there was no significant decrease in the resting potential (see Table 1).

TCX did not significantly alter myocardial cell automacity, although the spontaneous beating rate of heart cells in culture has already been shown to be dependent on calcium ions [14]. However, the membrane currents underlying cardiac pacemaker activity are supported by a time-dependent inward current (if), by a time-dependent potassium outward current (iK), and by a T-type inward calcium current (T-iCa). The iK and if currents are modulated by calcium ions [12, 19]. Therefore, on the one hand, the absence or minimal effect of TCX on cardiac automaticity seems to be consistent with a specific blockade of L-type calcium channels by the toxin. This is in agreement with the results of Gomez *et al.* [20]. On the other hand, we have shown that the acute addition of TCX evoked severe arrhythmias, which were caused by early after-depolarizations, and, which are favored by the action potential shortening. Arrhythmias may also be due to an altered spread of activation in the myocardial cell layer resulting from electrical uncoupling, decreased upstroke velocity, and/or phospholipase A_2 activity, as mentioned above.

In conclusion, we have shown that TCX has a complex range of adverse effects on cardiac cell physiology, including sustained changes in the action potential plateau, in action potential duration, and in contraction amplitude, which seem to be consistent with specific L-type calcium channel blockade. Other components, such as phospholipase A_2, may contribute to additional, mainly transitory effects, on cell communication, upstroke velocity and rhythm.

Acknowledgements

The skillful technical assistance of Simone Almanza and Martine Degois is gratefully acknowledged. This work was supported by grants from the Conseil Regional de Bourgogne; the Institut National de la Santé et de la Recherche Médicale; and the Foundation pour la Recherche Médicale; and by grants from Mr. and Mrs. Vidal-Madjar (Paris); Mr. Jean Ariet (Paris) in memory of his wife, Mrs. Rebecca Ariet; and Mrs. Flora Dabbah (Geneva) in memory of her husband, Mr. Yossef Dabbah.

References

1. Catterall WA: Molecular pharmacology of voltage sensitive sodium channels. Pharmacology 2: 190–195, 1988
2. Reynolds IJ, Snyder S: Calcium antagonist receptors. In: T Narahashi (ed). Ion channels. Plenum Publishing Corporation, New York, 1988: 1 pp 133–249
3. Possani LD, Martin BM, Yatani A, Mochca-Moarles J, Zamudio FZ, Gurrola GB, Brown AML: Isolation and physiological characterization of taicatoxin, a complex toxin with specific effects on calcium channels. Toxicon 30: 1343–1364, 1992
4. Grynberg A, Athias P, Degois ML: Effect of change in growth environment on cultured myocardial cells investigated in a standardized medium. *In Vitro* Cell Develop Biol 22: 44–50, 1986
5. Athias P, Grynberg A: Electrophysiological studies on heart cells in culture. In A Pinson (ed). The heart cell in culture. CRC Press, Boca Raton, Florida, 1987: 1 pp 125–158
6. Courtois M, Khatami S, Fantini E, Athias P, Mielle P, Grynberg AL: Polyunsaturated fatty acids in cultured cardiomyocytes: Effect on physiology and β-adrenoreceptor function. Am J Physiol 262: H451–H456, 1992
7. Bligh EG, Dyer WJ: A rapid method of total lipid extraction and purification. Can J Biochem Physiol 370: 911–917, 1954
8. Mayorek N, Bar-Tana J: Lipid synthesis in permeabilized cultured rat hepatocytes. J Biol Chem 264: 4450–4455, 1989
9. De Weille JR, Schweitz H, Maes P, Tartae A, Lazdunski M: Calciseptine, a peptide isolated from black mamba venom, is a specific blocker of the L-type calcium channel Proc Natl Acad Sci 88: 2437–2440, 1991

10. Kuroda H, Chen Y-N., Watanabe TX, Kimura T, Sakakibara SL: Solution synthesis of calciseptine, an L-type specific calcium channel blocker. Peptide Res 5: 265–268, 1992
11. Athias P, Groz B, Klepping J: Determination of the ionic basis of the spontaneous activity of cultured rat heart cells using microinjection technique. Biol cell 37: 183–188, 1980
12. Varro A, Papp JG: The impact of single cell voltage clamp on the understanding of the cardiac ventricular action potential. Cardioscience 3: 131–144, 1992
13. Fabiato A: Calcium-induced release of calcium from sarcoplasmic reticulum. J Gen Physiology 85: 189–320, 1985
14. De la Chapelle-Grosz B, Athias P: Gentamicin causes fasts depression of action potential and contraction in cultured cardiomyocytes. Eur J Pharmacol 152: 111–120, 1988
15. Severs NJ: The cardiac gap junction and intercalated disc. Intern J Cardiol 26: 137–173, 1990
16. Page E, Shibata Y: Permeable junctions between cardiac cells. Ann Rev Physiol 43: 431–441, 1981
17. Spray DC, Burt JM: Structure activity relations of the cardiac gap junction channel. Am J Physiol 258: 195–205, 1990
18. Munster PN, Weingart R: Effect of phorbolester on gap junctions of neonatal rat heart cells. Pflügers Archiv - Europ J Physiol 423: 181–188, 1993
19. Hagiwara N, Irisawara H: Modulation by intracellular Ca^{2+} of the hyperpolarization-activated inward current in rabbit sino-atrial node cells. J Physiol 409: 121–141, 1989
20. Gomez JP, Portreau D, Branka E, Raymand G: Developmental changes in Ca^{2+} currents from new born rat cardiomyocytes in primary culture. Pflüg Arch-Eur J Physiol 428: 241–249, 1994

Molecular and Cellular Biochemistry **160/161**: 67–74, 1996.

Pharmacological characteristics of endothelin receptors in the rabbit ventricular myocardium: The nonselective endothelin receptor antagonist PD 145065 antagonizes the positive inotropic effect of endothelin-3 but not of endothelin-1

Ikuo Norota and Masao Endoh
Department of Pharmacology, Yamagata University School of Medicine, Yamagata 990-23, Japan

Abstract

Endothelin-3 (ET-3) elicited a concentration-dependent positive inotropic effect on rabbit papillary muscle, the maximal response being approximately 65% of the maximal response to isoproterenol. ET-1 induced a positive inotropic effect over the concentration range below 10^{-9} M, at which ET-3 did not produce a positive inotropic effect, but the maximal response to ET-1 was equivalent to or slightly lower than that of ET-3. The nonselective ET receptor antagonist PD 145065 effectively antagonized the positive inotropic effect of ET-3 in a concentration-dependent manner and abolished it at 10^{-5} M. PD 145065 decreased the positive inotropic effect induced by ET1 at lower concentrations (< 10^{-9} M) but it did not affect the main portion of the concentration-response curve for the positive inotropic effect, i.e., the effect induced by high concentrations (> 10^{-9} M) of ET-1. PD 145065 antagonized also the positive inotropic effect of sarafotoxin S6c. PD 145065 inhibited the specific binding of $[^{125}I]$ET-1 and of $[^{125}I]$ET-3 with a high- and a low-affinity site for competition. ET_B selective ligands, RES-701-1 and sarafotoxin S6c, displaced $[^{125}I^{uc}]$ET-3 with high affinity but they scarcely affected the $[^{125}I]$ET-1 binding. These findings indicate that different subtypes of the ET receptor are responsible for the induction of the positive inotropic effect of ET-3 and ET-1. ET receptors involved in the production of the positive inotropic effect in the rabbit ventricular myocardium have pharmacological characteristics that are different from those of conventional ET receptors originally classified based on the pharmacological findings in noncardiac tissues. The positive inotropic effect of ET-3 in the rabbit ventricular muscle may be mediated predominantly by ET_{A1} receptors that are susceptible to PD 145065 as well as BQ-123 and FR139317, and partially mediated by ET_B receptors that are inhibitable with RES-701-1. ET_{A2} receptors that are resistant to ET_A selective as well as nonselective antagonists may mainly be responsible for the positive inotropic effect of ET-1 in the rabbit ventricular muscle. (Mol Cell Biochem **160/161**: 67–74, 1996)

Key words: endothelin-1, endothelin-3, positive inotropic effect, PD 145065, nonselective endothelin receptor antagonist, $[^{125}I]$endothelin-1, $[^{125}I]$endothelin-3, receptor binding, rabbit ventricular myocardium

Introduction

Endothelin (ET) is a potent vasoactive polypeptide of 21 amino acids, which was originally extracted from the culture medium of porcine aortic endothelial cells [1]. In the heart ET constricts coronary vessels, facilitates the secretion of atrial natriuretic peptide and induces cardiac hypertrophy. In addition, ET produces a pronounced positive inotropic effect on mammalian atrial and ventricular myocardium [2–9]. Specific, high-affinity binding sites for ET are densely distributed in the cardiac muscle of various mammalian species including rat, rabbit, guinea pig and dog [9, 10], and a autocrine/paracrine role for ET that is released from endothelial cells, as new type of cytokine in the regulation of contraction, has

Address for offprints: M. Endoh, Department of Pharmacology, Yamagata University School of Medicine, 2-2-2 Iida-nishi, Yamagata 990-23, Japan been proposed in the heart [11, 12].

Two subtypes of receptors have been cloned: ET_A receptor, which responds preferentially to ET-1 and less effectively to ET-3; and ET_B receptor, which responds to ET-1 and ET-3 nonselectively [13, 14]. A third subtype, ET_C, that reacts specifically with ET-3 has been identified by means of detection of immunoreactivity [15]. These three major classes of ET receptors are assumed to mediate a wide range of biological actions of the isopeptides of ET [16–19]. More recently, it has been proposed that ET_A receptors can be further classified as ET_{A1} and ET_{A2} subtypes on the basis of their susceptibility to a selective ET_A receptor antagonist, BQ-123; the ET_{A2} subtype is resistant to BQ-123 [20]. ET_B receptors have also been classified as ET_{B1} receptors, which mediate vasorelaxation through the release of NO and are inhibited by PD142893, bosentan and RES-701-1, and ET_{B2} receptors, which mediate direct vasoconstriction and are inhibited by SB2096701 and BQ-788 [21].

While it is evident that different subtypes of receptors are involved in the induction of positive inotropic effect of ET isopeptides in rabbit ventricular muscle [22], the pharmacological characteristics of subtypes responsible for the ET-induced positive inotropic effect have not yet been fully clarified. In the present study we investigated the influence of a nonselective ET receptor antagonist PD 145065 [23–25] on the positive inotropic effect of ET-1 and ET-3 in the rabbit ventricular myocardium.

Materials and methods

Isolated rabbit papillary muscle preparation

Experimental procedures employed in the present study are the same as those in the previous study described elsewhere [26,27]. Male albino rabbits (1.8–2.4 kg) were used in these experiments. Two or three papillary muscles were excised from the right ventricle of each rabbit after anesthesia with pentobarbital sodium (30 mg/kg, i.v.). Muscles were mounted in 20 ml organ baths that contained Krebs-Henseleit solution [with 57 μM ascorbic acid and 27 μM ethylenediamine-tetraacetic acid (EDTA), disodium salt, added to prevent autoxidation of the test compounds]. The solution was bubbled with 95% O_2 and 5% CO_2 at 37°C (pH 7.4). The concentrations of the various components of the solution were as follows (mM): Na^+, 142.9; K^+, 5.9; Mg^{2+}, 1.2; Ca^{2+}, 2.5; $H_2PO_4^-$, 1.2; HCO_3^-, 24.9; SO_4^{2-}, 1.2; Cl^-, 127.8; and glucose, 11.1. The muscles were stimulated electrically by square-wave pulses of 5 ms duration and a voltage about 20% above threshold, at a frequency of 1.0 Hz. The development of isometric tension was detected with strain gauge transducers (Shinkoh UL 10 GR; Minebea, Tokyo, Japan) and recorded on a thermal pen-writing oscillograph (Recti-Horiz-8K; San-ei Instruments, Tokyo, Japan). During an equilibration period of 1 h, the muscles were initially stretched under a tension of 5 mN and the length of each muscle was later adjusted to give 90% of the maximum developed tension.

(±)-Bupranolol (3×10^{-7} M) and prazosin (10^{-7} M) were allowed to act for 20 min before and during the administration of ET-3 and ET-1 to avoid any possible modulation of the response by norepinephrine that was released during the experiments. ET-3 and ET-1 were administered in a cumulative manner with stepwise increases of log 0.5 units in the concentration of the agonist. Once a steady contractile force had been achieved, the agonist was brought to the next higher concentration to be tested. Only one concentration-response curve for ET-3 or ET-1 was determined in each preparation.

The nonselective ET receptor antagonist PD 145065 (10^{-7}–10^{-5} M) was allowed to act 30 min before and during determinations of concentration-response curve for the ET isopeptides.

At the end of each experiment, the maximum contractile force was determined for each muscle by administration of isoproterenol (10^{-7}–10^{-4} M), after other drugs had been washed out for 120 min. The concentration of isoproterenol was raised in a stepwise manner until the maximum contractile force was achieved. The response to ET-3 or ET-1 was expressed as a percentage of the maximal response (maximum contractile force – basal force) to isoproterenol (ISO_{max}).

Binding assays with [^{125}I]ET-1 and [^{125}I]ET-3

Details of the experimental procedures used to prepare the membrane fraction and of the binding assay were essentially the same as those described previously [27, 28]. In brief, pieces of ventricular muscle, including free walls and septum, were excised and homogenized in 10 volumes of ice-cold buffer [0.25 M sucrose that contained 5 mM tris(hydroxymethyl)aminomethane-HCl (Tris-HCl) and 1 mM $MgCl_2$, pH 7.4] in a Polytron PT-10 (Kinematica). The homogenizer was activated three times for 15 sec each at a setting of 7. The homogenate was centrifuged at 500 × g for 15 min at 4°C. The supernatant was filtered through a single layer of cheese-cloth and centrifuged at 50,000 × g for 20 min at 4°C. The resulting pellet was washed twice with ice-cold incubation buffer (50 mM Tris-HCl, 10 mM $MgCl_2$, pH 7.5) by repeated resuspension and recentrifugation. The final pellet was resuspended in ice-cold incubation buffer at a protein concentration of about 1 mg/ml.

The binding assay was carried out as follows. The incubation mixture contained 150 μl of a suspension of membranes (approximately 150–300 μg of protein), 50 μl of a solution of [^{125}I]ET-3 or [^{125}I]ET-1, and 50 μl of incubation buffer without or with PD 145065 at various concentrations.

The reaction, started by the addition of [^{125}I]ET-3 or [^{125}I]ET-1, was incubated for 90 min at 25°C and it was terminated by the addition of 2 ml of ice-cold incubation buffer. Aliquots were taken and subjected to rapid filtration through a Whatman GF/C filter in a cell harvester (type M-24R; Brandel, Gaithersburg, MD). The filter was washed rapidly with 12 ml (3 × 4 ml) of ice-cold incubation buffer, dried for 60 min at 95°C, and placed in a scintillation vial. The radioactivity bound to the membrane was counted by means of Gamma Counter (1470 Wallac WIZARD, Turku, Finland) at an efficiency of 82%. Nonspecific binding of [^{125}I]ET-3 or [^{125}I]ET-1 was defined as the radioactivity that bound to the membrane fraction and was not inhibited by a high concentration of unlabeled ET-3 (10^{-6} M) or of ET-1 (10^{-6} M). The specific binding was calculated from the total bound radioactivity minus the nonspecific binding. The percentage of specific binding was 58% for [^{125}I]ET-3 or 88% for [^{125}I]ET-1 at the concentrations of 40 pM and 200 pM used for the competition analysis, respectively. All incubations were performed in duplicate. Protein concentrations were determined by the method of Lowry *et al.* [29] with bovine serum albumin as the standard.

Statistical analysis

Experimental values are presented as means ± S.E. Significance of differences between mean values was estimated by Student's *t*-test. A p value of less than 0.05 was considered to be significant. Analysis of the curve for the competition of [^{125}I]ET-3 or [^{125}I]ET-1 with PD 145065 by a non-linear curve-fitting method was performed by use of the computer programs that were originally made available by Mr. S. Nagatsuka (Daiichi Pure Chem. Co. Ltd., Tokyo) and partially modified by Tsuchihashi and Nagatomo [30]. Briefly, kinetic analysis was carried out on an NEC PC 9801F computer system that performs iterative non-linear regression. This program fits the binding data to equation based on the mass action for one or more classes of binding sites and determines whether the fit for the two-site model is statistically better than for the one-site model by using the theory of Munson and Rodbard [31]. Using the weighted sum of squares principle, the goodness of fit was evaluated between either a model having only one receptor subtype and a model having two receptor subtypes in the competition analysis. Competition curves were analyzed according to a general model for the interactions of several ligands with several classes of sites on the basis of the mass-action law [32]. The mathematical model of the competition curves were described previously [30]. In the comparison, statistical significance of two-site goodness of fit over one-site fit was judged by Akaike's information criterion (AIC), in which the model that gives a lower AIC value has better goodness-of-fit [30].

Peptides and drugs

The peptides and drugs used were obtained from the indicated sources: ET-1, ET-3 and sarafotoxin S6c (Peptide Institute, Osaka, Japan); PD 145065 (AcD-Bhg-L-Leu-L-Asp-L-Ile-L-Ile-L-Trp, Bhg = 5H-dibenzyl[a,d]cycloheptene10,11-dihydroglycine; Ciba-Geigy Japan, Takarazuka, Japan); RES-701-1 (Gly-Asn-Trp-His-Gly-Thr-Ala-Pro-Asp-Trp-Phr-Phe-Asn-Tyr-Tyr-Trp; Kyowa Hakko Kogyo Co. Ltd., Tokyo Research Laboratories, Tokyo, Japan); prazosin hydrochloride, (–)-isoproterenol hydrochloride (Sigma Chemical Co., St. Louis, MO); (±)-bupranolol hydrochloride (Kaken Pharmaceutical Co., Tokyo); pentobarbital sodium (Tokyo Kasei Kogyo Co., Tokyo); [^{125}I]ET-1 and [^{125}I]ET-3 (specific activity, 2,000 Ci/mmol, each) (Amersham, Buckinghamshire, UK). ET-1 and ET-3 were dissolved in phosphate-buffered saline that contained 0.05% BSA. A fresh solution of isoproterenol was prepared in 0.1% (w/v) ascorbic acid on each day that experiments were performed. It was diluted with a 0.9% (w/v) solution of NaCl and kept on ice prior to use.

Results

Characteristics of the positive inotropic effects of endothelin-3 and endothelin-1

ET-3 and ET-1, administered in a cumulative manner, increased the contractile force in a concentration-dependent manner in isolated rabbit papillary muscle as shown in Figs 1 and 2. The maximal responses to ET-3 and ET-1, respectively, in the absence of PD 145065 were achieved at 10^{-7} M and amounted to 65% (n = 4) and 54% (n = 9) of ISO_{max}. The EC_{50} values for ET-3 and ET-1 were 13.3 nM and 6.66 nM, respectively, indicating that there are not marked differences in the efficacy and potency of these two ET isopeptides in inducing the positive inotropic effect. However, there was a definite difference between the concentration-response curves for ET-3 and ET-1. ET-3 never produced a positive inotropic effect at concentrations of 10^{-9} M or lower (Fig. 1), while ET-1 at concentrations of 10^{-9} M and lower elicited a positive inotropic effect that reached a level equal to less than 20% of ISO_{max} (Fig. 2). These findings are essentially consistent with the previous observations in rabbit papillary muscle in this laboratory [22].

PD 145065, a nonselective antagonist of ET receptors, at concentrations of 10^{-7} M and higher, inhibited the positive inotropic effect of ET-3 in a concentration-dependent manner and at 10^{-5} M it abolished the positive inotropic effect of ET-3 at 3×10^{-7} M (the highest concentration examined) as shown in Fig. 1. The mode of antagonistic action of PD

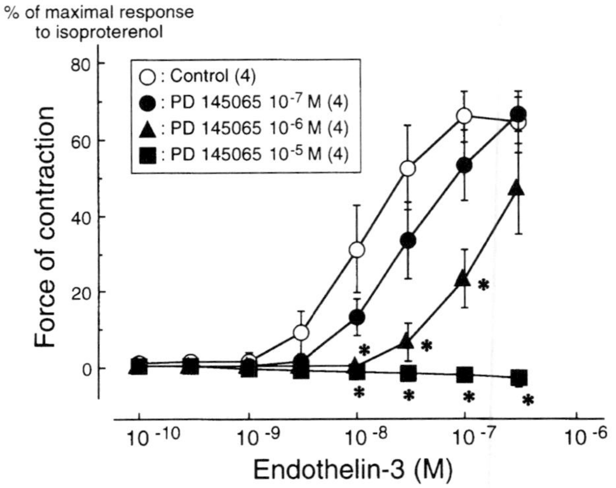

Fig. 1. Influence of a nonselective ET receptor antagonist, PD 145065, on the positive inotropic effects of ET-3 in isolated rabbit right ventricular papillary muscles that were electrically stimulated at 1 Hz at 37°C. Ordinate, force of contraction expressed as a percentage of the maximal response to isoproterenol; abscissa, molar concentration of ET-3 on a logarithmic scale. Vertical bars: S.E.M. Numbers in parentheses indicate numbers of muscle preparations examined. *$p < .05$ vs. corresponding values in the absence of PD 145065. Basal values: muscle length, 4.97 ± 0.15 mm; cross-sectional area, 0.55 + 0.07 mm^2; basal force, 12.7 ± 3.53 mN/mm^2; maximum force induced by isoproterenol, 36.7 ± 6.25 mN/mm^2 (n = 16).

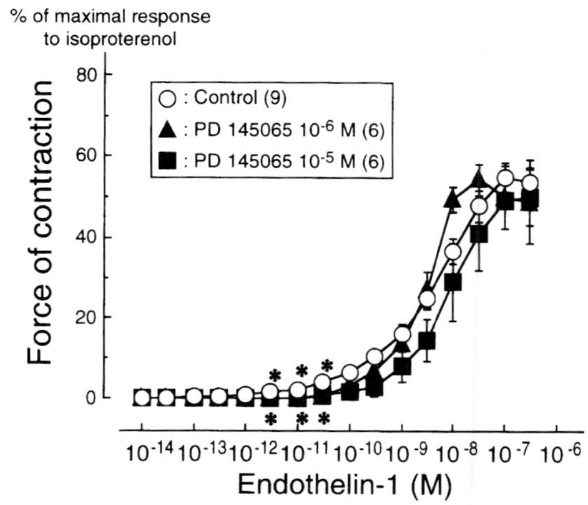

Fig. 2. Influence of a nonselective ET receptor antagonist, PD 145065, on the positive inotropic effects of ET-1 in isolated rabbit right ventricular papillary muscles that were electrically stimulated at 1 Hz at 37°C. Ordinate, force of contraction expressed as a percentage of the maximal response to isoproterenol; abscissa, molar concentration of ET-1 on a logarithmic scale. Vertical bars: S.E.M. Numbers in parentheses indicate numbers of muscle preparations examined. *$p < .05$ vs. corresponding values in the absence of PD 145065. Basal values: muscle length, 5.19 + 0.23 mm; cross-sectional area, 0.81 ± 0.09 mm^2; basal force, 7.59 ± 1.45 mN/mm^2; maximum force induced by isoproterenol, 33.1 ± 6.99 mN/mm^2 (n = 21).

145065 appears to be noncompetitive but it was not clear in the present study because of the limited availability of the agonist. The basal force of contraction was unaffected by PD 145065 up to 10^{-5} M (data not shown).

The positive inotropic effect of ET-1 was influenced by PD 145065 much less than that of ET-3 (Fig. 2). The main portion of the concentration-response curve for the positive inotropic effect of ET-1 was not shifted to the right by PD 145065 at concentrations up to 10^{-5} M and the slope of the concentration-response curve for ET-1 became steeper in the presence of PD 145065. The positive inotropic effect of ET-1 at low concentrations ($<10^{-9}$ M) was significantly decreased by PD 145065 at 10^{-6} M and 10^{-5} M (Fig. 2).

Influence of PD 145065 on the specific binding of [^{125}I]ET-3 and [^{125}I]ET-1

The specific binding of [^{125}I]ET-3 and of [^{125}I]ET-1 to the rabbit ventricular membrane fraction was saturable. Scatchard analysis of the data revealed the presence of a single class of binding sites. The B_{max} and the K_d values for the specific binding of [^{125}I]ET-3 and [^{125}I]ET-1 were 42 and 230 fmol/mg protein and 0.036 and 0.129 nM, respectively [22]. PD 145065 inhibited the specific binding of 40 pM [^{125}I]ET-3 (Fig. 3) and 200 pM [^{125}I]ET-1 (Fig. 4) in a concentration-dependent manner, with both high- and low-affinity binding. The analysis of the competition curves revealed that a better fit was obtained with a model in which there were two binding sites. For the binding of [^{125}I]ET-3, the AIC values were –76.5 for one-site and –65.1 for two-site model. The high-affinity binding site of PD 145065 for competition of [^{125}I]ET-3, with a K_i value of 4.48×10^{-8} M, accounted for 64.3% of the total binding sites; the K_i value for low-affinity site was

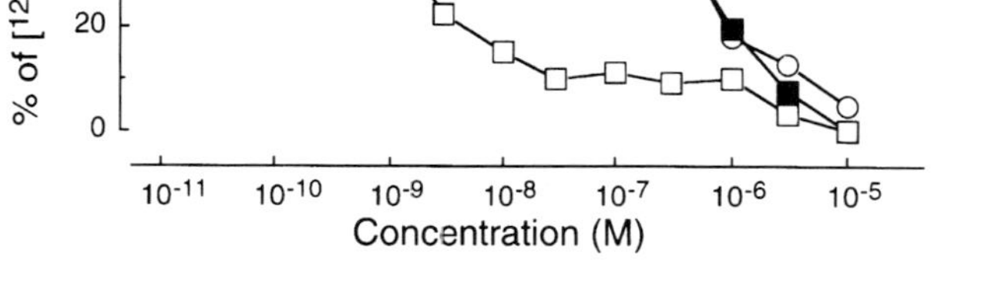

Fig. 3. Inhibition by PD 145065 of the specific binding of [^{125}I]ET-3 to membrane fractions derived from rabbit ventricular myocardium. Influence of RES-701-1 and sarafotoxin S6c is also presented for comparison. Values in parentheses are numbers of experiments determined in duplicate. Ordinate, binding of [^{125}I]ET-3 expressed as a percentage of the specific binding in the absence of unlabeled ligands; abscissa, molar concentration of PD 145065, RES701-1 and sarafotoxin on a logarithmic scale. The concentration of [^{125}I]ET-3 used in experiments was 40 μM. Levels of specific binding in this series was 57.6 ± 1.6% (n = 9).

6.90 × 10^{-7} M (r = 0.999; Fig. 3 and Table 1). An ET_B subtype-selective antagonist, RES-701-1, inhibited the [^{125}I]ET-3 binding with K_i value equivalent to that of PD 145065 (Fig. 3). An ET_B subtype-selective agonist, sarafotoxin S6c, displaced the [^{125}I]ET-3 with an extremely high affinity (K_i value = 2.3 × 10^{-10} M).

For the binding of [^{125}I]ET-1, the AIC values were –85.2 and –78.7 for one-site and two-site model, respectively. The high-affinity binding site of PD 145065 for competition of [^{125}I]ET-1, with a K_i value of 2.04 × 10^{-7} M, accounted for 57.3% of total binding sites; K_i for the low-affinity site was 4.20 × 10^{-7} M (r = 0.999; Fig. 4 and Table 1). RES-701-1 and sarafotoxin S6c scarcely affected the specific [^{125}I]ET-l binding (Fig. 4).

Effects of a selective antagonist and agonist of ET_B receptors

RES-701-1, at 10^{-6} M and 10^{-5} M shifted the concentration-response for the positive inotropic effect of ET-3 to the right in parallel, the pA_2 value (–log IC_{50}), determined at 30% of ISO_{max}, being 6.26 [27]. A selective agonist of ET_B receptors, sarafotoxin S6c, elicited a positive inotropic effect in a concentration-dependent manner at concentrations of 3 × 10^{-8} M and higher [27]. PD 145065 at 10^{-5} M abolished the positive inotropic effect of sarafotoxin S6c, while RES-701-1 at 10^{-5} M did not affect the positive inotropic effect of sarafotoxin S6c [27].

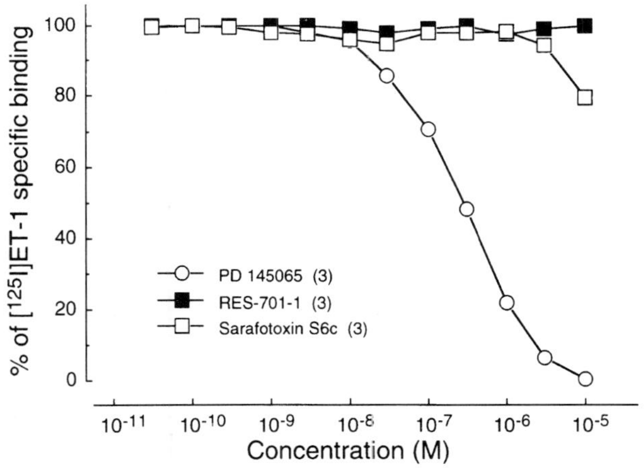

Fig. 4. Inhibition by PD 145065 of the specific binding of [^{125}I]ET-1 to membrane fractions derived from rabbit ventricular myocardium. Influence of RES-701-1 and sarafotoxin S6c is also presented for comparison. Values in parentheses are numbers of experiments determined in duplicate. Ordinate, binding of [^{125}I]ET-1 expressed as a percentage of the specific binding in the absence of unlabeled ligands; abscissa, molar concentration of PD 145065, RES-701-1 and sarafotoxin on a logarithmic scale. The concentration of [^{125}I]ET-1 used in experiments was 200 pM. Levels of specific binding in this series was 88.3 ± 1.3% (n = 9).

Discussion

Subtypes of endothelin receptor in the rabbit ventricular myocardium

An ET receptor subtype-nonselective antagonist, PD 145065, at 10^{-7}–10^{-5} M, antagonized the positive inotropic effect of ET-3. Since the inotropic effect of ET-3 is likewise antagonized by ET_A subtype-selective antagonists, BQ-123 [22, 33] and FR139317 [27, 34, 35], the subtype that mediates the positive inotropic effect of ET-3 may belong to a class of ET_{A1} receptor [20]. The positive inotropic effects of two derivatives of sarafotoxin, [Thr^2]SRTb and [Glu^9]SRTb [22], as well as sarafotoxin S6c [27] were also effectively antagonized by BQ-123 [22], FR139317 [27] and PD 145065. Thus, these agonists may mainly activate a subtype in one and the same class.

By contrast, the positive inotropic effect of ET-1, namely the main portion of the concentration-response curve for the positive inotropic effect of ET-1 was not shifted to the right but was shifted rather to the left (Fig. 2). The mode of antagonism induced by PD 145065 against the positive inotropic effect of ET-1 was very similar to the antagonism induced by ET_A subtype-selective antagonists, BQ-123 [22] and FR139317 [27]. Because the positive inotropic effect of ET-1 was consistently enhanced by the presence of these receptor antagonists [22, 27, present study], it is postulated that ET-1 elicits an inhibitory action via ET_A receptors in the rabbit ventricular muscle as has been described in other species [36, 37].

While the main portion of the positive inotropic effect of ET-1 was not antagonized by BQ-123, indicating that the effect may be mediated by ET_{A2} subtype (which is resistant to BQ-123 [20]), the subtype that mediates the positive inotropic effect of ET-1 (> 10^{-9} M) was also resistant to PD 145065. Therefore the ET receptor that mediates the positive inotropic effect of ET-1 in the rabbit ventricle may belong to a novel as yet undefined subclass.

Although the magnitude of the positive inotropic effect at lower concentrations (< 10^{-9} M) of ET-1 was small ([22] and Fig. 2), the effect is inhibitable with BQ-123 [22] and FR139317 [27], and therefore it may be mediated by ET_{A1} receptor [20].

PD 145065 inhibited the specific [^{125}I]ET-3 and [^{125}I]ET-1 binding with high affinity in membrane fractions derived from the rabbit ventricular muscle (Figs. 3 and 4; Table 1), The kinetic analysis of the competition curves implies that PD 145065 inhibits the specific binding of both ET isopeptides in biphasic manner with high- and low-affinity. However, the deference in K_i values of PD 145065 for high- and low-affinity sites is small, i.e., ten- and twofold in [^{125}I]ET-3 and [^{125}I]ET-1 binding, respectively, indicating that PD 145065 behaves as a nonselective ET receptor antagonist in

Table 1. Effects of PD 145065 on the [^{125}I]-3 or [^{125}I]ET-1 binding in membrane fractions derived from rabbit ventricular myocardium

[^{125}I]ET-3 binding

	n	K_i value (M)	R_H %	Slope factor
PD 145065	3	High $(4.48 \pm 1.80) \times 10^{-8}$ Low $(6.90 \pm 3.43) \times 10^{-7}$	64.28 ± 16.49	0.632 ± 0.108

[^{125}I]ET-1 binding

	n	K_i value (M)	R_H %	Slope factor
PD 145065	3	High $(2.04 \pm 1.28) \times 10^{-7}$ Low $(4.20 \pm 1.77) \times 10^{-7}$	57.34 ± 14.50	0.828 ± 0.132

Values are means ± S.E. n, number of rabbits, the concentration of [^{125}I]ET-3 or [^{125}I]ET-1 used was 40 pM or 200 pM.

rabbit ventricular muscle as in other tissues [23–25].

The K_i values of PD 145065 for high affinity binding sites in rabbit ventricular muscle were 4.48×10^{-8} M and 2.04×10^{-7} M when [^{125}I]ET-3 and [^{125}I]ET-1 were used as ligands (Table 1). Although PD 145065 was more potent to inhibit the [^{125}I]ET-3 binding than [^{125}I]ET-1 binding, the difference is not sufficient to explain the pronounced difference in the antagonistic action on the positive inotropic effect of ET isopeptides. Moreover, although PD 145065 inhibited completely the specific binding of [^{125}I]ET-1, it did not affect the main portion of the concentration-response curve for the positive inotropic effect of ET-1. The discrepancy between functional and binding characteristics has also been found with ET_A subtype-selective antagonists BQ-123 and FR139317 in rabbit ventricular muscle. Namely the specific binding of [^{125}I]ET-1 to membrane fractions derived from rabbit ventricular muscle was completely inhibited by BQ-123 [22] and FR139317 [27] with high affinity but the main portion of the concentration-response curve for the positive inotropic effect of ET-1 was not shifted to the right by these antagonists. Although we cannot explain the precise mechanism underlying these discrepancies, it is likely that the characteristics of ET binding assay may have affected to some extent the results obtained in the binding study: (1) while the receptor antagonist is generally employed as radioligand, the radio-labeled ET receptor agonist is used in the ET receptor binding assay; and (2) the ET agonist binding could be hardly displaced by washout or by cold receptor ligands once the binding had been established [27]. While it is generally considered that the agonist binding can be theoretically analyzed in the same way as the antagonist binding, it has not been so well characterized as the antagonist binding and further study will be required employing ET receptor antagonists as radioligands.

An ET_B subtype-selective antagonist RES-701-1 [38, 39] and an ET_B agonist sarafotoxin S6c inhibited the [^{125}I]ET-3 binding with no or little effects on the [^{125}I]ET-1 binding, indicating that ET-3 but not ET-1 binds preferentially to ET_B subtype. In the functional study, RES-701-1 produced a parallel shift of the concentration-response curve for the positive inotropic effect of ET-3 [27]. Although sarafotoxin S6c displaced the [^{125}I]ET-3 binding with extremely high affinity (K_i value = 2.3×10^{-10} M), it elicited a positive inotropic effect first at concentrations of 3×10^{-8} M and higher in isolated rabbit papillary muscle [27]. Furthermore, the positive inotropic effect of sarafotoxin S6c was not affected by RES-701-1 but abolished by BQ-123, FR139317 [27] as well as by PD 145065. Therefore, while ET_B subtype may partially contribute to the positive inotropic effect of ET-3, the pharmacological characteristics of the ET_B receptor in the rabbit ventricular muscle may also be different from those in noncardiac tissues, including the property that the subtype is susceptible to ET_A subtype-selective antagonists.

Relationship between phosphoinositide hydrolysis and the positive inotropic effects induced by isopeptides of endothelin

In rabbit ventricular muscle, we found a close correlation between the positive inotropic effect and the acceleration of PI hydrolysis in response to the alpha adrenoceptor agonist, phenylephrine, to angiotensin II or to ET-1 and ET-3 in the rabbit ventricular myocardium [26, 27, 40–42]. Since the positive inotropic effect of these agonists including endothelin isopeptides develops gradually even after the accumulation of IP_3 returned to the control level [26, 42], IP_3 may not play a crucial role in the receptor-mediated functional regulation, We consider therefore that it is more likely that the activation of protein kinase C induced by diacylglycerol, endogenously generated as another product of PI hydrolysis, mediates predominantly the functional regulation due to the receptor activation in the rabbit ventricular muscle (for review see [43]). However, in the rabbit ventricle, phorbol esters, such as phorbol 12-myristate 13-acetate [44] and phorbol 12,13-dibutyrate [45, 46], that activate protein kinase C *in vitro*, do not mimic but instead inhibit both the acceleration of PI hydrolysis and the positive inotropic effect induced by receptor activation. This indicates that the activation of protein kinase C induced by externally applied phorbol esters acts through the signal transduction process different from that activated by endogenously generated diacylglycerol, which hinders to examine the functional relevance of receptor-mediated activation of protein kinase C. Inhibitors of protein kinase C, such as staurosporine, NA 0345 and H-7, antagonized the positive inotropic effect mediated by alpha adrenoceptors selectively and consistently by approximately 20% of the maximum response to phenylephrine in the rabbit papillary muscle [47].

In summary, endothelin isopeptides, ET-3 and ET-1, elicit the positive inotropic effect associated with acceleration of PI hydrolysis on the rabbit ventricular muscle, essentially with equipotent and equi-effective activity. The ET receptor subtypes involved in the positive inotropic effect of both isopeptides, however, have different pharmacological characteristics from each other. The positive inotropic effect of ET-3 is abolished by PD 145065 but the effect of ET-1 is essentially unaffected by PD 145065. Pharmacological characteristics of ET receptors in the rabbit ventricular myocardium are atypical in many respects and all ET receptor antagonists examined up to the moment, including a nonselective antagonist PD 145065, failed to inhibit the main portion of the concentration-response curve for the positive inotropic effect of ET-1.

Acknowledgements

This study was supported in part by Grant-in-Aid (Nos. 06274202 and 06274201) for Scientific Research on Priority Areas and by a Grant-in-Aid for Scientific Research (B) (No. 06454155) from the Ministry of Education, Science and Culture, Japan, and the Mitsubishi Prize for Medicine, which was awarded to M.E. in 1994 and 1995. The authors thank Dr. T. Okada, International Research Laboratories, Ciba-Geigy Japan, Takarazuka, Japan, for the generous supply of PD 145065, Dr. Y. Matsuda, Kyowa Hakko Kogyo Co. Tokyo, Japan, for a gift of RES-701-1, to Kaken Pharmaceutical Co. Ltd., Tokyo, Japan, for a gift of (±)-bupranolol.

References

1. Yanagisawa M, Kurihara H, Kimura S, Tomobe Y, Kobayashi M, Mitsui Y, Yazaki Y, Goto K, Masaki T: A novel potent vasoconstrictor peptide produced by vascular endothelial cells. Nature 332: 411–415, 1988
2. Hu JR, Von Harsdorf R, Lang RE: Endothelin has potent inotropic effects in rat atria. Eur J Pharmacol 158: 275–278, 1988
3. Ishikawa T, Yanagisawa M, Kimura S, Goto K, Masaki T: Positive in otropic action of novel vasoconstrictor peptide endothelin on guinea pig atria. Am J Physiol 255: H970–H973, 1988
4. Moravec CS, Reynolds EE, Stewart RW, Bond M: Endothelin is a positive inotropic agent in human and rat heart *in vitro*. Biochem Biophys Res Commun 159: 14–18, 1989
5. Shah AM, Lewis MJ, Henderson AH: Inotropic effects of endothelin in ferret ventricular myocardium. Eur J Pharmacol 163: 365–367, 1989
6. Vigne P, Lazdunski M, Frelin C: The inotropic effect of endothelin-l on rat atria involves hydrolysis of phosphatidylinositol. FEBS Lett 249: 143–146, 1989
7. Watanabe T, Kusumoto K, Kitayoshi T, Shimamoto N: Positive inotropic and vasoconstrictive effects of endothelin-1 in *in vivo* and *in vitro* experiments. Characteristics and the role of L-type calcium channels. J Cardiovasc Pharmacol 13 (suppl 5): S108–S111, 1989
8. Li K, Stewart DJ, Rouleau J-L: Myocardial contractile actions of endothelin-1 in rat and rabbit papillary muscle. Circ Res 69: 301–312, 1991
9. Takanashi M, Endoh M: Characterization of positive inotropic effect of endothelin on mammalian ventricular myocardium. Am J Physiol 261: H611–H619, 1991
10. Gu XH, Ou RC, Casley DJ, Daly MJ, Nayler WG: Effect of age on endothelin-l binding sites in rat cardiac ventricular membranes. J Cardiovasc Pharmacol 19: 764–769, 1992
11. Krämer BK, Nishida M, Kelly RA, Smith TW: Endothelins. Myocardial actions of a new class of cytokines. Circulation 85: 350–356, 1992
12. Mebazaa A, Mayoux E, Maeda K, Martin LD, Lakatta EG, Robotham JL, Shah AM: Paracrine effects of endocardial endothelial cells on myocyte contraction mediated via endothelin. Am J Physiol 265: H1841–H1846, 1993
13. Arai H, Hori S, Aramori I, Ohkubo H, Nakanishi S: Cloning and expres sion of a cDNA encoding an endothelin receptor. Nature 348: 730–732, 1990
14. Sakurai T, Yanagisawa M, Takuwa Y, Miyazaki H, Kimura S, Goto K, Masaki T: Cloning of a cDNA encoding a non-isopeptide-selective subtype of the endothelin receptor. Nature 348: 732–735, 1990
15. Naruse M, Naruse K, Nishikawa T, Yoshihara I, Ohsumi K, Suzuki N, Demura R, Demura H: Endothelin-3 immunoreactivity in gonadotrophs of the human anterior pituitary. J Clin Endocrinol Metab 74: 968–972, 1992
16. Masaki T, Kimura S, Yanagisawa M, Goto K: Molecular and cellular mechanism of endothelin regulation. Implications for vascular function. Circulation 84: 1457–1468, 1991
17. Haynes WG, Davenport AP, Webb DJ: Endothelin: progress in pharmacology and physiology. Trends Pharmacol Sci 14: 225–228, 1993
18. Rubanyi GM, Polokoff MA: Endothelins: molecular biology, biochemistry, pharmacology, physiology, and pathophysiology. Pharmacol Revs 46: 325–415, 1994
19. Bax WA, Saxena PR: The current endothelin receptor classification: time for reconsideration? Trends Pharmacol Sci 15: 379–386, 1994
20. Sudjarwo SA, Hori M, Tanaka T, Matsuda Y, Okada T, Karaki H: Subtypes of endothelin ET_A and ET_B receptors mediating venous smooth muscle contraction. Biochem Biophys Res Commun 200: 627–633, 1994
21. Douglas SA, Meek TD, Ohlstein EH: Novel receptor antagonists welcome a new era in endothelin biology. Trends Pharmacol Sci 15: 313–316, 1994
22. Kasai H, Takasaki M, Takanashi M, Endoh M: Pharmacological properties of endothelin receptor subtypes mediating positive inotropic effects in rabbit heart. Am J Physiol 266: H2220–H2228, 1994
23. Cody WL, Doherty AM, He JX, Depue PL, Waite LA, Topliss JG, Haleen SJ, Ladouceur D, Flynn MA, Hill KE, Reynolds EE: The rational design of a highly potent combined ET_A and ET_B receptor antagonist (PD 145065) and related analogues. Med Chem Res 3: 154–162, 1993

24. Doherty AM, Cody WL, He JX, Depue PL, Cheng XM, Welch KM, Flynn MA, Reynolds EE, Ladouceur DM, Davis LS, Keiser JA, Haleen SJ: *In vitro* and *in vivo* studies with a series of hexapeptide antagonists. J Cardiovasc Pharmacol 22 (suppl 8): S98–S102
25. Warner TD, Allcock GH, Vane JR: Reversal of established responses to endothelin-l *in vivo* and *in vitro* by the endothelin receptor antagonists, BQ-123 and PD 145065. Br J Pharmacol 112: 207–213, 1994
26. Takanashi M, Endoh M: Concentration- and time-dependence of phosphoinositide hydrolysis induced by endothelin-l in relation to the positive inotropic effect in the rabbit ventricular myocardium. J Pharmacol Exp Ther 262: 1189–1194, 1992
27. Endoh M, Norota I, Yang H-T, Fujita S, Takanashi M: The positive in otropic effect and the hydrolysis of phosphoinositide induced by endothelin-3 in rabbit ventricular myocardium: inhibition by a selective antagonist of ET_A receptors, FR139317. J Pharmacol Exp Ther 1996 (in press)
28. Endoh M, Hiramoto T, Ishihata A, Takanashi M, Inui J: Myocardial α_1-adrenoceptors mediate positive inotropic effect and changes in phosphatidylinositol metabolism. Species differences in receptor distribution and the intracellular coupling processes in mammalian ventricular myocardium. Circ Res 68: 1179–1190, 1991
29. Lowry OH, Rosebrough NJ, Farr AL, Randall RJ: Protein measurement with the Folin phenol reagent, J Biol Chem 193: 265–275, 1951
30. Tsuchihashi H, Nagatomo T: Characterization of ^{3}H-dihydroalprenolol binding to β-adrenergic receptors of rat brain: two binding sites of racemic propranolol in displacement experiments. Chem Pharm Bull 35: 2979–2984, 1987
31. Munson PJ, Rodbard D: Ligand: a versatile computerized approach for characterization of ligand-binding systems. Anal Biochem 107: 220–239, 1980
32. Feldman HA: Mathematical theory of complex ligand-binding systems at equilibrium: some methods for parameter fitting. Anal Biochem 48: 317–338, 1972
33. Ihara M, Noguchi K, Saeki T, Fukuroda T, Tsuchida S, Kimura S, Fukami T, Ishikawa K, Nishikibe M, Yano M: Biological profiles of highly potent novel endothelin antagonists selective for the ET_A receptor. Life Sci 50: 247–255, 1991
34. Aramori C, Nirei H, Shoubo M, Sogabe K, Nakamura K, Kojo H, Notsu Y, Ono T, Nakanishi S: Subtype selectivity of a novel endothelin antagonist, FR139317, for the two endothelin receptors in transfected Chinese hamster ovary cells. Mol. Pharmacol 43: 127–131, 1993
35. Sogabe K, Nirei H, Shoubo M, Nomoto A, Ao S, Notsu Y, Ono T: Pharmacological profile of FR139317, a novel, potent endothelin ET_A receptor antagonist. J Pharmacol Exp Ther 264: 1040–1046, 1993
36. Hilal-Dandan R, Urasawa K, Brunton LL: Endothelin inhibits adenylate cyclase and stimulates phosphoinositide hydrolysis in adult cardiac myocytes. J Biol Chem 127: 10620–10624, 1992
37. Ono K, Tsujimoto G, Sakamoto A, Eto K, Masaki T, Ozaki Y, Satake M: Endothelin-A receptor mediates cardiac inhibition by regulating calcium and potassium currents. Nature 370: 301–304, 1994
38. Matsuda Y, Tanaka T, Morishita Y, Nozawa M, Ohno T, Yamada K: Pharmacological profile of RES-701-1, a novel, potent endothelin subtype B receptor antagonist of microbial origin. Circulation 88 (suppl Pt 2): I-281, 1993
39. Karaki H, Sudjarwo SA, Hori M, Tanaka T, Matsuda Y: Endothelin ET_B receptor antagonist, RES-701-1: effects on isolated blood vessels and small intestine. Eur J Pharmacol 262: 255–259, 1994
40. Takanashi M, Norota I, Endoh M: Potent inhibitory action of chlorethylclonidine on the positive inotropic effect and phosphoinositide hydrolysis mediated via myocardial alpha$_1$-adrenoceptors in the rabbit ventricular myocardium. Naunyn-Schmiedebergs Arch Pharmacol 343: 669–673, 1991
41. Ishihata A, Endoh M: Pharmacological characteristics of the positive inotropic effect of angiotensin II in the rabbit ventricular myocardium. Br J Pharmacol 108: 999–1005, 1993
42. Yang H-T, Endoh M: Dissociation of the positive inotropic effect of methoxamine from the hydrolysis of phosphoinositide in rabbit ventricular myocardium: a comparison with the effects of phenylephrine and the subtype of the *alpha*-l adrenoceptor involved. J Pharmacol Exp Ther 269: 732–742, 1994
43. Endoh M: The effects of various drugs on the myocardial inotropic response. Gen Pharmacol 26: 1–31, 1995
44. Kushida H, Hiramoto T, Satoh H, Endoh M: Phorbol ester does not mimic, but antagonizes the alpha-adrenoceptor-mediated positive inotropic effect in the rabbit papillary muscle. Naunyn-Schmiedebergs Arch Pharmacol 337: 169–176, 1988
45. Endoh M, Takanashi M: Differential inhibitory action of phorbol-12,13-dibutyrate on the positive inotropic effect of endothelin-l and Bay K 8644 in the isolated rabbit papillary muscle. J Cardiovasc Pharmacol 17(suppl 7): S165–S168, 1991
46. Endoh M, Otomo J, Norota I, Takanashi M: Selective inhibition by phorbol 12,13-dibutyrate of the α_1-receptor-mediated positive inotropic effect. Int J Cardiol 40: 191–201, 1993
47. Endoh M, Norota I, Takanashi M, Kasai H: Inotropic effects of staurosporine, NA 0345 and H-7, protein kinase C inhibitors, on rabbit ventricular myocardium: selective inhibition of the positive inotropic effect mediated by α_1-adrenoceptors. Jpn J Pharmacol 63: 17–26, 1993

Molecular and Cellular Biochemistry **160/161**: 75–82, 1996.

Involvement of nitric oxide synthase and protein kinase C activation on chagasic antibodies action upon cardiac contractility

Leonor Sterin-Borda, Graciela Cremaschi, Ana Maria Genaro, Agustina Vila Echagüe, Juan Carlos Goin and Enri Borda

Cefybo-Conicet and Department of Pharmacology School of Medicine and Dentistry, University of Buenos Aires, Serrano 669, 1414 Buenos Aires, Argentina

Abstract

We have already demonstrated the presence of antibodies in the sera of chagasic patients with the ability to interact with neurotransmitter receptors triggering several intracellular pathways of transduction signals. Here we show that, chagasic IgG induced protein kinase C (PKC) translocation to rat cardiac membranes and this effect was inhibited by muscarinic cholinergic blockers atropine and AF-DX 116 pointing to the participation of M_2 receptors in this effect. It was also able to stimulate nitric oxide synthase (NOS) activity and this action was blunted by phospholipase C (PLC) and PKC inhibitors indicating that the production of nitric oxide (NO) would be the consequence of the cascade of enzymatic pathways triggered by mAChR activation. PKC and NOS activities were involved in chagasic IgG negative inotropic actions on rat isolated myocardium as its effects were blunted by staurosporine and L-N-monomethyl arginine. Furthermore, low concentrations of chagasic IgG inhibited the cardiac mechanical action of carbachol in a non-competitive manner. These data suggested that PKC activation in myocardium by chagasic IgG would be involved in its physiological actions by modulating NOS activity. The participation of PKC-mediated phosphorylation of mAChR leading to receptor desensitization as one of the causes of dysautonomia is also discussed. (Mol Cell Biochem **160/161**: 75–82, 1996)

Key words: atria contractility, protein kinase C, nitric oxide, muscarinic cholinergic receptors, chagasic IgG

Introduction

Chagas disease, one of the most common determinants of congestive heart failure and sudden death in the world, is caused by a parasite, *Trypanosoma cruzi* (*T. cruzi*) which is widely distributed in south and central America [1]. Chronic chagasic cardiomyopathy involves parasite-induced autoimmune reactions including low levels of tissue parasites at this stage of the disease [2], the presence of anti-myocardial antibodies (Ab) [3, 4], the presence of myocardial, muscle and nerve lymphomononuclear cell infiltrates [5, 6] and abnormalities in circulating lymphocyte subpopulations [7, 8]. Among the antimyocardial Ab we have described the presence of different populations of specific Ab, that trigger neurotransmitter receptor-mediated biological effects in the sera of *T. cruzi* infected mice and human chagasic patients [9–13] and that would alter the physiology and biochemistry of normal myocardium. Furthermore, we have recently described that a monoclonal Ab anti-*T. cruzi* interacting with normal myocardium was able to modulate its physiology through the activation of cardiac β adrenoceptors (β AR) and muscarinic cholinergic receptors (mAChR) [14] mimicking the effects of circulating Ab from human and murine chagasic sera.

It is known that, chagasic heart disease has neurogenic nature and is caused by a poor regulation of autonomic control of heart activity, in which, denervation of both parasympathetic and sympathetic systems of the heart was proven to be higher in the form [15]. So, Chagas heart disease is a cardioneuromyopathy in which sympathetic and parasympathetic systems are affected [16]. This dysautonomia is

Address for offprints: L. Sterin-Borda, Cefybo-Conicet, Serrano 669, 1414 Buenos Airies, Argentina characterized by a slow and progressive blockade of cardiac neurotransmitter receptors prior to the development of the cardiomyopathy [16]. Analyzing the prevalence of antibodies against neurotransmitter receptors, we observed a strong association between them and seropositive chagasic patients with cardiac dysautonomic syndrome. Thus, the frequency of IgG with autonomic activity was higher on sera from seropositive patients with dysautonomic syndrome than those without dysautonomia [17]. We can speculate that antibodies fixation could have important functional implications in the neurogenic nature of chagasic heart disease.

Regarding chagasic Ab interaction with mAChR we have previously shown that these Ab were able to decrease cardiac contractility triggering mAChR coupled transduction signals, namely, inhibition of adenylate cyclase system, activation of guanylate cyclase with cyclic GMP (cGMP) production [18] and activation of phosphoinositide (PI) turnover [19]. Additionally, we have demonstrated that chagasic Ab directly recognize mAChR as judged by immunoprecipitation and immunoblotting studies [18] and as it was able to recognize and immunoprecipitate a fraction of soluble receptors similarly to an anti human M2 mAChR monoclonal Ab [20]. Recently we have also demonstrated, that chagasic Ab were able to induce phosphorylation and desensitization of human mAChR in a receptor transfected-cell line [21].

It was also demonstrated that purified cardiac mAChR are excellent substrates for protein kinase C (PKC) [22] and that PKC mediated phosphorylation of mAChR modulates receptor/G-protein interactions [23].

The purpose of the present study was to analyze if chagasic Ab were able to induce PKC activation and subsequent translocation to cardiac membranes and to study whether this association could play any role in the transduction signals triggered by chagasic IgG interaction with mAChR, such as nitric oxide synthase activation, and its functional effect upon myocardial mechanical behaviour.

Materials and methods

Serum selection

Sera from 10 patients with positive *T. cruzi* serology, absence of clinical symptoms with normal electrocardiogram and chest X ray were used. Ten normal sera (control) were from volunteers who had negative serology for Chagas disease and had no evidence of cardiovascular disease. Patients included in this study were between 25–60 years of age.

Purification of human IgG

Total IgG was isolated from sera of *T. cruzi*-infected patients and normal subjects by DEAE-cellulose (Sigma} chromatography. The eluted peaks were concentrated by ultrafiltration to about 9–10 mg protein/ml. The degree of IgG purification was tested by microimmunoelectrophoresis using goat anti human total serum protein and goat anti human total Ig. Only one line of precipitation was obtained in all cases.

Contractile assay

Rat were sacrificed by decapitation. The atria were carefully dissected from the ventricles, attached to a glass holder and immersed in a tissue bath containing Krebs-Ringer Bicarbonate (KRB) solution gassed with 5% CO_2 in oxygen and maintained at pH 7.4 and 30°C. KRB solution was composed as reported previously [17]. A preload tension of 750 mg was applied to the atria and tissues were allowed to equilibrate for 1 h. The initial control values for contractile tension of the isolated atria were recorded using a force transducer coupled to an ink writing oscillograph [17]. The preparations were paced with a bipolar electrode and an SK4 Grass Stimulator. The stimuli had a duration of 2 ms and their voltage was 10% above threshold. Inotropic effects (dF/dt) were assessed by recording the maximum rate of isometric force development during electrical stimulation at a fixed frequency of 150 beats/min. Control values (= 100%} refer to the dF/dt before the addition of drugs. The absolute value for dF/dt at the end of the equilibration period (60 min) was 7.3 ± 0.5 g/s. Cumulative dose-response curves to carbachol were made according to the method of Van Rossum [24]. A maximal effect was achieved within 6 min after each dose. Strips of right ventricle were treated as atria preparation. They were exposed to field stimulation with 1 ms duration at 45 beats/min of frequency. The absolute values for dF/dt at the end of equilibrium was 18.3 ± 1.2 g/s.

Determination of nitric oxide synthase activity

Nitric oxide synthase (NOS) activity was measured in tissues by production of (U-^{14}C) citrulline from (U-^{14}C) arginine according to the procedure described by Bredt and Snyder [25] for brain slices. Briefly, after 20 min preincubation in KRB solution, tissue were transferred to 500 μl of prewarmed KRB equilibrated with 5% CO_2 in oxygen in the presence of (U-^{14}C) arginine (0.5 μCi). Appropriate concentrations of drugs were added and the hearts were incubated for 20 min under 5% CO_2 in oxygen at 37°C. Tissues were then homogenized with a Ultraturrax in 1 ml of medium containing 20 mM HEPES pH 7.4, 0.5 mM EGTA, 0.5 mM EDTA, 1 mM dithiothreitol, 1 μM leupeptin and 0.2 mM phenylmethanesulfonyl fluoride at 4°C. After centrifugation at 20000 g for 10 min at 4°C, supernatants were applied to 2 ml columns of Dowex AG 50 WX-8 (sodium form); (^{14}C) citrulline was eluted with 3 ml of water and quantified by liquid scintillation counting. Measurement of basal NOS activity by the above mentioned procedure was inhibited 95% in the presence of 0.5 mM L-NMMA. The results (pmol/g/tissue wet weight) obtained were expressed as the difference between values in the absence and in the presence of L-NMMA.

Protein kinase C (PKC) activity assay

PKC activity was assayed by measuring the incorporation of ^{32}P from gamma-^{32}P-ATP into histone H_1. Incubations were conducted for 30 min at 30°C in a final volume of 85 μl. In final concentrations, the assay mixture contained 25 μM ATP (0.4 μCi), 10 mM Mg acetate, 5 mM β–mercaptoethanol, 50 μg of histone H_1, 20 mM HEPES, pH 7.5 and unless otherwise indicated, 0.2 mM $CaCl_2$ and 10 μg/ml of phophatidylserine vesicles. The incorporation of (^{32}P) phosphate into histone was linear for at least 20 min. The reaction was stopped by the addition of 2 ml of ice-cold 5% trichloroacetic acid, 10 mM H_3PO_4. The radioactivity retained on GF/C glass-fiber filters after filtration was determined by counting the filters in 2 ml of scintillation fluid. PKC activity was determined after substracting the incorporation in the absence of calcium and phospholipids. The data were expressed in picomol of phosphate into the substrate per minute and per milligram of protein (pmol/min/mg/prot).

Drugs

Carbachol, atropine, 2-nitro-4-carboxyphenyl-N,N-diphenylcarbamate (NCDC), L-arginine, N^G-monomethyl L-arginine (L-NMMA), staurosporine were purchased from Sigma Chemical Company and AF-DX 116 was kindly provided by Boehringer Ingelheim Pharmaceuticals Inc. Stock solutions were freshly prepared in the corresponding buffers. The drugs were diluted in the bath to achieve the final concentrations stated in the text.

Statistical analysis

Student's *t*-test for unpaired values was used to determine the levels of significance. When multiple comparison were necessary, after analysis of variance, the Student-Newman-Keuls test was applied. Differences between means were considered significant if $p < 0.05$.

Results

Stimulation of cardiac PKC activity by chagasic IgG

To analyze if chagasic IgG effects upon myocardium would lead to PKC activation, rat auricles and ventricles preparations were incubated for different times with chagasic antibody and PKC enzymatic activity was determined on both cytosolic and membrane fractions. As is shown in Fig. 1 chagasic IgG (1×10^{-7} M) induced PKC translocation within 30 min of incubation with both tissues, being the fold-stimulation in atria membranes greater than in ventricle membranes (2.82 vs 1.91 respectively). We also found that chagasic IgG stimulates atrial PKC activity in a dose dependent manner (Fig. 2) and that chagasic Ab induced-PKC activation was specific as IgG from non-chagasic (normal) patients had no effect on enzymatic activity (Fig. 3 panel A). Chagasic IgG induced-activation of PKC activity was similar to that obtained with the muscarinic cholinergic agonist carbachol (1×10^{-7} M) (Fig. 3 panel B) and both Ab and carbachol effects were blunted by the muscarinic cholinergic antagonist atropine (5×10^{-6} M) and the M_2 subtype antagonist AF-DX 116

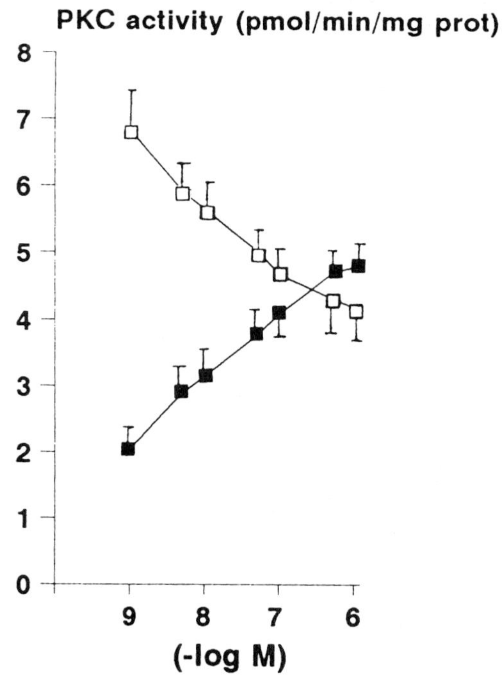

Fig. 1. Time course activation of PKC by chagasic IgG on both rat atria and ventricle preparations. Isolated atria or ventricles were incubated for different times with chagasic IgG (1×10^{-7} M) and cytosol and membrane fractions were obtained as indicated in Materials and methods. PKC activity was measured as described on both fractions, cytosol (□—□) and membrane (■—■) and results were expressed as percentage of total PKC activity of ten chagasic patients.

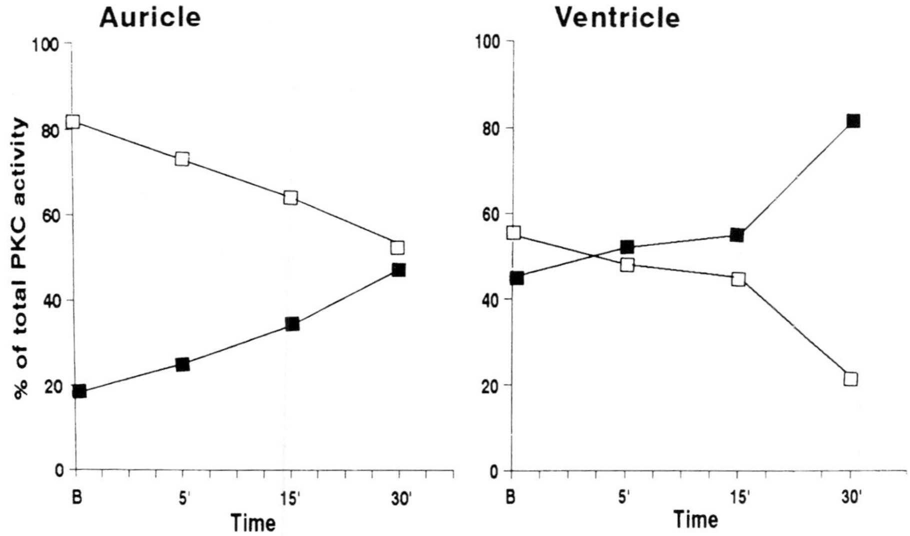

Fig. 2. Dose-response curve of chagasic IgG-induced activation of PKC activity. Atria were incubated with increasing concentrations of chagasic IgG for 30 min and PKC activity was determined on both cytosolic (□—□) and membrane (■—■) fractions as described. Enzymatic activity values shown are the mean ± S.E.M. of 5 independent experiments performed in duplicate.

(5×10^{-6} M) (Fig. 3 panel A and B); pointing to mAChR involvement in chagasic IgG stimulation of PKC activity. Furthermore, both chagasic IgG and carbachol effects on PKC activation were abrogated by the phospholipase C inhibitor NCDC (1×10^{-5} M), indicating a phospholipase C mediated activation of PKC by mAChR triggering through chagasic IgG and carbachol (Fig. 3 panel A and B). It is worth noting that none of the drugs used had effect upon basal PKC activity (data non shown).

To determine if the activation of PKC is involved in IgG stimulation on NOS activity, tissues were incubated with inhibitors of the different enzymatic pathways involved in muscarinic receptor-dependent activation of PI hydrolysis. Figure 4 shows that the inhibition of phospholipase C or PKC activities by NCDC (5×10^{-5} M) or staurosporine (1×10^{-9} M) respectively blocked the stimulatory action of chagasic IgG, indicating that PKC phosphorylation via phospholipase C-mediated hydrolysis was involved in this effect.

Chagasic IgG action on NOS activity

Chagasic IgG effects upon myocardium induces also the activation of NOS activity. As can be seen in Fig. 4 there was a dose-dependent increase in NOS activity when atria or ventricle were exposed to different concentrations of chagasic IgG. It is worth noting that the values of NOS activity in auricle were significantly higher than ventricles values ($P < 0.001$} coincident with that obtained on chagasic IgG-stimulated PKC activity, and with the major distribution of mAChRs and NOS immunoreactive neurons in ventricles [26]. L-NMMA (1×10^{-5} M) a reversible NOS inhibitor, blunted the stimulatory action of chagasic IgG while the natural substrate L-arginine (5×10^{-5} M) completely reversed the inhibitory action of L-NMMA on chagasic IgG effect (Table 1). Chagasic IgG effect mimicked carbachol stimulation of NOS activity and both carbachol and chagasic IgG effects could be blunted by atropine and AF-DX 116 (1×10^{-7} M) (Table 1), pointing to the participation of M_2 mAChR subtype in this effect. Normal IgG was not effective in the studied system (Table 1).

Inhibitory effect of chagasic IgG upon contractility

Chagasic IgG also exerted changes in the contractile behaviour of the heart. As can be seen in Fig. 5 there was a dose-dependent inhibition of contractility on atria and ventricle when they were exposed to different concentrations of chagasic IgG. Chagasic IgG effects resembled those obtained with carbachol and both, antibody and agonist actions, were blunted by 1×10^{-7} M atropine and AF-DX 116 (Table 2). Normal IgG did not result in changes in dF/dt.

To assess if the endogenous NO signalling system participates in the contractile action of chagasic IgG, atria or ventricle were incubated with L-NMMA (1×10^{-5} M). Table 2 also shows that L-NMMA, completely blocked the inhibitory action of both chagasic IgG and carbachol being this action reversed by L-arginine. Moreover, Fig. 5 shows that NCDC and staurosporine, also blunted the inotropic action of chagasic IgG on both auricle and ventricle; indicating that phospholipase C, PKC and NOS are enzymatic pathways involved in the muscarinic receptor-dependent activation of the inotropic negative effect of chagasic IgG. To determine

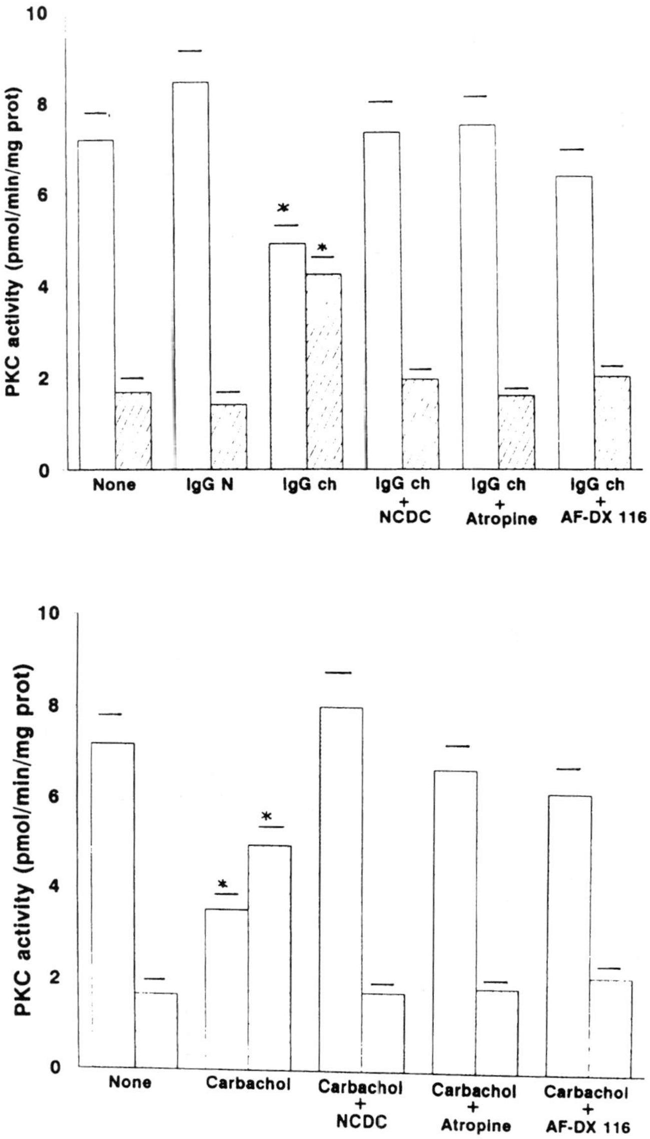

Fig. 3. Blockade of chagasic IgG and carbachol actions upon PKC activity by muscarinic cholinergic antagonists. Rat atria were incubated for 30 min alone or with the muscarinic cholinergic blockers atropine or AF-DX 116 or phospholipase C blocker NCDC. Incubations were continued for 30 min more in the presence of 1×10^{-7} M normal IgG (IgG N) or chagasic IgG (IgG ch) or for 15 min in the presence of carbachol. PKC activity was determined as indicated on cytosol (□) and membrane (▨) fractions. Results are the mean ± S.E.M. of at least four independent experiments performed in duplicate. *$p < 0.05$ with respect to control basal values (none) obtained in the absence of drugs or antibodies.

whether the biological effect of the IgG fraction from chagasic sera and carbachol is mediated by the same receptor site, the action of carbachol upon dF/dt in the presence of a low concentration of chagasic IgG (5×10^{-8} M) was tested. It can be seen in Table 3 that chagasic IgG, unlike normal IgG, inhibited the dose-response curve of carbachol in an uncompetitive fashion.

Discussion

Chronic Chagas' heart disease is an extraordinarily complex process with a poorly understood pathophysiology. In fact it has been described as a cardiac neuromyopathy in which the cardiac sympathetic and parasympathetic systems are affected [15].

This dysautonomia has been explained by a slow and progressive blockade of myocardial neurotransmitter receptors by autoantibodies directed against these receptors [27]. So, we demonstrated the presence of different populations of specific antibodies against cardiac β adrenergic and mAChRs in the sera of chagasic patients and that these antibodies were able to interact with those receptors, modifying normal myocardial behaviour [10, 13, 18].

In an attempt to elucidate the nature of the parasympathetic mechanism involved, here we characterized the participation of PKC and NO signalling system on cardiac mAChR activation by human chagasic IgG.

Our data show that both PKC and NOS activities are involved in the M_2 mAChR-mediated biological effects of chagasic IgG in the heart. It is possible that NO acting as an intracellular messenger, can modulate the muscarinic cholinergic dependent inhibition of contractility. So, chagasic IgG activation of mAChRs led to stimulation of NOS activity and triggered inotropic negative action on isolated myocardium. These effects were mimicked by carbachol and were blocked by a specific M_2 mAChR antagonist. In addition, NOS activation and inotropic negative action of chagasic IgG were blunted when the activity of NOS was inhibited. Also, the inhibitory effect of L-NMMA on the above IgG effects, could be reversed by the addition of the enzyme substrate (L-arginine).

It is known that some of the physiological effects of NO are mediated by the activation of guanylate cyclase by increasing production of cGMP [28]. Moreover, sodium nitroprusside (a NO source} and 8-bromo cyclic GMP (a cGMP analogue) are able to alter contractility of rat atria [29] and other mammalian heart [30]. Also the inotropic negative effects of low concentrations of muscarinic agonist on isolated rat atria are related to the production of NO and cGMP [29]. Thus, it is plausible that in the M_2 mAChR-mediated inhibition of contractility by chagasic IgG, NO acting as a messengers increased production of cGMP that in turn modulated the contractile behaviour of the heart.

The mechanism by which chagasic IgG activating M_2 mAChS induced NOS activation and inotropic negative action seems to involved phospholipase C and PKC activation, since agents known to inhibit both enzymatic activities efficiently interfered with |these effects. The fact that the stereospecific NOS inhibitor L-NMMA abrogate the short-time stimulation of NOS induced by chagasic IgG strongly suggest that the constitutive NOS is active in this system. Calcium/calmodulin dependent constitutive NOS has been reported in vascular

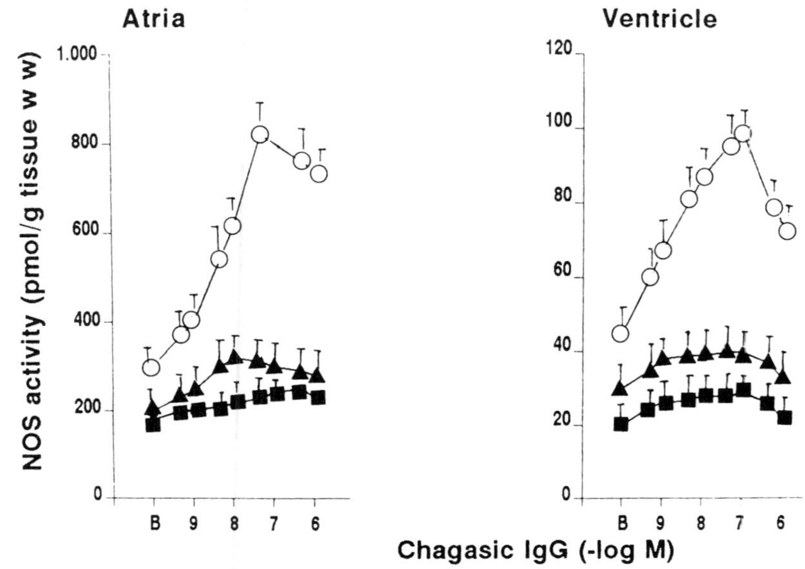

Fig. 4. Stimulation of NOS activity by increasing concentrations of chagasic IgG in atria and ventricle. Chagasic IgG effect alone (O—O) or in the presence of NCDC (■—■) or staurosporine (▲—▲) are shown. Values are mean ± S.E.M. of ten chagasic IgG in every concentrations of each group performed in duplicates.

Table 1. Stimulatory effect of chagasic IgG on NOS activity. Comparison with carbachol and normal IgG

Additions	NOS activity (pmol/g tissue w w)	
	Atria	Ventricle
None (basal)	300 ± 12	40 ± 3
Chagasic IgG	800 ± 20*	98 ± 8*
Chagasic IgG+L-NMMA	244 ± 10	38 ± 2
Chagasic IgG+L-NMMA+L-arg	815 ± 25*	105 ± 10*
Chagasic IgG+atropine	340 ± 14	42 ± 3
Chagasic IgG+AF-DX 116	330 ± 12	41 ± 2
Carbachol	830 ± 22*	89 ± 7*
Carbachol+L-NMMA	242 ± 11	37 ± 2
Carbachol+L-NMMA+L-arg	842 ± 20*	93 ± 6*
Carbachol+atropine	280 ± 15	41 ± 2
Carbachol+AF-DX 116	277 ± 13	39 ± 1
Normal IgG	296 ± 9	37 ± 2

Nitric oxide synthase (NOS) was measured by incubating atria with or without L-NMMA (1×10^{-5} M) or atropine and AF-DX 116 (1×10^{-7} M) for 20 min and then for an additional 15 min with 1×10^{-7} M chagasic IgG, normal IgG or carbachol. L-arginine (L-arg) (5×10^{-5} M) was added together with L-NMMA. Results are mean ± S.E.M. of ten different chagasic or normal patients or seven carbachol experiments performed in duplicate. *Significantly different from basal (none) values with $p < 0.001$.

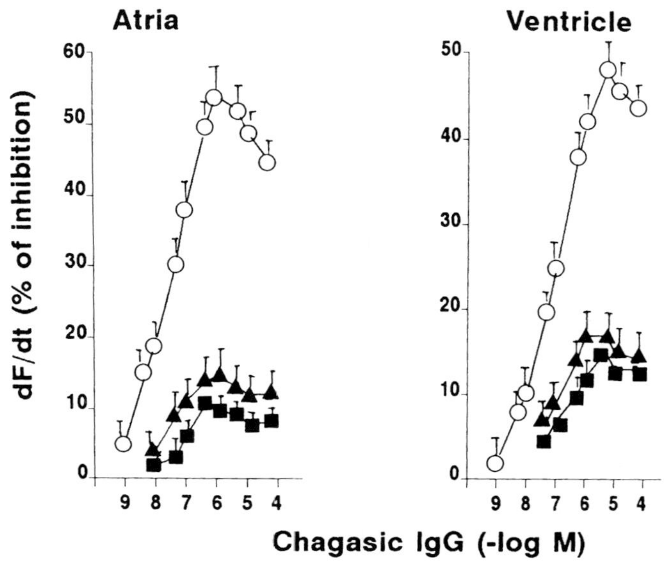

Fig. 5. Decrease in dF/dt of isolated rat atria and ventricle by increasing concentrations of chagasic IgG. Chagasic IgG effect alone (O—O) or in the presence of NCDC (■—■) or staurosporine (▲—▲) are shown. Values are mean ± S.E.M. of ten chagasic IgG in each group.

endothelial and smooth muscle cells [31] while reports on the presence of this enzyme in culture cardiac cells are conflictive [26, 32]. Recently we have demonstrated the existence of PKC-dependent constitutive NOS on isolated rat atria that is sensible to carbachol stimulation [29].

Additionally, we here show that the effect of chagasic IgG on the constitutive NOS activation is also modulated by PKC enzymatic activity. Furthermore, PKC is activated by chagasic IgG as it was able to translocate the enzymatic activity to rat atrial and ventricle membranes. These effects were similar to that obtained with carbachol and seemed to be related to mAChR mediated stimulation of PI turnover with the consequent activation of phospholipase C as both carbachol and chagasic IgG triggering of PKC translocation to cardiac membranes was blunted by muscarinic cholinergic and M_2 antagonist and phospholipase C inhibitor NCDC. It is worth noting that mAChRs of the M_2 subtype are coupled to PI

Table 2. Inhibitory effect of chagasic IgG on dF/dt. Comparison with carbachol and normal IgG

Additions	dF/dt (g/s) Atria	Ventricle
None (basal)	7.6 ± 0.6	18.5 ± 1.2
Chagasic IgG	4.6 ± 0.2*	13.2 ± 1.3*
Chagasic IgG+L-NMMA	7.8 ± 0.4	19.0 ± 1.4
Chagasic IgG+L-NMMA+L-arginine	3.9 ± 0.3*	13.5 ± 1.3*
Chagasic IgG+atropine	7.6 ± 0.4	18.2 ± 1.0
Chagasic IgG+AF-DX 116	8.0 ± 0.5	18.7 ± 1.4
Carbachol	1.8 ± 0.1*	10.2 ± 0.9*
Carbachol+L-NMMA	7.5 ± 0.4	18.4 ± 1.3
Carbachol+L-NMMA+L-arginine	1.7 ± 0.1*	10.3 ± 0.8*
Carbachol+atropine	7.3 ± 0.3	18.9 ± 1.3
Carbachol+AF-DX 116	7.4 ± 0.4	19.2 ± 1.5

Atrial and ventricle contractility (dF/dt) were measured by incubating tissues with or without L-NMMA (1×10^{-5} M) or atropine and AF-DX 116 (1×10^{-7} M) for 20 min and then for an additional 15 min with 1×10^{-7} M chagasic IgG, normal IgG or carbachol. L-arginine (5×10^{-5} M) was added together with L-NMMA. Results are mean ± S.E.M. of ten different chagasic or normal patients or of eight carbachol experiments performed in duplicate *Significantly different from basal (none) values with $p < 0.001$.

Table 3. Inhibitory effect of chagasic IgG on the negative inotropic effect of carbachol compared with normal IgG

Additions	Emax (g/s)	Kd (1×10^{-8} M)
Carbachol	0.22 ± 0.02	5.2 ± 0.2
Normal IgG + carbachol	0.25 ± 0.02	5.3 ± 0.1
Chagasic IgG + carbachol	4.36 ± 0.18*	3.5 ± 0.2

Values were obtained from Lineweaver-Burk plots get with the dF/dt dose-response curves of carbachol in absence or presence of 5×10^{-8} M chagasic or normal IgG. *Significantly different from carbachol alone or normal IgG+carbachol with $p < 0.001$. Emax: maximal effect of carbachol; Kd: effective concentration of carbachol causing 50% of the maximum response (basal values: 8.6 × 0.5 g/s>.

turnover in rat atria [29]. Chagasic IgG mediated-PKC activation via mAChR would have two effects on myocardium physiology. First, when being activated by chagasic IgG it would induce NOS activity leading to the negative inotropic effect of the heart. Second, it would induce mAChR phosphorylation leading to receptor desensitization and probably down-regulation. It is well known that mAChR belongs to the family of G-protein-coupled receptors that are regulated by protein phosphorylation [22, 33]. It has also been shown that agonist stimulation of cardiac mAChR results in a marked increase in the phosphorylation of the receptor on serine and threonine residues and that the process seems to be associated with receptor desensitization [34, 35]. Additionally, it was demonstrated *in vitro* phosphorylation of purified cardiac mAChR by PKC [11] and that chagasic IgG is able to phosphorylate mAChR [21].

So, chagasic IgG in an early step would activate mAChR and later on it might irreversibly bind to those receptors hiding them to the action of the true agonist and contributing to the neurotransmitter receptor blockade characteristic of the chronic stage of the disease. In fact, low concentrations of chagasic antibodies were able to interfere with carbachol actions on myocardial contractility, diminishing the Emax without affecting the Kd.

This paper and our previous observations [17, 18] strongly support the fact that neurotransmitter receptor antibodies from the sera of chagasic patients, would be responsible of the dysautonomia and the onset of the chronic development of Chagas disease.

Acknowledgements

This work has been supported by Grants 0352 from BID-CONICET and UBACYT 1995. The authors thank Mrs. Elvita Vanncchi for her technical assistance.

References

1. WHO Expert Committee: Control of Chagas Disease. Technical report series 811: 1–10, 1991
2. Koberle F: Pathology and pathological anatomy of Chagas disease. Bol. OPS 51: 404–428, 1961
3. Cossio PM, Diez C, Szarfman A, Kreutzer E, Candiolo B, Arana R.M: Chagasic cardiopathy: demonstration of a serum gamma globulin factor which react with endocardium and vascular structures. Circulation 49: 13–21, 1974
4. Acosta A, Santos Buch C. Autoimmune myocarditis induced by *T. cruzi*. Circulation 71: 1255–1261, 1985
5. Cossio PJ Laguens R, Diez C, Segal A, Arana R: Chagasic cardiomyopathy: Immunologic and morphological studies in myocardial biopsies. Am J Pathol 86: 533–538, 1977
6. Mirkin G, Jones M, Sanz 0, Rey RJ Sica R, Gonzalez Cappa S: Experimental Chagas disease: Electrophysiology and cell composition of the neuromyopathy inflammatory lesions in mice infected with the myotropic and pantropic strain of *T. cruzi*. Clin Immunol Immunopathol 73: 69–79, 1994
7. Bracco M, Sterin-Borda L, Fink S, Finiasz M, Borda ES: Stimulatory effect of lymphocytes of chagas patients on spontaneously beating rat atria. Clin Exp Immunol 55: 405–412, 1984
8. Gorelik G, Borda ES, Postan M, Gonzalez Cappa SM, Sterin Borda L: T lymphocytes from *T. cruzi* infected mice alter heart contractility: Participation of arachidonic acid metabolites. J Mol Cell Cardiol 24: 9–20, 1992
9. Sterin-Borda L. Borda ES: Participation of autonomic nerve system in the pathogenesis of Chagas disease. Acta Physiol Phar Ther Latinoan 44: 109–123, 1994
10. Borda ES, Pascual J, Cossio PM, Vega M, Arana RM, Sterin-Borda L: A circulating IgG in Chagas disease which binds to β adrenoceptors of myocardium and modulates its activity. Clin Exp Immunol 57: 679–686, 1984
11. Goin JC, Borda ES, Perez Leiros S, Storino R, Sterin-Borda L: Identification of antibodies with muscarinic cholinergic activity in human Chagas disease: Pathological implications. J Aut Nerv Sys 47: 45–52, 1994

12. Sterin-Borda L, Perez Leiros C, Wald M, Cremaschi G, Borda ES: Antibodies to β1 and β2 adrenoceptors in Chagas disease. Clin Exp Immunol 74: 349–354, 1988
13. Sterin-Borda L, Gorelik G, Borda ES: Chagasic IgG binding with cardiac muscarinic cholinergic receptors modifies cholinergic-mediated cellular transmembrane signals. Clin Immunol Immunopathol 61: 389–397, 1991
14. Cremaschi G, Zwirner N, Gorelik G, Malchiodi E, Charanonte M, Fossati C, Sterin-Borda L: Modulation of cardiac physiology by an anti-*Trypanosoma cruzi* monoclonal antibody after interaction with myocardium. FASEB J 9: 1995 (in press)
15. Koberle F: Pathogenesis of Chagas disease. In Ciba Foundation Simposium. Assoc Scientific Publishers, Amsterdam, 1974, pp 137–158
16. Iossa D, Dequatro Y, De-Ping-Lee D, Elkayan V, Caeiro T, Palmero H: Pathogenesis of cardiac neuropathy in Chagas disease and the role of autonomic nervous system. J Aut Nerv Sys 30: 583–588, 1990
17. Goin JC, Perez Leiros C, Borda ES, Sterin-Borda L. Human chagasic IgG and muscarinic cholinergic receptor interaction: Pharmacological and molecular evidence. Mol Neuropharm 3: 189–196, 1994
18. Goin J C, Perez Leiros C, Borda E S, Sterin-Borda L: Modification of cholinergic-mediated cellular transmembrane signals by the interaction of human chagasic IgG with cardiac muscarinic receptors. Nero Immuno Modulation 1: 284–291, 1994
19. Gorelik G, Borda ES, Bacman S, Cremaschi G, Sterin-Borda L: Chagasic IgG stimulates phosphoinositide hydrolysis via neurotransmitter receptors activation. Role of calcium. J Lipid Med 5: 249–259, 1992
20. Perez Leiros C, Sterin-Borda L, Borda ES, Goin JC, Hosey M: Regulacion de receptores muscarinicos de acetilcolina M_2 humanos por IgG de pacientes chagasicos. Acta Physiol Pharm Therap Latinoam 44: 145–146, 1994
21. Peres Leiros C, Sterin-Borda L, Vila Echague A, Goin JC, Genaro A, Cremaschi G, Borda ES: Paticipacion del oxido nitrico en el efecto muscarinico cardiaco de la IgG chagasic humana. Medicina 55: S561–S562, 1995
22. Richardson MM, Hosey M: Agonist-independent phosphorylation of purified cardiac muscarinic cholinergic receptor by protein kinase C. Biochemistry 29: 8555–8561, 1990
23. Richardson RM, Ptasienski J, Hosey M: Functional effects of protein kinase C-mediated phosphorylation of chick heart muscarinic cholinergic receptors. J Biol Chem 267: 10127–10132, 1992
24. Van Rossum JM: Cumulative dose-response curve. Arch Int Pharmacodyn Ther 143: 299–330, 1963
25. Bredt DS, Snyder SH: Nitric oxide mediates glutamate-linked enhancement of cyclic GMP levels in the cerebellum. Proc Natl Acad Sci USA 86: 9030–9033, 1989
26. Schulz R, Nava E, Moncada S: Induction and biological relevance of a calcium independent nitric oxide synthase in the myocardium. Br J Pharmacol 105: 575–580, 1992
27. Pascual J, Cossio PM, Borda ES, Arana RM, Sterin-Borda L: Modification of sarcolemmal enzymes by chagasic IgG and its effect on cardiac contractility. Biochem Pharmacol 35: 3839–3845, 1986
28. Balligan JL, Kelly RA, Marsden PA, Smith TW, Michel T: Control of cardiac muscle cell function by an endogenous nitric oxide signalling system. Proc Natl Acad Sci USA 90: 347–351, 1993
29. Sterin-Borda L, Vila Echague A, Perez Leiros C, Genaro A, Borda ES: Endogenous nitric oxide signalling system and the cardiac acetylcholine receptor-inotropic response. Br J Pharmacol 115: 1525–1531, 1995
30. Shah AM, Lewis MJ, Henderson AH: Effects of 8-bromo cyclic GMP on contraction and on inotropic response of ferret cardiac muscle. Mol Cell Cardiol 23: 55–64, 1991
31. Mulsch A, Busse R: N-nitro, L-arginine impairs nitric oxide synthesis from L-arginine. Naunyn Schmiedeberg's Arch Pharmacol 347: 143–147, 1990
32. Stein B, Drogemuller A, Mulsch A, Schmitz W, Scholz H: Calcium-dependent constitutive nitric oxide synthase is not involved in the cyclic GMP-increasing effects of carbachol in ventricular myocytes. J Pharm Exp Ther 266: 919–925, 1993
33. Sibley DR, Benovic JL, Caron MG, Lefkowitz RJ: Regulation of transmembrane signalling by receptor phosphorylation. Cell 48: 913–922, 1987
34. Kwatra MM, Hosey MM: Phosphorylation of the cardiac muscarinic receptors in intact chick heart and its regulation by a muscarinic agonist. J Biol Chem 261: 12429–12432, 1986
35. Kwatra MM, Ptasienski J, Hosey MM: The porcine heart M_2 muscarinic receptor agonist-induced phosphorylation and comparison properties with the chicken heart receptor. Mol Pharmacol 35: 553–558, 1989

Molecular and Cellular Biochemistry **160/161:** 83–87, 1996.

Kinetics of myocardial phospholipase D

Jian Dai, Song-Yan Liu and Vincenzo Panagia

Division of Cardiovascular Sciences, St. Boniface General Hospital Research Centre, and Departments of Anatomy and Physiology, Faculty of Medicine, University of Manitoba, Winnipeg, R2H 2A6 Canada

Abstract

Myocardial phospholipase D (PLD) is located in different subcellular membranes, including sarcolemma (SL) and sarcoplasmic reticulum (SR). In this study, the kinetics of PLD-dependent hydrolytic and transphosphatidylation activities were examined in SL and SR fractions isolated from rat heart by measuring the formation of phosphatidic acid and phosphatidylethanol, respectively. The results showed that, compared to SR PLD, SL PLD had a higher V_{max}, i.e. 373 vs. 70 nmol/mg protein/h for the hydrolytic activity and 415 vs. 60 nmol/mg protein/h for the transphosphatidylation activity. In comparison with the SR enzyme, SL PLD had a lower K_m value for the hydrolytic activity (0.46 vs. 0.65 mM), but a higher K_m for the transphosphatidylation activity (225 vs. 179 mM). These distinctive kinetic parameters suggest that SL PLD and SR PLD may be isoforms of the enzyme and/or have different membrane domain. Therefore, SL- and SR-localized PLD activities may be under independent control mechanism(s) and play distinct roles in normal conditions and pathological processes. (Mol Cell Biochem **160/161:** 83–87, 1996)

Key words: phospholipase D, hydrolysis, transphosphatidylation, sarcolemma, sarcoplasmic reticulum, heart

Introduction

Phospholipase D (PLD) catalyses the hydrolysis of several phospholipids, including phosphatidylcholine (PtdCho), an optimal substrate for PLD, to form phosphatidic acid (PtdOH) and a non-phosphorylated base. In addition to the hydrolytic activity, PLD also catalyzes a transphosphatidylation reaction in which the phosphatidyl moiety of substrate phospholipids is transferred to primary alcohols producing phosphatidyl-alcohols. PtdOH formed from PLD hydrolysis may be further dephosphorylated to *sn*-1,2-diacylglycerol (DAG) by a functionally coupled PtdOH phosphohydrolase. Both PtdOH and DAG play a role as second messenger for the regulation of cellular functions [1–3].

Several studies have shown that PLD may be activated by a wide variety of agonists and is regulated by G proteins and protein kinases [1–3]. PLD hydrolytic activity and subsequent formation of PtdOH have been reported to regulate the Ca^{2+} movements in myocardial cells [4]. Therefore, this reaction may influence the heart function in normal conditions [4] as well as in disease states characterized by abnormal Ca^{2+} homeostasis such as the ischemia and reperfusion injury [5].

Our previous studies have shown that PLD is located in different subcellular membranes of the cardiac cells [7]; while the major localization was found at the surface membrane (sarcolemma, SL), a considerable activity was present in the sarcoplasmic reticular (SR) membrane [7]. The activities of PLD were detected in myocardial membranes by other investigators [6, 8]; however, information on the biochemical properties of cardiac PLD is very limited. In this paper, we studied the kinetics of the hydrolytic and transphosphatidylation activities of SL and SR PLD. The evidence provided herein indicates that the two localizations of the enzyme have different kinetic parameters for either the hydrolytic or transphosphatidylation reactions.

Materials and methods

Adult male Sprague-Dawley rats weighing 250–300 g were sacrificed by decapitation and the hearts were quickly excised. Purified SL and SR preparations were obtained as previously indicated [7].

The hydrolytic activity of PLD was assayed by measuring the formation of labeled PtdOH from [^{14}C]-PtdCho [7]. The transphosphatidylation activity of PLD was tested by in-

Address for offprints: V. Panagia, Division of Cardiovascular Sciences, St. Boniface General Hospital Research Centre, 351 Tache Avenue, Winnipeg, Manitoba, R2H 2A6 Canada cluding ethanol in the assay medium and by measuring the formation of labeled phosphatidylethanol (PtdEtOH) [9]. The kinetic analysis was performed by plotting the rate of formation of the reaction products against substrate concentration. In order to completely fit in the hyperbolic curve of the Michaelis-Menten equation when the substrate was below a so called 'critical concentration' [10, 11], a phase-separation model was used [12]. This model assumes that in aqueous microdispersions of phospholipid substrates under a critical concentration, the physical state of the phospholipid is a monomer. Above the critical concentration, however, two substrate species are present; monomers and aggregates. For kinetic calculations, only the contribution from the aggregated form was taken into consideration. Therefore, each substrate concentration was corrected by subtracting the critical concentration value. This value was obtained by assuming various values to fit a Lineweaver-Burk plot with the experimental data at a graphically feasible range of substrate concentrations [10, 11]. The best fit (correlation coefficient: 0.992 ± 0.002, $n = 30$) was achieved with a critical concentration of 0.283 ± 0.008 mM. The V_{max} and K_m values of the PLD hydrolytic activity were directly obtained from Lineweaver-Burk plots for a PtdCho concentration range of 0.1–3.0 mM, and the values of the transphosphatidylation activity were obtained for an ethanol concentration range of 10–600 mM. The membrane proteins were determined as indicated elsewhere [7].

Data points are the average of three determinations (variance < 10%). Each set of experiments was performed at least three times and a representative experiment is presented.

L-α-1-Palmitoyl-2-oleoyl-[oleoyl-1-^{14}C]PtdCho (specific activity, 58.0 mCi/mmol) was purchased from New England Nuclear, Mississauga, Ontario (Canada). PtdCho, PtdOH, and oleic acid (sodium salt) were obtained from Serdary Research Laboratories, London, Ontario (Canada). Silica gel 60A F-254 TLC plates were purchased from Whatman International Ltd., Maidston (U.K.). The organic solvents were bought from Mallinckrodt, Montreal, Quebec (Canada). All other chemicals were from Sigma Chemical Company, St. Louis, Missouri (U.S.A.). PtdEtOH was synthesized as previously indicated [9].

Results

PLD is a bifunctional enzyme that catalyzes two reactions, i.e. hydrolysis and transphosphatidylation [1, 3, 7]. As the basal activities of myocardial PLD are not detectable in the presence of exogenous [^{14}C]-PtdCho [9], the oleate-stimulated activities [9] were analyzed in this study. The data in Fig. 1 show the SL PLD-dependent formation of PtdOH and PtdEtOH in the presence of a fixed concentration of [^{14}C]-PtdCho and increasing concentrations of the acceptor alcohol, ethanol. The augmented availability of ethanol enhanced the synthesis of PtdEtOH and reduced that of PtdOH. This finding suggests that the two PLD-catalyzed reactions are competitive, as observed in other tissues [10].

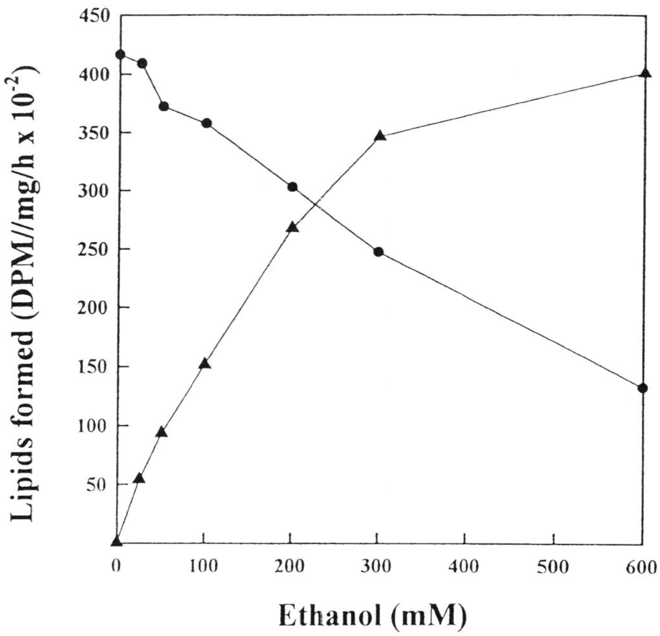

Fig. 1. Effect of scalar increases of the ethanol concentration on phosphatidic acid and phosphatidylethanol production by sarcolemmal phospholipase D. Values are the average of three determinations (variance < 10%) from a representative experiment. PLD activity was assayed as indicated in Materials and Methods. PtdOH (●), PtdEtOH (▲).

The effect of varying PtdCho concentrations on PtdOH formation in myocardial SL and SR is shown in Fig. 2. The kinetic patterns are typical plots of velocity versus substrate concentrations for an enzyme-catalyzed reaction. The K_m and V_{max} values of SL PLD were 0.46 mM and 373 nmol/mg protein/h, respectively. The corresponding values for SR PLD were 0.65 mM and 70 nmol/mg protein/h. Apparently, this means that SL PLD has a higher hydrolytic activity and a higher affinity for the phospholipid substrate than the SR enzyme. The effect of varying the ethanol concentrations on PtdEtOH synthesis in myocardial SL and SR is shown in Fig. 3. As seen in the case of the hydrolytic reaction, SL PLD transphosphatidylation showed a V_{max} value higher than that of SR PLD (415 vs. 60 nmol/mg protein/h). Conversely, SR PLD had a lower K_m value and, therefore a higher affinity for ethanol than SL PLD (179 vs. 225 mM). The kinetic values of the SR transphosphatidylation reaction were similar to those reported for the brain synaptosomal enzyme [10].

Discussion

In an earlier report, we provided the first evidence for the presence of PLD in different membrane systems of the car-

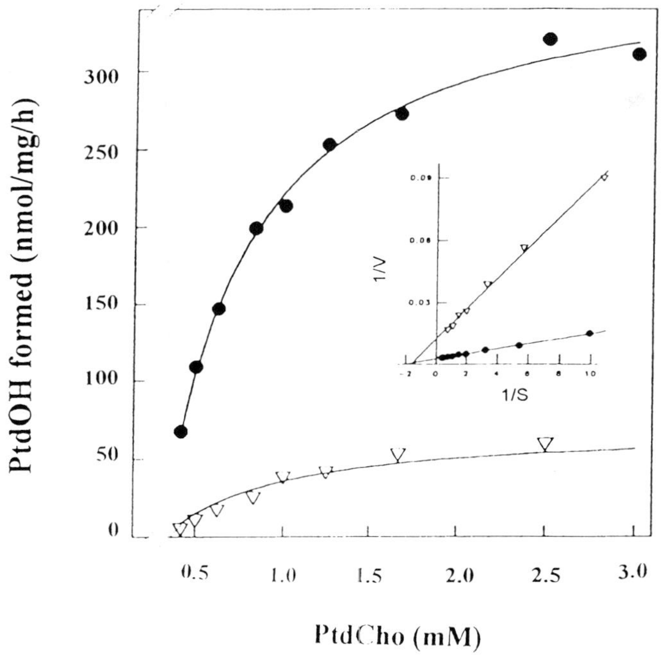

Fig. 2. Effect of exogenous phosphatidylcholine on phosphatidic acid formation in sarcolemmal and sarcoplasmic reticular membranes. Values are the average of three determinations (variance < 10%) from a representative experiment. The inset represents Lineweaver-Burk plots which were constructed from the values of the enzyme specific activity at different concentrations of the substrate. The hydrolytic activity of SL (●) and SR (∇) PLD was assayed according to the Materials and Methods. V_{max} was 373 and 70 nmol PtdOH formed/mg protein/hr for SL and SR respectively. K_m was 0.46 and 0.65 mM for SL and SR, respectively.

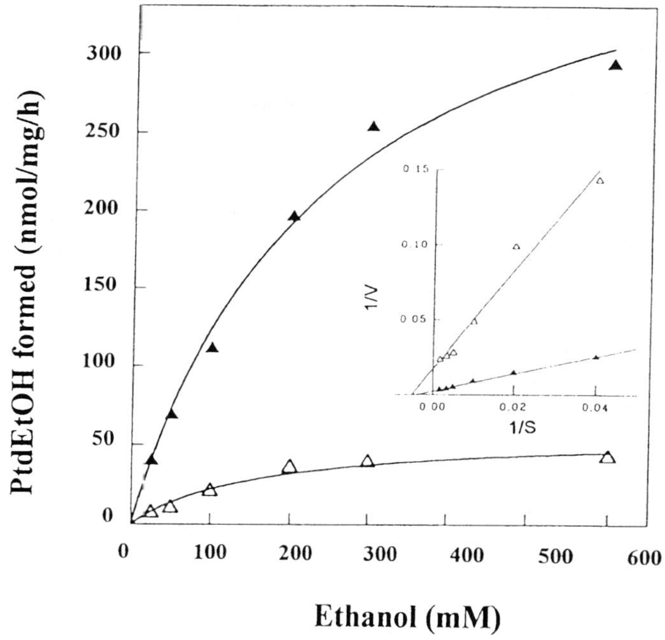

Fig. 3. Effect of increasing ethanol concentrations on phosphatidylethanol formation by sarcolemmal and sarcoplasmic reticular membranes. Values are the average of three determinations (variance < 10%) from a representative experiment. The transphosphatidylation activity of SL (▲) and SR (Δ) PLD was assayed in the presence of 2.5 mM [^{14}C]-PtdCho and different concentrations of ethanol according to Materials and Methods. V_{max} was 415 and 60 nmol PtdEtOH formed/mg protein/h for SL and SR, respectively. K_m was 225 and 179 mM for SL and SR, respectively.

diac cell, including SL and SR [7]. The biochemical properties of the SL PLD [7] and the possibility that its activity is affected during cardiomyocyte stimulation [13] suggest that the role of this enzyme may be that of actively generating important lipid molecules for signal transduction. The functional significance of PLD-dependent increase of PtdOH in cardiac SL membranes is indicated by some studies. For example, the treatment with exogenous PLD increased the cardiac SL Na^+-Ca^{2+} exchange activity [14], Ca^{2+} binding and force of contraction of the heart [15]. The observation that exogenous addition of PtdOH or PLD generates Ca^{2+}-dependent slow action potentials in depolarized rat atrium supports the role of PtdOH in mediating the increase of Ca^{2+} influx into the cardiac cells [16]. PtdOH has also been reported to stimulate the SL Ca^{2-} pump [17] which is involved in the efflux of Ca^{2+} from the cell during relaxation of the myocardium [18]. These results suggested that SL PLD hydrolytic activity may be related to Ca^{2+} movements within the cardiomyocyte and may influence the heart function in normal conditions as well as in disease states characterized by abnormal Ca^{2+} homeostasis. In fact, we have found that exogenously added PtdOH releases Ca^{2+} from the SR of adult cardiomyocytes to increase $[Ca]_i$, which may contribute to PtdOH-related positive inotropic effect [4]. Furthermore, a rapid formation of PtdOH due to stimulation of the heart by agonists [13], its potential action in enhancing phosphoinositide synthesis [19] and hydrolysis [20], and stimulating the phosphorylation of certain proteins [21] indicate that this phospholipid may be intimately involved in signal transduction [21, 22]. In contrast to the numerous knowledge on PLD hydrolytic activity, the information about the transphosphatidylation activity is still limited [23]. Phosphatidylethanol, which is an anionic phospholipid at physiological pH [24] and normally amounts to 1–2% of the total cellular phospholipids [25], can alter the physicochemical properties of the membrane by increasing its fluidity [25]. Minor quantities of PtdEtOH activate 5′-nucleotidase [25], the enzyme responsible for adenosine synthesis which, in the heart, has been found at both the SL and SR level [26]. PtdEtOH can also substitute phosphatidylserine as a cofactor for the activation of PKCγ [27], thus influencing its normal activation pattern.

SL and SR PLD share some common features in their regulatory mechanisms, such as sulphydryl group dependence [28, 29]. However, they display significantly different responses to Fe^{2+} ions and ischemia-reperfusion injury [5, 29]. These findings suggested that SL and SR PLD might have dissimilar biochemical properties and/or regulatory mechanisms. Indeed, differences in PLD kinetics have been observed in this study between the two subcellular localizations of the enzyme. As already reported [7], the pattern for PLD purification in the two fractions follows that seen for the respective positive marker enzymes, indicating that the PLD activities detected in the present study are inherent to each membrane. SL PLD hydrolytic activity exhibited a higher V_{max} and higher affinity for the substrate than SR PLD. As discussed above, the product of this reaction, PtdOH, is involved in the signalling processes of the cardiomyocytes.

Therefore, the higher V_{max} and lower K_m values for the PLD hydrolytic activity in SL may result in a prompt formation of the signal molecule upon agonist stimulation, thus favoring a rapid cell response. On the other hand, the SR transphosphatidylation reaction showed a K_m value lower than that seen in SL membranes. Therefore, an augmented availability of ethanol or other acceptor alcohols may competitively reduce the formation of PtdOH and its dephosphorylated product DAG at the SR level, thus compromising the PLD hydrolytic pathway at this level.

The mechanism underlying the kinetic diversity between SL PLD and SR PLD in the heart remains unknown. The present findings showed that both K_m and V_{max} values were different in the two subcellular localizations of the enzyme. That may suggest the existence of more than one PLD isoenzyme, as it has been recently reported for the rat brain membranes [30]. However, the possibility that specific lipid domains may surround and influence the activity of the same enzyme molecule in SL and SR membranes cannot be excluded. At any rate, the different kinetic properties of SL and SR PLD suggest that these activities may have a distinct role in the myocardial cell function. Given that, it seems important to further examine the regulatory mechanisms and biochemical characteristics of the PLD activities localized in the SL and in the SR membranes of the cardiomyocyte.

Acknowledgements

This research was supported by a grant from the Heart and Stroke Foundation of Manitoba (V.P.). J.D. was the recipient of a Studentship from the Manitoba Health Research Council.

References

1. Exton JH: Phosphatidylcholine breakdown in signal transduction. Biochim Biophys Acta 1212: 26–42, 1994
2. Lamers JM, De Jonge HW, Panagia V, Van Heugten HAA: Receptor-mediated signalling pathways acting through hydrolysis of membrane phospholipids in cardiomyocytes. Cardioscience 4: 121—31, 1993
3. Liscovitch M, Chalifa V: Signal-activated phospholipase D. In: M Liscovitch (ed). Signal-activated phospholipases. R.G. Landes Company, Austin, 1994, pp 31–63
4. Xu Y-J, Panagia V, Shao Q, Wang X, Dhalla NS: Phosphatidic acid increases intracellular free Ca^{2+} and cardiac contractile force. Am J Phyisol (Heart Circ Physiol), 1996, in press
5. Dai J, Panagia V: Distinct changes of sarcolemmal and sarcoplasmic reticular phospholipase D activity in ischemic-reperfused heart. FASEB J 7: A121, 1993
6. Moraru II, Popescu LM, Maulik N, Liu X, Das DK: Phospholipase D signalling in ischemic heart. Biochim Biophys Acta 1139: 148–154, 1992
7. Panagia V, Ou C, Taira Y, Dai J, Dhalla NS: Phospholipase D activity in subcellular membranes of rat ventricular myocardium. Biochim Biophys Acta 1064: 242–250, 1991
8. Baldini PM, Incerpi S, Zannetti A, De Vito P, Luly P: Selective activation by atrial natriuretic factor of phosphatidylcholine-specific phospholipase activities in 14 purified heart muscle plasma membranes. J Mol Cell Cardiol 26: 1691–1700, 1994
9. Dai J, Williams SA, Ziegelhoffer A, Panagia V: Structure activity relationship of the effect of cis-unsaturated fatty acids on heart sarcolemmal phospholipase D activity. Prostagland Leuk Essent Fatty Acids. 52: 167–171, 1995
10. Kobayashi M, Kanfer JN: Phosphatidylethanol formation via transphosphatidylation by rat brain synaptosomal phospholipase D. J Neurochem 48: 1597–1603, 1987
11. Allgyer TT, Wells MA: Phospholipase D from savoy cabbage: purification and preliminary kinetic characterization. Biochemistry 18: 5348–5353, 1979
12. Wells MA: The mechanism of interfacial activation of phospholipase A2. Biochemistry 13: 2248–2257, 1974
13. Ye H, Wolf RA, Kurz T, Corr PB: Phosphatidic acid increases in response to noradrenaline and endothelin-l in adult rabbit ventricular myocytes. Cardiovasc Res 28: 1828–1834, 1994
14. Philipson KD, Nishimoto AY: Stimulation of Na+-Ca2+ exchange in cardiac sarcolemmal vesicles by phospholipase D. J Biol Chem 259: 16–19, 1984
15. Langer GA, Rich TL: Phospholipase D produces increased contractile force in rabbit ventricular muscle. Circ Res 56: 146–149, 1985
16. Knabb MT, Rubio R, Berne RM: Calcium-dependent atrial slow action potentials generated with phosphatidic acid or phospholipase D. Pflugers Arch 401: 435–437, 1984
17. Carafoli E: How calcium crosses plasma membranes including sarcolemma. In: L.H. Opie (ed). Calcium antagonists and cardiovascular disease. Raven Press Publishing, Boston, 1984, pp 29–41
18. Dhalla NS, Pierce GN, Panagia V, Singal PK, Beamish RE: Calcium movements in relation to heart function. Basic Res Cardiol 77: 117–139, 1982
19. Moritz A, De Graan PNE, Gispen WH, Wirtz KWA: Phosphatidic acid is a specific activator of phosphatidylinositol-4-phosphate kinase. J Biol Chem 267: 7207–7210, 1992
20. Ross AH, Boyce SY, Kurz T, Wolf RA: Stimulation and binding of myocardial phospholipase C by phosphatidic acid. Am J Physiol 269: C349–C358, 1995
21. Bocokino SB, Wilson PB, Exton JH: Phosphatidate-dependent protein phosphorylation. Proc Natl Acad Sci USA 88: 6210–6213, 1991
22. Meij JTA, Panagia V: Phospholipase D: a new avenue to the phospholipid signalling pathway in the myocardium. In: B. Ostadal and N.S. Dhalla (eds). Heart function in health and disease. Kluwer Academic Publishing, Boston, 1993, pp 79–90
23. Yu C-H, Liu S-Y, Panagia V: The transphosphatidylation activity of phospholipase D. Mol Cell Biochem 157: 101–105, 1996
24. Hoek JB, Rubin E: Alcohol and membrane associated signal transduction. Alcohol and Alcoholism 25: 143–156, 1990
25. Omodeo Salé F, Lindi C, Palestini P, Masserini M: Role of phosphatidylethanol in membranes. Effect on membrane fluidity, tol-

erance to ethanol and activity of membrane-bound enzymes. Biochemisty 30: 2477–2482, 1991
26. Panagia V, Heyliger CE, Singal PK, Dhalla NS: Subcellular distribution of cardiac 5′-nucleotidase: alteration of microsomal pool in hypertrophied pig heart. J Mol Cell Cardiol 18: 815–826, 1986
27. Asaoka Y, Kikkawa V, Sekiguchi K, Shearman MS, Kosaka Y, Nakano Y, Satoh T, Nishizuka Y: Activation of brain-specific protein kinase C subspecies in the presence of phosphatidylethanol. FEBS Lett 231: 221–224, 1988
28. Dai J, Meij JTA, Panagia V: Depression of cardiac sarcolemmal phospholipase D activity by oxidant-induced thiol modification. Circ Res 71: 970–977, 1992
29. Dai J, Meij JTA, Dhalla V, Panagia V: Involvement of thiol groups in the impairment of cardiac sarcoplasmic reticular phospholipase D activity by oxidants. J Lipid Mediators Cell Signalling 11: 107–118, 1995
30. Massenburg D, Han J-S, Liyanaga M, Patton WA, Rhee SG, Moss J, Vaughan M: Activation of rat brain phospholipase D by ADP-ribosylation factors 1,5, and 6: Separation of ADP-ribosylation factor-dependent and oleate-dependent enzymes. Proc Natl Acad Sci USA 91: 11718–11722, 1994

Molecular and Cellular Biochemistry **160/161**: 89–93, 1996.

Is cysteine residue important in FITC-sensitive ATP-binding site of P-type ATPases? A commentary to the state of the art

Albert Breier,[1] Attila Ziegelhöffer,[2] Konrad Famulsky,[3] Marek Michalak[3] and Ján Slezák[2]

[1]*Institute of Molecular Physiology and Genetics, Slovak Academy of Sciences, Vlárska 5, 833 34 Bratislava, Slovak Republic;* [2]*Institute for Heart Research, Slovak Academy of Sciences, Dúbravská cesta 9, 842 33 Bratislava, Slovak Republic;* [3]*Department of Biochemistry, Cardiovascular Disease Research Group, University of Alberta, Edmonton, 424 Heritage Medical Research Center, Edmonton, Alberta, T6G 2S2, Canada*

Abstract

Treatment of P-type ATPases (from mammalian sources) by fluorescein isothiocyanate (ITC) revealed the ITC label on a lysine residue that was than considered as essential for binding of ATP in the ATP-binding site of these enzymes. On the other hand, experiments with site directed mutagenesis excluded the presence of an essential lysine residue that would be localized in the ATP binding sites of ATPases. Other previous studies, including those of ourselves, indicated that the primary site of isothiocyanate interaction may be the sulflhydryl group of a cysteine residue and this may be essential for binding of ATP. In addition considerable knowledge accumulated since yet also about the differences in stability of reaction product of isothiocyanates with SH- or NH_2- groups. Based upon evaluation of the data available up to now, in present paper the following tentative roles for lysine and cysteine residues located in the ATP-binding site of P-type ATPases are proposed: The positively charged micro-domain of the lysine residue may probably attract the negatively charged phosphate moiety of the ATP molecule whereas the cysteine residue may probably be responsible for recognition and binding of ATP by creation of a proton bridge with the amino group in position 6 on the adenosine ring of ATP. (Mol Cell Biochem **160/161**: 89–93, 1996)

Key words: isothiocyanates, lysine and cysteine in the ATP-binding site, P-type ATPases, ($Na^+ + K^+$)-ATPase, Ca^{2+} ATPase

Introduction

Fluorescein isothiocyanate (FITC) was first time used for labeling the ATP binding site of purified sarcoplasmic reticulum Ca^{2+} -ATPase and of plasma membrane ($Na^+ + K^+$)-ATPase by Pick and Basilian [1] and Carilli *et al.* [2], respectively. The labeled enzyme was treated subsequently with trypsin, the tryptic fragment carrying the FITC label was subsequently purified and its amino acid sequence was determined [3–6]. The label was found attached to the NH_2 -group of a lysine residue. Similar result was also obtained when investigating the Ca^{2+} -ATPase of plasma membrane and of the ($H^+ + K^+$)-ATPase from stomach [7, 8]. In contrast to these findings we have demonstrated that p-bromophenyl isothiocyanate (PBITC), FITC and ATP-ITC ([4-isothiocyanatophenyl-(thioureidohexyl)carbamoylmethyl]-ATP)* were inhibiting the activity of heart sarcolemmal ($Na^+ + K^+$)-ATPase by reacting with an SH-group located likely in the proximity of ATP-binding site of the enzyme [9, 10]. The latter SH- group proved to be essential for binding of ATP to the enzyme and it seemed be identical with the SH-groups revealed in the active site of ($Na^+ + K^+$)-ATPase by Patzelt-Wenczler *et al.* [11] and Schoner *et al.* [12]. By means of disulfide analogues of ATP, these authors could clearly demonstrate the existence of two types of reactive sulfhydryl groups that were located inside or close to the ATP-binding site of ($Na^+ + K^+$)-ATPase and that may fulfill the role to recognize and bind ATP by interacting with the 6-amino group

*The ATP analogue was synthesized in the Laboratory of Dr. P. Gemeiner, Institute of Chemistry, Slovak Academy of Sciences, Bratislava, Slovak Republic, as it was already described elsewhere [18].

of the purine ring on the ATP molecule.

From data available about the reactivity of isothiocyanates [12] it may be assumed that all isothiocyanates applied, i.e. PBITC [9] and FITC [1–3] and ATP-ITC [9, 10] are fully capable to inhibit the ATPases by interacting with the same SH-groups on the enzyme molecule. Hence, the question arose why in enzyme processed by labeling, tryptic digestion and sequencing [3–6] always a lysine NH_2-group and never a cysteine SH-group were identified as the recipient of isothiocyanate label. A plausible answer to the above question seems to be offered by peculiarities in the reactivity of isothiocyanates at different pH values. Isothiocyanates react easily with NH_2-groups of amino acids providing N,N'-disubstituted derivatives of thiourea (IV, in Fig. 1). The resulting compounds are stable in the presence of reactive nucleophilic substances such as β-mercaptoethanol or dithiotreitol (III in Scheme I) as well as in alkaline milieu [13, 14]. Nevertheless, isothiocyanates can react also with SH-groups and this reaction may occur in a broad range of pH far more readily than that with the NH_2-groups [14]. Moreover, it was demonstrated on the example of several dehydrogenases that in neutral or slightly acidic conditions, isothiocyanates are reacting selectively and exclusively with the SH-groups of enzymes only [15]. However, the resulting S-esters of dithiocarbamic acid are less stable compounds than the N,N'-derivatives of thiourea and in contrast to the latter they may dissociate already in slight alkaline milieu [13, 14]. Therefore, if an SH-group would represent the site of primary isothiocyanate interaction, the position of the label will be very vulnerable to shifts of pH towards the alkaline side. In respect to the above facts the following comment may be made about the events involved in labeling the active site of P-type ATPases (from mammalian sources) by isothiocyanates and identifying the final recipient of the label: (i) In experiments of Farley *et al.* [3] and others [4–8] with labeling the active sites of ATPases, the reaction with FITC was performed at pH = 9.2. This milieu was not securing a selective interaction of the inhibitor and enabled the isothiocyanate to react with both the NH_2- as well as the SH-groups in the active site of the enzyme (II in Fig. 1). However, in experiments with PBITC, FITC and ATP-ITC [9, 10] that were carried out at pH = 6 (in the aim to restrict the interaction of isothiocyanate exclusively to SH-groups) a similar inhibition of the ATPase could be achieved. This indicated, that in both cases rather the essential SH- than some NH_2-groups may represent the primary site of isothiocyanate interaction in the ATP-binding centers of the ATPases. (ii) The removal of unbound, excess FITC in the presence of β-mercaptoethanol or dithiothreitol as well as the consequently following, relatively long procedure of tryptic digestion and isolation of the peptide fragment, were carried out at alkaline pH [3–8]. Hence, the alkali-labile S-esters of dithiocarbamic acid that at the beginning might be carrying probably the majority, if not all of the FITC label bound in ATP- binding site of the ATPases investigated, could not survive the processes sub (ii) and dissociated. (iii) The free isothiocyanate label that became liberated in the process of dissociation of the respective S-ester of dithiocarbamic acid (V in Fig. 1.) could than jump over and react in turn with the NH_2-group of a close lying lysine residue (IV in Fig. 1). Such a residue was always found to be present in the tryptic fragments described by Farley *et al.* [3] and others [4–8]. (iv) An evidence that in alkaline pH, FITC may be relocated from an SH-group to an NH_2-group was demonstrated on mercaptoethylamine as a model molecule by Wilderspin and Green [16] already in 1983. Later, Swoboda and Hasselbach [17] also considered the possibility of relocation of FITC from a cysteine to a lysine even in the active site of the Ca^{2+}-ATPase from sarcoplasmic reticulum. However, in the papers of Farley *et al.* [3] and others [4–8] such a possibility was not discussed.

When considering all the facts and possibilities indicated sub (i–iv) it may be assumed that it is little probable that the isothiocyanate label may persist to be bound to an SH-group also after the procedures tryptic digestion and isolation of the tryptic fragments. Hence, the conception [3–8] that the tryptic fragments with FITC attached to an NH_2-group may represent the complete structure of ATP binding sites of the investigated ATPases, still may remain a matter of discussion. Nevertheless, there are also further arguments that make the essentiality of lysine NH_2-group in the ATP-binding sites of ATPases questionable.

Data in Table 1 demonstrate that the inhibition of (Na^+ + K^+)-ATPase activity by PBITC, FITC or ATP-ITC at pH 7.0 can be reversed by dithiothreitol. It is known that only the S-esters but not the N-esters of dithiocarbamic acid may be considered sensitive to reactive nucleophilic agents [14]. Thus, the recovery of the isothiocyanates-inhibited (Na^+ +

Table 1. Recovery from isothiocyanates inflicted inhibition of (Na^+ + K^+)-ATPase with dithiohiothreitol treatment.

Group	Isothiocyanates		
	PBITC	FITC	ATP-ITC
A	10.4 ± 0.2	10.4 ± 0.2	10.4 ± 0.2
B	1.3 ± 0.1	3.7 ± 0.1	2.8 ± 0.1
C	9.2 ± 0.2	9.4 ± 0.2	9.1 ± 0.3

Results represent the activities (mmol P_i/min.mg) of differently treated (Na^+ + K^+)-ATPase preparations isolated from dog kidney [23]. (A) Control experiment. Samples were treated similarly as in the groups B and C but in the absence of isothiocyanates and dithiothreitol. (B) Samples were incubated with the respective isothiocyanate (I mmol/mg protein) in 10 mmol/l Tris-HCl buffer pH = 7.0 for 2 h at 4°C. Two times centrifuged at 96.000 × g and resuspended in the same medium in the absence of dithiothreitol. (C) Samples were handled similarly as sub B but after the second centrifugation they were treated additionally with dithiothreitol (5 mmol/1) in the same bufter for 60 min at 4°C. After reaction with dithiothreitol the membranes were spunned down at 96.000 × g, again resuspended in 10 mmol/l Tris-HCl bufter pH 7.2 and used for measurement of the (Na^+ + K^+)-ATPase activity according to [9].

K$^+$)-ATPase activity may result from displacement of the inhibitor from an essential thiol group only.

An important evidence speaking against the presence of an essential lysine residue in the ATP-binding site of the (Na^+ + K^+)-ATPase was brought in studies with site directed mutagenesis of the latter enzyme [19] as well as of Ca^{2+}-ATPase from the sarcoplasmic reticulum [20]. In these studies the lysine residue considered as essential was replaced by either alanine, or arginine or glutamic acid respectively. However, the complete loss in ATPase activity expected consequently was missing. On the other hand, the FITC-sensitive lysine residue that was found in ATP-binding site of ATPases also seems to exhibit functional importance. The replacement of this lysine residue by glutamic acid (negatively charged amino acid residue), strongly decreased the affinity of the (Na^+ + K^+)-ATPase to ATP [19] and the transport of calcium by the Ca^{2+}-ATPase of sarcoplasmic reticulum [20]. This indicates that the above lysine residue is not essential for binding of ATP but with its positive charge may participate in creation of the ATP-binding site, i.e. it should be at least close to the amino acid residue essential for binding of ATP. Close to the FITC-sensitive lysine residue involved in binding of ATP exists also a cysteine residue (for references see Table 2). This cystein residue may be considered as highly conserved because in each mammalian P-type ATPase (Table 2) it was found to be localized in the same position in respect to the FITC-sensitive lysine residue. The primary structures given in Table 2. represent our concept of putative structures of the ATP-binding sites of P-type ATPases. The structure starts with a stretch of five charged amino acid residues followed by a short hydrophobic region and it is flanked by another charged region of four amino acid residues (Fig. 2). The charged region situated closer to N-terminus contains a lysine residue (*) whereas a cysteine residue (#) was identified to be present in the charged region on the side of the C-terminus. In all mammalian ionic pumps studied so far the

Table 2. Primary structure of FITC sensitive peptide fragments

ATPase	Sequence	pI
(H^+ + K^+)-ATPase 1	LVM **K**GA PER VLE R**C**S S	8.899
(H^+ + K^+)-ATPase 2	VVM **K**GA PER ILE K**C**S T	8.730
(Na^+ + K^+)-ATPase	LVM **K**GA PER ILD R**C**S S	8.899
$SRCa^{2+}$-ATPase	MFV **K**GA PEG VID R**C**T H	7.003
$PMCA^{2+}$-ATPase (1)	IFS **K**GA SEI ILK K**C**F K	9.807
$PMCA^{2+}$-ATPase (2)	MYS **K**GA SEI VLK K**C**C K	9.687

K and **C** represent the respective lysine and cysteine residues that were found to be labeled by FITC [3–8]. The sequences were adopted from the primary structure of cDNA of the following enzymes: (H^+ + K^+)-ATPase 1 – from gastric mucosa [24]; (H^+ + K^+)-ATPase 2 – putative distal colon (H^+ + K^+)-ATPase [25]; (Na^+ + K^+)-ATPase [26]; $SRCa^{2+}$-ATPase [27]. Similar structure of this tragment was also found in the cases of SERCA1, SERCA2a, SERCA2b and SERCA3 isoenzymes of the sarcoplasmic calcium pump [28]; $PMCa^{2+}$-ATPase (1) [29] and $PMCa^{2+}$-ATPase (2) [30]. Values of pI were deduced from the corresponding primary structures.

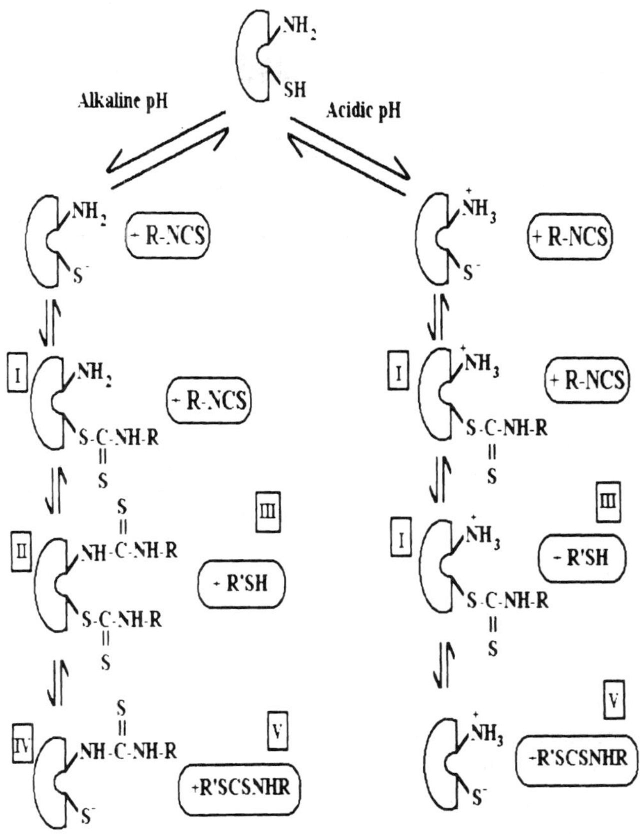

Fig. 1. Reactivity of isothiocyanate with SH- and NH_2-groups of the enzymes. In alcaline medium when the thiol groups are in S^- form and the amino groups in NH_2 form, isothiocyanates are interacting preferably with thiol groups. The reaction is fully reversible and provides S-esthers of dithiocarbamic acid (I) that are sensitive to reactive nucleophilic substances (III) such as dithiothreitol or β-mercaptoethanol. The removal of reactants may induce a shift in equilibrium of the reaction (I). Consequently, the reaction product may dissociate easily and if the isothiocyanate is in excess, this may react in addition also with NH_2 groups by forming stable N,N'-disubstituted derivatives of thiourea (II) that are fully insensitive to NH_2 groups. However, if added in this stage, dithiothreitol or β-mercaptoethanol may induce the dissociation of a further amount from the existing S-esthers of dithiocarbamic acid. Eventually, this may even more enhance the amount of enzyme modified selectively on the NH_2 groups (IV). In acidic or neutral medium, when thiol groups may be partially dissociated and NH_2 groups are predominantly in the NH_3^+ form, isothiocyanates attack selectively the sulfhydryl groups of proteins only (I). However, a change in pH to the alcaline side may cause the dissotiation of a certain portion of S-esthers of dithiocarbamic acid, enabling thus an additional reaction of isothiocyanates with NH_2 groups (II) or relocation of the bound isothiocyanate from SH- to NH_2-groups (IV).

above cysteine and lysine residues are separated from each other by 9 amino acids.

From pI values deduced from the primary structure in Table 2 it may be assumed that at pH = 7.0-7.4 in contrast to ATP-binding sites of calcium pumps from the sarcoplasmic reticulum the ATP-binding sites of plasma membrane ATPases are in a more ionized form. In the mean time we have no explanation for this phenomenon. Nevertheless, it may be at

K* G A P(S) E X V(I) L(I) Z R(K) C# B(F) Z(B)

Fig. 2. General scheme of putative ATP binding site of P-type ATPases. The cysteine and lysine residue are signed by # and *, respectively. Z-charged aminoacid, B-polar aminoacid. ■ – charged region; ▒ – hydrophobic region.

least speculated that this difference may reflect the diversities in micro-environments of the sarcolemmal and sarcoplasmic ionic pumps. Such speculations could be based on the well known content of acidic phospholipids and cholesterol in the plasma membrane of mammalian cells which is exceeding that in the membranes of sarcoplasmic reticulum.

Inhibition of the ATPase activity was also achieved by application of the pyridoxal derivative of ATP [22] or pyridoxal phosphate itself [23]. But, unfortunately, since the role of the FITC-sensitive lysine residue in binding of ATP has been taken as granted, the procedures applied in experiments with pyridoxal exhibited similar shortcomings as in experiments with FITC. In addition, it can not be excluded that the Shiff base formed by reaction of lysine residue and pyridoxal in the ATP-binding site are not altering considerably its fine structure.

When considering all 'pro and contra,' from experiments with modification of lysine [19, 20 22, 23] and of cysteine [9–12] it may be tentatively assumed the following: (i) Forming a positively charged microenvironment, the lysine residue may fulfill an important function, most probably by accommodating the negatively charged phosphate moiety of the ATP molecule in the micro-domain of ATP-binding sites of P-type ATPases; (ii) As concerns the cysteine residue, it may be proposed that this residue may play the role of recognition and primary binding of ATP by forming a proton bridge with the amino group in position 6 on the adenosine moiety of the ATP molecule. A future experiment replacing the respective cysteine (see Table 2) with the aid of site directed mutagenesis may provide the last information needed to complete the definitive picture about the composition and function of the ATP-binding site of P-type ATPases.

Acknowledgement

This work was supported in part by the Slovak Grant Agency for Science 9, grants No. 2/571/93, 2/1258/95 and 2/1001/95.

References

1. Pick U, Bassilian S: Modification of the ATP-binding site of the Ca^{2+}-ATPase from sarcoplasmic reticulum by fluorescein isothiocyanate. FEBS Lett 123: 127–130, 1981
2. Carilli CT, Farley RA, Perlman DM, Cantley LC: The active site structure of Na^+ and K^+ stimulated ATPase. Location of a specific isothiocyanate reactive site. J Biol Chem 257: 5601–5606, 1981
3. Farley RA, Tran CM, Carilli CT, Hawke D, Shively JE: The amino acid sequence of a fluorescein labeled peptide from the active site of (Na^+ + K^+)-ATPase. J Biol Chem 259: 9532–9535, 1984
4. Kirley KL, Wallick ET, Lane LK: The amino acid sequence of the fluorescein isothiocyanate reactive site of lamb and rat kidney Na^+- and K^+-dependent ATPase. Biochim Biophys Res Commun 125: 767–773, 1984
5. Ohta T, Morohashi M, Kawamura M, Yoshida M: The amino acid sequence of the fluorescein-labeled peptides of electric ray and brime shirm (Na,K)-ATPase. Biochem Biophys Res Commun 130: 221–228, 1985
6. Mitchinson C, Wilderspin AF, Trinnaman BJ, Green NM: Identification of a labeled peptide after stoichiometric reaction of fluorescein isothiocyanate with Ca^{2+} dependent adenosine triphosphatase of sarcoplasmic reticulum. FEBS Lett 146: 87–93, 1982
7. Filoteo AG, Gorski JP, Penniston JT: The ATP binding site of the erythrocyte membrane calcium pump. Amino acid sequence of the fluorescein isothiocyanate-reactive region. J Biol Chem 262: 6526–6530, 1987
8. Farley RA, Faller LD: The amino acid sequence of an active site peptide from the proton-potassium ATPase of gastric mucosa. J Biol Chem 260: 3899–3901, 1985
9. Ziegelhöffer A, Breier A, Džurba A, Vrbjar N: Selective and reversible inhibition of heart sarcolemmal (Na^+ + K^+)-ATPase by p-bromophenylisothiocyanate. Evidence for a sulfhydryl group in the ATP binding site of the enzyme. Gen Physiol Biophys 2: 447–456, 1983
10. Breier A, Ziegelhöffer A, Stankovičová T, Dočolomanský P, Gemeiner P, Vrbanová A: Inhibition of (Na/K)-ATPase by electrophilic substances: Functional implications. Mol Cell Biochem 147: 187–192, 1995
11. Patzelt-Wenczler R, Schoner W: Evidence for two different reactive sulfhydryl groups in the ATP binding sites of (Na^+ + K^+)-ATPase. Eur J Biochem 114: 79–87, 1981
12. Schoner W, Sepersu EH, Pauls H, Patzelt-Wenczler R, Kreickman H, Rempeter G: Comparative studies on the ATP-binding sites in Ca^{2+}-ATPase and (Na^+ + K^+)-ATPase by the use of ATP analogues. Z Naturforsch 37c: 692–705, 1982
13. Drobnica L, Gemeiner P: In: JL Fox, Z. Deyl and A. Blazej (eds). Use of isothiocyanate as 'reporter' groups in modification of enzymes. Protein Structure and Evolution. Marcel Dekker Inc, New York, 1976, pp 105–115
14. Drobnica L, Kristián P, Augustín J: The chemistry of isothiocyanates, In: S. Patai (ed.) The Chemistry of Cyanates and their Thio Derivatives. Part 2. Wiley and Sons, Chichester, 1977, pp 1003–1221
15. Gemeiner P, Drobnica L: Selective and reversible modification of essential thiol groups of D-glyceraldehyd-3-phosphate dehydrogenase by isothiocyanates. Experientia 35: 857–858, 1979
16. Wilderspin AF, Green M: The reaction of fluorescein isothiocyanate with thiols: A method for assay of isothiocyanates. Analyt Biochem 132: 449–455, 1983
17. Swoboda G, Hasselbach W: Reaction of fluorescein isothiocyanate with thiol and amino group of sarcoplasmic ATPases. Z Naturforsch 40c: 863–875, 1985
18. Gemeiner P, Bíliková Z, Uhrín D, Šoltés L, Mosbach K: ATP derivatives for biorecognition technology: High-performance liquid chromatography and nuclear magnetic resonance spectra. Biotech Appl Biochem 11: 176–183, 1989
19. Wang K, Farley RA: Lysine 480 is not an essential residue for ATP binding or hydrolysis by Na,K-ATPase. J Biol Chem 267: 3577–3580, 1992

20. Maruyama K, Mac Lennan DH: Mutation of aspartic acid 351, lysine 352 and lysine 515 alters the calcium transport activity of the calcium ATPase expressed in COS-1 cells. Proc Natl Acad Sci USA 85: 3314–3318, 1988
21. Breier A, Turi Nagy L, Ziegelhöffer A, Monošíková R, Džurba A: Hypothetical structure of the ATP binding site of (Na^+ + K^+)-ATPase. Gen Physiol Biophys 8: 283–286, 1989
22. Yamamoto H, Imamura Y, Tagaya M, Fukui T, Kawakita M: Calcium dependent conformational change of the ATP binding site of calcium transporting ATPase of sarcoplasmic reticulum as revealed by an alteration of target-site specificity of adenosine triphosphate pyridoxal. J Biochem (Tokyo) 106: 1121–1125, 1989
23. Tamura S, Tagaya M, Madea M, Futai M: Pig gastric (H^+ + K^+)-ATPase. Lys-492 conserved in cation transporting ATPase is modified with pyridoxal 5'-phosphate. J Biol Chem 264: 8580–8584, 1989
24. Jorgensen PL: Purification of Na^+, K^+-ATPase: Enzyme sources, preparative problems, and preparation from mammalian kidney, In: S Fleischer and B Fleischer (eds). Methods in Enzymology, Vol 156. Academic Press, London, 1988, pp 29–43
24. Shull GE, Lingrel JB: Molecular cloning of the rat stomach proton potassium ATPase. J Biol Chem 261: 16788– 16791, 1986
25. Crownson MS, Shull GE: Isolation and characterization of a cDNA encoding the putative distal colon H^+, K^+-ATPase. Similarity of deduced amino acid sequence to gastric H^+, K^+-ATPase and Na^+, K^+ ATPase and mRNA expression in distal colon, kidney and uterus. J Biol Chem 267: 13740–13748, 1992
26. Shull GE, Schwartz A, Lingrel JB: Amino-acid sequence of the catalytic subunit of the (Na^+ + K^+)-ATPase deduced from a complementary DNA. Nature 316: 691–695, 1985
27. Mac Lennan DH, Brandl CJ, Korozak B, Green MN: Aminoacid sequence of a Ca^{2+}-ATPase from rabbit muscle sarcoplasmic reticulum deduced from its complementary DNA sequence. Nature. 316: 696–700,1985
28. Inesi G, Kirtley MR: Structural features of cation transport ATPases. J Bioenerg Biomembran 24: 271–283, 1992
29. Verma AK, Filoteo AG, Stanford DR, Wiebnen ED, Penniston JT, Strehler EE, Fischer R, Heim R, Vogel G, Mathews S, Strehler-Page MA, James P, Vorherr T, Krebs J, Carafoli E: Complete primary structure of a human plasma membrane Ca^{2+} pump. J Biol Chem 263: 4152–14159, 1988
30. Shull GE, Greeb J: Molecular cloning of the plasma membrane Ca^{2+} transporting ATPase from rat brain. J Biol Chem 263: 8646-8657, 1988

Molecular and Cellular Biochemistry **160/161**: 95–99, 1996.

Reinstatement of the ATP high energy paradigm

Kurt R. H. Repke
Max Delbrück Centre for Molecular Medicine, Robert-Rössle-Str. 10, D-13122 Berlin-Buch, Germany

Abstract

The paradigm that the hydrolysis of ATP releases high Gibbs energy able to perform work has increasingly been questioned over the last two decades. Results from theoretical and experimental studies have been interpreted to indicate that the synthesis of ATP from ADP and P_i does not require energy supply and that binding of ATP per se can transmit utilizable energy to an enzyme. As has recently been concluded, all this has led to a change of the ATP high energy paradigm in bioenergetics. Starting from this challenge, the present review singles out the striking sources of the apparent dichotomy in bioenergetics, and endeavours to eliminate the apparent contradictions by the application of the prior knowledge on both the participation of the enzyme protein in energy exchange processes and the particular reactivities of phosphorus that make it an outstanding element for functionally variable work assignments in enzymatic systems. (Mol Cell Biochem **160/161**: 95–99, 1996)

Key words: ATP synthesis, ATP hydrolysis, phosphoryl transfer, ion-motive force, energy transfer, coupling mechanism

Introduction

Sources of doubt about energy-transfer in ATP hydrolysis

The paradigm on the existence and function of energy-rich phosphate compounds as high Gibbs energy carrier through phosphoryl transfer, founded by Lipmann in 1941 [15] and reinforced by him in 1960 [16] and 1975 [17], had undisputedly determined the way of bioenergetics during the three decenniae since 1941. However, as already stated by Lipmann [16], we have remained pretty much in the dark about the events, subsequent to an initial ATP binding step, which are sadly needed for the conversion of phosphoryl potential into osmotic as well as into mechanical work. This helps to explain that over the years criticisms have repeatedly been voiced. So, Nicholls and Ferguson [18] declared in 1992 the existence and role of the high-energy phosphate bond even as a myth. In 1993, de Meis [19] summarized his review on the concept of energy-rich phosphate compounds as follows: 'The energy level of the phosphate compound varies during the catalytic cycle of the enzyme before it is hydrolyzed. Work is coupled with this transition and not with the hydrolysis of the compound.'

As I review in the following three paragraphs, the sources of doubt about Lipmann's ATP high energy paradigm include the uncertainty about the energetics of ATP hydrolysis and synthesis, the lack of agreement on the chemistry of ATP hydrolysis and synthesis, and the contradictions concerning the fundamentals of the coupling of ion-motive forces with ATP synthesis and hydrolysis. When I nevertheless here dare to rationalize elder and newer thoughts and data stemming from biochemical and physical knowledge, I have felt invited for it by Racker's insight [20]: 'What we can learn from a study of the history of bioenergetics is that we should keep an open mind and be susceptible to novel ideas regardless of how foreign they are to current concepts.'

Energetics of ATP synthesis and hydrolysis

As stated in 1993 by Pedersen and Amzel [21], the mechanism by which H^+-transporting ATP synthase makes, releases, and turns over ATP remains unclear. According to Boyer *et al.* (1973) [22], Grubmeyer *et al.* (1982) [23] and Slater (1987) [24] it appears that ATP can be made from ADP and P_i bound to the catalytic centre with little or no energy input such that the energy for ATP synthesis is required not for the esterification of ADP by phosphate per se, but for the dissociation of ATP from the enzyme. In 1994, Cross [25] reconfirmed that the primary use of energy supplied is not to make ATP, but rather to promote its release from catalytic sites where it forms spontaneously from tightly bound ADP and P_i.

Address for offprints: K.R.H. Repke, Max Delbrück Centre for Molecular Medicine, Robert-Rössle-Str. 10, D-13122 Berlin-Buch, Germany

As documented by de Meis in 1989 [26] and 1993 [19], the energy of ATP hydrolysis varies greatly depending on whether it occurs in solution or in the catalytic centre. Various theoretic and experimental data have suggested to him that enzymes can perform work without hydrolysis of ATP. So, in the enzyme's conformational transition induced by ATP binding, work should be performed prior to ATP hydrolysis which appeared indispensable only to prepare the product release. In the reverse process, ATP was suggested, also by de Meis, to be formed from ADP and P_i in the catalytic centre without energy supply. Energy would be required for the conversion of ATP from a 'low energy' into a 'high energy' state which he thought to result from an alteration of water activity in the catalytic centre due to a conformation change of the enzyme. According to de Meis [19], the pertinent studies, selected by him up to 1993, have led to a rejection of the ATP high energy paradigm in bioenergetics.

The interpretations, reviewed above, all did not take into account the prior knowledge, expounded by Lumry and Biltonen [27] in 1969, that the most common and most important role, played by the protein in specific biochemical processes, is the provision of a sink or source for mechanical and chemical Gibbs energy. In general, Lumry [28] found in estimates of the Gibbs energy stored conformationally that something like -10 kcal/mole is the upper limit. The defined participation of protein conformation in energy exchange processes accounts for the seeming paradox that in phosphoryl transfer reactions, catalyzed by various enzymes, the equilibrium constant for the bound substrate-product couple is generally found near one. According to Knowles [29] this means that the forces of differential binding, that distinguish between substrate and product, do ensure that the transition state for the catalytic step is flanked by complexes of essentially equal Gibbs energy.

In ATP hydrolysis by the catalytic F_1 sector of H^+-transporting ATP synthase, the plainly measured small $\Delta G^{o\prime}$ change has been suggested to result from large compensating enthalpy and entropy changes in the protein (cf. [27, 28]). As derived by Jencks [30], the ready reversibility of the hydrolysis of ATP bound to the enzyme indicates that the great Gibbs energy change of ATP hydrolysis, which is observed in solution, is largely conserved as mechanical Gibbs energy in the enzyme's protein fabric so that it can be utilized in the resynthesis of ATP from enzyme-bound ADP and P_i. In line with this reasoning, Caplow *et al.* [31] observed that the Gibbs energy change in the hydrolysis of microtubule-bound nucleotide triphosphate is near zero since all the Gibbs energy from hydrolysis is stored in the microtubule lattice until useful work can be done.

Chemistry of ATP synthesis and hydrolysis

As stated by Pedersen and Carafoli [32] in 1987, reaction cycles have been proposed for several ion motive ATPases, but alone they do not provide molecular/chemical details of energy coupling events. In particular, the chemistry of the condensation reaction of ADP and P_i to form ATP has not conclusively been elucidated. Most probably, ATP synthesis and hydrolysis occur at the same catalytic site of F_1 [33].

In 1974, Mitchell [34] initiated a novel approach to the mechanism of the reversible F_1 ATPase.* Starting from the knowledge that nucleophilic substitution reactions at phosphate phosphorus generally proceed by way of pentavalent transitional intermediates, he developed a nucleophilic substitution mechanism for ATP hydrolysis and ADP phosphorylation which included pseudorotation of the ligands of pentavalent phosphorus. Boyer [35], however, immediately rejected this approach by stating that ATP hydrolysis very likely proceeds by a metaphosphate mechanism not involving pentacovalency. Nevertheless, Kozlov and Skulachev [36] resumed in 1977 Mitchell's approach including the postulate on intermediary pseudorotation of the ligands of pentavalent phosphorus intermediates. However, the involvement of such pseudorotations in these processes has soon be excluded by two independent physico-chemical techniques [37, 38]. So, the phosphoric residue transfer catalyzed by mitochondrial F_1 ATPase was found to proceed with inversion of configuration at the γ-phosphorus atom which is incompatible with pseudorotation of phosphorus ligands [37]. Studies on mitochondria-catalyzed ATP hydrolysis in ^{18}O-labelled water showed that the four oxygen atoms of P_i were equivalent during the time enzyme-bound P_i underwent exchange [38]. Hence, alternatively a turnstile rotation mechanism, in which no oxygen is steadily liganded to a side chain in the catalytic centre (cf. [39]), appeared us to underly the unimpeded exchange [14].

Since the evidence for a pentacoordinated transition state in ATP hydrolysis and synthesis appeared to be strong [40, 41], we decided to design the adjacent associative phosphorane mechanism of ATP processing by F_1 (cf. Fig. 1) which accounts for the facility of the reversible phosphoryl transfer by turnstile rotation of phosphorus ligands [14]. The favourable features of this mechanism include the provision of a single path for both the minimum energy consuming cleavage and formation of the phosphoric acid anhydride bond, as well as the inherent coincidence of phosphoryl and energy transfer on ATP hydrolysis and ATP synthesis (cf. below).

*F_0F_1, H^+-transporting ATP synthase (EC 3. 6. 1. 34)

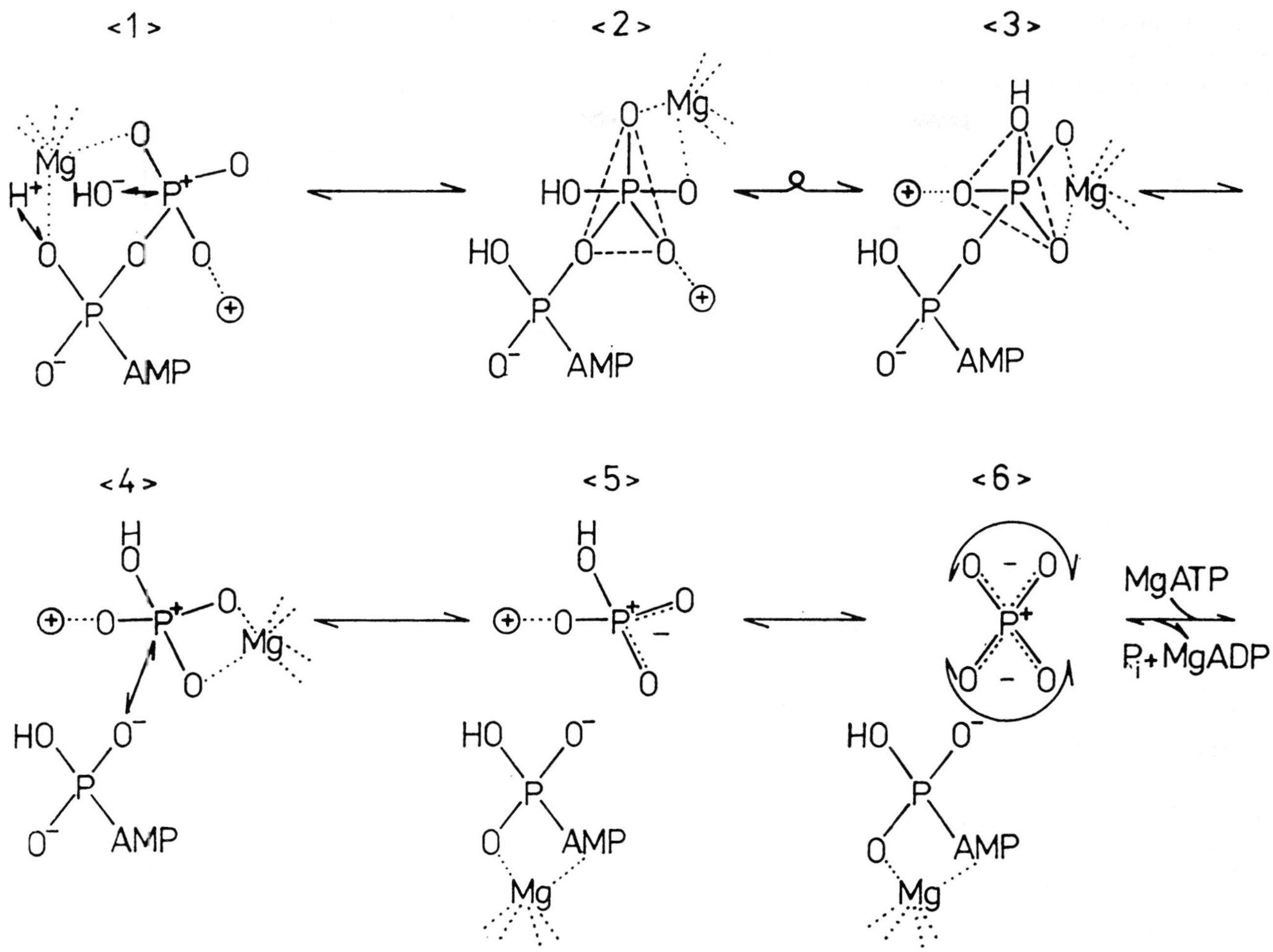

Fig. 1. Hypothetical, adjacent-associative mechanism of ATP hydrolysis (<1> → <6>) and ATP formation (reverse direction) by H^+-transporting ATP synthase via turnstile rotation of the ligands of intermediary phosphorane states (<2> ⇄ <3>). As indicated in <6>, bound P_i can rotate at the catalytic site so that the four oxygens are spatially randomized. The circumstantial evidence in favour of this mechanism has been described elsewhere [14]. For details of phosphorus chemistry underlying this scheme cf. refs. [10, 11,13,141].

Coupling of proton-motive forces with ATP synthesis and hydrolysis

The real heart of the mechanism – how the proton flux through F_0F_1 effects the formation of the phosphoric anhydride bond between ADP and P_i – is not yet understood [42]. As stated by Pedersen and Carafoli [32], one of the central and most fundamental questions is to establish whether the proton fluxes are directly or indirectly coupled with the chemical reactions.

Mitchell [43] proposed still in 1985 that ATP and its hydrolysis products may participate relatively directly in the carriage of the protons by H^+-transporting ATP synthase. Such mechanism has become untenable through the more recent information provided by Boyer [44] that the realization of the proton-motive force across mitochondrial inner membrane proceeds indirectly and requires protein-mediated conformational coupling between the protonophoric structures in F_0 and the peptide sequences forming the catalytic centre in F_1. In Boyer's binding change mechanism for ATP synthase [44] conformational changes driven by proton translocation across F_0 are transmitted to the catalytic site region in F_1 where they promote P_i and ADP binding and ATP release after its 'spontaneous' formation. Hence, this concept does not explain mechanistically the well-known coupling between proton 'downhill' movement and ATP synthesis as well as between ATP hydrolysis and 'uphill' proton transport (for references cf. [14]).

This deficiency has been proposed by us to be made up by the introduction of the adjacent-associative phosphorane mechanism of phosphoryl transfer (Fig. 1) from ATP to water (<1> → <6>) and from orthophosphate to ADP forming

ATP (reverse direction) [14]. In terms of this mechanism, the Gibbs energy-transferring step is the relaxation of the high-energy arrangement of ATP terminal phosphorus ligands to a low-energy arrangement (<1> → <4>), and the Gibbs energy-conserving step is the stretch of the low-energy arrangement of the orthophosphate phosphorus ligands to the high-energy arrangement of ATP terminal phosphorus ligands (<4> → <1>). These polytopal ligand rearrangements through their turnstile rotations are accordingly assumed to provide the vectorial momenta which, via communicating peptide sequences, are transmitted to or elicited by the coupled protonophoric centres. In this way, the chemical partial processes produce or utilize high protonic chemical potentials and thus realize the inherent coincidence of phosphoryl and energy transfer.

Conclusion

The outlined mechanism of ATP hydrolysis and formation via intermediary pentacovalent phosphorane states shows its feasibility by its capability to explain: the storage and transfer of Gibbs energy, the interconversion of scalar and vectorial Gibbs energies, and the inherent coincidence of phosphoryl and energy transfer. The diverse lines of reasoning brought forward here provide detailed physicochemical support for Lipmann's paradigm that ATP serves as high Gibbs energy carrier and that the energy release takes off from the terminal phosphoryl of ATP.

Dedication

The utilization of high energy phosphates in myocardial contraction and the impact of cardiac glycosides on the energetics of cardiac contractility were the central research topics of Albert Wollenberger when I first met him in the fifties and sixties. His superior insight in these intricate problems, documented in several painstaking reviews [1–3] had great influence on the choice and elaboration of my own research objectives [4–7]. His foresight that cardiac glycosides will prove to be most valuable weapons in the attack on one of the major problems of cell biology, the utilization of the energy of ATP for osmotic work, has been shown to come true by our studies [8–14]. Hence, it is a pleasure for me to dedicate the present review to Albert Wollenberger on the occasion of the fiftieth anniversary of his doctoral thesis at Harvard University.

References

1. Wollenberger A: The energy metabolism of the failing heart and the metabolic action of the cardiac glycosides. J Pharmacol Exp Ther 97: 311–352, 1949
2. Wollenberger A: Digitalis: action on metabolism and the contractile system. In: JL Mongar and AVS de Reuck (eds). Ciba Foundation Symposium on Enzymes and Drug Action. Churchill, London, 1962, pp 127–150
3. Wollenberger A: Agents influencing myocardial contraction: Cardiac glycosides. In: RD Tanz, F Kavaler and J Roberts (eds). Factors Influencing Myocardial Contractility. Academic Press, New York and London, 1967, pp 505–518
4. Repke KRH, Portius HJ: Über die Identität der Ionenpumpen-ATPase in der Zellmembran des Herzmuskels mit einem Digitalis-Rezeptorenzym. Experientia 19: 452–458, 1963
5. Repke KRH: Über den biochemischen Wirkungsmodus von Digitalis. Klin Wschr 42: 157–165, 1964
6. Repke KRH: Effect of digitalis on membrane adenosine triphosphatase of cardiac muscle. In: BB Brodie, JR Gillette and R Capek (eds). Proceedings of the Second International Pharmacological Meeting, Prague 1963, Vol. 4: Drugs and Enzymes. Pergamon Press, Oxford, 1965, pp 65–87
7. Repke KRH, Est M, Portius HJ: Über die Ursache der Speziesunterschiede in der Digitalisempfindlichkeit. Biochem Pharmacol 14: 1785–1802, 1965
8. Repke KRH, Weiland J, Megges R, Schön R: Approach to the chemotopography of the digitalis recognition matrix in Na^+/K^+-transporting ATPase as a step in the rational design of new inotropic steroids. Prog Med Chem 30: 135–202, 1993
9. Repke KRH, Megges R, Weiland J, Schön R: Location and properties of the digitalis receptor site in Na^+/K^+-ATPase. FEBS Lett 359: 107–109, 1995
10. Repke KRH: Concept on the link between chemical and electro-chemical free energies in (NaK)-ATPase transport function. In: G Semenza and E Carafoli (eds). FEBS-Symposium No. 42: Biochemistry of Membrane Transport. Springer-Verlag, Berlin, 1977, pp 363–373
11. Repke KRH: On the mechanism of energy release, transfer, and utilization in Na,K-ATPase transport work: old ideas and new findings. Ann N Y Acad Sci 402: 272–286, 1982
12. Repke KRH, Schön R: Role of protein conformation changes and transphosphorylations in the function of Na^+/K^+- transporting adenosine triphosphatase: an attempt at an integration into the Na^+/K^+ pump mechanism. Biol Rev 67: 31–78, 1992
13. Repke KRH, Schön R: Chemistry and energetics of transphosphorylations in the mechanism of Na^+/K^+-transporting ATPase: an attempt at a unifying model. Biochim Biophys Acta 1154: 1–16, 1992
14. Repke KRH, Schön R: Synthesis of a self-contained concept of the molecular mechanism of energy interconversion by H^+-transporting ATP synthase. Biol Rev 69: 119–145, 1994
15. Lipmann F: Metabolic generation and utilization of phosphate bond energy. Adv Enzymol 1: 99–162, 1941
16. Lipmann F: Attempts toward a formulation of biological use of energy in terms of chemical potentials. In: D Nachmansohn (ed.) Molecular Biology. Academic Press, New York and London, 1960, pp 37–47
17. Lipmann F: The roots of bioenergetics. In: Ciba Foundation Symposium 31 (new series): Energy Transformation in Biological Systems. Elsevier, Excerpta Medica, North Holland, Amsterdam, Oxford, New York, 1975, pp 3–22
18. Nicholls DG, Ferguson SJ: Bioenergetics 2. Academic Press, London, 1992, p 48
19. de Meis L: The concept of energy-rich phosphate compounds: water,

transport ATPases, and entropic energy. Arch Biochem Biophys 306: 287–296, 1993

20. Racker E: From Pasteur to Mitchell: a hundred years of bioenergetics. Fed. Proc. 39: 210–215, 1980
21. Pedersen PL, Amzel LM: ATP synthases. Structure, reaction center, mechanism, and regulation of one of nature's most unique machines. J Biol Chem 268: 9937–9940, 1993
22. Boyer PD, Cross RL, Momsen W: A new concept for energy coupling in oxidative phosphorylation based on a molecular explanation of the oxygen exchange reactions. Proc Natl Acad Sci USA 70: 2837–2839, 1973
23. Grubmeyer C, Cross RL, Penefsky HS: Mechanism of ATP hydrolysis by beef heart mitochondrial ATPase. Rate constants for elementary steps in catalysis at a single site. J Biol Chem 257: 12101–12105, 1982
24. Slater EC: The mechanism of the conservation of energy of biological oxidations. Eur J Biochem 166: 489–504, 1987
25. Cross RL: Our primary source of ATP. Nature 370: 594–595, 1994
26. de Meis L: Role of water in the energy of hydrolysis of phosphate compounds – energy transduction in biological membranes. Biochim Biophys Acta 973: 333–349, 1989
27. Lumry R, Biltonen R: Thermodynamic and kinetic aspects of protein conformations in relation to physiological function. In: SN Timasheff and GD Fasman (eds). Structure and Stability of Biological Macromolecules. Marcel Dekker, New York, 1969, p 200
28. Lumry R: Conformational mechanisms for free energy transduction in protein systems, old ideas and new facts. Ann N Y Acad Sci 227: 46–76, 1974
29. Knowles JR: Enzyme-catalyzed phosphoryl transfer reactions. Ann Rev Biochem 49: 877–919, 1980
30. Jencks WP: Utilization of binding energy and coupling rules for active transport and other coupled vectorial processes. Methods in Enzymology 171: 145–164, 1989
31. Caplow M, Ruhlen RL, Shanks J: The free energy for hydrolysis of a microtubule-bound nucleotide triphosphate is near zero: all of the free energy for hydrolysis is stored in the microtubule lattice. J Cell Biol 127: 779–788, 1994
32. Pedersen PL, Carafoli E: Ion motive ATPases. II. Energy coupling and work output. Trends in Biochem Sci 12: 186–189, 1987
33. Vignais PV, Sartre M: Recent developments on structural and functional aspects of the F_1 sector of H^+-linked ATPases. Mol Cell Biochem 60: 33–70, 1984
34. Mitchell P: A chemiosmotic molecular mechanism for proton-translocating adenosine triphosphatases. FEBS Lett 43: 189–194, 1974
35. Boyer PD: Energy transduction and proton translocation by adenosine triphosphatases. FEBS Lett 50: 91–94, 1975
36. Kozlov IA, Skulachev VP: H^+-adenosine triphosphatase and membrane energy coupling. Biochim Biophys Acta 463: 29–89, 1977
37. Webb MR, Grubmeyer C, Penefsky HS, Trentham DR: The stereochemical course of phosphoric residue transfer catalyzed by beef heart mitochondrial ATPase. J Biol Chem 255: 11637–11639, 1980
38. Mitchell RA, Lamos CM, Russo JA: Mitochondrial-catalyzed ATP hydrolysis in highly enriched $[^{18}0]\ H_20$. Frequency distributions of ^{18}O-labelled P_i species. Biochim Biophys Acta 592: 406–414, 1980
39. Luckenbach R: Dynamic Stereochemistry of Pentacoordinated Phosphorus and Related Elements. Thieme, Stuttgart, 1973, p 225
40. Wimmer MJ, Rose IA: Mechanism for oxygen exchange in the chloroplast photophosphorylation system. J Biol Chem 252: 6769–6775, 1977
41. Cross RL: The mechanism and regulation of ATP synthesis by F_1-ATPases. Ann Rev Biochem 50: 681–714, 1981
42. Gräber P: Kinetics of proton-transport coupled ATP synthesis in chloroplasts. In: G Milazzo and M Blank (eds). Bioelectrochemistry III. Plenum Press, New York, 1990, p 279
43. Mitchell P: The correlation of chemical and osmotic forces in biochemistry. J Biochem 97: 1–18, 1985
44. Boyer PD: The binding change mechanism for ATP synthase – some probabilities and possibilities. Biochim Biophys Acta 1140: 215–250, 1993

Molecular and Cellular Biochemistry **160/161**: 101–109, 1996.

Regulation of and intervention into the oxidative pentose phosphate pathway and adenine nucleotide metabolism in the heart

Heinz-Gerd Zimmer

Carl-Ludwig-Institute of Physiology, University of Leipzig, Liebigstr. 27, D-04103 Leipzig, Germany

Abstract

The capacity of the oxidative pentose phosphate pathway (PPP) in the heart is limited, since the activity of glucose-6-phosphate dehydrogenase (G-6-PD), the first and regulating enzyme of this pathway, is very low. Two mechanisms are involved in the regulation of this pathway. Under normal conditions, G-6-PD is inhibited by NADPH. This can be overcome in the isolated perfused rat heart by increasing the oxidized glutathione and by elevating the $NADP^+$/NADPH ratio. Besides this rapid control mechanism, there is a long-term regulation which involves the synthesis of G-6-PD. The activity of G-6-PD was elevated in the rat heart during the development of cardiac hypertrophy due to constriction of the abdominal aorta and in the non-ischemic part of the rat heart subsequent to myocardial infarction. The catecholamines isoproterenol and norepinephrine stimulated the activity of myocardial G-6-PD in a time- and dose-dependent manner. The isoproterenol-induced stimulation was cAMP-dependent and due to increased new synthesis of enzyme protein. The G-6-PD mRNA was elevated by norepinephrine. As a consequence of the stimulation of the oxidative PPP, the available pool of 5-phosphoribosyl-l-pyrophosphate (PRPP) was expanded. PRPP is an important precursor substrate for purine and pyrimidine nucleotide synthesis. The limiting step in the oxidative PPP, the G-6-PD reaction, can be bypassed with ribose. This leads to an elevation of the cardiac PRPP pool. The decline in ATP that is induced in many pathophysiological conditions was attenuated or even entirely prevented by i.v. infusion of ribose. In two *in vivo* rat models, the overloaded and catecholamine-stimulated heart and the infarcted heart, the normalization of the cardiac adenine nucleotide pool by ribose was accompanied by an improvement of global heart function. Combination of ribose with adenine or inosine in isoproterenol-treated rats was more effective to restore completely the cardiac ATP level within a short period of time than either intervention alone. (Mol Cell Biochem **160/161**: 101–109, 1996)

Key words: adenine nucleotide synthesis, catecholamines, cardiac hypertrophy, ischemic heart disease, 5-phosphoribosyl-1-pyrophosphate, ribose

Introduction

It is typical for the heart that it restores its adenine nucleotide pool only very slowly once it has been depleted by a brief period of ischemia. Upon reperfusion it takes several days both in the dog and rat heart until normal ATP levels are regained [1, 2]. This phenomenon is related to the fact that the degradation products of ATP such as adenosine, inosine, and hypoxanthine that are produced during ischemia can permeate the cell membrane. They are washed out and lost from the heart [3, 4] and thus are not available for reutilization via the 'salvage pathways'. The repletion of the ATP pool can therefore be attained only via the de novo synthesis (biosynthesis) of adenine nucleotides (Fig. 1). This process, however, is very slow [5], since the available pool of 5-phosphoribosyl-pyrophosphate (PRPP), an important precursor substrate for the synthesis of purine and pyrimidine nucleotides, is limited in the heart [6]. PRPP originates from ribose-5-phosphate (Fig. 2) which is generated in the oxidative pentose phosphate pathway (PPP).

The oxidative PPP [7] is the link between carbohydrate and fatty acid as well as purine and pyrimidine nucleotide metabolism (Fig. 2). Glucose-6-phosphate (G-6-P) originating from glycogenolysis or from glucose taken up by the myo-

Address for offprints: H.-G. Zimmer, Carl-Ludwig-Institute of Physiology, University of Leipzig, Liebigstr. 27, D-04103 Leipzig, Germany

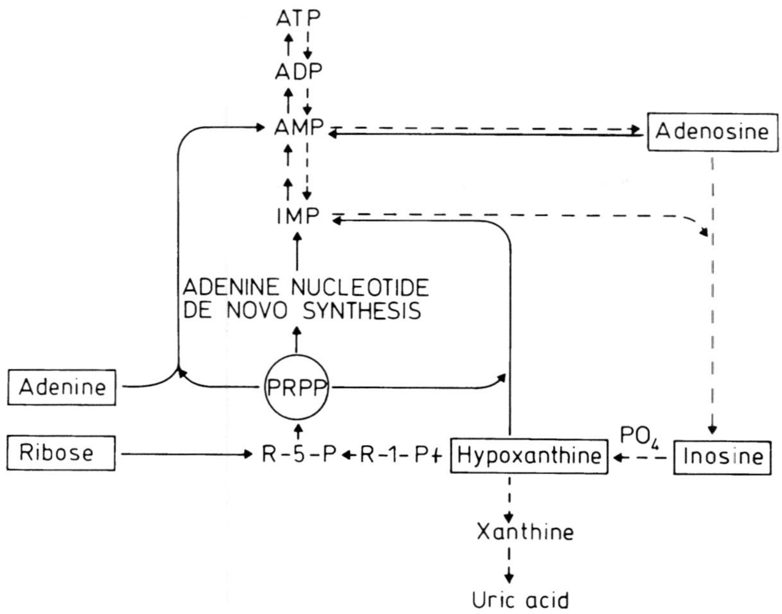

Fig. 1. Schematic representation of the central role of 5-phosphoribosyl-1-pyrophosphate (PRPP) for purine nucleotide metabolism. PRPP is the starting point for adenine nucleotide *de novo* synthesis and can be used up by adenine and hypoxanthine for AMP and IMP formation, respectively. The degradation pathways are indicated by the broken arrows. The routes of synthesis are symbolized by the continuous arrows. The precursor substances that can be used in the various synthetic processes are in boxes. IMP: Inosine monophosphate; AMP: Adenosine monophosphate; ADP: Adenosine diphosphate; ATP: Adenosine triphosphate.

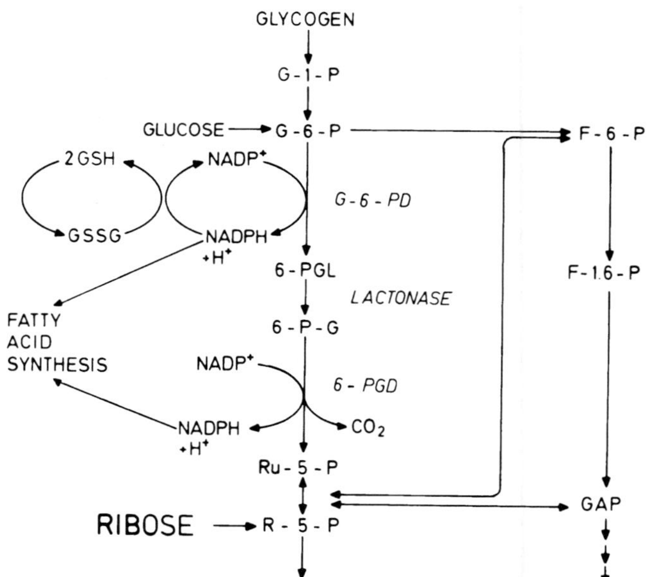

Fig. 2. The oxidative pentose phosphate pathway (center), and its connections to glycolysis (right hand side) via the transaldolase and transketolase reactions. G-1-P: Glucose-1-phospate; G-6-P: Glucose-6-phosphate; F-6-P: Fructose-6-phosphate; F-1,6-P: Fructose-1,6-biphosphate; GAP: Glyceraldehyde-3-phosphate; 6-PGL: 6-Phosphogluconolactone; 6-PG: 6-Phosphogluconate; Ru-5-P: Ribulose-5-phosphate; R-5-P: Ribose-5-phosphate; PRPP: 5-Phosphoribosyl-1-pyrophosphate; $NADP^+$: Nicotinamide Adenine Dinucleotide Phosphate; GSH: Reduced glutathione; GSSG: Oxidized glutathione; G-6-PD: Glucose-6-phosphate dehydrogenase; 6-PGD: 6-Phosphogluconate dehydrogease.

cardial cell is metabolized predominantly via glycolysis. A small portion of G-6-P, however, enters the oxidative PPP of which G-6-PD is the first and rate-limiting enzyme. This pathway serves mainly two functions: (1) It provides reducing equivalents in the form of NADPH [7] which can be used for the synthesis of free fatty acids and for the conversion of oxidized (GSSG) to reduced glutathione (GSH). This is important for the detoxification of reactive oxygen species. (2) Ribose-5-phosphate is generated which is transformed to 5-phosphoribosyl-1-pyrophosphate (PRPP), and this is an essential precursor substance for the de novo synthesis of purine nucleotides, for the utilization of orotic acid for UTP synthesis as well as for the salvage of the purine bases adenine and hypoxanthine to form AMP and IMP, respectively (Fig. 1).

In comparative studies the capacity of the oxidative PPP has been assessed in several rat organs. To do this, the activity of G-6-PD [8, 9], the available pool of PRPP [10], and the rate of purine nucleotide biosynthesis [5] were measured. There was a parallel behaviour of these three parameters in the organs examined. They were all highest in kidney, followed by liver, heart and skeletal muscle [6]. Thus, the capacity of the oxidative PPP is very low in muscular organs. In the heart, this becomes of critical importance when there is a need for restoration of the ATP pool as during the reperfusion following severe ischemia. In this situation, the PRPP pool is limiting for the biosynthesis of adenine nucleotides as well as for the conversion of adenine to AMP and of hypoxanthine to IMP.

The low capacity of the oxidative PPP is not confined to the rat heart. Comparative studies on a variety of animal species including man have shown that the activity of G-6-PD is always lower than the activity of 6-PGD, another enzyme in the pathway, and that G-6-PD activity in the human heart is on the same order of magnitude as in the rat heart on which most of the studies were done [11]. Thus, the cardiac PPP is poorly developed in all animal species.

There are two possibilities to elevate the available pool of PRPP in the heart: (1) Stimulation of the activity of cardiac G-6-PD, the first and regulating enzyme of the oxidative PPP. (2) Bypass of this rate-limiting step with ribose. Both approaches have been utilized in experimental studies. As to the regulation directed at the G-6-PD, there is a short-term and long-term control.

Short-term control of the oxidative pentose phosphate pathway

Studies on rat liver have shown that G-6-PD is always inhibited by NADPH [7, 12]. This inhibition was competitive with $NADP^+$. Since the enzyme was almost completely inhibited at physiological concentrations of free $NADP^+$ and free NADPH, it followed that the regulation of G-6-PD is a matter of de-inhibition. It turned out that among over hundred cell constituents tested, only two of these counteracted the inhibition by NADPH. These were AMP and GSSG. Only GSSG was highly effective at concentrations that may occur physiologically [7].

On the basis of these results, studies were carried out on the isolated perfused Langendorff rat heart. A 40-fold elevation of the GSSG concentration was induced by perfusion with an oxidizing agent, tert-butyl hydroperoxide. In addition, the $NADP^+$/NADPH ratio was increased 2-fold under these conditions. So if GSSG is able to overcome the inhibition of G-6-PD exerted by NADPH, then the flow through the oxidative PPP should be stimulated. This should result in an elevation of the available pool of PRPP. In fact, the incorporation of ^{14}C-adenine into myocardial adenine nucleotides which was used as a relative measure of the PRPP pool was increased 3-fold within the first 15 min after tert-butyl hydroperoxide administration [13].

Since both an increased GSSG and $NADP^+$ concentration can overcome the NADPH-mediated inhibition of G-6-PD, it was difficult to decide which of these two factors may be more relevant. Since GSSG was elevated to a greater extent than the $NADP^+$/NADPH ratio and since the latter may still be inhibitory, it was concluded that GSSG is more important for the regulation of G-6-PD [13]. This appears to be a rapid control mechanism that becomes effective within minutes. Such a mechanism may be activated when oxygen free radicals are produced in the myocardium. In the neutrophils, the activation of the oxidative PPP has been shown to be associated with the respiratory burst and with the process of phagocytosis [14].

Long-term effects on the oxidative pentose phosphate pathway

Apart from the 'fine' control mechanism exerted by NADPH that is involved in the short-term regulation of G-6-PD, there is a long-term mechanism operating in the heart. This is also directed at G-6-PD. It involves the increase in enzyme protein synthesis and has been shown to be activated in the rat heart during the development of cardiac hypertrophy and by catecholamines which stimulate α- and β-adrenergic receptors.

Cardiac hypertrophy

Figure 3 shows the changes in the activity of G-6-P D, in the rate of adenine nucleotide *de novo* synthesis, and in protein synthesis in the rat heart subsequent to constriction of the abdominal aorta. This time-course study revealed that the first increase in cardiac G-6-PD activity occurred after 48 h and reached its maximum after 3 and 4 days. The synthesis of both cardiac adenine nucleotide and protein synthesis decreased within the first few hours after aortic banding and then started to increase. Thus, the stimulation of the oxidative pentose phosphate pathway correlated very well with the elevation of adenine nucleotide and protein synthesis [15].

A similar picture evolved in another model of compensatory cardiac hypertrophy. After ligation of the descending branch of the left coronary artery in rats, the non-ischemic part of the heart develops hypertrophy. Enzymehistochemical studies have revealed that G-6-PD activity started to increase in the non-ischemic area of the infarcted heart after 10 h, and this increase was maintained until the fifth day following coronary artery ligation [16]. Interestingly, the rates of adenine nucleotide and protein synthesis were also found to be elevated in the non-infarcted rat heart after 48 h [17].

Catecholamines

Pure stimulation of cardiac β-adrenergic receptors with isoproterenol also resulted in the development of cardiac hypertrophy. This was characterized by an elevation of G-6-PD activity, adenine nucleotide and protein synthesis (Fig. 4). However, the sequential increase was different. Cardiac adenine nucleotide *de novo* synthesis was maximally stimulated already after 5 h, at a time when there was no change in G-6-PD activity. Protein synthesis reached its peak after 12 h, at a time when G-6-PD activity had just started to be elevated.

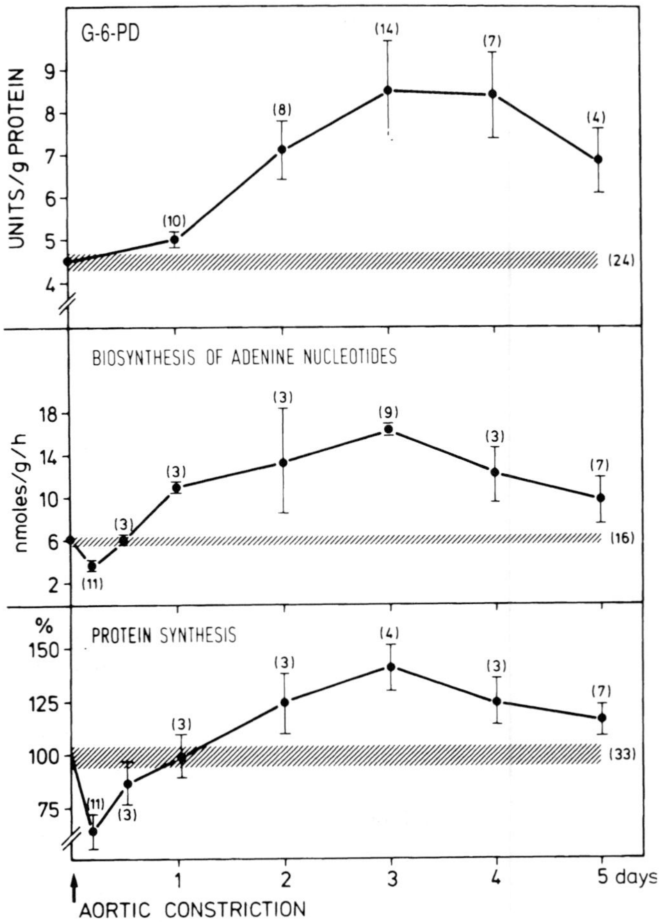

Fig. 3. Changes in the activity of glucose-6-phosphate dehydrogenase (G-6-PD), in the rate of adenine nucleotide *de novo* synthesis, and in protein synthesis in rat hearts subsequent to constriction of the abdominal aorta. This was induced by placing a metal rod of 0.65 mm diameter alongside the abdominal aorta beneath the diaphragm. A thread was placed around the aorta and the metal rod and tightly fixed with a nod. The metal rod was then pulled out leaving the aorta constricted to a diameter corresponding to the diameter of the rod. Mean values ± S.E.M.; number of experiments in parentheses.

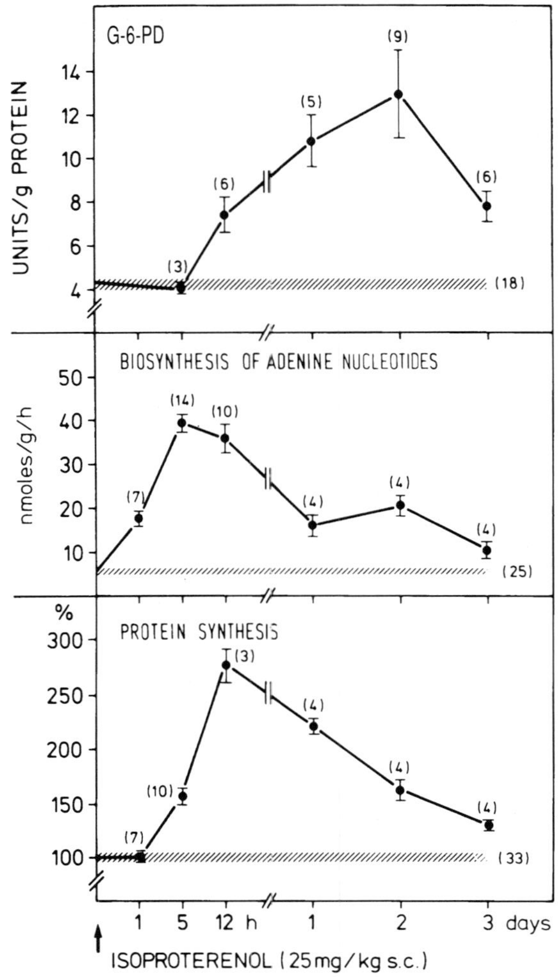

Fig. 4. Effect of isoproterenol (25 mg/kg, s.c.) on the activity of glucose-6-phosphate dehydrogenase (G-6-PD), adenine nucleotide and protein synthesis in rat hearts. Mean values ± S.E.M.; number of experiments in parentheses.

It thus appears that there are two phases that can be distinguished subsequent to isoproterenol application: The first phase was characterized by an immediate increase in the cAMP and G-6-P contents. Parallel to these metabolic changes, heart rate and contractility were also elevated [6]. After 12 and 24 h, these parameters had returned to the respective control levels. However, myocardial adenine nucleotide biosynthesis which was also enhanced very early remained elevated, though at a somewhat lower level (Fig. 4). In the second phase which started at about 12 h subsequent to isoproterenol injection, protein synthesis began to increase and when this had achieved the maximum, G-6-PD activity became stimulated.

From this time course it was inferred (1) that cAMP may be the trigger for the stimulation of cardiac G-6-PD activity and (2) that G-6-PD activation is dependent on protein synthesis and thus may reflect an increased enzyme protein synthesis. Studies were therefore performed to eliminate the isoproterenol-induced increase in the cAMP signal and to prevent the increase in protein synthesis. It was then examined which effect these interventions have on cardiac G-6-PD activity. In fact, β-receptor blockade with atenolol completely prevented the stimulation of enzyme activity [6]. Likewise, cycloheximide which inhibits translation, and actinomycin D which interferes with transcription blunted the stimulating isoproterenol effect on G-6-PD activity [6]. These results strongly indicated that cardiac β-adrenergic receptors and enzyme protein synthesis are involved in the isoproterenol-mediated increase in G-6-PD activity.

In this context, recent molecular biological studies are of interest. Continuous i.v. infusion of norepinephrine for three days which induced cardiac hypertrophy [18] increased both the mRNA level and the activity of cardiac G-6-PD in a time-

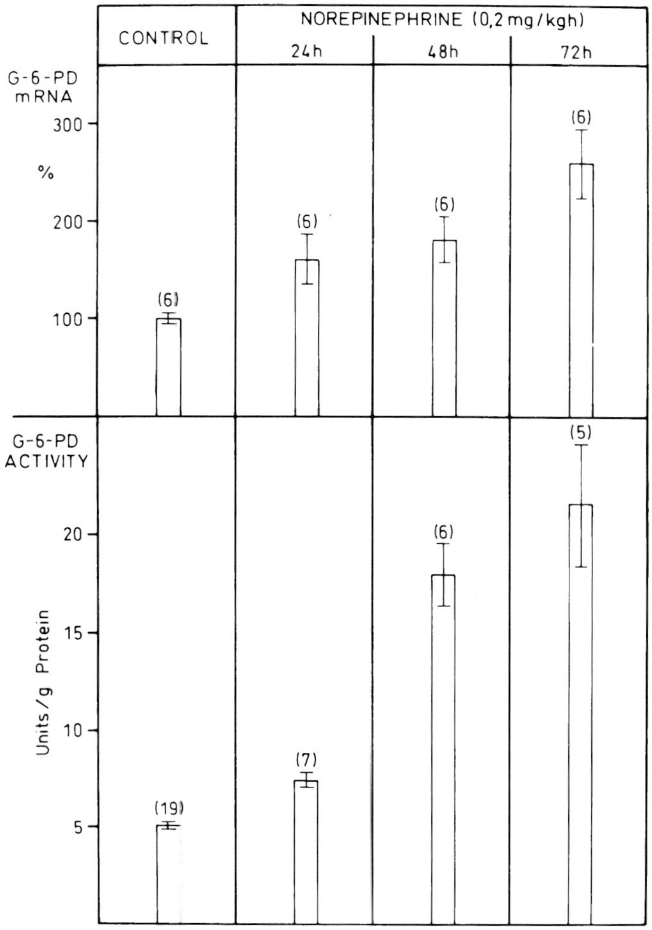

Fig. 5. Changes in the mRNA and activity of myocardial glucose-6-phosphate dehydrogenase (G-6-PD) within the first three days during continuous i.v. infusion of norepinephrine (0.2 mg/kg/h). Data are mean values ± S.E.M.; number of experiments in parentheses.

dependent fashion (Fig. 5). This increase was antagonized by simultaneous administration of carvedilol, a β-adrenergic blocker and vasodilator with α_1-blocking activity [19]. The mRNA and activity of 6-phosphogluconate dehydrogenase (6-PGD), one of the following enzymes of the oxidative PPP, was only slightly and unspecifically enhanced. It thus appears that both α- and β-adrenergic agonists promote specifically gene expression of cardiac G-6-PD.

Since the effects of α- and β-adrenergic stimulation on the oxidative PPP are similar, it is interesting to explore the mechanisms whereby different second messengers trigger the same metabolic process. In regard to isoproterenol it is known that cAMP is the second messenger for a number of metabolic and mechanical processes such as stimulation of glycogenolysis, lipolysis, and the inotropic effect. β-Adrenergic blockade prevented the isoproterenol- [6] and attenuated the norepinephrine-induced increase in cardiac G-6-PD activity [19]. It has been speculated that cAMP may induce transcription of the respective genes by activating a cAMP-dependent protein kinase and by phosphorylation of protein factor(s) that may act as transcriptional activators of a cAMP-sensitive promoter [20, 21]. It is not quite clear how the second messengers of α-adrenergic stimulation, i.e. inositol-1,4,5-trisphosphate (IP_3) and diacylglycerol [22] are related to the stimulation of the oxidative PPP. However, it is known that diacylglycerol activates protein kinase C which in turn phosphorylates a range of cellular proteins [23]. Moreover, protein kinase C may be involved in the increase of c-*myc* mRNA levels as well as of c-*fos*, c-*jun*, and egr-1 mRNAs [24] that has been shown to occur as a result of α_1-adrenergic stimulation in neonatal rat cardiac myocytes in culture. Also in the isolated perfused working rat heart it has been demonstrated recently that norepinephrine increased the mRNA of c-*fos* and c-*myc* [25]. These members of the immediate early gene program, alone or together with other transcription factors, may potentially regulate gene expression such as that of G-6-PD.

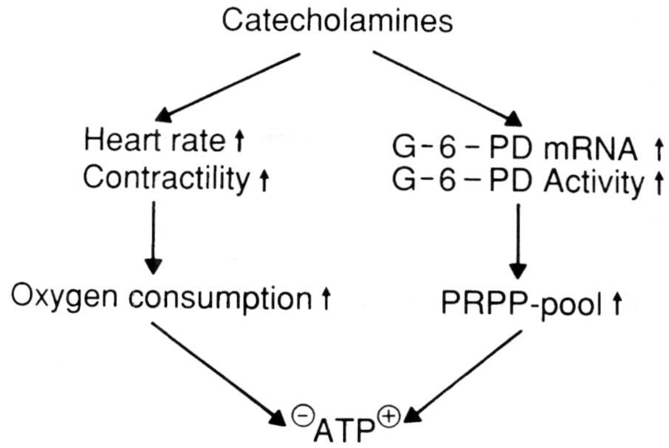

Fig. 6. Proposed homeostatic mechanism regarding the cardiac effects of catecholamines.

The stimulation of the oxidative PPP by catecholamines may represent part of a long-term homeostatic mechanism (Fig. 6). This concept is based on the fact that catecholamines increase heart rate, contractility, and oxygen consumption of the heart. Thus, there is a tendency for the decline in the ATP level in the catecholamine-stimulated heart. In the long run, this can be counteracted by the α- and β-receptor-mediated increase in the capacity of the oxidative PPP which leads to the elevation of the available PRPP pool. This can then be used up for the *de novo* synthesis and for the salvage pathways that are involved in ATP resynthesis (Fig. 1).

Thyroid hormones

Triiodothyronine (T_3) also induces cardiac hypertrophy [15]. However, the effect on the oxidative PPP is quite different depending on the rat model. Fig. 7 shows that daily injections of T_3 for 5 days have no effect whatsoever on the activity of

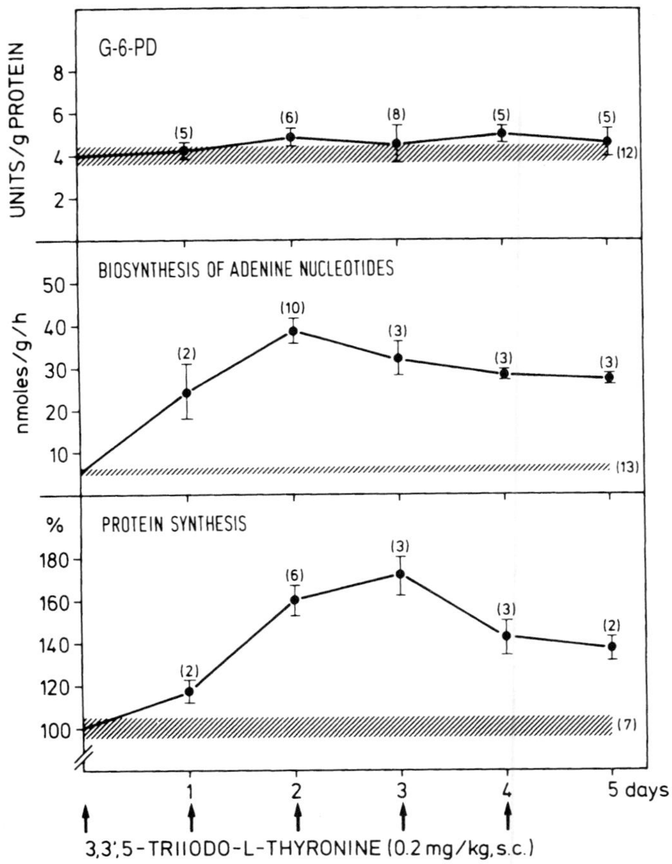

Fig. 7. Effect of daily s.c. injection of 3,3',5-triiodo-L-thyronine (T_3) on the activity of glucose-6-phosphate dehydrogenase (G-6-PD), and on adenine nucleotide and protein synthesis in rat hearts. Data are mean values ± S.E.M.; number of experiments in parentheses.

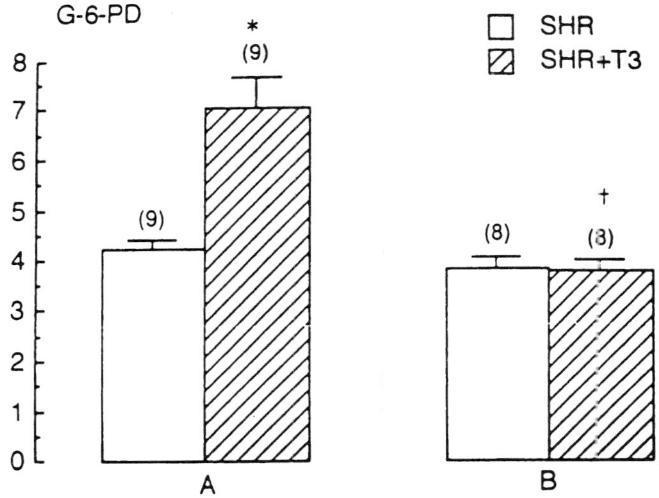

Fig. 8. Activity of cardiac glucose-6-phosphate dehydrogenase (G-6PD) in spontaneously hypertensive rats (SHR) with daily injections of 0.9% NaCl (blank bars) and with injections of triiodothyronine (T_3) at the end of the treatment period (hatched bars) of 14 days' duration (A) and at the end of 14 days after discontinuation of treatment (B). Mean values ± S.E.M.; number of experiments in parentheses. * $p < 0.05$ vs. control; $^+$ $p < 0.01$ vs. T_3 in A.

cardiac G-6-PD, although there is a substantial stimulation of both adenine nucleotide and protein synthesis. In spontaneously hypertensive rats, however, daily applications of T_3 for 14 days do increase cardiac G-6-PD activity. When T_3-treatment is discontinued for 14 days, cardiac G-6-PD activity is normal again (Fig. 8).

From the results obtained in the studies with catecholamines and T_3 one can deduce a particular time course of metabolic events that is of general interest. After administration of both catecholamines and T_3, there is an immediate stimulation of glycogenolysis and glycolysis [26] and a positive chronotropic and inotropic effect as well as a concomitant elevation in the PRPP pool, in adenine nucleotide and protein synthesis. During this immediate phase, there was no change in G-6-PD activity [6]. It was only in the subsequent second phase that cardiac G-6-PD activity was enhanced. Since G-6-PD activity was unchanged during the first phase, a mechanism other than an enhanced flow through the oxidative PPP may be responsible for the elevation of the PRPP pool and for the stimulation of adenine nucleotide biosynthesis. The alternative is that the PRPP may be elevated via the non-oxidative PPP through the transaldolase and transketolase reactions. This is a very likely mechanism, since the activities of the enzymes of the non-oxidative PPP were shown to be much higher than those of G-6-PD and 6-PGD in muscle [27]. Furthermore, the large amounts of NAD^+ present in the heart [28] may favor glycolysis, whereas the small quantity of $NADP^+$ may be limiting rather than stimulating the oxidative PPP.

Ribose

Apart from affecting the activity or the mRNA of cardiac G-6-PD, there is another intervention into the oxidative PPP that ultimately results in an elevation of the myocardial PRPP pool and consequently of adenine nucleotide *de novo* synthesis. This is the administration of ribose. Ribose bypasses the critical and rate-limiting step in the oxidative PPP, the G-6-PD reaction (Fig. 2), and leads via ribose-5-phosphate to the formation of PRPP and to the enhancement of adenine nucleotide biosynthesis [10].

This metabolic approach has been utilized in many experimental models, e.g. in isolated cardiac myocytes [29], after isoproterenol stimulation of the intact rat heart [10, 30], during development of cardiac hypertrophy [31], during recovery from temporary oxygen deficiency in the isolated rat [32, 33] and guinea pig heart [34], in the *in vivo* rat [2] and dog heart [35–38], after permanent coronary artery ligation [39], in chronic alcoholic cardiomyopathy as assessed in the isolated working rat heart [40], and in primary rat muscle cultures [41].

In all these studies, ribose proved to be an intervention that either attenuated or prevented entirely the decline of the myocardial ATP pool. For instance, after 15 min of reversible regional ischemia the ATP fell by about 40%. During the reperfusion period, the ATP pool recovered to some extent, most probably due to the rephosphorylation of ADP and AMP. However, even after three days ATP was still lower than the respective control value of sham-operated rats [2]. This corresponded well with the results obtained in dogs [1]. However, when ribose had been administered as continuous i.v. infusion, ATP had been normalized already after 12 h. So the metabolic recovery period was speeded up by at least 60 h [2].

In two pathophysiological situations in the intact rat there was a good correlation between the restoration of the cardiac ATP pool achieved by ribose and heart function. To evaluate heart function *in vivo*, a sensitive method that is applicable in small laboratory animals was developed. Ultraminiature catheter pressure transducers (Millar Instruments, Inc., Houston, Texas) were inserted into the left ventricle via the right carotid artery to measure heart rate, left ventricular pressure, and contractility as the maximal rate of rise in pressure, LV dp/dt_{max} [42]. Using this method it was examined whether a depressed heart function may be improved by normalizing the cardiac ATP pool with ribose.

In the first model, depression of all hemodynamic parameters was induced in rats by severe constriction of the abdominal aorta in combination with a single s.c. dose of isoproterenol. 24 h after this combined intervention, both the ATP level and heart function had deteriorated in animals that had received continuous i.v. infusion of 0.9% NaCl. When ribose had been administered for 24 h, the biosynthesis of cardiac adenine nucleotides was stimulated to such an extent that the depression of ATP and of the total adenine nucleotide pool was prevented. This metabolic normalization was accompanied by an elevation of the depressed left ventricular pressure and LV dp/dt_{max}. The pressure-rate-product had also returned to a near normal value [43]. Thus, ribose was able to normalize an impaired global heart function concomitantly with the restoration of the cardiac adenine nucleotide pool.

A marked impairment of heart function was also obtained by experimental myocardial infarction which was induced by ligation of the descending branch of the left coronary artery in rats. There was a progressive decline in left ventricular systolic pressure, in LV dp/dt_{max}, and in the pressure-rate-product [39]. Left ventricular enddiastolic pressure was elevated markedly. Cardiac output and stroke volume were depressed. The ATP content in the nonischemic region was lower than control after 24 h and recovered spontaneously toward the control within the following three days. Continuous i.v. infusion of ribose attenuated the fall and promoted the restoration of the ATP level in the nonischemic myocardium within 4 days after coronary artery ligation. The elevation of left ventricular enddiastolic pressure was attenuated with ribose after 2 and 4 days [39]

Combination of ribose with verapamil, metoprolol, adenine, and inosine

Ribose has the advantage that it has purely metabolic effects and is entirely neutral in functional terms as far as the heart and circulatory system is concerned [36, 44]. This is in contrast to the purine nucleosides adenosine [45] and inosine [46]. Another characteristic feature of ribose is that is can be combined with drugs that are used in conventional therapy of heart diseases. Ribose did not alter the hemodynamic effects of the calcium antagonist verapamil and of the β-adrenergic receptor blocker metoprolol in the intact rat. The negative chronotropic and inotropic effects of metoprolol and verapamil were not affected by ribose. Neither the verapamil-induced decrease in total peripheral resistance nor the metoprolol-induced reduction of cardiac output were influenced by ribose. On the other hand, ribose retained its typical metabolic effect, i.e. the stimulation of cardiac adenine nucleotide biosynthesis, despite the pronounced hemodynamic changes brought about by these drugs [44]. Thus ribose appears to be a suitable and appropriate adjunct in the medical therapy of certain heart diseases.

In another study, the cardiac ATP pool in intact rats was reduced by isoproterenol. Continuous i.v. infusion of ribose for 5 h did not affect the isoproterenol-induced ATP decline. Likewise, adenine administration for the same period of time had no effect, while inosine attenuated the ATP fall to some extent. Thus, the exposure time was too short to affect substantially the cardiac ATP pool. However, when ribose was given in combination with adenine (Fig. 9) or inosine (Fig. 10) for the same period of time, 5 h, the cardiac ATP level was entirely normal. These results indicate that neither the biosynthesis of adenine nucleotides (ribose application) nor the salvage pathways (adenine, inosine administration) by themselves are capable to replenish the cardiac ATP within 5 h. Only when the available PRPP pool is elevated by ribose (Fig. 1) can the ATP be normalized within this period of time [47]. This is an important aspect, when one considers ribose for clinical application.

The combination of ribose with adenine also proved to be beneficial in liver and kidney. Assessment of ATP by means of nuclear magnetic resonance spectroscopy in a mouse liver perfusion model detected a faster and more complete recovery within the first minutes of reperfusion after a period of exhaustive ischemia [48]. ATP levels and total adenine nucleotide contents were also higher in slices of dog kidneys that were obtained after 5 days of hypothermic preservation with adenine in combination with ribose compared to control values and with values obtained after adenosine treatment [49].

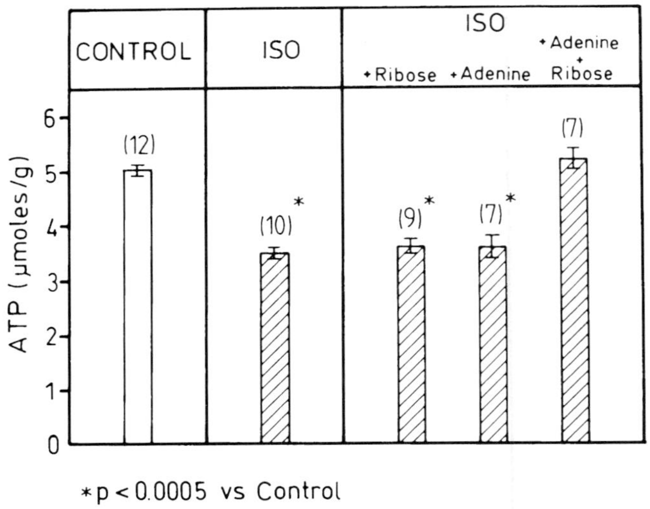

Fig. 9. Myocardial ATP values in rat hearts under control conditions, 5 h after s.c. application of isoproterenol (ISO, 25 mg/kg) alone or together with continuous i.v. infusion of ribose (200 mg/kg/h), or adenine (50 mg/kg/h) alone and in combination. Mean values ± S.E.M.; number of experiments in parentheses.

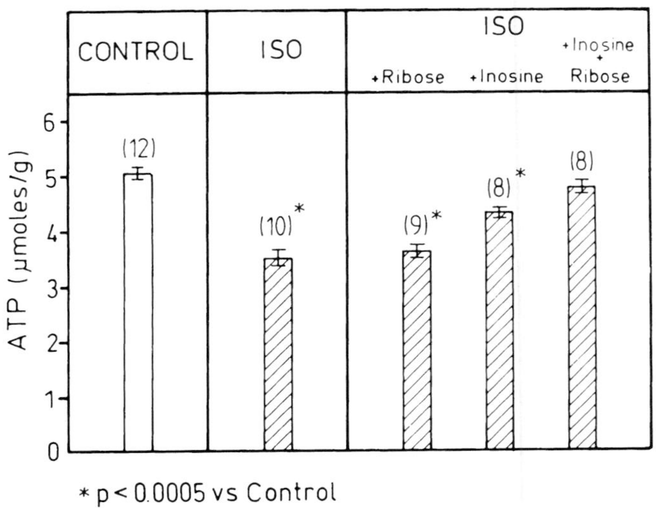

Fig. 10. Myocardial ATP values in rat hearts under control conditions, 5 h after s.c. application of isoproterenol (ISO, 25 mg/kg) alone or together with continuous i.v. infusion of ribose (200 mg/kg/h), or inosine (200 mg/kg/h) alone or in combination. Mean values ± S.E.M.; number of experiments in parentheses.

Conclusion

It is a characteristic feature of cardiac carbohydrate metabolism that the capacity of the oxidative PPP is very limited. This metabolic deficiency has two consequences. (1) The heart cannot replenish immediately and completely its ATP pool once it has been depleted. This creates many problems in the setting of myocardial ischemia and reperfusion. (2) The heart has no efficient system to detoxify oxygen free radicals. There are several possibilities for intervention into the oxidative PPP in the heart. (1) Immediated 'fine' control of G-6-PD activity by GSSG and the $NADP^+$/NADPH ratio, (2) long-term increase in G-6-PD synthesis by catecholamines, (3) administration of ribose to bypass the reaction catalyzed by G-6-PD. From these possibilities the last two are of physiological and therapeutic interest.

Acknowledgement

The experimental research referred to in this article was supported by Deutsche Forschungsgemeinschaft (Zi 199/8-1-4) and carried out in the Department of Physiology, University of Munich, Germany.

References

1. Reimer KA, Hill ML, Jennings RB: Prolonged depletion of ATP and of the adenine nucleotide pool due to delayed resynthesis of adenine nucleotides following reversible myocardial ischemic injury in dogs. J Mol Cell Cardiol 13: 229–239, 1981
2. Zimmer H-G, Ibel H: Ribose accelerates the repletion of the ATP pool during recovery from reversible ischemia of the rat myocardium. J Mol Cell Cardiol 16: 863–866, 1984
3. Gerlach E, Deuticke B, Dreisbach: Der Nucleotid-Abbau im Herzen bei Sauerstoffmangel und seine mögliche Bedeutung für die Coronardurchblutung. Naturwiss 50: 228–229, 1963
4. Imai S, Riley AL, Berne RM: Effect of ischemia on adenine nucleotides in cardiac and skeletal muscle. Circ Res 15: 443–450, 1964
5. Zimmer H-G, Trendelenburg C, Kammermeier H, Gerlach E: *De novo* synthesis of myocardial adenine nucleotides in the rat. Acceleration during recovery from oxygen deficiency. Circ Res 32; 635–642, 1973
6. Zimmer H-G, Ibel H, Suchner U: β-Adrenergic agonists stimulate the oxidative pentose phosphate pathway in the rat heart. Circ Res 67: 1525–1534, 1990
7. Eggleston LV, Krebs HA: Regulation of the pentose phosphate cycle. Biochem J 138: 425–435, 1974
8. Glock GE, McLean P: Further studies on the properties and assay of glucose-6-phosphate dehydrogenase and 6-phosphogluconate dehydrogenase of rat liver. Biochem J 55: 400–408, 1953
9. Glock GE, McLean P: Levels of enzymes of the direct oxidative pathway of carbohydrate metabolism in mammalian tissues and tumours. Biochem J 56: 171–175, 1954
10. Zimmer H-G, Gerlach E: Stimulation of myocardial adenine nucleotide biosynthesis by pentoses and pentitols. Pflügers Arch 376: 223–227, 1978
11. Zimmer H-G, Ibel H, Suchner U, Schad H: Ribose intervention in the cardiac pentose phosphate pathway is not species-specific. Science 223: 712–714, 1984
12. Negelein E, Haas E: Über die Wirkungsweise des Zwischenferments. Biochem Z 282: 206–220, 1935
13. Zimmer H-G, Bünger R, Koschine H, Steinkopff G: Rapid stimulation of the hexose monophosphate shunt in the isolated perfused rat heart: Possible involvement of oxidized glutathione. J Mol Cell Cardiol 13: 531–535, 1981
14. Borregaard N, Schwartz JH, Tauber AI: Proton secretion by stimu-

lated neutrophils. Significance of hexose monophosphate shunt activity as source of electrons and proteins for the respiratory burst. J Clin Invest 74: 455–459, 1984

15. Zimmer H-G, Ibel H, Gerlach E: Significance of the hexose monophosphate shunt in experimentally induced cardiac hypertrophy. Basic Res Cardiol 75: 207–213, 1980
16. Bechtelsheimer H, Zimmer H-G, Gedigk P: Histologische und histochemische Untersuchungen am nicht-infarzierten Myokardanteil beim experimentellen Herzinfarkt der Ratte. Virchows Arch. Abt. B Zellpath. 9: 115–124, 1971
17. Zimmer H-G, Gerlach E: Some metabolic features of the development of experimentally induced cardiac hypertrophy. Eur Heart J 3 (Suppl A): 83–92, 1982
18. Zierhut W, Zimmer H-G: Significance of myocardial α-and β-adrenoceptors in catecholamine-induced cardiac hypertrophy. Circ Res 65: 1417–1425, 1989
19. Zimmer H-G, Lankat-Buttgereit B, Kolbeck-Rühmkorff C, Nagano T, Zierhut W: Effects of norepinephrine on the oxidative pentose phosphate pathway in the rat heart. Circ Res 71: 451–459, 1992
20. Riabowol KT, Fink JS, Gilman MZ, Walsh DA, Goodman RH, Feramisco JR: The catalytic subunit of cAMP-dependent protein kinase induces expression of genes containing cAMP-responsive enhancer elements. Nature 336: 83–86, 1988
21. Sassone-Corsi P, Visvader J, Ferland L, Mellon PL, Verma IM: Induction of proto-oncogene fos transcription through the adenylate cyclase pathway: characterization of a cAMP-responsive element. Genes and Development 2: 1529–1538, 1988
22. Berridge MJ, Irvine RF: Inositol trisphosphate, a novel second messenger in cellular signal transduction . Nature 312: 315–321, 1984
23. Nishizuka Y: The molecular heterogeneity of protein kinase C and its implications for cellular regulation. Nature 334: 661–665, 1988
24. Iwaki K, Sukhatme VP, Shubeita HE, Chien KR: α- and β-Adrenergic stimulation induces distinct patterns of immediate early gene expression in neonatal rat myocardial cells. J Biol Chem 265: 13809–13817, 1990
25. Kolbeck-Rühmkorff C, Horban A, Zimmer H-G: Effect of pressure and volume overload on proto-oncogene expression in the isolated working rat heart. Cardiovasc Res 27: 1998–2004, 1993
26. Seymour A-ML, Eldar H, Radda GK: Hyperthyroidism results in increased glycolytic capacity in the rat heart. A ^{31}P-NMR study. Biochim Biophys Acta 1055: 107–116, 1990
27. Wagner KR, Kauffman FC, Max SR: The pentose phosphate pathway in regenerating skeletal muscle. Biochem J 170: 17–22, 1978
28. Glock GE, McLean P: Levels of oxidized and reduced diphosphopyridine nucleotide and triphosphopyridine nucleotide in animals tissues. Biochem J 61: 388–390, 1955
29. Dow JW, Nigdikar S, Bowditch J: Adenine nucleotide synthesis *de novo* in mature rat cardiac myocytes. Biochim Biophys Acta 847: 223–227, 1985
30. Zimmer H-G, Ibel H, Steinkopff G, Korb G: Reduction of the isoproterenol-induced alterations in cardiac adenine nucleotides and morphology by ribose. Science 207: 319–321, 1980
31. Zimmer H-G, Peffer H: Metabolic aspects of the development of experimental cardiac hypertrophy. Basic Res Cardiol 81 (Suppl. 1): 127–137, 1986
32. Chatham JC, Challiss RAJ, Radda GK, Seymour A-ML: Studies on the protective effect of ribose in myocardial ischaemia by using ^{31}P-nuclear-magnetic-resonance spectroscopy. Biochem Soc Trans 13: 885–886, 1985
33. Pasque MK, Spray TL, Pellom GL, van Trigt P, Peyton RB, Currie W, Wechsler AS: Ribose-enhanced myocardial recovery following ischemia in the isolated working rat heart. J Thorac Cardiovac Surg 83: 390–398, 1982
34. Seifart HI, Delabar U, Siess M: The influence of various precursors on the concentration of energy-rich phosphates and pyridine nucleotides in cardiac tissue and its possible meaning for anoxic survival. Basic Res Cardiol 75: 57–61, 1980
35. Haas GS, DeBoer LWV, O'Keefe DD, Bodenhamer RM, Geffin GA, Drop LJ, Teplick RS, Daggett WM: Reduction of postischemic myocardial dysfunction by substrate repletion during reperfusion. Circulation 70 (suppl I): I65–I74, 1984
36. Mauser M, Hoffmeister HM, Nienaber C, Schaper W: Influence of ribose, adenosine, and 'AICAR' on the rate of myocardial adenosine triphosphate synthesis during reperfusion after coronary artery occlusion in the dog. Circ Res 56: 220–230, 1985
37. Ward HB, St.Cyr JA, Cogordan JA, Alyono D, Bioanco RW, Kriett JM, Foker JE: Recovery of adenine nucleotide levels after global ischemia in dogs. Surgery 96: 248–253, 1984
38. Wyatt DA, Ely SW, Lasley RD, Walsh R, Mainwaring R, Berne RM, Mentzer RM: Purine-enriched asanguineous cardioplegia retards adenosine triphosphate degradation during ischemia and improves postischemic ventricular function. J Thorac Cardiovasc Surg 97: 771–778, 1989
39. Zimmer H-G, Martius PA, Marschner G: Myocardial infarction in rats: Effects of metabolic and pharmacologic interventions. Basic Res Cardiol 84: 332–343, 1989
40. Clay MA, Stewart-Richardson P, Tasset DM, Williams JF: Chronic alcoholic cardiomyopathy. Protection of the isolated ischaemic working heart by ribose. Biochem Internat 17: 791–800, 1988
41. Zoref-Shani E, Shainberg A, Sperling O: Characterization of purine nucleotide metabolism in primary rat muscle cultures. Biochim Biophys Acta 716: 324–330, 1982
42. Zimmer H-G: Measurement of left ventricular hemodynamic parameters in closed-chest rats under control and various pathophysiologic conditions. Basic Res Cardiol 78: 77–84, 1983
43. Zimmer H-G: Normalization of depressed heart function in rats by ribose. Science 220: 81–82, 1983
44. Zimmer H-G, Zierhut W, Marschner G: Combination of ribose with calcium antagonist and β-blocker treatment in closed-chest rats. J Mol Cell Cardiol 19: 635–639, 1987
45. Isselhard W, Hamaji M, Mäurer W, Erkens H, Welter H: Adenosine-induced increase in myocardial adenine nucleotides without adenosine-induced systemic hypotension. Basic Res Cardiol 80: 47–61, 1985
46. Seesko RC, Zimmer H-G: Hemodynamic effects of inosine in combination with positive and negative inotropic drugs: Studies on rats *in vivo*. J Cardiovasc Pharmacol 16: 249–256, 1990
47. Zimmer H-G, Schneider A: Nucleotide precursors modify the effects of isoproterenol. Studies on heart function and cardiac adenine nucleotide content in intact rats. Circ Res 69: 1575–1582, 1991
48. Lee RGL, Springer C, Kasulis P, Lanir A, Frazer J, Clouse ME: Nuclear magnetic resonance assessment of adenosine triphosphate (ATP) dynamics in ischemic mouse livers perfused with adenine and ribose. Invest Radiol 22: 685–687, 1987
49. McAnulty JF, Southard JH, Belzer FO: Improved maintenance of adenosine triphosphate in five-day perfused kidneys with adenine and ribose. Transplant Proc 19: 1376–1379, 1987

Molecular and Cellular Biochemistry **160/161:** 111–115, 1996.

The endogenous digitalis-like factor

František Kölbel and Vratislav Schreiber[1]

1st Department of Medicine, Charles University 2nd School of Medicine, Charles University 2nd School of Medicine;
[1]Laboratory for Endocrinology and Metabolism, 3rd Department of Medicine, Charles University 1st School of Medicine, Prague, Czech Republic

Abstract

Intensive search into the presence of endogenous digitalis-like factor (EDLF) started shortly after identification of the alpha subunit of the Na,K,-ATPase as being receptor for digitalis glycosides. After years of skepticism, present data testify EDLF really exists. Most probably, the EDLF has chemical structure of either ouabain or of one of its isomers. It is secreted by the adrenal cortex, and, under conditions of stress, it's secretion is regulated differently from the secretion of both gluco- and mineralocorticoids. The physiological role of the EDLF has not been fully understood yet. In the newborn's kidneys, the inhibition of the Na,K-ATPase may assist to increase elimination of surplus sodum from the organism. In individuals of any age, the inhibitory influence of EDLF upon Na,K-ATPase in the arterial wall smooth muscle cells increases peripheral vascular resistance and thus, blood pressure. In the tissue culture, direct positive inotropic influences of EDLF upon rat cardiomyocytes was observed. However, the importance of positive inotropic effect of the EDLF upon the heart in clinical medicine remains to be elucidated. (Mol Cell Biochem **160/161:** 111–115, 1996)

Key words: digitalis-like factor, ouabain, Na,K-ATPase, adrenal cortex

Introduction

Since the discovery of endogenous ligands for opioid receptors, the endorphins, the idea of endogenous ligands of humoral receptors became feasible. In this review, we present some of the fundamental facts regarding the discovery of endogenous digitalis-like factor (DLF), and review data on its possible source, chemical structure, as well as data on its potential role in the organism.

History of the DLF

Hypothesis on the potential existence of a DLF was most probably put forward by Ringer in 1885, and was reviewed again by Szent-Györgyi in 1953 (see [1]). First experimental data describing presence in blood serum of a substance with digoxin-like immunoreactivity appeared in 1979 and 1980 and originated from two independent laboratories: Gruber *et al.* [2, 3] described the presence of digoxin-like immunoreactivity in the serum of animals and men with hypertension and blood volume-expansion. Another observation of the presence in blood serum of a material capable to react upon enzymoimmunoanalysis like digoxin was published by Schreiber *et al.* [4, 5] in rats with pressure-induced cardiac hypertrophy and with hyperthyroidism. In both cases, the presence of the substance with digitalis-like reactivity was identified by means of commercially available kits designed for determination of serum digoxin levels. In parallel, also the ability of that material to inhibit the ^{86}Rb influx into human erythrocytes in a way similar to digitalis glycosides was proved. During the next years attempts started to answer three main questions: (1) identification of the chemical formula of DLF; (2) identification of its source in the organism; (3) identification of its physiological role.

Identification of the chemical formula of DLF

The chemical formula of aglycons, or genins of most substances inhibiting Na,K-ATPase in the way digoxin does is based on cyclopentanoperhydrophenanthrene nucleus with

Address for offprints: F. Kölbel, 1st Department of Medicine, Charles University 2nd School of Medicine, V úvalu 84, 150 18 Praha 5 – Motol, Czech Republic the lactone ring positioned on C 17 (Fig. 1). With respect to that, potential candidates for the endogenous DLF were searched mainly among lipids and among the steroid compounds. Of steroids, the candidates for the DLF were progesteron derivatives [6, 7] and dehydroepiandrosterone sulphate [8]. The possibility that dehydroepiandrosterone could be responsible for endogenous digitalis-like activity was critically evaluated later [9]. As digoxin-like immunoreactivity of 14-OH oestrone, and for 14-OH oestradiol were demonstrated, the number of steroid compounds candidating for the position of DLF was great. Besides that, also several nonsteroid compounds were listed as potential DLF, namely ascorbic acid, non-esterified fatty acids, bile acids, lysophospholipids and an unidentified amino-glyco-steroid (see [10]). It became evident later that in the human plasma was present a substance with properties very closely resembling ouabain and the precise identification of the structure of DLF was given [11]. The active fraction was isolated from 85 liters of human plasma of healthy individuals not receiving any digitalis compound. It was analysed by mass spectrometry, its ability to inhibit Na,K-ATPase was tested in two independent ways as well as the ability of this fraction to react with polyclonal antiserum against ouabain. All tests performed showed the fraction of human plasma was structurally, biologically, as well as immunologically indistinguishable from ouabain. The ability of that fraction to react with antibodies against ouabain leaves open the question whether this substance, endogeneous ouabain, is identical to substances detected in the past by means of their ability to cross-react with antidigoxin antibodies, and isolated from material other than blood serum. E.g., detailed analysis of a substance with DLF activity extracted from bovine hypothalamus disclosed this substance is an isomer of ouabain in which two hydroxyl groups are altered in comparison to ouabain [12]. This structural difference may explain differences of physiological effects of the hypothalmic DLF from those of ouabain. Mainly, this DLF had higher positive inotropic effect on cultured rat myocytes than ouabain without toxic component, and caused vasoconstriction of rat aorta and pulmonary artery at concentrations in which ouabain was ineffective [12].

Another question requiring further elucidation is, whether endogenous ouabain is the only one substance with digitalis-like properties that is synthesized in the organism. The analysis of DLF isolated from patients with chronic renal failure disclosed presence of two different substances. One of them had properties of ouabain, the other those of digoxin [13]. The synthesis of more than one substance with digitalis like properties would not be exceptional – e.g. the Bufo marinus toad, a producer of bufalin synthesizes, besides bufalin at least another one digitalis-like steroid with significant vasoconstrictory activity [14]. Nevertheless, it seems to be fairly clear the mammals including man are able to synthesize at least one substance identical with ouabain.

Identification of the source of the DLF

The DLF was isolated from a wide variety of tissues, e.g. bovine brain, porcine myocardium, liver, the adrenals, as well as biological fluids – human milk, colostrum, human plasma

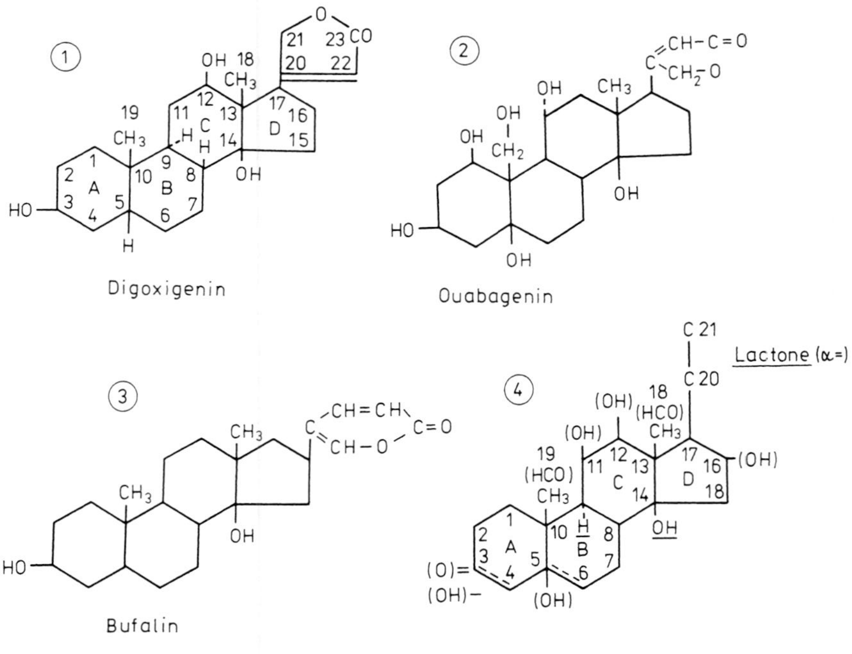

Fig. 1. Structure of digoxigenin (1), ouabagenin (2), of bufalin (3). General structure of genins in which formula is based on cyclopentanoperhydrophenanthrene nucleus (4). Underlined are substitutes essential for the cardiotonic action.

(Table 1) and, curiously, even from normal as well as from cataract-affected eye lenses [15]. From the very beginning of research of DLF, attempts were made to identify the site of its origin (see [10]). It was the opinion of first investigators that DLF originated in the brain [16], more precisely, in the hypothalamus [17–19] and adenohypophysis [20].

Potential interconnection between the development of cardiac hypertrophy and adrenal function was proposed in the past, at different times and at different levels by Liebegott – in humans [21], and by Hort [22], in experimental animals. Also the finding that in adrenalectomized rats, isoprenaline was not able to induce cardiac hypertrophy [23] was relevant and stressed the role of adrenal cortex in adaptive changes of the heart. When a good correlation between cardiac and adrenal weights was observed in animals with experimental cardiac hypertrophy of different origin it was hypothesized the DLF originated in adrenal cortex [5]. In rat adrenal extract the presence of a fraction was identified with chromatographic mobility close to aldosterone that exhibited strong digoxin-like immunoreactivity [24, 25]. Interesting in this respect was also the observation, in rats with cardiac pressure overload and hypertrophy, of decreased adrenal weight after long-term digitoxin treatment [26]. Pernollet and collaborators [27], Dorris [28], Dorris and Stocco [29] and Shaikh *et al.* [30] presented data indicating adrenal cortex was a source of a DLF. Presence of the endogenous ouabain-like substance was finally demonstrated in human, beef and rat adrenals in concentrations exceeding approximately 500 times that in the blood serum [11]. Likewise it was shown the adrenocortical cells in tissue culture secreted a compound possessing ouabain-like immunoreactivity. Their secretion was negatively influenced by raised concentration of potassium in the cultivation medium [11]. In humans, administration of dexamethasone suppressed secretion of both cortisol and of DLF, whereas administration of adrenocorticotrophic hormone (ACTH) increased plasma levels of both cortisol and of DLF [31]. Identical laboratory experiments has shown at the same time that both production and renal clearance play an important role in determination of the plasma level of DLF [31].

Doubts were thrown on these data by some other clinical observations. Although plasma level of immunoreactive ouabain in a patient with primary hyperaldosteronism decreased after unilateral adrenalectomy, in another two patients with bilateral adrenalectomy, the serum level of immunoreactive ouabain-like material remained normal [32]. This corresponds to older observation of digoxin-like immunoreactivity present in serum of adrenalectomized rats for which adrenocortical hormone was, most probably responsible [33]. Still another question is however, the specificity of the laboratory method or methods used to detect the presence of DLF. A logic requirement for future research in this field is standardisation of laboratory methods enabling us to compare data obtained at different laboratories. The key position of adrenal cortex in the secretion of DLF remains, in spite of all mentioned controversies, fairly clear.

Table 1. DLF content of different tissues and of biological fluids. Tissues were obtained from 300 g freshly killed male rat as well as from human cadaver

Tissue source	DLF (ng/g tissue) rat	DLF (ng/g tissue) human	Reference
Adrenal	1.98	0.95	[30]
Liver	0.39	0.57	ibid.
Kidney	0.09	0.22	ibid.
Heart	< 0.04	ND	ibid .
Plasma (adult)	ND	0.1[1]	[50]
Plasma (neonatal)	ND	0.5[1]–1.0	ibid.
Milk	ND	35.6[1]	[41]
Colostrum	ND	61.3[1]	ibid.
Peritoneal dilysis fluid	ND	0.09[1]	[13]

[1]values in ng/ml; ND not determined

Another question is whether adrenal cortex is the only one source of DLF, at least in humans. The presence of DLF in human colostrum and human milk in concentrations exceeding those in blood make it probable the mammary gland either itself secretes the DLF or extracts it from blood, concentrates it and secretes it concentrated. Also the origin of DLF isolated from beef hypothalamus remains obscure as yet [12]. It is evident that further analysis of this problem is necessary.

Physiological role of DLF

Elevated levels of plasma DLF were found in different pathologic states, e.g. arterial hypertension in nondiabetic as well as in diabetic patients [34], primary aldosteronism, pheochromocytoma, acromegaly, and chronic renal failure (see [32]), in congestive heart failure [35], and in obese persons [36]. Further, elevated plasma levels of DLF were also found in patients with acute myocardial infarction [37], in pregnant women [38, 39], in pre-term and full-term neonates [40], in umbilical cord blood of newborns and in mothers colostrum [41]. Rare are observations of the coincidence of adrenal tumor, elevated plasma level of ouabain in the plasma of both peripheral and tumor corresponding adrenal veins, and with arterial hypertension ('ouabainoma'). From the broad spectrum of scattered clinical data, it is impossible to define the most probable physiological role of the DLF . Most probably, DLF is active primarily in inducing natriuresis. The presence in the urine of healthy salt loaded subjects of a natriuretic factor binding to specific digoxin antibodies was demonstrated during early stage of the investigation of the DLF [42]. Participation of the DLF in different experimental models of DOCA-high salt-intake hypertension without partial nephrectomy was demonstrated [43, 44] as well as in models of hypertension in which DOCA-high salt regimen was supplemented by partial nephrectomy [45–47]. Further, in a way not yet precisely identified, the DLF may rise blood pressure, most probably via its action on vascular smooth muscle Na^+-Ca^{2+} exchange [47]. Reports on the direct action of DLF upon the heart and heart muscle cells are scarce, but exist. In tissue culture, the DLF isolated from bovine hypothalamus exhibited positive inotropic effect at doses three orders of magnitude lower than ouabain [12]. Other data are less specific and, consequently less persuasive. Kaye *et al.* [48], and Bagrov *et al.* [37] both hypothesized on potential arrhythmogenic effects of DLF. As a curiosity, signs of digitalis intoxication with Mobitz type I atrioventricular block were described as a result of intoxication caused by endogenous DLF of the toad [49]. The physiological role of the appearance in the blood sera DLF both in experimental animals and in patients exposed to hemodynamic overload remains still to be elucidated. In connection with these attempts, we feel obliged to again underline the necessity of standardization of methods used for detection of the presence of DLF at different conditions, both in clinical research, and in experiments using laboratory animals.

Conclusions

Intensive research on the existence of DLF lasts approximately 15 years. In spite of initial general skepticism, relatively very small number of laboratories in several countries of the world collected data that are perceived critically [52], but cannot be overlooked. Mainly, it was demonstrated the organisms of a wide variety of species including humans are able to synthesize a substance meeting criteria of DLF. This substance is either ouabain or a compound very closely related to it. Further, it must be admitted that besides endogenous ouabain some other compound or compounds exist that belong into the larger family of endogenous digitalis-like compounds. The tissue most probably synthesizing DLF *de novo* (or, maybe, assembling it from precursors) is adrenal cortex. The physiological role of the DLF remains still to be mostly elucidated. Most probably, this substance (or, substances) participate in regulation of blood pressure and, perhaps also in regulation of the inotropic state of the heart. Standardization of laboratory methods used by different working groups for the detection of DLF in biological material under various experimental as well as clinical conditions seems to be one of important prerequisites of successfull research in this field in the future.

References

1. Hinson PJ, Dawnay AB, Raven PW: Why we should give a qualified welcome to ouabain: A whole new family of adrenal steroid hormones? J Endocrinol 146: 369–372, 1995
2. Gruber KA, Whitaker JN, Plunkett WC, Buckalew VM, Jr: Detection of an endogenous digoxin-like substance (endoxin) in plasma of dog, sheep and man. Kidney Intl 16: 817, 1979
3. Gruber KA, Whitaker JM, Buckalew VM, Jr: Endogenous digitalis-like substance in plasma of volume-expanded dogs. Nature 287: 743–745, 1980
4. Schreiber V, Kölbel F, Štěpán J: Apparent (?) digoxin immunoreactivity in serum of rats with cardiac overload. To the problem of endogenous cardiotonic (endocardin). Czech text. Čas lék čes 119: 768–770, 1980
5. Schreiber V, Kölbel F, Štěpán J, Gregorová I, Přibyl T: Digoxin-like immunoreactivity in the serum of rats with cardiac overload. J Mol Cell Cardiol 13: 107–110, 1981
6. Kim RS, LaBella FS, Zunza H, Templeton JF: Progesterone derivatives that bind to the digitalis receptor: Structure-activity relationships. Mol Pharmacol 18: 402–405, 1980
7. LaBella FS, Bihler I, Templeton J, Kim RS, Hnatowich M, Rohrer D: Progesterone derivatives that bind to the digitalis receptor: effects on Na^+,K^+-ATPase and isolated tissues. Fed Proc 44: 2806–2811, 1985
8. Vasdev S, Longerich L, Johnson E: Dehydroepiandrosterone sulphate as a digitalis-like factor in plasma of healthy human adults. Res Commun Chem Pathol Pharmacol 49: 387–399, 1985
9. Goto A, Yamada K, Ishii M, Yoshika M, Sugimoto T: Is dehydroepiandrosterone sulphate a natriuretic digitalis-like factor in human serum? Clin Chim Acta 192: 145–151, 1990
10. Schreiber V: The endogenous dgitalis-like factor. Physiol Res 40: 207–211, 1991
11. Hamlyn JM, Blaustein MP, Bova S, DuCharme DW, Harris DW, Mandel F, Mathews WR, Ludens JH: Proc Natl Acad Sci USA 88: 6259–6263, 1991
12. Tymiak AA, Norman JA, Bolgar M, DiDonato GC, Lee H, Parker WL, Lo L-Ch, Berova N, Nakanishi K, Haber E, Haupert GT, Jr: Physicochemical characterization of a ouabain isomer isolated from bovine hypothalamus. Proc Natl Acad Sci USA 90: 8189–8193, 1993
13. Weinberg U, Dolev S, Werber MM, Shapiro MS, Shenkman L: Identification and preliminary characterization of two human digitalis-like substances that are structurally related to digoxin and ouabain. Biochem Biophys Res Commun 188: 1024–1029, 1992
14. Bagrov AY, Roukoyatkina NI, Fedorova OV, Pinaev AG, Ukhanova MV: Digitalis-like and vasoconstrictor effects of endogenous digoxin-like factor(s) from the venom of Bufo marinus toad. Eur J pharmacol 234: 165–172, 1993
15. Lichtstein D, Gati I, Samuelov S, Berson D, Rozenman Y, Landau L, Deutsch J: Identification of digitalis-like compounds in human cataractous lenses. Eur J Biochem 216: 261–268, 1993
16. Fishman MC: Endogenous digitalis-like activity in mammalian brain. Proc Natl Acad Sci USA 76: 4661–4663, 1979
17. Haupert CT, Jr, Sancho JM: Sodium transport inhibitor from bovine hypothalamus. Proc Natl Acad Sci USA 76: 4651–4660, 1979
18. Wijdicks EFM, Vermeulen M, van Brummelen P, den Boer NC, van Gijn J: Digoxin-like immunoreactive substance in patients with aneurysmal subarachnoid haemorrhage. Brit Med J 294: 729–732, 1987
19. Yamada K, Goto A, Hui C, Yagi N, Sugimoto T: Effects of the Fab fragment of digoxin antibody on the natriuresis and increase in blood pressure induced by intracerebroventricular infusion of hypertonic saline solution in rats. Clin Sci 82: 625–630, 1992
20. Takahashi H, Iyoda I, Matsuzawa M, Okabayashi H, Kaubara S, Yoshimura M, Ijichi H: Evidence for a digitalis-like substance in the hypothalamopituitary axis in rat: Implications in the central cardiovasclar regulation associated with an excessive salt intake (Abstr). Abstr 11th Sci Meeting Int Soc Hypertension, Heidelberg 1986
21. Liebegott G: Studien zur Orthogonie und Pathologie der Nebennieren. Beitr Pathol Anat 109: 93–178, 1944

22. Hort W: Morphologische und physiologische Untersuchungen an Ratten während eines Lauftrainings und nach dem Training. Virchows Arch 320: 197–237, 1951
23. Kölbel F, Kapitola J, Schreiberová O, Kölblová V: The influence of adrenalectomy on effect of isoproterenol in the rat. Physiol bohemoslov 19: 281–285, 1970
24. Schreiber V, Kölbel F, Štěpán J, Přibyl T, Jahodova J, Kubova V: Correlations bertween adrenal weight and heart weight in rats with a cardiac overload. Physiol bohemoslov 30: 289–294, 1981
25. Schreiber V, Štěpán J, Gregorová I, Krejčíková J: Crossed digoxin immunoreactivity in chromatographic fractions of rat adrenal extract. Biochem Pharmacol 30: 805–806, 1981
26. Kölbel F, Schreiber V, Kubová V, Jahodová J: Long term digitoxin administration to normal rats and to rats with experimental cardiac hypertrophy (Abstr). J Mol Cell Cardiol 15 (Suppl I): 142, 1983
27. Pernollet MG, Mohammad R, Meyer P, deVynck MA: Are the circulating digitalis-like compounds of adrenal origin (Abstr). Abstr 11th Sci Meeting Int Soc Hypertension Heidelberg, 1986
28. Doris PA: Immunological evidence that the adrenal gland is a surce of an endogenous digitalis-like factor. Endocrinology 123: 2440–2444, 1988
29. Doris PA, Stocco DM: An endogenous digitalis-like factor derived from the adrenal gland: Studies of adrenal tissue from various sources. Endocrinology 125: 2573–2579, 1989
30. Shaikh IM, Lav BWC, Valdes SR: Isolation of digoxin-like immunoreactive factors from mammalian adrenal cortex. J Biol Chem 266: 13672–13673, 1991
31. Vinge E, Erfurth EM, Lundin S: Effects of adrenal functional tests on the levels of endogenous digitalis-like substances and some pituitary hormones. Acta Endocrinol Copenh 128: 29–34, 1993
32. Naruse K, Naruse M, Tanabe A, Yoshimoto T, Watanabe Y, Kurimoto F, Horiba N, Tamura M, Inagami T, Demura H: Does plasma immunoreactive ouabain originate from the adrenal gland? Hypertension 23, (Suppl I): I102–I105, 1994
33. Schreiber V, Přibyl T, Štěpán J, Horký K: Digitalis-like biological activity and immunoreactivity of synthetic $^{1-24}$corticotropin and its natriuretic action. Physiol bohemoslov 31: 83–85, 1982
34. Okamoto S, Ikeda M, Morise T, Miyamori I, Takeda R: Elevated endogenous Na,K-ATPase inhibitor activity in hypertensive diabetic patients with a family history of hypertension. J Diabet Complications 5: 146–147, 1991
35. Gottlieb SS, Rogowski AC, Weiberg M, Krichten CM, Hasmilton BP, Hamlyn JM: Elevated concentrations of endogenous ouabain in patients with congestive heart failure. Circulation 86: 420–425, 1992
36. Giampietro O, Clerico A, Penno G, Gregori G, Del-Chicca MG, Cionini R, Volpe L, Navalesi R: Endogenous digitalis-like factors (EDLF) in obese individuals: Preliminary results. J Nucl Biol Med 36: 41–45, 1992
37. Bagrov AZ, Fedorova OV, Maslova MN, Rovkoyatkina NI, Ukhanova MV, Zhdabko EP: Endogenous plasma Na,K-ATPase inhibitory activity and digoxin-like immunoreactivity after acute myocardial infarction. Cardiovasc Res 25: 371–377, 1991
38. Pudek MR, Whitfield MF, Jacobson BE, Wittmann BK, King JR: Perinatal changes in digoxin-like immunoreactive substance. Pediat Res 18: 1097–1099, 1984
39. Krivoy N, Jakobi P, Paldi E, Alroy G: Total digoxin-like immunoreactive factor(s) in healthy population, uncomplicated term pregnancies and neonates. J Endocrinol Invest 13: 9–12, 1990
40. Bertrand J-M, Langhendries, JP, Gras A, Battisti O: Digoxin-like immunoreactive substance in serum of preterm and full-term neonates. Eur J Pediat 146: 145–146, 1987
41. Clerico A, Paci A, Del Chicca MG, Biver P, Giampietro O: Endogenous digitalis-like factors in human milk. Clin Chem 38: 504–506, 1992
42. Klingmüller D, Weiler E, Kramer HJ: Digoxin-like natriuretic activity in the urine of salt loaded healthy subjects. Klin Wochenschr 60: 1249–1253, 1982
43. Zicha J, Štolba P, Pohlová I,Kuneš J, Jelínek J: A different role of digoxin-like factor in the maintenance of elevated blood pressure in rats treated with DOCA and saline in youth or only in adulthood. J Hypertension 2 (suppl 2): 481–483, 1984
44. Zicha J, Kuneš J, Štolba P: Endogenous digoxin-like factor contributes to the elevation of systemic resistance in rats exposed to high salt intake from prepuberty. J Hypertension 3 (suppl 3): S17–S19, 1985
45. Kojima I, Yoshihara S, Ogata E: Involvement of endogenous digitalis-like substance in genesis of deoxycorticosterone-salt hypertension. Life Sci 30: 1775–1781, 1982
46. Nakamura T, Ichikawa S, Sato K, Tajima Y, Fuzita H, Murata K: Time-related alterations in an endogenous digitalis-like factor in the development of DOCA-salt hypertension in rats. Clin and Exper 9: 1733–1744, 1987
47. Yamada K, Goto A, Nagoshi H, Hui C, Yagi N, Sasabe M, Omata M: Hypertension 23 (Suppl 1): I110–I114, 1994
48. Kaye GC, Williams A, Camm AJ: Digoxin-like immunoreactivity during atrial arrhythmias The Lancet 848, vol I: 689, 1986
49. Chern M-S, Ray C-Y, Wu D: Biologic intoxication due to digitalis-like substance after ingestion of cooked toad soup. Am J Cardiol 67: 443–444, 1991
50. Ludens JH, Clark MA, DuCharme DW, Harris DW, Lutzke BS, Mandel F, Mathews WR, Sutter DM, Hamlyn JM: Hypertension 17: 923–929, 1991
51. Nicholls MG, Richards AM, Lewis LK, Yandle TG: Ouabain: A new steroid hormone? The Lancet 346: 1381–1382, 1995

Molecular and Cellular Biochemistry **160/161**: 117–120, 1996.

Functional antagonism by a monoclonal antibody to digoxin in a test system of cultured rat heart myocytes

G. Wallukat, H.-U. Simon,[1] W.-D.Müller,[1] I. Wolf[1]

Max Delbrück Centre for Molecular Medicine, Robert-Rössle-Str. 10, 13125 Berlin-Buch, Germany; [1]Institute of Clinical Immunology, Friedrich-Schiller Universität Jena, Am Johannesfriedhof 3, 07740 Jena, Germany

Abstract

A monoclonal antibody with a high affinity for digitoxin (K_A = 0.50 nM) and digoxin (K_A = 0.55 nM) was produced by somatic cell fusion. This antibody, designated 2A3(47), displayed little cross reactivity with other glycosides. In cultured rat heart myocytes, 2A3(47), antagonized the positive chronotropic effect exerted by digitoxin but did not alter that of ouabain. Our results suggest that this monoclonal antibody may prove to be useful in treating digoxin and digitoxin intoxication. (Mol Cell Biochem **160/161**: 117–120, 1996)

Key words: glycosides, monoclonal antibodies, cardiac myocytes, cell culture, digitoxin intoxication

Introduction

Digitoxin, one of the large group of cardiac glycosides, is the drug most frequently prescribed in Europe for patients with congestive heart failure. Polyclonal Fab fragments of digitoxin and digoxin antibodies are of proven value in the treatment of life-threatening poisoning with glycosides [1–5]. Some monoclonal antibodies directed against digoxin [6–10] and digitoxin [3, 11] also have been described. These antibodies can be used for the immunoassay for cardiac glycosides, for the detection of endogeneus digoxin-like substances, and for the therapy of glycoside intoxications.

A prerequisite for successful treatment is a higher affinity of the glycoside to the monoclonal antibody in comparison with the receptor on heart muscle cells (i.e. Na, K-ATPase). The Na,K-ATPase serves as the receptor for cardiac glycosides [12]. Binding of the glycosides to the receptor inhibits the Na,K-ATPase resulting in increased cytosolic Na^+ concentration, followed by an elevation of intracellular Ca^{++} [13]. The Na,K-ATPase is a heterodimeric protein composed of a 111 kD catalytic α-subunit and a 55 kD β-subunit. The α-subunits can exist in three isoforms and in the adult rat ventricular myocardium the (αl- and α2- isoforms are detectable [14]. However, in the neonatal rat heart ventricle the α1- and the α3- isoform are present [15]. Each of the isoforms of the Na,K-ATPase possesses a different sensitivity to cardiac glycosides. The α1- isoform, with a low affinity, is ouabain resistent. The α2- and α3- isoforms possess a high affinity to this glycoside. The α3- enzyme represent the fetal isoform detectable in the failing human [16] and in the hypertrophied heart [17].

Some attempts were made to develop test models for proving functional antagonism of monoclonal antibodies directed against cardiac glycosides [6, 7]. As early as in 1959 the group of Wollenberger showed that cultured cardiomyocytes were capable to respond functionally to cardiac glycosides [18–20] and thus represent a suitable experimental system to investigate pharmacologically the action of drugs [21]. We reasoned that characterization of such a monoclonal antibody in terms of its affinity to digitoxin and digoxin, cross reactivity to other glycosides and related steroids, and ability to antagonize the functional effects of digitoxin in cultured rat heart myocytes prepared from newborn rats would be a worthwhile undertaking.

Address for offprints: G. Wallukat, Max Delbrück Centre for Molecular Medicine, Robert-Rössle-Str. 10, 13125 Berlin-Buch, Germany

Materials and methods

Antibody preparation and ELISA

Digoxin-bovine serum albumin conjugate preparation, immunization, and fusion were carried out as described previously by Simon *et al.* [8]. Culture supernatants were tested in a standard ELISA assay [8]. Initial screening was done on plates coated with digoxin-BSA and BSA to identify hapten-specific clones. One particular antibody, designated 2A3(47), was chosen for further study. Antibody specificity was assessed from the binding inhibition curves. An ELISA was employed using different dilutions (0.5–40 ng/ml) of glycosides and structure-related steroids. NUNC-ELISA plates were coated with 100 µl of the purified monoclonal antibody (20 µg/ml) and 50 µl glycoside samples were added into the wells. The detection of the free binding sites were performed by the addition of 50 µl digoxin-peroxidase-conjugate in a dilution of 1:5000. Affinity constants were determined graphically from Scatchard plots. To determine the cross reactivity 10 ng/ml of the glycosides or the structure-related steroids were used.

Cell cultures

For the functional characterization of the monoclonal antibody cultured rat heart myocytes were used as a functional test system [22, 23].

As described previously in detail [24], single cells were dissociated aseptically from the minced heart ventricles of 1–3 day old Wistar rats with a 0.2% solution of trypsin and were cultured for 4 days at 37°C as monolayers on the bottom of 45 ml Müller flasks (1.2×10^6 seeded cells in 3 ml Medium) on a slowly moving rocker apparatus. Halle medium SM 20-I [25] containing 10% heat-inactivated calf serum and 2 µM fluorodeoxyuridine was used as culture fluid. On the day of the experiment, the medium was replaced and the cells were incubated at 37°C for 2 h in 2 ml SM 20-I on the rocker apparatus.

The pulsations of spontaneously beating single myocytes or clusters of myocytes were counted at 37°C on a heated stage of an inverted microscope, on which 10 small circular fields of the cell layer were inspected through the perforation of a metal template. Dose-response curves were obtained by cumulative addition of the glycosides. The effects were measured 5 min after the addition of the antibodies or the drugs. The number of beats of the cardiomyocytes in each of the 10 fields was counted for 15 sec each time. As a rule, this procedure was repeated twice in different cultures to yield results representing a total of 30 cells or cell clusters. The basal rate of beating was in the range of 134 ± 8 beats/min.

Drugs

Digoxin, digoxigenin, digitoxigenin, ouabain, lanatoside C, cortisol, and prednisolone were obtained from Sigma Chemie, Deisenhofen, Germany; Digitoxin from Labor Chemie Apolda, Germany; Fluorodeoxyuridine from Serva, Heidelberg, Germany; and trypsin from Leidholdt Biochemica, Kleinmachnow, Germany. Acetyldigoxin and digoxigenin-bis-digitoxoside were kindly given by Dr. W. Hubl, Dresden, Germany.

Statistical analysis

All experimental data are means ± S.E.M. and were analysed using the Student *t*-test. Differences were considered significant when probability values (p) were less than 0.05.

Results

A monoclonal antibody against digoxin was produced as described by Simon *et al.* [8]. This monoclonal antibody 2A3(47) which cross reacts with the glycoside digitoxin was classified as an immunoglobulin of the IgG1 class. It exhibited a high affinity for digitoxin (0.50 nM) and digoxin (0.55 nM).

Percent cross-reactivities with other glycosides determined by ELISA are shown in Table 1. The anti-digoxin monoclonal antibody shows a cross reactivity of 83% to digitoxin.

Digitoxigenin-bis-digitoxoside, lanotoside C, acetyldigoxin, digitoxigenin, digoxigenin, cortisol, and prednisolone were weakly or not recognized by the antibody 2A3(47).

The incubation of the cultured rat heart myocytes with the glycosides digitoxin, digoxin, and ouabain produced a positive chronotropic effect. The stimulatory effect was dose-dependent and a maximal stimulation of the myocytes was observed at glycoside concentrations of 10 µM. The preincubation of the glycosides digitoxin and digoxin with the

Table 1. Specificity of monoclonal antidigoxin antibody 2A3(47): Percentage of cross reactivity with glycosides and structure-related steroids measured in the ELISA

	Percentage of cross reactivity (%)
digoxin	100
digitoxin	83
digoxigenin-bis-digitoxoside	10
lanotoside C	3
acetyldigoxin	3
digitoxigenin	1
digoxigenin	1
cortisol	< 0.001
prednisolone	< 0.001

Table 2. Influence of digitoxin, digoxin, and ouabain on the pulsation rate of cultured neonatal rat heart myocytes. Effect of pretreatment with the glycosides with the mAB 2A3(47) in a dilution of 1:1000

Concentration	Increase in number of beats/min					
	Digitoxin		Digoxin		Ouabain	
μM	–mAB	+mAB	–mAB	+mAB	–mAB	+mAB
0.1	8.6 ± 0.8	0.9 ± 0.8[1]	7.4 ± 1.1	1.9 ± 1.2[1]	5.5 ± 1.0	7.0 ± 1.1[2]
1	17.1 ± 1.6	4.4 ± 0.8[1]	18.4 ± 1.9	6.5 ± 1.6[1]	12.8 ± 1.1	13.2 ± 1.5[2]
6	25.5 ± 1.6	9.9 ± 1.3[1]	27.8 ± 1.7	10.0 ± 1.8[1]	21.1 ± 1.5	20.3 ± 1.4[2]

mAB = monoclonal antibody, [1]$p<0.05$ [2]not significant

monoclonal antibody 2A3(47) at a dilution of 1:1000 for 1 h leads to neutralization of these glycosides (Table 2). Under these conditions the glycosides were not able to increase the beating rate of the myocytes. Pretreatment of ouabain with the monoclonal antibody did not influence the positive chronotropic effect of this glycoside (Table 2), suggesting that the monoclonal anti-digoxin antibody 2A3(47) did not recognize the glycoside ouabain.

In a concentration of 6 μM digitoxin and ouabain caused a similar increase in the pulsation rate of the rat heart myocytes. Addition of the monoclonal antibody 2A3(47) (1:2000) to the glycoside-stimulated myocytes led in the case of digitoxin to a loss of activity. After incubation for 5 min the monoclonal antibody 2A3(47) reduced the positive chronotropic response to digitoxin by 50%. After an incubation time of 30 min the digitoxin effect was completely abolished. In contrast, the antibody in the same dilution did not alter the positive chronotropic effect of ouabain (Fig. 1).

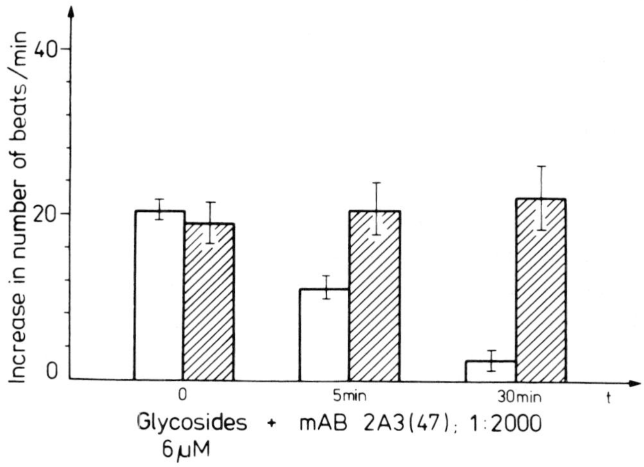

Fig. 1. Functional antagonism of the monoclonal antibody 2A3(47) (mAB) on digitoxin (open bars) prestimulated cultured neonatal rat cardiomyocytes. The positive chronotropic effect of ouabain (hatched bars) was not influenced.

Discussion

The results of this study indicate that the monoclonal antibody 2A3(47) displays high affinity for digitoxin and digoxin and limited or no cross reactivity with other glycosides and structure-related steroids. Its affinity for digoxin (0.55 nM) and digitoxin (0.50 nM) is similar or little higher than that of monoclonal antibodies reported by others (from 0.25–1.14 nM) [7, 26]. In cultured rat heart myocytes, expressing the α3 isoform of the Na,K-ATPase and responding to digitoxin like embryonic chick heart myocytes with a positive chronotropic effect [18, 19], the monoclonal antibody 2A3(47) antagonized the positive chronotropic effect of this cardiac glycoside. On the other hand the effect of oubain was not influenced. These findings suggest that the monoclonal antibody is able to neutralize and to dissociate fairly specifically digitoxin and digoxin from their binding sites.

Although periodic determination of digitoxin serum levels has reduced the incidence of intoxication, management of patients with advanced glycoside intoxication remains a major clinical problem. Because the mechanism of action of the digitalis glycosides consist in the inhibition of the membrane Na,K-ATPase system [12, 13] it was anticipated that antibodies to glycosides could be used to reverse its biologic effect [6]. In fact, Fab fragments prepared from polyclonal affinity purified sheep anti-glycoside antibodies have been successfully used to rescue several patients from severe intoxication resulting from overdoses of digitoxin and digoxin [5]. Similar results were observed in animal experiments by Lechat *et al.* [27]. In these experiments digoxin intoxication in guinea pigs was reversed by monoclonal antibodies directed against digoxin. Urtizberea *et al.* [11] used Fab fragments of a monoclonal anti-digitoxin antibody to overcome acute digitoxin poisoning in rabbits. Functional antagonism of cardiac glycoside effects by a specific antibody has been demonstrated in a number of systems, including *in vitro* erythrocyte cation transport, isolated papillary muscles, and atrial strips [6, 7]. The results of our study indicate that spontaneously beating cultured rat heart myocytes represent a novel *in vitro* system for testing reversal of cardiac glycoside effects by monoclonal antibodies directed against these glycosides. Due to their homogeneity monoclonal antibodies can serve as reagents that circumvent many of the limitations encountered with polyclonal preparations. This renders them more suitable for therapeutic purposes. Our results obtained with the monoclonal antibody 2A3(47) suggest that this reagent satisfies the necessary criteria to qualify as a additional therapeutic agent.

Acknowledgements

The authors are grateful to Mrs. M. Wegener, Mrs. H. Schmidt, Mrs. K. Karczewski and Mrs. E. Walther for skillful technical assistance and to Prof. A. Wollenberger for critically reading of the manuscript.

References

1. Butler VP Jr, Chen JP: Digitoxin-specific antibodies. Proc Nat Acad Sci 57: 71–78, 1967
2. Leikin J, Vogel S, Graff J, Goldman-Leikin R, Butler VP, Haber E, Smith TW: Use of Fab fragments of digoxin-specific antibodies in the therapy of massive digoxin poisoning. Ann Emerg Med 14: 172–178, 1985
3. Schaumann W, Kaufmann B, Neubert P, Smolarz A: Kinetics of the Fab fragments of digoxin antibodies and of bound digoxin in patients with severe digoxin intoxication. Eur J Clin Pharmacol 30: 527–533, 1986
4 Smolarz A, Abshagen U: Digitalis-Antikörper-Fragmente bei 90 schweren Glykosidvergiftungen. Klinische Erfahrungen aus einer multizentrischen Studie. Herz/Kreislauf 6: 261–266, 1986
5. Antman EM, Wenger TL, Butler VP, Haber E, Smith TW: Treatment of 150 cases of life-threatening digitalis intoxication with digoxin-specific Fab antibody fragments. Final report of a multicenter study. Circul 81: 1744–1752, 1990
6. Mudgett-Hunter M, Margolies MN, Ju A, Haber E: High affinity monoclonal antibodies to the cardiac glycoside, digoxin. J Immunol 129: 1165–1172, 1982
7. Wong PC, Reilly TM, Karol RA, Timmermans PBMWM: Functional antagonism by monoclonal antibody to digoxin in the guinea-pig atria. Eur J Pharmacol 136: 437–440, 1987
8. Simon H-U, Hubl W, Müller W-D, Wolf I, Schlenvoigt G, Jäger L: Produktion und Charakterisierung eines hochaffinen monoklonalen Antikörpers mit Digoxin- und Digitoxin-Spezifität. Allerg Immunol Leipz 36: 343–350, 1990
9. Sawada J, Janejai N, Terao T: Production and characterization of monoclonal and polyclonal antibodies against digoxin. Eisei Shikenjo Hokoku 108: 29–33, 1990
10. Kehayov IR, Tyutyulkova SN, Lolov S, Kyurkchiev SD: Generation of monoclonal antidigoxin antibodies. Hybridoma 9: 493–509, 1990
11. Urtizberea U, Latour M, Edelman L, Carcagne J, Armand J, Scherrmann JM: Immunopharmacological activity of monoclonal antidigitoxin. Dev Biol Stand 71: 147–157, 1990
12. Smith TW: Digitalis: mechanisms of action and clinical use. N Engl J Med 318: 358–365, 1988
13. Smith TW: The fundamental mechanism of inotropic action of digitalis. Therapie 44. 431–435, 1989
14. Shyjan A W, Levenson R: Antisera specific for the $\alpha 1$, $\alpha 2$, $\alpha 3$, and β-subunits of the Na,K-ATP ase: Differential expression of α and β subunits in rat tissue membranes. Biochemistry 28: 4531–4535, 1989
15. Lucchesi PA, Sweadner KJ: Postnatal changes in Na,K-ATPase isoform expression in rat cardiac ventricle. J Biol Chem 266: 9327–9331, 1991
16. Zahler R, Gilmore-Hebert M, Baldwin JC, Franco K, Benz EJ: Expression of α isoforms of Na,KvATPase on human heart. Biochim Biophys Acta 1149: 189–194, 1993
17. Charlemagne D, Swynghedauw B: Myocardial phenotypic changes in Na,K-ATPase on left ventricular hypertrophy: Pharmacological consequences. Europ Heart J 16 Suppl C: 20–23, 1995
18. Halle W: Über die Züchtung von Herzmuskelzellen in einem semisynthetischen Medium und ihr Verhalten gegenüber Digitoxin. Naturwiss 46: 267, 1959
19. Wollenberger A, Halle W: Einfluß von Reserpin und Dichlorisoproterenol auf die durch Adrenalin und Digitoxin hervorgerufenen Wirkungen an Kulturen spontan schlagender isolierter Zellen des embryonalen Hühnerherzens. Monatsberichte der Deutschen Akademie der Wissenschaften 5: 38–42, 1963
20. Halle W, Wollenberger A: Myocardial and other cell cultures. In: A Schwartz (ed.). Methods in Parmacology. Appleton-Century Crofts, New York, 1971, 1: 191–246
21. Wollenberger A: Isolated heart cells as a model of the myocardium. Basic Res Cardiol 80 Suppl 2: 9–13, 1985
22. Wallukat G, Wollenberger A: Involvement of β_2-adernergic receptors in the potentiation of the chronotropic action of isoprenaline evoked in rocker-cultured neonatal rat cells by pyruvate and L(+) lactate. In: R.E. Beamish, U. Panagia, and N.S. Dhalla (eds). Pharmacological aspects of heart disease. Martinus Nijhoff, Boston, 1987, pp 217–231
23. Wallukat G, Wollenberger A: Die kultivierte Herzmuskelzelle – ein funktionelles Testsystem zum Nachweis von Auto-Antikörpern gegen den beta-adrenergen Rezeptor. Acta Histochem Suppl.- Band XXXV: 145–149, 1988
24. Wallukat G, Wollenberger A: Supersensitivity to β-adrenoceptor stimulation evoked in cultured neonatal rat heart myocytes by L(+)-lactate and pyruvate. J Auton Pharmacol 13: 1–14, 1993
25. Halle W, Wollenberger A: Differentiation and behavior of isolated embryonic and neonatal heart cells in a chemically defined medium. Am J Cardiol 25: 292–299, 1970
26. Collignon A, Geniteau-Legendre M, Sandre C, Quero A-M, Labarre C: Specific binding characteristics of high affinity monoclonal antidigitoxin antibodies. Hybridoma 7: 355–366, 1988
27. Lechat P, Mudgett-Hunter M, Margolies MN, Haber E, Smith TW: Reversal of lethal digoxin toxicity in guinea pigs using monoclonal antibodies and Fab fragments. J Pharmacol Exp Ther 229: 210–213, 1984

PART II

ISCHEMIA-REPERFUSION AND CARDIOPROTECTION

Molecular and Cellular Biochemistry **160/161**: 123–128, 1996.

Biphasic changes in relaxation following reperfusion after myocardial ischemia

Susana M. Mosca, Martín Carriquiriborde and Horacio E. Cingolani

Centro de Investigaciones Cardiovasculares, Universidad Nacional de La Plata, 60 y 120, 1900 La Plata, Argentina

Abstract

The present study provides evidences of left ventricular diastolic alterations following reperfusion in a model of global ischemia. Isolated perfused rabbit and rat hearts, were subjected to ischemia for 15 and 20 min respectively, followed by 30 min of reperfusion. In rabbit heart at the end of the reperfusion period, isovolumic left ventricular developed pressure (LVDP) and $+dP/dt_{max}$ stabilized at 55 ± 3% and 60 ± 2% of preischemic values respectively and, in rat heart LVDP = 61 ± 8% and $+dP/dt_{max}$ = 57 ± 9% of preischemic values. Stunned heart was then obtained from both species. Left ventricular end diastolic pressure (LVEDP) values stabilized at the end of reperfusion period at values higher than preischemic conditions in both species (38.9 ± 4.4 mmHg and 30.3 ± 3.1 mmHg in rabbit and rat respectively). The time constant of relaxation (***T***) increased early in reperfusion in both species, but then decreased and stabilized at the end of reperfusion period at values lower than preischemic values. The ratio between both maximal velocities ($+\dot{P}/-\dot{P}$), also showed a transitory impairment in relaxation, followed by normalization and stabilization at values lower than preischemic values. This biphasic pattern in relaxation was detected in both species. The changes in relaxation were dissociated from the diastolic compliance and could be the result of a transitory calcium overload and/or sarcoplasmic reticulum dysfunction. The faster myocardial relaxation at the end of reperfusion period is consistent with the decreased myofilament sensitivity, which characterizes the stunned myocardium. (Mol Cell Biochem **160/161**: 123–128, 1996)

Key words: myocardial stunning, diastolic stiffness, myocardial relaxation

Introduction

Reperfusion after myocardial ischemia of brief duration does not induce necrosis, but results in a contractile dysfunction [1, 2]. This phenomenon has been called myocardial stunning [3]. The contractile dysfunction in stunned myocardium could be the result of a decrease in either the intracellular free Ca^{2+} concentration during the cardiac cycle ($[Ca^{2+}]i$ transient) or the myofilament responsiveness to Ca^{2+}. Ca^{2+} myofilament responsiveness is a function of two terms: Ca^{2+} myofilament sensitivity and maximal Ca^{2+}-activated force. Since it was described that during myocardial stunning the Ca^{2+} transient does not decrease or even increase [4–6], a decrease in Ca^{2+} myofilament responsiveness seems to be the mechanism responsible for the contractile dysfunction. Furthermore, changes in Ca^{2+} myofilament responsiveness either by a decrease in maximal contractile force [7] or myofilament sensitivity [6] have been reported. A decrease in Ca^{2+} myofilament sensitivity is expected to induce a faster relaxation through a decreased Ca^{2+} buffering by the 'Ca^{2+} sink' troponin C [8]. However experimental results about the diastolic alterations during stunning are controversial. Whereas most of studies are in agreement about an increase in diastolic stiffness [9–12], an impaired [13–15] or accelerated myocardial relaxation [12, 16] has been reported during stunning. In the present study we have evaluated in two different species, rabbit and rat, the changes in stiffness and myocardial relaxation following reperfusion after ischemia. The data provide evidences that reperfusion following ischemia induces a monotonic increase in diastolic stiffness and biphasic changes in relaxation: an early transient impairment followed by a faster relaxation when stunning is established.

Address for offprints: H.E. Cingolani, Centro de Investigaciones Cardiovasculares, Universidad Nacional de La Plata, 60 y 120, 1900 La Plata, Argentina

Materials and methods

Isolated heart preparation

The isolated rabbit and rat hearts were perfused with Ringer's solution (118 mM NaCl, 2 mM $CaCl_2$, 5,9 mM KCl, 1,2 mM $MgSO_4$, 20 mM $NaHCO_3$, and 11.1 mM dextrose, equilibrated with 95% O_2-5% CO_2 at pH = 7.38, 37°C) by the non-recirculating Langendorff technique. Rabbit hearts were paced at 160 ± 4 beats/min and rat hearts at 298 ± 2 beats/min by atrial pacing. Global ischemia was induced by abolishing coronary flow. Hearts were maintained at 37°C during the ischemic period by immersion in a water-jacketed beaker containing perfusate. A latex balloon tied to the end of a polyethylene tube was passed into the left ventricle through the mitral valve; the opposite end of the tube was then connected to a Statham P23XL pressure transducer. The balloon was filled with aqueous solution to an left ventricular end-diastolic pressure (LVEDP) of 9–13 mmHg, and then kept isovolumic throughout the experiment. Coronary perfusion pressure (CPP) was monitored at the cannulation point of the aorta. Left ventricular (LV) pressure and CPP were recorded with a direct writing recorder. After 10 min of stabilization, the coronary flow rate, controlled with a peristaltic pump, was 43 ± 2 ml/min and 10 ± 1 ml/min for rabbit and rat hearts, respectively.

Protocol

After 10 min during which pressure development was allowed to stabilize, two experimental groups were performed:

Rabbit (n = 20): hearts were submitted to a 15 min period of global ischemia, followed by a 30 min period of reperfusion.

Rat (n = 12): hearts were submitted to a 20 min period of global ischemia, followed by a 30 min period of reperfusion. In rat hearts a 20 min period of global ischemia was selected because no stunning was observed with shorter periods of ischemia (data not shown).

Under these experimental conditions, in rabbit and rat hearts myocardial stunning rather than infarction predominates [17, 18]. Although myocardial stunning results from the interaction in which the coronary vasculature and myocardium are involved, we think that isolated buffer-perfused heart provides a good physiological model to study stunning. A similar model for studies on postischemic dysfunction have been used by some investigators [19–21].

Systolic function

The left ventricular developed pressure (LVDP) values were obtained by subtracting LVEDP from the peak systolic pressure. By electronic differentiation the first derivative of left ventricular (dP/dt) was obtained. Maximal rate of rise of left ventricular ($+dP/dt_{max}$) was measured. Myocardial contractility was assessed by $+dP/dt_{max}$ and LVDP.

Diastolic stiffness

Since our experimental model is an isovolumic preparation, LVEDP was used as an index of ventricular diastolic stiffness [22].

Myocardial relaxation

To study the myocardial relaxation, the following measurements were performed:

- Ratio between maximal rate of rise ($+dP/dt_{max} = +\dot{P}$) and maximal rate of decay of left ventricular pressure ($-dP/dt_{max} = -\dot{P}$), ($+\dot{P}/-\dot{P}$);
- Half time of relaxation ($t_{1/2}$): time for pressure fall from peak pressure to half peak pressure.
- time constant of relaxation (***T***). It was calculated by using a monoexponential equation with asymptotic offset:

$$P = P_o \cdot e^{-t/T} + P_b$$

Where P is left ventricular pressure measured from the $-dP/dt_{max}$ until LVEDP reach + 5 mmHg. P_b is the asymptotic value, P_o is the value of left ventricular pressure at time t = 0 during relaxation, t is elapsed time during the relaxation period, and ***T*** is the relaxation time constant.

Analysis of data

Data are expressed as mean ± standard error of the mean. Multiple comparison of parameters between the two experimental groups was performed by one-way ANOVA. When a significant F value was found, the data were compared using Student-Newman-Keuls test [23].

Results

The time course of LVDP and $+dP/dt_{max}$ in rabbit and rat hearts, during ischemia and reperfusion are shown in Fig. 1. LVDP and $+dP/dt_{max}$ showed an early and almost complete recovery during reperfusion. After 30 min of reperfusion, LVDP (upper panel) and $+dP/dt_{max}$ (lower panel) stabilized at 55 ± 3% and 60 ± 2% in rabbit and 61 ± 8% and 57 ± 9% of preischemic values in rat. This pattern indicates that a stunned myocardium

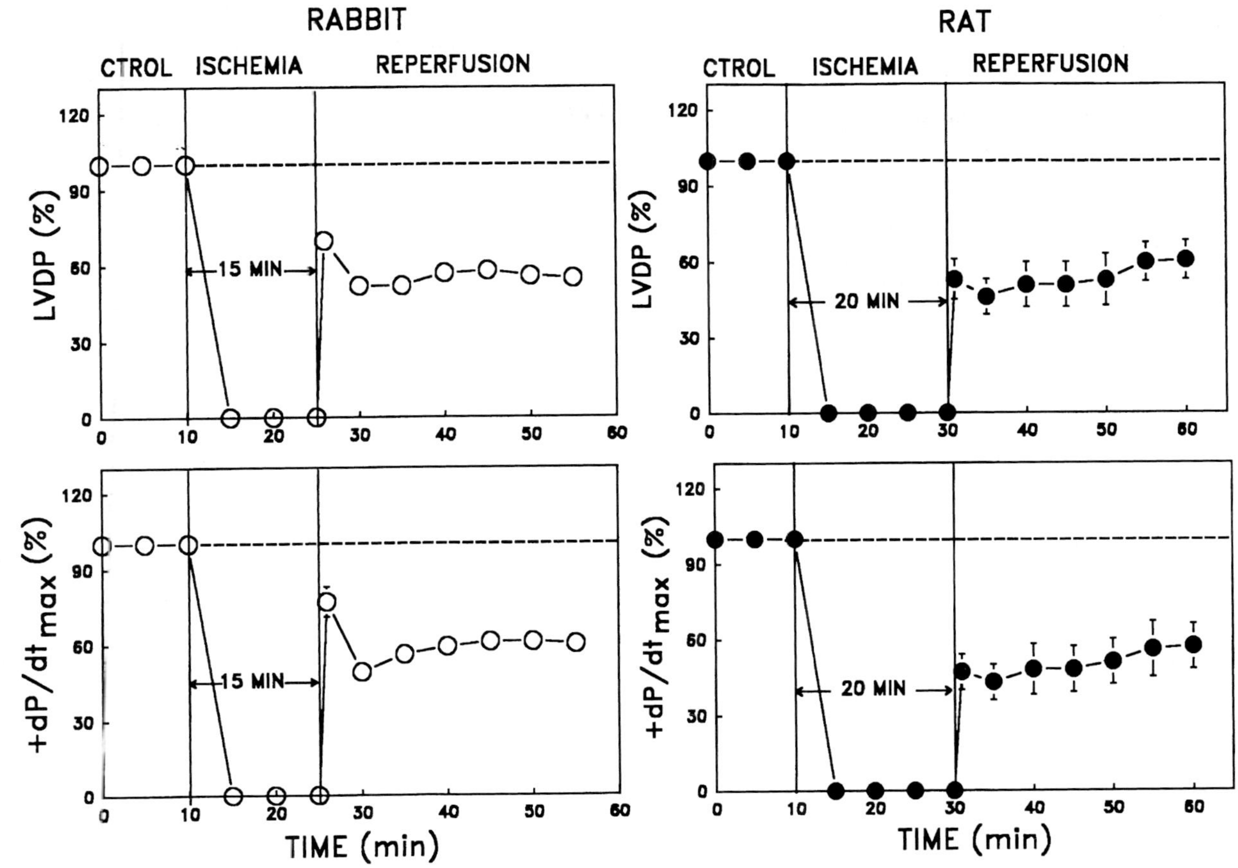

Fig. 1. Time course of changes in left ventricular developed pressure (LVDP, upper panel) and maximal rate of rise of left ventricular pressure ($+dP/dt_{max}$, lower panel), expressed as percentage of preischemic values, in isolated perfused rabbit and rat hearts are plotted.

have been induced [16–18]. Although the duration of ischemic period in the rat was longer than in the rabbit (see Materials and methods) the recovery during reperfusion was not significantly different between both species.

The changes in passive diastolic stiffness of rabbit and rat hearts, assessed by LVEDP at constant volume are depicted in Fig. 2. LVEDP increased monotonically in rabbit hearts during the reperfusion from a value of 18.5 ± 1.9 mmHg at the onset of reperfusion (preischemic value = 11.3 ± 0.7 mmHg) to 38.9 ± 4.4 mmHg at the end of reperfusion period. As reported previously a contracture during the ischemic period in the rat heart was observed [20, 24, 25]. LVEDP reached 34.3 ± 6.2 mmHg at the end of ischemia (preischemic value = 10.4 ± 0.4 mmHg). LVEDP tended to decrease during reperfusion from a value of 40.8 ± 4.4 mmHg at the onset of reperfusion but stabilized at 30.3 ± 3 mmHg at the end of reperfusion. Nevertheless, in both species LVEDP stabilized at values higher than preischemic values at the end of reperfusion, when stunning was established. Statistical analysis showed that there were no differences between the increments of LVEDP at the end of reperfusion between both species. The reperfusion following ischemia is inducing a stiff myocardium.

Changes in the time constant of relaxation, ***T***, from rabbit

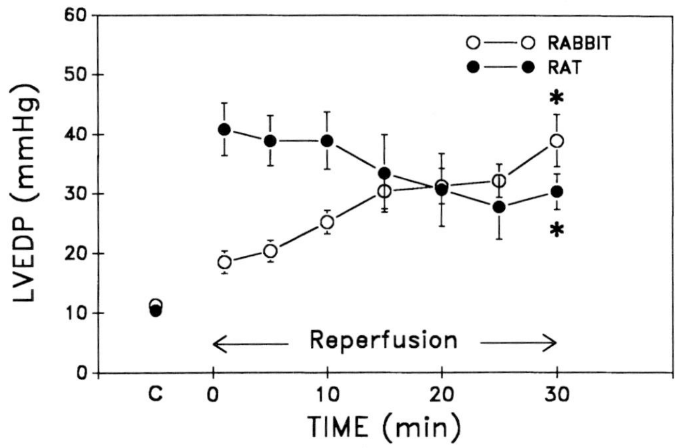

Fig. 2. Values of left ventricular end-diastolic pressure (LVEDP) during reperfusion in rabbit and rat hearts. Note that whereas in the rabbit hearts there was a monotonic increase in LVEDP during reperfusion, in the rat hearts a transient recovery followed by stabilization at higher than preischemic values was observed. *$p < 0.05$ with respect preischemic values.

and rat hearts during reperfusion, are shown in Fig. 3. In both species, ***T*** increased from a control value of 42 ± 2 to 69 ± 8 ms (rabbit) and from 31 ± 2 to 51 ± 7 ms (rat), at approxi-

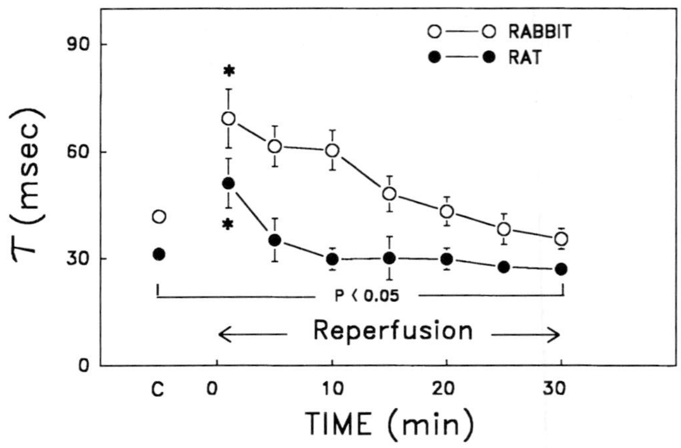

Fig. 3. Absolute values of the time constant of relaxation (***T***) during reperfusion in rabbit and rat hearts. Note the biphasic behaviour of this parameter during reperfusion. It showed higher values early in reperfusion and lower values than preischemic values when the stunning was established in both species. $*p < 0.05$ with respect preischemic values.

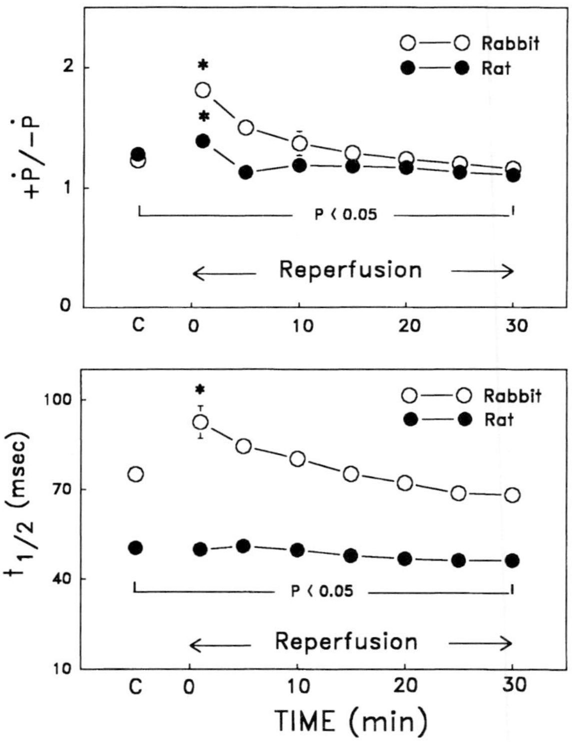

Fig. 4. Absolute values of ratio $+\dot{P}/-\dot{P}$ (upper panel) and half time of relaxation ($t_{1/2}$, lower panel) during reperfusion in rabbit and rat hearts. $+\dot{P}/-\dot{P}$ showed a biphasic behaviour in both species. $t_{1/2}$ showed a biphasic behaviour in rabbit but not in rat. In both species $t_{1/2}$ stabilized at lower than preischemic values at the end of reperfusion. $*p < 0.05$ with respect preischemic values.

mately 1 min of reperfusion. After 30 min of reperfusion the time constant of relaxation in rabbit and rat hearts stabilized at lower values than in preischemic situation. The ratio $+\dot{P}/-\dot{P}$ (Fig. 4, upper panel) increased from preischemic values of 1.23 ± 0.02 to 1.81 ± 0.07 and from 1.28 ± 0.03 to 1.39 ± 0.05 in rabbit and rat hearts, respectively at approximately 1 min after the onset of reperfusion. This proportionally greater increase in $+\dot{P}$ than in $-\dot{P}$ reflected a negative lusitropic effect [26–29]. However, this increase in $+\dot{P}/-\dot{P}$ is followed by a progressive decrease in the ratio and at the end of reperfusion values of 1.16 ± 0.03 (rabbit) and 1.11 ± 0.04 (rat) were detected. The present results indicate a transitorily impairment of relaxation early in the reperfusion. However, late in reperfusion an acceleration in relaxation was detected in both species. Figure 4, lower panel, shows absolute values of $t_{1/2}$ during reperfusion, in rabbit and rat hearts. $t_{1/2}$ showed the same pattern than ***T*** and $+\dot{P}/-\dot{P}$ ratio in rabbit hearts. This parameter increased early in reperfusion and decreased at the end of reperfusion. In rat heart, $t_{1/2}$ did not change early in reperfusion and reached values lower than control at the end of reperfusion. In summary, the diastolic alterations detected during reperfusion in the two species were a decrease in diastolic compliance and biphasic changes in relaxation: early impairment followed by a faster relaxation when stunning was established.

Discussion

In our experiments the left ventricular diastolic function exhibits different changes during reperfusion, depending on whether myocardial relaxation or diastolic stiffness was been evaluated. The diastolic stiffness assessed by the left ventricular end-diastolic pressure, at constant volume, increased in the stunned myocardium. This finding in stunned condition, is in agreement with previous reports [9–13]. An increase in diastolic intracellular Ca^{2+} with concomitant increase in resting force was recently demonstrated in rat stunned myocardium [12, 30]. This finding may explain the increase of diastolic stiffness detected in our and others studies. Myocardial relaxation, however, exhibited a behavior independent of diastolic stiffness. The dissociation between relaxation and diastolic stiffness is not new. It was previously reported by us [31] during reperfusion and was showed by Momomura *et al.* [22] after several pharmacological interventions. In previous experiments performed by us [31] a transitory impairment in relaxation (as measured by ***T***) with a return to preischemic values at the end of reperfusion was detected. In our previous report we were unable to detect values of ***T*** lower than preischemic values at the end of reperfusion period [31]. Increasing the number of experiments and using other parameters of relaxation to characterize lusitropism ($+\dot{P}/-\dot{P}$ and $t_{1/2}$), the faster relaxation at the end of reperfusion became evident (Figs 3 and 4). Although our data do not provide evidence on the mechanisms involved in the biphasic changes in relaxation detected by us, some possibilities could be considered. Cytosolic Ca^{2+} overload has been implicated as a primary factor in the pathogenesis of 'stunned' myocardium [32–34]. Myocardial ischemia induces a progressive rise in free intracellular Ca^{2+} concentration, that persists during the early stages of reperfusion [11, 35–38]. This Ca^{2+} overload is produced by complex mechanisms that include the alteration of function of sarcoplasmic reticulum (SR) as a major determinant [39, 40]. Other possible mechanism could be a decrease in Ca^{2+} efflux by Na^+/Ca^{2+} exchanger due to the inhibitory effect of ischemic acidosis [41, 42] or as a consequence of elevated intracellular Na^+ concentration [43–45]. Myocardial relaxation is the result of several cellular processes, including SR function and myofilament Ca^{2+} responsiveness. Altered myofibrillar Ca^{2+} responsiveness is a well-documented feature in reperfused myocardium [37, 6, 7] and has been attributed to either changes in maximal Ca^{2+}-activated force [6], Ca^{2+} myofilament sensitivity [7], or both. Both effects can explain the acceleration in relaxation observed in our experimental preparation when the stunning was established. In any case, either an impaired or an accelerated relaxation can be detected during reperfusion depending on how long after the onset of reperfusion the measurements are performed.

References

1. Heydrinckx G, Millard RW, McRitchie RJ, Maroko PR, Vatner SF: Regional myocardial function and electrophysiologic alterations after brief coronary artery occlusions in conscious dogs. J Clin Invest 56: 978–985, 1975
2. Kloner RA, DeBoer LWV, Darsee JR, Ingwall JS, Hale S, Tumas J, Braunwald E: Prolonged abnormalities of myocardium salvaged by reperfusion. Am J Physiol 241 (Heart Circ Physiol 10): H591–H599, 1981
3. Braunwald E, Kloner R: The stunned myocardium: prolonged, postischemic ventricular dysfunction. Circulation 66: 1146–1149, 1982
4. Kusuoka H, Koretsune Y, Chacko VP, Weisfeldt ML, Marban E: Excitation-contraction coupling in postischemic myocardium. Does failure of activator Ca^{2+} transient underlie stunning? Circ Res 66: 1268–1276, 1990
5. Lee JA, Allen DG: Changes in intracellular free calcium concentration during long exposures to stimulated ischemia in isolated mammalian ventricular muscle. Circ Res 71: 58–69, 1992
6. Carrozza JP, Bentivegna LA, Williams ChP *et al.*: Decreased myofilament responsiveness in myocardial stunning follows transient calcium overload during ischemia and reperfusion. Circ Res 71: 1334–1340, 1992
7. Hofmann PA, Miller WP, Moss RL: Altered calcium sensitivity of isometric tension in myocyte-sized preparations of porcine postischemic stunned myocardium. Circ Res 72: 50–56, 1993
8. Blinks JR, Endoh M: Modification of myofibrillar responsiveness to Ca^{2+} as an inotropic mechanism. Circulation 73, Suppl III: III-85–III-98, 1986
9. Jeremy RW, Koretsune Y, Marban E, Becker LC: Relation between glycolisis and calcium homeostasis in postischemic myocardium. Circ Res 70: 1180–1190, 1992
10. Eberli FR, Weinberg EO, Grice WN, Horowitz GL, Apstein CS: Protective effect of increased glycolitic substrate against systolic and diastolic dysfunction and increased coronary resistance from prolonged global underperfusion and reperfusion in isolated rabbit hearts perfused with erythrocyte suspensions. Circ Res 68: 466–481, 1991
11. Marban E, Kitakaze M, Koretsune Y, Yue DT, Chacko VP, Pike MM: Quantification of $[Ca^{2+}]_i$ in perfused hearts: Critical evaluation of the 5F-BAPTA and nuclear magnetic resonance method as applied to the study of ischemia and reperfusion. Circ Res 66: 1255–1267, 1990
12. Gao WD, Atar D, Backx PH, Marban E: Relationship between intracellular calcium and contractile force in stunned myocardium. Direct evidence for decreased myofilament Ca^{2+} responsiveness and altered diastolic function in intact ventricular muscle. Circ Res 76: 1036–1048, 1995
13. Charlat M, O'Neill P, Hartley C, Roberts R, Bolli R: Prolonged abnormalities of left ventricular diastolic wall thinning in the 'stunned' myocardium in conscious dogs: time course and relation to systolic function. J Am Coll Cardiol 13: 185–194, 1989
14. Przylenk K, Patel B, Kloner RA: Diastolic abnormalities of postischemic 'stunned' myocardium. Am J Cardiol 60: 1211–1213, 1987
15. Apstein CS, Bing OHL, Levine HJ: Acute cardiac ischemia and reperfusion: contractility, relaxation, and glycolisis. Am J Physiol 235: H637–H643, 1978
16. Lew WYW, Nishikawa Y, Su H: Cardiac myocyte function and left strains after brief ischemia and reperfusion in rabbits. Circulation 90: 1942–1950, 1994
17. Jennings R, Schaper J, Hill M: Effect of reperfusion late in the phase of reversible ischemic injury: changes in cell volume, electrolytes, metabolites and ultrastructure. Circ Res 56: 262–268, 1985
18. Walker DM, Marber MS, Walker JM, Yellon DM: Preconditioning in isolated superfused rabbit papillary muscles. Am J Physiol 266 (Heart Circ Physiol 35): H1534–H1540, 1994
19. Weiss RG, Kalil-Filho R, Herskowitz A, Chacko VP, Litt M, Stern MD, Gerstenblith G: Tricarboxylic acid cycle activity in postischemic rat hearts. Circulation 87: 270–282, 1993
20. Asimakis GK, Inners-McBride K, Medellin G, Conti VR: Ischemic preconditioning attenuates acidosis and post-ischemic dysfunction in isolated rat heart. Am J Physiol 263: H887–H894, 1992
21. Lasley RD, Anderson GM, Mentzer RM: Ischemic and hypoxic preconditioning enhance post-ischemic recovery of function in the rat heart. Cardiovasc Res 27: 565–570, 1993
22. Momomura S, Izuka T, Serizawa T, Sugimoto I: Separation of rate of left ventricular relaxation from changes stiffness in rats. Am J Physiol 24: 1468–1476, 1988
23. Wallenstein S, Zucker CL, Fleiss JL: Some statistical methods useful in circulation research. Circ Res 47: 1–9, 1980
24. Kingsley PB, Sako EY, Yang MQ, Zimmer SD, Ugurbil K, Foker JE, From AHL: Ischemic contracture begins when anaerobic glycolisis stops: a ^{31}P-NMR study of isolated rat hearts. Am J Physiol 261: H469–H478, 1991
25. Griffiths EJ, Halestrap AP: Protection by ciclosporin A of ischemia/reperfusion-induced damage in isolated rat hearts. J Mol Cell Cardiol 25: 1461–1469, 1993
26. Grassi de Gende AO, Pérez Alzueta AD, Cingolani HE: Effect of isoproterenol on relation between maximal rate of contraction and maximal rate of relaxation. Am J Physiol 2: H404–H409, 1977
27. Vittone L, Cingolani HE, Mattiazzi RA: The link between myocardial contraction and relaxation: The effects of calcium antagonists. J Mol Cell Cardiol 17: 255–263, 1981
28. Mattiazzi RA, Garay A, Cingolani HE: Critical evaluation of isometric indexes of relaxation in rat and cat papillary muscles and toad ventricular strips. J Mol Cell Cardiol 18: 749–758, 1986
29. Vittone L, Mundiña-Weilenmann C., Mattiazzi RA, Cingolani HE: Physiologic and pharmacologic factors that affect myocardial relaxation. JPM 32(1): 7–18, 1994

30. Meissner A, Morgan JP: Contractile dysfunction and abnormal Ca^{2+} modulation during postischemic reperfusion in rat heart. Am J Physiol 268 (Heart Circ Physiol 37): H100–H 111, 1995
31. Mosca SM, Gelpi RJ, Cingolani HE: Dissociation between myocardial relaxation and diastolic stiffness in the stunned heart: its prevention by ischemic preconditioning. Mol Cell Biochem 129: 171–178, 1993
32. Marban E, Koretsune Y, Corretti MC, Chacko VP, Kusuoka H: Calcium and its role in myocardial cell injury during ischemia and reperfusion. Circulation 80 (suppl IV): 17–22, 1989
33. Kusuoka H, Marban E: Cellular mechanisms of myocardial stunning. Ann Rev Physiol 54: 243–56, 1992
34. Kusuoka H, Kitakaze M, Koretsune Y, Inoue M, Marban E: Pathophysiology and pathogenesis of contractile failure in stunned myocardium. Jpn Circ J 55: 878–884, 1991
35. Kihara Y, Grossman W, Morgan JP: Direct measurement of changes in intracellular calcium transients during hipoxia, ischemia and reperfusion of the intact mammalian heart. Circ Res 65: 1029–1044, 1989
36. Steenbergen C, Murphy E, Levy L, London RE: Elevation in cytosolic free calcium concentration early in myocardial ischemia in perfused rat heart. Circ Res 60: 700–707, 1987
37. Mohabir R, Lee HK, Kurz RW, Clusin WT: Effects of ischemia and hypercarbic acidosis on myocyte calcium transients, contraction, and pH_i, in perfused rabbit hearts. Circ Res 69: 1525–1537, 1991
38. Steenbergen C, Murphy E, Watts JA, London RE: Correlation between cytosolic free calcium, contracture, ATP, and irreversible ischemic injury in perfused rat heart. Circ Res 66: 135–146, 1990
39. Krause SM, Jacobus WE, Becker LC: Alterations in cardiac sarcoplasmic reticulum calcium transport in the postischemic 'stunned' myocardium. Circ Res 65: 526–530, 1989
40. Davis MD, Lebolt W, Feher JJ: Reversibility of the effects of normothermic global ischemia on the ryanodine-sensitive and ryanodine-insensitive calcium uptake of cardiac sarcoplasmic reticulum. Circ Res 70: 163–71, 1992
41. Philipson KD, Bersohn MM, Nishimoto AY: Effects of pH on Na^+-Ca^{2+} exchange and Ca^{2+} permeability of cardiac sarcolemmal vesicles. J Biol Chem 258: 5905–5910, 1982
42. Philipson KD, Nishimoto AY: Na^+-Ca^{2+}-exchange in inside-out cardiac sarcolemmal vesicles. J Biol Chem 257: 5111–5117, 1982
43. Tani M: Mechanisms of Ca^{2+} overload in reperfused ischemic myocardium. Ann Rev Physiol 52: 543–59, 1990
44. Kusuoka H, Camilión de Hurtado MC, Marban E: Role of Na^+/Ca^{2+} exchanger in the mechanism of stunning: Protective effect of reperfusion with high sodium solution. J Am Coll Cardiol 21: 240–248, 1993
45. Pike M, Kitakaze M, Marban E: Na-NMR measurements of intracellular sodium in intact ferret hearts during ischemia and reperfusion. Am J Physiol 259: H1767–H1773, 1990

Molecular and Cellular Biochemistry **160/161**: 129–136, 1996.

Signal transduction in myocardial ischaemia and reperfusion

A. Lochner, E. Tromp and R. Mouton

Department of Medical Physiology and Biochemistry, University of Stellenbosch Faculty of Medicine; MRC Programme for Experimental Biology, Tygerberg 7505; Republic of South Africa

Abstract

Recent studies in the non-ischaemic myocardium indicated that drugs stimulating cAMP formation inhibit α_1-mediated inositol phosphate generation, while α_1-adrenergic stimulation lowered tissue cAMP levels, implicating cross-talk between α_1- and β–adrenergic signalling pathways in normal physiological conditions. Massive amounts of endogenous catecholamines, predominantly noradrenaline, are released during myocardial ischaemia and reperfusion, causing stimulation of both α_1- and β-adrenergic receptors which, in turn, may contribute to intracellular Ca^{2+} overload and subsequent cell damage. Since no information is available regarding cross-talk in pathophysiological conditions, the aim of this study was to evaluate the interactions between α_1- and β-adrenergic signalling pathways during different periods of ischaemia and reperfusion.

Isolated rat hearts were perfused retrogradely for 30 min before being subjected to (i) 5–25 min global ischaemia and (ii) 1–5 min of reperfusion after 20 min global ischaemia. Drugs (prazosin, 10^{-7} M; propranolol, 10^{-6} M; phenylephrine 3×10^{-5} M; isoproterenol 10^{-9} M) were added 10 min before the onset of ischaemia and were present during reperfusion.

Increasing periods of ischaemia caused an immediate rise and progressive lowering in tissue cAMP and Ins(1,4,5)P_3 levels respectively. In contrast, reperfusion caused an elevation in Ins(1,4,5)P_3 levels and reduced cAMP. Prazosin elevated cAMP levels during both ischaemia and reperfusion, while propranolol had no effects on tissue Ins(1,4,5)P_3. The activity of the α_1-adrenergic signal transduction pathway appears to have an inhibitory effect on the activity of the β-adrenergic system during ischaemia and reperfusion. (Mol Cell Biochem **160/161**: 129–136, 1996)

Key words: signal transduction, adrenergic receptors, ischaemic myocardium, ischaemia-reperfusion injury

Introduction

Several recent studies indicated that inhibitory cross-talk occurs between the α_1- and β-adrenergic signalling pathways under physiological conditions [1–3]. For example, using cultured myocytes, Guse and coworkers [1] showed that formation of cAMP in response to β-adrenergic stimulation transiently inhibited inositol phosphate production, whereas α_1-adrenergic stimulation in turn, has been demonstrated to inhibit β-mediated increased cAMP levels [3].

However, as far as we know, no information is available regarding the possibility of cross-talk in myocardial ischaemia and reperfusion, pathological conditions which are associated with the release of large amounts of endogenous noradrenaline [4–6] and thus simultaneous stimulation of the α_1- and β-receptors. The effects of ischaemia on both the α_1- and β-adrenergic response of the myocardium have been thoroughly investigated. Acute myocardial ischaemia causes sensitization of the adenylyl cyclase system, an increase in the number of β-adrenergic receptors [7] and impaired function of the inhibitory G protein [8], while reperfusion reversed the latter [8]. Myocardial ischaemia has also been reported to double the α_1-adrenergic receptor density and to increase sensitivity to noradrenaline stimulation [9, 10].

Although release of noradrenaline during reperfusion may contribute to the resumption of contractile activity, the unimpeded stimulation of both α_1- and β-receptors may be potentially harmful, since both processes can contribute to intracellular Ca^{2+} overloading [11, 12]. In view of the above, the aim of the present study was to evaluate the possibility of cross-talk between the α_1- and β-adrenergic signalling pathways on post-receptor events in the ischaemic-reperfused

Address for offprints: A. Lochner, Department of Medical Physiology and Biochemistry, PO Box 19063, Tygerberg 7505, Republic of South Africa rat heart. Tissue cAMP and inositol (1,4,5) trisphosphate ($InsP_3$) levels were used as indicators of the activity of the β- and α_1-adrenergic pathways respectively. In addition, incorporation of myo-[^{3}H]-inositol into tissue inositolphosphates was studied in a number of selected experiments. The respective roles of the α_1- and β-adrenergic signal transduction pathways were evaluated by the use of appropriate blockers and stimulants.

Materials and methods

Animals

Male Wistar rats (230–250 g) were used in all experiments. The animals were fed *ad libitum* until experimentation. Institutional and international ethical guide lines were observed in the handling of the experimental animals. The rats were anaesthetized with pentobarbital (50 mg/kg) intraperitoneally.

Perfusion technique

Immediately after removal of the heart, it was arrested in ice-cold saline and mounted onto the aortic cannula for retrograde perfusion. Hearts were perfused at 37°C with oxygenated (95% O_2, 5% CO_2) Krebs-Henseleit buffer (KHB) containing (in mM): NaCl 124; $NaHCO_3$ 26.1; KCl 4.93; KH_2PO_4 1.23; $MgSO_4$ 1.62; Na_2SO_4 0.6; $CaCl_2$ 1.25; glucose 10. The perfusion systems were filled with 50 ml KHB. Hearts were perfused with 20 ml KHB in a non-recirculating manner to remove all traces of blood. Thereafter the hearts were perfused with the remaining 30 ml in a recirculating manner (perfusion pressure 80 mmHg) for 30 min to stabilize. Hearts were then subjected to different periods (5–25 min) of normothermic global arrest by clamping of the aortic cannula. Cooling was prevented by a tightly-stoppered double-walled perfusion jacket kept at 40°C. Reperfusion was introduced by unclamping of the aortic cannula and perfusion continued at 80 mmHg as before. Ventricles were freeze-clamped at the end of the stabilization period, at the end of ischaemia or at the end of the reperfusion period. Tissues were stored in liquid N_2 until extraction.

Drugs (propranolol 10^{-6} M, prazosin 10^{-7} M, phenylephrine $10^{-7} - 3 \times 10^{-5}$ M or isoproterenol 10^{-9} M) were added after 20 min of stabilization, for a period of 10 min before the onset of sustained ischaemia. The drugs were also present during reperfusion. In a separate series propranolol and prazosin were administered during the reperfusion period only.

In a number of studies hearts were perfused, after the initial washout of blood with 20 ml KHB, for 2 h in a recirculating manner with 30 ml KHB containing 1.5 μCi/ml myo-[^{3}H]-inositol for labelling of the phosphatidylinositol stores. At the end of this period, the extracellular label was washed out with 20 ml non-recirculating KHB. From thereon the following different protocols were used: *Controls*: Hearts were perfused with recirculating KHB containing 10 mM LiCl for 35 min. *Global ischaemia and reperfusion*: After perfusion with KHB containing 10 mM LiCl for 15 min, hearts were subjected to global ischaemia for 20 min. In a number of studies the hearts were reperfused for 1 min.

In these studies the drugs (phenylephrine 3×10^{-5} M; propranolol 10^{-6} M) were added during the 15 min perfusion period with 10 mM LiCl, before the onset of global ischaemia. Propranolol was present for 15 min, whereas phenylephrine was added 5 min later.

Biochemical analyses

Inositol phosphates from ventricular tissue were extracted by homogenizing tissue (± 300 mg) at 4°C in 10 ml 10% trichloro-acetic acid containing 1 mM EDTA, using a Polytron PT 10 homogenizer (3 × 5 sec, setting 7). Samples were incubated on ice for 10 min before centrifugation for 15 min at 1000 g. The supernatants were washed three times with equal volumes of diethyl ether to remove traces of trichloro-acetic acid. The washed extracts were neutralized to pH 7.5 with 1 M $NaHCO_3$ and diluted to 20 ml with distilled water and 100 μl aliquots were assayed directly using Amersham's [^{3}H] inositol-1,4,5-trisphosphate assay system.

For separation of the [^{3}H] inositol phosphates, the washed neutralized extracts were applied to Dowex 1 × 8 columns (formate form) and the $InsP_1$, $InsP_2$ and $InsP_3$ fractions eluted as described previously [13].

For measurement of tissue cAMP, ventricular tissue (± 300 mg) was extracted with 1.2 ml 6% perchloric acid according to the method of Lamprecht and Trautschold [14]. After neutralization, cAMP levels were determined using Amersham's cyclic AMP[^{3}H] assay kit. For this assay, a standard curve ranging from 0.125–8 picomoles cAMP/sample was included.

Experimental programme

(A) Effect of ischaemia and reperfusion on tissue cAMP and $InsP_3$: (1) Control hearts, perfused for either 30 or 50 min. (2) Hearts subjected to 5, 10, 20 and 25 min global ischaemia. (3) Hearts subjected to 20 min global ischaemia and reperfused for 0.5, 1, 2 or 5 min.

(B) Effects of α_1-stimulation, α_1-blockade, β-stimulation and β-blockade on tissue cAMP and $InsP_3$: (1) Control hearts, perfused with propranolol, prazosin, isoproterenol or phenylephrine for 10 min after an initial stabilization period of

20 min. (2) Hearts perfused with propranolol, prazosin, isoproterenol or phenylephrine for 10 min before being subjected to 20 min global ischaemia. (3) Hearts perfused with propranolol, prazosin, isoproterenol or phenylephrine for 10 min before being subjected to 20 min global ischaemia, followed by 0.5, 1, 2, or 5 min reperfusion. (Drugs present during reperfusion as well). (4) Hearts perfused for a 30 min stabilization period in the absence of drugs, subjected to 20 min global ischaemia and 1 min reperfusion. Prazosin or propranolol present during reperfusion only.

(C) Effect of ischaemia and reperfusion on incorporation of myo-[³H]-inositol in tissue inositolphosphates: (1) Control hearts, perfused for 35 min in the presence of 10 mM LiCl with or without drugs (propranolol, phenylephrine). (2) Hearts were perfused for 15 min with or without drugs in the presence of 10 mM LiCl and then subjected to 20 min global ischaemia. (3) Hearts were perfused as above and reperfused for 1 min. (Drugs present during reperfusion as well).

Statistical analysis

Results are expressed as mean ± S.E.M. P-values were determined by the non-paired Students *t*-test and P values < 0.05 were regarded as statistically significant.

Materials

The Ins(1,4,5)P_3[^{3}H] and cyclic AMP-[^{3}H] assay systems were obtained from Amersham, England. Prazosin and phenylephrine were obtained from Sigma Chemical Company, St Louis, Mo. Routine laboratory chemicals were of Analar grade and supplied by Merck, Darmstadt.

Table 1. Effects of global ischaemia and reperfusion on tissue Ins(1,4,5)P_3 and cAMP levels

	cAMP (pmoles/g wet weight)	Ins(1,4,5)P_3 (pmoles/g wet weight)
Controls		
(20)	354.43 ± 18.48	2940 ± 167
Ischaemia		
5 min (10)	560.33 ± 50.1*	2026 ± 239*
10 min (5)	521.55 ± 19.90*	1889 ± 209*
20 min (20)	625.65 ± 26.10*	1804 ± 106*
25 min (6)	654.00 ± 27.50*	1098 ± 108*
Reperfusion after 20 min ischaemia		
0.5 min (4)	504.39 ± 16.50**	1126 ± 197*
1 min (20)	473.94 ± 16.64**	2207 ± 148***
2 min (6)	523.20 ± 18.24	2330 ± 244
5 min (4)	343.21 ± 22.88	3480 ± 368***

Control hearts were perfused for either 30 or 50 min (80 mmHg; 95% O_2, 5% CO_2). The values obtained at the two time intervals were similar and therefore pooled. Number of hearts indicated in brackets. *p < 0.001 vs controls; **p < 0.005 vs 20 min ischaemia; ***p < 0.05 vs 20 min ischaemia

Results

Effects of global ischaemia and reperfusion on tissue $InsP_3$ and cAMP levels (Table 1)

The effects of increasing periods (5–25 min) of global ischaemia were studied on tissue $InsP_3$ and cAMP levels, whereas the effects of different periods of reperfusion (0.5–5 min) were evaluated after exposure of the heart to 20 min of ischaemia only.

Myocardial ischaemia caused a significant increase in tissue cAMP within 5 min (p < 0.001). Increasing the period of ischaemia to 25 min did not cause a significant further elevation. Reperfusion after exposure of the heart to 20 min ischaemia, caused a significant lowering in tissue cAMP content within 30 sec. Increasing the period of reperfusion caused a progressive lowering in cAMP approaching those of control hearts after 5 min.

Tissue $InsP_3$ levels, on the other hand, showed a significant reduction within 5 min of ischaemia, becoming more pronounced with increasing periods of ischaemia. For example after 25 min of ischaemia, tissue $InsP_3$ levels were reduced by 63% when compared to controls. Reperfusion after 20 min of ischaemia caused a significant increase in tissue $InsP_3$ within 1 min and values normalized within 5 min.

Effects of β-adrenergic stimulation or blockade (Table 2)

For comparison purposes, the values obtained in control hearts, perfused without drugs, as shown in Table 1, are included.

β-blockade by propranolol (10^{-6} M) when added to control perfused hearts had no effect on either tissue cAMP or $InsP_3$ contents, while stimulation with isoproterenol (10^{-9} M) caused the expected significant increase in tissue cAMP, while having no effect on $InsP_3$ levels.

β-blockade before the onset of ischaemia caused a significant lowering of tissue cAMP at the end of 20 min global ischaemia (p < 0.01 when compared to the response in the absence of drugs), while having no effect on tissue $InsP_3$. The presence of β-blockade before the onset of ischaemia also significantly reduced tissue cAMP levels after 1 min of reperfusion (p < 0.02). However, when added at the onset of

Table 2. Effect of β-blockade (propranolol 10^{-6} M) or stimulation (isoproterenol 10^{-9} M) on tissue Ins(1,4,5)P_3 and cAMP levels during 20 min ischaemia and reperfusion.

	cAMP (pmoles/g wet weight)			Ins(1,4,5)P_3 (pmoles/g wet weight)
Control				
No drugs (20)	354.43 ± 18.48			2940 ± 167
Propranolol (10)	343.39 ± 18.28	$p < 0.001$		2423 ± 266
Isoproterenol (11)	566.97 ± 43.08			2960 ± 358
Ischaemia				
No drugs (20)	625.65 ± 26.10	$p < 0.01$		1804 ± 106
Propranolol* (6)	506.58 ± 28.37		NS	1564 ± 135
Isoproterenol* (15)	581.35 ± 33.05			1776 ± 250
Reperfusion (1min)				
No drugs (20)	473.94 ± 16.64	$p < 0.02$		2207 ± 148
Propranolol* (8)	414.30 ± 13.68		$p < 0.002$	2260 ± 240
Propranolol** (6)	572.08 ± 33.70			1778 ± 256
Isoproterenol* (10)	648.68 ± 15.37			2136 ± 220

Control hearts were perfused for 30 min and freeze-clamped. Propranolol or isoproterenol was added during the last 10 min of perfusion. *Propranolol or isoproterenol added during the stabilization phase 10 min before onset of global ischaemia and present during reperfusion. **Propranolol added for 1 min during reperfusion only. NS = not significant

reperfusion only, it was ineffective and cAMP levels were similar to those obtained during ischaemia alone. β-blockade had no effect on the pattern of $InsP_3$ changes characteristic of ischaemia and reperfusion.

β-stimulation with isoproterenol before onset of ischaemia did not cause a further rise in cAMP during ischaemia, while its presence prevented the reduction in cAMP occurring with 1 min reperfusion in the absence of drugs and values were obtained similar to those at the end of 20 min ischaemia. As was observed with β-blockade, stimulation of the β-adrenergic pathway did not affect the $InsP_3$ changes during ischaemia and reperfusion.

Effects of a_1-adrenergic stimulation or blockade (Table 3)

As before, control values for tissue cAMP and $InsP_3$ (Table 1) were included for comparison purposes. Alpha-1-blockade with prazosin (10^{-7} M) had no effect on tissue $InsP_3$ and cAMP values, while stimulation with phenylephrine at a concentration of 3×10^{-5} M, caused a significant elevation in both second messenger levels.

Blocking of the α_1-adrenergic pathway with prazosin before the onset of global ischaemia, caused a significant elevation in tissue cAMP at the end of the ischaemic period, while not affecting the characteristic reduction in tissue $InsP_3$ levels during ischaemia. Prazosin, regardless of the time of administration, also caused a significant increase in tissue cAMP after 1 min of reperfusion ($p < 0.005$), while preventing the rise in $InsP_3$ occurring in the absence of drugs.

Alpha-1-stimulation with phenylephrine caused a significant decline in tissue $InsP_3$ levels both at the end of ischaemia and after 1 min of reperfusion, when compared to their respective controls perfused in the absence of drugs. cAMP remained high under both conditions.

Simultaneous stimulation of α_1- and β-receptors

Addition of phenylephrine and isoproterenol to control perfused hearts, while causing a significant increase in tissue cAMP, prevented the characteristic rise in $InsP_3$ induced by phenylephrine alone. Tissue cAMP and $InsP_3$ at the end of ischaemia and reperfusion were similar to those obtained with phenylephrine alone.

Elimination of the β-effects of phenylephrine (Table 3)

According to the increase in tissue cAMP elicited by phenylephrine (3×10^{-5} M), this α_1-stimulant has significant β-effects at this concentration. Elimination of the β-effects by addition of propranolol (10^{-6} M), caused a significant lowering of tissue cAMP in control hearts, although not to the levels observed in hearts perfused without drugs. Propranolol, as before, had no effect of tissue $InsP_3$ levels. Addition of propranolol significantly lowered tissue cAMP at the end of ischaemia as well as after reperfusion.

Effects of ischaemia, reperfusion and α_1-stimulation on incorporation of myo-[^{3}-H]-inositol into tissue inositolphosphates (Table 4)

In an attempt to find an explanation for the observation that phenylephrine at a concentration of 3×10^{-5} M caused a reduction in tissue $InsP_3$ values after ischaemia and reperfusion, an additional series of experiments were performed in which the phosphoinositide stores were labelled with myo-[^{3}H]-inositol and the rate of incorporation into $InsP_1$, $InsP_2$ and $InsP_3$ determined. Myocardial ischaemia caused a reduction in [^{3}H]-labelled $InsP_2$ and $InsP_3$, concomitant with an increase in [^{3}H]-labelled $InsP_1$ (the total [^{3}H]InsP accumulation was unchanged).

Table 3. Effects of α_1-blockade (prazosin 10^{-7} M) or stimulation (phenylephrine 3×10^{-5} M) on tissue Ins(1,4,5)P_3 and cAMP levels during 20 min ischaemia and reperfusion

	cAMP (pmoles/g wet weight)			$Ins(1,4,5)P_3$ (pmoles/g wet weight)		
Controls						
No drugs (20)	354.43 ± 18.48			2940 ± 167		
Prazosin (10)	418.45 ± 28.45			3166.5 ± 292	p < 0.001	
			p < 0.001			
Phenylephrine (10)	641.93 ± 25.64			4725 ± 360		
			p < 0.001			
Phenylephrine + propranolol (10)	489.30 ± 22.05			4136 ± 344		< 0.005
Phenylephrine + isoproterenol (6)	662.13 ± 23.40			2663 ± 476		
Ischaemia						
No drugs (20)	625.65 ± 26.10			1804 ± 106		
		p < 0.001				
Prazosin (10)	853.99 ± 37.52			2058 ± 243		
			p < 0.001			
Phenylephrine (15)	808.87 ± 19.19			1487 ± 137		< 0.05
		p < 0.001				
Phenylephrine + propranolol (10)	546.18 ± 22.95			1359 ± 159		
Phenylephrine + isoproterenol (6)	732.17 ± 44.55			1775 ± 175		
Reperfusion (1 min)						
No drugs (20)	473.94 ± 16.64			2207 ± 148		
		p < 0.005			p < 0.01	
Prazosin* (10)	599.83 ± 32.4			1618 ± 143		
			p 0.005			
						p < 0.05
Prazosin** (5)	718.55 ± 28.89			1868 ± 183		
Phenylephrine (15)	577.18 ± 17.51			1666 ± 183		
Phenylephrine + propranolol (10)	428.86 ± 28.74			1898 ± 434		
Phenylephrine + Isoproterenol (5)	616.11 ± 78.0			1573 ± 268		

Control hearts were perfused for 30 min and freeze-clamped. Drugs were added during last 10 min of perfusion. Prazosin or phenylephrine were added 10 min before the onset of ischaemia and was present during reperfusion. **Prazosin added for 1 min during reperfusion only. Propanolol and Isoproterenol added in same concentrations as in Table 2.

Reperfusion for 1 min caused a significant increase in [^{3}H]InsP_3 when compared with the values obtained at the end of 20 min ischaemia. The [^{3}H]InsP_1 values at the end of reperfusion was significantly higher than those observed in control or ischaemic hearts, indicating a stimulation of PI breakdown.

Phenylephrine (with propranolol added to block its β-effects) caused a significant increase in [^{3}H]InsP_2 and [^{3}H]InsP_1 of control hearts, whereas in the case of ischaemic and reperfused hearts [^{3}H]InsP_1 was increased. These results indicated significant stimulation of phosphoinositide pathway, despite the lowering of or unchanged InsP_3.

Discussion

The main findings of this study can be summarized as follows: (i) simultaneous α_1- and β-receptor stimulation of control perfused hearts showed attenuation of the α_1-response, while the β-adrenergic response remained unaltered; (ii) myocardial ischaemia causes a significant concomitant increase in tissue cAMP and a reduction in tissue InsP_3 levels, while reperfusion for 1 min elicited the opposite effects; (iii) α_1-blockade increased tissue cAMP both during ischaemia and reperfusion, indicating an inhibitory effect of α_1-stimulation on the β-adrenergic pathway; (iv) β-stimulation or -blockade had no effect on the activity of the phosphatidylinositol pathway, as indicated by tissue InsP_3 levels.

Table 4. [³H]-myo-inositol incorporation into ventricular inositolphosphates: Effects of ischaemia, reperfusion, phenylephrine (3×10^{-5} M) and propranolol (10^{-6} M)

		$InsP_1$		$InsP_2$		$InsP_3$	
Control							
No drugs (5)		10143 ± 927		8939 ± 963		8678 ± 529	
			< 0.001		< 0.01		
Phenlephrine + propranolol (6)		34699 ± 3364		14113 ± 1000		10268 ± 542	
				< 0.05		< 0.005	0.005
					< 0.001		
Ischaemia							
No drugs (4)		13047 ± 1603		5800 ± 878		5707 ± 402	
			< 0.005				
Phenylephrine + propranolol	< 0.02	72654 ± 11008		7844 ± 606		5737 ± 746	< 0.05
Reperfusion							
No drugs (4)		20656 ± 3160		9192 ± 793		9119 ± 1303	
			< 0.005				
Phenylephrine + propranolol (4)		60603 ± 9152		10335 ± 2039		6482 ± 1405	

Incorporation of [³H]-myo-inositol into $InsP_1$, $InsP_2$ and $InsP_3$ expressed as dpm/g wet weight.

Experimental models

The well-characterized model of the globally ischaemic-reperfused rat heart [15] was used for this study. Although the possibility of cross-talk from other cell types is recognized, we were of opinion that results obtained in ventricular tissue will mainly reflect metabolic changes occurring in myocytes, being the predominant cell type. Except for the study using myo-[³H]-inositol incorporation for evaluation of the phosphatidylinositol pathway, perfusion with 10 mM lithium was omitted from the experimental protocol of all other studies. Lithium which inhibits dephosphorylation of inositol monophosphate ($InsP_1$), yields higher values for both the incorporation of [³H]inositol into inositol phospholipids and inositolphosphates as well as for $InsP_3$ [16]. However, it has been shown that lithium inhibits agonist-induced stimulation of cAMP [17], which renders it unsuitable for studies on cross-talk. Apart from the absolute values being lower in the absence of lithium [16], the effects of ischaemia and reperfusion on tissue inositolphosphates as obtained in this study were the same as observed previously [13]. In this study tissue $InsP_3$ levels were used as indicator of the activity of the phosphatidylinositol pathway. Previous studies from our laboratory showed that measurements of both $InsP_3$ mass and incorporation of myo-[³H]-inositol could be used to this effect [6, 13].

Effects of ischaemia and reperfusion

Although the effects of ischaemia and reperfusion on the activity of the α_1- and β-adrenergic signalling pathways have been studied by many researchers, this study (as far as we know) is the first to emphasize that the changes are exactly the opposite *viz.* an immediate and maintained increase in cAMP during ischaemia and an immediate (within 30 sec) reduction upon reperfusion, whereas tissue $InsP_3$ showed the opposite (Table 1).

The exact reasons for the characteristic increase in tissue cAMP during ischaemia have not been fully elucidated, but has been attributed to release of endogenous catecholamines, dual sensitization of the adrenergic system [7] and inactivation of phosphodiesterases [18]. Although the ability of β-adrenergic blockers to reduce the effects of ischaemia on cAMP is contentious, our results with propranolol (Table 2) confirm those of Corr *et al.* [19] who also observed a partial reduction of the ischaemia-induced increase in cAMP by β-blockade in cats. The rapid reduction observed within 30 sec of reperfusion is probably due to washout as well as a reversal of the inhibition of phosphodiesterase activity [18]. The increase in intracellular Ca^{2+}, characteristic of reperfusion [20, 21], has been shown to effectively inhibit adenylyl cyclase activity [22–24] and therefore may also contribute to the lowering of tissue cAMP upon reperfusion.

Measurement of tissue $InsP_3$ levels using Amersham's assay system, showed a progressive reduction with increasing periods of ischaemia, confirming previous observations from our laboratory [6, 13]. Results obtained using incorporation of myo-[³H]-inositol into inositolphosphates were very similar to those reported recently by Anderson *et al.* [25], *viz.* an increase in [³H]-$InsP_1$ and concomitant decreases in [³H]$InsP_2$ and [³H]$InsP_3$ (Table 4). These changes were not caused by noradrenaline release, since α_1-blockade by prazosin [6, 25] or prior reserpination [25] had no effect on the results obtained. The inhibition of the PI pathway during ischaemia has been ascribed to (i) decrease in the phosphoinositide degrading enzymes [26]; (ii) the ATP dependence of the PI kinases [27] and (iii) inhibition of PI kinases by increasing ADP concentrations [28].

Interaction between α_1- and β-adrenergic signalling pathways

The possibility of interaction between the α_1- and β-adrenergic signalling pathways at a post-receptor level was investigated by determining the pattern of tissue cAMP and $InsP_3$ changes during normoxia, global ischaemia and reperfusion and also by the use of appropriate blockers or stimulants.

In contrast to observations made in myocyte cultures [1–3], we could not obtain clear evidence for cross-talk in the perfused rat heart under normal physiological conditions. Beta-stimulation with isoproterenol, with its significant increase in tissue cAMP, had no effect on basal tissue $InsP_3$ values and conversely, α_1-stimulation with phenylephrine (in the presence of propranolol to eliminate its β-effects) did not reduce tissue cAMP levels. However, simultaneous administration of isoproterenol and phenylephrine to control perfused hearts caused an attenuation of the α_1-response to phenylephrine (4136 ± 344 vs 2664 ± 467 nmoles/g wet weight), while not reducing cAMP levels, indicating an inhibitory effect of β-stimulation on the α_1-response to phenylephrine (Table 3).

The latter findings are in agreement with those obtained by Guse and coworkers [1]. Using cultured cardiac myocytes, the inhibition of the α-response by β-stimulation was transient (30 sec) and disappeared after 15 min, while in our experiments attenuation was still present after 10 min of exposure to the stimulants. The finding of Guse *et al.* [1] that the effect of β-stimulation can be mimicked by forskolin suggested that cAMP or a cAMP-activated second step might be the inhibitory agent. Edes and coworkers [29] also reported that isoproterenol mediated a decrease in basal $InsP_3$ levels in guinea pig hearts, due to a reduction in phosphoinositidase C.

The finding that α_1-stimulation alone, or in combination with isoproterenol had no inhibitory effect on the β-response in control perfused hearts, was surprising in view of the convincing evidence to the contrary reported in both myocytes [2, 3, 30], and isolated perfused hearts [31, 32]. For example, adenylyl cyclase activity stimulated by isoprenaline was reduced by noradrenaline, phenylephrine and methoxamine [2]. The observation that α_1-stimulation attenuated the β-response was attributed to either phosphodiesterase activation [3] or inhibition of the cAMP synthetic pathway via coupling to a Gi protein [2]. The β-effects of phenylephrine at the high concentration used in our study (3×10^{-5} M) could have masked any inhibitory effects. However, using lower concentrations of phenylephrine (0.1–1 μM) which significantly increased tissue $InsP_3$ values (results not shown) or including propranolol (Table 3), did not elicit attenuation of the β-response. An explanation for these discrepancies is not readily available and further work is required.

In contrast to the results obtained in control perfused hearts, inhibitory cross-talk by the β-adrenergic pathway on the α_1-response could not be demonstrated in global ischaemia and manipulation of tissue cAMP by either β-blockade or stimulation had no effect on the ischaemia-induced reduction in $Ins(1,4,5)P_3$ (Table 2).

To determine whether the significant reduction in cAMP occurring within 1 min of reperfusion was involved in the concomitant stimulation of the phosphatidylinositol pathway during reperfusion, propranolol or isoproterenol was added to the perfusate. Both interventions were without effect, indicating that also during reperfusion the activity of the β-adrenergic signalling system had no effect on the α_1-adrenergic pathway. Also simultaneous stimulation of both receptor types by phenylephrine and isoproterenol had no effect on tissue $Ins(1,4,5)P_3$ levels during both ischaemia and reperfusion (Table 3).

During both ischaemia and reperfusion, α_1-receptor blockade by prazosin markedly increased tissue cAMP levels (Table 3), suggesting attenuation of the β-response by α_1-receptor stimulation. Efforts to directly demonstrate that α_1-receptor stimulation by phenylephrine inhibited the ischaemia-induced cAMP accumulation yielded two problems: (i) prior exposure to phenylephrine did not further increase tissue $InsP_3$ values in both ischaemia and reperfusion; (ii) phenylephrine at 3×10^{-5} M had marked β-effects which masked the possibility of observing inhibitory cross-talk.

However, evaluation of the effect of phenylephrine on the activity of the phosphatidylinositol pathway by using myo-[^{3}H]-inositol incorporation into tissue inositolphosphates as indicator, showed marked stimulation of $InsP_1$ accumulation during both ischaemia and reperfusion indicating that phenylephrine did stimulate the phosphoinositide pathway. The latter problem was addressed by simultaneous perfusion with phenylephrine and propranolol. Although propranolol reduced the phenylephrine effects on tissue cAMP, propranolol alone caused a larger reduction in cAMP (Table 2). It was therefore impossible to determine whether the reduction in cAMP observed with phenylephrine and propranolol was due to inhibitory cross-talk or due to β-blockade only. However, despite the latter results, the very significant stimulation in cAMP accumulation induced by α_1-blockade (even when administered for 1 min only) suggests attenuation of the β-response by α_1-adrenergic stimulation in both ischaemia and reperfusion, as was reported by several others in the normal myocardium [2, 30–32].

Attenuation of the β-response during reperfusion may possibly be due to intracellular Ca^{2+} overloading: α_1-stimulation as well as upregulation of the $InsP_3$ receptor population [33] may contribute to the characteristic rise in intracellular Ca^{2+} during early reperfusion, which in turn, can effectively inhibit adenylyl cyclase activity [22–24]. The inhibitory effects of α_1-stimulation during ischaemia are more difficult to explain, in view of the lowered $InsP_3$ levels, which were not affected by prazosin.

Finally, it has been suggested that the physiological significance of inhibitory cross-talk between the cAMP and inositolphosphate systems is a cardioprotective effect reciprocally lowering the positive inotropic reaction of α_1- and β-stimulation and thereby preventing overstimulation of the myofibrillar ATPase [1]. The results obtained in the present study on inhibitory cross-talk during ischaemia and reperfusion, suggest that it may be a protective mechanism aimed at reducing the combined effects of α_1- and β-stimulation.

References

1. Guse AH, Berg I, Gercken G: Inhibition of α_1-adrenoceptor mediated inositol phosphate accumulation in cultured cardiac myocytes by cyclic AMP-generating compounds. J Mol Cell Cardiol 23: 1375–1382, 1991
2. Barret S, Honbo N, Karliner JS: Alpha-l-adrenoceptor-mediated inhibition of cellular cAMP accumulation in neonatal rat ventricular myocytes. Arch Pharmacol 347: 384–393, 1993
3. Brunton LL, Buxton ILO: α-Adrenergic receptors on rat ventricular myocytes: characteristics and linkage to cAMP metabolism. Am J Physiol 251: H307–H313, 1986
4. Schömig A, Dart AM, Dietz R, Mayer E, Kübler W: Release of endogenous catecholamines in the ischemic myocardium of the rat. Circ Res 55: 689–701, 1984
5. Schömig H: Catecholamines in myocardial ischemia, systemic and cardiac release. Circulation 82 (suppl. II): II-13–II-22, 1990
6. Mouton R, Genade S, Boschmans SA, Perkins MF, Lochner A: The role of α_1-adrenergic stimulation in inositol phosphate metabolism during post-ischaemic reperfusion. Life Sciences 51: 2033–2040, 1992
7. Strasser RH, Krimmer J, Braun-Dullaeus R, Marquetant R, Kübler W: Dual sensitization of the adrenergic system in early myocardial ischemia: Independent regulation of the β-adrenergic receptors and the adenylyl cyclase. J Mol Cell Cardiol 22: 1405–1423, 1990
8. Niroomand F, Weinbrenner C, Weis A, Bangert M, Schwencke C, Marquetant R, Beyer T., Strasser RH, Kübler W, Rauch B: Impaired function of inhibitory G proteins during acute myocardial ischemia of canine hearts and its reversal during reperfusion and a second period of ischemia. Circ Res 76: 861–870, 1995
9. Butterfield MC, Chess-Williams R: Enhanced alpha-adrenoceptor responsiveness and receptor number during global ischemia in the Langendorff perfused rat heart. Br J Pharmacol 100: 641–645, 1990
10. Heathers GP, Evers AS, Corr PB: Enhanced inositol trisphosphate response to alpha-1-adrenergic stimulation in cardiac myocytes exposed to hypoxia. J Clin Invest 83: 1409–1413, 1988
11. Sharma AD, Saffitz J E, Lee BI, Sobel BE: Alpha-adrenergic mediated accumulation of calcium in reperfused myocardium. J Clin Invest 72: 802–818, 1983
12. Beresewicz A: Anti-ischemic and membrane stabilizing activity of calmodulin inhibitors. Bas Res Cardiol 84: 631–645, 1989
13. Mouton R, Huisamen B, Lochner A: Effects of ischaemia and reperfusion on sarcolemmal inositol phospholipid and cytosolic inositol phoshate metabolism in the isolated perfused rat heart. Mol Cell Biochem 105: 127–135, 1991
14. Lamprecht W, Trautschold I: Determination with hexokinase and glucose-6-phosphate dehydrogenase. In: H.U. Bergmeyer (Ed.) Methods of enzymatic analysis, Academic Press, New York, 1974, p 2105
15. Edoute Y, Van der Merwe E, Sanan D, Kotzé JCN, Steinmann C, Lochner A: Normothermic ischemic cardiac arrest of the isolated rat heart. Effects of time and reperfusion on myocardial ultrastructure, mitochondrial oxidative function and mechanical recovery. Circ Res 53: 663–678, 1983
16. Bester EP, Mouton R, Malan JF, Lochner A: The effect of lithium on the heart. Cardiovasc. J of S A 4: 210–214, 1993
17. Belmaker RM: Receptors, adenylate cyclase, depression and lithium. Biol Psychiatry 16: 333–350, 1981
18. Podzuweit T, Nennstiel P, Bader R, Muller A: Ischaemia causes inhibition of cyclic nucleotide phosphodiesterases. J Mol Cell Cardiol 26, CXVI, 1994.
19. Corr PB, Witkowski FX, Sobel BE: Mechanisms contributing to malignant dysrhythmias induced by ischemia in the cat. J Clin Invest 61: 109–119, 1978
20. Bourdillon PDV, Poole-Wilson PA: The effects of ischaemia and reperfusion on calcium exchange and mechanical function in isolated rabbit myocardium. Cardiovasc Res 15: 121–130, 1981
21. Tani M, Neely JR: Role of intracellular Na^+ and Ca^{2+} overload and depressed recovery of ventricular function of reperfused ischemic rat hearts: possible involvement of H^+/Na^+ and Ca^{2+} exchange. Circ Res 65: 1045–1056, 1989
22. Boyajian CL, Garritsen A, Cooper DM: Bradykinin stimulates Ca^{2+} mobilization in NCB-20 cells leading to direct inhibition of adenylyl cyclase. J Biol Chem 266: 4995–5003, 1991
23. De Bernardi MA, Seki T, Brooker G: Inhibition of cAMP accumulation by intracellular calcium mobilization in C6-2B cells stably transfected with substance K receptor cDNA. Proc Nat Acad Sci USA 88: 9257–9261, 1991
24. Cooper DMF, Mons N, Karpen JW: Adenylyl cyclases and the interaction between calcium and cAMP signalling. Nature 374: 421–424, 1995
25. Anderson KE, Dart AM, Woodkock EA: Inositol phosphate release and metabolism during myocardial ischemia and reperfusion in rat heart. Circ Res 76: 261–268, 1995
26. Schwertz DW, Halverson J, Isaacson T, Feinberg H, Palmer JW: Alterations in phospholipid metabolism in the globally ischemic rat heart: Emphasis on phosphoinositide specific phospholipase C activity. J Mol Cell Cardiol 19: 685–693, 1987
27. Quist E, Satumtira N, Powell P: Regulation of polyphosphoinositide synthesis in cardiac muscle. Arch Biochem Biophys 271: 21–32, 1989
28. Kasinathan C, Xu ZC, Kirchberger MA: Polyphosphoinositide formation in the isolated cardiac plasma membranes. Lipids 24: 818–823, 1989
29. Edes J, Solaro RJ, Kranias EG: Changes in phosphoinositide turnover in isolated guinea pig hearts stimulated with isoproterenol. Circ Res 65: 989–996, 1989
30. Hazeki O, Ui M: Modification by islet-activating protein of receptor-mediated regulation of cAMP accumulation in isolated rat heart cells. J Biol Chem 256: 2856–2863, 1981
31. Watanabe AM, Hathaway DR, Besch HR, Farmer BB, Harris RA: α-Adrenergic reduction of cyclic adenosine monophosphate concentrations in rat myocardium. Circ Res 40: 596–602, 1977
32. Keely SL, Corbin JD, Lincoln T: Alpha-adrenergic involvement in heart metabolism: effects on adenosine cyclic 3'5'-monophosphate dependent protein kinase, guanosine cyclic 3'5'-monophosphate and glucose transport. Mol Pharmacol 13: 965–975, 1977
33. Huisamen B, Lochner A: Regulation of $InsP_3$ receptor population in rat atria and ventricles. Mol Cell Biochem 140: 23–30, 1994

Molecular and Cellular Biochemistry **160/161**: 137–141, 1996.

Reversible and irreversible damage in reoxygenated 'ischemic' ventricular myocytes in culture

Arié Pinson and Régine Tirosh

Laboratory for Myocardial Research, The Hebrew University – Hadassah Medical School, Jerusalem, Israel

Abstract

The LDH release pattern from cardiomyocytes under 'ischemia-like' conditions shows two phases. In the initial slow phase, reoxygenation immediately stops further enzyme release. Accelerated LDH release, which occurs concomitantly with lysosomal enzyme release, characterizes the second phase of 'ischemia.' Reoxygenation at this stage does not put a stop to further enzyme release.

Reoxygenation during the first phase of 'ischemia' rapidly restored ATP level, while in the second phase, ATP levels remained low even after 6 h of reoxygenation.

This study as well as previous data seem to suggest that irreversible cellular damage leading to cell death, occurs by synergistic action of many effectors, each of which does not necessarily cause irreversible damage. (Mol Cell Biochem **160/161**: 137–141, 1996)

Key words: cardiomyocytes, ischemia, heart, reoxygenation, ATP, LDH, hexosaminidase

Introduction

The functioning of the heart, which is essentially an oxidative organ, predominantly depends on oxidative phosphorylation [1, 2]. During oxygen restriction, high-energy phosphate (HEP) levels are maintained through increased carbohydrate metabolism via anaerobic glycolysis [3, 4]. Both ATP production and levels subsequently rapidly decline with a concomitant increase in the levels of degradation products, such as ADP, AMP and adenosine, which are degraded further to the purine bases [5–7]. Indeed, most of the ATP is converted to purine bases within 90 min of the onset of ischaemia [7].

Cultured heart cells have been used by several researchers to study oxygen deprivation at the cellular level [8–10]. However, enzyme release could only be detected after the cells had been kept under anoxic conditions for very long periods of up to 16 h in their studies. In addition, various agents, such as dinitrophenol, cyanide, deoxyglucose and other metabolic inhibitors, were employed in attempts to impair the respiratory chain or glycolysis [10, 11], thereby introducing variables that are not present under 'true' anoxic conditions.

We proposed a different approach for simulating anoxic injury in cultured heart cells that does not require artificial agents [12]. We showed that reducing the volume of glucose-depleted, anoxic extracellular fluid so that the height of the liquid above the cells approximates to the diameter of a single cultured cell (a total volume of 0.2 ml per 35 mm-diameter Petri dish) corresponds, at the cellular level, to partial or total restriction of the renewal of intercellular fluid. In the same system, extracellular volumes ranging from 1.0–2.0 ml gave rise to anoxic conditions *per se*. We also devised a suitable incubation unit and a procedure for achieving proper preequilibration of the medium with N_2/CO_2. Under these conditions, cellular damage, as reflected by enzyme release, was inversely proportional to the volume of the medium, and presumably attributable to the effects of the catabolic metabolite (i.e. lactic acid) remaining in the vicinity of the cells [12]. Moreover, we have shown that the onset of cellular damage, as reflected by electron microscopic studies, [13], enzyme release [12], HEP depletion [14] occurred in cultured heart cells at similar rates to those observed in other experimental *in vivo* and perfused heart systems [5, 15]. It certainly resembled such situations more closely than other previously described studies with cultured cells.

Reoxygenation of ischemic heart leads to free radical formation, which enhances cardiac injury [1]. This also occurs in reoxygenated cultured cardiomyocytes [16, 17], or in the presence of a free radical-generating system [18, 19].

This study was designed to provide more information about the transition from reversible to irreversible cellular damage in cultured ventricular myocytes.

Materials and methods

Cell cultures

Heart cells cultured from 1-day-old Wistar rats were prepared as previously described [20, 21], with seeding densities of 1.8–2.0 × 10^6 cells per Petri dishes of 35 mm in diameter (Falcon 3001). Under these conditions the starting percentage of 80–90% muscle cells is maintained for at least 6 days in culture.

Oxygen deprivation and volume restriction

These methods have been described previously [12, 22]. Cells cultured for 5 days were used for these experiments. The growth medium was discarded and the cells immediately washed with Ham-F10 culture medium. Anoxia was achieved by maintaining the cells in Petri dishes with 2.0 ml medium, preequilibrated by bubbling through a mixture of 95% N_2 and 5% CO_2 for at least 30 min. For anoxia combined with volume restriction, the same conditions were used except that the volume of extracellular medium was 0.4 ml. In this paper, the term anoxia refers to cultures incubated with 2 ml oxygen-deprived medium, and 'simulated mild ischemia' to plates with 0.4 ml medium.

Reoxygenation

Reoxygenation following anoxia or 'ischemia' for various periods, was carried out together with reperfusion. This was done by adjusting the volume of the extracellular medium to 2 ml with Ham F10 medium w/o sera. Enzyme release during reoxygenation was considered to be a marker of continuing cellular injury.

Enzyme determinations

Lactate dehydrogenase (LDH) was determined according to the method of Wroblewski and La Due [23] and hexosaminidase, as described by Galjaar [25]. ATP was determined by the chemiluminescence method of Holmsen *et al.* [25].

Results

During simulated ischemia (Fig. 1) LDH follows a two-phase: Moderate release during the first phase of oxygen deprivation, and accelerated LDH release during the second phase. Reoxygenation during the first phase of anoxic injury, as indicated by cellular enzyme release, immediately came to a halt, however, during the second phase, reoxygenation no longer put a stop to further enzyme release. This biphasic pattern is even clearer with hexosaminidase release (Fig. 2). Indeed, there is hardly any release of lysosomal enzyme during the initial phase of oxygen deprivation, but its release is sharply accelerated during the second phase. During this phase, hexosaminidase continues to be exported in spite of restoration of the oxygen supply. The cellular ATP content during the critical phases is shown in Fig. 3. During the initial phase of 'ischemia,' ATP is less than 20% of the controls, but returns to 60% and 80% of the controls after the first and second hours of reoxygenation, respectively. During longer periods, cellular ATP shows some fluctuations. During the second phase, ATP content falls to approximately 3–4% of the controls, barely reaching 30% after 6 h of reoxygenation.

Discussion

Ischemia is a circulatory event involving partial (mild ischemia) or total occlusion of a coronary artery. Myocardial ischemia is the result of an imbalance between oxygen supply and demand.

Since myocardial cell damage is the major manifestation of ischemic injury, we attempted to simulate, in culture, the conditions existing in the immediate environment of the cardiac cell during such circulatory events. The cardiac cell rapidly becomes anoxic, and mobility of the intercellular fluid is greatly impaired with the result that substrates fail to reach the cell, and released metabolites, such as lactate, are not removed from its vicinity. In order to approach these conditions, in previous studies, the smallest possible volume of the extracellular fluid of 0.2 ml was used with cellular enzyme release serving as a marker of cell damage [12, 22].

In this research, we sought to further confirm that the injury to cardiac cells caused by reducing the volume of extracellular medium was similar to that occurring in other experimental model systems with respect to reversible and irreversible damage during the reoxygenation period. To avoid 'drastic' injury, a volume of 0.4 ml extracellular fluid per 35 mm-diameter Petri dish was used and termed 'simulated mild ischemia.'

As previously described [13, 26], a classical picture of ischemic injury was found in 'simulated mild ischemia.' The sarcolemmal structure is disorganized, displaying swelling

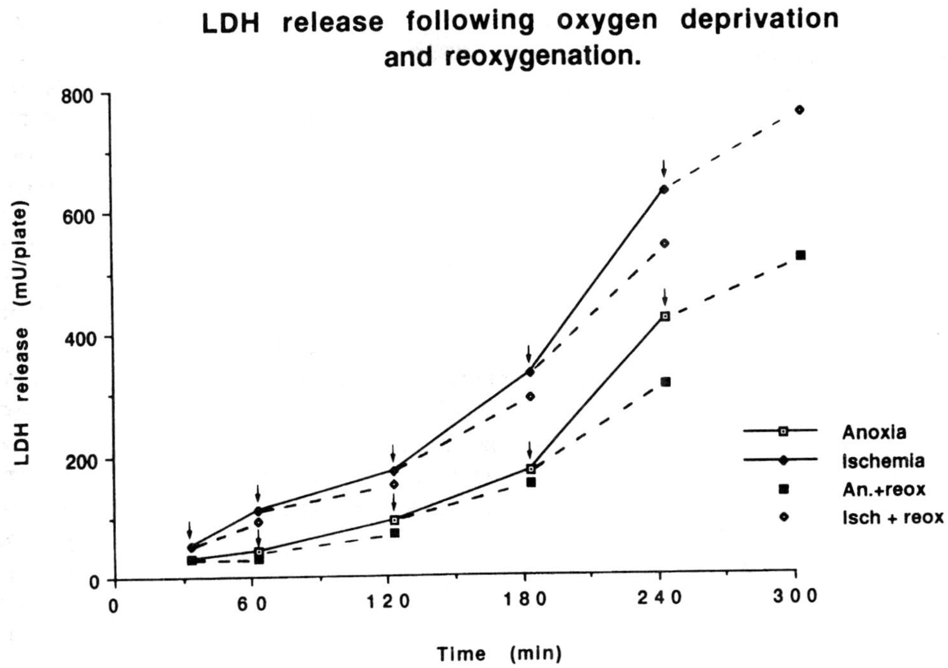

Fig. 1. LDH release during a 60 min period of reoxygenation (broken lines) following the oxygen deprivation period (solid lines). The arrows indicate the point at which reoxygenation was initiated.

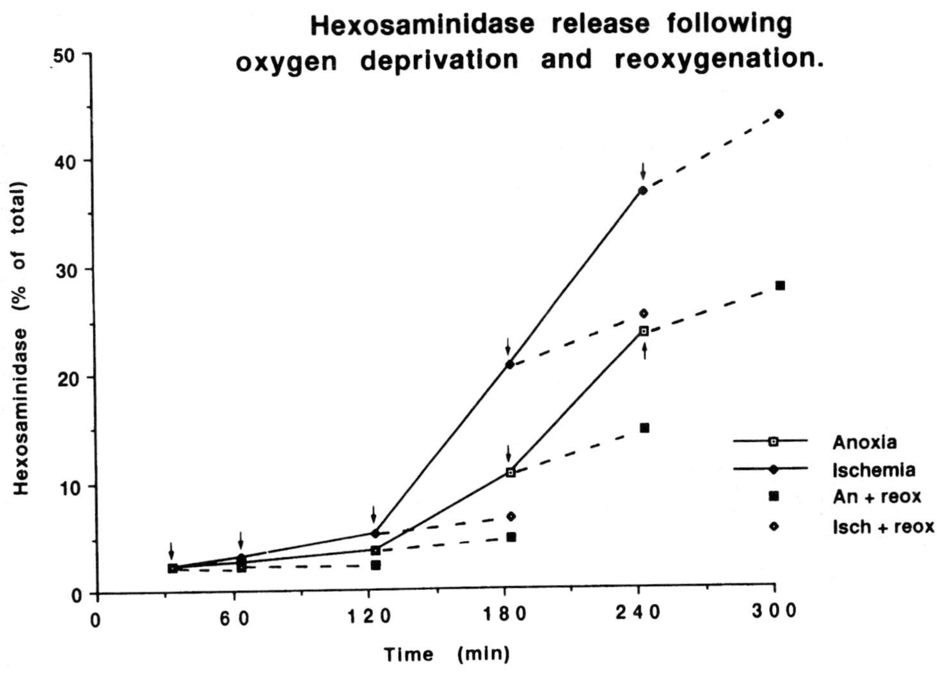

Fig. 2. Hexosaminidase release during a 60 min period of reoxygenation (broken lines) following the oxygen deprivation period (solid lines). The arrows indicate the point at which reoxygenation was initiated.

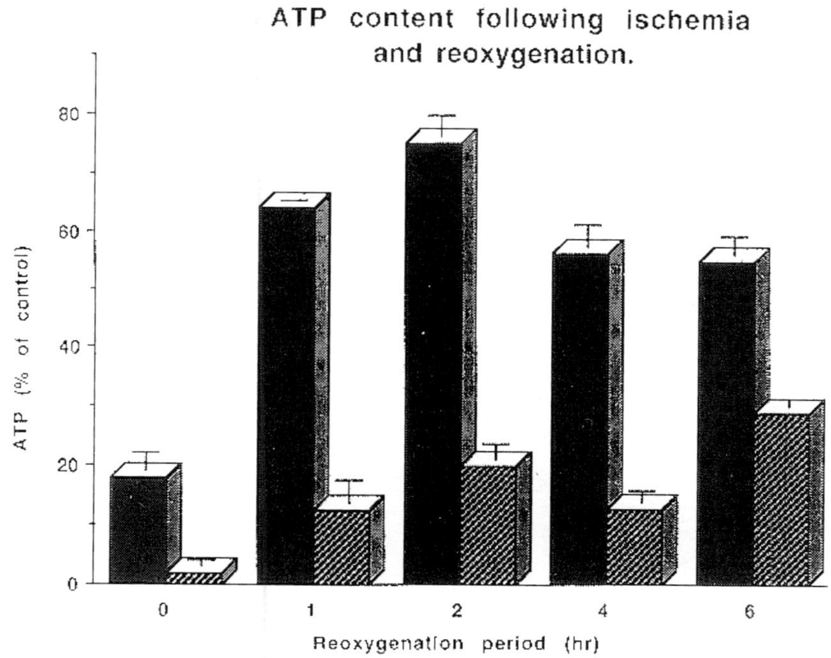

Fig. 3. Cellular ATP content (per cent of controls) during various reoxygenation periods (1–6 h) following ■ 75 and ▨ 120 min of simulated ischemia.

and giant mitochondria with disoriented cristae [13]. Membrane blebbing in freeze-fracture studies [26] and changes in plasma membrane topology [27, 28] are among the other manifestations supporting disorganization of the sarcolemmal bilayer, which presumably underlies enzyme release.

In the present paper as well as in an earlier study [14], we showed that under similar anoxic conditions, HEP content fell dramatically although a certain amount of ATP and even some CrP [14] remained, whereas during 'simulated-mild ischemia,' ATP levels were reduced even further and CrP was virtually exhausted. Lysosomal enzyme release only started at this stage of anoxic or 'ischemic' injury (Fig. 2) . However, due to the larger volume of extracellular fluid in anoxia, the 'washout' of metabolites, such as lactate, continues, thereby presumably maintaining normal cellular pH for longer periods. This is not so in 'ischemia-like' conditions since the metabolites remain in the vicinity of the cell.

The 'ischemia-like' conditions are clearly more damaging to the cells and study of various events may indicate the stage at which the cells enter the irreversible phase. Early enzyme release is not necessarily an indication of irreversible cellular damage since the release stops immediately upon reoxygenation, and the cells regain their beating function. ATP also returns to near normal values. Thus, early enzyme release is possibly an outcome of sarcolemmal reorganization during oxygen deprivation [27, 28] when PE, which alone does not form bilayers, is translocated to the outer leaflet and may form some monolayer domains there. PE in such domains may rotate freely and can also readily interact with proteins. Since PE translocation is reversible in the early stages of 'ischemia,' this might explain the early protein release during the reversible stage of ischemia.

On the other hand, exhaustion of ATP alone, as occurs with metabolic inhibitors, did not significantly increase enzyme release, as compared to 'simulated ischemia' [29]. The cells also recovered quickly when the inhibitor was washed out.

Lysosomal enzymes, however, are released at a later stage of 'simulated ischemia,' during which reoxygenation no longer puts a stop to further cellular damage. While lysosomal enzyme release marks the onset of irreversible cell damage, it does not necessarily mean that activity of these enzymes is the primary or sole cause of cell death. Indeed, based on previous findings, we put forward the hypothesis that cell killing is a synergistic series of events [30]. During ischemia, these include the release of lysosomal enzymes, combined with lack of HEP, hyperosmotic shock, low pH and ion imbalance, which act concomitantly to cause irreversible damage. At this stage, the production of active oxygen species during reoxygenation may further increase cellular damage.

Acknowledgements

This research was supported by grants from Mr. and Mrs. Vidal-Madjar (Paris); Mr. Jean Ariet (Paris) in memory of his wife Mrs. Rebecca Ariet; and Mrs. Flora Dabbah (Geneva) in memory of her husband Mr. Yossef Dabbah.

References

1. Hearse DJ: Reperfusion of the ischemic myocardium. J Mol Cell Cardiol 9: 605–616, 1984
2. Doorey AJ, Barry WH: The effects of inhibition of oxidative phosphorylation and glycolysis in cultured chick heart cells. Circ Res 53: 192–201, 1983
3. Taylor SH: Insulin and heart failure. Br Heart J 33: 329–333, 1971
4. Haider W, Eckersberger F, Wolner E: Preventive insulin administration for myocardial protection in cardiac surgery. Anesthiology 60: 422–429, 1984
5. Jennings RB, Reimer KA, Hill ML, Mayer SA: Total myocardial ischemia *in vitro*: I. Comparison of high energy phosphate production, utilization and repletion and adenine nucleotide catabolism in total ischemia *in vitro* vs. severe ischemia *in vivo*. Circ Res 49: 992–899, 1981
6. Reibel DK, Rovetto MJ: Myocardial adenosine salvage rates and restoration of ATP content following ischemia. Am J Physiol. 237: H247–H252, 1979
7. Schoutsen B, De Jong JW, Harmsen E, De Tombe PP, Achterberg PW: Myocardial xanthine oxidase/dehydrogenase. Biochim Biophys Acta 762: 519–524, 1983
8. Van der Laarse A, Hollaar L, Van der Valk LJM, Witteveen SAGJ: Enzyme release from and enzyme depletion in rat heart cell cultures during anoxia. J Mol Med 3: 123–131, 1978
9. Higgins TJC, Allsopp D, Bailey PJ: The effect of extracellular calcium concentration and Ca-antagonist drugs on enzyme release and lactate production by anoxic heart cell cultures. J Mol Cell Cardiol 12: 909–927, 1980
10. Wenzel DG, Acosta D: Labilization of lysosomes and mitochondria *in situ* by hypoxia and hypoxia-related factors. Res Commun Chem Pathol Pharmacol 12: 173–176, 1975
11. Higgins TJC, Allsopp D, Bailey PJ, D'Souza EDA: The relationship between glycolysis, fatty acid metabolism and membrane integrity in neonatal myocytes. J Mol Cell Cardiol 13: 599–615, 1981
12. Vemuri R, Yagev S, Heller M, Pinson A: Studies on oxygen and volume restrictions in cultured cardiac cells I. A model for ischemia and anoxia with a new approach. *In Vitro* 21: 521–525, 1985
13. Ne'eman Z, Pinson A: Oxygen and extracellular fluid restriction in cultured heart cells: Electron microscopy studies. Cardiovas Res 24: 555–559, 1990
14. Vemuri R, De Jong JW, Hegge JAJ, Huizer T, Heller M, Pinson A: Studies on oxygen and extracellular fluid restriction in cultured heart cells: High energy phosphate metabolism. Cardiovasc Res 23: 254–261, 1989
15. Jennings RB, Hawkins HK, Lowe JE, Hill ML, Klotman S, Reimer KA: Relation between high energy phosphate and lethal injury in myocardial ischemia in the dog. Am J Pathol 92: 187–202, 1978
16. Qualife RA, Kohomoto O, Barry WH: Mechanism of reoxygenation injury in cultured ventricular myocytes. Circulation 83: 566–577, 1991
17. Bengt EK, Hallberg C, Sjorgen KG, Hjalmarson A: Reoxygenation-induced cell damage of isolated neonatal rat ventricular myocytes can be reduced by chair-breaking antioxidants. Free Rad Biol Med 16: 117–121, 1994
18. Samuni A, Winkelsberg D, Pinson A, Mitchell JB, Russo A: Nitroxide stable radicals protect beating cardiomyocytes against oxidative damage. J Clin Invest 87: 1526–1530, 1991
19. Mohsen M, Pinson A, Samuni A: Do nitroxide radicals protect beating cardiomyocytes from oxidative damage. Mol Cell Biochem 145: 103–110, 1995
20. Yagev S, Heller M, Pinson A: Changes in cytoplasmic and lysosomal enzyme activities in cultured heart cells: The relationship to cell differentiation and cell population in culture. *In Vitro* 20: 893–898, 1984
21. Pinson A, Padieu P, Harary I: Techniques for culturing heart cells. In: A Pinson (ed.) The Heart Cell in Culture. CRC Press, Boca Raton, 1987: Vol I, pp 7–22
22. Pinson A: Heart cells in culture as a model for studying anoxia and ischemia in the myocardium. In: A Pinson (ed.) The Heart Cell in Culture. CRC Press, Boca Raton, 1987, Vol III, pp 35–48
23. Wroblewski F, La Due JS: Lactic dehydrogenase dehydrogenase activity in blood. Proc Soc Exp Biol Med 90: 210–213, 1955
24. Galjaard H: Genetic Metabolic Diseases: Early Diagnosis and Prenatal Analysis. Elsevier/North Holland, Amsterdam, 1980, p 327
25. Holmsen H, Storm E, Day HJ: Determination of ATP and ADP in blood platelets. A modification of the firefly luciferase assay for plasma. Anal Biochem 46: 489–509, 1972
26. Musters RJP, Post JA, Verklej AJ: The isolated neonatal rat cardiomyocyte used in an *in vitro* model for 'ischemia.' I. A Morphological Study. Biochim Biophys Acta 1019: 270–277, 1991
27. Vemuri R, Mersel M, Heller M, Pinson A: Studies on oxygen and volume restriction in cultured cardiac cells: Possible rearrangement of sarcolemmal lipid moieties during anoxia and ischemia-like states. Mol Cell Biochem 79: 39–46, 1988
28. Musters RJP, Otten E, Biegelman E, Bijvelt J, Keijzer JJH, Post JH, Op den Kamp JAF, Verkleij AJ: Loss of asymmetric distribution of sarcolemmal phosphatidylethanolamine during simulated ischemia in the isolated neonatal rat cardiomyocyte. Circ Res 73: 514–523, 1993
29. Pinson A, Zilberman Y, Tirosh R, Tremblover V, Shohami E: High oligomycin concentrations augment 6 keto-PGF1α production in ventricular cardiomyocytes. Biochem Biophys Acta 1211: 283–288, 1994
30. Ginsburg I, Misgav R, Pinson A, Varuni J, Ward PA, Kohen R: Synergism among oxidants, proteinases, phospholipase, microbial hemolysins, cationic proteins and cytokines: A possible major cause of cell and tissue destruction in inflammation (a working hypothesis). Inflammation 16: 519–538, 1992

Molecular and Cellular Biochemistry **160/161:** 143–151, 1996.

Effects of hypoxia, simulated ischemia and reoxygenation on the contractile function of human atrial trabeculae

Andreas Lammerich, Jürgen Bohm,[1] Ingolf Schimke,[2] Kay-Dietrich Wagner, Eberhard Storch and Joachim Günther

Institute of Physiology, Humboldt-University, Tücholsky Str. 2, D-10117; [1]Clinic of Cardiac Surgery, Humboldt-University, Schumannstr. 20–21, D-10098; [2]Clinic of Internal Medicine I, Humboldt-University, Schumannstr. 20–21, D-10098 Berlin, Germany

Abstract

Hypoxia, ischemia and reoxygenation cause contractile dysfunction which will be characterized by the time course of isometric contraction of human atrial trabeculae. Post-rest potentiation (PRP) and postextrastimulatory potentiation (PEP) were elicited to obtain indirect information about the role of the sarcoplasmic reticulum (SR) in excitation-contraction coupling. As lipid peroxidation could cause SR dysfunction, thiobarbituric acid reactive substances (TBARS) were measured. After 30 min of hypoxia (H) or simulated ischemia (H combined with acidosis-SI), contractile force decreased to 15% and 6%, respectively, of control ($p \leq 0.05$), whereas the normalized rate of both contraction and relaxation increased. In group H, rapid reoxygenation produced a recovery of contractile force to about 60%. After post-hypoxic reoxygenation the TBARS concentration was increased. In group Sl, rapid reoxygenation and a rather gradual correction of acidosis produced complete recovery of contractile force. PRP and PEP were maintained during H and Sl. Particularly post-ischemic reoxygenation caused a marked depression of PRP and partly of PEP. Thus, alteration of SR Ca^{2+} handling occurs predominantly during reoxygenation rather than during H or Sl, probably associated with the damaging effect of increased oxygen radicals. The depression of potentiation occured along with delayed relaxation, temporary increased resting force, mechanical alternans, and spontaneous activity which are further characteristics for SR dysfunction. Thus, for a possibly beneficial effect of low pH during Sl combined with its gradual correction during reoxygenation on the recovery of contractile function, developed force should not be the only index. (Mol Cell Biochem **160/161:** 143–151, 1996)

Key words: hypoxia, simulated ischemia, reoxygenation, contractile function, human atrial trabeculae

Introduction

Effects of ischemia and reperfusion on the contractile function of the human heart ventricle are of great clinical interest and have been studied in various experimental animal models [1–5]. The ischemic depression of right atrial contractility plays an important hemodynamic role, as well [6, 7]. Contractile dysfunction induced by ischemia/reperfusion is associated with disturbances in the cellular Ca^{2+} homeostasis [1, 3]. Thus, transsarcolemmal Ca^{2+} movement, the function of the sarcoplasmic reticulum (SR), as well as the myofilament responsiveness to Ca^{2+} could be affected [5, 8]. As these mechanisms differ between species, investigations on human myocardium are also necessary [5, 9–11]. In the study isometrically contracting human atrial trabeculae were used as a model.

In isolated trabeculae, hypoxia, as a key component of ischemia, is mimicked by the gradual replacement of O_2 by N_2. Particular attributes of ischemia are extracellular acidosis and enhanced extracellular K^+ concentration. In addition, acidosis has been shown to modify the recovery of myocardial function during reperfusion [12, 13]. Therefore, we combined hypoxia with acidosis to address the implications of acidosis and to simulate ischemia more closely by replac-

Address for offprints: A. Lammerich, Institute of Physiology, Humboldt-University, Tücholsky Str. 2, D-10117 Berlin, FRG ing O_2 by a N_2/CO_2 mixture at high extracellular K^+ concentration [11].

Primarily, we intended to evaluate the contractile function of the human atrial trabeculae during hypoxia, simulated ischemia as well as in the respective recovery during reoxygenation. The comprehensive characterization of contractile dysfunction includes different parameters for the time course of rhythmically developed force in order to address pathophysiological mechanisms being involved. Furthermore, we tested the ability of the muscle to respond to perturbations in stimulation rhythm with potentiation of force. This could provide additional insight into the Ca^{2+} homeostasis, especially into those steps of the excitation-contraction coupling which are related to the role of the SR [8, 14, 15]. Developing intracellular acidosis, as during hypoxia and early ischemia, normally causes a rise in Na_i due to activation of Na^+-H^+ exchange [4] and a secondary rise in Ca_i via Na^+-Ca^{2+} exchange [1,11], which is probably the major route of Ca^{2+} overload [3–5]. There is a close correlation between cellular gain in Na^+ and Ca^{2+} and myocardial injury during ischemia/reoxygenation [1 ,3] related to excessive myofilament activation and impaired mitochondrial ATP synthesis as well as activation of proteases and lipases, which could damage cellular structures [1, 3, 5].

In contrast to hypoxia and early ischemia, our experimental variant in human myocardium resembles a severe ischemia. The very low extracellular pH (pH_e) may depress Na^+-H^+ exchange and Na^+-Ca^{2+} exchange, thereby probably reducing the accumulation of Na_i and Ca_i respectively [1, 5, 9, 12, 13]. Therefore, we assumed that very low pH_e could attenuate irreversible injury of contractile function during the simulated severe ischemia and thereby exert a beneficial effect on subsequent reoxygenation as reported by others [13, 16].

Reperfusion associated with reoxygenation and restoration of normal pH does not necessarily restore contractile function. Instead, rapid correction of pH_e may accelerate Ca^{2+} overload and limit the recovery of contractile function [17], whereas acidemic reperfusion has been shown to ameliorate postischemic contractile dysfunction by preventing a rapid increase in Na_i and Ca_i [3, 12]. Hence, in the proposed model of reoxygenation injury, pO_2 was more rapidly increased than pHe. Thus, the recovery of oxidative metabolism would precede Ca^{2+} overload and promote sufficient ATP supply for ion transports and contraction. In this way, the simulated ischemia itself and that mode of reoxygenation could improve the recovery of contractile function.

In recent studies, the beneficial effect of low pH_e either during severe ischemia or reperfusion / reoxygenation has been addressed [3, 9, 12, 16]. In the study, these approaches are combined in order to examine, how Na^+-H^+ exchange in cooperation with Na^+-Ca^{2+} exchange may modulate the contractile function during simulated ischemia and during post-ischemic recovery in comparison to hypoxia and post-hypoxic recovery.

As the SR plays a critical and important role in the excitation-contraction coupling [8], its function should be characterized. Ca^{2+} handling by the SR is susceptible to hypoxia, ischemia and reperfusion damage, among others, to that form of damage caused by reactive O_2 radicals [18]. This type of injury is related to lipid peroxidation, which we characterize by measurement of thiobarbituric acid reactive substances (TBARS) [19]. Concerning SR function after ischemia/reoxygenation, SR Ca^{2+} uptake was found to be reduced in human myocardium, as well [20, 21]. Therefore, the physiological role of the SR will be addressed under various aspects: Developed force reflecting the SR Ca^{2+} content, relaxation rate and the maintenance of stable diastolic force reflecting an adequate SR Ca^{2+} pump activity [8], post-extrasystolic potentiation of force indicating the ability of SR to accumulate additional Ca^{2+} that has entered the cell during the action potential of the preceding extra contraction (Ref. see [14]), post-rest potentiation of force determined by the rest-dependent SR Ca^{2+} content, thereby indicating the ability of the SR to store Ca^{2+} [8], the occurrence of spontaneous contractile activity and oscillations in the amplitude of the rhythmically developed force (mechanical altemans) reflecting SR dysfunction [10, 22].

The results demonstrate that the contractile function of the human cardiac preparations responds to hypoxia, simulated ischemia and reoxygenation in a manner comparable to the main characteristics of true ischemia/reperfusion. During the simulated ischemia, acidosis, being combined with hypoxia, has a small additive depressing effect on contractile function, whereas the improved recovery of developed force during rapid reoxygenation but delayed correction of pH_e is strongly influenced by the removal of H^+. The concomitant occurrence of characteristic pathophysiological signs for SR dysfunction despite complete recovery of developed force suggests, however, that peak force should not be used as the only index of contractile function.

Materials and methods

Cardiac preparations

Specimens of right atrial appendages were removed from patients during the course of cannulation for open-heart surgery. Ethical approval was obtained from the hospital ethical committee. Dissection of trabeculae and measurement of isometric force were performed in an oxygenated solution of the following composition (mmol/l): 140 NaCI, 5.4 KCI, 1.5 $CaCI_2$, 1.1 $MgCI_2$, 10 Tris/HCI, 11.1 glucose; pH 7.4. The atrial trabeculae (length > 2.5 mm and diameter < 1 mm) were horizontally attached via clips to a fixed hook and to the force transducer in an organ bath with a temperature of 31°C.

Measurement of isometric twitch

Field stimulation and measurement of isometric force were achieved using the stimulator module, the force transducer and the bridge amplifier of the Plugsys system 603 (Hugo Sachs Elektronik, FRG). Biphasic impulses with a duration of 7 ms and a current of 30% above threshold were applied. Data acquisition, analysis of the contraction curve and statistics were performed using a software developed in our laboratory. Peak force (PF), time to peak force (TPF) and the relaxation time for the decline of force by 50% of PF (RT50) were determined from the force signal. From the dF/dt signal, maximum dF/dt (dF/dt_{max}) and minimum dF/dt (dF/dt_{min}) were obtained to calculate the ratios (dF/dt_{max})/PF and (dF/dt_{min})/PF.

The preparations were equilibrated for about 45 min at a stimulation frequency of 0.25 Hz and at low preload. After that, the extracellular K^+ concentration was increased to 30 mM, and the trabeculae were progressively lengthened until maximum contractile force was reached. Subsequently, this load was reduced by 50% to maintain a stable contractile force at a stimulation frequency of 0.5 Hz during the experimental protocol.

Solutions were bubbled in a pre-organ bath with either pure O_2 to obtain a pO_2 of about 80 kPa as aerobic control, or with pure N_2 for 30 min to lower the pO_2 successively to a constant value of about 4 kPa to mimic hypoxia (group H) (Fig. 1). In the group simulated ischemia (group SI), the solutions were bubbled with a mixture of 95% N_2/5% CO_2, which resulted in a similar pO_2 as in hypoxia and in a pH of about 6.65. In both groups, the trabeculae were then allowed to recover during 30 min of reoxygenation with pure O_2 which increased pO_2 rapidly and the pH more slowly under the respective condition (Fig. 1).

The maximum postextrasystolic potentiated contraction (PECmax) was found by various extrasystolic intervals in the range of 150–1000 ms. Postrest potentiated contractions (PRC) were elicited when the rhythmic stimulation was interrupted by pauses in the range of 7.5–120 sec.

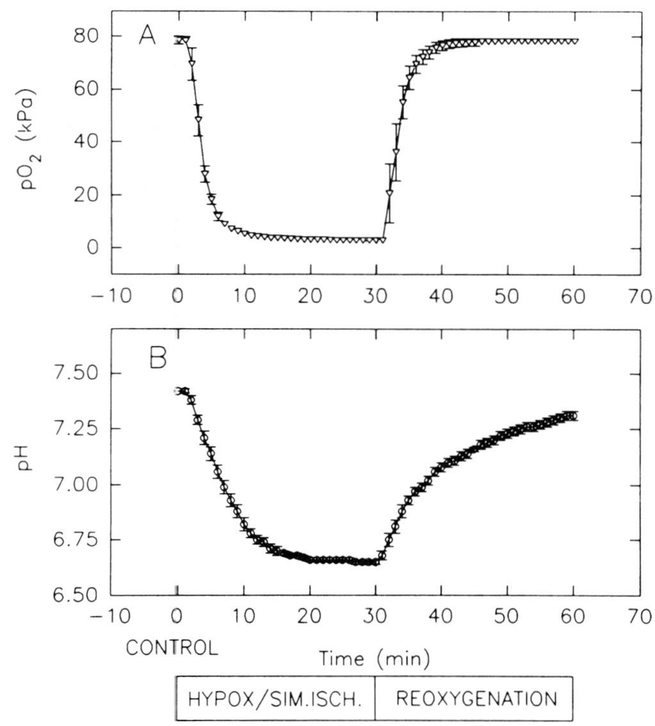

Fig. 1. Time course of pO_2 and pH in the organ bath during the experimental protocol (A) Switching from pure O_2 to 100% N_2 led to a rapid decrease in pO_2 which completely is reversed when O_2 reappears for reoxygenation. (B) Decrease in pH due to the application of 95% N_2/5% CO_2 mixture to simulate ischemic conditions and recovery due to readministration of pure O_2 after 30 min, n = 6.

Measurement of myocardial concentration of thiobarbituric acid reactive substances (TBARS)

Tissue samples of human atrial trabeculae were homogenized 1/150 vol. % in ice-cooled 150 mM KCI solution containing 0.01% butylated hydroxyanisole by means of a glass-teflon homogenizer. The measurements of the TBARS concentration was based on the method of Ohkawa *et al.* [19]. Briefly, 0.15 ml homogenate was incubated in a reaction mixture (final concentration: 1.35 M acetic acid, 0.15 mM sodium dodecyl sulfate, 20 mM 1-thiobarbituric acid, with NaOH to pH 3.5; final volume: 0.8 ml) at 90°C for 60 min. After cooling in ice water, 1 ml n-butanol/pyridine mixture (15/1) was added. This mixture was shaken vigorously, centrifuged and the organic layer taken for fluorimetric measurement of the TBARS fluorescence (excitation 515 nm, emission 553 nm). The TBARS concentration was calculated using 1.1.3.3.-tetraethoxypropane dissolved in acetic acid as standard.

Statistical analysis

Values are given as mean ± S.E.M. Significances were determined by the Wilcoxon test for unpaired and paired samples. Significance was assumed at $p \leq 0.05$.

Results

Human atrial trabeculae were exposed either to hypoxia (group H) or to hypoxia combined with acidosis (simulated ischemia, group SI) for 30 min and subsequently reoxygenated. The 31 trabeculae of the two groups were similar in length, diameter as well as in baseline twitch characteristics (Table 1). The functional changes were therefore normalized with respect to these baseline values.

Table 1. Characteristics of human atrial trabeculae and of their rhythmic isometric twitches under aerobic control conditions

Group	H (n = 15)	Sl (n = 16)
Age of patients (years)	57.07 ± 3.52	55.95 ± 2.60
Length of trabeculae (mm)	5.49 ± 0.30	5.09 ± 0.30
Diameter of trabeculae (mm)	0.65 ± 0.04	0.65 ± 0.03
Isometric twitch characteristics (aerobic control)		
PF (mN)	6.89 ± 0.91	9.43 ± 1.39
$(dF/dt_{max})/PF$ (1/s)	9.49 ± 0.32	9.97 ± 0.24
TPF (ms)	199.38 ± 5.26	195.47 ± 3.27
$(dF/dt_{min})/PF$ (1/s)	3.87 ± 0.21	3.98 ± 0.24
RT50 (ms)	169.63 ± 9.54	162.00 ± 6.72

PF = peak force; (dF/dtmax)/PF = maximal rate of contraction normalized to peak force; TPF = time to peak force; (dF/dtmin)/PF = maximal rate of relaxation normalized to peak force; RT50 = time to 50% of relaxation.

Effects of hypoxia and reoxygenation on contractile properties (group H)

In group H, the rapid fall in pO_2 in the bath solution (cf. Fig. 1) caused an initial small increase in peak force (PF), which was followed by a strong decrease to about 15% of aerobic control (Fig. 2). An increase in $(dF/dt_{max})/PF$ was found which indicates that PF fell more than the maximum rate of contraction. This marked fall in PF may be partially due to an abbreviation in force development during the twitches as reflected by the shorter time to peak force (TPF).

The relaxation was not markedly influenced at the end of the hypoxic period (Fig. 2D, E, F). The maximum relaxation rate (dF/dt_{min}) was reduced almost proportionally to PF so that half time of relaxation (RT50) was nearly equal to control. However, around the 15th min we found a temporary minimum of relaxation rate and consequently a maximum of RT50. The increase in TPF may also reflect a delayed onset of relaxation, i.e. a prolonged and more tonic rise in force appeared, and consequently, $(dF/dt_{max})/PF$ was diminished.

Reoxygenation produced a nearly constant recovery of PF to about 64% of pre-hypoxic control within 15 min. This was paralleled by a progressive increase in TPF and a decrease in $dF/dt_{max}/PF$, e.g. by a gradual deceleration in contraction. In contrast, the significant deceleration in relaxation, indicated by decreasing $(dF/dt_{min})/PF$ and increasing RT50, occured rapidly in a few minutes.

Effects of simulated ischemia and reoxygenation on contractile properties (group Sl)

Replacing O_2 by a mixture of 95% N_2 and 5% CO_2 led to a rapid fall in pO_2 combined with a decrease in pH_e for simulation of ischemia (Sl) (Fig. 1). In group Sl, PF decreased considerably but without the initial slight increase observed in H. After 30 min of Sl, PF was depressed to 6.5% of control, and thus, to lower values than in group H. Similar to H, the twitches in group Sl are characterised by accelerated relaxation (increased $(dF/dt_{min})/PF$ and decreased RT50) and abbreviated contraction (shortened TPF and increased $(dF/dt_{max})/PF$).

Post-ischemic reoxygenation resulted in an initially delayed but complete recovery of developed PF (102% of pre-ischemic control). Similar to H, the very early period of reoxygenation induced a permanent deceleration of contraction, indicated by a decrease in contraction rate and increase in TPF, which means that the force of twitches rose more tonically.

In contrast to the rapid changes in relaxation observed in group H, the relaxation in Sl decelerated only gradually (e.g. decrease in relaxation rate $(dF/dt_{min})/PF$, increase in RT50). Finally, the values were not different from those of H.

Postextrasystolic potentiated contractions (PEC)

In order to assess the role of the sarcoplasmic reticulum (SR) for reuptake, storage and release of Ca^{2+} the degree of twitch potentiation in response to perturbations of the regular stimulation was tested. The maximal postextrasystolic potentiated contraction (PECmax) was attained by an interval between an extrastimulus and the preceding regular stimulus of 500 ms in aerobic control, Reox H and Reox Sl, of 300 ms in H and of 200 ms in Sl (cf. Fig. 3, bottom line). PECmax for aerobic control conditions amounted to 1.25. In group H, extrastimulation had a stronger potentiating effect, and reoxygenation restored it to pre-hypoxic values. In Sl, we found the greatest potentiation but after post ischemic reoxygenation, PECmax was even significantly reduced compared to PECmax of control.

Post-rest potentiated contractions (PRC)

Post-rest potentiated contractions (PRC) were measured after pauses ranging from 7.5–120 sec. Under aerobic control conditions, there was a post-rest potentiation of about 1.35, which was maintained up to pauses of 60 sec and then decreased slowly after longer pauses. The PRC's in hypoxia amounted to 3.4, whereas after reoxygenation, long pauses led to depression of force development compared with rhythmic activity. In Sl, we observed progressive post-rest potentiation along with prolongation of pauses (8-fold increase in PRC120). Instead, the post-ischemic reoxygenation resulted in post-rest depression already after short pauses.

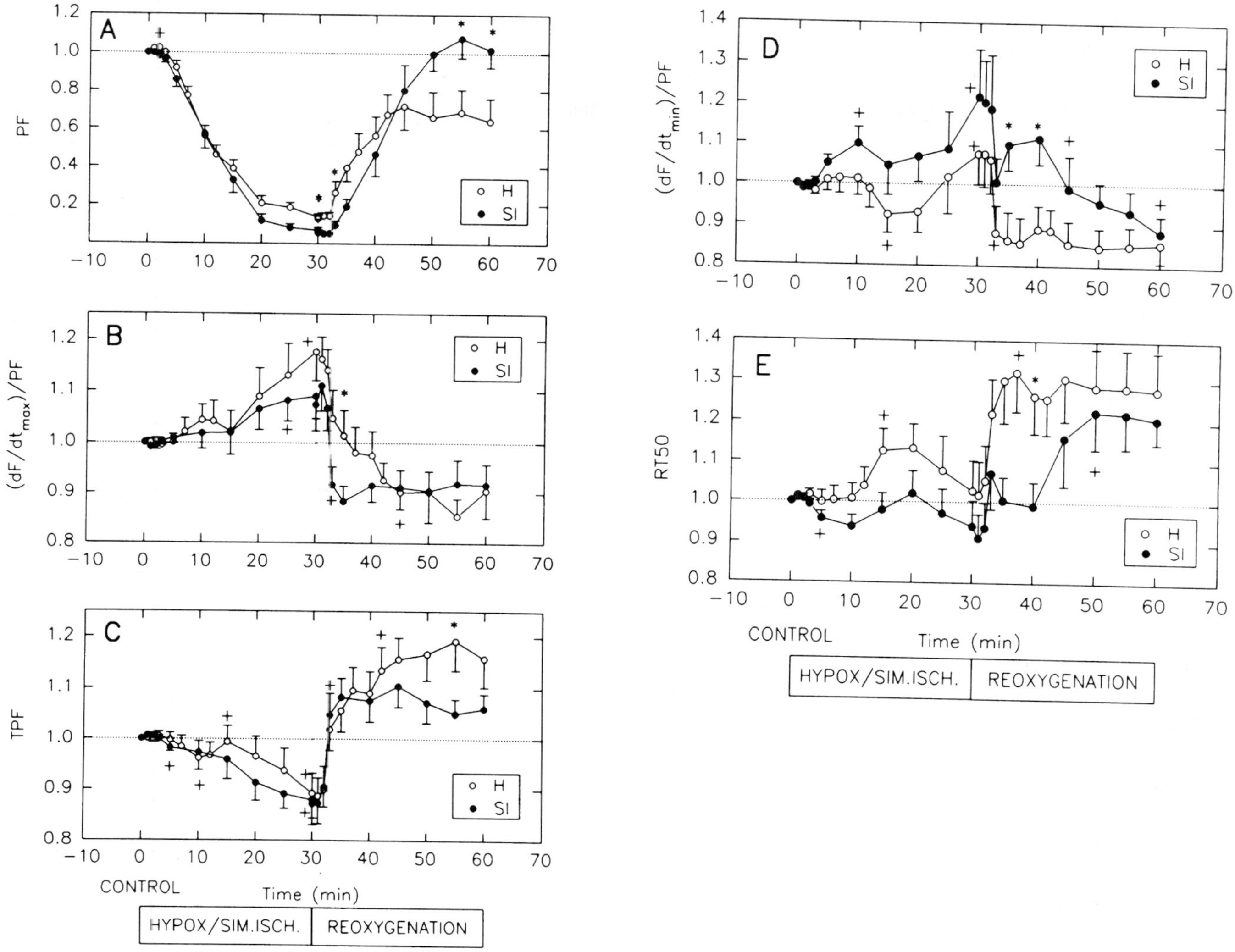

Fig. 2. Characteristics of isometric twitches of human trabeculae during hypoxia (H)/simulated ischemia (SI) and subsequent reoxygenation. The parameter values are normalized with respect to the respective aerobic control values at 0 min (dashed lines) (Table 1). Parameters are: (A) Peak force (PF), (B) maximum rate of contraction normalized to respective PF ($(dF/dt_{max})/PF$), (C) Time to PF (TPF), (D) maximum rate of relaxation normalized to respective PF ($(dF/dt_{min})/PF$), (E) Time for 50% of relaxation (RT50). $^*p < 0.05$ for differences between the two groups at an individual time point; $^+p < 0.05$ for differences within one group with respect to preceding values or control.

Mechanical alternans and spontaneous activity

An alternation of peak force of rhythmic contractions (mechanical alternans) and spontaneous activity occured during the early post-hypoxic and late post-ischemic period of reoxygenation (Fig. 5) along with delayed relaxation (cf. Fig. 2) and a temporary increase in resting force.

Oxygen radical formation during hypoxia/reoxygenation

The concentration of TBARS in the trabeculae was measured after post-hypoxic reoxygenation. Compared to control, we found a significantly higher TBARS concentration after post-hypoxic reoxygenation (0.57 ± 0.05 nmol/mg prot, n = 11 vs. 0.39 ± 0.05 nmol/mg prot, n = 14, $p < 0.05$).

Discussion

In the experimental approach, the aerobic control conditions were transformed into hypoxia (group H) or simulated ischemia (group SI), whereby the decrease in pH_e by about 0.8 pH units could have decreased pH_i by 30–60% of pH_e [23]. After 30 min, these conditions were reversed by reoxygenating with pure O_2 (Fig. 1). Our results demonstrate the typical myocardial contractile response to hypoxia, ischemia and reoxygenation [1–3, 5, 8, 10, 18]. Obviously,

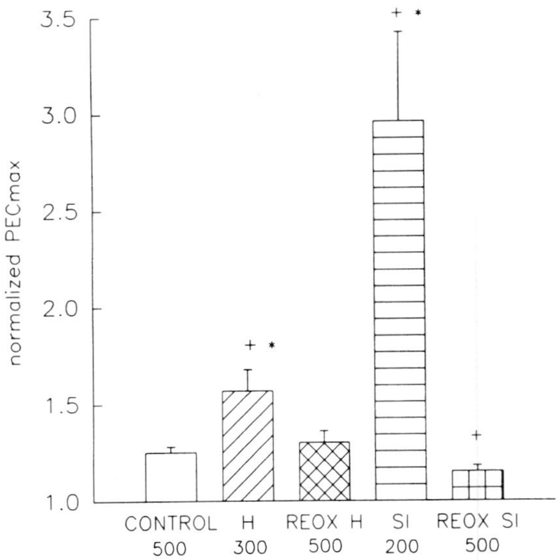

Fig. 3. Maximum postextrasystolic potentiations under aerobic control conditions (control), after 30 min of hypoxia (H) or simulated ischemia (SI) and of reoxygenation (REOX H and REOX SI, respectively) expressed as the ratio between the amplitude of the postextrasystolic potentiated twitch and of rhythmic twitch preceding the extrastimulus. The optimum extrastimulus intervals, which were required for maximal potentiation under the respective experimental conditions, are listed in the bottom row. $+p < 0.05$ with respect to control; $*p < 0.05$ with respect to H.

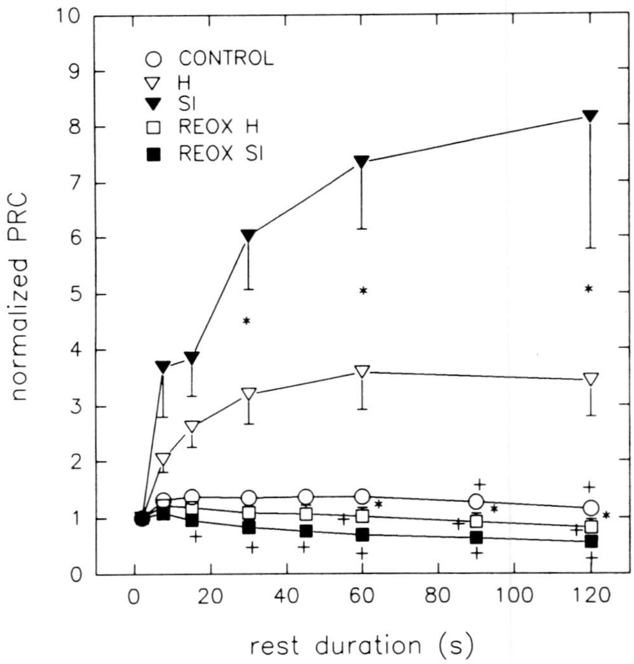

Fig. 4. Postrest potentiations at various rest duration under aerobic control conditions (control), after 30 min of hypoxia (H) or simulated ischemia (SI) and of reoxygenation (REOX H and REOX SI, respectively) expressed as the ratio between the amplitude of the postrest twitch and of the preceding rhythmic twitch. $+p < 0.05$ with respect to the preceding value of the same group at the shorter pause; $*p < 0.05$ between groups at equal rest duration.

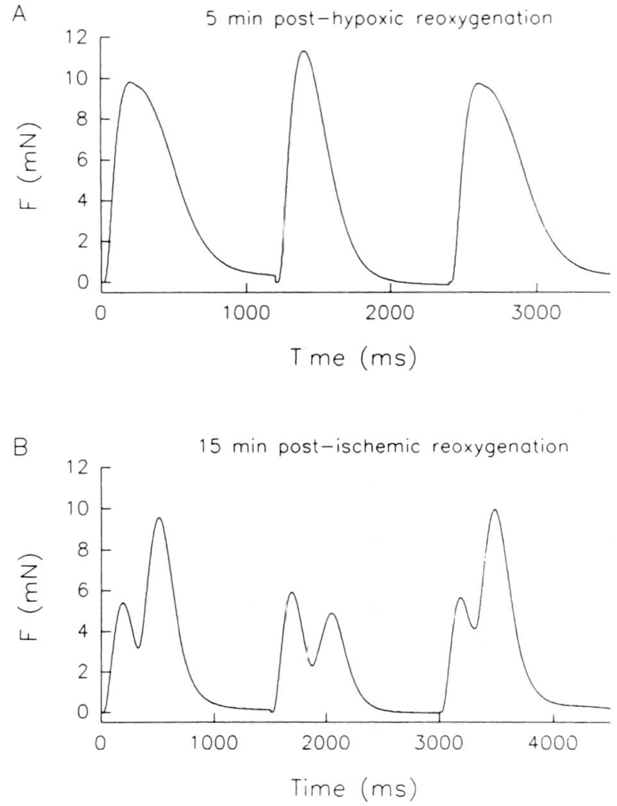

Fig. 5. Typical records demonstrating spontaneous activation and mechanical alternans in human atrial trabeculae. Contractions of a preparation after 5 min of post-hypoxic reoxygenation (A) and of another preparation after 15 min of post-ischemic reoxygenation (B). The regular electric stimulus was followed by a spontaneous activation resulting in twitches with double peaks.

the experimental procedures were suitable to reproduce these essential properties of the contractile function of the human myocardium. As very complex processes and mechanisms occur during these interventions, the discussion will include only those aspects which are conclusively consistent with the results.

Contractile function during hypoxia (H) and simulated ischemia (SI)

In both groups, peak force (PF) fell similarly during the respective initial period (Fig. 2). However, in group H, a slight temporary gain in PF is observed during the first few minutes. For this phase, a mild intracellular alkalization, due to a rapid decrease in creatine phosphate, is reported to cause the positive inotropic effect [1, 2]. In group SI, that effect was absent which suggests that the initial decrease in pH_e induced also a decrease in pH_i, preventing an alkalization as in H. During 15 min of H and SI, a very similar decrease in force occured, although in SI the decrease in pO_2 is accompanied by acidosis. The initial fall of PF is attributed primarily to a decrease in the sensitivity of contractile myofilaments to Ca^{2+} which is suggested to result from the increase in inorganic phosphate originating from rapid breakdown of energy-rich phosphates [1, 8]. In group SI, the stronger decrease in PF is likely on account of this strong effect combined with a rather slight desensitizing effect of H^+ [1, 8].

The increase in $(dF/dt_{max})/PF$ in both groups indicates a decrease in myofilament sensitivity to Ca^{2+} [24]. Accordingly, the short TPF displays how the rise in force steadily ceases earlier. In hypoxia, TPF is temporarily increased after 15 min, which also could indicate a slow onset of relaxation.

In both groups, the relaxation rate is differently influenced. In group H, the temporary decrease in relaxation rate may be caused by slower removal of cytosolic Ca^{2+}. An explanation can be provided by the glycolytic H^+ production. Accordingly, the transsarcolemmal H^+ gradient would drive the Na^+-H^+ exchange followed by the continuous hypoxic accumulation of Na_i. This, in turn, would inhibit Ca^{2+} extrusion via the Na^+-Ca^{2+} exchange and thus twitch relaxation [4, 5]. During the late hypoxic period, growing intracellular acidosis would then lower the Ca^{2+} sensitivity of the myofilaments, thereby promoting the relaxation as indicated by enhanced $(dF/dt_{min})/PF$. In group SI, the strong acidosis is a conclusive reason for a lower Ca^{2+} sensitivity of the myofilaments compared to group H, which could have produced, as is typical for this [1, 8, 24], higher relaxation rate.

Contractile function during reoxygenation

In group H, the recovery of the contractile force during reoxygenation followed the rapid increase in the pO_2 more closely than the recovery in SI did (Figs 1 and 2). Thus, for the post-ischemic recovery, the additional effect of the only gradual increase in pHe has to be taken into account.

Post-hypoxic and post-ischemic recovery of PF development during reoxygenation require an adequate sensitivity of the myofilaments to Ca^{2+} together with a substantial Ca^{2+} delivery.

In group H, the onset of the oxidative metabolism is thought to stop the glycolytic production of H^+ and to normalize the concentration of ATP and thereby that of P_i which would restore the sensitivity of the myofilaments to Ca^{2+}. Furthermore, excessive transsarcolemmal Ca^{2+} entry is suggested to occur for the following reasons: (1) Extrusion of excessive intracellular H^+ in exchange for Na^+ would activate to Ca^{2+} entry via Na^+-Ca^{2+} exchange; (2) recovery of the Ca^{2+} current through the Ca^{2+} channels and (3) hyperpermeability of the sarcolemma for Ca^{2+} due to reoxygenation damage were reported [2, 4, 5, 18]. The latter may result from peroxidation by oxygen radicals [18], which we could also detect as TBARS [19], and from the attack of Ca^{2+} activated lipases and proteases on membrane structure [5]. These proposed changes during the reoxygenation-mediated recovery should affect contractile function in a characteristic manner. The re-increase in Ca^{2+} sensitivity should decelerate both twitch contraction and relaxation [24]. Also, the modified role of sarcolemmal Ca^{2+} movements, mainly via Na^+-Ca^{2+} exchange due to increased Na_i, would prolong Ca^{2+} entry and diminish Ca^{2+} extrusion and, therefore, decelerate contraction and relaxation [1, 4, 5]. This is exactly what is observed. The rate of contraction $(dF/dt_{max})/PF$ declined continuously below the pre-hypoxic control value. This decline in the ratio $(dF/dt_{max})/PF$ would mean that PF increases more than dF/dt_{max} during reoxygenation. Thus, the fast increase in force is followed by a further tonic increase in force indicated by the continuous prolongation of TPF. Such a biphasic rise in force is consistent with enhanced tonic transsarcolemmal Ca^{2+} entry [25].

Relaxation was extremely delayed after 5 min of reoxygenation, while the recovery of PF development continued. The decelerated relaxation is also consistent with a combined effect of increasing sensitivity of the myofilaments to Ca^{2+} and delayed Ca^{2+} removal. Accordingly, the above postulated increase in Na_i, could have inhibited the Ca^{2+} extrusion via Na^+-Ca^{2+} exchange, and the reactive oxygen radicals could have damaged the SR membranes, thereby depressing the uptake of Ca^{2+} for relaxation as well as the storage of Ca^{2+} for release [4, 5, 8, 18].

The recovery of PF development in group H reached only 60% of control after about 15 minutes without further improvement in PF development or in disturbed contraction and relaxation rate.

Different effects of post-ischemic and post-hypoxic reoxygenation on contractile function are considered to depend on the delayed restoration of pH_e in group SI. So, reoxygenation is divided in an early and a late post-ischemic period. In the early period of about 10 min, the existing still low pH_e (Fig. 1) would inhibit Na^+-H^+ exchange and Na^+-Ca^{2+} exchange[1, 3, 4, 9, 10]. The low pH_e would maintain the low Ca^{2+} sensitivity of the myofilaments [3, 23]. This should delay the onset of the recovery of PF development despite the preceding rapid increase in pO_2 which, in fact, was observed in comparison to group H.

In this period, the ratio $(dF/dt_{max})/PF$ decreased markedly. The inhibition of the sarcolemmal exchange processes due to extracellular acidosis would contradict an intensified Ca^{2+} influx via Na^+-Ca^{2+} exchange. So, it is unlikely that the steep decrease in (dF/dtmax)/PF was caused by a Na^+-Ca^{2+} exchange-mediated tonic rise in force as proposed for group H, because TPF was not different between the two groups and PF was smaller than in H. Therefore, we suggest that the main cause for the decrease in $(dF/dt_{max})/PF$ in SI is the decrease in dF/dt_{max}, which may be rather attributed to diminished Ca^{2+} release from a possibly damaged SR.

The relaxation in the early post-ischemic period was similar to that at the end of preceding ischemia but clearly faster than in H. This suggests that extracellular acidosis maintained a high relaxation rate, probably due to a low Ca^{2+} sensitivity of the myofilaments.

In the late post-ischemic reoxygenation period, pH_e reached nearly normal values. Thus, the further recovery of PF, attaining even the level of aerobic control, should correspond to the increase in pH_e. However, the abnormal time course of the twitches signalizes a still disturbed Ca^{2+} handling. A rapid increase in pH_e can cause a noticeable positive inotropic effect [17]. The resulting transsarcolemmal pH gradient and the disinhibition of Na^+-H^+ exchange and Na^+-Ca^{2+} exchange would induce Na_i accumulation and subsequently enhance the transsarcolemmal Ca^{2+} entry. This could lead to a relative increase in the contribution of transsarcolemmal Ca^{2+} entry to activator Ca^{2+} resulting in the tonic component of force development.

The relaxation process in the late post-ischemic period, associated with increasing pH_e should be influenced by Na_i accumulation via Na^+-H^+ exchange, which would inhibit Ca^{2+} extrusion via Na^+-Ca^{2+} exchange[5]. Indeed, we found a progressive deceleration of relaxation. Additionally, this delay of the relaxation associated with the transient increase in the diastolic tension is a typical feature for Ca^{2+} overloading [8]. Possible damaging effects on the membrane integrity of the sarcolemma as well as of the SR by Ca^{2+} activated proteases and lipases and by reactive O_2 species could have supported the process of Ca^{2+} overloading [5].

Role of the sarcoplasmic reticulum

The substantial contribution of the SR to the Ca^{2+} homeostasis in human atrial myocardium as, e.g. in rat myocardium [26], is expressed by the relatively high degree of post rest potentiation (PRP) in control conditions maintained also after long pauses. PRP may reflect the balance between SR Ca^{2+} uptake into the SR for release on the one hand and spontaneous SR Ca^{2+} leakage on the other (Ref. see [14]).

During H and SI, the greatest amplitudes of normalized PRP occured. An explanation can be given by a shift of the Ca^{2+}-tension relationship to a lower Ca^{2+} sensitivity [1]. Under the experimental conditions, at relatively low extracellular Ca^{2+} concentration and low stimulation frequency, activation of regular contractions would involve a lower but activation of potentiated contractions the steeper part of the sigmoid rightward shifted Ca^{2+}-tension relationship curve. Assuming a similar potentiation of Ca^{2+} release as in normal conditions, this, in agreement with other [14], would explain the observed increase in the ratio between PRC and regular contraction. In SI, the additional rightward shift of the Ca^{2+}-tension relationship curve due to the more severe acidosis could have caused the highest relative increase in PRC. Obviously, the ability of the SR to accumulate and to store additional Ca^{2+} is preserved during H and SI.

During reoxygenation, the capacity of the SR Ca^{2+}-ATPase is presumably too low to reduce Ca^{2+} promptly during increased influx, as seen by the delay of the relaxation and by the transient increase in diastolic force. Concomitant mechanical alternans and spontaneous activity (Fig. 5) were also observed by others [15, 22]. Mechanical alternans is caused by alternating Ca^{2+} release from a SR with high Ca^{2+} load at rhythmic contractile activity [10, 22]. Arrhythmia signalizes spontaneous SR Ca^{2+} release which hence triggers transsarcolemmal Ca^{2+} entry [10].

After 30 min of reoxygenation, PRP changed to post-rest depression of force, i.e. SR Ca^{2+} uptake could not compensate for a possibly increased leakage of SR Ca^{2+} during the pauses, as reported by others [14]. Electrophysiological studies suggest that oxygen radicals, which were also produced in the experiments, could have damaged the SR Ca^{2+} release channels [27].

The potentiation by extra stimulation is explained by a second Ca^{2+} influx during the relaxation of the extra contraction. This Ca^{2+}, taken up into the SR together with the Ca^{2+} released during the regular contraction, enhances the amount of releasable Ca^{2+} for the activation of the postextrastimulatory potentiated contraction (PEC) (Ref. see [14]). The great amplitudes of PEC in H and particularly in SI (Fig. 3), as also reported by others [14], may be explained by the same causes as discussed above for PRC. The increase in PF of PEC after H and SI suggests that intensified SR Ca^{2+} uptake due to extra stimulation is preserved, and concomitant enhanced ATP turnover may also not limit the potentiation. The small post-ischemic decrease in PF of PEC can not be conclusively explained. Possibly, a striking depression of SR Ca^{2+} uptake was caused by the combined effect of ischemia and reoxygenation [1, 28].

In conclusion, rapid correction of hypoxia by reoxygenation produced a partial recovery of force development in group H. In group SI, however, this and a delayed correction of acidosis resulted in complete recovery of force development. The significant post-rest potentiation in control conditions indicates an important role of the SR for the Ca^{2+} homeostasis in human atrial myocardium. Postextrastimulatory and post-rest potentiation appeared during hypoxia and simulated ischemia, but were depressed in particular after post-ischemic reoxygenation. Thus, the SR seems to be more susceptible to reoxygenation probably associated with the damaging effect of increased oxygen radicals. The depression of potentiation during reoxygenation occured along with further characteristics of SR dysfunction. Thus, for a possibly beneficial effect of low pH during ischemia combined with its gradual correction during reoxygenation on the recovery of contractile function, developed peak force should not be the only index.

Acknowledgment

We are grateful to Prof. Gabriele Pfitzer for very helpful advices on the manuscript, to Prof. Wolfgang Konertz for continuing support and to Miss B. Schröter for careful technical assistance.

References

1. Allen DG, Orchard CH: Myocardial contractile function during ischemia and hypoxia. Circ Res 60: 153–168, 1987
2. Hearse DJ, Bolli R: Reperfusion induced injury: manifestations, mechanisms, and clinical relevance. Cardiovasc Res 26: 101–108, 1992
3. Pierce GN, Czubryt MP: The contribution of ionic imbalance to ischemia/reperfusion-induced injury. J Mol Cell Cardiol 27: 53–63, 1995
4. Silverman HS, Stern MD: Ionic basic of ischaemic cardiac injury: insights from cellular studies. Cardiovasc Res 28: 581–597, 1994
5. Tani M: Mechanisms of Ca^{2+} overload in reperfused ischemic myocardium. Ann Rev Physiol 52: 543–559, 1990
6. Goldstein JA, Tweddell J, Barzilai B, Yagi Y, Jaffe AS, Cox JL: Right atrial ischemia exacerbates hemodynamic compromise associated with experimental right ventricular dysfunction. JACC 18: 1564–1572, 1991
7. Kesler DF, Galyanov M, Zhukov KG: Diastolic function in patients with coronary artery disease: Effect of angina and heart failure functional class. Int J Cardiol 47: 211–215, 1995
8. Mubagwa K: Sarcoplasmic reticulum function during myocardial ischaemia and reperfusion. Cardiovasc Res 30: 166–175,1995
9. Lazdunski M, Frelin C, Vigne P: The sodium/hydrogen exchange system in cardiac cells: Its biochemical and pharmacological properties and its role in regulating internal concentrations of sodium and internal pH. J Mol Cell Cardiol 17: 1029–1042, 1985
10. Orchard CH, Cingolani HE: Acidosis and arrhythmias in cardiac muscle. Cardiovasc Res 28: 1312–1319, 1994
11. Wilde MM, Aksnes G: Myocardial potassium loss and cell depolarisation in ischaemia and hypoxia. Cardiovasc Res 29: 1–15, 1995
12. Harada K, Franklin A, Johnson RG, Grossman W, Morgan JP: Acidemia and hypernatremia enhance postischemic recovery of excitation-contraction coupling. Circ Res 74: 1197–1209, 1994
13. Poole-Wilson PA: Regulation of intracellular pH in the myocardium; relevance to pathology. Mol Cell Biochem 89: 151–155, 1989
14. Mattheussen M, Mubagwa K, Rusy BF, Van Aken H, Flameng W: Potentiated state contractions in isolated hearts: Effects of ischemia and reperfusion. Am J Physiol 264: H1663–1673, 1993
15. Tatsumi T, Asayama J, Yamahara Y, Mijazaki H, Inoue M, Omori I, Inoue D, Nakagawa M: Effects of hypoxia and reoxygenation on regular contractions and postrest contraction in rat papillary muscles. Jpn Heart J 23: 363–371, 1991
16. Van Hardefeld C, Schouten VJA, Muller A, Van Der Meulen ET, Elzinga G: Exposure of energy depleted rat trabeculae to low pH improves contractile recovery: role of Calcium. Am J Physiol 268: H1510–H1520, 1995
17. Hata K, Takasago T, Saeki A, Nishioka T, Goto Y: Stunned myocardium after rapid correction of acidosis. Circ Res 74: 79–805, 1994
18. Goldhaber JI, Weiss JN: Oxygen free radicals and cardiac reperfusion abnormalities. Hypertension 20: 118–127, 1992
19. Ohkawa H, Ohishi N, Yagi K: Assay for lipid peroxides in animal tissue by thiobarbituric acid reaction. J Anal Biochem 95: 351–358, 1979
20. Luciani GB, D'Agnolo A, Mazzuco A, Galucci V, Salviati G: Effects of ischemia on sarcoplasmic reticulum and contractile myofilament activity in human myocardium. Am J Physiol 265: H1334–1341, 1993
21. Zucchi R, Testoni S, Ronca G, Calafiore AM, Barsotti A, Mariani M: Sarcoplasmic reticulum Calcium uptake in reperfused human myocardium. Proc Internat Soc Heart Res (Mondüzzi Editore): 493–495, XV Eürop Section Meeting (Denmark) 1994
22. Nwasokwa O N: Mechanism of mechanical alternans in ischemia-reperfusion: Role of deficient relaxation of the strong twitch. Am J Physiol 269: H169–H175, 1995
23. Orchard CH, Kentish JC: Effects of changes of pH on the contractile function of cardiac muscle. Am J Physiol 258: C967–C981, 1990
24. Lammerich A, Günther J, Storch E: Interpretation of the inotropic effect of 2,3-butanedione monoxime on the isometric twitch of guinea pig papillary muscle. Gen Physiol Biophys 13: 377–392, 1994
25. Gross TH, Günther J, Storch E: The isometric twitch of rabbit papillary muscle: Reflection of the cellular Calcium movements? Gen Physiol Biophys 6: 521–538, 1989
26. Lammerich A, Günther J, Pfitzer G, Storch E, Vetter R: Alterations of cardiac contractile function are related to changes in membrane Calcium transport in spontaneously hypertensive rats. J Hypertension 13: 1313–1324, 1995
27. Holmberg SRM, Cumming DVE, Kusama Y, Hearse DJ, Poole-Wilson PA, Shattock MJ, Williams AJ: Reactive oxygen species modify the structure and function of the cardiac sarcoplasmic reticulum calcium-release channel. Cardioscience 2: 19–25, 1991
28. Hohl CM, Garleb M; Altschuld RA: Effects of simulated ischaemia and reperfusion on the sarcoplasmic reticulum of digitonin-lysed cardiomyocytes. Circ Res 70: 716–723, 1992

Molecular and Cellular Biochemistry **160/161:** 153–158, 1996.

The guanine nucleotide-binding regulatory proteins (G proteins) in myocardium with ischemia

Mitsumasa Ohyanagi and Tadaaki Iwasaki
First Department of Internal Medicine, Hyogo College of Medicine, Nishinomiya, Japan

Abstract

Guanine nucleotide-binding regulatory proteins (G proteins) play a major role in the regulation of a number of physiological processes, such as stimulation or Inhibition of adenylate cyclase activity or gaiting of ionic channels. Myocardial ischemia could induce the changes in receptor-G protein signal transduction system in the heart. Therefore, this article will focus on the role and alterations of G proteins (especially, Gs and Gi) in myocardial ischemia. The Gi protein rapidly loses functional activity during very early myocardial ischemia. In contrast to Gi protein, the function of Gs protein during this phase has not been evaluated. Moreover, the changes in Gs protein after 30 min of ischemia are contradictory. However, the sensitization of the adenylate cyclase activity in the very early phase of acute ischemia is gradually replaced by a decrease in adenylate cyclase activity with prolonged ischemia. The decrease in the function and amount of Gs protein may be one of the factors that induce these changes. The function of Gs protein was also decreased in the canine hearts with ischemia and reperfusion. In contrast to ischemia and reperfusion, there are no significant alterations in G proteins and modulation of adenylate cyclase in the stunned myocardium. It has become increasingly evident that Gi protein may play an important role in the cardioprotective effects of preconditioning. When β-adrenoceptor densities are reduced in chronic myocardial ischemia, decreased in the amount and function of Gi protein and increased amount of Gs protein may play the role in preservation of the adenylate cyclase activity. These alterations in G proteins may play the important role in the myocardial function during myocardial ischemia. (Mol Cell Biochem **160/161:** 153–158, 1996)

Key words: G protein, myocardial ischemia

Introduction

Guanine nucleotide-binding regulatory proteins (G proteins) are polymorphic proteins that play a major role in the regulation of a number of physiological processes, such as stimulation or inhibition of adenylate cyclase activity or gaiting of ionic channels (for review, references [1–4]). Three types of G protein appear to be important in the heart: (1) Gs protein mediates stimulation of adenylate cyclase and production of cyclic AMP (cAMP) via β-adrenoceptors and activates cAMP-dependent protein kinase (PKA); this may promote influx of extracellular Ca^{2+} through phosphorylation of calcium channels. (2) Gi protein mediates inhibition of adenylate cyclase and formation of cAMP via muscarinic acetylcholine receptors or adenosine A_1 receptors and activates certain potassium channels via muscarinic acetylcholine receptors. (3) Gq/11 acts independently of adenylate cyclase by stimulating a phospholipase C via α1-adrenoceptors [4].

Acute myocardial ischemia could induce the changes in receptor-G protein signal transduction system in the heart. Alterations induced by ischemia could occur at the level of G proteins. However, the reports on changes in G proteins are contradictory. Therefore, this article will focus on the role and alterations of G proteins (especially, Gs and Gi) in myocardial ischemia.

Acute myocardial ischemia

Sustained myocardial ischemia

Alterations in G proteins during early myocardial ischemia. β-adrenoceptor stimulation leads to a positive inotropic response via the receptor-G protein-adenylate cyclase complex

Address for offprints: M. Ohyanagi, First Department of Internal Medicine, Hyogo College of Medicine, 1-1, Mukogawa-cho, Nishinomiya, Hyogo, 663, Japan

[3]. Acute myocardial ischemia induces a release of endogenous catecholamine [5, 6]. The density of β-adrenoceptor is unaltered during very early myocardial ischemia [7]. Although Strasser and Marquetant proposed a dual sensitization of the β-receptor-adenylate cyclase system with the increase in enzyme activity possibly due to activation of protein kinase C, the Gi protein rapidly loses functional activity during very early myocardial ischemia. Even within 5 min after ischemia, the carbachol-mediated GTPase activity, as a functional parameter for Gi activity, is greatly reduced [9]. This impaired function of muscarinic receptor linked signal transduction during early myocardial ischemia could play one of the role in the sensitization of the adenylate cyclase system by a reduced ability to counteract adenylate cyclase activation. These sensitization of adenylate cyclase system could contribute to the occurrence of ischemia induced tachyarrhythmias and to the deterioration of myocardial function. These impaired function of Gi protein are still present after 15 min of ischemia [10]. In contrast to the function, the amount of Gi α is unaltered during 15 min of ischemia [9, 11]. Wolff *et al.* [12] showed that the function of Gi protein after 30 min of ischemia was unaltered. However, the alterations in the amount of Gi after 30 min of ischemia is not significant [11] or decreased [13]. These discrepancies may depend largely on species used and the methods by which ischemia is introduced.

While the function of Gi protein during very early phase (i.e. 5 min after ischemic insult) was investigated, the function of Gs protein during this phase has not been evaluated. While Maisel *et al.* [14] reported that the amount of Gs α determined by an ELISA technique markedly decreased within 15 min after induction of ischemia, van den Ende *et al.* [13] have shown no alterations in the amount of immunodetectable Gs protein determined by Western blot analysis within 30 min after induction of ischemia. These discrepancies may depend on duration of ischemia and the methods of analysis for Gs protein. However, Wollf *et al.* [12] have shown the decrease in the function of Gs protein after 30 min of ischemia. The sensitization of the adenylate cyclase activity in the very early phase of acute ischemia is gradually replaced by a decrease in adenylate cyclase activity with prolonged ischemia [15]. The decrease in the function of Gs protein may be one of the factors that induce these changes. This decrease in Gs protein might contribute to the decreased contractility during ischemia.

Alterations in G proteins during late myocardial ischemia. In general, the density of β-adrenoceptors is increased during acute late myocardial ischemia [16]. However, no significant elevation in cAMP levels in 1 h ischemic myocardium was observed in our study [17]. Ischemia-induced increases in the density of β-adrenoceptors and decreases in adenylate cyclase activity have also been reported by Susanni *et al.* [18] and Strasser *et al.* [15], although different species of animals and types of ischemia were studied. An abnormality in post-receptor signal transduction for cAMP production in tissues appears to be responsible for this finding. As same as early myocardial ischemia, the function of Gs protein is decreased during late (1 or 3 h [18, 19]) ischemia. The amount of Gs protein was also decreased, whereas there are no significant changes in the amount of Gs protein during early (30 min) ischemia [13]. The steady-state mRNA for Gs protein is also decreased [19]. These data suggest that decreased steady-state mRNA is a contributor to the decreased Gs protein levels, although other mechanisms of decreased protein levels, such as post-translational modification or phosphorylation of the enzyme, cannot be ruled. These decreases in the function and amount of Gs protein may help explain the decrease in adenylate cyclase activity during late ischemia, even when catecholamine levels and the density of β-adrenoceptors are increased.

While the function of Gi protein during early myocardial ischemia is investigated in a few studies [9–11], there are very few studies for the amount of Gi protein during late myocardial ischemia. Tobise *et al.* [20] showed no significant alterations in the immunodetectable Gi protein after 40 min of ischemia, whereas Will-Shahab *et al.* demonstrated the decrease in the immunodetectable Gi protein after 30 min of ischemia [11]. We also showed the decrease in the mRNA for Gi protein and immunodetectable Gi after 1 or 3 h ischemia [19]. These data suggest that decreased steady-state mRNA is a contributor to the decreased protein levels of Gi as same as Gs protein.

The mechanism for the decrease in Gs and Gi after late myocardial ischemia remains unclear. However, since β-actin mRNA showed no changes even in the ischemic region in our study [19], the decrease in Gs and Gi is not attributable to ischemia-induced nonspecific changes. It is known that cAMP levels decrease in ischemic myocardial cells [7], and this decrease is one factor responsible for decreased myocardial contractility. A decrease in Gs could cause a reduction in myocardial cell cAMP levels. A decrease in Gi, on the other hand, may prevent a decrease in myocardial contractility by suppressing the decrease in cellular cAMP levels.

Ischemia and reperfusion

Myocardial ischemia and reperfusion is observed in patients with coronary spasms or myocardial infarction accompanied with recanalization. However, there are few reports of the alterations in G protein of myocardium with ischemia followed by reperfusion [13, 21]. Van den Ende *et al.* [13] reported that adenylate cyclase responsiveness was impaired during ischemia (30 min), and that this responsiveness recovered during reperfusion (30 min), whereas inotropic responses to forskolin and dibutylyl cAMP were absent in reperfused rat hearts. They also showed that cAMP content decreased time-dependently during ischemia and did not recover after reperfusion, indicating ATP-depletion. However, they showed that immunodetectable Gs and Gi were not altered by ischemia and reperfusion. Therefore, they hypothesized that an additional mechanism downstream of cAMP formation appears to be defective during reperfusion, which prevent recovery of inotropic responses to hormonal stimulation. However, in contrast the these studies, Kiuch *et al.* [21] reported that the function of Gs protein determined by reconstitution assay using mouse lymphoma S49 adenylyl cyclase-deficient (cyc-) membrane was decreased in the canine hearts with ischemia (45 min) and reperfusion (45 min). They also showed that immunodetectable Gi was decreased. These discrepancies may depend largely on the duration of ischemia and reperfusion (i.e. shorter periods of ischemia (10–30 min) may not affect G proteins, but after prolonged ischemia (45 min) may induce a decrease in G proteins) and species used, and the methods by which ischemia is introduced.

Table 1. The effects of ischemia on G proteins

	Gs protein		Gi protein	
	Function	Amount	Function	Amount
I. Acute phase				
a. sustained ischemia				
5 min			↓9)	
15 min	↓14)		↓10)	→9,11)
30 min	↓12)	→11,13)	→12)	→13) ↓ 11)
40 min		→20)		→20)
60 min	↓18,19)	↓18,19)		↓19)
180 min	↓19)	↓19)		↓19)
b. ischemia (I) and reperfusion (R)				
I; 30 min, R; 30 min		→13)		→13)
I;45min, R; 45min	↓21)			↓21)
c. stunned myocardium		→22)	→23)	→23)
d. preconditioning		→29)	↑28)	→28)
II. Chronic phase				
5 weeks	→?34)	↑34)	↓34)	↓34)
6 months				↓36)

↑: increase, →: no significant change, ↓: decrease,) see references.

Myocardial stunning

Myocardial stunning is defined as a transient reversible myocardial dysfunction induced by brief periods of ischemia, which do not lead to necrosis. This phenomenon is likely to be of importance in the myocardial dysfunction observed in patients after periods of unstable angina or coronary spasms and after therapeutic thrombolysis and percutaneous transluminal coronary angioplasty. There are only two studies for the alterations in G proteins in stunned myocardium. Fu *et al.* [22, 23] demonstrated no significant alterations in the density of β-adrenoceptors, the content of mRNA and protein of G proteins, and modulation of adenylate cyclase in biopsies from stunned myocardium. They speculated that the normal β-adrenergic signaling system in the stunned myocardium 1 h after the last ischemic period might be due to fast recovery of signal transduction as a result of reperfusion. As shown above chapter of ischemia and reperfusion, reperfusion may reverse the alteration of adenylate cyclase system in ischemic myocardium. They proposed that this is probably extremely important for the ability of the stunned myocardium to response to β adrenergic stimulation.

Ischemic preconditioning

Single or repeated episodes of cardiac ischemia and reperfusion causes the myocardium to become very resistant to infarction. This phenomenon, termed ischemic preconditioning, has been demonstrated by various investigators [24, 25]. Although the precise mechanism for preconditioning remain controversial, it has become increasingly evident that adenosine may play an important role in the cardioprotective effects of preconditioning. Liu *et al.* [26] reported that intracoronary infusion of adenosine or an A1 selective adenosine receptor agonist was as effective as preconditioning in limiting infarct size by prolonged myocardial ischemia, and that the infarcted size-limiting effect of preconditioning was lost after pretreatment with adenosine receptor antagonists. The A1 selective adenosine receptor couples to Gi protein. Thornton *et al.* [27] showed that pertusis toxin, which blocked Gi proteins, greatly attenuates preconditioning. Therefore, Gi proteins are likely involved in the ischemic preconditioning. During early myocardial ischemia, activation of Gi proteins is expected to be predominantly induced by activation of adenosine A1 receptors, due to the increased myocardial release of adenosine. Niroomand *et al.* [28] demonstrated that although during the first 5 min of myocardial ischemia the function of Gi protein was significantly reduced, during a following 15 min period of reperfusion this decreased responsiveness was reversed completely exceeding control activities. They also showed that the amount of Gi protein did not alter during 15 min of ischemia. This suggests that the underlying mechanism of ischemic preconditioning is the increased responsiveness of Gi proteins after a brief period of ischemia and reperfusion. A continued activation of adenosine A1 receptors is required for the cardioprotective mechanisms. This adenosine production may be enhanced by increased 5′-nucleotidase activity caused by protein kinase C [29]. Fu *et al.* [30] also showed no significant changes in the amount of Gi protein in biopsies from preconditioned myocardium 30 min after the last coronary occlusion. They evaluated the amount of Gi protein by ELISA technique using specific Gi α antipeptide antibody. However, they showed no alterations in the activity of adenylate cyclase. The reason for the discrepancy in the changes in adenylate cyclase activity between Niroomand's and Fu's studies is not clear. It may be ascribed to differences in the duration of ischemia and reperfusion and species differences. Gi proteins mediate activation of ATP sensitive potassium channels. Gross and Auchampach [31] showed that ischemic preconditioning was lost after pretreatment with glibenclamide, an inhibitor of these channels. However, in rabbits blockade of the ATP-sensitive potassium channels did not prevent ischemic preconditioning [32]. Further study is needed to clarify whether the ATP-sensitive potassium channels involve the mechanism of ischemic preconditioning.

Although there are a lot of studies for the role of Gi protein in ischemic preconditioning, very few studies have evaluated the effects of preconditioning on Gs protein. Iwase *et al.* [33] reported that functional activity of Gs determined by S49 lymphoma cyc^- cell reconstituted assay was greater in the preconditioned heart than in the control heart and that the activity of the catalytic unit of the adenylate cyclase was enhanced. Although it has become increasingly evident that adenosine may play an important role in the cardioprotective effects of preconditioning, they proposed that inhibition of adenylate cyclase activity is not the target action of the cardioprotective effect of preconditioning potentially mediated by the A1-adenosine receptor Gi protein pathway. Fu *et al.* [30] showed no significant changes in the Gs proteins and the activity of adenylate cyclase in biopsies from preconditioned myocardium 30 min after the last coronary occlusion. Whereas Iwase *et al.* [33] evaluated the function of Gs protein, Fu *et al.* [30] studied the amount of Gs protein. However, the reason for the discrepancy in the changes in adenylate cyclase activity between two studies is not clear. It may be ascribed to differences in the duration of ischemia and reperfusion and species differences as same as Gi proteins.

Chronic myocardial ischemia

Whereas the alterations in G proteins during acute myocardial ischemia are evaluated in many studies as shown above, the effects of chronic myocardial ischemia on G proteins have been addressed in very few studies. Hammond *et al.* [34] demonstrated that regional myocardial downregulation of the Giα2 and β-adrenoceptors in porcine model of chronic episodic myocardial ischemia (5 weeks) without infarction or heart failure. They showed that the decrease in Giα2, shown by quantitative immunoblotting, was associated with decreased ability of oxotremorine to inhibit adenylate cyclase in homogenates of myocardium obtained from the ischemic bed. Longabaugh *et al.* [35] reported that teatment of adipocytes with adenosine receptor agonists resulted in decreased pertussis toxin substrate and Gi. Moreover, adenosine is released abundantly during myocardial ischemia. Therefore, Hammond *et al.* [34] proposed that prolonged adenosine receptor activation may be important in the regulation of Giα2. They also showed that Gsα in ischemic endocardium was increased.

Meggs *et al.* [36] also showed that the amount of Gi protein determined by ADP-ribosylation were reduced in rat myocardium with moderate ventricular hypertrophy and ventricular dysfunction after 6 months of coronary arterial stenosis. Their model was associated with a reduction in α-1 adrenoceptor density and norepinephrine stimulated phosphoinositol turnover. Although the identity of the regulatory protein that couples the α-1 adrenoceptors with phospholipase C has not been demonstrated with certainty, they speculated that the decrease in pertussis toxin-labeled substrates may reflect the reduction in phosphoinositol turnover in myocytes with prolonged coronary constriction.

When β-adrenoceptor densities are reduced in chronic myocardial ischemia, decreased amount and function of Gi protein and increased amount of Gs protein may play the role in preservation of the adenylate cyclase activity. These compensation mechanism might contribute to maintain the myocardial contractility during chronic ischemia.

G protein in coronary microvascular tone during myocardial ischemia

Previous studies have shown that coronary arterioles dilated in response to the reduction in perfusion pressure [37]. This response may play the important role in the defense mechanism that preserves coronary flow in the presence of coronary stenosis. Recent studies have shown that glibenclamide, which is a potent and selective blocker of ATP sensitive potassium channel, abolished the microvascular dilation in response to a reduction in perfusion pressure [38, 39]. Although ATP sensitive potassium channel is regulated by the intracellular ATP level, Gi proteins is one of factors that affect the activity of this channel. Komaru *et al.* [40] showed that whereas coronary microvessels dilated in response to the reduction in perfusion pressure in the vehicle group, microvessels did not dilate during the reduction in perfusion pressure in the pertussis toxin pretreatment group. Therefore, they proposed that pertussis toxin sensitive G proteins is crucially involved in microvascular control during autoregulation and ischemia. Gawler *et al.* [41] showed abolition of the expression of Gi protein in hepatocytes of animal models with diabetes mellitus. Komaru *et al.* [40] suggested that if abolition of the expression of Gi protein occurs at the coronary microvascular level in diabetes mellitus, the dysfunction of pertussis toxin sensitive G protein may be responsible for an impaired defense mechanism during ischemia.

References

1. Gilman AG: G proteins and dual control of adenylate cyclase. Cell 36: 577–579, 1984
2. Gilman AG: G proteins: Transducers of receptor-generated signals. Ann Rev Biochem 56: 615–649, 1987
3. Fleming JW, Wister PL, Watanabe AM: Signal transduction by G proteins in cardiac tissues. Circulation 85: 420–433, 1992
4. Brodde O-E, Michel MC, Zerkowski H-R: Signal transduction mechanisms controlling cardiac contractility and their alterations in chronic heart failure. Cardiovasc Res 30: 570–584, 1995
5. Wollenberger A, Shahab L: Anoxia-induced releases of noradrenaline from isolated perfused heart. Nature 207: 88–89, 1965
6. Muntz K, Hagler HK, Boulas HJ, Willerson JT, Buja LM: Redistribution of catecholamines in the ischemic zone of the dog heart. Am J Pathol 114: 64–78, 1984
7. Will-Shahab L, Schubert B: Functional status of sympathetic system in cardiovascular disease: myocardial infarction. In: P.K. Ganguly (ed.). Cathecholamines and heart disease. Boca Raton, FL: CRC Press, 1991, pp 123–143
8. Strasser RH, Marquetant R: Sensitization of the β-adrenergic system in acute myocardial ischemia by a protein kinase C-dependent mechanism. Eur Heart J 12(suppl F): 48–53, 1991
9. Niroomand F, Bangert M, Beyer T, Rauch B: Reduced adenylyl cyclase inhibition by carbachol and GTP during acute myocardial ischemia. J Mol Cell Cardiol 24: 471–475, 1992
10. Strasser RH, Marquetant R: Supersensitivity of adenylyl cyclase system in acute myocardial ischemia: Evaluation of three independent mechanisms. Basic Res Cardiol 85 (Suppl 1): 67–78, 1990
11. Will-Shahab L, Rosenthal W, Schulze W, Kuttner I: G protein function in the ischaemic myocardium. Eur Heart J 12 (Suppl F): 135–138, 1991
12. Wolff AA, Hines DK, Karliner JS: Preserved β-adrenoceptor-mediated adenylyl cyclase activity despite receptor and postreceptor dysfunction in acute myocardial ischemia. Am Heart J 128: 542–550, 1994
13. Van den Ende R, Batink HD, Michel MC, van Zwieten PA: Influence of ischemia and reperfusion on cardiac signal transduction. G protein content, adenylyl cyclase activity, cyclic AMP content, and forskolin and dibutyryl cyclic AMP-induced inotropy in the rat Langendorff heart. Fundam Clin Pharmacol 8: 408–416, 1994
14. Maisel AS, Ransnas LA, Insel PA: β-adrenergic receptors and Gs protein in myocardial ischemia and injury. Basic Res Cardiol 85 (Suppl 1): 47–56, 1990
15. Strasser RH, Krimmer J, Braun-Dullaeus R, Marquetant R, Kubler W: Dual sensitization of the adrenergic sysytem in early myocardial ischemia: independent regulation of β-adrenergic receptors and the adenylyl cyclase. J Mol Cell Cardiol 22: 1405–1423, 1990
16. Thandroyen FT, Muntz KH, Buja LM, Willerson: Alterations in β-adrenergic receptors, adenyate cyclase, and cyclic AMP concentrations during acute myocardial ischemia and reperfusion. Circulation 82 (Suppl II): II-30–II-37, 1990
17. Ohyanagi M, Matsumori Y, Iwasaki T: β-adrenergic receptors in ischemic and nonischemic canine myocardium: Relation to ventricular fibrillation and effects of pretreatment with propranolol and hexamethonium. J Cardiovasc Pharmacol 11: 107–114, 1988
18. Susanni EE, Manders WT, Knight DR, Vatner DE, Vatner SF, Homoy CJ: One hour of myocardial ischemia decreases the activity of the stimulatory guanine-nucleotide regulatory protein Gs. Circ Res 65: 1145–1150, 1989
19. Ohyanagi M, Yamamoto J, Nakamura K, Shibuya J, Morita M, Masutani M, Arii T, Iwasaki T: Messenger RNA for the guanine nucleotide-binding regulatory protein (G protein) is reduced in the acute ischemic myocardium. J Mol Cell Cardiol 27: 1131–1139, 1995
20. Tobise K, Abe M, Onodera S: Prolonged period of global ischemia causes no change in the GTP-binding proteins in the isolated perfused rat heart. Eur J Pharmacol 208: 183–187, 1991
21. Kiuchi K, Shen Y-T, Vatner SF, Vatner DE: Mechanisms mediating responsiveness to β-adrenergic stimulation after coronary reperfusion in conscious dogs. Am J Physiol 267: H1578–H1588, 1994
22. Fu L-X, llebekk A, Kirkeben KA, Aksnes G, Waagstein F, Bergh C-H, Hoebeke J, Liang Q-M, Hjalmarson A: Oxygen free radical injury and Gs mediated signal transduction in the stunned porcine myocardium. Cardiovasc Res 26: 449–455, 1992
23. Fu MLX, llebekk A, Xu X, Pergola G, Eriksson P, Kirkeben KA, Aksnes G, Hoebeke J, Hjalmarson A: Gi-mediated muscarinic adenylyl cyclase inhibition in timolol-treated stunned porcine myocardium. Acta Physiol Scand 151: 291–299, 1994
24. Murry CE, Jennings RB, Reimer KA: Preconditioning with ischemia: a delay of lethal cell injury in ischemic myocardium. Circulation 66: 1150–1153, 1982
25. Murry CE, Richard VJ, Reimer KA, Jennings RB: Ischemic preconditioning slows energy metabolism and delays ultrastructural damage during a sustained ischaemic episode. Circ Res 66: 913–931, 1990
26. Liu GS, Thornton J, Van Winkle DM, Stanley AWH, Olsson RA, Downey JM: Protection against infarction afforded by preconditioning is mediated by A1 adenosine receptors in rabbit heart. Circulation 84: 350–356, 1991
27. Thornton JD, Liu GS, Downey JM: Pretreatment with pertussis toxin blocked the protective effects of preconditioning: evidence for a G-protein mechanism. J Mol Cell Cardiol 25: 311–320, 1993
28. Niroomand F, Weinbrenner C, Weis A, Bangert M, Schwencke C, Marquetant R, Beyer T, Strasser RH, Kubler W, Rauch B: Impaired function of inhibitory G proteins during acute myocardial ischemia of canine heart and its reversal during reperfusion and a second period of ischemia. Possible implications for the protective mechanism of ischemic preconditioning. Circ Res 76: 861–870, 1995
29. Kitakaze M, Kitabatake A: Increased 5′-nucleotidase activity caused by protein kinase C enhances adenosine production in hypoxic cardiomyocytes. Circulation 84 (Suppl II): II-620, 1991
30. Fu L-X, Kirkeboen KA, Liang Q-M, Sjogren K-G, Hjalmarson A, llebekk A: Free radical scavenging enzymes and G protein mediated receptor signaling systems in ischaemically preconditioned porcine myocardium. Cardiovasc Res 27: 612–616, 1993
31. Gross GJ, Auchampach JA: Blockade of ATP-sensitive potassium channels prevents myocardial preconditioning in dogs. Circ Res 70: 223–233, 1992
32. Thornton JD, Thornton CS, Sterling DL: Blockade of ATP-sensitive potassium channels increases infarct size but does not prevent preconditioning in rabbit hearts. Circ Res 72: 44–49, 1993
33. Iwase T, Murakami T, Tomita T, Miki S, Nagai K, Sasayama S: Ischemic preconditioning is associated with a delay in ischemia-induced reduction of β-adrenergic signal transduction in rabbit hearts. Circulation 88: 2827–2837, 1993
34. Hammond HK, Roth DA, McKiman MD, Ping P: Regional myocardial downregulation of the inhibitory guanine triphosphate-binding protein (Giα2) and β-adrenergic receptors in a porcine model of chronic episodic myocardial ischemia. J Clin Invest 92: 2644–2652, 1993
35. Longabaugh JP, Didsbury J, Spiegel A, Stiles GL: Modification of the

rat adipocyte A1 adenosine receptor-adenylate cyclase system during chronic exposure to an A1 adenosine receptor agonist: Alterations in the quantity of Gsα and Giα are not associated with changes in their mRNAs. J Biol Chem 262: 841–847, 1987

36. Meggs LG, Huang H, Li P, Capasso JM, Anversa P: Chronic coronary arterial stenosis impairs α1-adrenoceptor signaling and cardiac performance in rats. Am J Physiol 263: H929–H938, 1992
37. Chilian WM, Layne SM: Coronary microvascular responses to reductions in perfusion pressure: Evidence for persistent arteriolar vasomotor tone during coronary hypoperfusion. Circ Res 66: 1227–1238, 1990
38. Komaru T, Lamping KG, Eastham CL, Dellsperger KC: Role of ATP-sensitive potassium channels in coronary microvascular autoregulatory responses. Circ Res 66: 1227–1238, 1990
39. Narishige T, Egashira K, Akatsuka Y, Katsuda Y, Numaguchi K, Skata M. Takeshita A: Glibenclamide, a putative ATP-sensitive K^+ channel blocker, inhibits coronary autoregulation in anesthetized dogs. Circ Res 73: 771–776, 1993
40. Komaru T, Wang Y, Akai K, Sato K, Sekiguchi N, Sugimura A, Kumagai T, Kanatuka H, Shirato K: Pertussis toxin-sensitive G protein mediates coronary microvascular control during autoregulation and ischemia in canine heart. Circ Res 75: 556–566, 1994
41. Gawler D, Milligan G, Spiegel AM, Unson CG, Houselay MD: Abolition of the expression of inhibitory guanine nucleotide regulatory protein Gi activity in diabetes. Nature 327: 229–232, 1987

Molecular and Cellular Biochemistry **160/161:** 159–166, 1996.

Role of reactive oxygen species in cardiovascular aging

Claudio Muscari, Antonella Giaccari, Emanuele Giordano, Carlo Clô, Carlo Guarnieri and Claudio Marcello Caldarera
Center of Research on Heart Metabolism, Department of Biochemistry 'G. Moruzzi,' University of Bologna, Bologna, Italy

Abstract

Biochemical and structural changes occurring in the myocardium with aging are mainly resulting from the association of a general tissue atrophy with the hypertrophy of the remaining myocytes. Whilst hypertrophy seems to be a compensatory process to the loss of cardiomyocytes and to a mild systolic hypertensive condition that accompanies elderly people, atrophy should be the modification more closely related to aging '*per se.*' In support to the free radical theory of aging, several signs of oxidative damage have been shown in the aged heart, such as lipofuscin accumulation, decreased phospholipid unsaturation index, greater formation of both hydrogen peroxide and 8-hydroxy-2'deoxyguanosine. As a compensatory reaction, the activities of the main oxygen-radical scavenger enzymes are stimulated in the mitochondria of aged rat heart. Endothelium-mediated vasoregulation is more susceptible to oxidative stress in aged with respect to young rats, suggesting that also the vasculature can be negatively influenced by the oxygen free radicals generated during aging. The possible primary role of oxygen free radicals in the development of myocardial atrophy is also discussed. Mol Cell Biochem **160/161**: 159–166, 1996)

Key words: reactive oxygen species, 8-hydroxy-2'-deoxyguanosine, heart, nitric oxide, aging, rat

Introduction

Cardiovascular functions considerably change with age, not only for the increasing incidence of coronary diseases, but also for the several modifications in life-style of elderly subjects, i.e. a decreased physical activity. Therefore, it is extremely difficult to identify structural and functional features of heart and vessels exclusively dependent on the natural process of aging.

The general increase in systolic blood pressure that accompanies senescence [1, 2], although not usually consistent, can partly explain the hypertrophic structure of the aged heart in mammals. In fact, a variety of biochemical and cellular modifications observed in the heart of aged mammals, including humans, have been described also in young subjects whose cardiac muscle becomes hypertrophied as a consequence of an hemodynamic overload [3].

Most changes found both in aged and hypertrophied cardiomyocytes affect the excitation-contraction coupling, leading to a prolongation of both contraction and relaxation times. This is mainly due to an increased duration of the action potential [4] and to a reduced biosynthesis of the Ca^{2+}-stimulated ATPase pump of sarcoplasmic reticulum [5, 6] both causing a prolongation of the cytosolic Ca^{2+} transient without affecting the developed peak-force [7].

An altered contractility has also been shown in hearts of elderly men. Clinical studies report both a prolongation in left ventricular ejection time and a decrease in the diastolic filling rate [8–11].

In spite of these modifications, cardiac performance (i.e. cardiac output, ejection fraction) seems to be adequate at rest in most elderly subjects without evident signs of heart disease. By contrast, older subjects can not reach the maximal heart rate obtained by younger people under a physical effort [12, 13]. This seems to be linked to an age-dependent reduction in contractile response of cardiac muscle to catecholamine stimulation [14, 15]. Nevertheless, cardiac output is maintained because of an increased diastolic enlargement of the left ventricular chamber, according to the Frank-Starling's law [12].

Since the myocardial adaptations described above may be largely attributed to the hypertrophic process, we will focus

Address for offprints: C.M. Caldarera, Dipartimento di Biochimica 'G. Moruzzi,' Universita' di Bologna, Via Irnerio, 48, 40126 Bologna, Italy our attention on other biochemical and structural changes that are more closely connected to aging '*per se*.' Aging is also a cause of atrophy of myocardial tissue because the total volume occupied by myocytes decreases with age [16]. This is accompanied by a general decrease in protein synthesis [17, 18], loss of myocytes [19] and increase in the amount of collagen and elastin [20]. Thus the hypertrophy of the remaining myocytes could be considered as a compensative response to systolic hypertension and/or myocardial atrophy that naturally occur in elderly people. Conversely, the heart of a young subject with a moderate degree of afterload-induced hypertrophy, which leads to an enhancement of heart weight as little as that of aged subjects, does not develop signs of atrophy.

In order to investigate the myocardial modifications that are intrinsically related to the aging process, many researchers have looked at the free radical theory of aging [21] as the most invoked hypothesis of cellular senescence, especially for tissues with elevated oxidative metabolism. Free radicals generated by oxygen metabolism are among the most reactive and therefore harmful substances naturally produced in living organisms. The possibility that free radicals mediate the age-related biochemical changes of the heart is attractive; however the relationship between some typical alterations of the aged heart and oxidative stress has not been well elucidated. Here we try to speculate whether the most documented free radical-driven alterations observed in heart and vessels of aged animals may contribute to the natural process of cardiovascular aging.

Oxidative damage in the aged myocardium

A large amount of lipofuscin is detectable in cytosol, lysosomes and mitochondria of senescent mammalian cells [23, 24]. This fluorescent pigment represents an end-product of the conjugation of malondialdehyde, a stable result of lipid peroxidation, with primary amino groups present in the cell. Lipofuscin accumulates during aging because it is not degraded by cells. Its formation is closely dependent on oxygen; in fact, an increase of lipid fluorescence in cultured cardiomyocytes is related to either increasing oxygen tension or prolonged exposure to molecular oxygen [24]. It has been recently proposed that lipofuscin formation is due to an interplay between the production of hydrogen peroxide by mitochondria and the iron delivered from cellular structures undergoing intralysosomal degradation [25]. The resulting generation of hydroxyl free radicals via Fenton-type mechanisms would induce lipid peroxidation and intermolecular cross-linking, with the formation of lipofuscin. In the rat heart the cytosolic and mitochondrial levels of fluorescent pigments are significantly increased at 14–18 months of age [26], a period considered in the mature life of the animal (Table 1). In the same age period, the production of mitochondrial hydrogen peroxide and the concentration of oxidized glutathione increase in comparison to younger rats and this is paralleled by a decrease in the reduced form of glutathione [27]. No changes in the production of superoxide radicals (O_2-) were found in the submitochondrial particles at the level of the rotenone-inhibited region of the respiratory chain during the lifespan of the rat. On the contrary, a marked increase in O_2- formation was observed at the level of the antimycin-inhibited region of the electron transport chain in 14-month and 18-month-old rats [26] (Fig. 1). These data indicate that the mitochondrial production of oxygen free radicals in the heart muscle peaks at middle age in the rat. This elevated flux of oxygen radicals is probably not thoroughly tolerated, since it corresponds to a marked decline in the survival curve of a typical aging colony of rats. One may speculate that the oldest animals represent a 'survivor' group able to overcome the oxidative stress at middle age. However, also the 24-month-old rats showed signs of increased production of oxygen-reactive species at myocardial level, as evidenced by a 25% increase in mitochondrial hydrogen peroxide generation with respect to 3-month-old rats [27, 28].

The parallel increase in the activity of mitochondrial superoxide dismutase [29], glutathione peroxidase [30] and catalase [31], which all are enzymes neutralizing free radical toxicity, might be interpreted as a compensatory response of the cell to the increased oxidative stress. On the contrary, the cytosolic activities of these enzymes decline or are not modified in senescent rat hearts. Glutathione-S-transferase activity also decreases with age [32].

When hearts from rats aged 3 months and 24 months were isolated and subjected to the same periods of ischemia and reperfusion, conditions that rise the production of oxygen free radicals, the greatest levels of malondialdehyde were observed in the younger group [33]. These findings seem to indicate that more efficient antioxidant defences are present in the mitochondria of the aged heart.

Other antioxidative mechanisms are probably activated by the myocardial cell during aging to contrast mitochondrial oxygen toxicity. For instance, it has been reported that ubiquinone biosynthesis is stimulated in heart mitochondrial membranes of 24-month-old rats [34]. Since the mitochondrial content of coenzyme Q_9, the most abundant form of ubi-

Table 1. Changes in fluorescent product levels of rat heart during aging (% fluorescence/mg protein).

Age	Tissue homogenate	Mitochondria
3 months	4.5 ± 0.3	56.4 ± 4.7
14 months	4.9 ± 0.6	57.4 ± 3.8
18 months	5.8 ± 0.4*	87.2 ± 5.5*
24 months	9.1 ± 1.1*	98.3 ± 6.1*

Values are expressed as mean ± S.D. of 6 separate experiments. * $p < 0.05$ vs the corresponding 3-month-old group. Adapted from Muscari *et al.* [26]

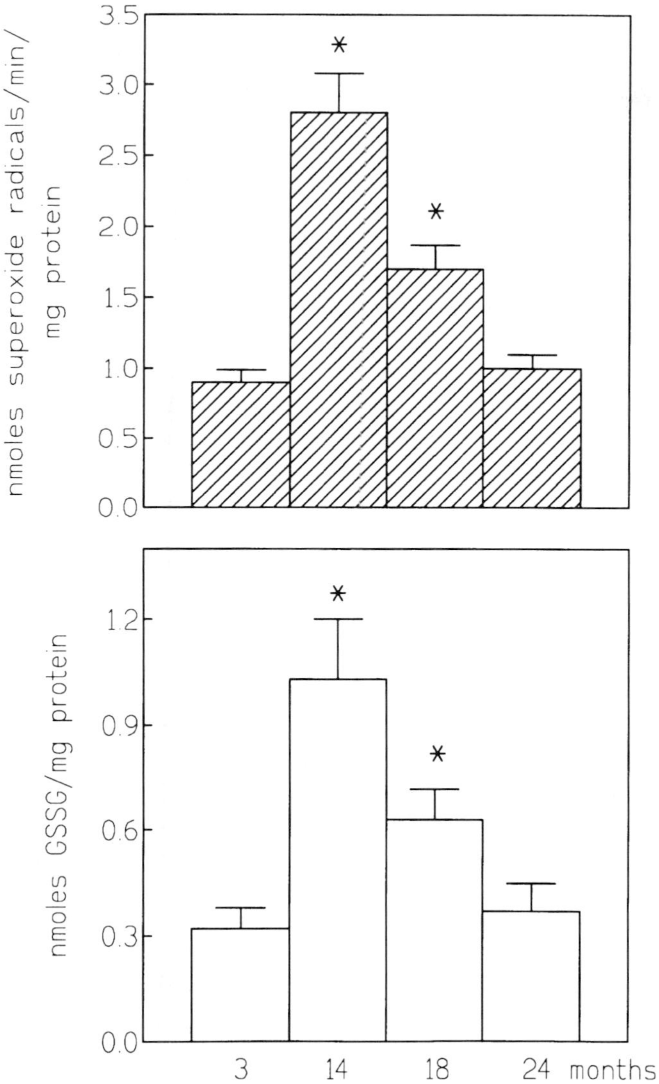

Fig. 1. Similar behaviour of myocardial superoxide radical production and GSSG levels in rats at different months of age. Upper panel: Production of O_2- by submitochondrial particles at the level of antimycin-inhibited region of the respiratory chain. Lower panel: Myocardial GSSG levels. *$p < 0.05$ vs 3-month-old group. Values are expressed as mean ± S.D. of 4–6 separate experiments. Adapted from Muscari *et al.* [26].

quinone in the rat, was not modified in the aged myocardium, it has been supposed that the greater rate of ubiquinone biosynthesis could compensate an increased rate of its degradation. It is interesting to underline that the reduced form of coenzyme Q is a powerful scavenger of superoxide radicals. The fact that enhanced ubiquinone synthesis has been observed only in mitochondria and not in microsomes suggests that this event may be triggered by the elevated presence of radical species in the former organelles.

The study of mitochondrial membrane composition shows a decreased unsaturation index for most phospholipids [35] and an increased conjugated diene concentration [28]. These alterations cause a functional perturbation of membranes that mirrors the progressive loss of fluidity with age. A rise in superoxide radical formation, paralleled by a decrease in membrane fluidity, has been observed in brain, heart and liver plasma membranes of aged rats [36]. This local superoxide radical generation is likely enzyme-mediated since it is Ca^{2+} sensitive and inhibited by both indomethacin, a ciclooxygenase inhibitor, and bromophenocyl bromide, a phospholipase A_2 inhibitor. On the contrary, in old rat hearts, no modifications were observed in fatty acid composition at microsome level, neither a consistent loss of delta-6-desaturase activity [37].

Table 2. Effect of malondialdehyde (MDA) on adenylate cyclase activity in heart membranes from young and aged rats

	% of basal activity			
	1 month		24 months	
Treatment	-MDA	+MDA	-MDA	+MDA
None	100 (80)*	100 (56)*	100 (40)*	100 (34)*
Isoprenalin (100 μM)	615	338	208	166
Forskolin (100 μM)	2540	1270	1183	946
Spermine (100 μM)	58	76	82	96

*pmoles cAMP/mg protein/min

The response *in vitro* of adenylate cyclase to stimulants, such as isoprenalin, a beta-adrenergic agonist, or forskolin, a diterpene which directly activates the catalytic component of the transduction system, is strongly lower with the enzyme from aged (24 month-old) rat heart than with that from young (1 month-old) one. Also spermine, a biogenic amine able to inhibit adenylate cyclase activity, is less effective with the enzyme from aged hearts. Preincubation (10 min) of heart membranes from young rats with 0.5 μM MDA leads to a similar reduction of responsiveness to both stimulants and inhibitors.

In both animal models and human the contractile response to beta-adrenergic stimulation declines with advancing age. Together with the evidence of a lower density of beta-adrenergic receptors in sarcolemmal membranes [38], a progressive impairment of functionality of membrane-bound adenylate cyclase has been reported [39]. Preincubation of heart membranes from young rats with 0.5 μM malondialdehyde caused a reduction in basal adenylate cyclase activity of the same magnitude of that induced by aging (Table 2). Besides, the presence of malondialdehyde impaired the response of adenylate cyclase to both stimulating and inhibiting effectors. The attenuation of the enzymatic response to various modulators was also observed in cardiomyocytes obtained from aged rats [40].

Age-related oxidation of myocardial DNA

Most cellular components are known to be susceptible to damage by free radicals. However, while combined processes of degradation and biosynthesis operate in concert to maintain, as long as possible, a functional pool of intact proteins, lipids and carbohydrates, damaged DNA is not substituted and then accumulates, particularly in non-replicating cells [41]. Although DNA is subjected to numerous repair processes, it has been estimated that the average number of oxidative damages to DNA per cell per day is about 10,000 in human and about tenfold higher in rat, due to the more intense oxidative metabolism in the latter species [42]. The concentration of an oxidative adduct of guanosine, 8-hydroxy-2'-deoxyguanosine (8-OH-dG), is generally increased in aged rat tissues. [43]. Also myocardial 8-OH-dG content is twofold higher in 27-month-old than in 8-month-old mice [44].

The amount of this oxidative adduct of guanosine is about hundred fold greater in mitochondrial DNA (mtDNA) than in nuclear DNA [45]. As a matter of fact mtDNA, which is histone free, is more susceptible to various chemical attacks and more frequently damaged by oxygen free radicals, in view of its close spatial relation to the inner mitochondrial membrane. Moreover, DNA repair is much less efficient in mitochondria than in the nucleus [46]. The mammalian organelles are not able to provide specific recombinational substitutions of the modified bases, but excise the altered nucleosides. Thus, mitochondrial endonucleases initiate the degradation of damaged mtDNA, contributing to select the unaffected genome that will assure the survival of the wild population of mitochondria. However, these repair mechanisms appear to be almost inadequate in mitochondria, allowing the accumulation of a variety of damages to mtDNA.

In human hearts, 8-OH-dG has been correlated to the presence of a particular kind of mtDNA mutation that produces the detaching of a double stranded fragment, generating a molecule of deleted mtDNA [47]. This molecule is smaller than normal mtDNA, so it should duplicate faster during mitochondrial neogenesis.

It has been postulated that the accumulation of deleted molecules of mtDNA over the lifetime of an organism may be a causative factor for the aging process. In support of this view, an increasing number of studies have documented age-dependent increases in the level of deleted mtDNA in human heart and in other non-dividing or slow-dividing cells [47, 48].

The most frequent deletion related to aging occurs in human tissues between nucleotide 8470 and 13459, a portion of the mitochondrial genome coding for some subunits of ATP synthase, cytochrome oxidase and NADH dehydrogenase [48]. Therefore, a possible consequence of the accumulation of this deletion is a progressive decrease of the bioenergetic capacity with age. Indeed, the amount of cytochrome-oxidase has been found significantly reduced in the heart of aged subjects [49]. Besides this deletion of 4977 base pair, the so called 'common deletion,' since it was frequently detected in a number of mitochondrial neurodegenerative disease [50], a 7.4 Kb deletion is specifically produced in the heart [51]. Hayakawa *et al.* [47] reported an age-associated increase in the human myocardial content of deleted mtDNA up to 7% with respect to the normal wild type of mtDNA. The same group found a parallel rise of the 8-OH-dG level that accumulated with age up to 1.5%. It has been proposed that 8-OH-dG can induce strand breaks in mtDNA, leading to the generation of long regions of single-stranded DNA, which are a prerequisite for large deletions. Moreover, the production of reactive oxygen species would be enhanced in the mitochondria with deleted mtDNA, because of the absence of some key-proteins of the respiratory chain.

Ozawa's group has also recently associated the fragmentation of cardiac mtDNA with premature aging in a 19-year-old patient using a very extensive approach, since 180 kinds of primer pair for Polymerase Chain Reaction were employed [52]. This study revealed up to 235 different types of deletion of various size, accounting for 84% of the total mtDNA. In this patient the amount of 8-OH-dG corresponded to that of a normal subject aged 78 years.

The study of cardiac energetics in the aged myocardium shows that ATP and phosphocreatine levels are reduced by some 20% [53, 54]. However, a causal link between oxidative damage to mtDNA and defective mitochondrial energy output has not been demonstrated; moreover, another mitochondrial metabolic alterations have been related to the aging process, such as the impairment of fatty acid beta-oxidation, Krebs cycle and ATP translocation from matrix to intermembrane space, and the reduced response to Ca^{2+} increase of Ca^{2+}-dependent mitochondrial enzymes [55, 56].

Free radicals and vascular tone

The vascular tree is subjected with age to several modifications leading to marked structural changes in the vessel wall, e.g. a reduced compliance of the aorta which is among the causes of systolic hypertension in elderly people, since it provokes a blood wave reflecting from the periphery to the heart [57].

Beside the structural modification of the vessels arising with age, functional mechanisms have also been evoked to explain the impaired regulation of its smooth muscle layer. In particular, endothelium-mediated vasodilation seems to be blunted by the aging process [58, 59]. Oxygen free radicals can influence the vascular function by a direct effect on the vascular smooth muscle cells (VSMC) or through the paracrine release of endothelial factors [60, 61]. Another free radical molecule has recently been shown to play a critical role in vascular homeostasis, namely the nitric oxide radical (•NO), a powerful vasorelaxant agent generated in the endothelium by the enzyme NO synthase in response to a variety of chemical and physical stimulations [62]. These two radical forms, O_2- and •NO, readily interact each other to

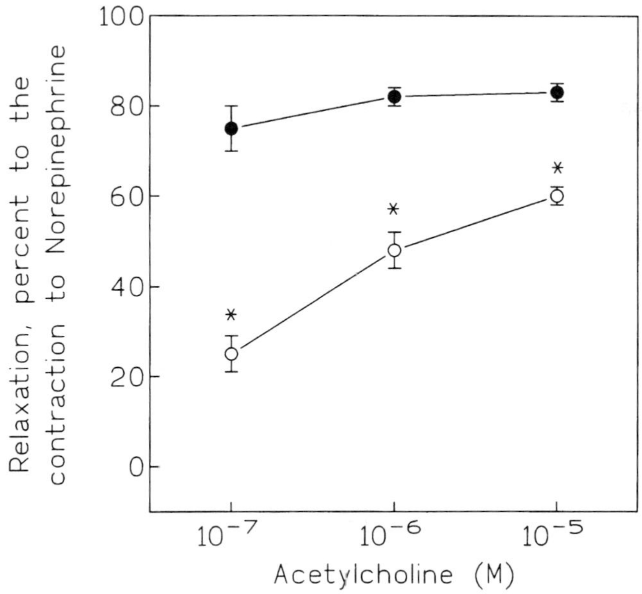

Fig. 2. Dose-response curve of relaxation to Acetylcholine by aortic strips from 3-month-old (filled) and 24-month-old (unfilled) rats. Relaxation values are expressed as percentage of the contraction to 0.1 μM Norepinephrine. Values are expressed as mean ± S.E. of 3–5 separate experiments. *$p < 0.01$ vs the corresponding relaxation value obtained with aortic strips from 3-month-old rats. Adapted from Guarnieri *et al.* [64] .

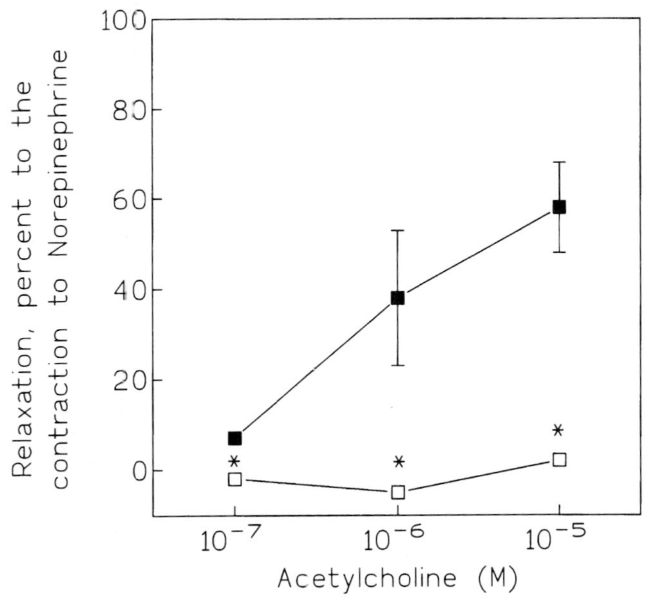

Fig. 3. Dose-response curve of relaxation to Acetylcholine by aortic strips from 3-month-old (filled) and 24-month-old (unfilled) rats previously exposed to oxygen radical species. Before norepinephrine contraction, the strips were subject for 30 min to a flux of oxygen radicals generated by adding 0.5 mM hypoxanthine and 4 mIU xanthine oxidase into the organ bath. Values are expressed as mean ± S.E. of 3–5 separate experiments. *$p < 0.05$ vs the corresponding relaxation value obtained with aortic strips from 3-month-old rats. Adapted from Guarnieri *et al.* [64] .

generate peroxynitrite, a strong oxidant agent which rapidly decomposes under physiological conditions, forming the more harmful reactive oxygen species hydroxyl radical. Therefore, an increased O_2- production, not only results in the impairment of the physiological roles of nitric oxide in the vascular wall, but also leads to the formation of more damaging chemical compounds [63].

Our recent experiments demonstrated that the vasorelaxant, •NO-driven effect of acetylcholine (ACh) on rat aortic strips, was markedly impaired in 24-month-old with respect to 3-month-old animals (Fig. 2) and vasodilating response to ACh was lowered after exposure of the strips to an oxygen-radical flow. This effect was more pronounced in the tissue of 24-month-old rats (Fig. 3) [64]. Vascular function of aged rats is thus particularly susceptible to oxidative stress and endothelial cells seem to be a main target of oxygen free radicals.

Conclusion

The cardiovascular system is influenced by the natural presence of oxygen free radicals, especially during aging. So far it is still impossible to define whether oxidative damage to cellular components may be a cause or only a parallel phenomenon of senescence. Myocardial atrophy is related to cell death, a process that may be induced by reactive oxygen species [65, 66] (Fig. 4). This deleterious event could be triggered in the aged heart by the increase in mitochondrial hydrogen peroxide generation. Recently it has been reported that oxygen stress reduced cell viability and induced the formation of several mtDNA deletions in cultured fibroblasts carrying normal mitochondrial genome. By contrast, a fibroblast cell line lacking mtDNA exhibited a marked resistance to cell death, indicating that fragmentation of mtDNA is an important modulator in eliminating damaged cells [67]. Moreover, a cell death regulating factor, the Bcl-2 proto-oncogene product, is located within mitochondria and is closely involved in the antioxidant defence system operated by the cell. Thus, damaged mitochondria could contribute to regressive processes such as aging and degenerative disease, by controlling some regulating factors that modulate cell death machinery.

Other works indicate that food restriction might be a tool to prolong life-span by reducing calory production and oxidative stress [44, 68]. Mice fed with a diet that is 40% lower in calories exhibited an extension in average life span of the same magnitude. The rates of mitochondrial superoxide and hydrogen peroxide generation from heart, brain and kidney were lower with respect to the *ad libitum* fed animals. This effect was connected with a decrease in protein carbonyl content and 8-OH-dG concentration that was significant in all the three

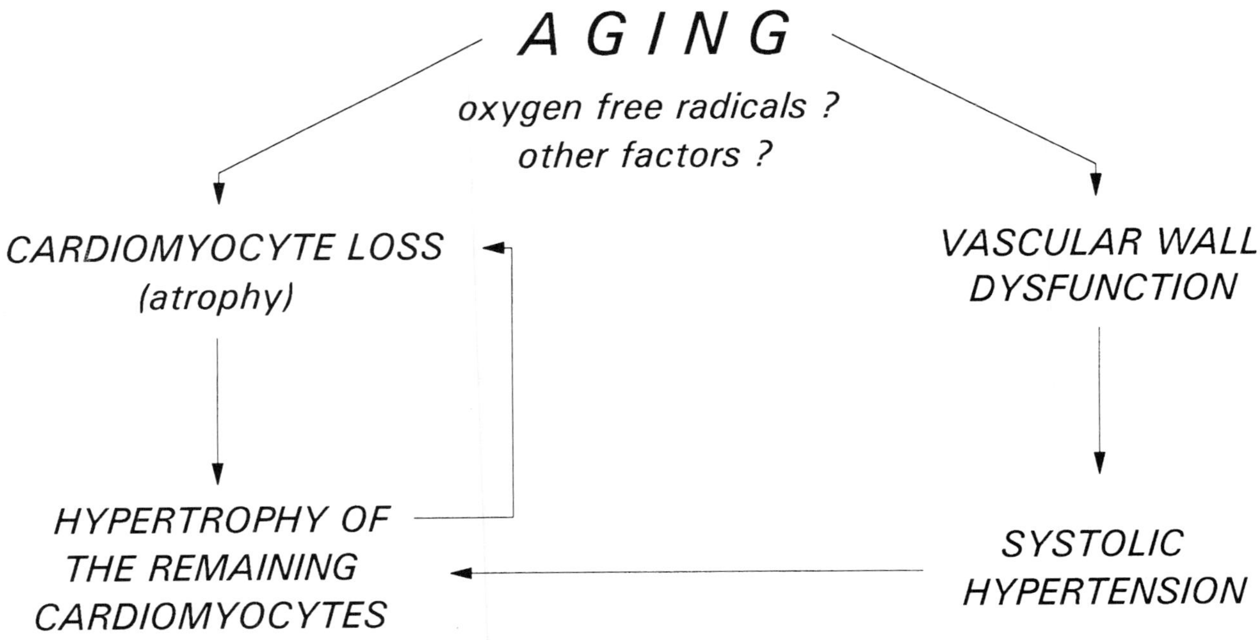

Fig. 4. Hypothetic causative relationship between oxygen free radicals and cardiovascular aging.

organs. In contrast, in all tissues examined there were no modifications in the scavenger-enzyme activity.

These results, taken together, suggest that one primary mechanism of aging in the heart and other long-lived postmitotic cells is associated with oxidative damage arising from the spontaneous generation of oxygen reactive species [69].

The main sites of O_2- formation in the respiratory chain are located at the level of complex I and III [28]. About 2% of molecular oxygen is continuously converted in oxygen reactive species. It is conceivable that the basal production of oxy-radicals, if we assume that they have a causal role in the aging process, may be sufficient to determine a constant death of myocardial cells. However, it is significant that the major amount of fluorescent products is formed in rat cardiac mitochondria beginning from 14 months of age, when the mitochondrial O_2- production is more evident and well correlated with the parallel elevation of H_2O_2 production and the peak level of Mn^{2+} -SOD activity [29]. This corresponds to an increased oxidative damage to cell components [26, 27] that probably trigger some signals for accentuated protein turnover [70] and mitochondrial biogenesis [71]. These events might favour the adaptation of the heart muscle toward a condition of hypertrophy [72] or non-mitotic division of cardiac cells [19], compensating for both decreased myocyte mass and number.

In conclusion, oxygen free radicals can be considered as factors that influence myocardial aging and senescence. More studies are needed to better define the specific role of these reactive species in the process of hypertrophy and atrophy that both characterize the aged heart.

References

1. Rosentahal J: Aging and the cardiovascular system. Gerontology 33 (suppl 1): 3–8, 1987
2. Walsh R: Cardiovascular effects of the aging process. Am J Med 82 (suppl B): 34–40, 1987
3. Lakatta EG: Myocardial adaptations in advanced age. Basic Res Cardiol 88 (suppl 2): 125–133, 1993
4. Jullien T, Verdetti J: Comparative electrophysiological response of young and old rat myocardium to pharmacological agents. Gen Pharmacol 19: 759–766, 1988
5. Besse S, Assayag P, Delcayre C, Carre F, Cheav SL, Lecarpentier Y, Swynghedauw B: Normal and hypertrophied senescent rat heart: mechanical and molecular characteristics. Am J Physiol 265 (1 Pt 2): H183–H190, 1993
6. Maciel LMZ, Polikar R, Rohrer D, Popovich BK, Dillmann WH: Age induced decreases in the messenger RNA coding for the sarcoplasmic reticulum Ca2+ -ATPase of the rat heart. Circ Res 67: 230–234, 1990
7. Lakatta EG, Yin CP: Myocardial aging: functional alterations and related cellular mechanisms. Am J Physiol 242: H927–H941, 1982
8. Rengo F, Ferrara N, Leosco D: Ventricular function in the elderly. Aging 3: 9–17, 1991
9. Merino A, Alegria E, Castello R, Martinez-Caro D: Influence of age on left ventricular contractility. Am J Cardiol 62: 1103–1108, 1988
10. Willems JL, Roelandt H, De Greest H, Kesteloot H, Joossems JV: The left ventricular ejection time in elderly subjects. Circulation 42: 37–42, 1970
11. Spirito P, Maron B: Influence of aging on Doppler echocardiographic indices of left ventricular diastolic function. Br Heart J 59: 672–679, 1988
12. Fleg JL: Alterations in cardiovascular structure and function with advancing age. Am J Cardiol 57: 33C–44C, 1986
13. Safar M: Ageing and its effects on the cardiovascular system. Drugs 39 (suppl 1): 1–8, 1990
14. Lakatta EG, Gerstenblith G, Angell CS, Shock NW, Weisfeldt ML: Diminished inotropic response of aged myocardium to catecholamines.

Cir Res 36: 262–269, 1975
15. Amerini S, Fusi F, Piazzesi G, Mantelli L, Ledda F, Mugelli A: Influence of age on the positive inotropic effect mediated by alpha-and beta-adrenoceptors in rat ventricular strips. Dev Pharmacol Ther 8: 34–42, 1985
16. Carrier L, Chassagne C, Boheler KR, Schwartz K: Molecular basis of cardiac aging. Presse Med 21: 1196–1198, 1992
17. Crie JS, Milliward DJ, Bates PC, Griffin EE, Wildenthal K: Age-related alterations in cardiac protein turnover. J Mol Cell Cardiol 13: 589–598, 1981
18. Starnes JW, Edington DW, Beyer RE: Myocardial protein synthesis during ageing and endurance exercise in rats. J Gerontol 38: 660–665, 1983
19. Anversa P, Palankal T, Sonnenblick EH, Olivetti G, Meggs LG, Capasso JM: Myocyte cell loss and myocyte cellular hyperplasia in the hypertrophied aging rat heart. Circ Res 67: 871–885, 1990
20. Eghbali Me, Eghbali Ma, Robinson TF, Seiter S, Blumenfeld OO, Cohen-Gould L, Geraci M, Valentine M: Collagen accumulation in heart ventricles as a function of growth and aging. Cardiov Res 23: 723–729, 1989
21. Harman D: Prolongation of life: role of free radical reaction in aging. Am Geriat Soc 8: 721–735, 1969
22. Ikeda H, Tauchi H, Shimasaki H, Veta N, Sato T: Age and organ difference in amount and distribution of autofluorescent granules in rats. Mech Ageing Dev 31: 139–146, 1985
23. Shimasaki H, Nozawa T, Privett OS, Anderson WR: Detection of age related fluorescent substance in rat tissue. Arch Biochem Biophys 183: 443–451, 1977
24. Gao G, Johansson U, Rundquist I, Ollinger K: Lipofuscin-induced autofluorescence of living neonatal rat cardiomyocytes in culture. Mech Ageing Dev 73: 79–86, 1994
25. Brunk UT, Jones CB, Sohal RS: A novel hypothesis of lipofuscinogenesis and cellular aging based on interactions between oxidative stress and autophagocytosis. Mutat Res 275: 395–403, 1992
26. Muscari C, Frascaro M, Guarnieri C, Caldarera CM: Mitochondrial function and superoxide generation from submitochondrial particles of aged rat hearts. Biochim Biophys Acta 1015: 200–204, 1990
27. Muscari C, Caldarera CM, Guarnieri C: Age-dependent production of mitochondrial hydrogen peroxide, lipid peroxides and fluorescent pigments in the rat heart. Bas Res Cardiol 85: 172–178, 1990
28. Nohl H, Hegner D: Do mitochondria produce oxygen radicals *in vivo*? Eur J Biochem 82: 563–567, 1878
29. Lammi-Keefe CJ, Swan PB, Hegarty PVJ: Copper-zinc and manganese superoxide dismutase activities in cardiac and skeletal muscles during aging in male rats. Gerontology 30: 153–158, 1984
30. Hazelton GA, Lang CA: Glutathione peroxidase and reductase activities in the aging mouse. Mech Ageing Dev 29: 71–81, 1985
31. Nohl H, Hegner D, Summer KH: Responses of mitochondrial superoxide dismutase, catalase and glutathione peroxidase activities to aging. Mech Ageing Dev 11: 141–151, 1984
32. Ji LL, Dillon D, Wu E: Myocardial aging: Antioxidant enzyme systems and related biochemical properties. Am J Physiol 261 (2 Pt 2): R368–R392, 1991
33. Muscari C, Guarnieri C, Biagetti L, Finelli C, Caldarera CM: Influence of age on oxidative damage in mitochondria of ischemic and reperfused rat hearts. Cardioscience 1: 275–278, 1990
34. Muscari C, Biagetti L, Stefanelli C, Giordano E, Guarnieri C, Caldarera CM: Adaptive changes in Coenzyme Q biosynthesis to myocardial reperfusion in young and aged rats. J Mol Cell Cardiol 27: 283–289, 1995
35. Lewin MB, Timiras PS: Lipid changes with aging in cardiac mitochondrial membranes. Mech Ageing Dev 24: 343–351, 1984
36. Sawada M, Sester U, Carlson JC: Superoxide radical formation and associated biochemical alterations in plasma membrane of brain, heart, and liver during the lifetime of the rat. J Cell Biochem 48: 296–304, 1992
37. Lopez-Jimenez JA, Bordoni A, Hrelia S, Rossi CA, Turchetto E, Zamora-Navarro S, Biagi PL: Evidence for a detectable delta-6-desaturase activity in rat heart microsomes: aging influence on enzyme activity. Biochem Biophys Res Commun 192: 1037–1041, 1993
38. Gudmundsdottir E, Benediktsdottir VE, Gudbjarnason S: Combined effects of age and dietary fat on beta-receptors and Ca^{2+} channels in rat hearts. Am J Physiol 260 (1 Pt 2): H66–H72, 1991
39. Pignatti C, Tantini B, Zanfanti ML, Sacchi P, Clo' C: Influence of Mg^{2+} on the *in vitro* responsiveness of adenylate cyclase from hearts of aging rats. Cardioscience 4: 105–109, 1993
40. Tantini B, Pignatti C, Sacchi P, Zanfanti ML, Mariani F, Clo' C: Influence of spermine on adenylate cyclase responsiveness of cultured cardiomyocytes from aging rats. In: C.M. Caldarera, C. Clo' and M.S. Moruzzi (eds). Polyamines, Biological and Clinical Aspects, CLUEB, Bologna, Italy, 1994, pp 287–294
41. Holmes GE, Bernstein C, Bernstein H: Oxidative and other DNA damages as the basis of aging: A review. Mutation Res 275: 305–315, 1992
42. Ames BN, Gold LS: Endogenous mutagens and the causes of aging and cancer. Mutation Res 250: 3–16, 1991
43. Fraga CG, Shigenaga MK, Park JW, Degan P, Ames BN: Oxidative damage to DNA during aging: 8-hydroxy-2'-deoxyguanosine in rat organ DNA and urine. Proc Natl Acad Sci (USA) 87: 4533–4537, 1990
44. Sohal RS, Agarwal S, Candas M, Forster MJ, Lal H: Effect of age and caloric restriction on DNA oxidative damage in different tissues of C57BL/6 mice. Mech Ageing Dev 76: 215–224, 1994
45. Richter C: Reactive oxygen and DNA damage in mitochondria. Mutation Res 275: 249–255, 1992
46. Richter C: Do mitochondrial DNA fragments promote cancer and aging? FEBS Lett 241: 1–5, 1988
47. Hayakawa M, Hattori K, Sugiyama S, Ozawa T: Age-associated oxygen damage and mutations in mitochondrial DNA in human hearts. Biochem Biophys Res Commun 189: 979–985, 1992
48. Yen TC, Su JH, King KL, Wei YH: Ageing-associated 5Kb deletion in human liver mitochondrial DNA. Biochem Biophys Res Commun 178: 124–131, 1991
49. Muller-Hoker J: Cytochrome-c-oxidase deficient cardiomyocytes in the human heart. An age-related phenomenon. Am J Pathol 134: 1167–1173, 1989
50. Holt IJ, Harding AE, Morgan-Huges JA: Deletions of muscle mitochondrial DNA in patients with mitochondrial myopathies. Nature 331: 717–719, 1978
51. Hattori K, Tanaka M, Sujiyama S, Oayashi T, Ito T, Sataket T, Hanaki Y, Asai J, Nagano M, Ozawa T: Age-dependent increase in deleted mitochondrial DNA in the human heart: possible contributory factor to presbycardia. Am Heart J 21 (6 Pt 1): 1735–1742, 1991
52. Katsumata K, Hayakawa M, Tanaka M, Sujiyama S, Ozawa T: Fragmentation of human heart mitochondrial DNA associated with premature aging. Biochem Biophys Res Commun 202: 102–110, 1994
53. Muscari C, Finelli C, Stefanelli C, Flamigni F, Guarnieri C, Caldarera CM: Age-dependent differences of ATP breakdown and ATP-catabolite release in ischemic and reperfused hearts. Mech Ageing Dev 67: 1–11, 1993
54. Finelli C, Guarnieri C, Muscari C, Ventura C, Caldarera CM: Incorporation of [14C]hypoxanthine into cardiac adenine nucleosides: Effect of aging and post-ischemic reperfusion. Biochim Biophys Acta 1180, 262–266, 1993
55. Muscari C, Caldarera I, Rapezzi C, Branzi A, Caldarera CM: Biochemical correlates with myocardial aging. Cardioscience 3: 67–78, 1992

56. Guarnieri C, Muscari C, Finelli C, Caldarera CM: Age-related changes in cardiac mitochondrial energetics under the influence of calcium in rat. Cardioscience 4: 117–120, 1993
57. Yin FC: The aging vasculature and its effects on the heart. In: M.L. Weisfeldt (ed.). The Aging Heart. Raven Press, New York, 1980, pp 137–213
58. De Mey C, Vanhoutte PM: Effect of age and spontaneous hypertension on the tachyphylaxis to 5-hydroxytryptamine and angiontensin II in the isolated rat kidney. Hypertension 3: 718–724, 1981
59. Moritoki H, Matsugi T, Takase H, Ueda H, Tanioka A: Evidence for the involvement of cyclic GMP in adenosine-induced, age-dependent vasodilation. Br J Pharmacol 100: 569–575, 1990
60. Katusic ZS, Vanhoutte PM: Superoxide anion is an endothelium-derived contracting factor. Am J Physiol 257: H33–H37, 1989
61. Tesfamarian B, Cohen RA: Role of superoxide anion and endothelium in vasoconstrictor action of prostaglandin endoperoxide. Am J Physiol 262: H1915–H1919, 1992
62. Palmer RMJ, Ferrige AG, Moncada S: Nitric oxide release accounts for the biological activity of endothelium-derived relaxing factor. Nature 327: 524–526, 1987
63. Darley-Usmar V, Radomski M: Free radicals in the vasculature: the good, the bad and the ugly. The Biochemist 16(5): 15–18, 1994
64. Guarnieri C, Giordano E, Muscari C, Grossi L, Caldarera CM: Alphatocopherol pretreatment improves endothelium-dependent vasodilatation in aortic strips of young and aging rats exposed to oxidative stress. Mol Cell Biochem, in press
65. Persoon-Rothert M, Egas-Kenniphaas JM, van der Valk-Kokshoorm EJ, Buys JP, van der Laarse A: Oxidative stress-induced perturbations of calcium homeostasis and cell death in cultured myocytes: role of extracellular calcium. Mol Cell Biochem 136: 1–9, 1994
66. Bhatnagar A: Biochemical mechanisms of irreversible cell injury caused by free radical-initiated reactions. Mol Cell Biochem 137: 9–16, 1994
67. Yoneda M, Katsumata K, Hayakawa M, Tanaka M, Ozawa T: Oxygen stress induces an apoptotic death associated with fragmentation of mitochondrial genome. Biochem Biophys Res Commun 209: 723–729, 1995
68. Sohal RS, Ku HH, Agarwal S, Forster MJ, Lal H: Oxidative damage, mitochondrial oxidant generation and antioxidant defenses during aging and in response to food restriction in the mouse. Mech Ageing Dev 74: 121–133, 1994
69. Sohal RS, Brunk UT: Mitochondrial production of pro-oxidants and cellular senescence. Mutation Res 275, 295–304, 1992
70. Davies KJA: Protein damage and degradation by oxygen radicals. J Biol Chem 262: 9895–9901, 1987
71. Davies KJA, Quintanilha AT, Brooks GA, Parker L: Free radicals and tissue damage produced by exercise. Biochem Biophys Res Commun 107: 1198–1205, 1982
72. Meerson FZ: Development of modern components of the mechanism of cardiac hypertrophy. Circ Res 34 (suppl 2): 58–63, 1974

Molecular and Cellular Biochemistry **160/161**: 167–177, 1996.

Uncoupling of mitochondrial oxidative phosphorylation alters lipid peroxidation-derived free radical production but not recovery of postischemic rat hearts and post-hypoxic endothelial cells

I.E. Blasig,[1] B.F. Dickens[2], W.B. Weglicki[2] and J.H. Kramer[2]

[1]*Forschungsinstitut für Molekulare Pharmakologie, Forschungsverbund Berlin e.V., Alfred-Kowalke-Str. 4, D-10315 Berlin, Germany;* [2]*Departments of Medicine and Physiology, Division of Experimental Medicine, The George Washington University Medical Center, 2300 Eye St, N.W., Washington, D.C. 20037, USA*

Abstract

The contribution of mitochondrial free radical production towards the initiation of lipid peroxidation (LPO) and functional injury in the post-ischemic heart is unclear. Using the isolated rat heart model, the effects of the uncoupler of mitochondrial oxidative phosphorylation dinitrophenol (DNP, 50 μM final) on post-ischemic lipid peroxidation-derived free radical production and functional recovery were assessed. Hearts were subjected to 30 min total global ischemia followed by 15 min of reperfusion in the presence of DNP. As expected, DNP enhanced oxygen consumption before (11.3 ± 0.9 μmol/min, $p < 0.001$) and during reperfusion (at 10 min: 7.9 ± 0.7 μumol/min), compared to the heart with control treatment (8.2 ± 0.5 and 6.7 ± 0.3, respectively). This effect was only associated with a higher incidence of ventricular tachycardia during reperfusion (80 vs. 50% for control treatment, $p < 0.05$). Electron spin resonance spectroscopy (ESR) and spin trapping with α-phenyl-tert-butylnitrone (PBN, 3 mM final) were used to monitor free radical generation during reperfusion. The vascular concentration of PBN-radical adducts (untreated: 6.4 ± 1.0 nM, at 10 min) decreased in the presence of DNP (1.7 ± 0.4 nM, $p < 0.01$). The radical concentration inversely correlated with myocardial oxygen consumption. Total liberation of free radical adducts during the initial 10 min of reperfusion was reduced by DNP (0.59 ± 0.09 nmol, $p < 0.01$) compared to the respective control treatment (1.26 ± 0.16 nmol). Similar effects, prevention of PBN adduct formation and unchanged viability in the presence of DNP, were obtained with endothelial cells during post-hypoxic reoxygenation. Since inhibition of mitochondrial phosphorylation can inhibit the formation of LPO-derived free radicals after an ischemic/hypoxic interval, mitochondria may represent an important source of free radicals capable of initiating lipid peroxidative injury during reperfusion/reoxygenation. (Mol Cell Biochem **160/161**: 167–177, 1996)

Key words: ESR spin trapping, lipid radicals, mitochondrial uncoupling, oxygen consumption, myocardial reperfusion, endothelial cell reoxygenation

Introduction

The generation of free radicals during post-ischemic reperfusion injury is now a generally accepted occurrence [1–8]. Under normal circumstances, a major pathway of free radical production in cells involves electron 'leakage' from the mitochondrial electron transport chain to molecular oxygen, resulting in reduced species, in particular $\cdot O_2$- and H_2O_2 [9]

Address for offprints: I.E. Blasig, Forschungsgruppe Radikal- und Zellbiochemie, Forschungsinstitut für Molekulare Pharmakologie, Forschungsverbund Berlin e.V., Aflred-Kowalke-Str. 4, D-10315 Berlin, Germany

It has been proposed that normally about 2% of oxygen reduced by the respiratory chain is converted to reactive oxygen species [10]. This percentage is thought to be higher in the oxidatively-stressed, post-ischemic myocardium [11] where conditions (tissue acidosis [12], iron mobilisation [13]) may exist to foster the conversion of $\cdot O_2$- to more potent primary oxidants ($\cdot O_2H$, $\cdot OH$).

The potential contribution of mitochondrial-derived free radicals to reperfusion injury, is further suggested by electron spin resonance (ESR) spectroscopy studies which identified signals from frozen post-ischemic myocardium as being ubiquinone-type radicals associated with the mitochondrial respiratory chain [11, 14]. Furthermore, submitochondrial particles [15] or mitochondria prepared from post-ischemic [16] myocardium have been shown to produce higher amounts of free radicals than those from non-ischemic tissue. Moreover, reperfusion of isolated ischemic rat hearts with an anoxic perfusate decreased post-ischemic radical production [17], presumably by limiting the availability of the electron acceptor (O_2) and maintaining a reduced mitochondrial electron transport chain.

In spite of the evidence that mitochondria are a potential source of primary oxygen free radicals in the reperfused myocardium [12], little evidence is available demonstrating that mitochondrial-derived oxidants contribute to lipid peroxidative injury and functional alterations of post-ischemic hearts. Moreover, recent findings [5, 18, 19] suggest that the extent of primary oxygen free radical production in reperfused tissue is not always proportional to the severity of oxidative injury. This can be partly attributed to the indiscriminant nature of primary radical attack on cellular components which may be critical or non-critical to the functional and structural integrity of the heart. Alternatively, we have recently demonstrated that lipid peroxidation (LPO)-derived free radical detection provides a more reliable index of oxidative injury to the post-ischemic myocardium [20], since these radicals are generated as by-products of primary radical attack on membrane lipids resulting in cellular and, hence, myocardial disturbances. In the present study, ESR spin trapping (α-phenyl-tert-butylnitrone) was used to monitor alterations in LPO-derived free radical production in post-ischemic rat hearts exposed to the oxidative phosphorylation uncoupler 2,4-dinitrophenol (DNP). As endothelial cells, in addition to myocytes [20, 21], has been thought to be an important cell type generating free radicals the effect of DNP on post-hypoxic endothelial cell injury has also been studied. With DNP it is possible to determine the relative contribution of mitochondrial primary free radical production towards oxidative injury of endothelial cells during reoxygenation damage and myocardial dysfunction during reperfusion injury.

Materials and methods

Materials

Reagents and solvents were from Sigma/St. Louis, MO or Fisher/Pittsburgh, PA, 2,4-dinitrophenol (DNP) from Eastman Kodak/Albany, NY, α-phenyl-tert-butylnitrone (PBN), 2,2,6,6-tetramethylpiperidine-1-oxyl (TEMPO) from Aldrich/ Milwaukee, WI. All buffers, reagents, and spin trap solutions were prepared from 0.8 μm pore filtered, double-deionized, distilled water; none of the prepared solutions contained detectable levels of trace metals (atomic absorption). Bovine aortic endothelial cells (AG 07684, Coriell Inst. Med. Res., Camden, NY) and male Sprague-Dawley rats (230–320 g) were used. All animal experiments were guided by the principles for the care and use of laboratory animals as recommended by the US Department of Health and Human Services and approved by the George Washington University Animal Care and Use Committee.

Heart perfusion technique, time course, and treatment

The Langendorff and working heart preparation perfusion models (non-paced), and global ischemia/reperfusion have been described previously [20]. Briefly, hearts were excised and cannulated by the aorta to a working heart perfusion apparatus within 1 min. Hearts were perfused at 37°C in the Langendorff mode for a 5 min wash-out period at a constant perfusion pressure of 75 mmHg, and an incision was made in the pulmonary artery to ensure the ejection of coronary effluent. The perfusate was Krebs-Henseleit buffer (KHB) supplemented with 5 mM glucose, and gassed with Carbogen to maintain a pH of 7.4. The working mode was prepared by cannulating the left atrium via the left pulmonary vein (preload 10.5 mmHg); the buffer entered the atrium, passed to the left ventricle, from which it was ejected through the aortic cannula against a pressure of 75 mmHg. Pressure measurements were obtained using a Bentley model 800 transducer connected to a 20 g cannula which was inserted into the left ventricle (left ventricular pressure), or via a side arm of the aortic cannula (aortic pressure). After the wash-out period, hearts were placed in 37°C jacketed chamber and perfused in the working mode for 20 min. Any heart that did not achieve a steady level of function during this time was excluded. At the end of this stabilization period in the working mode, baseline measurements (pre-ischemic control) for functional parameters, oxygen consumption and electrocardiogram (ECG) were taken. Thereafter, the heart was converted back to Langendorff perfusion for 10 min and pre-ischemic control values were obtained in this mode. Normothermic total global ischemia was achieved by cross clamping above the aortic cannula for 30 min, followed by normoxic Langendorff reperfusion for the initial 15 min. Infusion of DNP and PBN (begin 5 min before ischemia) were conducted only during the pre-ischemic Langendorff control and reperfusion phase. After Langendorff reperfusion, hearts were converted to the working mode for an additional 25 min, during which post-ischemic recovery of functional parameters, oxygen consumption and ECG were recorded. Isosmotic 120 mM PBN/saline solution was prepared prior to use [4, 22]. Hearts were infused with 3 mM PBN solution (final) via the side arm of the aortic cannula. Coronary flow (CF) was monitored to permit adjustment of the PBN infusion rate (~2.5% of CF) as needed. Effluent was collected sequentially for 30 s durations before and after ischemia. Effluent sample volume was estimated, and then stored in liquid N_2 [20]. DNP (50 μM final) was infused simultaneously with PBN, beginning 5 min before ischemia, and during the 15 min Langendorff reper-fusion phase.

Heart function, rhythm disturbances, and oxygen consumption

The following measurements were taken during control perfusion and reperfusion: heart rate (HR); CF and aortic flow (AF); left ventricular end-diastolic pressure (LVEDP); left ventricular developed pressure (LVDP = left ventricular systolic peak pressure - LVEDP); first derivative of left ventricular pressure ($LVdp/dt_{max}$); systolic (AOP_{syst}) and diastolic aortic pressure (AOP_{diast}); pressure volume work index [(AF + CF) × AOPsyst]. Continuous epicardial ECG recordings were made using two silver electrodes which were attached to the apex of the heart and to the cannula near the aortic root. The ECG was analyzed for HR, incidence and duration of ventricular fibrillation (VF), ventricular tachycardia (VT) and normal sinus rhythm (SR), and whether VF or VT were sustained (persisting throughout the study) or nonsustained (spontaneously reverting to regular rhythm). ECG, aortic and left ventricular pressures were recorded by using a precalibrated 6-channel physiograph (MFE CP-6). AF was measured by a calibrated flow meter (Brooks, tube size R-2-15-B) and CF by timed collection of the coronary effluent. Coronary arteriovenous O_2-difference ($AVDO_2$ = perfusate O_2-concentration – effluent O_2-concentration) was monitored with a Yellow Springs Instrument 5300 monitor and Clark electrode placed in a thermostated (37 °C) chamber and these values were used to calculate myocardial O_2-consumption ($MVO_2 = AVDO_2 \times CF$) and O_2-efficiency ($MVO_2/LVdp/dt_{max}$). The electrode was calibrated with O_2 and argon and perfusate (arterial) O_2-concentration was measured before and after each experiment.

Endothelial cell preparation and incubation

Cells were cultured in MEM and grown in 150 cm^2 culture flasks (Corning/Wexford, PA) as described [23]. Following trypsination and addition of growth medium suspensions were washed twice (400 × g, 4 min, 24°C) with an iso-osmotic 10 mM potassium phosphate buffer (pH 7.4) containing (mM): NaCl, 145; $MgCl_2$, 1.2; $CaCl_2$, 1.25; and glucose, 5, (incubation solution). The final pellet was resuspended in the incubation buffer and cellular viabilities (trypan blue dye, 0.1% final, exclusion method [23]) of > 90% were obtained prior to initiating incubation in the presence or absence of 50 μM DNP. Cell suspensions were diluted ($2.5 \cdot 10^6$ cells/ml final) into either N_2-gassed (hypoxic) or Carbogen-gassed control incubation buffer; suspensions were then gassed again (10 s) to verify oxygen content before initiating cell incubation [22]. For the hypoxia/reoxygenation studies, cell suspensions were incubated (37°C shaking water bath) for 100 min in anaerobic-sealed test-tubes, and subsequently reoxygenated with Carbogen for 15 min. At the end of hypoxia and at various times of reoxygenation, cells ($5 \cdot 10^6$/sample) were incubated with 10 mM PBN for 2 min. Then, the samples were rapidly filtered through a vacuum filtration manifold (Whatman GF/C filters) and frozen in liquid N_2. Endothelial cell viability was determined, at the end of reoxygenation and after 115 min of control incubation.

Sample extraction and radical detection

Post-ischemic [20] and post-hypoxic [24] free radical production was monitored using a PBN spin-trapping procedure and ESR spectroscopy. Subdued room light was maintained during all preparations involving the spin trap to avoid photolytic degradation. Effluent samples (6 ml) and cell filtrates (2 ml), containing spin trap adducts from control treated and DNP-treated hearts and cells, respectively, were thawed and extracted with 2 ml toluene (HPLC-grade, N_2-saturated) as described [22]. The toluene-extracted volumes were reduced under N_2-gas (24°C) to 0.6 ml before ESR spectroscopy. PBN extraction efficiency was 90% [24] and no significant ESR signal artefact was noted due to extraction. N_2-gassing was used devoid extracted samples of O_2 after being placed in 5 mm (i.d.) quartz tubes. PBN radical adduct (PBN-R) signal intensities were measured (24°C) with a Bruker ER 100 X-band spectrometer, equipped with 100 kHz modulation, and using the following settings: Modulation amplitude 0.16 mT, microwave power 20 mW, scan range 8 mT, scan time 200 s, and time constant 200 ms. Signal intensities, which are proportional to the concentration of PBN-R, were measured directly from the field scan [4, 22]; values were adjusted to the corresponding volumes of coronary effluent and cell solution sample, respectively. The values were expressed as vascular free radical and cell solution concentrations, respectively, (nM) by double integration of the experimental spectra using TEMPO nitroxide radical as an integration standard [24]. Signal averaging was applied (as needed) to improve signal resolution using an EPR Data Acquisition System (Scientific Software Services/Bloomington, IL). The free radicals detected in this study were consistent with the LPO-derived radical species observed in our earlier reports involving the post-ischemic rat heart model and post-hypoxic cell model. The radicals were identified as secondary radicals occurring during the radical-induced lipid peroxidation [4, 20]. The amount of myocardial PBN-R liberation (pmol/min) at each perfusion time was obtained by multiplying the adduct concentration by the respective coronary flow [1].

Statistics

Data were expressed as mean ± S.E.M. A one-way analysis of variance was carried out to test any difference between the mean values of all groups. If differences were established, treated group values were compared to those of the control treatment using the *t*-test. Where no normal distribution exist for a group, the parameter free Wilcoxon-test was used. An analogue procedure was performed for distribution of discrete variables such as the incidence of VF and VT. An overall chi-squared test for 2 × n tables was constructed, followed by a sequence of 2 × 2 chi-squared test in order to compare individual groups. When $p < 0.05$ was obtained, the differences were considered significant. Correlation analyses were assessed by the Spearman's rank correlation coefficient (r_S, stochastic independent) which was tested against zero (two-sided, $p < 0.05$).

Results

Effect of DNP on functional recovery and arrhythmias during reperfusion of the heart

Since DNP was infused simultaneously with PBN during Langendorff perfusion, Table 1 shows the effect on pre-ischemic and reperfusion parameters measured during this mode of perfusion. During the pre-ischemic period of the untreated heart, PBN increased CF ($p < 0.001$) and contractility ($p < 0.05$) compared to the pre-infusion values. These effects are consistent with earlier observations [25–27]. During reperfusion typical disturbances were observed; recovery of LVDP and LVdp/dt_{max} were 36 and 43%, respectively, lower ($p < 0.01$ each) than the pre-ischemic values (with PBN infusion) and LVEDP increased more than 3-fold. DNP infusion reduced the pre-ischemic LVDP ($p < 0.01$) and HR, and

Table 1. The effect of 50 μM dinitrophenol (DNP) on pre-ischemic function and postischemic functional recovery of isolated Langendorff perfused rat hearts.

Function	Before ischemia 5 min	0 min	Reperfusion 15 min
		Control treatment	
LVDP (mmHg)	83.0 ± 4	91.0 ± 4	58.0 ± 12
LVEDP (mmHg)	4.0 ± 1	4.0 ± 1	13.0 ± 4
LVdp/dt_{max} (mmHg/s)	2027.0 ± 100	2606.0 ± 188	1477.0 ± 330
HR (min^{-1})	212.0 ± 5	211.0 ± 3	229.0 ± 20
CF (ml/min)	9.5 ± 0.5	14.4 ± 0.8	15.5 ± 0.6
		DNP treatment	
LVDP (mmHg)	81.0 ± 3	75.0 ± 3**	28.0 ± 11
LVEDP (mmHg)	5.0 ± 1	5.0 ± 1	38.0 ± 8*
LVdp/dt_{max} (mmHg/s)	1933.0 ± 67	2320.0 ± 212	729.0 ± 301
HR (min^{-1})	232.0 ± 7	193.0 ± 20	61.0 ± 29**
CF (ml/min)	12.0 ± 2.3	18.0 ± 0.8**	14.0 ± 1.3

DNP was infused simultaneously with PBN, for 5 min before the ischemic period (0 min values) and during the initial 15 min of reperfusion following 30 min total global ischemia. Functional parameters: Left ventricular developed pressure (LVDP); left ventricular end-diastolic pressure (LVEDP); contractility (LVdp/dt_{max}); heart rate (HR); and coronary flow (CF). Values are means ± S.E.M. n = 8 per group; *, **denote $p < 0.05$ and 0.01, respectively, compared to the respective pre-ischemic or post-ischemic values with control treatment.

increased LVdp/dt_{max}, CF was significantly elevated ($p < 0.01$), compared to the pre-ischemic control treatment. During reperfusion, DNP treatment significantly affected LVEDP ($p < 0.05$) and HR ($p < 0.01$) only, when compared to the control treatment.

Table 2 demonstrates the effect of prior DNP treatment on post-ischemic recovery of isolated working rat hearts. Measurements were made 25 min after ending exposure (Langendorff perfusion) to DNP and PBN. In the group receiving control treatment, functional alterations typical of reperfusion injury were seen: LVDP and LVdp/dt_{max} were 34 and 32% lower ($p < 0.001$, $p < 0.01$), respectively, than the pre-ischemic values. HR, AF, AOP_{syst} were also reduced (13, 61, 12%; $p < 0.05$, $p < 0.001$, $p < 0.01$) and LVEDP was enhanced (2.6-fold; $p < 0.001$), respectively. DNP-treated working rat hearts experienced additional losses of functional properties during reperfusion, but no significant differences were found compared to the control treatment. Post-ischemic recovery of cardiac work (calculated as % recovery of pressure volume work) was similarly affected (Fig. 1). Compared to recovery of hearts with control treatment (47.8 ± 7.5% of pre-ischemic value), the DNP-treated group (32.9 ± 7.2%) was found not to be significantly different.

Figure 2 shows the effect of DNP on the incidence of VF, VT and sustained arrhythmias during the entire reperfusion phase (40 min) following 30 min of ischemia. DNP increased the incidence of VF to 50, of VT to 80, and of sustained arrhythmias to 10% (compared to the group with control

Table 2. The effect of 50 μM dinitrophenol (DNP) on post-ischemic recovery of functional properties of isolated working rat hearts

Function	Pre-ischemic control (–10 min)	Recovery (40 min)
		After control treatment
LVDP (mmHg)	138.0 ± 8	91.0 ± 6
LVEDP (mmHg)	8.0 ± 1	21.0 ± 3
$LVdp/dt_{max}$ (mmHg/s)	5653.0 ± 364	3855.0 ± 358
HR(min^{-1})	314.0 ± 14	272.0 ± 14
CF (ml/min)	18.3 ± 0.7	16.2 ± 1.3
AF (ml/min)	49.4 ± 2.5	19.1 ± 5.2
AOP_{syst} (mmHg)	106.0 ± 3	93.0 ± 2
AOP_{diast} (mmHg)	68.0 ± 2	72.0 ± 2
		After DNP treatment
LVDP (mmHg)	122.0 ± 5	71.0 ± 12^{ns}
LVEDP (mmHg)	10.0 ± 2	33.0 ± 7^{ns}
$LVdp/dt_{max}$ (mmHg/s)	5381.0 ± 248	2740.0 ± 499^{ns}
HR (min^{-1})	323.0 ± 18	229.0 ± 39^{ns}
CF (ml/min)	21.7 ± 1.5	17.2 ± 1.7^{ns}
AF (ml/min)	49.7 ± 4.1	9.3 ± 5.7^{ns}
AOP_{syst} (mmHg)	102.0 ± 3	90.0 ± 4^{ns}
AOP_{diast}(mmHg)	65.0 ± 3	69.0 ± 2^{ns}

DNP was infused simultaneously with PBN, for 5 min before the ischemic period and during the initial 15 min of reperfusion following 30 min total global ischemia. Functional parameters: Left ventricular developed pressure (LVDP); left ventricular end-diastolic pressure (LVEDP); contractility ($LVdp/dt_{max}$); heart rate (HR); coronary (CF) and aortic flow (AF); aortic systolic (AOP_{syst}) and diastolic pressure (AOP_{diast}). Values are means ± S.E.M. n = 8 per group; ns denotes that functional recovery was not significantly different when compared to the post-ischemic values (recovery) after control treatment.

treatment; 40, 50, and 0%, respectively). However, only the effect on VT was significantly different ($p < 0.05$).

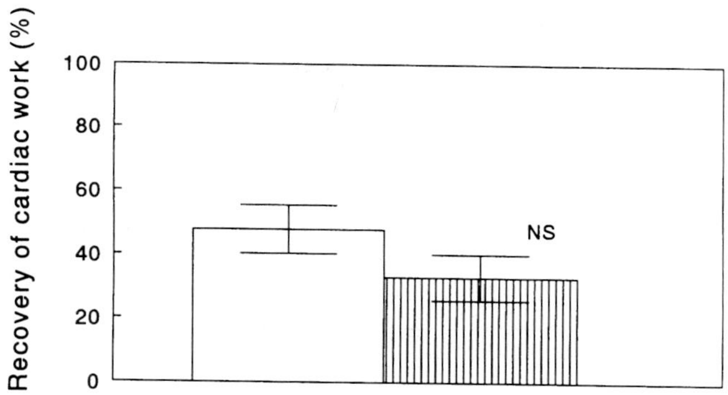

Fig. 1. Effect of 50 μM dinitrophenol (DNP, ▥) on post-ischemic recovery of cardiac work (pressure-volume work index) of isolated perfused rat hearts. 5 min before and during the initial 15 min of reperfusion following 30 min of total global ischemia, hearts were infused (including PBN) with DNP. During that time, hearts were perfused in Langendorff mode, and thereafter they were converted to working mode reperfusion for an additional 25 min (40 min total recovery); percent recovery of cardiac work was assessed at this time. Values are means ± S.E.M. of 8 hearts in each group; NS denotes for not significantly different from the group after receiving control treatment (□).

Effect of DNP on myocardial oxygen consumption

During the pre-ischemic period of the control treated group, PBN infusion caused a significantly ($p < 0.001$) higher MVO_2 (8.2 ± 0.5 μmol/min) compared to values prior to infusion (5.4 ± 0.2 μmol/min). Additional infusion of DNP caused significantly higher MVO_2 (11.3 ± 0.9 μmol/min; $p < 0.01$). During Langendorff reperfusion (10th min) of the control treated group, the MVO_2 (6.6 ± 0.3 μmol/min) was lower compared to the respective pre-ischemic interval ($p < 0.001$). DNP induced a nonsignificantly increased MVO_2 (7.9 ± 0.7 μmol/min) compared to the control infused reperfusion group, and it was significantly ($p < 0.01$) different from the respective DNP-treated pre-ischemic value (Fig. 3, Langendorff mode). The post-ischemic recovery of MVO_2 during working reperfusion was not significantly different after DNP treatment compared to the control treatment group (Fig. 3, working mode). The efficiency of O_2 consumption (Langendorff perfusion) did not show great variation during the pre-ischemic period among the non-infused (45.2 ± 3.3 pmol/mmHg) and the PBN-infused (52.7 ± 4.7 pmol/mmHg) groups. However, the administration of DNP diminished (i.e. higher ratio) O_2-efficiency during the pre-ischemic perfusion period (88.8 ± 5.6 pmol/mmHg, p < 001) compared to the control treated group (PBN only). In the working mode, the post-ischemic recovery of O_2-efficiency was nearly complete in the control treated group (35.2 ± 1.9 pmol/mmHg) and in the DNP group (42.9 ± 5.1 pmol/mmHg) compared to the pre-ischemic value (38.8 ± 2.2 pmol/mmHg, control treated group).

Effect of DNP on myocardial radical metabolism

Prior to ischemia, DNP did not significantly influence the vascular concentration of PBN-R or myocardial liberation of PBN-R (Figs. 4 and 6). Reperfusion of the control treated group led to significant increases of free radical concentration, with maximum levels occurring within the first 5 min of reperfusion, followed by a return towards pre-ischemic levels by 15 min. DNP inhibited post-ischemic enhancement of free radical concentration completely (Fig. 4). Post-ischemic PBN-R concentration was found to be inversely correlated with MVO_2. That means hearts which consumed lower amounts of O_2, as in the case of the control treatment, displayed higher free radical concentrations, and those consuming higher O_2 such as the DNP group, were associated with lower free radical levels (Fig. 5). The total myocardial liberation rate of PBN-R increased nearly 3 times ($p < 0.001$) in the control treated group after the onset of reperfusion compared to the pre-ischemic period, showing maxima after 5 min and normalisation after about 15 min. Of interest, how-

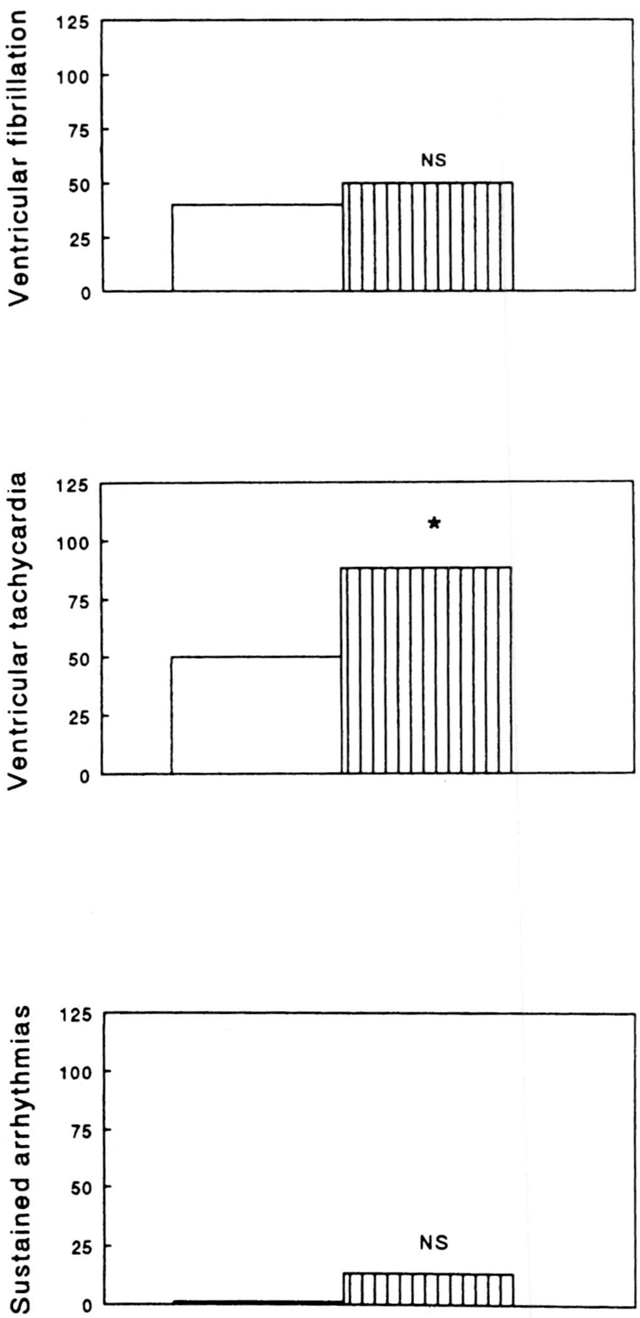

Fig. 2. Effect of 50 μM dinitrophenol (DNP, ▥) on the incidence of arrhythmias (as %) in isolated perfused rat hearts during the entire reperfusion period (40 min), after 30 min of ischemia. DNP infusion and the reperfusion methods are described in Fig. 1. Values are means of 10 hearts in each group; *denotes $p < 0.05$, NS for not significantly different from the post-ischemic group with control treatment (□).

ever, were our findings that the amount of myocardial PBN-R released during reperfusion from DNP-treated hearts were considerably less than that observed for the control treated group (Fig. 6). These results highlight the importance of considering free radical production in both coronary flow-dependent (concentration) and -independent (liberation) terms, when evaluating the effects of agents possessing vasoactive properties.

Effect of DNP on viability and radical metabolism of endothelial cells

Endothelial cells subjected to 100 min of hypoxia did not display a significant loss of viability; however, following 15 min of reoxygenation, the cell viability decreased to 67.4 ± 3.2% compared to 90.3 ± 2.6% after 115 min of untreated control incubation ($p < 0.001$). DNP treatment did not show significant effects either under control condition (78.1 ± 6.3%) or during hypoxia/reoxygenation (64.1 ± 5.7%) when compared to the respective untreated control group (Fig. 7).

Figure 8 shows the effect of DNP on the time course of PBN-R concentration in the incubation solution during endothelial cell reoxygenation following 100 min of hypoxia. Reoxygenation of the untreated group led to significant increases of free radical concentration, with maximum levels occurring within the first 5 min of reoxygenation, followed by a return towards pre-hypoxic levels by 15 min (latter data not shown). DNP inhibited posthypoxic enhancement of free radical concentration completely.

Discussion

The results of this investigation agree with previous reports [1, 4, 20, 28] which demonstrated transient bursts of lipid peroxidation (LPO)-derived free radicals in the myocardium during the early minutes of reperfusion following ischemia. The data obtained in the endothelial cell experiments show similar patterns of free radical production and lipid peroxidation, respectively, following hypoxia as reported earlier [24]. In the present study, it has to be recognized that the PBN spin trapping agent may itself possess concentration-dependent cardioactive effects [29]. The concentration (3 mM) used in this study for detecting the radicals caused an increase in coronary flow during the pre-ischemic perfusion phase. Similar effects have been observed earlier [17, 25–27] and appeared to be due to a direct vasodilatory action of the trap. The mechanism of this vasodilatory action may involve an EDRF(endothelium-derived relaxing factor)-like activity [26]. Nitric oxide (NO), a proposed form of EDRF, is known to be inactivated by $\cdot O_2$-. By trapping $\cdot O_2$-, PBN might indirectly potentiate the vasodilating effects of NO. In addition to its vasoactive influence, PBN also induced a positive inotropic effect and increased myocardial oxygen consumption; however its mode of action is unclear. PBN-mediated protection to the post-ischemic heart has been largely attributed

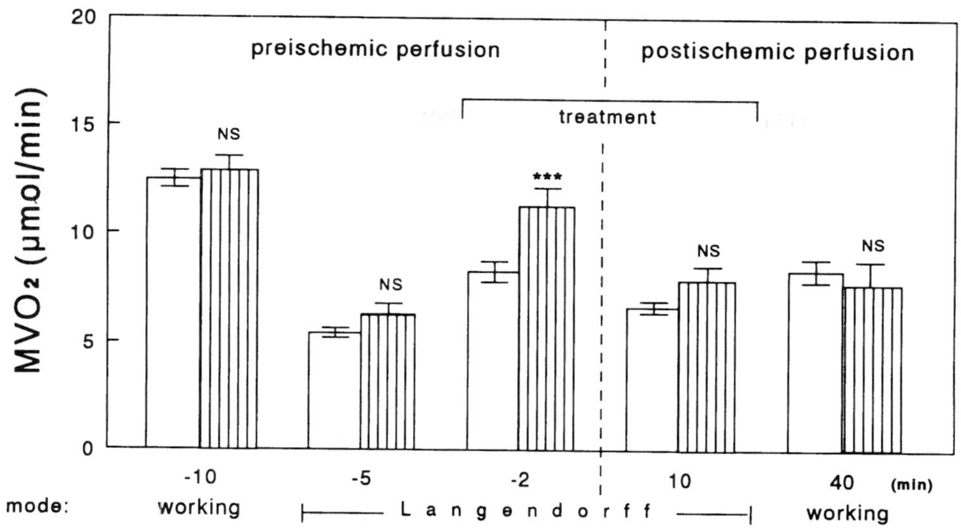

Fig. 3. Effect of 50 µM dinitrophenol (DNP, ▥) on O_2-consumption (MVO_2) of isolated perfused rat hearts before and after 30 min of ischemia. DNP infusion and the reperfusion methods are described in Fig. 1. Values are means ± S.E.M. of 8 hearts in each group. ***denotes $p < 0.001$, respectively, and NS for not significantly different from the preischemic or post-ischemic group at the comparable time (control treatment, □).

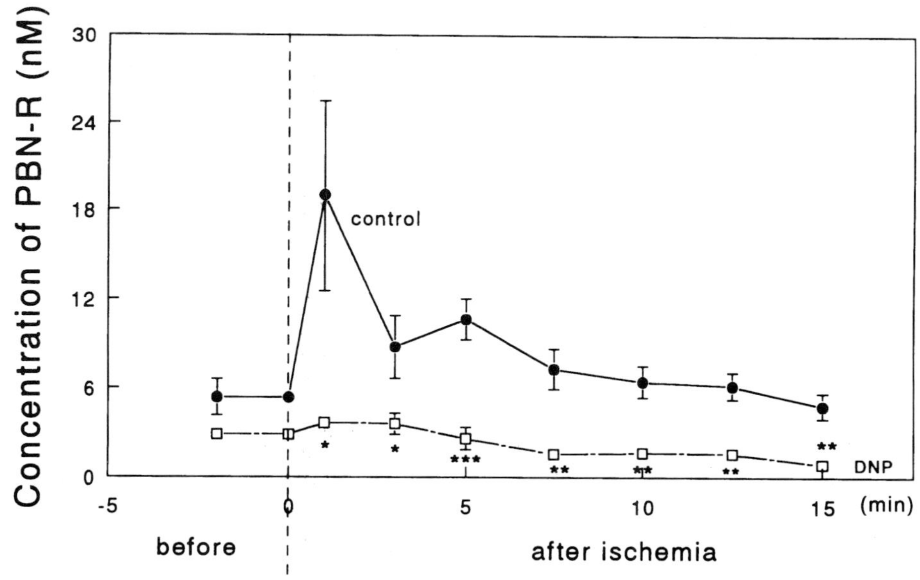

Fig. 4. Effect of 50 µM dinitrophenol (DNP, n = 8) on the effluent concentration of PBN free radical adduct (PBN-R) from 30 min ischemic/reperfused hearts. Isolated rat hearts were infused with the spin trap in the presence or absence of DNP during the pre-ischemic and post-ischemic Langendorff perfusion period (further details see Fig. 1). Sample extraction and ESR spectroscopy parameters have been described in Materials and method. Values are means ± S.E.M. *, **, and ***denote $p < 0.05$, 0.01, and 0.001, respectively, compared to the post-ischemic group with control treatment (n = 10) at comparable times.

to the free radical scavenging properties of nitrone spin traps [30, 31]. PBN decreased reperfusion-induced arrhythmias [29], and improved contractility and cellular integrity of post-ischemic hearts [32]. That PBN may afford protection directly to mitochondria is suggested by evidence of mitochondrial uptake of this trap [33] and the finding that PBN preserved phosphocreatine content of the post-ischemic heart [32]. Thus, when assessing the effects of a mitochondrial inhibitor on post-ischemic free radical generation and functional recovery (Langendorff mode), the influence of PBN-infusion was always considered when using the control treated group for comparison.

DNP treatment of perfused hearts induced responses typical of a mitochondrial uncoupling agent: mitochondrial respiration was raised as shown by the elevation of myocardial oxygen consumption. Uncoupling of respiration and phos-

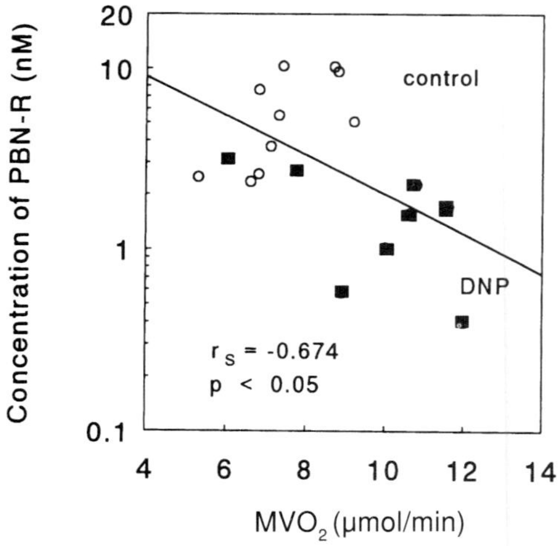

Fig. 5. Inverse correlation between vascular concentration of PBN free radical adduct (PBN-R) and myocardial O_2-consumption (MVO_2) after 10 min of reperfusion of DNP-treated or control treated Langendorff reperfusion (Details for infusion see Fig. 1). Isolated rat hearts were subjected to 30 min total global ischemia prior to reperfusion. The rank correlation coefficient (r_S) according to Spearman was determined

phorylation leads to a substantial decline in mitochondrial ATP synthesis. ATP synthesis in the presence of DNP can, in a greater amount, occur via glycolysis because mitochondrial oxidation of NADH is maintained, if not enhanced, as indicated by the oxygen consumption measured. Moreover, the increase of coronary flow associated with DNP insures that glycolysis will not be inhibited by tissue accumulation of lactate or NADH and substrate deficiency. This concept receives support from the fact that recovery of energy-dependent contractile events was similar in DNP-treated post-ischemic hearts compared to the control treatment. In total, the energy supply seems to be limited as indicated by a slight reduction of contractile parameters both before and after ischemia. DNP treatment also dramatically reduced vascular LPO-derived free radical concentration during reperfusion. This may be an indirect response to DNP, since it is presumed that uncoupling mitochondrial phosphorylation will reduce production of reactive oxygen species (protonated superoxide anion, hydroxyl radical) capable of initiating lipid peroxidative injury (i.e. LPO-derived radicals) during reperfusion. When compared to the control treated group, it is curious that the DNP-induced reduction of free radical concentration occurred without a more dramatic improvement in post-ischemic functional recovery. This observation suggests that DNP influences multiple metabolic pathways which are independent of each other: (1) it reduces LPO injury by inhibiting mitochondrial primary free radical production; and (2) it worsens contractile recovery by interrupting mitochondrial ATP synthesis.

Changes in post-ischemic free radical concentration were inversely correlated to myocardial oxygen consumption. DNP stimulated oxygen consumption and reduced LPO-derived free radical concentration. This relationship does not suggest the lack of involvement of molecular oxygen in processes leading to free radical production. Indeed ischemic hearts reperfused with anoxic perfusate have been shown to generate substantially lower amounts of free radicals than normoxically reperfused hearts [17]. This relationship may also indicate that high vascular radical concentrations lead to re-

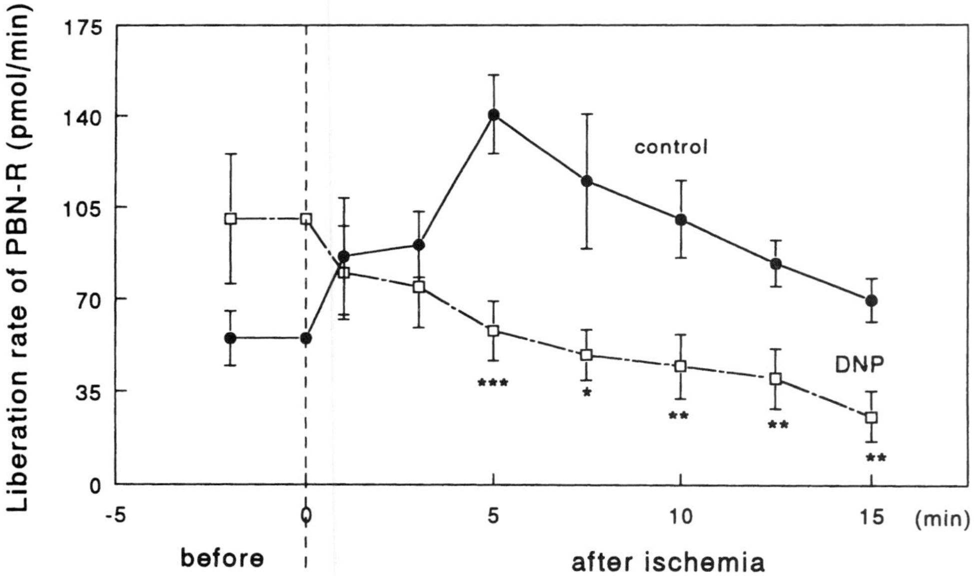

Fig. 6. Effect of 50 μM dinitrophenol (DNP, n = 8) on the myocardial liberation rate of PBN free radical adduct (PBN-R) during Langendorff reperfusion of isolated rat hearts. PBN spin trapping was conducted for 5 min before and during the initially 15 min after 30 min of total global ischemia, and values which are means ± S.E.M. were corrected for coronary flow. *, **, and ***denote $p < 0.05$, 0.01 and 0.001, respectively, compared to the post-ischemic values with control treatment (n = 10) at comparable times.

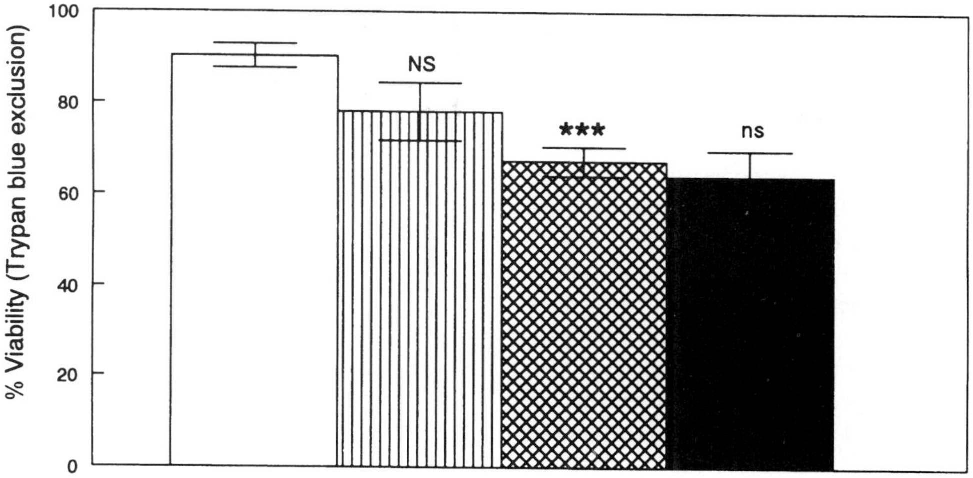

Fig. 7. Effect of 50 μM dinitrophenol (DNP) on cellular viability of endothelial cell suspensions exposed to 115 min control incubation (▥, n = 7) and to 100 min hypoxia/15 min reoxygenation (■, n = 7), respectively. DNP was incubated throughout the experiment. PBN was added during the 2 min before filtration of the medium. Values are means ± S.E.M. *** and NS denote $p < 0.001$ and not significantly different, respectively, compared to the control incubation group without DNP (□, n = 8). ns denotes not significantly different either compared to the DNP-treated control group or compared to the untreated reoxygenation group (▩, n = 8).

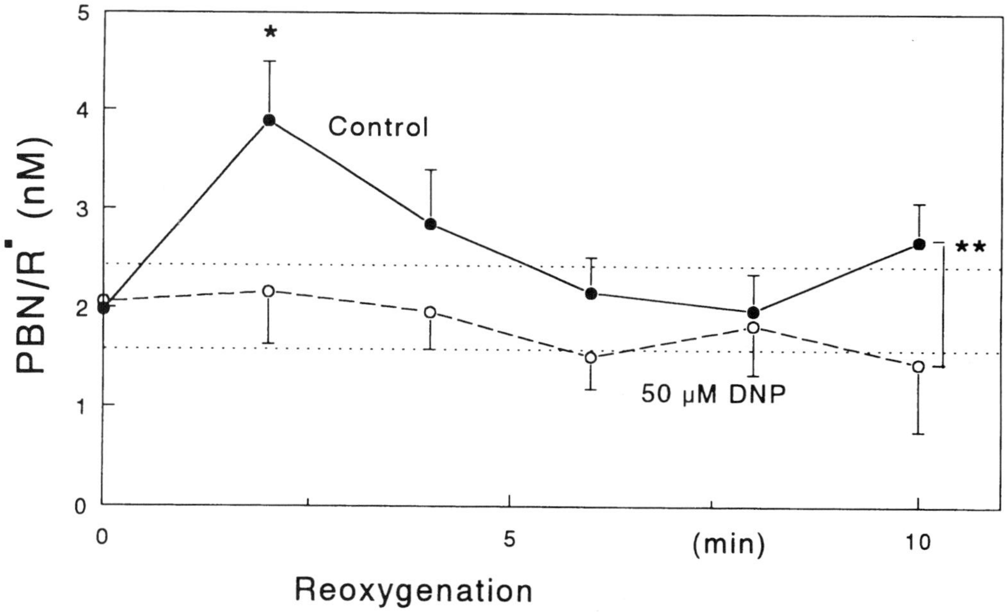

Fig. 8. Effect of 50 μM dinitrophenol (DNP, n = 7) on the concentration of PBN free radical adduct (PBN-R) in the incubation solution of endothelial cells during reoxygenation following 100 min of hypoxia. Endothelial cell suspensions were incubated with DNP throughout the experiment. PBN was added 2 min before filtration of the medium. Extraction of the incubation solution and ESR spectroscopy parameters have been described in Materials and methods. Values are means ± S.E.M. * and **denote $p < 0.05$ and 0.01, respectively, compared to the untreated reoxygenation group (n = 8) at comparable times.

duced myocardial function and a resultant decrease in the myocardial oxygen demand. This interpretation is underscored by recent findings demonstrating a significant: (1) inverse correlation between vascular free radical concentration and recovery of both post-ischemic contractility or left ventricular developed pressure [1, 20, 35]; and (2) direct correlation between myocardial leakage of cellular proteins and free radical concentration [20]; this latter relationship suggests the occurrence of free radical-induced disturbances of membrane functional and structural integrity during post-ischemic stress. DNP totally abolished free radical production from the reperfused heart and reoxygenated endothelial cells. It is also likely that the transfer of single electrons to molecular oxygen is diminished. This suggestion has been supported by studies using different uncoupling agents which inhibit the generation of H_2O_2 in isolated mitochondrial preparations [34].

Although we demonstrate a prominent role for mitochondria in the mechanism(s) of tissue oxidative stress, our data do not preclude the contribution of other free radical generating pathways [36] towards the oxidative injury associated with reperfusion injury. Moreover, the cellular source(s) of free radical generation in post-ischemic hearts has been addressed by the present study. Earlier, it has been shown that oxidatively-stressed vascular endothelial cells [37] or ventricular myocytes [20, 22, 38] have the potential of contributing to the free radical production seen in the post-ischemic heart tissue. In fact, myocytes, which have a high mitochondrial content and negligible xanthine oxidase activity compared to endothelial cells appear also to be more sensitive to anoxia/reoxygenation induced injury [20, 21]; this cell model also displayed a time course of LPO-derived free radicals which was similar to that of the post-ischemic rat heart [20, 22]. This suggestive evidence has led us to the assumption that mitochondria of ventricular myocytes may be a principle source of free radicals observed in the post-ischemic tissue model. However, it cannot be excluded that mitochondria in endothelium may also contribute to radical liberation of the whole heart during post-hypoxic and post-ischemic circumstances, respectively. They also produce lipid-derived free radicals which are completely abolished by DNP.

In general, uncoupling reduced radical concentrations in hearts and endothelial cells and supports the assumption that mitochondria are an important source of post-ischemic/post-hypoxic radical generation. Moreover, these and earlier [39] results suggest that phosphorylation related processes or degradation reactions of high energy phosphates may also be involved in radical formation in addition to respiratory chain reactions. However, mitochondrial radical production seems to be of limited relevance for the reperfusion/reoxygenation injury as prevention of mitochondrial radical production was not accompanied by improvement of post-ischemic/post-hypoxic recovery of hearts or endothelial cells, respectively.

In summary, our findings suggest that mitochondria are a potential source of myocardial free radicals which can induce lipid peroxidation injury (as LPO-derived free radicals) in the isolated perfused post-ischemic heart. Our conclusions are supported by previous ESR investigations using submitochondrial particles [15] or mitochondria isolated from ischemic and non-ischemic myocardium [16], and by mitochondrial inhibitor studies using uncoupling agents [34]. Studies with DNP also demonstrate that contractile dysfunction during reperfusion can be partially dissociated from oxidative tissue injury, and can be explained by treatment-mediated imbalances between myocardial high energy phosphate production and utilisation.

Acknowledgements

This work was supported by NIH RO1-HL36418, RO1-HL49232, DFG BI 308/1-1 and BMBF BEO 03110015A/B.

References

1. Blasig IE, Shuter S, Garlick P, Slater TF: Relative time-profiles for free radical trapping, coronary flow, enzyme leakage, arrhythmias and function during myocardial reperfusion. Free Rad Biol Med 16: 35–41, 1994
2. Damerau W, Ibel J, Thürich T, Assadnazari A, Zimmer G: Generation of free radicals in Langendorff and working hearts during normoxia, hypoxia, and reoxygenation. Basic Res Cardiol 88: 141–119, 1993
3. Kumar C, Okuda M, Ikai I, Chance B: Luminol enhanced chemiluminescence of the perfused rat heart during ischemia and reperfusion. FEBS Lett 272: 121–124, 1990
4. Mak IT, Kramer JH, Weglicki WB: Antioxidant properties of active and inactive isomers of nicardipin in cardiac membranes, endothelial cells, and perfused rat hearts. Coronary Artery Disease 3: 1095–1103, 1992
5. Nohl H, Stolze K, Napetschnig S, Ishikawa T: Is oxidative stress primarily involved in reperfusion injury of the ischemic heart? Free Rad Biol Med 11: 581–588, 1991
6. Pietri S, Culcasi M, Stella L, Cozzone PJ: Ascorbyl free radical as a reliable indicator of free radical-mediated myocardial ischemic and postischemic injury. A real-time continuous flow ESR study. Eur J Biochem 193: 845–854, 1990
7. Powell RS, Hall D: Use of salicylate as a probe for ·OH formation in isolated ischemic rat hearts. Free Rad Biol Med 9: 133–141, 1990
8. Bolli R, Jeroudi MO, Patel BS, Aruoma Oi, Halliwell B, Lai EK, McCay PB: Marked reduction of free radical generation and contractile dysfunction by antioxidant therapy begun at the time of reperfusion: Evidence that myocardial 'stunning' is manifestation of reperfusion injury. Circ Res 65: 607–622, 1989
9. Paraidathanthiu T, DeGroot H, Kehrer JP: Production of reactive oxygen by mitochondria from normoxic and hypoxic rat heart tissue. Free Rad Biol Med 13: 289–297, 1992
10. Rashba YE, Vartanyan LS, Bayder LM, Krinitskaya LA: Quantitative estimation of the rate of superoxide radical generation in mitochondrial membranes by ESR-technique. Biofizika USSR 34: 57–62, 1989
11. Baker JE, Kalyanaraman B: Ischemia-induced changes in myocardial paramagnetic metabolites: Implications for intracellular oxy-radical generation. FEBS Lett 244: 311–314, 1989
12. Ambrosio G, Zweier JL, Duilio C, Santoro G, Elia PP, Tritto I, Chiariello M, Flaherty JT: Inhibition of mitochondrial respiration and lipid peroxidation in reperfused hearts. Circulation 80(4): 11–31 (Abstract), 1989
13. Prasad MR, Engelman RM, Jones RM, Das DK: Effects of oxyradicals on oxymyoglobin. Deoxygenation, haem removal and iron release. Biochem J 263: 731–736, 1989
14. Zweier JL, Kuppusami P, Williams R, Rayburn BK, Smith D, Weisfeldt ML, Flaherty JT: Measurement and characterization of postischemic free radical generation in the isolated perfused heart. J Biol Chem 264: 18890–18895, 1989
15. Ueta H, Ogura R, Sugiyama M, Kagiyama A, Shin G: $\cdot O_2$- spin trapping on cardiac submitochondrial particles isolated from ischemic and non-ischemic myocardium. J Mol Cell Cardiol 22: 893–899, 1990
16. Otani H, Tanaka H, Inoue T, Umemoto M, Omoto K, Tanaka K, Sato T, Osako T, Masuda A, Nonoyama A, Kagawa T: *In vitro* study on

contribution of oxidative metabolism of isolated rabbit heart mitochondria to myocardial reperfusion injury. Circulation Res 55: 168–175, 1984
17. Davies MJ, Garlick PB, Slater TF, Hearse DJ: Detection of free radical production during myocardial reperfusion injury using electron spin resonance spin trapping. In: C. Rice-Evans and T. Dormandy (eds). Free radicals: Chemistry, pathology and medicine. Richelieu Press, London, 1988, pp 303–319
18. Atiq Khalid M, Samara ZO, Ashraf M: Association of endogenously produced OH radicals with anoxia reoxygenation injury in isolated rat myocytes. FASEB J 5(5): A1282 (Abstract), 1991
19. Henry TD, Archer SL, Nelson D, Weir EK, From AHL: Postischemic oxygen radical production varies with duration of ischemia. Am J Physiol 264: H1478–H1484, 1993
20. Kramer JH, Misik V, Weglicki WB: Lipid peroxidation-derived free radical production and postischemic myocardial reperfusion injury. Ann N Y Acad Sci 723: 180–196, 1994
21. Freedman AM, Kramer JH, Mak IT, Cassidy MM, Weglicki WB: Propranolol preserves ultrastructure in adult cardiocytes exposed to anoxia/reoxygenation: A morphometric analysis. Free Rad Biol Med 11: 197–206, 1991
22. Kramer JH, Mak IT, Freedman AM, Weglicki WB: Propranolol reduces anoxia/reoxygenation-mediated injury of adult myocytes through an antiradical mechanism. J Mol Cell Cardiol 23: 1231–1244, 1991
23. Dickens BF, Weglicki WB, Li Y-S, Mak IT: Magnesium deficiency in vitro enhances free radical-induced intracellular oxidation and cytotoxicity in endothelial cells. FEBS Lett 311: 187–191, 1992
24. Kramer JH, Dickens BF, Misik V, Weglicki WB: Phospholipid hydroperoxides are precursors of lipid alkoxyl radicals produced from anoxia/reoxygenated endothelial cells. J Mol Cell Cardiol 27: 371–381, 1995
25. Blasig IE, Schönheit K: Improvement of heart function by the spin trap alpha-phenyl-t-butyl nitrone possibly caused by its nitroxyl radical adduct. In. K.J.A. Davies (ed.) Oxidative damage and repair: Chemical, biological and medical aspects. Pergamon Press, New York, 1992, pp 421–425
26. Li X-Y, Sun J-Z, Bradamante S, Piccinini F, Bolli R: Effects of the spin trap alpha-phenyl N-tert-butyl nitrone on myocardial function and flow: a dose-response study in the open chest dog and in the isolated rat heart. Free Rad Biol Med 14: 277–285, 1993
27. Konorev EA, Baker JE, Joseph J, Kalyanaraman B: Vasodilatory and toxic effects of spin traps on aerobic cardiac function. Free Rad Biol Med 14: 127–137, 1993
28. Pissarek M, Jänichen F, Blasig IB, Haseloff RF, Keller T, Tapp E, Krause E-G: Cardioprotective potency of the radical scavenger S-2-(3 aminopropionylamino) ethylphosphorothioic acid in the post-ischemic rat heart. Mol Cell Biochem 145: 121–129, 1995
29. Hearse DJ, Tosaki A: Free radicals and reperfusion-induced arrhythmias: protection by spin trap agent PBN in the rat heart. Circ Res 60: 375–383, 1987
30. Tosaki A, Haseloff RF, Hellegouarch A, Schonheit K, Martin VV, Das DK, Blasig IE: Does the antiarrhythmic effect of DMPO originate from its oxygen radical trapping property or the structure of the molecule itself? Basic Res Cardiol 87: 536–547, 1992
31. Biasig IE, Volodarski LB, Tosaki: A. Nitrone spin trap compounds. Mode of cardioprotective action. Pharm Pharmacol Lett 3: 135–138, 1993
32. Bradamante S, Monti E, Paracchini L, Lazzarini E, Piccini F: Protective activity of the spin trap tert-butyl-alpha-phenyl nitrone (PBN) in reperfused rat heart. J Mol Cell Cardiol 24: 375–386,1992
33. Cova D, De Angelis L, Monti E, Piccinini F: Subcellular distribution of two spin trapping agents in rat heart: Possible explanation for their different protective effects against doxorubicin-induced cardiotoxicity Free Rad Res Commun 15: 353–360, 1991
34. Loschen G, Azzi A, Flohe L: Mitochondrial H_2O_2 formation and energy conservation. In. L. Flohe (ed.) Glutathione. Thieme-Verlag, Stuttgart, 1974, pp 228–236
35. Blasig IE, Ebert B, Hennig C, Pali T, Tosaki A: Inverse relationship between ESR spin trapping of oxyradicals and degree of functional recovery during myocardial reperfusion in isolated working rat heart. Cardiovasc Res 24: 263–270, 1990
36. Kloner RA, Przyklenk K, Whittaker P: Deleterious effects of oxygen radicals in ischemia/reperfusion. Resolved and unresolved issues. Circulation 80: 1115–1127, 1989
37. Zweier JL, Kuppusamy P Lutty GA: Measurement of endothelial cell free radical generation: Evidence for a central mechanism of free radical injury in postischemic tissues. Proc Natl Acad Sci USA 85: 4046–4050, 1988
38. Dickens BF, Weglicki WB, Li Y-S, Kramer JH: Rapid alkoxyl radical production during endothelial cell hypoxia/reoxygenation. FASEB J 5(5): A1283 (Abstract), 1991
39. Blasig IE, Steinschneider AY, Lakomkin VL, Ledenev AN, Korchazkina OV, Ruuge EK: ESR spin trapping and NMR spectroscopy of the same heart shows correlation between energy depression and radical formation during post-ischemic reperfusion. FEBS Lett 267: 29–32, 1990

Molecular and Cellular Biochemistry **160/161:** 179–185, 1996.

Relationship between mechanical dysfunction and depression of sarcolemmal Ca^{2+}-pump activity in hearts perfused with oxygen free radicals

Taku Matsubara and Naranjan S. Dhalla

Division of Cardiovascular Sciences, St. Boniface General Hospital Research Centre and Department of Physiology, Faculty of Medicine, University of Manitoba, Winnipeg, Manitoba, Canada R2H 2A6

Abstract

Although *in vitro* studies have shown that oxygen free radicals depress the sarcolemmal Ca^{2+}-pump activity and thereby may cause the occurrence of intracellular Ca^{2+} overload for the genesis of contractile failure, the exact relationship between changes in sarcolemmal Ca^{2+}-pump activity and cardiac function due to these radicals is not clear. In this study we examined the effects of oxygen radicals on sarcolemmal Ca^{2+} uptake and Ca^{2+}-stimulated ATPase activities as well as contractile force development by employing isolated rat heart preparations. When hearts were perfused with medium containing xanthine plus xanthine oxidase, the sarcolemmal Ca^{2+}-stimulated ATPase activity and ATP-dependent Ca^{2+} accumulation were depressed within 1 min whereas the developed contractile force, rate of contraction and rate of relaxation were increased at 1 min and decreased over 3–20 min of perfusion. The resting tension started increasing at 2 min of perfusion with xanthine plus xanthine oxidase. Catalase showed protective effects against these alterations in heart function and sarcolemmal Ca^{2+}-pump activities upon perfusion with xanthine plus xanthine oxidase whereas superoxide dismutase did not exert such effects. The combination of catalase and superoxide dismutase did not produce greater effects in comparison to catalase alone. These results are consistent with the view that the depression of heart sarcolemmal Ca^{2+} pump activities may result in myocardial dysfunction due to the formation of hydrogen peroxide and/or hydroxyl radicals upon perfusing the hearts with xanthine plus xanthine oxidase. (Mol Cell Biochem **160/161**: 179–185, 1996)

Key words: oxygen free radicals, cardiac contractile failure, sarcolemmal Ca^{2+}-pump activity, sarcolemmal Ca^{2+}-stimulated ATPase, rat heart

Introduction

Several studies have suggested that oxygen free radicals such as superoxide and hydroxyl radicals as well as active species of oxygen including H_2O_2 are involved in cardiac cell injury and contractile failure under some pathophysiological conditions like myocardial ischemia-reperfusion [1–5]. Although cellular membranes are considered to be the site of attack by oxyradicals, the exact mechanisms of myocardial dysfunction induced by oxygen radicals are still unclear. Recently, some investigators have demonstrated greater susceptibility of myocardial sarcolemma to oxygen free radical injury or ischemia-reperfusion-induced peroxidative injury compared to sarcoplasmic reticulum and mitochondrial membranes [6, 7]. Studies in our laboratory [8–10], which were undertaken to investigate the effects of oxygen free radicals on heart sarcolemmal Ca^{2+}-pump and Na^+-Ca^{2+} exchange systems *in vitro*, have led us to conclude that the sarcolemmal Ca^{2+}-transport activities are altered upon exposure of cardiac membranes to oxygen free radicals. These alterations caused by oxygen radicals were considered to result in the occurrence of intracellular Ca^{2+} overload and functional abnormalities in the myocardium. The purpose of this study was to examine the effects of oxygen free radicals generated by xanthine plus xanthine oxidase (X + XO) system on the isolated perfused rat heart in order to characterize the relationship between changes in cardiac contractile function and the sarcolemmal Ca^{2+}-pump activity.

Address for offprints: N.S. Dhalla, Division of Cardiovascular Sciences, St. Boniface General Hospital Research Centre, 351 Tache Avenue, Winnipeg, Manitoba, Canada R2H 2A6

Materials and methods

Perfusion of the isolated heart

Adult male Sprague-Dawley rats (250–300 g) were decapitated and their hearts were rapidly excised and washed in ice-cold buffer. The aorta was attached to a cannula and the heart was perfused immediately by the Langendorff technique with Krebs-Henseleit solution containing (in mM) NaCl 120.0, $NaHCO_3$ 25.0, KCl 4.7, KH_2PO_4 1.2, $MgSO_4$ 1.2, $CaCl_2$ 1.25 and glucose 11.0. This perfusion solution was continuously oxygenated with a mixture of 95% O_2 and 5% CO_2 (pH 7.4) and its temperature maintained at 37°C. The heart was electrically stimulated (Phipps and Bird stimulator) at 300 beats/min via a square wave of 1.5 ms duration at twice the threshold voltage. The coronary flow was maintained at 10 ml/min by a Harvard peristaltic pump. Cardiac contractile force, the rate of contraction (+dF/dt) and the rate of relaxation (-dF/dt) were recorded on a Beckman recorder via a force displacement transducer (Grass FT.03). A resting tension of 2 g was applied to the heart at the beginning of each experiment.

To generate oxygen radicals, X (2 mM, Sigma) and XO (10 U/1, Sigma) were dissolved in the above perfusion medium and mixed for 60 min. All hearts were allowed to equilibrate for 20–30 min with normal oxygenated Krebs-Henseleit medium before starting any experimental intervention. In the first set of experiments, hearts were perfused with the perfusion medium containing X + XO for 20 min to examine changes in contractile parameters. In some hearts, the period of perfusion with this solution was for l, 2, 3, 5, 10 and 20 min to study changes in the sarcolemmal Ca^{2+}-pump activities (Ca^{2+}-stimulated ATPase and ATP-dependent Ca^{2+} accumulation) with respect to contractile dysfunction. The second set of experiments was performed to examine the effects of superoxide dismutase (SOD: 1.5×10^5 U/1, Sigma) and catalase (CAT: 1.0×10^5 U/1, Sigma) on changes induced by X + XO. Immediately after the experimental period, the hearts were rapidly removed from the perfusion apparatus; the ventricles were frozen using the Wollenberger clamp precooled in liquid nitrogen. All hearts were stored at -90°C for the isolation of sarcolemmal membranes. No change was observed in the Ca^{2+}-pump activities in the sarcolemmal membrane isolated from the frozen heart stored for 6 weeks.

Isolation of sarcolemmal membrane

Sarcolemmal membranes were isolated from a pool of 3–4 hearts according to the method described earlier [8–11] . The frozen hearts were crushed to a fine powder in a stainless steel percussion mortar precooled in liquid nitrogen. The frozen tissue powder was allowed to thaw on the ice and then homogenized (3.5 ml/g tissue) in ice-cold 0.6 M sucrose, 10 mM imidazole-HCl, pH 7.0 using a Polytron PT-20 (5 × 20 s, setting 5). The resulting homogenate was centrifuged at 12,000 g for 30 min and the pellet was discarded. After diluting (5 ml/g tissue) with 140 mM KCl, 20 mM 3-(N-morpholino)propanesulfonic acid (MOPS), pH 7.4 (KCl-MOPS buffer), the supernatant was centrifuged at 95,000 g for 60 min. The resulting pellet was suspended in the KCl-MOPS buffer and layered over a 30% sucrose solution containing 0.3 M KCl, 50 mM $Na_4P_2O_7$ and 0.1 M tris(hydroxymethyl)aminomethane (Tris)-HCl, pH 8.3. After centrifugation at 95,000 g for 90 min (utilizing a Beckman swinging bucket rotor), the band at the sucrose-buffer interface was taken and diluted with 3 volumes of KCl-MOPS buffer. A final centrifugation at 95,000 g for 30 min resulted in a pellet rich in sarcolemma. All these steps for the isolation of sarcolemma were carried out at 0–4°C. The pellet was resuspended in 0.25 M sucrose-10 mM histidine buffer, pH 7.0, and quickly frozen in liquid nitrogen. Samples were then stored at -90°C for up to 3 weeks before the biochemical studies were carried out. The freezing period had a minimal effect (< 5% reduction) on the sarcolemmal Ca^{2+}-pump activities when compared with membranes assayed immediately after the isolation [12]. The protein concentration was measured using the method of Lowry *et al.* [13]. The purity of the membrane preparations was examined by measuring the activities of marker enzymes such as Na^+-K^+-ATPase (sarcolemmal marker), rotenone-insensitive NADPH (sarcoplasmic reticular marker), and cytochrome *c* oxidase (mitochondrial marker) in both homogenate and sarcolemmal membrane according to the procedures used in our laboratory [8, 14]. Sarcolemmal Na^+-K^+-ATPase activity was enriched about 18-fold with respect to the homogenate values and the other marker enzyme activities showed negligible contamination in the sarcolemmal fraction with sarcoplasmic reticular and mitochondrial membranes.

Measurements of Mg^{2+}-ATPase and Ca^{2+}-stimulated ATPase activities

The experimental conditions for the determination of ATPase activities were the same as reported elsewhere [8, 9]. For the estimation of Mg^{2+}-ATPase, sarcolemmal vesicles (20–40 μg protein) were preincubated at 37°C for 5 min in 0.5 ml of medium containing 140 mM KCl-10 mM MOPS-Tris (pH 7.4), 2 mM $MgCl_2$, 5 mM sodium azide, and 0.1 mM ethylene glycol-bis (β-aminoethyl ether)-N,N,N',N'-tetraacetic acid (EGTA). The reaction was started by the addition of 4 mM Tris-ATP, pH7.4, and terminated 5 min later with 0.5 ml of ice-cold 12% trichloroacetic acid; the liberated phosphate was measured by the method of Taussky and Shorr [15]. Estimation of total (Ca^{2+}-Mg^{2+})-ATPase was made in medium containing 140 mM KCl-10 mM MOPS-Tris, pH 7.4, 2 mM $MgCl_2$, 5 mM sodium azide, and $1 \times 10^{-}$

^{5} M free Ca^{2+}. The concentration of free Ca^{2+} in the medium was adjusted by using EGTA [16]. The Ca^{2+}-stimulated ATPase activity was the difference between the total ATPase and Mg^{2+}-ATPase activities.

Determination of ATP-dependent Ca^{2+} accumulation

ATP-dependent Ca^{2+} uptake was carried out by a method described elsewhere [8, 9] with some modifications. Sarcolemmal vesicles (50 μg protein) were preincubated at 37°C for 5 min in 0.25 ml of medium containing 140 mM KCl-10 mM MOPS-Tris, pH 7.4, 2 mM $MgCl_2$, and $^{45}CaCl_2$-EGTA, which contained 1×10^{-5} M free Ca^{2+} [16]. Ca^{2+} uptake was initiated by adding 4 mM Tris-ATP, pH 7.4. After 5 min of incubation at 37°C, the reaction was stopped by the addition of 0.03 ml ice-cold stopping solution containing KCl-MOPS and 1 mM $LaCl_3$, pH 7.4. Samples (0.25 ml from 0.28 ml of the total reaction mixture) were filtered through Millipore filters (pore size 0.45 μm), washed twice with 2.5 ml ice-cold KCl-MOPS and 1mM $LaCl_3$, and the radioactivity was determined for calculating the total Ca^{2+} uptake. Nonspecific Ca^{2+} binding was also measured in the absence of ATP for each set of experiments. The ATP-dependent Ca^{2-} uptake was calculated by subtracting nonspecific Ca^{2+} binding from the total Ca^{2+} accumulation.

Statistical analysis

Results are presented as mean ± S.E. All data were subjected to analysis of variance followed by the Bonferroni corrected '*t*' test for multiple comparison. A P value of less than 0.05 was considered to be statistically significant.

Results

Changes in contractile force development and resting tension

Figure 1 shows the time course of changes in developed tension, resting tension, +dF/dt, and -dF/dt upon perfusing the heart with X + XO. An initial increase in developed tension was observed within 1 min but continued perfusion with X + XO resulted in a marked depression in the developed tension and an increase in the resting tension. The initial increase and then decline of the developed tension was accompanied by similar biphasic changes in +dF/dt as well as -dF/dt. Developed tension as well as +dF/dt and -dF/dt declined within 3 min whereas the increase in resting tension was seen in 2 min; complete contractile failure was seen at 20 min of perfusion with X + XO.

Alterations in Sarcolemmal ATP-dependent Ca^{2+}-pump activities

To examine the effects of X + XO on sarcolemmal Ca^{2+}-pump activity, Ca^{2+}-stimulated ATPase and ATP-dependent Ca^{2+} uptake were assayed in the rat hearts perfused for different time intervals (Table 1). Sarcolemmal vesicles from hearts subjected to 1 min perfusion with X + XO showed a significant decrease in Ca^{2+}-stimulated ATPase activity and ATP-dependent Ca^{2+} accumulation compared to the control group. The decrease of Ca^{2+}-stimulated ATPase and ATP-dependent Ca^{2+} accumulation due to perfusion with X + XO was time-dependent. Mg^{2+}-ATPase activity also showed a time-dependent change; significant differences were seen at 5 min of perfusion with X + XO. These results indicate that ATP-dependent Ca^{2+} accumulation and Ca^{2+}-stimulated ATPase were more sensitive than Mg^{2+}-ATPase to oxygen radicals generated in perfused hearts by X + XO. A previous study from our laboratory [14] has shown that the sarcolemmal vesicles used for these experiments released the accumulated Ca^{2+} in the presence of extravesicular Na^{+}, a phenomenon which is not observed for the purified sarcoplasmic reticulum vesicles. Furthermore, unlike the sarcoplasmic reticulum, the sarcolemmal Ca^{2+}-stimulated ATPase activities were inhibited by low concentrations (2–4 μM) of vanadate. Thus we can rule out the possibility that ATP-dependent Ca^{2+} accumulation and Ca^{2+}-stimulated ATPase activities in the sarcolemmal vesicles are due to contamination by fragments of the sarcoplasmic reticulum. It should also be noted that the control values for ATP-dependent Ca^{2+} accumulation and Ca^{2+}-stimulated ATPase activity reported here compare favourably with those reported elsewhere [8, 14, 17].

Effects of SOD and CAT on sarcolemmal Ca^{2+}-pump activities and contractile function

Another series of experiments was performed to assess the effects of SOD and CAT on X + XO-induced depression in sarcolemmal ATP-dependent Ca^{2+} pump activities and cardiac contractile function. As shown in Table 2, SOD showed no protective effect on the depression in Ca^{2+}-stimulated ATPase activity and ATP-dependent Ca^{2+} accumulation. On the other hand, CAT showed a protective effect with respect to changes in both activities. The combination of CAT and SOD did not exert effects greater than those observed with CAT alone on the X + XO-induced depression in the ATP-dependent Ca^{2+} pump activities. SOD had no protective effect against the X + XO-induced contractile failure and increase of resting tension whereas CAT showed a protective effect (Table 2). Sarcolemmal ATP dependent Ca^{2+} accumulation was also examined in control and X + XO-perfused hearts at varying concentrations of Ca^{2+} in the incubation medium (Fig. 2). A signifi-

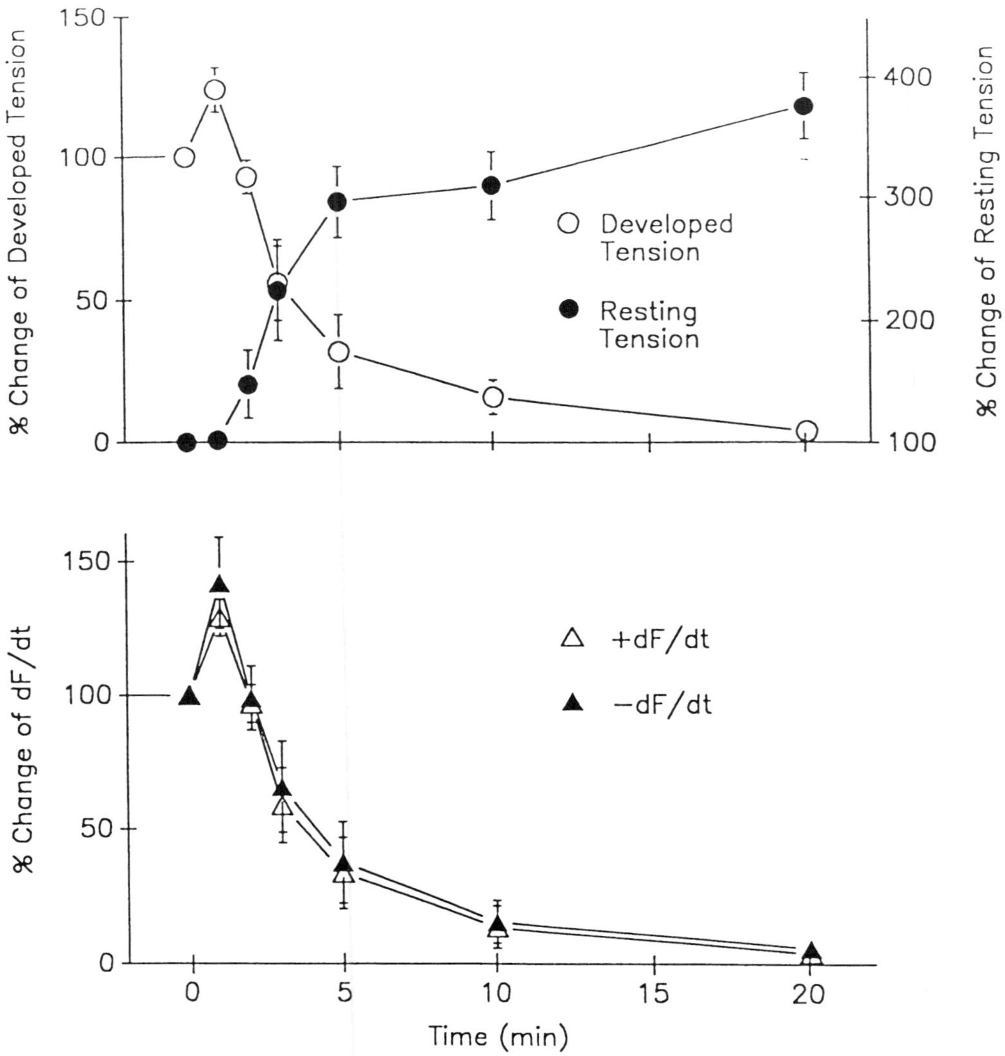

Fig. 1. Effects of xanthine plus xanthine oxidase on developed tension, resting tension, rate of contraction (+dF/dt) and rate of relaxation (-dF/dt) in isolated perfused rat hearts. Values are mean ± S.E. of 6 experiments and are expressed as percent changes of control values recorded just before X + XO exposure. Control values for developed tension and resting tension were 7.2 ± 0.6 g and 2 g whereas those for +dF/dt and -dF/dt were 137 ± 12 g/sec and 131 ± 16 g/sec, respectively.

cant depression of ATP-dependent Ca^{2+} accumulation at different concentrations of Ca^{2+} was observed in sarcolemmal vesicles from X + XO-perfused hearts in comparison to the control values. Representation of these data on a double-reciprocal plot illustrates the noncompetitive nature of the X + XO-induced depression of ATP-dependent Ca^{2+} accumulation. Table 3 shows the values, calculated from the double-reciprocal plot, for the apparent constant (Ka) and maximal velocity (Vmax) of ATP-dependent Ca^{2+} accumulation. The Vmax value (26.6 ± 1.4 and 8.1 ± 1.7 nmol Ca^{2+}/mg protein/5 min for control and X + XO-treated, respectively) was decreased by 70% in the sarcolemmal samples isolated from the X + XO-perfused hearts; this change was prevented by addition of SOD plus CAT in the perfusion medium. The Ka value for the ATP-dependent Ca^{2+} accumulation was not affected in the experimental samples indicating no changes in the affinity of the sarcolemmal Ca^{2+}-pump system for Ca^{2+} upon perfusing the heart with X + XO.

Discussion

This study demonstrated that perfusion of the isolated perfused rat heart with a medium containing X + XO, a well known oxygen free radical generating system, resulted in a depression of sarcolemmal Ca^{2+}-pump activities and produced a biphasic response of the cardiac contractile function. The degree of changes in heart sarcolemmal Ca^{2+} pump activities were dependent upon the duration of perfusion time with the medium containing X + XO. Following an initial increase in

Table 1. Effects of xanthine plus xanthine oxidase on sarcolemmal Ca^{2+}-pump activities in the rat hearts perfused for different time intervals

Perfusion time with X + XO	ATP-dependent Ca^{2+} accumulation (nmol Ca^{2+}/mg protein/5 min)	ATPase activities (µmol Pi/mg protein/h)	
		Mg^{2+}-ATPase	Ca^{2+}-stimulated ATPase
Control	24.4 ± 1.0	100.4 ± 2.8	11.8 ± 0 9
min	15.0 ± 1.9*	94.9 ± 7.0	6.5 ± 1.1*
2min	12.0 ± 1.8*	92.2 ± 4.4	6.1 ± 1.2*
3min	8.8 ± 1.7*	90.8 ± 7.5	5.8 ± 1.1*
5min	7.2 ± 1.1*	84.8 ± 4.1*	5.4 ± 1.4*
10min	4.3 ± 1.3*	85.5 ± 6.1*	4.8 ± 1.3*
20 min	3.8 ± 1 .0*	80.3 ± 6.4*	4.1 ± 0.7*

Each value is a mean ± S.E. of 4–6 experiments. Control values are the result of perfusion of hearts for 2–20 min with normal Krebs-Henseleit solution. Since values for control hearts were overlapping, the data were grouped together. *Significantly different from control value ($p < 0.05$). X: xanthine; XO: xanthine oxidase.

the developed tension, +dF/dt and -dF/dt, these parameters declined gradually, but this depression was preceded by an increase of resting tension. While the sarcolemmal Ca^{2+}-pump activities in the X + XO-perfused hearts were decreased significantly at 1 min, a significant depression of the developed tension or +dF/dt and -dF/dt was seen at 3 min and a significant increase in resting tension was observed at 2 min. These findings are consistent with the view that the heart sarcolemmal Ca^{2+} pump activities are affected by oxygen free radicals and confirm our previous observations concerning the depressant effects of oxyradicals on the sarcolemmal Ca^{2+}-stimulated ATPase and ATP-dependent Ca^{2+} uptake activities under *in vitro* conditions [8–10]. Furthermore, the results regarding the biphasic changes in the developed force by the isolated heart upon perfusion with X + XO are similar to those reported earlier [18].

It is now well known that a defect in the ability of myocardial cell to maintain normal Ca^{2+} homeostasis may lead to the occurrence of intracellular Ca^{2+} overload and subsequent irreversible cellular injury and contractile failure [19, 20]. Sarcolemmal Ca^{2+} transport systems, such as Na^+-Ca^{2+} exchanger and ATP-dependent Ca^{2+} pump, are considered to effect Ca^{2+} extrusion from the myocardial cell [17, 20, 21]. Since Na^+-Ca^{2+} exchanger in the sarcolemmal membrane is considered to have a high pumping rate for Ca^{2+} extrusion from the myocardial cell in comparison to the sarcolemmal ATP-dependent Ca^{2+} pump [17, 20–22], the sarcolemmal Na^+-Ca^{2+} exchanger is thought to be the major route for Ca^{2+} efflux from the myocardial cell whereas the sarcolemmal Ca^{2+} pump is considered to have an ancillary role in the Ca^{2+} efflux process. However, it has been suggested that the sarcolemmal Ca^{2+}-pump may be more important for Ca^{2+} extrusion from the myocardial cell because this system is active enough to respond to the lower cytosolic Ca^{2+} concentration normally found in diastole (3×10^{-7} M) where the Ca^{2+} efflux is carried one-third by the Ca^{2+}-pump and two-thirds by the Na^+-Ca^{2+} exchange [23]. Philipson and Ward [24] have reported that the *in vivo* capacity of the Ca^{2+} pump is roughly equal to that of the Na^+-Ca^{2+} exchanger, considering normal intracellular physiologic conditions. Accordingly, the depression of heart sarcolemmal Ca^{2+} pump activity observed in this study can be seen to cause an increase in the intracellular Ca^{2+} concentration, which will result in the initial increase of contractile force development, and subsequent occurrence of intracellular Ca^{2+} overload accompanied by an increase of resting tension and loss of the myocardium to generate contractile force.

In the present study, we used X + XO for the generation of oxygen free radicals for studying the effects in the perfused heart. Although superoxide is the primary radical produced with XO and its substrates, it has been considered that contractile dysfunction induced with this oxyradical generating system in isolated perfused hearts is mediated by hydroxyl radical and/or H_2O_2 instead of superoxide radicals [25, 26]. Also in this study, SOD was not effective in protecting changes in the mechanical function and sarcolemmal Ca^{2+}

Table 2. Effects of superoxide dismutase and catalase on xanthine plus xanthine oxidase induced changes in Ca^{2+}-pump activities and mechanical function

Perfusion Groups	Developed tension (% of control)	Resting tension (% of control)	ATP-dependent Ca^{2+} accumulation (nmol Ca^{2+}/mg protein/5 min)	ATPase activities (µmol Pi/mg protein/h)	
				Mg^{2+} -ATPase	Ca^{2+} -stimulated ATPase
Control	102.0 ± 4	103 ± 2	24.7 ± 1.0	100.9 ± 3.5	11.7 ± 1.0
X+XO	1.0 ± 1*	390 ± 18*	3.6 ± 1.2*	81.5 ± 6.5*	4.1 ± 0.9*
X + XO + SOD	3.0 ± 2*	403 ± 42*	3.7 ± 0.7*	84.1 ± 6.2*	4.5 ± 0.5*
X+XO+CAT	94.0 ± 2	108 ± 3	23.3 ± 1.8	93.6 ± 7.3	9.0 ± 1.0
X+XO+SOD+CAT	96.0 ± 3	110 ± 4	22.1 ± 1.4	96.2 ± 3.5	9.2 ± 1.2

Each value is a mean ± of 4–6 experiments. X: xanthine; XO: xanthine oxidase; SOD; superoxide dismutase; CAT: catalase. Control values are the result of perfusion of heart with Krebs-Hensleit solution for 20 min. All other hearts were perfused for 20 min in the presence of X + XO with or without different scavengers. Developed tension and resting tension are expressed as percent of values obtained just before starting the perfusion with X + XO. Initial values for the developed tension for control, X + XO, X + XO + SOD, X + XO + CAT and X + XO + SOD + CAT groups were 7.1 ± 1.0, 7.4 ± 0.8, 6.8 ± 0.9, 7.0 ± 1.0 and 6.7 ± 0.8 g, respectively. *Significantly different from control value ($p < 0.05$).

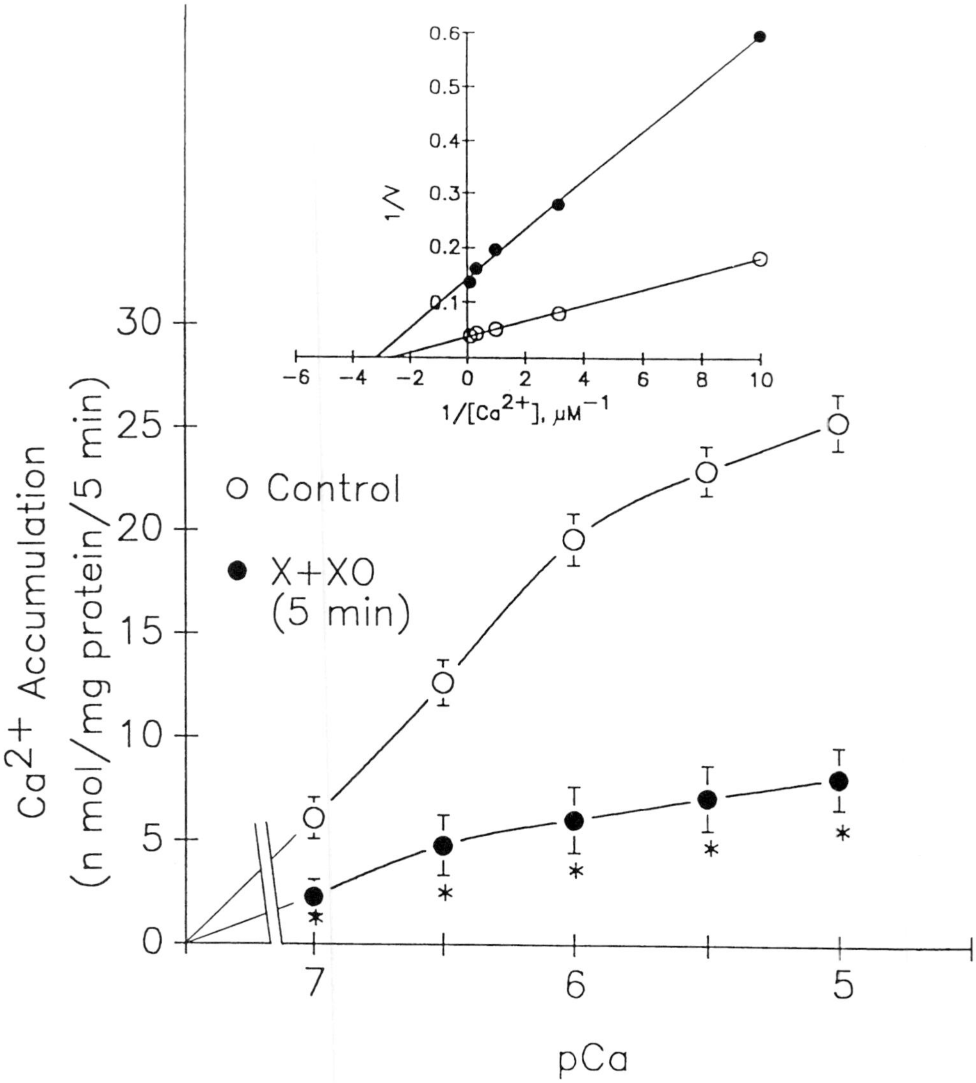

Fig. 2. Effect of 5 min perfusion of the isolated heart with xanthine plus xanthine oxidase (X + XO) on sarcolemmal ATP-dependent Ca^{2+} accumulation at different concentrations of Ca^{2+} in the incubation medium. The assay was carried out by using 0.1–10 μM Ca^{2+} for 5 min. The upper section of the figure shows the Lineweaver-Burk plot of the data. Each value is a mean ± S.E. of 5 experiments. *P < 0.05, compared to the control values.

Table 3. Effects of xanthine plus xanthine oxidase on kinetic parameters of ATP dependent Ca^{2+} accumulation

Perfusion	Ka (μmol Ca^{2+})	Vmax (nmol Ca^{2+}/mg protein/5 min)
Control	0.40 ± 0.13	26.6 ± 1.4
X + XO (5 min)	0.31 ± 0.07	8.1 ± 1.7*
X+XO+SOD+CAT(5min)	0.42 ± 0.11	25.7 ± 2.6

Each value is a mean ± S.E. of 4–5 experiment. X: xanthine; XO: xanthine oxidase; SOD: superoxide dismutase; CAT: catalase. The values for Ka and Vmax were obtained from data according to the Lineweaver-Burk plot analysis. *Significantly different from control value (p < 0.05).

pump activities, suggesting that superoxide radicals may not be directly participating in toxic effects of X + XO. Although H_2O_2 is considered to be toxic at high concentrations, the major danger of H_2O_2 accumulation is the production of hydroxyl radicals by the Habar-Weiss reaction or the Fenton reaction [3]. In fact, hydroxyl radicals are known to be more toxic than superoxide radicals or H_2O_2 [27, 28]. Previous *in vitro* studies [8–10] have reported that SOD had protective effects on the depression of sarcolemmal Ca^{2+} pump activities by X + XO. In these studies, SOD was used in the incubation medium upon treating the sarcolemmal membrane with X + XO whereas in the present study SOD was used in a buffer solution for perfusing the isolated hearts with X + XO. This difference in experimental methods may be the reason why

SOD was ineffective in protecting the sarcolemmal Ca^{2+}-pump activities in this study. Nonetheless, the results reported here indicate that CAT was capable of preventing the X + XO induced changes in contractile parameter and sarcolemmal Ca^{2+}-pump activities which observation can be taken to suggest the involvement of H_2O_2 and/or hydroxyl radicals in the genesis of mechanical and biochemical changes in the heart perfused with X + XO.

Acknowledgments

This research was supported by a grant from the Medical Research Council of Canada (MRC Group in Experimental Cardiology). Dr. Matsubara was a postdoctoral fellow of the Medical Research Council of Canada during the tenure of this study.

References

1. Jolly SR, Kane WJ, Bailie MB, Abrams GD, Lucchesi BR: Canine myocardial reperfusion injury: Its reduction by the combined administration of superoxide dismutase and catalase. Circ Res 54: 277–285, 1984
2. Ambrosio G, Becker LC, Hutchins GM, Weisman HF, Weisfeldt ML: Reduction in experimental infarct size by recombinant human superoxide dismutase: Insights into the pathophysiology of reperfusion injury. Circulation 74: 1424–1433, 1986
3. Hammond B, Hess ML: The oxygen free radical system: Potential mediator of myocardial injury. J Am Coll Cardiol 6: 215–220, 1985
4. Gauduel Y, Duvelleroy MA: Role of oxygen radicals in cardiac injury due to reoxygenation. J Mol Cell Cardiol 16: 459–470, 1984
5. Shlafer M, Kane PF, Kirsh MM: Superoxide dismutase plus catalase enhances the efficacy of hypothermic cardioplegia to protect the globally ischemic, reperfused heart. J Thorac Cardiovasc Surg 83: 830–839, 1982
6. Kramer JH, Mak IT, Weglicki WB: Differential sensitivity of canine cardiac sarcolemmal and microsomal enzymes to inhibition by free radical-induced lipid peroxidation. Circ Res 55: 120–124, 1984
7. Romaschin AD, Wilson GJ, Thomas U, Feitler DA, Tumiati L, Mickle DAG: Subcellular distribution of peroxidized lipids in myocardial reperfusion injury. Am J Physiol 259: H116–H123, 1990
8. Kaneko M, Beamish RE, Dhalla NS: Depression of heart sarcolemmal Ca^{2+}-pump activity by oxygen free radicals. Am J Physiol 256: H368–H374, 1989
9. Dixon IMC, Kaneko M, Hata T, Panagia V, Dhalla NS: Alterations in cardiac membrane Ca^{2+} transport during oxidative stress. Mol Cell Biochem 99: 125–133, 1990
10. Kaneko M, Elimban V, Dhalla NS: Mechanism for depression in heart sarcolemmal Ca^{2+}-pump by oxygen free radicals. Am J Physiol 257: H804–H811, 1989
11. Hata T, Kaneko M, Beamish RE, Dhalla NS: Influence of oxygen free radicals on heart sarcolemmal Na^+-Ca^{2+} exchange. Coronary Artery Dis 2: 397–407, 1991
12. Dixon IMC, Hata T, Dhalla NS: Sarcolemmal calcium transport in congestive heart failure due to myocardial infarction in rats. Am J Physiol 262: H1387–H1394, 1992
13. Lowry OH, Rosebrough NJ, Farr AL, Randall RJ: Protein measurement with the Folin phenol reagent. J Biol Chem 193: 265–275, 1951
14. Dixon IMC, Eyolfson DS, Dhalla NS: Sarcolemmal Na^+-Ca^{2+} exchange activity in hearts subjected to hypoxia reoxygenation. Am J Physiol 253: H1026–H1034, 1987
15. Taussky HH, Shorr E: A microcolorimetric method for the determination of inorganic phosphorus. J Biol Chem 202: 675–685, 1953
16. Ganguly PK, Pierce GN, Dhalla KS, Dhalla NS: Defective sarcoplasmic reticular calcium transport in diabetic cardiomyopathy. Am J Physiol 244: E528–E535, 1983
17. Caroni P, Carafoli E: The Ca^{2+}-pumping ATPase of heart sarcolemma. J Biol Chem 256: 3263–3270, 1981
18. Matsubara T, Musat-Marcu S, Misra HP, Dhalla NS. Protective effect of vanadate on oxyradical-induced changes in isolated perfused heart. Mol Cell Biochem 153: 79–85, 1995
19. Shen AC, Jennings RB: Kinetics of calcium accumulation in acute myocardial ischemic injury. Am J Pathol 67: 441–452, 1972
20. Dhalla NS, Pierce GN, Panagia V, Singal PK, Beamish RE: Calcium movements in relation to heart function. Basic Res Cardiol 77: 117–139, 1982
21. Mechmann S, Pott L: Identification of Na-Ca exchange current in single cardiac myocytes. Nature 319: 597–599, 1986
22. Reuter H: Exchange of calcium ions in the mammalian myocardium. Circ Res 34: 599–605, 1974
23. Chapman RA: Control of cardiac contractility at the cellular level. Am J Physiol 245: H535–H552, 1983
24. Philipson KD, Ward R: Ca^{2+} transport capacity of sarcolemmal Na^+-Ca^{2+} exchange. Extrapolation of vesicle data to *in vivo* conditions. J Mol Cell Cardiol 18: 943–951, 1986
25. Burton KP, McCord JM, Ghai G: Myocardial alterations due to free-radical generation. Am J Physiol 246: H776–H783, 1984
26. Blaustein AS, Schine L, Brooks WW, Fanburg BL, Bing OHL: Influence of exogenously generated oxidant species on myocardial function. Am J Physiol 250: H595–H599, 1986
27. Del Maestro RF: An approach to free radicals in medicine and biology. Acta Physiol Scand Suppl. 492: 153–168, 1980
28. Fridovich I: The biology of oxygen radicals. Science 201: 875–880, 1978

Molecular and Cellular Biochemistry **160/161:** 187–194, 1996.

Relation between energy metabolism, glycolysis, noradrenaline release and duration of ischemia

Anna Cargnoni, Claudio Ceconi, Salvatore Curello, Massimo Benigno, Jan Willem de Jong[1] and Roberto Ferrari[2]

Fondazione Clinica del Lavoro, Centro di Fisiopatologia Cardiovascolare 'Salvatore Maugeri,' Gussago, Brescia, Italy; [1]*Thoraxcentre, Erasmus University Rotterdam, The Netherlands;* [2]*Cattedra di Cardiologia, Universita' degli Studi di Brescia, Brescia, Italy*

Abstract

We studied the effect of 12–36 min of global ischemia followed by 36 min of reperfusion in Langendorff perfused rabbit hearts (n = 26). Metabolism was determined in terms of peak and total release of purines (adenosine, inosine, hypoxanthine), lactate and noradrenaline during reperfusion; and myocardial content of nucleotides (ATP, ADP, AMP), glycogen and noradrenaline at the end of reperfusion. An inverse relationship ($r = -0.79$) existed between duration of ischemia and developed pressure post-ischemia. Early during reperfusion, after 12 min of ischemia, the purine concentration (peak release) increased 100× ($p < 0.01$), that of lactate and noradrenaline 10× ($p < 0.05$) . Total purine release rose with progression of the ischemic period (30× after 36 min of ischemia; $p < 0.01$), concomitant with a reduction in nucleotide content. Lactate release was independent from the duration of ischemia, although glycogen had declined by 30% ($p < 0.01$) after 36 min of ischemia. The acid insoluble glycogen fraction, which presumably contains proglycogen, increased substantially during short-term ischemia. Peak noradrenaline increased 100× and 200× ($p < 0.05$) after 24 and 36 min of ischemia, respectively. Total noradrenaline release due to various periods of ischemia mirrored its peak release. Function recovery was inversely related to total purine and noradrenaline efflux (both $r = -0.81$); it correlated with tissue nucleotide content ($r = 0.84$). In conclusion, larger amounts of noradrenaline are released only after a substantial drop in myocardial ATP. During severe ischemia ATP consumption more than limited ATP production by anaerobic glycolysis, is a key factor affecting recovery on subsequent reperfusion. In contrast to lactate efflux, purine and noradrenaline release are useful markers of ischemic and reperfusion damage. (Mol Cell Biochem **160/161**: 187–194, 1996)

Key words: ATP breakdown. catecholamine, glycogen, ischemia

Introduction

Many critical factors are involved in the onset of severe myocardial ischemic damage. The most important are: (1) loss of tissue adenine nucleotide with consequent failure to restore energy metabolism with reperfusion [1–6]; (2) accumulation of toxic metabolites [7–9]; (3) calcium overload [10]; (4) release of noradrenaline [11–15]. Data about these metabolic alterations, however, derive from a combination of several individual studies employing a variety of different experimental preparations and ischemic protocols. In addition, the analytical techniques utilized for enzymatic determination of catecholamine and of adenine nucleotide often are of low sensitivity [4, 8, 12, 16, 17].

There is therefore the need of systematic data analyzing in details the temporal sequences of the metabolic alterations which occur as ischemia progresses in the light of finding out the more suitable index of ischemic and reperfusion damage.

Isolated perfused heart preparation was selected for the present study because it allows the control of many variables that *in vivo* influence the severity of ischemic damage such as: myocardial perfusion, work load, substrate availability, neurohumoral factors.

Address for offprints: R. Ferrari, Cattedra di Cardiologia, Universita' degli Studi di Brescia, c/o Spedali Civili, P.le Spedali Civili, 1, 25123 Brescia, Italy

Materials and methods

Heart perfusion

Male New Zealand white rabbits (2.0–2.3 kg), maintained on a standard diet, were used. They were treated in conformity with the guiding principles of the American Physiological Society. The animals were stunned by a blow on the head; the hearts were quickly removed and perfused as previously described [18]. The perfusion solution was heated to 37°C, bubbled with 95% O_2 and 5% CO_2 and transported at a rate of 22 ml/min to the aortic cannula with a Gilson Minipuls 2 rotary pump. The hearts were jacketed (40–44°C) to provide a constant myocardial temperature of 37°C, independent of coronary flow, checked by an Ellab thermometric probe (model CTD 85) in the pulmonary artery. They were paced using suprathreshold rectangular pulses at 0.1 ms duration at a rate of 180 beats/min.

After a 30-min period of aerobic equilibration (22 ml/min), the hearts were randomly divided into four groups: Control, these hearts (n = 7) were aerobically perfused for a further period of 60 min (no ischemia); the other hearts were made ischemic by stopping the coronary flow for: (A) 12 min (n = 7); (B) 24 min (n = 6); (C) 36 min (n = 6). At the end of ischemia, coronary flow was reestablished and all ischemic hearts were reperfused for 36 min. Left ventricular pressure was determined from a fluid-filled balloon inserted into the ventricular cavity as previously described [18].

Biochemical assays

Analysis of coronary effluent

Coronary effluent fractions were collected in cooled vials (0°C) at regular intervals during the entire period of reperfusion for purine, catecholamine and lactate determinations. Two ml of perfusate were added to 30 µl 10 mM $Na_2S_2O_5$ for catecholamine assay. Another aliquot (0.5 ml) was added to 1.0 ml $HClO_4$ (6%) for the assay of lactate. Treated and untreated samples were stored at –80°C. Purines were determined with our reverse-phase high-pressure liquid chromatography (HPLC) method with detection at 254 nm [19, 20]. Noradrenaline was determined by reverse-phase HPLC with electrochemical detection on perfusate extracted with Al_2O_3 as previously described by us [21]. L-lactate was enzymatically determined by the method of Noll [22].

Tissue determinations

At the end of the experiments, the hearts were clamped with precooled Wollenberger tongs and stored in liquid nitrogen. The frozen tissue was homogenized with $HClO_4$. The mixture was thawed and centrifuged; an aliquot of the supernatant fraction was neutralized with KOH, and analyzed for nucleotide and catecholamine content. Nucleotides were determined by HPLC using the method of Harmsen *et al.* [23]. After Al_2O_3 extraction, noradrenaline was assayed as described above. Another aliquot was neutralized with NaOH, and analyzed enzymatically for glycogen content [24, 25], using rabbit-liver glycogen as the standard.

In the acid precipitate, protein and glycogen were assayed. Protein was measured according to Bradford [26] using bovine serum albumine as the standard. A factor of 70 mg protein/g wet weight was used to express release data, based on heart weight, per gram protein. Glycogen was determined as described before [24, 25].

Statistics

Data are reported as mean ± S.E.M. A one-way analysis of variance was first carried out to test for any differences between all groups. If a difference was established, each of the groups was compared with the control group (aerobic control), using the unpaired *t*-test with Bonferroni correction. $p < 0.05$ was considered significant.

Results

Mechanical changes

During 60 min of control aerobic perfusion, mechanical function was well maintained since developed pressure declined only by 15% (n.s.). The recovered pressures, measured at the end of reperfusion after different ischemic periods are depicted in Fig. 1. Compared with the control group, there was a complete recovery of developed pressure in group A, while, with prolongation of ischemia, it significantly decreased in groups B and C (to 66% and 38% of control, respectively; $p < 0.01$, see Fig. 1). There was an inverse relationship ($y = 61.69–1.04x$; $r = –0.79$, $p < 0.001$) between the duration of ischemic period and the recovered pressure during reperfusion.

Diastolic pressure in the control hearts remained 0 mmHg during the whole time of perfusion. The same was true for the hearts in group A. However, in groups B and C it rose to 24 ± 11 and 44 ± 14 mmHg, respectively ($p < 0.001$ vs control, at the end of reperfusion) (Fig.1).

Metabolite release

When the hearts were perfused under aerobic conditions, small amounts of purine (adenosine, inosine and hypoxanthine), noradrenaline and lactate were present in coronary

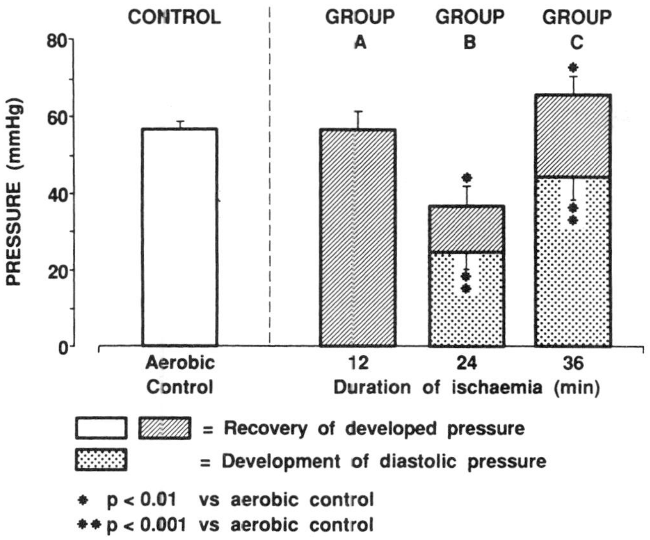

Fig. 1. Recovery of developed pressure and development of diastolic pressure after various periods of ischemia: decline in mechanical recovery and increment in diastolic tension are only evident after prolonged ischemia. For the non-ischemic controls (aerobic perfusion), the 60 min perfusion values were used. Mean values ± S.E.M. (n = 6–7) are depicted. Pre-intervention left ventricular systolic pressure was 63.1 ± 2.5 mmHg and diastolic pressure was 0 mmHg (n = 26).

effluent. Peak release of these metabolites was observed during the first minute of reperfusion (groups A, B and C, Fig. 2). Purine release, which was very low in control hearts (0.044 ± 0.004 µmol/min/g protein), became 100-fold higher after 12 min of ischemia (group A; $p < 0.01$). If the ischemic period was prolonged from 12 to 24 min (group B) it increased and it doubled after 36 min of ischemia (group C, Fig. 2A). The peak releases of individual purines are shown in Table 1. Inosine represented the major component of total purine released in the aerobic period (42%). After various ischemic periods (groups A, B and C), release of all purines increased. In particular, during reperfusion, inosine rose to 63% in group C, at the expense of adenosine.

Peak noradrenaline release (Fig. 2B) rose 10 fold in group A ($p < 0.05$ vs Control) and by an extra two orders of magnitude in groups B and C. Peak lactate release reached already a plateau in group A ($p < 0.01$ vs Control, see Fig. 2C).

Noradrenaline and lactate release declined to baseline in a few minutes (4 min for group B) whereas purine release persisted substantially longer (7–8 min for group B; Fig. 3).

Figure 4 shows the total release of the compounds determined in perfusate collected during 36 min of reperfusion. The behaviour observed for purines was quite similar to that described for their peak release (compare Figs. 2A and 4A). Total noradrenaline release in group A was not different from that in control (Fig. 4B), whereas peak release in group A exceeded that in control 10-fold (Fig. 2B). No increment in total lactate release was observed after ischemia in groups A, B and C; all values were comparable to those obtained for

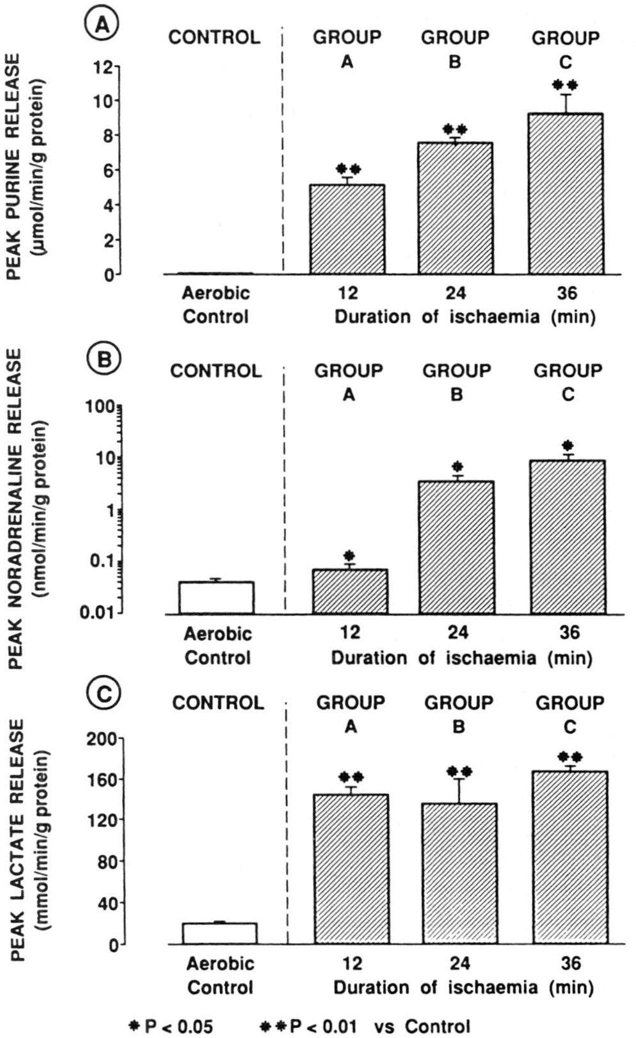

Fig. 2. Differences in peak release pattern of purines, noradrenaline and lactate after various periods of ischemia. Purines comprise adenosine, inosine and hypoxanthine. Please, note that the catecholamine data are given on a logarithmic scale. For other details, see legend to Fig. 1.

aerobic control perfusion (Fig. 4C), contrasting the behaviour of peak lactate release (Fig. 2C).

Tissue changes

The myocardial content of nucleotides (sum of ATP, ADP, and AMP), noradrenaline and glycogen is shown in Fig. 5. In groups A, B and C, total nucleotide content decreased ($p < 0.01$) compared with aerobic control (to 76%, 43% and 21%, respectively, see Fig. 5A). The ATP, ADP and AMP levels are specified in Table 1. There was a significant decline in ATP and ADP levels, proportional with the duration of the ischemic period. In contrast, AMP content decreased in group A ($p < 0.05$), and increased in groups B ($p < 0.01$) and C (n.s.).

Table 1. Purine release as well as nucleotide content after various periods of ischemia

Variable	Aerobic control	Ischemic period (min)		
		12 (Group A)	24 (Group B)	36 (Group C)
Peak release (μmol/min/g protein)				
Adenosine	0.015 ± 0.002	0.89 ± 0.12	0.92 ± 0.28	1.30 ± 0.70
Inosine	0.020 ± 0.002	3.09 ± 0.26**	4.82 ± 0.53**	5.84 ± 0.84**
Hypoxanthine	0.009 ± 0.001	0.84 ± 0.06**	1.87 ± 0.15**	2.14 ± 0.11 **
Total release (μmol/36′/g protein)				
Adenosine	0.49 ± 0.07	1.29 ± 0.15*	3.14 ± 0.71**	3.26 ± 0.63**
Inosine	0.76 ± 0.09	6.19 ± 0.34**	15.11 ± 1.76**	19.08 ± 2.55*
Hypoxanthine	0.37 ± 0.03	1.55 ± 0.08**	4.75 ± 0.41**	5.87 ± 0.52**
Tissue content (μmol/g protein)				
ATP	35.2 ± 1.5	27.2 ± 1.3	13.3 ± 2.2**	5.12 ± 1.04**
ADP	5.94 ± 0.31	4.35 ± 0.12**	3.07 ± 0.13**	2.22 ± 0.21 **
AMP	1.01 ± 0.11	0.52 ± 0.06*	1.81 ± 0.12**	1.16 ± 0.23
Energy charge	0.905 ± 0.005	0.915 ± 0.004	0.804 ± 0.021**	0.675 ± 0.044**

The values pertain to the 36-min reperfusion period, either early (peak release), over the whole period (total release), or to the end (myocardial content). The data show major ATP breakdown after the longer periods of ischemia. Energy charge is [(ATP) + 0.5 (ADP)]/[(ATP)+(ADP)+(AMP)]. Mean ± S.E.M.; n = 5–7. *p < 0.05, **p < 0.01 vs control values

Besides myocardial nucleotide content, the phosphocreatine (PCr) tissue level was determined. In the control group, it amounted to 56.3 ± 3.4 μmol/g protein at the end of experiment. No significant differences were observed due to ischemia/reperfusion.

Energy-charge data are useful to complete the view on myocardial energetic status in the reperfused hearts after various ischemic periods. The aerobic control value was 0.905 ± 0.005. Only in groups B and C energy charge decreased, to 88% and 75% with respect to control ($p < 0.01$, Table 1).

Panels B and C of Fig. 5 depict the noradrenaline and glycogen content. They show a decrease with the prolongation of the ischemic period from 12 min to 24 and 36 min. The decrements in noradrenaline content were not statistically significant. Total glycogen decreased to 49% and 30% in groups B and C, respectively ($p < 0.001$ vs Control, Fig . 5C). Acid-insoluble glycogen increased from 7.3 ± 0.4 mg/g protein in the control group to 17.1 ± 1.6 (group A, $p < 0.001$), 12.3 ± 1.9 (group B, $p < 0.05$) and 7.7 ± 1.3 mg/g protein (group C, n.s.)

Correlation between function and metabolism

The linear correlations between the recovery of developed pressure upon reperfusion and metabolite release (peak and total) as well as tissue content are reported in Table 2. A linear inverse and significant relationship existed between recovery of function and peak release of purines, noradrenaline and lactate, although in the last case, the correlation coefficient is low. The myocardial nucleotide and glycogen contents correlated strongly with developed pressure. Total lactate release and noradrenaline tissue content exhibited no linear relation with recovered mechanical function.

Discussion

Energy metabolism

We observed, in line with literature [6, 27–29], that reperfusion was unable to restore basal levels of cardiac adenine nucleotide content (Fig. 5A). This is especially true for ATP stores (see Table 1). In contrast to ATP, phosphocreatine was resynthetized during reperfusion. The finding of a declined AMP content in the hearts reperfused after 12 min of ischemia could be explained with the increased AMP dephosphorylation which occurs in anoxic and ischemic conditions. On the other hand, more severe ischemia causes large ATP catabolism with a probable congestion of the breakdown pattern and consequent AMP accumulation. More specifically, this accumulation could be due to increased cytosolic Pi which inhibits the activity of 5′-nucleotidase, the enzyme responsible for cardiac adenosine production [30].

Our data reveal that the lacking recovery in tissue adenine nucleotides, compared with aerobic levels, was balanced with purines lost during the reperfusion period (Table 1). This indicates that a portion of ATP, consumed during ischemia,

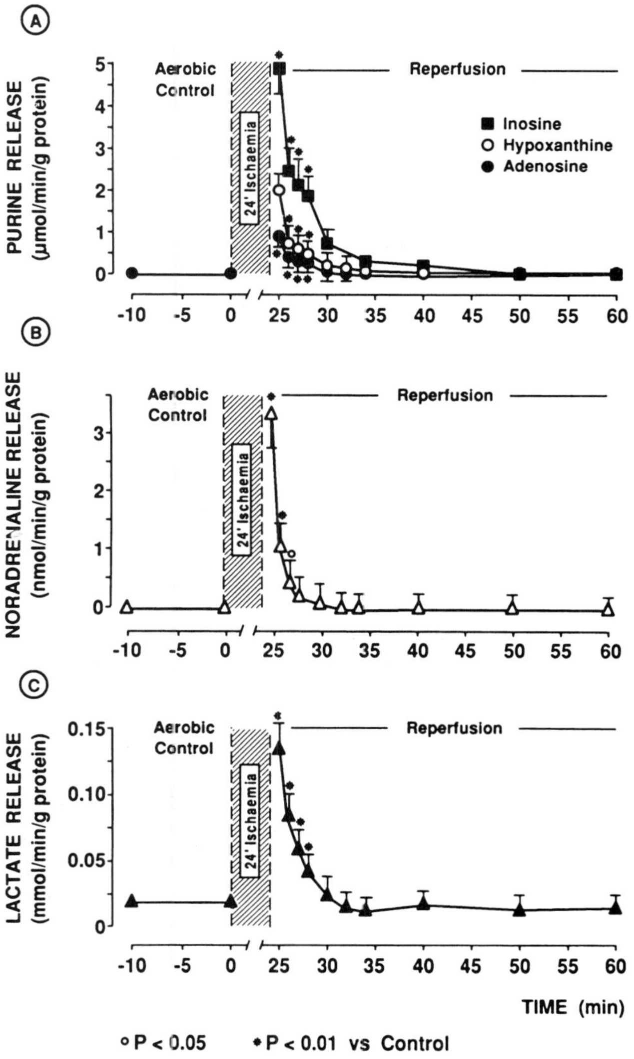

Fig. 3. Time course of metabolite release after 24 min of ischemia. The efflux of noradrenaline and lactate is already back to baseline in 4 min, when purine release is still elevated. For other details, see legend to Fig. 1.

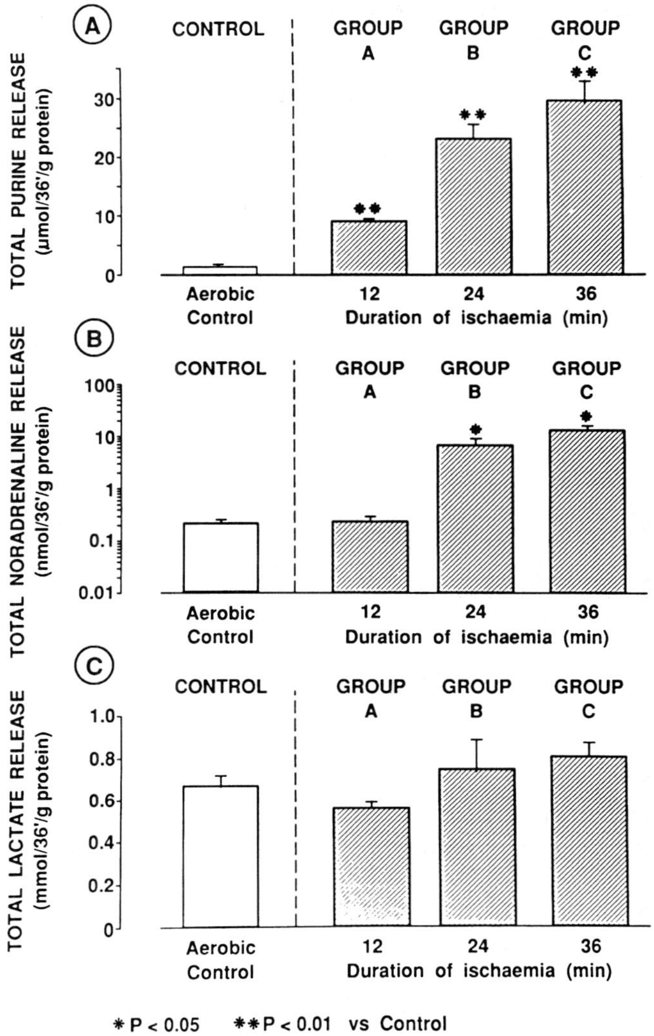

Fig. 4. Differences in total release pattern of purines, noradrenaline and lactate after various periods of ischemia. Depicted are the amounts released during 36 min of reperfusion. Purines are released in proportion with duration of ischemia. Major noradrenaline release takes place only after the longer periods of ischemia, but no-flow does not affect total lactate release. For other details, see legends to Figs. 1 and 2.

was not only degradated to ADP and AMP, but catabolized to adenosine, inosine and hypoxanthine which were immediately released from myocytes. Further degradation of hypoxanthine does not take place in rabbit heart [31].

After exhaustive ATP depletion during total ischemia [32], ATP levels can be partly restored during reperfusion from ADP by oxidative phosphorylation. But ATP resynthesis by the purine salvage pathway is too slow to play an important role in a relatively brief reperfusion period [6, 28, 29]. These data suggest that ATP degradation to purines during ischemia could slow high-energy phosphate restoration during reperfusion with delayed mechanical and metabolic cardiac recovery (stunned myocardium; ref. [27]). Our findings also emphasize the specificity of purine release to characterize the occurrence of myocardial ischemic conditions [33], where lactate release seems to be inadequate as a marker (Figs. 2 and 4).

Noradrenaline release and tissue content

The study of noradrenaline release rate is of relevance in oxygen deprivation conditions as already evidenced by the pionieristic work of Wollenberger A. *et al.* [34]. We found a large noradrenaline release at the first minute of reperfusion (peak release) after 12, 24 or 36 min of ischemia (Figs. 2B and 3B). This noradrenaline efflux represents washout of the neurotransmitter accumulated within the extracellular space during the ischemic period [14, 35, 36]. We observed also an increment of noradrenaline release during the whole reperfusion period (total release), after 24 and 36 min of

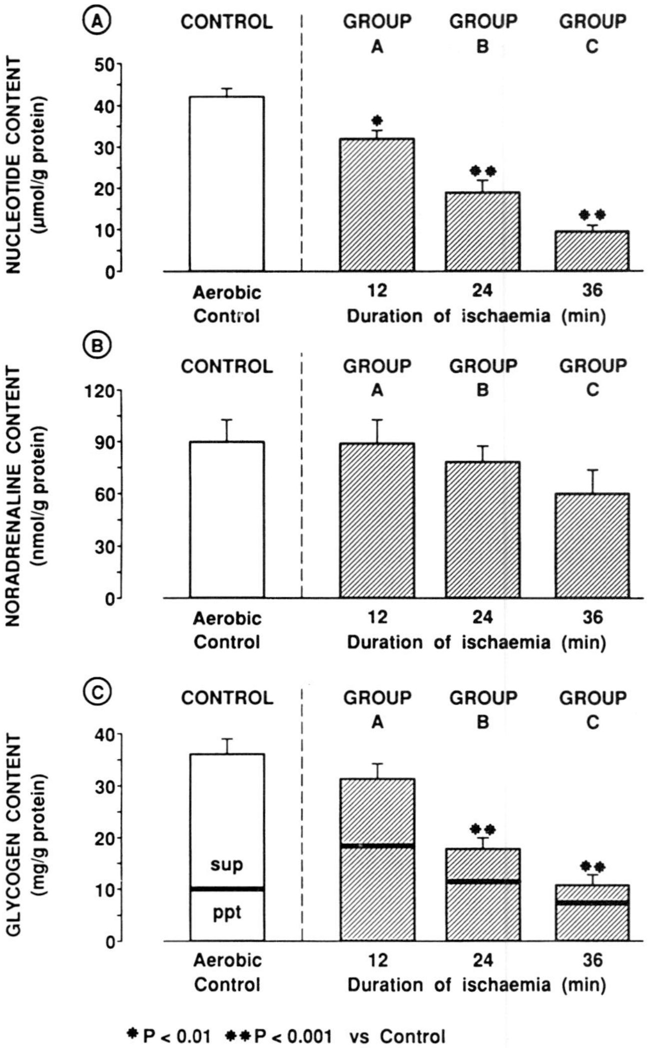

Fig. 5. Decline in tissue nucleotide, noradrenaline, and glycogen content after various periods of ischemia. Adenine nucleotides comprise ATP, ADP and AMP. Glycogen found in the acid-soluble (sup) and acid-insoluble (ppt) fractions is given in the upper and lower portions of the bars, respectively. The ratio sup/ppt in the post-ischemic groups differed from Control ($p < 0.001$). The data show substantial nucleotide and glycogen (acid-soluble) breakdown with increased duration of ischemia, with minor changes in noradrenaline content. The shorter period of ischemia induces an increase in the acid-insoluble glycogen fraction, which presumably comprises proglycogen. For other details, see legend to Fig. 1.

ischemia (Fig. 4B), and a balanced, progressive, but not significant, decline in myocardial noradrenaline content (Fig. 5B). On the other hand, it should be recalled that noradrenaline released during reperfusion represents only a small portion, from 0.1–14%, of the basal tissue content.

Based on the positive relationship between total noradrenaline release and total purine release ($r = 0.80$, $p < 0.001$) and on the inverse correlation with ATP myocardial levels ($r = -0.75$, $p < 0.001$) and glycogen content ($r = -0.71$, $p < 0.001$), we propose the following influence of endogenous noradrenaline on energetic and carbohydrate metabolism during myocardial ischemia. Released noradrenaline stimulates glycogenolysis by cyclic-AMP formation and activation of phosphorylases, as earlier suggested [13, 37]. Metz and Bernauer [16] hypothesized that noradrenaline release is not essential for the activation of anaerobic glycogenolysis, but is important for stimulating anaerobic glycolysis [14]. However, these authors found no evidence for an alternative regulating mechanism during ischemia [16].

Our data also suggest that cardiac noradrenaline, e.g. by activation of a phosphatase such as mitochondrial ATPase, could affect breakdown of ATP stores, exacerbating the ischemic consumption of high-energy phosphates. However, little is known about the effect of noradrenaline on metabolism during myocardial ischemia, in contrast to the great number of studies on its arrhythmogenic and inotropic effects upon reperfusion [11, 13–15). In the light of these novel data and observation, the metabolic role of noradrenaline released during ischemia deserves more attention. As a marker of ischemia, it is perhaps only useful after longer periods of ischemia (cf. Figs. 2B and 4B).

Glycolytic metabolism

The glycolytic pathway is profoundly changed by ischemia. In the early phase of ischemia, glycogenolysis is stimulated by local catecholamine release [13]; phosphofructokinase is activated by an increase in ADP, AMP and Pi levels and a contemporary reduction in ATP content [37]. As a result, anaerobic glycolysis is accelerated, with ATP production and lactate accumulation [5]. We have studied alterations in gly-

Table 2. Correlation between function recovery and metabolic variables after various periods of ischemia

	Peak Concentration (per g protein)			Release (per 36′/g protein)			Tissue Content (per g protein)		
	Purine (µmol/)	NorA (nmol/)	Lactate (mmol/)	Purine (µmol/)	NorA (nmol/)	Lactate (mmol/)	Nucl (µmol/)	NorA (nmol/)	Glycogen (mg/)
vs Developed pressure	r –0.66	–0.81	–0.41	–0.81	–0.81	–0.32	0.84	0.21	0.84
	p < 0.001	< 0.001	< 0.05	< 0.001	< 0.001	n.s.	< 0.001	n.s.	< 0.001

Calculated are the correlation coefficients and their p-values for developed pressure vs peak and total release of metabolites, and vs tissue compounds. Function data, obtained at the end of normoxic control perfusion and after 36 min of reperfusion were used. Mean ± S.E.M. (n = 24–26). NorA = noradrenaline, Nucl = adenine nucleotides colytic metabolism, determining myocardial glycogen content and lactate release upon reperfusion. In agreement with the literature [17], our results show that, after reperfusion, the glycogen content was reduced proportionally with the duration of ischemia. Furthermore, myocardial glycogen levels were closely correlated with the recovery of developed pressure after different ischemic periods (Table 2). We expected lactate release during reperfusion to be correlated with glycogen consumption. During 36 min of reperfusion, we observed that all reperfused hearts maintained a total lactate release similar to control hearts (Fig. 4C). However, at the first minute of reperfusion, all treated groups showed a high lactate release, compared with aerobic release. These peak releases were independent of the period of ischemia (12–36 min, Fig. 2C). Increased peak release is in apparent contrast with correspondent unchanged total release (cf. Fig. 2C and 4C), but is explained by lower-than-baseline lactate release after the early minutes of reperfusion (Fig. 3C).

We want to underline that lactate washout is independent from the duration of ischemia. Possible mechanisms include either a saturable lactate export from the myocardial cells or an inhibition of anaerobic glycolysis, without a stop in glycogenolysis. The first hypothesis is unproved because lactate accumulating in tissue was released approximately in proportion to the tissue concentration [38]. The latter hypothesis is supported by several studies: when ischemia is prolonged, glyceraldehyde-3-phosphate dehydrogenase and phosphofructokinase are inhibited by excess NADH, acidosis and lactate accumulation [39–42]. Therefore, lactate production stops while other glycolytic products accumulate in myocytes, such as glucose-1-phosphate and glucose-6-phosphate [5, 41, 43, 44]. For this reason lactate as a quantitative marker for myocardial ischemia is unacceptable (see also Figs. 2C and 4C). Apstein *et al.* [45] already pointed at the limitations of lactate production as an index of myocardial ischemia.

A novel observation is the substantial increase in acid-insoluble glycogen due to 12 or 24 min of ischemia, followed by reperfusion. This fraction presumably comprises of proglycogen, a low-molecular-weight form of glycogen with the protein glycogenin attached to it [46]. Glycogenin primes glycogen biogenesis [47]. Further investigations have to substantiate that proglycogenin, described for skeletal muscle [46], does indeed occur in myocardial tissue, where it may be affected by ischemia/reperfusion.

Conclusion

Our data demonstrate that: (1) impaired restoration of myocardial function during reperfusion is correlated with a large loss of purines and consequent delay to reestablish adequate myocardial energy metabolism; (2) ATP production by anaerobic glycolysis is independent from the duration of ischemia; (3) cardiac noradrenaline release is possibly involved in aggravating energy waste during ischemia. These findings suggest that manipulations which increase energy saving during ischemia [5] or that inhibit purine release, with accelerated repletion of ATP stores on reperfusion [19, 29], are useful to improve mechanical function during post-ischemic reperfusion. In contrast to lactate release, purine and possibly noradrenaline release are useful markers of ischemic damage.

Acknowledgements

This work has been supported by the National Research Council (C.N.R.) target project 'Prevention and control of disease factors' # 91.00156PS41. We thank Miss Roberta Ardesi, Miss Patrizia Martina, Miss Selma Nieukoop and Miss Michela Palmieri for their expert technical assistance, and Miss Roberta Bonetti for secretarial help.

References

1. Asimakis GK, Sandhu GS, Conti VR, Sordahl LA, Zwischenberger JB: Intermittent ischemia produces a cumulative depletion of mitochondrial adenine nucleotides in the isolated perfused rat heart. Circ Res 66: 302–310, 1990
2. Clarke K, O'Connor AJ, Willis RJ: Temporal relation between energy metabolism and myocardial function during ischemia and reperfusion. Am J Physiol 253: H412–H421, 1987
3. De Scheerder IK, Maas AAM, Nieukoop AS, Van der Meer P, Huizer T, Roelandt JRTC, De Jong JW: Cardiac ATP breakdown and mechanical function during recurrent periods of anoxia. Cardioscience 3: 189–195, 1987
4. Jennings RB, Reimer KA, Hill ML, Mayer SE: Total ischemia in dog hearts, *in vitro*. Circ Res 49: 892–900, 1981
5. Murry CE, Richard VJ, Reimer KA, Jennings RB: Ischemic preconditioning slows energy metabolism and delays ultrastructural damage during a sustained ischemic episode. Circ Res 66: 913–931, 1990
6. Reibel DK, Rovetto MJ: Myocardial adenosine salvage rates and restoration of ATP content following ischemia. Am J Physiol 237: H247–H252, 1978
7. Katz AM, Messineo FC: Lipid membrane interactions and the pathogenesis of ischemic damage in the myocardium. Circ Res 48: 1-16, 1981
8. Neely JR, Grotyohann LW: Role of glycolytic products in damage to ischemic myocardium. Circ Res 55: 816–824, 1984
9. Van der Vusse GJ, Stam H: Lipid and carbohydrate metabolism in the ischaemic heart. Basic Res Cardiol 82 (suppl 1): 149–153, 1987
10. Nayler WG: The role of calcium in the ischemic myocardium. Am J Pathol 102: 262–270, 1981
11. Carlsson L, Abrahamsson T, Almgre O, Svensson: Ischemia-induced release of noradrenaline and creatine kinase in the isolated, working rat heart. Acta Physiol Scand 130: 83–93, 1987
12. Dart AM, Riemersma RA, Schömig A, Ungar A: Metabolic requirements for release of endogenous noradrenaline during myocardial ischemia and anoxia. Br J Pharmacol 90: 43–50, 1987
13. Mayer SE, Williams BJ, Smith JM: Adrenergic mechanisms in glyco-

gen metabolism. Ann NY Acad Sci 39: 682–702, 1967
14. Schömig A, Dart AM, Dietz R, Mayer E, Kübler W: Release of endogenous catecholamines in the ischemic myocardium of the rat. Part A: locally mediated release. Circ Res 55: 689–701, 1984
15. Schömig A, Fischer S, Kurz T, Richardt G, Schömig E: Nonexocytotic release of endogenous noradrenaline in the ischemic and anoxic rat heart: Mechanism and metabolic requirements. Circ Res 60: 194–205, 1987
16. Metz V, Bernauer N: The effect of reserpine and guanethidine on carbohydrate metabolism in ischaemic rat myocardium. Cardiovasc Res 23: 385–389, 1989
17. Taegtmeyer H, Roberts AFC, Raine AEG: Energy metabolism in reperfused heart muscle: Metabolic correlates to return of function. J Am Coll Cardiol 6: 864–870, 1985
18. Ferrari R, Ceconi C, Curello S, Guarnieri C, Caldarera CM, Albertini A, Visioli O: Oxygen mediated myocardial damage during ischemia and reperfusion. Role of cellular defence against oxygen toxicity. J Mol Cell Cardiol 17: 937–945, 1985
19. De Jong JW, Harmsen E, De Tombe PP, Keijzer E: Nifedipine reduces adenine nucleotide breakdown in ischemic rat heart. Eur J Pharmacol 81: 89–96, 1987
20. Harmsen E, De Jong JW, Serruys PW: Hypoxanthine production by ischemic heart demonstrated by high pressure liquid chromatography of blood purine nucleosides and oxypurines. Clin Chim Acta 115: 73–84, 1987
21. Ceconi C, Condorelli E, Quinzanini M, Rodella A, Ferrari R, Harris P: Noradrenaline, atrial nutriuretic peptide, bombesin and neurotensin in myocardium and blood of rats in congestive cardiac failure. Cardiovasc Res 23: 674–682, 1989
22. Noll F: L-(+)-Lactate. In: HU Bergmeyer (ed.) Methods of Enzymatic Analysis, Vol 6, 3rd ed., Verlag Chemie, Weinheim, 1984, pp 582–588
23. Harmsen E, De Tombe PP, De Jong JW: Simultaneous determination of myocardial adenine nucleotides and creatine phosphate by high pressure liquid chromatography. J Chromatogr 230: 131–136, 1987
24. Bergmeyer HU, Bernt E: Glycogen-determination with glucose oxidase and peroxidase. In: HU Bergmeyer (ed.) Methods of Enzymatic Analysis, Vol 3, 2nd ed., Verlag Chemie, Weinheim, 1974, pp 1205–1215
25. Keppler D, Decker K: Glycogen-determination with amyloglucosidase. In: HU Bergmeyer (ed.) Methods of Enzymatic Analysis, Vol. 3, 2nd ed., Verlag Chemie, Weinheim, 1974, pp 1127–1131
26. Bradford MM: A rapid and sensitive method for the quantification of microgram quantities of protein utilizing the principle for protein binding. Anal Biochem 72: 248–254, 1978
27. Flameng W, Sukekiro S, Molihoff T, Van Belle H, Janssen P: A new concept of long-term donor heart preservation: Nucleotide transport inhibition. J Heart Lung Transplant 10 (suppl 6): 990–998, 1987
28. Harmsen E, De Tombe PP, De Jong JW, Achterberg PW: Enhanced ATP and GTP synthesis from hypoxanthine or inosine after myocardial ischemia. Am J Physiol 246: H37–H43, 1984)
29. Swain JL, Heines JJ, Sabrina RL, Holmes EW: Accelerated repletion of ATP and GTP pools in post-ischaemic canine myocardium using a precursor of purine *de novo* synthesis. Circ Res 51(1): 102–105, 1982
30. Itoh R, Oka J, Ozasa H: Regulation of rat heart cytosol 5′-nucleotidase by adenylute energy charge. Biochem J 235:847-851, 1984
31. De Jong JW, Van der Meer P, Nieukoop AS, Huizer T, Stroeve RJ, Bos E. Xanthine oxidoreductase activity in perfused hearts of various species, including humans: Circ Res 67: 770–773, 1987
32. Steenbergen C, Murphy E, Watts JA, London RE: Correlation between cytosolic free calcium, contracture, ATP and irreversible ischemic injury in perfused rat heart. Circ Res 66: 135–146, 1990
33. Remme WJ, De Jong JW, Verdouw PD: Effects of pacing-induced myocardial ischemia on hypoxanthine efflux from the human heart. Am J Cardiol 40: 55–62, 1977
34. Wollenberger A, Shahab L: Anoxia-induced release of noradrenaline from the isolated perfused heart. Nature 207 (992): 88–89, 1965
35. Abrahamsson T, Almgren O, Carlsson L: Wash-out of noradrenaline and its metabolites by calcium-free reperfusion after ischemia: Support for the concept of ischemia induced noradrenaline release. Br J Pharmacol 81 (suppl 1): 22–24, 1984
36. Carlsson L, Abrahamsson T: Characterization of the inhibitory effect of some antidepressant drugs on the outward transport of norepinephrine in the ischemic myocardium. J Pharmacol Exp Ther 247: 715–720, 1988
37. Wollenberger A, Krause E: Metabolic control characteristics of the acutely ischemic myocardium. Am J Cardiol 22: 349–359, 1968
38. Makdessi SA, Andrieu JL, Chah QT, Tuduri AJ, Faucon G: Relationship between lactate arteriovenous gradient and tissue lactate content in subendocardial and subepicardial layers of the ischaemic *in situ* heart. Cardiovasc Res 16: 552–558, 1982
39. Gelet TR, Altschuld RA, Weissler AM: Effects of acidosis on the performance and metabolism of the anoxic heart. Circulation 39/40 (suppl. IV): 60–69, 1987
40. Kübler W, Spieckermann PG: Regulation of glycolysis in the ischemic and anoxic myocardium. J Mol Cell Cardiol 1 :351–377, 1970
41. Rovetto MJ, Lamberton WF, Neely JR: Mechanism of glycolytic inhibition in ischemic rat hearts. Circ Res 37: 742–751, 1975
42. Williamson JR, Schaffer S, Ford C, Safer B: Contribution of tissue acidosis to ischemic injury in the perfused rat heart. Circulation 53 (suppl 1): 3–14, 1976
43. Neely JR, Morgan HE: Relationship between carbohydrate and lipid metabolism and the energy balance of heart muscle. Annu Rev Physiol 36: 413–459, 1974
44. Opie LH: Effects of regional ischemia on metabolism of glucose and fatty acids: Relative rates of aerobic and anaerobic energy production during myocardial infarction and comparison with effects of anoxia. Circ Res 38: 52–74, 1976
45. Apstein CS, Gravino F, Hood JW Jr: Limitations of lactate production as an index of myocardial ischemia. Circulation 60: 877–888, 1979
46. Lomako J, Lomako WM, Whelan WJ: Proglycogen: A low-molecular weight form of muscle glycogen. FEBS Lett 279: 223–227, 1991
47. Smythe C, Cohen P: The discovery of glycogenin and the priming mechanism for glycogen biogenesis. Eur J Biochem 200: 625–631, 1991

Molecular and Cellular Biochemistry **160/161**: 195–208, 1996.

On the regulation of cellular energetics in health and disease

V.A. Saks, T. Tiivel, L. Kay, V. Novel-Chaté, Z. Daneshrad, A. Rossi, E. Fontaine, C. Keriel, X. Leverve, R. Ventura-Clapier,[1] K. Anflous,[1] J.-L. Samuel[2] and L. Rappaport[2]

Laboratories of Bioenergetics of Institute of Biological and Chemical Physics, Tallinn, Estonia, and of Joseph Fourier University, Grenoble; [1]Laboratoire de Cardiologie Cellulaire et Moleculaire INSERM Unité 446, Université Paris-Sud, Chatenay-Malabry; [2]INSERM Unité 127, Hôpital Lariboisiere, Paris, France

Abstract

Very recent experimental data, obtained by using the permeabilized cell technique or tissue homogenates for investigation of the mechanisms of regulation of respiration in the cells *in vivo*, are shortly summarized. In these studies, surprisingly high values of apparent Km for ADP, exceeding that for isolated mitochondria *in vitro* by more than order of magnitude, were recorded for heart, slow twitch skeletal muscle, hepatocytes, brain tissue homogenates but not for fast twitch skeletal muscle. Mitochondrial swelling in the hypo-osmotic medium resulted in the sharp decrease of the value of Km for ADP in correlation with the degree of rupture of mitochondrial outer membrane, as determined by the cytochrome c test. Very similar effect was observed when trypsin was used for treatment of skinned fibers, permeabilized cells or homogenates. It is concluded that, in many but not all types of cells, the permeability of the mitochondrial outer membrane for ADP is controlled by some cytoplasmic protein factor(s). Since colchicine and taxol were not found to change high values of the apparent Km for ADP, the participation of microtubular system seems to be excluded in this kind of control of respiration but studies of the roles of other cytoskeletal structures seem to be of high interest.

In acute ischemia we observed rapid increase of the permeability of the mitochondrial outer membrane for ADP due to mitochondrial swelling and concomitant loss of creatine control of respiration as a result of dissociation of creatine kinase from the inner mitochondrial membrane. The extent of these damages was decreased by use of proper procedures of myocardial protection showing that outer mitochondrial membrane permeability and creatine control of respiration are valuable indices of myocardial preservation. In contrast to acute ischemia, chronic hypoxia seems to improve the cardiac cell energetics as seen from better postischemic recovery of phosphocreatine, and phosphocreatine overshoot after inotropic stimulation.

In general, adaptational possibilities and pathophysiological changes in the mitochondrial outer membrane system point to the central role such a system may play in regulation of cellular energetics *in vivo*. (Mol Cell Biochem **160/161**: 195–208, 1996)

Key words: respiration, ADP diffusion, heart, skeletal muscle, liver, brain, *in vivo* regulation

Introduction

Professor Albert Wollenberger has always been deeply interested in the mechanisms of regulation of cardiac cell energetics in health and disease [1]. Therefore, we decided to dedicate to his anniversary this short review of our own experimental data collected in recent years in studies of permeabilized cells, skinned fibers or tissue homogenates, which were carried out in different countries. We hope that these results may be helpful in clarifying some very important but controversial issues in the field of cellular bioenergetics. This controversy concerns in fact the central problem of this area of science: We still do not know exactly by which feedback mechanism cardiac work (or any kind of cellular work) controls cellular respiration in spite of the fact that this phenomenon itself – linear relationship between work and

Address for offprints V.A. Saks, Laboratory of Bioenergetics, Joseph Fourier University, BP 53X-38041, Grenoble, Cedex, France oxygen consumption – has been known for long time and studied extensively [2]. The textbook explanation, that the answer to this question is given by simple Michaelis-Menten type kinetics on the basis of phenomenon of respiratory control by ADP and apparent Km for ADP determined in vitro and equal to 17–20 μM, is not satisfactory since many authors have shown that with elevation of workload and respiration rate, the cellular ADP concentration calculated from creatine kinase equilibrium does not increase and sometimes even decreases significantly ([3], for review see ref. [4]). The use of permeabilized cells and skinned fibers allows us to determine the affinities of mitochondrial systems to ADP in the cells *in situ*, and the results of these studies described below show that cellular structures may control the mitochondrial respiration much more strongly that it has been anticipated. This kind of control results in more significant compartmentation of adenine nucleotides in the cells [5]. It also allows channeling of energy from mitochondria to specialized cytosolic metabolic pathways by different kinases associated with mitochondria, such as creatine kinase, hexokinase, myokinase [4–7] and probably may participate in transduction of metabolic feedback signal into mitochondria (for details see ref. [4]).

Materials and methods

Preparations

Tissues studied. Rat heart, skeletal muscles (soleus and gastrocnemius), liver and brain were excised from narcotized Wistar rats and immediately placed into cooled physiological salt solution A [8].

Isolated cardiomyocytes and hepatocytes were prepared by using collagenase technique as described [9–12] and permeabilized by using saponin, 50 μg/mg or digitonin, 6 μg per mg dry weight, correspondingly.

Skinned fibers from heart and skeletal muscles were prepared by using the saponin as described earlier [8, 10].

To prepare rat brain homogenate, the tissue was carefully minced by scissors and mildly homogenized by hand in loosely fitted Teflon-glass homogenizer in solution B [8]. The homogenate was filtered to remove blood vessels and nonhomogenized tissue, and centrifuged for 5 min at 3000 g. Supernatant was taken for determination of enzyme activities to estimate the extent of soluble enzyme extraction from the cells, the pellet was resuspended in solution B and used in the experiments.

Determinations

Oxygen uptake was determined by using Yellow Spring oxygraph in physiological salt solution B as previously described [4, 8, 10].

NMR studies of chronically hypoxic rat hearts were performed as described earlier [11]. Hearts from control and chronically hypoxic rats (3 weeks at 10% oxygen) were perfused and subjected to jump in workload by changing calcium concentration in perfusate from 0.5 to 2.5 mM and by administration of isoproterenol (1 μM) and phosphorus NMR spectra were recorded.

Results

Skinned muscle fibers.

Figure 1 shows very clearly the differences in kinetics of regulation of respiration by ADP in isolated heart mitochondria and in permeabilized cardiomyocytes, where mitochondria are located in the cells *in situ* (in their almost normal cellular surrounding) but due to dissolution of the sarcolemma became accessible to externally added ADP [12, 13]. According to the classical data by Chance *et al.* [14], in isolated mitochondrial preparation respiration is very rapidly increased after addition of small amounts of ADP into the medium (dotted line), and the value of apparent Km for ADP is around 17 μM. However, in permeabilized cardiomyocytes apparent Km for ADP is increased by more than order of magnitude, up to 300 μM, and for maximal activation of respiration almost millimolar concentration of ADP is necessary (solid line). This kinetics is significantly changed when creatine is present and mitochondrial creatine kinase in the intermembrane space, at the outer surface of the inner mitochondrial membrane, is functionally coupled to the adenine nucleotide translocase and, due to this coupling, manifold increases the turnover of adenine nucleotides in mitochondria – an amplification effect at low ADP concentrations resulting in apparently decreased Km for ADP, now equal to 50–70 μM (see also ref. [4, 5, 12, 13]).

Earlier, we have shown that apparent Km for ADP, in absence of creatine, is equally high in permeabilized isolated cardiomyocytes and saponin-skinned cardiac fibers consisting of tens or even hundreds of myocytes [15]. This indicates that in cardiac cells apparent Km for ADP is high due to specific intracellular structures surrounding mitochondria. Figure 2 confirms the earlier observation by Veksler *et al.* [16] that decreased affinity of mitochondria to ADP (high apparent Km for ADP) in the cells *in situ* is tissue specific: Both in cardiac and red skeletal muscles (soleus) apparent Km for ADP is high, 300 μM and 400 μM, correspondingly, but very clearly, it is not so in fast-twitch skeletal muscle (gastrocne-

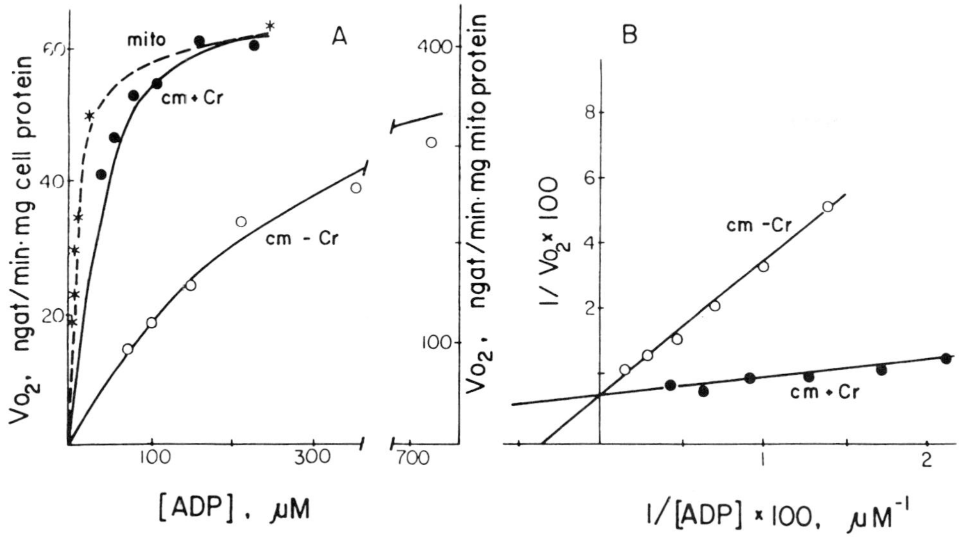

Fig. 1. A. The dependences of the respiration rate of skinned cardiomyocytes (left ordinate axis) and isolated rat heart mitochondria (right ordinate axis) on ADP concentration in medium. Dotted line - mitochondria, solid lines - cardiomyocytes (cm), open circles - without creatine, closed circles - with 25 mM creatine. B. Linearization of dependences for cardiomyocytes in double reciprocal plots. (From ref. [12] with permission).

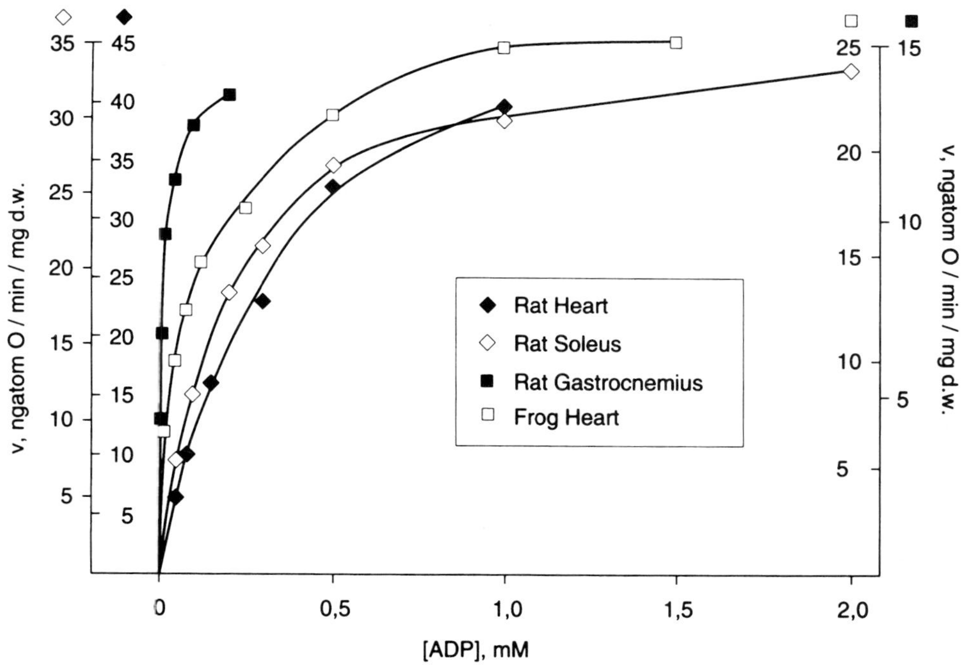

Fig. 2. The dependences of the respiration rates of skinned cardiac and red and white skeletal muscle fibers on the ADP concentration in the medium.

mius) in which apparent Km for ADP in regulation of respiration is 10–20 μM, perfectly equal to the value of this parameter for mitochondria *in vitro*. The conclusion from this fact is that intracellular diffusion of ADP is retarded as indicated by high apparent Km value for ADP not because of highly organized myofibrillar structures which may bind ADP – they are more numerous in white muscles which also have larger cell diameter. This conclusion is directly confirmed by Fig. 3, which shows that if so called 'ghost' fibers are produced from skinned cardiac or red skeletal muscle fibers by complete extraction of contractile proteins by 800 mM KCl solution, high apparent Km values are not changed [13].

However, this parameter can by very significantly decreased by treatment of both cardiac or red muscle fibers by trypsin (0,125 mg/ml) – pretreatment of the fibers in this so-

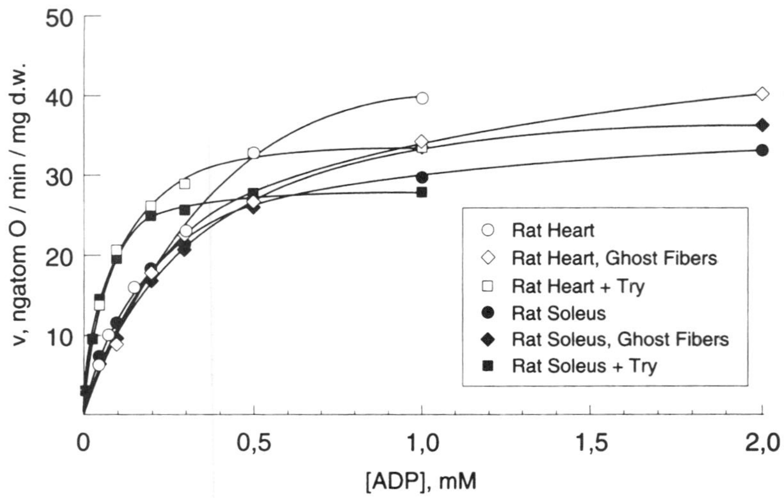

Fig. 3. The effects of treatment of skinned cardiac and skeletal muscle fibers with trypsin (125 μg/ml, 15 min) and of myosin extraction ('ghost' fibers) on the dependence of respiration rate on ADP concentration.

lution for 15 min at 4°C decreases apparent Km values to 40–100 μM (Fig. 3, ref. [5]). Also, the affinity of mitochondria to ADP is significantly increased if the outer mitochondrial membrane is disrupted by hypoosmotic treatment (Table 1, ref. [5, 13]).

Permeabilized hepatocytes

It is interesting to note here that while there are remarkable tissue specificities in the values of apparent Km for ADP among different muscles, isolated and permeabilized liver cells show the characteristics which are very similar to those of cardiac (or red muscle) cells. Our experimental data for permeabilized hepatocytes are summarized in Figs. 4–6 to describe in more details the effect of trypsin on the respiration regulation by ADP in the cells *in vivo*. Figure 4 shows that after trypsin treatment there is only minor decrease of cytoplasmic density but the mitochondria completely preserve their morphological appearance, and Fig. 5 shows that after trypsin treatment the mitochondrial outer membrane stays intact since the respiration is not dependent on the added cytochrome c (that means that the endogenous cytochrome c stays in the intermembrane space and interacts with the respiratory chain, in spite of dissociation from the membrane in the solution with high ionic strength) [19]. However, Fig. 6 shows that after trypsin treatment there is very significant change in the kinetics of respiration regulation by ADP in permeabilized hepatocytes. While before trypsin treatment the apparent Km for ADP is high and equal to about 300 μM (as in cardiomyocytes), after trypsin treatment there is practically no difference between permeabilized hepatocytes and isolated mitochondria *in vitro*: In both cases the Km value is close to 20–50 μM. Similar effect can be seen also in liver

Table 1. Apparent Km values for ADP in regulation of respirat on of permeabilized cells (skinned fibers) and tissue homogenates from rat

Preparation	Apparent Km (ADP) μM	Vmax ngntom/min/mg
1. Skinned cardiac fibers		
control	297 ± 35	29 ± 1.1
with creatine	85 ± 5	28 ± 4
'ghost' fibers	351 ± 32	46 ± 1.5
after trypsin	98 ± 8	26 ± 1
after hypoosmotic shock:		
– phase I	315 ± 23	
– phaseII	32 ± 5	
2. Skinned skeletal muscle		
a) soleus		
control	354 ± 46	14 ± 1
with creatine	105 ± 15	16 ± 4
'ghost' fibers	320 ± 22	17 ± 1
after trypsin	59 ± 6	22 ± 1
trypsin with trypsin inhibitor	304 ± 25	11 ± 1
after hypoosmotic shock:		
– phase I	304	
– phase II	41	
b) gastrocnemius		
control	10 ± 1	6 ± 0.3
with creatine after trypsin	25 ± 4	9 ± 0.5
3. Permeabilized hepatocytes		
control	275 ± 35	
after trypsin	48 ± 8	
after hypoosmotic shock	32 ± 8	
4. Brain homogenates		
control		
– phase I	400	
– phase II	40	
after trypsin	28	
with creatine	38	

Mean values and standard deviations for 5–7 experiments are given. 'Ghost' fibers were obtained by treatment with 800 mM KCl solution for 30 min. Trypsin treatment: 125 μg/ml, 15 min. Hypoosmotic shock: 30 min in 35–50 mOsM solution [13, 17].

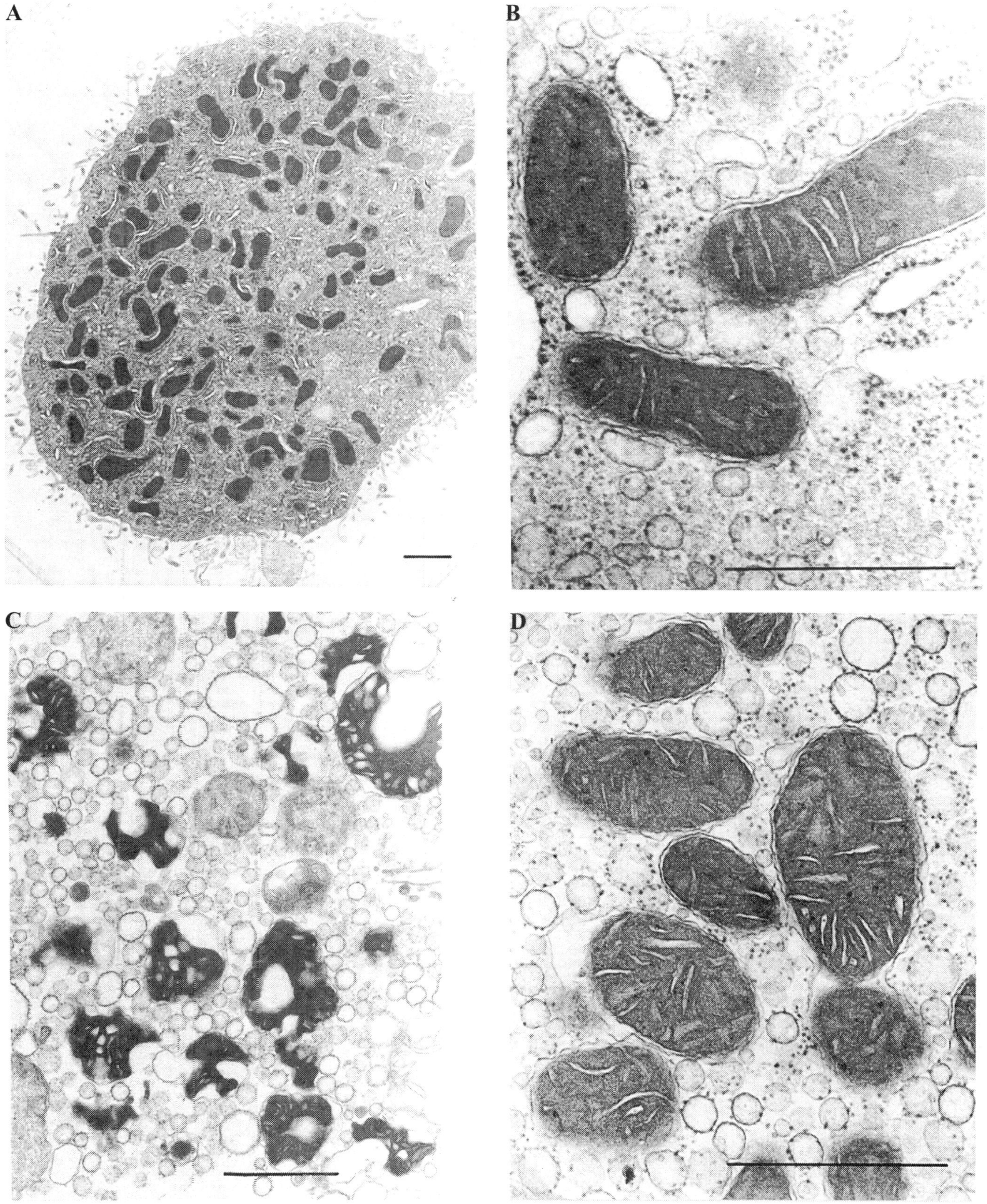

Fig. 4. Electron microscopy photographs of permeabilized hepatocytes before (A, B) or after treatment in hypoosmotic solution (C) or with trypsin (125 μg/ml, 15 min) (D). The bar length represents 1 μm (From ref. [17 and 18] with permission).

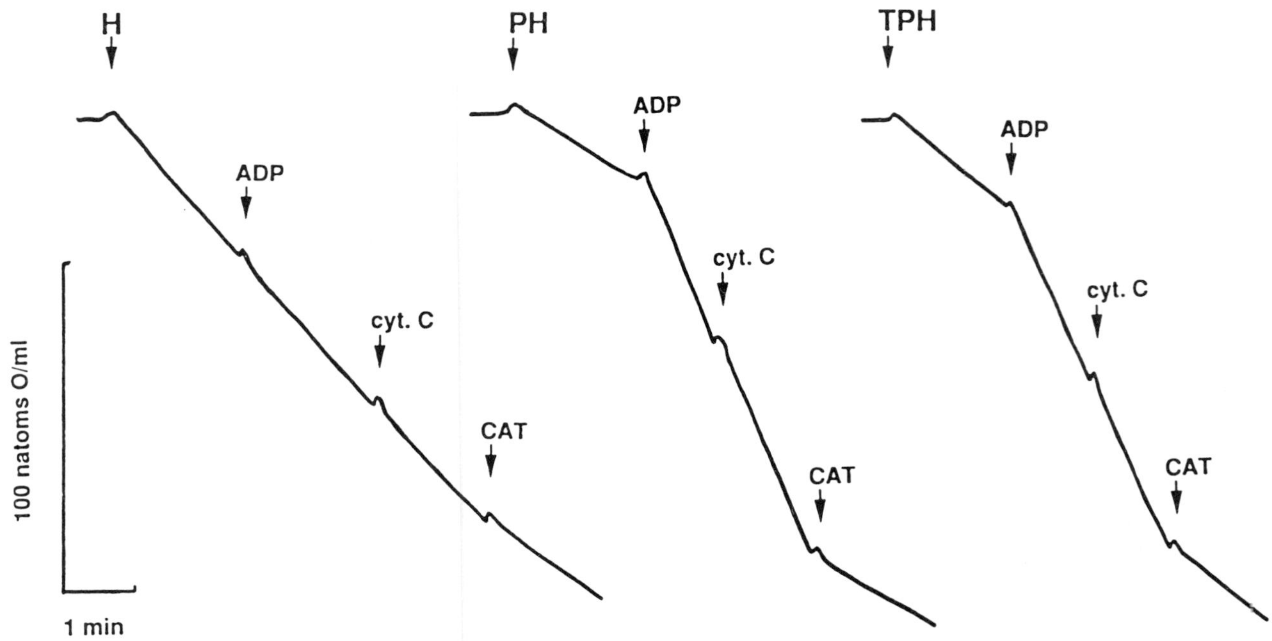

Fig. 5. Recording of respiration of isolated hepatocytes (H), permeabilized hepatocytes (PH) and trypsin-treated permeabilized hepatocytes (TPH). The recordings were made in the medium containing 125 mM KCl. ADP was added to 1 mM, cytochrome c to 8 μM, final concentration.

tissue homogenates (Figs. 6C and 6D) as well as in the cardiac tissue homogenates [5, 18]. These sharp changes, at the background of normal morphological appearance of permeabilized cells after trypsin treatment, exclude the possible explanation for high apparent Km for ADP by mitochondrial clustering (see ref. [18]). Also, it has been shown that inhibition of adenylate kinase, which may rapidly convert ADP into AMP and ATP, has no effect on apparent Km for ADP, this excluding the trivial explanation of high values of this parameter by rapid ADP consumption [12, 17, 18].

However, very significant decrease of apparent Km for ADP may be achieved also in hepatocytes by mitochondrial swelling in the hypotonic medium which disrupts the outer mitochondrial membrane (Fig. 4C, Table 1, ref. [12, 18]).

Brain tissue homogenates

Brain tissue was mildly homogenized, filtered, centrifuged for 5 min at 3000 g and supernatant was taken for determination of hexokinase and lactate dehydrogenase activities to estimate the extent of soluble enzyme extraction, and pellet was resuspended and used for oxygraphic determination of mitochondrial respiratory activities. Figures 7A and 7B show that this mild homogenization is enough to release practically all soluble glycolytic enzymes that means that practically all cells were mechanically broken and mitochondria were made accessible for externally added ADP. And Figs. 7C and 7D show that in this system one observes biphasic kinetics of respiration regulation by ADP most probably because of two populations of mitochondria: one population is characterized by high apparent Km for ADP in regulation of respiration equal to 400 μM and the second – by low Km for ADP – 40 μM. By analogy with cardiac muscle, population of mitochondria with low affinity for ADP represents intact population with controlled permeability for ADP, and the population of mitochondria with high affinity for ADP is most probably due to homogenization and some kind of damage to the membrane of mitochondria. However, in the presence of creatine, due to activation of creatine kinase reaction coupled to oxidative phosphorylation via translocase, the apparent Km for both populations was low. Also, the apparent Km for ADP for mitochondrial population with low affinity to ADP was decreased by treatment of the tissue homogenate by trypsin, and both in the presence of creatine and after trypsin the kinetics of respiration regulation by ADP was monophasic and close to that for isolated mitochondria. (Figs. 7C and 7D, Table 1).

Thus, some interesting conclusions can be made from these experiments. First, in cardiac, red skeletal muscle, liver and also brain cells *in vivo*; but not in fast skeletal muscle, apparent Km for ADP in regulation of mitochondrial respiration is high, and there are at least three ways to decrease this parameter:(1) isolation of mitochondria; (2) treatment of tissue by trypsin which do not change mitochondrial properties; (3) hypoosmotic treatment of permeabilized cells that results in mitochondrial swelling and selective disruption of mitochondrial outer membrane. Most probably, these results show that in these cells *in vivo* the permeability of mitochondrial outer membrane for ADP is very low and controlled by

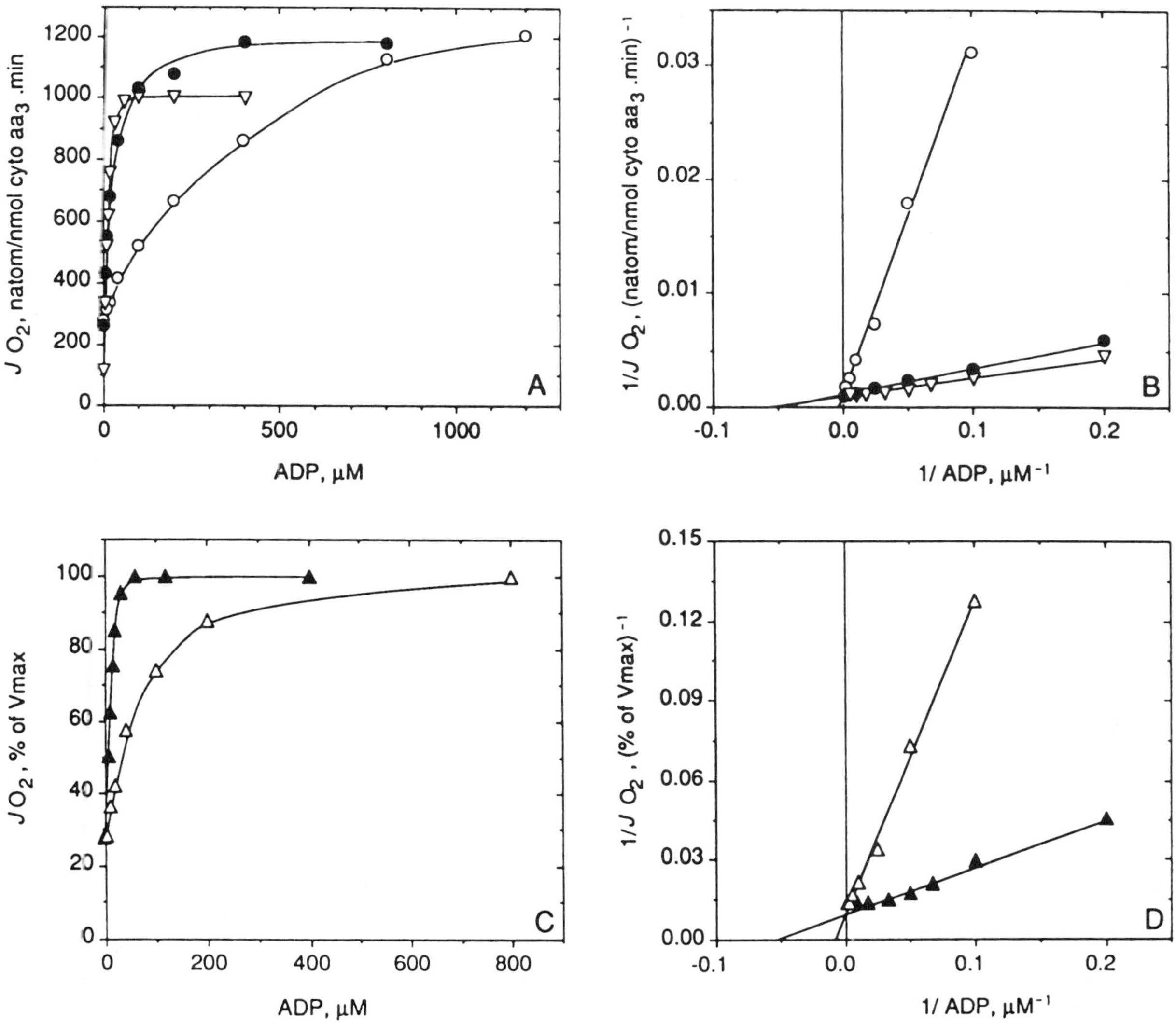

Fig. 6. The dependence of respiration rate of isolated liver mitochondria Δ and permeabilized hepatocytes before (O) or after (●) trypsin treatment on external ADP concentration (A) and their linearization (B), C, D – respiration rates of rat liver homogenates before (Δ) and after (▲) trypsin treatment. (From ref. [18] with permission).

some cytoplasmic proteins, or protein structures.

Since this kind of control is seen after removal of soluble fraction of cytoplasmic proteins but is lost after isolation of mitochondria, we may speculate that the unknown protein(s) which control the permeability of the outer mitochondrial membrane is (are) related to cytoskeleton structures [5]. To investigate if these structures may be represented by microtubular system we treated skinned cardiac fibers by colchicine and taxol which dissociate and stabilize the microtubular system, correspondhlgly, but did not find any effect of these treatments on apparent Km values for ADP (Table 2). Also, use of cytochalasine which prevents formation of actin microfilaments did not change these parameters (Table 2). Thus, most probably, if cytoskeletal structures are involved in control of mitochondrial function in the cell, their components others than tubulin or actin microfilaments may play this role.

Very low permeability of mitochondrial outer membrane for ADP probably controlled by cytoplasmic proteins obviously results in further compartmentation of adenine nucleotides in mitochondria including the intermembrane space, in cells *in vivo*. Consequently, ADP (and ATP) concentrations in the intermembrane space and cytoplasm may be very different. In this case ADP and ATP concentrations in the intermembrane space will depend on functioning of coupled creatine kinase and other peripheral kinases [5–7].In particular, because of its localization on the outer surface of the inner mitochondrial membrane and tight coupling to adenine nucleotide translocase by a mechanism of direct metabolic channeling mitochondrial creatine kinase may acquire effective control over respiration, due to amplifying effect on the turnover of adenine nucleotides [4, 5]. This may also be an important component of the intracellular mechanism for

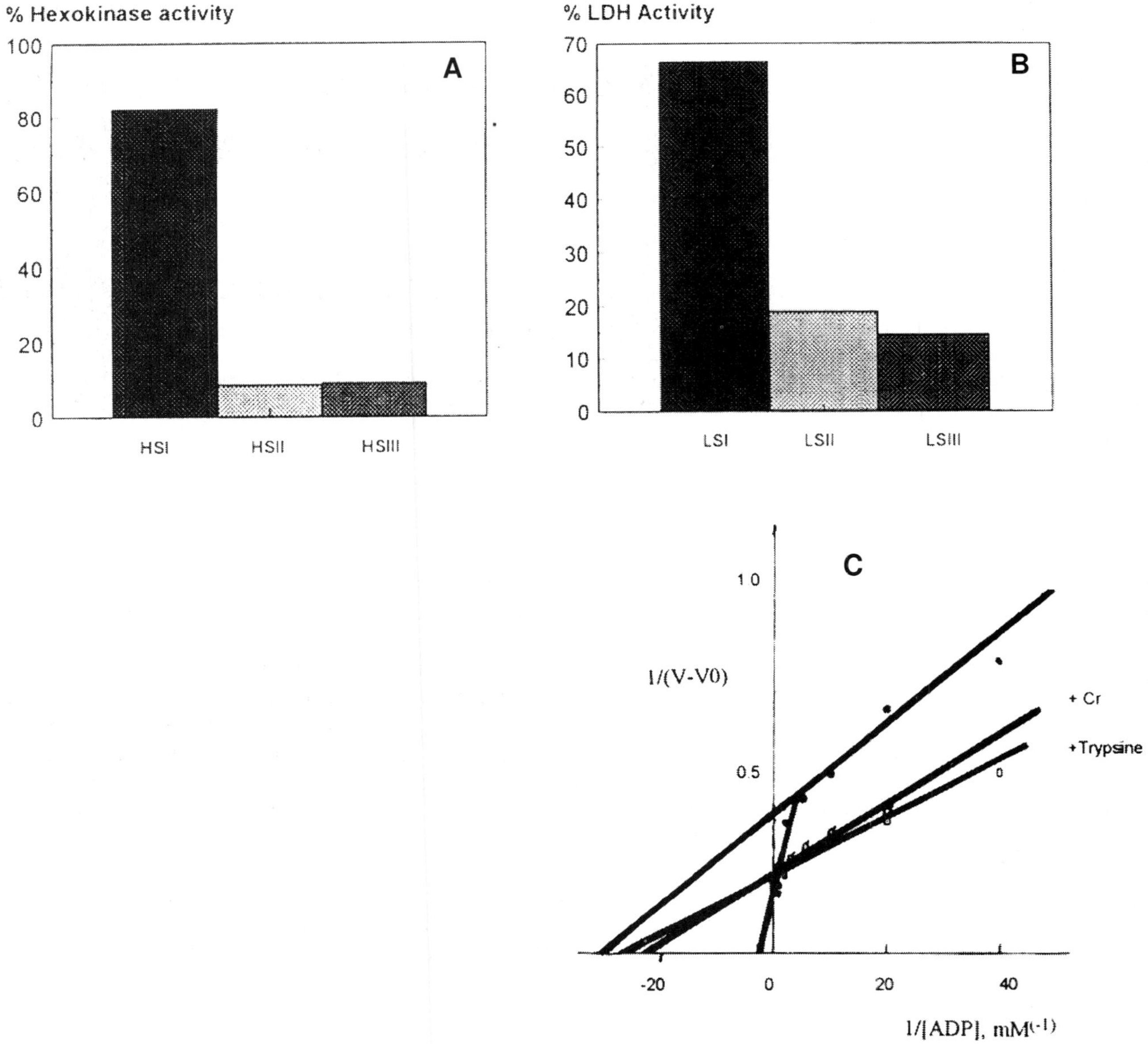

Fig. 7. Release of hexokinase (A) and lactate dehydrogenase (B) from rat brain tissue during homogenization, and the dependence of respiration rate of rat brain tissue homogenates on the external ADP concentration before (●) and after trypsin treatment (□) or in the presence of creatine, 25 mM (∅), (C). In A and B: HSI – supernatant after centrifugation of brain homogenate for 5 min at 3000 × g; HSII – supernatant after centrifugation for 5 min at 3000 × g of a suspension obtained after resuspension of pellet after first centrifugation; HSIII – supernatant after resuspension of the pellet from the second step by Polytron and centrifugation. These data show that both hexokinase and lactate dehydrogenase are extracted after first very mild homogenization, but the factor controlling affinity to ADP is still present in the pellets.

vectorial feedback metabolic signal transduction into mitochondria (for explanation, se ref. [4]).

Alternative explanations for low affinity of mitochondria for ADP in the cardiac, slow skeletal muscle, liver and brain cells in vivo by limited ADP diffusion from outside into the cells and by mitochondrial clustering are most probably excluded by the following experimental data. (1) Low affinity (high apparent Km) for ADP is tissue specific and is not observed in fast twitch skeletal muscle with large size of cells; also, this affinity is not influenced by modification of sarcomere structures after extraction of thick filaments by 800 mM KCl [13] or after inhibition of myokinase reaction which may modify intracellular ADP concentration [12, 16, 17]. It is important that isolated cardiomyocytes and skinned fibers consisting of many muscle cells display very similar characteristics [15]. (2) Treatment of the cells with trypsin does not change mitochondrial localization (clustering) and any diffusion distance but significantly decreases the value of ap-

Table 2. Values of appalent Km for ADP in regulation of respiration of skinned cardiac fibers treated by colchicine, taxol and cytochalasine

Treatment	Apparent Km (ADP), μM
1. Control	297 ± 58
2. Colchicine 3 mM	234 ± 102
3. Taxol/DMSO 10 μM (100 μM DMSO)	240 ± 107
4. Cytochalasine 50 μM, 90 min	258 ± 40

Mean values and standard deviations for 5–7 experiments are shown. Colchicine and taxol were present during all experiment in concentrations shown. DMSO: (dimethylsulphoxide).

parent Km for ADP [18]. (3) The affinity of mitochondria to ADP *in vivo* is very sensitive to adaptive changes and increases (apparent Km for ADP decreases) in response to decreased energy state or knockout of MM creatine kinase that may be explained by altered expression of proteins controlling membrane permeability but not by alteration of diffusion [16, 19].

The last section of this review shows that both outer mitochondrial membrane and coupled creatine kinase system may change during heart pathology, especially in acute ischemia.

Changes in the mitochondrial outer membrane and coupled creatine kinase system in ischemia and during heart preservation.

In these experiments with isolated rat hearts several protocols were used to study the effects of ischemia and heart preservation on mitochondrial function *in situ*. Control hearts were perfused with cold cardioplegic solution. In the second group total ischemia of the heart was induced for 1 h at 37°C by incubation in humid atmosphere in closed vessel. In the third and fourth groups hearts were preserved for 15 h at 4°C in St. Thomas cardioplegic solution without perfusion (simple immersion) or with low flow (0.3 ml per min) perfusion. From all these hearts skinned fibers were prepared and the respiration rates determined. The results of typical experiment of this type is shown on Fig. 8. While in control or cardioplegic hearts (Figs. 8A and 8B), respiration is at low nonsaturating ADP concentration of 0.1 mM strongly activated by creatine (20 mM) due to tight functional coupling of mitochondrial creatine kinase and ATP-ADP translocase (see the lower scheme), this stimulation is completely lost after 1

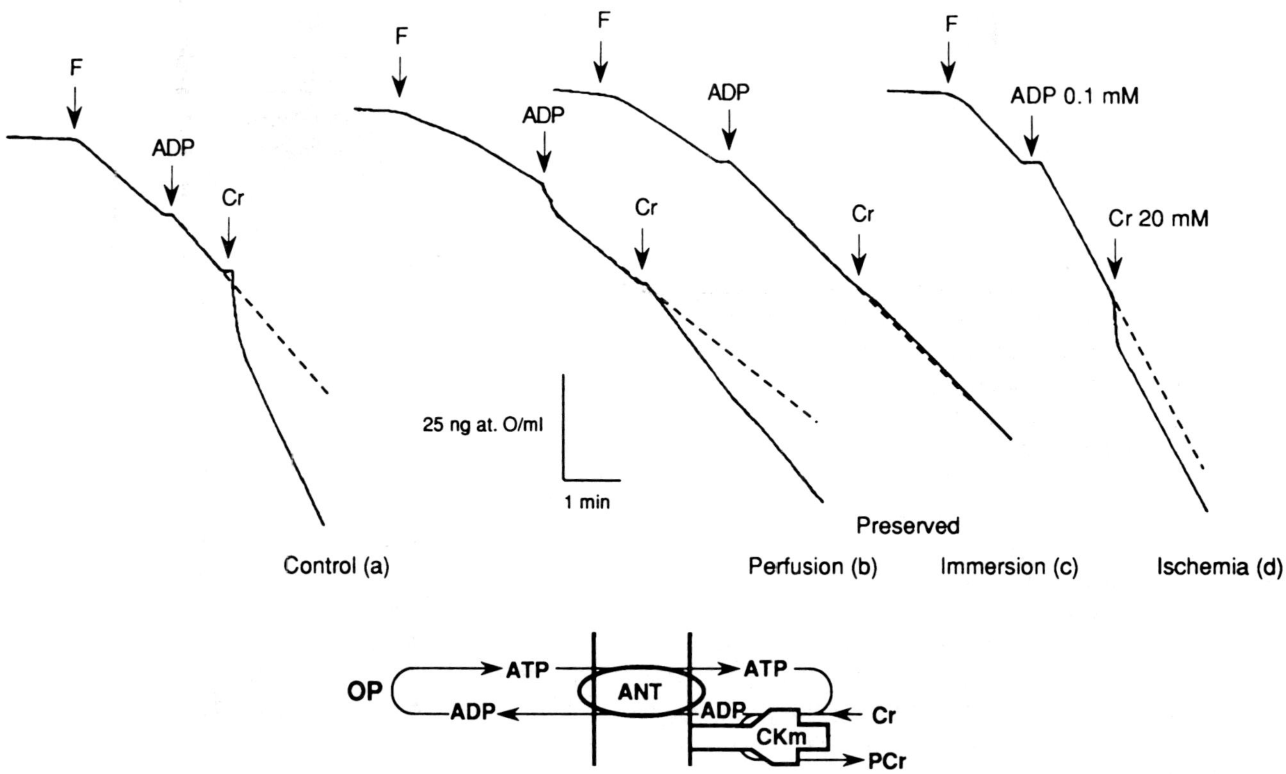

Fig. 8. Recordings of the respiration rates of skinned fbers from isolated rat heart in control (a), after 15 h of preservation in St. Thomas solution at 4°C by low-flow perfusion (b) or simple immersion (c), and after 1 h of normothermic ischemia (d). ADP concentration – 0.1 mM, creatine – 20 mM. Lower panel explains the stimulating effect of creatine on respiration by activation of mitochondrial creatine kinase (CK) reaction coupled to oxidative phosphorylation (OP) via adenine nucleotide translocase (ANT).

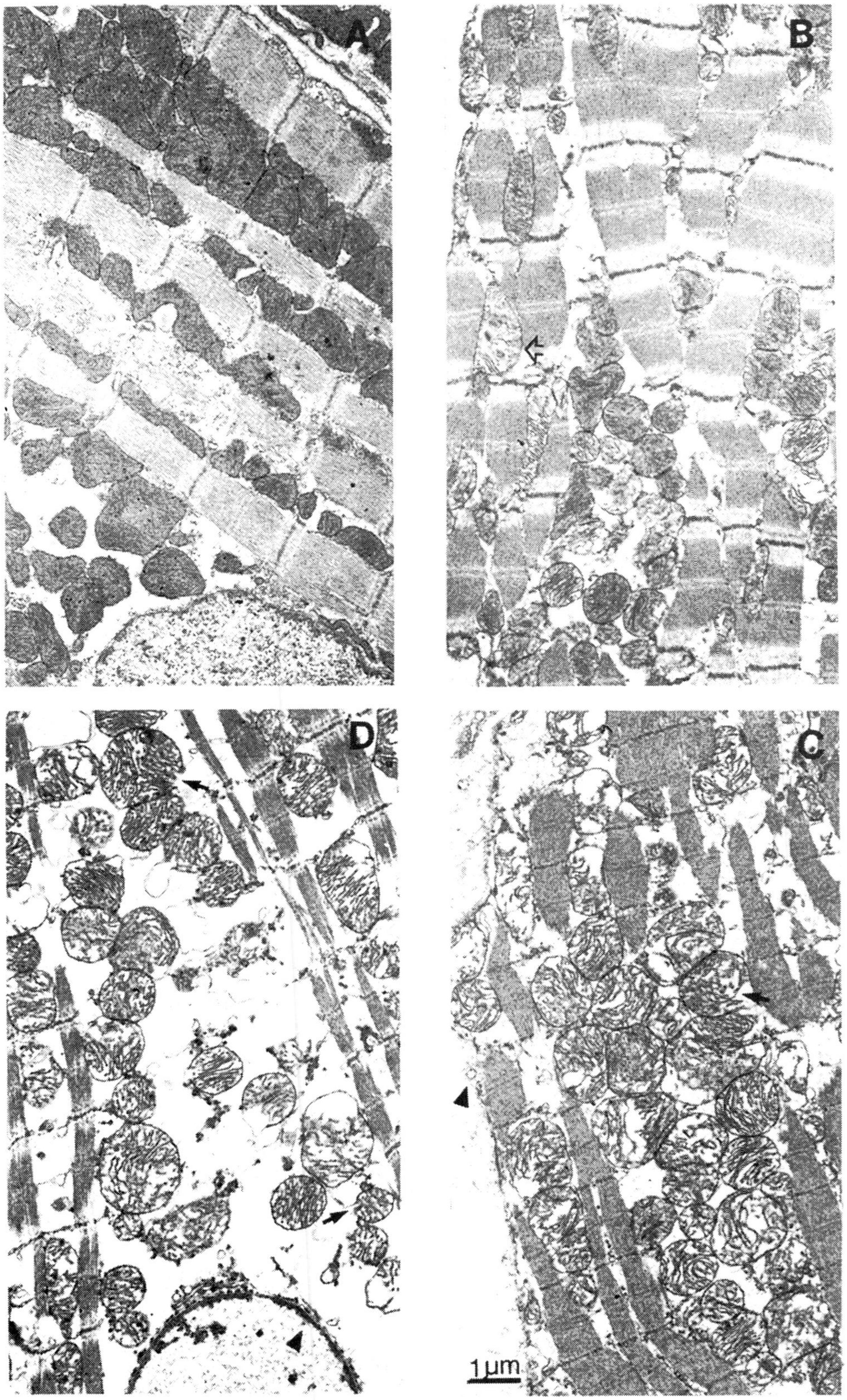

Fig. 9. Electron microscopy photographs of cardiac cells in skinned fibers in control (A), after 15 h of preservation with low-flow perfusion (B) or by immersion (C), and after 1 h of normothermic total ischemia. Very different extents of cellular damage including those for mitochondrial outer membrane (arrows) are evident.

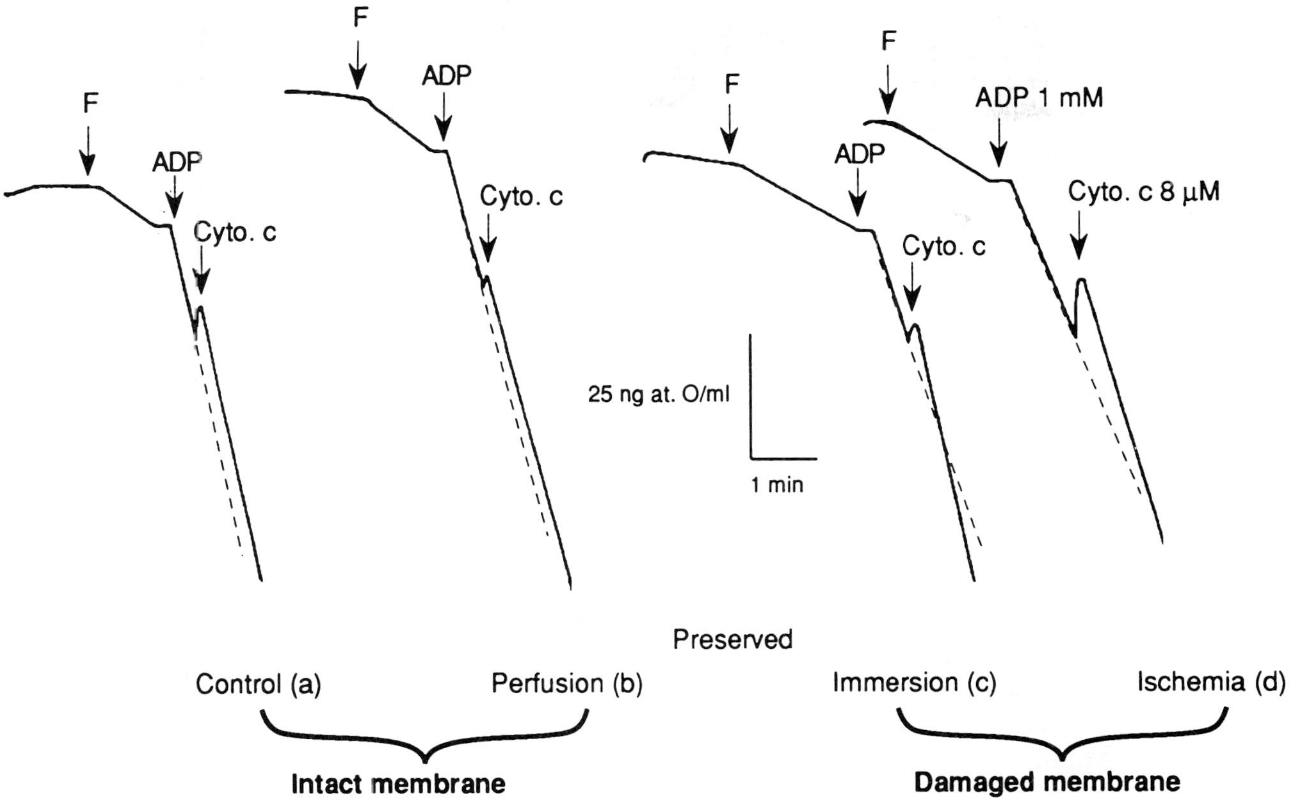

Fig. 10. The effect of cytochrome c (8 μM) on respiration of skinned cardiac fibers in presence of 1 mM ADP (state 3, St 3) in control (a), preserved by low-flow perfusion (b), or immersion (c), or after 1 h of ischemia (d).

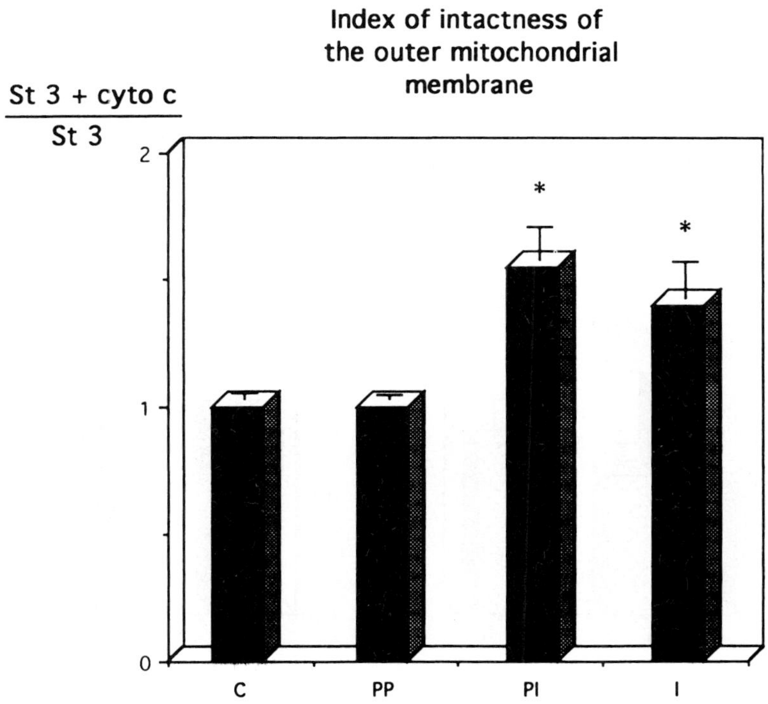

Fig. 11. Quantitative expression of data shown in Fig. 10 to calculate the index of intactness of outer mitochondrial membrane in skinned fibers from isolated hearts after different protocols of heart preservation.

h of total ischemia (Fig. 8D) or 15 h of preservation by immersion (Fig. 8C), but still observed after 15 h of preservation with low-rate perfusion (Fig. 8B). Comparison of the results given in Fig. 8 and those of morphological studies (Fig. 9) clearly shows that the loss of creatine control correlates with mitochondrial swelling in the cells most probably because of inorganic phosphate accumulation.

Similar conclusion can be made from experiments in which the outer mitochondrial membrane intactness was determined by cytochrome c test [17]. After total ischemia or preservation by immersion, respiration rate at 1 mM ADP decreased but could be restored by exogenous cytochrome c (Figs. 10 and 11) – that means that respiratory chain itself was not damaged but mitochondrial outer membrane was disrupted resulting in leakage of endogenous cytochrome c. Detailed kinetic analysis of the respiration of all these fibers in absence and presence of creatine is given in Fig. 12. In the perfusion preserved hearts where outer membrane is not damaged and apparent Km for ADP is still high, creatine significantly decreases this parameter value due to only moderate swelling (Fig. 9) and preservation of coupled creatine kinase. However, in the cases when outer mitochondrial membrane was damaged due to swelling (ischemia and preservation with immersion) apparent Km for ADP decreased and stimulatory effect of creatine was completely lost (Figs. 11B and 11D).

Very clear correlation between mitochondrial swelling, loss of the control of permeability of outer mitochondrial membrane for adenine nucleotides and decrease of the effect of creatine on respiration or on the apparent Km for ADP permits us to suggest that determination of the creatine effects on respiration of skinned cardiac fibers may be an easy and precise test for the evaluation of extent of cardiac cell damage and efficiency of preservation.

Chronic hypoxia, however, seems to improve the energy metabolism

It is interesting to note that in contrast to acute ischemia, chronic hypoxia induces adaptive changes which result in apparent increasing the efficiency of energy metabolism of the heart. This is illustrated in Fig. 13. The phosphocreatine level was lower in the hearts from hypoxic hearts but its restoration rate following isoproterenol administration was increased when compared to the controls. Symmetrically, the decrease in inorganic phosphate concentration was accelerated in these hearts. Increased rate of restoration of phosphocreatine finally resulted in its overshoot [11].Cellular mechanisms of this kind of increased efficiency of cellular ener getics are still unclear and interesting problems for further studies.

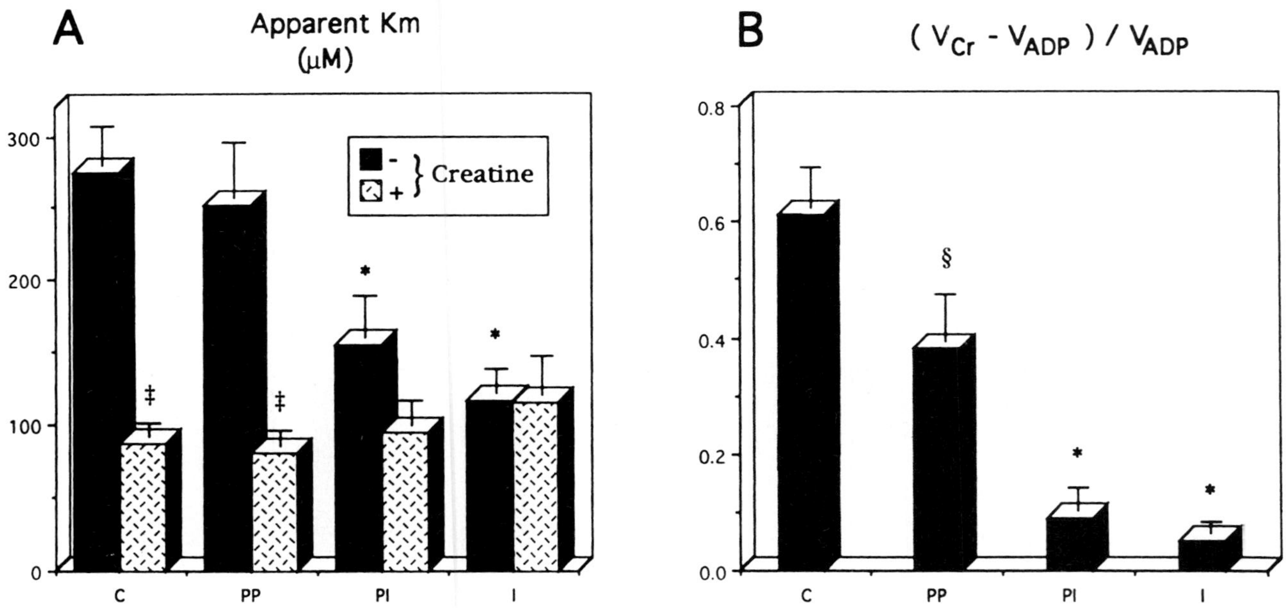

Fig. 12. Quantitative indices of control of mitochondrial respiration by creatine kinase reaction in skinned cardiac fibers taken from isolated hearts in control (C), after preservation with perfusion (PP) or immersion (PI) and after 1 h of normothermic ischemia (I). A: Apparent Km for ADP in absence and in presence of creatine, 20 mM. B: relative increase of respiration rate after addition of 20 mM (V_{CR}) in the presence of 0.1 mM ADP (ADP) (for more details see Fig. 8).

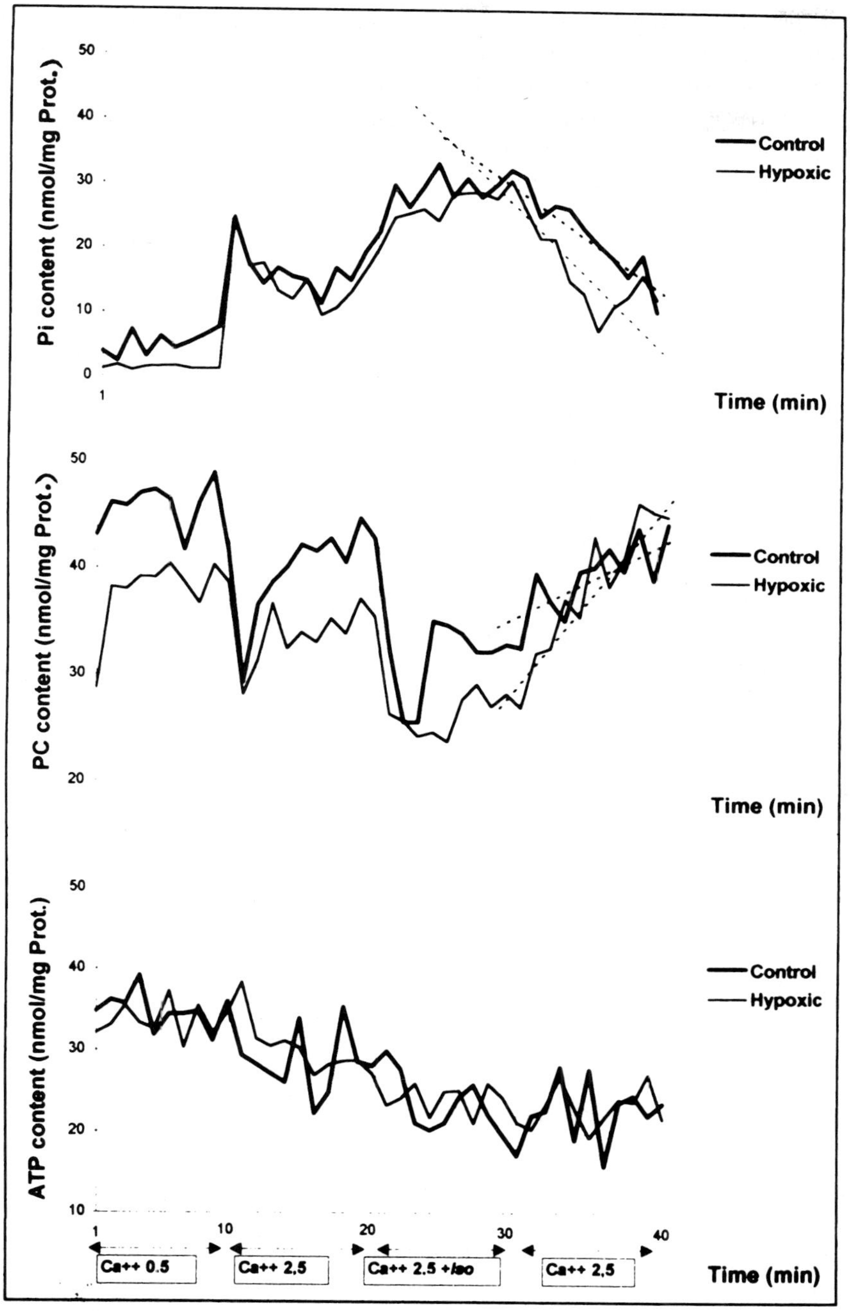

Fig. 13. Time course of changes in contents of phosphorylated compounds in isolated rat hearts from control and chronically hypoxic animals subjected to jumps in workloads. Phosphorus spectra corresponding to the sum of signals from 5 hearts were used for calculations. The data were normalized relative to the ATP content biochemically determined for hearts freeze-clamped after 10 min of perfusion with solution containing 0.5 mM calcium.

Conclusion

It has become clear that intracellular diffusion of ADP in the cardiac, red skeletal muscle, liver and brain cells but not in fast skeletal muscle is strongly controlled at the level of outer mitochondrial membrane by some still unknown cytoplasmic protein factors. To completely understand the feedback mechanism of regulation of cellular respiration, identification of these factors and clarification of the mechanism of manifold decrease of mitochondrial affinity to cytoplasmic (exogenous) ADP in these cells are necessary. A low permeability of outer mitochondrial membrane to cytosolic ADP together with numerous kinases associated with mammalian cell mitochondria allow targeted channeling of energy from mitochondria towards specialized cytosolic pathways.

In addition, damages to the mitochondrial outer membrane and coupled creatine kinase in acute ischemia appear to be among the earliest cellular injuries in this kind of pathology. In chronic hypoxia, however, there seems to be an increase in the efficiency of energy metabolism of the cell.

References

1. Wollenberger A, Krause EG: Metabolic control characteristics of the acutely ischemic myocardium. Am J Cardiol 22: 349–359, 1969
2. Neely JR, Liebermeister H, Battersby EJ, Morgan HE. Effect of pressure development on oxygen consumption by isolated rat heart. Am J Physiol 212: 804–814, 1967
3. Van B, Dounen C, Duszynsky J, Salama G, Vary TC, Lanoue KF: Effect of cardiac work on electrical potential gradient across mitochondrial membrane in perfused hearts. Am J Physiol 265: H453–H460, 1993
4. Saks VA, Khucllua ZA, Vasilyeva EV, Belikova YuO, Kuznetsov AV: Metabolic compartmentation and substrate channelling in muscle cells. Role of coupled creatine kinases in *in vivo* regulation of cellular respiration. J Mol Cell Biochem 133/134: 155–192, 1994
5. Saks VA, Kuznetsov AV, Khuchua ZA, Vasilyeva EV, Belokova YuO, Kesvatera T, Tiivel T: Control of cellular respiration *in vivo* by mitochondrial outer membrane and by creatine kinase. A new speculative hypothesis: Possible involvement of mitochondrialcytoskeleton interactions. J Mol Cell Cardiol 27: 625–645, 1995
6. Brdiczka D, Wallimann T: The impoltance of the outer mitochondrial compartment in regulation of energy metabolism. J Mol Cell Biochem 133/134: 69–84, 1994
7. Zeleznikar R, Dzeja P, Goldberg ND: Adenylate kinase-catalyzed phosphoryl transfer couples ATP utilization with its generation by glycolysis in intact muscle. J Biol Chem 270: 7311–7319, 1995
8. Saks VA, Kapelko VI, Kupriyanov VV, Kuznetsov AV, Lakomkin VL, Veksler VI, Sharov VG, Javadov SA, Seppet EK, Kairane C: Quantitative evaluation of relationship between cardiac energy metabolism and post-ischemic recovery of contractile function. J Mol Cell Cardiol 21: 67–78, 1989
9. Saks VA, Vasilyeva EV, Khuchua ZA, Belikova YO, Aliev MK, Kesvatera T, Tiivel T, Anflous K, Ventura-Clapier R, Keriel Ch, Fontaine E, Leverve X: ADP diffusion gradients and metabolic oscillatory signals to mitochondria in heart and liver cells. In: E Gnaiger, FN Gellerich, M Wyss (Eds). What is controlling life? Modern trends in Biothermokinetics, vol.3, 1994, pp 178–180
10. Veksler VI, Kuznetsov AV, Sharov VG, Kapelko VI, Saks VA: Mitochondrial respiratory parameters in cardiac tissue: novel method for assessement using saponin - skinned fibers. Biochim Biophys Acta 892: 191-196, 1987
11. Novel-Chate V, Aussedat J, Saks VA, Rossi A: Adaptation to chronic hypoxia alters cardiac metabolic response to beta stimulation: novel phase of phosphocreatine overshoot phenomenon. J Mol Cell Card 27: 1679–1687, 1995
12. Saks VA, Belikova YO, Kuznetsov AV: *In vivo* regulation of mitochondrial respiration in cardiomyocytes: specific restrictions for intracellular diffusion of ADP. Biochim Biophys Acta 1074: 302–311, 1991
13. Saks VA, Vasilyeva EV, Belikova YuO, Kuznetsov AV, Lyapina SA, Petrova L, Perov NA: Retarded diffusion of ADP in cardiomyocytes: possible role of mitochondrial outer membrane and creatine kinase in cellular regulation of oxidative phosphorylation. Biochim Biophys Acta 144: 46–53, 1993
14. Chance B, Williams GR: The respiratory chain and oxidative phosphorylation. Adv Enzymol 17: 65–134, 1956 15. Khuchua ZA, Vasilyeva EV, Clark JF, Korchazllkilla OV, Branishte T, Kapelko VI, Kuznetsov AV, Ventura-Clapier R, Steinschneider AYa, Lakomkin VL, Ruuge EK, Saks VA: The creatine kinase system and cardiomyopathy. Am J Cardiovasc Pathol 4: 223–234, 1992
16. Veksler VI. Kuznetsov AV, Anflous K, Mateo P, van Deursen J, Bé Wieringa, Ventura-Clapier R: Muscle creatine kinase-deficient mice. II. Cardiac and skeletal muscles exhibit tissue-specific adaptation of the mitochondrial function. J Bol Chem 270: 19921–19929, 1995
17. Saks VA, Belikova YuO, Vasilyeva EV, Kuznetsov AV, Fontaine E, Keriel C, Leverve X: Correlation between degree of rupture of outer mitochondrial membrane and changes of kinetics of regulation of respiration by ADP in permeabilized heart and liver cells. Biochem. Biophys Res Comm 208: 919–926, 1995
18. Fontaine EM, Keriel C, Lantuejoul S, Rigoulet M, Leverve X, Saks VA: Cytoplasmic cellular structures control permeability of outer mitochondrial membrane for ADP and oxidative phosphorylation in rat liver cells. Biochem Biophys Res Comm 213: 138–146, 1995
19. Clark JF, Khuchua ZA, Kuznetsov AV, Vasilyeva EV, Boehm E, Radda GK, Saks VA. Actions of the creatine analogue beta-guanidinopropionic acid on rat heart mitochondria. Biochem J 300: 211–216, 1994

Molecular and Cellular Biochemistry **160/161**: 209–215, 1996.

Salvage of jeopardized myocardium by ischemic preconditioning: Is the quest over?

Rudolf Strasser,[1] Patrik Htun[2] and Wolfgang Schaper[2]

[1]*Max-Planck-Institute for Physiological and Clinical Research, Kerckhoff-Klinik, Benekestrasse 4-8, 61231 Bad Nauheim;* [2]*Max-Planck-Institute for Physiological and Clinical Research, Department of Experimental Cardiology, Benekestrasse 2, D-61231 Bad Nauheim, Germany*

Abstract

Helmholtz is quoted to have said that if he'd had any influence in creation he would have returned the human eye to its maker for revisions. The same could be said of the heart with its only very rudimentary ability to defend itself against ischemia. Ischemia was obviously not a problem during evolution: Early man did not live much longer than prime time for reproduction and no selection bias existed to prevent vascular diseases, an affliction of later life. In spite of this natural disadvantage of aged males the number of existing although not very efficient defense mechanisms is surprisingly large. It is the general belief that the knowledge of these mechanisms may lead to the development of new therapies that hopefully improve the imperfect product of natural selection. (Mol Cell Biochem **160/161**: 209–215, 1996)

Key words: myocardial ischemia, stunned myocardium, preconditioning, angiogenic response

Introduction

The anti-ischemic defense mechanisms which the heart employs, albeit in an insufficient way, can be classified into short-term and long-term adaptations. They are: a) the transient increase in ischemia tolerance following brief occlusions, i.e. ischemic preconditioning, b) the downregulation of oxygen consumption and the upregulation of coronary-vascular tone in stunned myocardium, c) the de-differentiation (or markedly slowed degeneration) of myocytes in hibernating myocardium, d) the microvascular angiogenic response to progredient arterial occlusion (collateral circulation).

In this chapter we will concentrate on the molecular events that confer increased tolerance against ischemia and that are apparently activated by brief ischemic episodes.

Following its first description 10 years ago by Murry and Reimer [1, 2] many groups have tried to unravel its mechanism. Several hypotheses have been forwarded and many findings are still controversial or inconclusive or are restricted to certain species and hence lack universality.

From the pharmacological point of view interesting observations had been made with agonist/antagonist experiments around the adenosine A_1 receptor, the K^+ channel and around the Na^+/H^+ exchanger. These drugs were usually given instead of the brief coronary occlusions and the results were compared with groups of preconditioned animals. In essence, these pharmacological experiments had no direct relation to ischemic preconditioning, they were mainly tests to find out whether a pharmacological agent was able to reduce infarct size. This dates back to a period about 20 years ago when a mass screening was started to empirically search for active anti-ischemic compounds that ended somewhat disappointingly. However, it ended also because of the advent of rapid thrombolytic reperfusion therapy that seemed to make other more conservative therapies obsolete. We know now that thrombolysis is not always salvaging ischemic myocardium – and the renewed search for anti-ischemic drugs seems justified. Ischemic preconditioning provides now a better theoretical background, a testable hypothesis, for these new endeavours.

It is clear from all these previous research efforts that if blood flow cannot be restored within a certain time frame (that is modulateable) myocardium is destined to die in spite of all the efforts by ischemic preconditioning and by drugs.

Although 'life without oxygen' [3] is possible for extended times in lower animals and even for specialized mammals,

Address for offprints: W. Schaper, Max-Planck-Institute for Physiological and Clinical Research, Department of Experimental Cardiology, Benekestrasse 2, D-61231 Bad Nauheim, Germany these mechanisms are difficult to tap but gene therapy may provide solutions in the future.

The popularity of research in ischemic preconditioning stems to a large part from the fact that finally an intervention (apart from early reperfusion) was found that showed an unequivocally positive result in the long quest to salvage ischemic myocardium in the presence of a long lasting coronary occlusion. Henceforth brief coronary occlusion served as a positive control in the search for potentially active drugs. Although this was certainly a progress, the standard so set was a modest one because the protection reached by ischemic preconditioning was short and not immediately renewable and it is, on the basis of physiological and biochemical knowledge unrealistic to assume that the protection caused by ischemic preconditioning can be much prolonged beyond that achieved already now by new drugs without some residual energy supply and at normal body temperature.

The limits of tolerance towards ischemia

The quest for procedures and for pharmacological agents that increase the ischemia tolerance of tissues that are sensitive to oxygen deprivation is an old and long one. Although nature shows interesting cases for adaptation to environments of low oxygen tension or to long periods of breath-holding like in the habor seal or of the embryo in utero with its perfusion with mixed venous blood, none of these interesting adaptations could be 'bottled' into a drug with a meaningful application. Research into ischemia tolerance of the heart enjoyed success when, as in cardioplegia, the temperature can be brought close to zero centigrade and when contractile function can be completely abolished. However, even under these extreme conditions the heart survives not longer than perhaps 7 h when generation of a normal cardiac output after reperfusion is the endpoint and ultrastructural morphology shows alterations much sooner than 7 h [4]. This means that expectations from research into the molecular mechanisms of ischemic preconditioning should not be too high. Even if cardiac myocytes can be induced to completely downregulate contractile function, the life without oxygen at body temperature will not last long.

Years ago [5] we and others [6] have defined the constraints of survival following acute coronary occlusion in a number of mammals. The determinants of infarcts are (1) the size of the occluded artery and the size of its risk region, (2) the amount of residual (collateral) blood flow, (3) the duration of the ischemic period, (4) the prevailing myocardial oxygen consumption and (5) the variable tolerance against ischemia, a finding emanating from the new research in ischemic preconditioning. The quantitative contributions of these determinants vary from one species to the other. In the hearts of rats, rabbits, pigs and sheep collateral flow is so low (and completely invariant within the time frame of an acute experiment) that it is not a determinant. In the guinea pig collateral flow is so high that practically no significant infarcts develop upon acute coronary occlusion and is, hence, also without influence. Without collateral flow, and hence without oxygen supply, the prevailing myocardial oxygen consumption at the moment of coronary occlusion is also of no importance. Ischemia tolerance proper is therefore best studied in a species with no collateral flow. It is quite clear that in such a model increased ischemic tolerance by preconditioning must soon reach the limit laid down by the laws of physiology and biochemistry and that the range of modulation cannot be very large. We reported some time ago [7] experimental results and calculations assuming complete shutdown of some dispensable energy consuming functions that predicted practically unlimited structural survival of canine myocardium thanks to its significant residual blood flow. The outcome for the porcine heart was predictably much less favorable but the ischemic damage can theoretically be contained for parts of the ventricular wall for about 90 min.

Our prospective experiments in the porcine model showed that ischemia tolerance can indeed be modulated but a similar result as with the 'infarct-size-reduction' efforts of a decade ago was repeated: *It is indeed possible to delay infarction but not to prevent it.* In the case of ischemic preconditioning the delay is about 45 min, i.e. the infarct size that one would expect from coronary occlusion in a 'naive' pig heart at 45–60 min is delayed to about 90 min. This is impressive to experimentalists who know that this has not been achieved with drugs *in vivo* (positive drug studies with A1-agonists were performed in isolated hearts) at tolerable arterial pressures and following systemic injection, but it is unremarkable in the clinical setting.

The time relationships

We have tried to define the time frame within which ischemic preconditioning is active in the porcine model, i.e. (1) the minimal duration of the conditioning coronary occlusions, (2) the duration of reperfusion periods, (3) the duration of the protective effect, (4) the (im)possibility of its immediate renewability, and (5) the question of the existence of a period of late protection (second window of protection).

In these experiments [8], we found (Figs 1 and 2) that a cumulative period of 20 min of ischemia (2×10 or 4×5 min) is necessary for immediate protection, and at least 10 min of reperfusion between occlusions are needed. The protective effect lasts for about 40 min and cannot be immediately renewed and there is no second window of protection in the porcine model. However, protection can be renewed after about 24 h. These findings differ from those obtained in other species in that the minimum duration of the conditioning

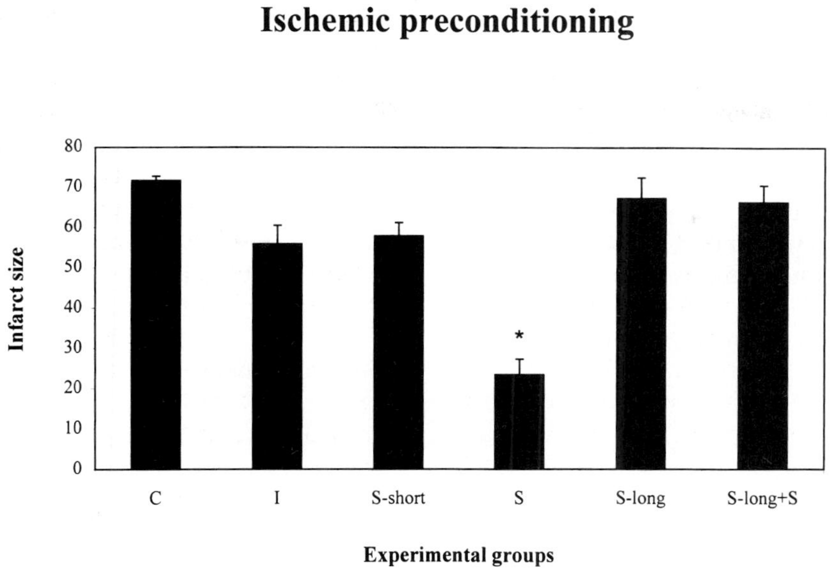

Fig. 1. Ischemic preconditioning: Time course of protection in anesthetized pigs. After the preconditioning ischemia all groups had a long occlusion of the mid left anterior descending coronary artery for 60 min followed by 90 min of reperfusion. Group C = only long occlusion; Group I =10 min of ischemia followed by 30 min of reperfusion; Group S-short = two times 10 min of ischemia separated by 10 min of reperfusion followed 10 min later by infarct induction; Group S (standard protocol) = two times 10 min of ischemia separated by 30 min of reperfusion followed by the long occlusion 30 min later; Group S-long = two times 10 min of ischemia separated by 30 min of reperfusion followed by the long occlusion 60 min later; Group S-long+S = the same protocol as group S-long followed by a new preconditioning protocol as in group S. Depicted is the infarct size as percent of the risk region. * = $p < 0.05$ vs. C, given are mean values and S.E.M. More than 10 min of preconditioning ischemia and more than two times 10 min of reperfusion are required for induction of protection. The protection is lost after 1 h and a renewal is not possible at that time. (Data from Strasser [8] and Sack [31]).

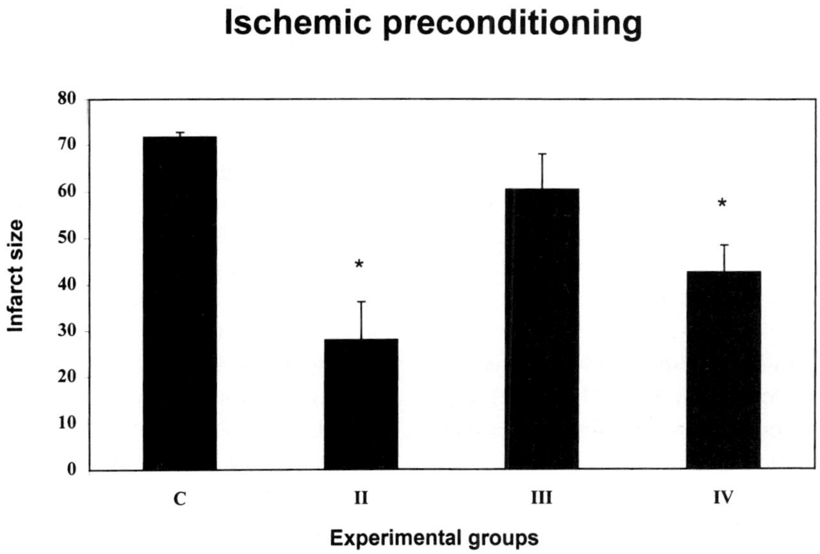

Fig. 2. Ischemic preconditioning: renewal of effect and second window of protection in anesthetized pigs. All groups had finally a long occlusion of the mid left anterior descending coronary artery for 60 min followed by 90 min of reperfusion. Group C underwent only the long occlusion. Groups II to IV had all the same preconditioning protocol with four times 5 min of ischemia induced with a PTCA-balloon separated by 5 min of reperfusion. Group II = infarct induction 30 min after the last ischemic cycle; Group III = infarct induction 24 h after the preconditioning ischemia; Group IV = second preconditioning ischemia 24 h after the first followed by the long occlusion 30 min later. Depicted are the infarct size as percent of the risk region. * = $p < 0.05$ vs. C, given are mean values and S.E.M. 24 h after the preconditioning ischemia the protection is completely lost in porcine myocardium. With a new preconditioning cycle a renewal of protection is possible at that time (Data from Strasser [8]).

stimulus is much shorter in the dog [9] and rat [10] and only the limited duration of the protective effect agrees well in all species. Differences exist also with regard to the immediate renewability of increased tolerance by another brief occlusion. This seems possible in the rat and rabbit but not in pigs. The rat differs in several other aspects in that the protective effect is apparently not mediated via Gi-proteins. Our finding of the absence of late protection (24 h after the conditioning stimulus) in the pig is supported by a preliminary report by the Bolli group [11].

The role of adenosine

It is generally accepted that adenosine plays a prominent role in the protective effect of brief occlusions and that its effect is mediated via the A_1-receptor. However, it is uncertain whether the protective effect is exclusively due to the actions of adenosine. The presence of adenosine in ischemic-reperfused tissue following brief occlusions is limited due to the rapid breakdown of the metabolite and it is difficult to explain the protective effect that persists much longer than the presence of the assumed transmitter. Although we found a protective effect using a synthetic A_1-agonist that resists enzymatic breakdown by tissue deaminases, the systemic dose necessary was tolerated only when using a cardiac pacemaker because of excessive bradycardia. Local administration of an A_1-agonist directly into the myocardium (where much higher concentrations can be reached) was clearly protective in our experiments (Fig. 3). An argument against the role of adenosine is the observation that 4×5 min of occlusion (or 2×10) are needed but tissue adenosine concentration rises only during the first occlusion, a finding which was known for some time [12].

The PKC-Hypothesis

An interesting hypothesis to explain the relatively long action produced by a short-lived transmitter (adenosine) was forwarded by the Downey group [13]. They hypothesized that occupation of the A_1-receptor leads to activation of G_i-proteins and translocation of protein kinase C (via PLC-activation) from the cytosol to the sarcolemma where it phosphorylates a so far unknown protein with a so far unknown protective action. Although the translocation of PKC was known since quite some time to occur in ischemia a protective role had not been ascribed to that. On the contrary: the phosphorylation of the Na^+/H^+-exchanger that occurs with PKC translocation favors sodium – and subsequently calcium overload of the ischemic cell. PKC activation in the brain is usually a sign of irreversible injury. We could show that stimulation of PKC with phorbolesters worsened the effects of ischemia and that inhibition of PKC with both staurosporine and bisindolylmaleimide reduced the damage by ischemia [14]. Furthermore we could show that the protective effects of A_1-agonists did not lead to PKC translocation [15]. The pathway from PLC to PKC remains furthermore contradictory. Although the PKC hypothesis would provide an elegant explanation for the memory effect because it takes time for the translocation in both directions, it clashes with more than one observed fact and is therefore not a valid model.

Other pathways

K^+-channels
We and others have shown that K^+-channel openers have an anti-ischemic effect and inhibitors of these channels exhibit opposite effects. It is furthermore known that the beneficial effects of A_1-agonists are inhibited by K^+-channel blockers and that A_1-antagonism can be neutralized by K^+-channel openers [16–19]. From these results it can be imagined that a connection exists between the A_1-receptor via G_i-protein to a K^+-channel. It is indeed known that G-proteins exhibit also ion channel activities [20]. This signal pathway appears to be the most probable as of this writing but it has several drawbacks, i.e. it cannot explain the memory and it is based almost exclusively on the availability of tool drugs that are not absolutely specific and do not exhibit their full effect *in vivo* because, as in the case of the K^+-channel openers, they decrease blood pressure.

Bradykinin and inhibitors of neutral peptidases
Bradykinin is reported to exert protective effects by mimicking ischemic preconditioning. The beneficial effect of converting enzyme inhibitors is also partially ascribed to the preservation of bradykinin. Neutral peptidase inhibitors reduce the damage inflicted by high doses of isoproterenol in the rat heart [21]. We were unable to reproduce these effects in our model of direct intramyocardial microinfusion. Neither with bradykinin nor with phosphoramidon, even in amounts that spilled over into the circulation and reduced blood pressure (in the case of bradykinin), a protective effect was observed. At present we do not know whether this is a peculiarity of our model or of the species utilized (Htun and Schaper, unpublished).

Trophic factors

Brief periods of ischemia upregulate a large number of genes among which the growth factor genes VEGF and IGF-2. The latter is known as a trophic factor for neurons with a survival value in ischemia. Our experiments show that local infusion of the IGF-2 peptide into the LV-myocardium has a local

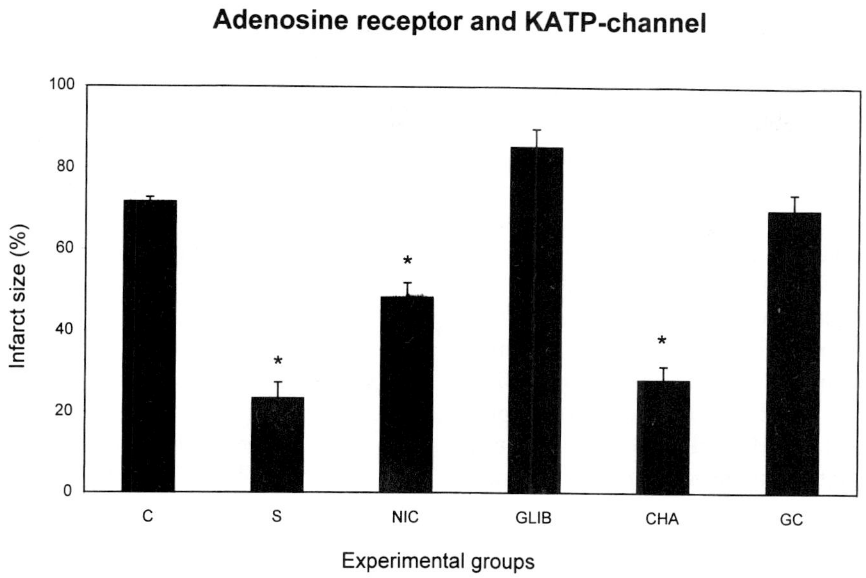

Fig. 3. Preconditioning in porcine myocardium: The interplay between the adenosine A_1-receptor and the K^+ATP-channel. All groups had a long occlusion for 60 min followed by 90 min of reperfusion after ischemic preconditioning or after the infusion of drugs. The control group C had only the long 60 min infarct occlusion. The ischemic preconditioning group S underwent 2 times 10 min of ischemia separated by 30 min of reperfusion, group NIC received an infusion of nicorandil (K^+_{ATP}-channel opener) 10 µg/kgBW bolus followed by 2.5 µg/kgBW/min. for 10 min ic, group GLIB Glibenclamide (inhibitor of the K^+_{ATP}-channel) 0.5 mg/kgBW over 10 min iv, group CHA cyclohexyladenosine (adenosine Al-agonist) CHA 2.5 µg/kgBW over 10 min ic and group GC Glibenclamide 0.5 mg/kgBW iv followed by cyclohexyladenosine 2.5 µg/kgBW ic. Depicted are the infarct size as percent of the risk region. * = $p < 0.05$ vs. C, given are mean values and S.E.M. The intracoronary infusion of the adenosine agonist CHA or the K^+_{ATP} -channel opener nicorandil is able to replace ischemic preconditioning in reducing infarct size in porcine myocardium. Glibenclamide, an inhibitor of the K^+_{ATP} -channel, completely blocks the protective effect of CHA. The ability to activate the K^+_{ATP} - channel seems to be mandatory for induction of the protective effect (Data from Strasser [19]).

protective effect that can be neutralized by simultaneous infusion of one of the IGF binding proteins (IGFBP-3). The latter is also upregulated on the mRNA level following brief occlusion but with a latency of about 30 min following that of IGF-2. The IGF-system is therefore an ideal candidate to explain the memory effect and the subsequent refractoriness of ischemic preconditioning. The protective effect is mediated via the insulin receptor, a receptor of the tyrosine kinase type that can be blocked by the TKR- antagonist Lavendustin which indeed inhibits the protective effect of IGF-2 (Vogt and Schaper, unpublished). The only reservation is that allegedly no new protein synthesis is needed for the protective effect of ischemic preconditioning [22], an isolated observation in need of confirmation.

Other G_i-coupled receptors

All receptors coupled to inhibitory G-proteins should exhibit the same protective effect as adenosine. Intuitively it is difficult to imagine that angiotensin-II, endothelin, muscarinic agonists and epinephrine exert the same beneficial effect. In our experience they don't, with the possible exeption of muscarinic agonists in the dog heart. Years ago we found that morphine derivatives with a strong muscarinic action that produce bradycardia markedly retarded the spreading of an infarct from the subendocardium toward the subepicardium in the canine heart, which we originally ascribed to the ensuing decrease of the myocardial oxygen consumption. However, the activation of inhibitory G-proteins with subsequent K^+-channel opening could have contributed.

Direct myocardial infusion of endothelin, angiotensin-II and epinephrine had no protective effects in our experiments (Htun, unpublished).

Are heatshock proteins relevant?

We showed [23] that heat shock proteins are quickly upregulated following brief ischemic episodes, thereby enlarging upon an observation by Dillmann who had shown heatshock protein upregulation in the canine heart after long (and close to irreversible) ischemic periods. Within the relatively short duration of our experiment we did not observe an upregulation of the proteins proper, i.e. at a time when the protective effect of ischemic preconditioning was already over, translation had not yet started. We concluded therefore that heatshock proteins are not involved. They may have a place in the late effect of short occlusions ('second window of protection') which can be observed in the rabbit and the dog but not in the pig (8, 24, 25).

Clinical importance

Prospective and retrospective studies have been carried out to show whether ischemic preconditioning exists also in the clinical setting and whether it has a protective effect in patients. Under the acute conditions of PTCA repeated occlusions were indeed able to reduce the signs of ischemia [26], but metaanalysis of several large clinical trials for thrombolysis with the question whether anginal attacks before acute coronary occlusions resulted in smaller infarctions remain inconclusive [27–31]. In conclusion one can say that ischemic preconditioning has probably no clinical relevance with therapeutical consequences. In the animal laboratory it provided a useful model (next to early reperfusion) as a positive standard against which anti-ischemic drugs can be tested. It provided also interesting insights into intracellular signal transduction albeit more on the side of what protein kinases cannot do. The existence of significant species differences is confusing and may lead to the conclusion that if molecular pathways are evolutionary not well preserved across species of mammals they may not be very important.

Is the Quest over?

Biomedical research often proceeds in waves dictated by the availability of new methods and new ideas. To make the heart more resistant against ischemia was a serious and very successful effort of cardiac surgeons who developed cardioplegia to perform the complicated corrections on a non-perfused and non-beating heart. The effects of temperature and of ions became firmly established. This was the first wave of research into reversible ischemia. The protective effect of short occlusions fascinated researchers who had tried in vain to find pharmacological principles to counteract the consequences of ischemia during the era that is still known for its zeal to 'reduce infarct size' (the second wave). The difficulty of that time was that there was no hint as to whether the envisioned goal was realistic. Only the scantiest of working hypotheses existed (reduce $M\dot{V}O_2$) for a mass screening of compounds. The finding of the protective effect of repeated occlusions gave meaning to these efforts because it showed that it was possible to delay infarcts via an 'endogenous route.' Together with Downey's hypothesis the discovery of ischemic preconditioning produced a third wave of research into the manipulation of ischemia that continues to increase our knowledge on signal transmission in cardiac myocytes and on cardiac metabolism. With regard to progress in the basic sciences the quest of ischemic preconditioning is certainly not over, with regard to new drug development the situation looks different. Here the quest may definitely be over particularly in view of the therapeutic options that presently exist. When the search for anti-ischemic drugs and principles started in 1972 very little could be done for patients with acute coronary occlusion: Thrombolysis and PTCA were years to come, calcium antagonists and beta blockers were on the market and effective in stable angina pectoris but inactive in acute coronary occlusion, bypass surgery and lipid lowering drugs were in its infancy. Today's topics for tomorrow cardiac therapy are research on angiogenesis, gene therapy for atherosclerosis, and the genetic reprogramming of the cardiac myocyte to undo terminal differentiation.

References

1. Murry CE, Jennings RB, Reimer KA: Preconditioning with ischemia: A delay of lethal cell injury in ischemic myocardium. Circulation 74: 1124–1136, 1986
2. Reimer KA, Murry CE, Yamasawa I: Four brief periods of myocardial ischemia cause no cumulative ATP loss or necrosis. Am J Physiol 251: H1306–H1315, 1986
3. Hochachka P: Living without oxygen. Harvard University Press; Cambridge, London. 1980
4. Schaper J, Schwarz F, Kittstein H, Staemmler G, Winkler B, Scheld H, Hehrlein F: The effects of global ischemia and reperfusion on human myocardium: Quantitative evaluation by electron microscopic morphometry. Ann Thorac Surg 33: 116–122, 1982
5. Schaper W: Therapeutic approaches to myocardial infarct size limitation. In: D. Hearse and D. Yellon (eds). Experimental infarcts and the microcirculation. Raven press. New York, 1984, pp 79–90
6. Reimer KA, Lowe JE, Rasmussen MM, Jennings RB: The wavefront phenomenon of ischemic cell death 1. Myocardial infarct size vs duration of coronary occlusions in dogs. Circulation 56: 786–794, 1977
7. Schaper W, Binz K, Sass S, Winkler B: Influence of collateral blood flow and of variations in $M\dot{V}O_2$ on tissue-ATP content in ischemic and infarcted myocardium. J Mol Cell Cardiol 19: 19–37, 1987
8. Strasser R, Arras M, Vogt A, Elsaesser A, Schwarz ER, Schlepper M, Schaper W: Preconditioning of porcine myocardium: How much ischemia is required for induction? What is its duration? Is a renewal of effect possible? Circulation 90: I-109, 1994
9. Li GC, Vasquez JA, Gallagher KP, Lucchesi BR: Myocardial protection with preconditioning. Circulation 82: 609–619, 1990
10. Yellon DM, Alkhulaifi AM, Browne EE, Pugsley WB: Ischaemic preconditioning limits infarct size in the rat heart. Cardiovasc Res 26: 983–987, 1992
11. Qiu Y, Tang X, Sun J, *et al*: Does the late phase of ischemic preconditioning protect against infarction and reperfusion arrhythmias in conscious pigs. Circulation 92: I-389, 1995
12. Hoffmeister HM, Mauser M, Schaper W: Repeated short periods of regional myocardial ischemia: Effect on local function and high energy phosphate levels. Bas Res Cardiol 81: 361–372, 1986
13. Liu GS, Thornton J, Van-Winkle DM, Stanley AW, Olsson RA, Downey JM: Protection against infarction afforded by preconditioning is mediated by A1 adenosine receptors in rabbit heart. Circulation 84: 350–356, 1991
14. Vogt A, Barancik M, Weihrauch D, Arras M, Podzuweit T, Schaper W: Inhibitoren von Proteinkinase C reduzieren die Infarktgrösse in Schweineherzen *in vivo*. Z Kardiol 84: 100, 1995
15. Vogt A, Barancik M, Weihrauch D, Arras M, Podzuweit T, Schaper W: Protection of ischemic porcine myocardium from infarction by an A1-receptor agonist is not mediated by protein kinase C. Circulation 90: I-371, 1994
16. Grover GJ, Sleph PG, Dzwonczyk S: Role of myocardial ATP-sensi-

tive potassium channels in mediating preconditioning in the dog heart and their possible interaction with adenosine A1-receptors. Circulation 86: 1310–1316, 1992
17. Auchampach JA, Gross GJ: Adenosine-A(1) receptors, K(ATP) channels, and ischemic preconditioning in dogs. Am J Physiol 264: H1327–H1336, 1993
18. Van Winkle DM, Chien GL, Wolff RA, Soifer BE, Kuzume K, Davis RF: Cardioprotection provided by adenosine receptor activation is abolished by blockade of the K-ATP channel. Am J Physiol 266: H829–H839, 1994
19. Strasser R, Sprengel U, Arras M, Schlepper M, Schaper W: Blockade des KATP Kanals verhindert die Infarktgrössenreduktion durch Adenosinagonisten am Schwein. Z Kardiol 84: 846, 1995
20. Kirsch GE, Codina J, Birnbaumer L, Brown AM: Coupling of ATP-sensitive K^+ channels to A1 receptors by G proteins in rat ventricular myocytes. Am J Physiol 259: H820–H826, 1990
21. Piedimonte G, Nadel J, Long C, Hoffman J: Neural endopeptidase in the heart: Neutral endopeptidase inhibition prevents isoproterenol-induced myocardial hypoperfusion in rats by reducing bradykinin degradation. Circ Res 75: 770–779, 1994
22. Thornton J, Striplin S, Liu GS, Swafford A, Stanley AW, Van-Winkle DM, Downey JM: Inhibition of protein synthesis does not block myocardial protection afforded by preconditioning. Am J Physiol 259: H1822–H1825, 1990
23. Andres J, Sharma HS, Knöll R, Stahl J, Sassen LMA, Verdouw PD, Schaper W: Expression of heat shock proteins in the normal and stunned porcine myocardium. Cardiovasc Res 27: 1421–1429, 1993
24. Marber MS, Latchman DS, Walker JM, Yellon DM: Cardiac stress protein elevation 24 h after brief ischemia or heat stress is associated with resistance to myocardial infarction. Circulation 88: 1264–1272, 1993
25. Kuzuya T, Hoshida S, Yamashita N, Fuji H, Oe H, Hori M, Kamada T, Tada M: Delayed effects of sublethal ischemia on the acquisition of tolerance to ischemia. Circ Res 72: 1293–1299, 1993
26. Deutsch E, Berger M, Kussmaul W, Hirshfeld J, Jr., Herrmann H, Laskey W: Adaptation to ischemia during percutaneous transluminal coronary angioplasty. Clinical, hemodynamic, and metabolic features. Circulation 82: 2044–2051, 1990
27. Muller DWM, Topol EJ, Califf RM, Sigmon KN, Gorman L, George BS, Kereiakes DJ, Lee KL, Ellis SG: Relationship between antecedent angina pectoris and short term prognosis after thrombolytic therapy for acute myocardial infarction. Am Heart J 119: 224–231, 1990
28. Barbash GI, White HD, Modan M, Van de Werf F: Antecedent angina pectoris predicts worse outcome after myocardial infarction in patients receiving thrombolytic therapy: Experience gleaned from the international tissue plasminogen activator/streptokinase mortality trial. J Am Coll Cardiol 20: 36–41, 1992
29. Kloner RA, Shook T, Przyklenk K, Davis VG, Junio L, Matthews RV, Burstein S, Gibson M, Poole WK, Cannon CP, *et al.* Previous Angina Alters In-Hospital Outcome in TIMI 4. A Clinical Correlate to Preconditioning?. Circulation 91: 37–47, 1995
30. Ottani F, Galvani M, Ferrini D, Sorbello F, Limonetti P, Pantoli D, Ruszicali F: Prodromal angina limits infarct size. A role for ischemic preconditioning. Circulation 91: 291–297, 1995
31. Sack S, Mohri M, Arras M, Schwarz E, Schaper W: Ischaemic preconditioning time course of renewal in the pig. Cardiovasc Res 27: 551–555, 1993

Molecular and Cellular Biochemistry **160/161**: 217–224, 1996.

Cytoprotective mechanisms in cultured cardiomyocytes

Hari S. Sharma,[1] Joachim Stahl,[2] Dirk Weisensee[3] and Iris Löw-Friedrich[3]

[1]*Cardiovascular and Molecular Biology Laboratory, Institute of Pharmacology, Erasmus University, Rotterdam, The Netherlands;* [2]*Max-Delbrück-Center for Molecular Medicine, Berlin, Germany;* [3]*Center for Internal Medicine, J. W. Goethe University Hospital, Frankfurt am Main, Germany*

Abstract

Tumor necrosis factor-α (TNF-α), a potent cytokine mainly secreted by macrophages exerts pleiotropic effects on different cell types. However, the intracellular mediators of its action are not yet well characterized. To get an insight into endogenous cytoprotective mechanisms, we developed an *in vitro* model based on cultured cardiomyocytes treated with TNF-α at which we examined gene expression of heat shock proteins (HSP-27, HSP-70 and ubiquitin). Cardiomyocytes were isolated from the hearts of 18 day old fetal mice by enzymatic dissociation and grown in minimum essential medium containing 10% fetal calf serum. Spontaneously contractile cells were serum deprived for 24 h and treated with TNF-α (25 ng/ml) for 1, 2, 4, 6, 8, 12, and 24 h After each incubation, cells were processed to extract total proteins for Western and total RNA for Northern blot analyses. TNF-α induced arrhythmias and cessation of spontaneous contractions in a concentration and time dependent manner. Steady state (ubiquitin) or undetectable mRNA levels (HSP-27, HSP-70) were drastically induced (> 4 fold for all three genes vs untreated control cells) by TNF-α, reaching maximal values between 6–8 h of stimulation. Thereafter, the expression of these stress genes declined but remained elevated as compared to control. By Western blot analysis, we found increased multiple bands of ubiquitin protein conjugates in TNF-α treated cells whereas no significant change in HSP-27 protein accumulation until 12 h was observed as compared to control. 24 h of TNF-α incubation resulted in partial cellular necrosis. Our results indicate that TNF-α induces in cardiomyocytes transiently gene expression for cytoprotective molecules like HSP-27, HSP-70 and ubiquitin, suggesting these stress proteins to participate in subsequent defense mechanisms, for example in postischemic myocardial recovery. (Mol Cell Biochem **160/161**: 217–224, 1996)

Key words: TNF-α, heat shock proteins, cardiomyocytes, cytoprotection

Introduction

Dysfunction of the coronary endothelium as well as injury to cardiac muscle cells are the consequences of myocardial ischemia and reperfusion. A number of factors including various cytokines like TNF-α are released in the postischemic myocardium [1–3]. These ischemia induced cytokines may mimic the local cellular injury and may contribute in altering molecular phenotype of the postischemic myocardium. Increase in local expression as well as in circulating TNF-α has been reported in experimental animals and in patients with myocardial ischemia and infarction [1, 3]. In a recent study, pretreatment of rat hearts with TNF-α was found to be protective from ischemia and reperfusion injury [4]. However, in the normal heart, TNF-α may exert negative inotropic effects by directly altering intracellular calcium homeostasis in a concentration and time dependent manner [5]. We have shown earlier that perfusion of spontaneously contracting cultures of cardiomyocytes with a high dose of TNF-α (10,000 units/ml) led to arrhythmias with time and complete cessation of spontaneous contractions followed by severe loss of myocyte inotropy [6]. More recently, we have reported that TNF-α as well as interleukins (IL-2, IL-3, IL-6) induce formation of stress proteins in cultured cardiomyocytes [7, 8].

Address for offprints: H.S. Sharma, Institute of Pharmacology, Erasmus University, Faculty of Medicine and Health Sciences, PO Box 1738, 3000 DR Rotterdam, The Netherlands.

We hypothesize that TNF-α stimulates cytoprotective mechanisms in cardiomyocytes, and such mechanisms can be examined by investigating the expression pattern of various heat shock/stress protein genes.

TNF-α, a multipotent cytokine, originally known for its anti-tumor activity, has now been implicated in several biologic processes including inflammation, immunoregulation, cytotoxicity, cytoprotection and angiogenesis [9–11]. TNF-α is predominantly secreted by activated monocytes and it exerts pleiotropic effects on different cell types [10, 12]. However, the intracellular mediators of its action are not yet well known. Biological effects of TNF-α in the target cell are initiated by its binding to high affinity cell surface receptors [13–15]. TNF receptors are expressed on the membranes of virtually all somatic cell types [13, 14]. TNF receptors channel signals to cytoplasm and nucleus, and thereby initiate profound alterations in the metabolic pathway and nuclear transcription [13–16]. Two immunologically distinct TNF receptors of 55 kDa and 75 kDa apparent molecular mass have been identified and characterized [13, 14]. Many studies have demonstrated that TNF-α induces phosphorylation of a stress protein of about 28 kDa in different cell types [7, 18–20] and this phosphorylation results from stimulation of G protein coupled signal transduction involving the mitogen activated protein (MAP) kinase cascade [17, 21]. Furthermore, TNF-α may function as indirect angiogenic growth factor because it is a potent noncytotoxic growth inhibitor for endothelial cells *in vitro*, and enhances neovascularization *in vivo* [11], probably by inducing a potent angiogenic polypeptide such as vascular endothelial growth factor [22].

Heat shock proteins (HSPs) are a group of highly conserved proteins that can be stimulated by heat shock and other environmental and pathophysiological stresses [23–29]. Also glucose regulated proteins (GRPs), ubiquitin, αβ-crystalline and heme oxygenase [24–26, 28–33] are included in the group of HSPs. These proteins participate in cellular defense mechanisms and enable cells to survive and recover from stressful conditions [23, 26, 27, 34]. The proteins of the HSP-70 family are relatively well characterized; they bind to ATP and help in posttranslational import of proteins into endoplasmic reticulum and mitochondria [23, 26, 35]. The small heat shock protein HSP-27 migrates to the nucleus upon stress, can act as a molecular chaperone and plays an important role in signal transduction and drug resistance [20, 24–27]. Ubiquitin is a stress protein of 8 kDa that also belongs to the small heat shock family of proteins and contributes to ATP dependent non-lysosomal protein degradation [30, 33, 36].

It is believed that the heart has its own endogenous system(s) for protecting itself against ischemia-reperfusion injury, and a number of HSPs that may act as chaperones in saving vital cellular proteins from degradation have been proposed [24–29, 34, 35]. Over the last many years, we have been interested in characterizing molecules in the post-ischemic myocardium that may play important roles in cellular defense, in particular against ischemia [8, 22, 25, 29, 33, 37]. In the present study we attempted to evaluate TNF-α induced endogenous cytoprotective mechanisms in cultured cardiomyocytes. We developed an *in vitro* model based on cardiomyocytes treated with an adequate amount of TNF-α and examined the expression patterns of HSP-27, HSP-70 and ubiquitin at mRNA and protein levels.

Materials and methods

Cell culture

Cardiomyocytes were isolated from hearts of 18 day old fetal mice by the method described elsewhere [6, 7]. Briefly, small pieces of fetal cardiac tissue were subjected to trypsin (Biochrom KG, Berlin, FRG) digestion (final concentration of trypsin 0.1%) in calcium-magnesium free phosphate buffered saline (PBS, 140 mM NaCl, 2.6 mM KCl, 1.4 mM KH_2PO_4, 8 mM $Na_2HPO_4.2H_2O$). The supernatants of the first two digestion steps were discarded and thereafter supernatants of two trypsinization steps were collected. Trypsin was inactivated by dilution of the supernatants into equal volumes of ice-cold alpha medium supplemented with 20% fetal calf serum (FCS), penicillin (25 U/ml), and streptomycin (25 μg/ml). All chemicals, antibiotics, and media were of tissue culture grade and procured from Sigma Chemie, Deisenhofen, FRG, or from Life technologies GmbH, Eggenstein, FRG. Cells were collected by centrifugation at 4°C in a Sorvall SS34 rotor at 1000 rpm for 10 min and seeded in petri-dishes with a coverslip at the bottom. The cell number was determined in a Neubauer chamber and adjusted to 1.5 $\times 10^6$ cardiomyocytes per dish (Nunc, Denmark, diameter 6 cm). The adherent cardiomyocytes were incubated for 2 days in alpha medium with 20% FCS and another 2 days in Dulbecco's minimum essential medium (DMEM) with 10% FCS. Later on the cultures were transferred to L-valine deficient DMEM to prevent the growth of any fibroblasts. The cardiomyocytes which were spontaneously contractile (average frequency 86 ± 16 cycles/min) in culture were incubated for 5 days at 37°C in a humidified atmosphere with 5% CO_2 prior to start of serum deprivation and incubations. The contractility of cardiomyocytes was followed in a microscope perfusion system at constant temperature of 37 ± 0.1°C according to the method described earlier [6].

Immunofluorescence staining

In order to check the purity of cardiomyocytes, culture dishes were fixed in ice-old methanol/acetone 1:1 for 10 min, washed with PBS and followed by incubation with anti-titin antibody (Boehringer Mannheim GmbH, Mannheim, FRG) diluted in PBS with 0.5% bovine serum albumin (BSA) for 60 min at room temperature. After washing twice with PBS, slides were incubated with TRITC-conjugated secondary antibody against mouse IgG (1:10) for another 60 min at room temperature. The preparations were washed again, mounted and photographed under a fluorescence microscope (Axioskop, Zeiss, Oberkochen, FRG) using a Kodak Tri-X film (400 ASA). In addition, we also stained cardiomyocytes grown for 5 days in culture with anti-actin antibodies (Boehringer Mannheim GmbH, Mannheim, FRG) using rhodamine phalloidin. Photographs were taken by using a fluorescence microscope equipped with epi-illumination.

TNF-α incubation

Cultures of cardiomyocytes were serum deprived for 24 h. and incubated with the recombinant TNF-α (25 ng/ml, British Biotechnology Ltd., UK) for 1, 2, 4, 6, 8, 12 and 24 h. TNF-α was directly added to the culture dishes without exchanging the medium to avoid any culture shock. We used 3×10^6 cardiac myocytes for each time point and repeated the experiments for at least three times. Cells were harvested after each incubation directly either in guanidinium isothiocyanate buffer [38] for isolation of total cellular RNA or in protein solubilization buffer [39] for protein assay.

RNA isolation and Northern blot analysis

Total cellular RNA was isolated from cardiomyocytes treated with TNF-α for different time periods by the method of Chomczynski and Sacchi [38]. RNA concentration was measured by optical density and the quality of RNA was tested on a denatured formaldehyde agarose gel. For Northern hybridization, 15 μg of total RNA was denatured at 65°C in buffer containing formamide and ethidium bromide and electrophoresed on 1% agarose gel containing 2.2 M formaldehyde [25, 33, 37]. RNA was transferred to Hybond-N membrane (Amersham Nederland B.V., Den Bosch, Holland) by vacuum blotting. Thereafter, filters were air-dried, UV crosslinked in a gene linker (Bio-Rad Laboratories B.V., Holland) and ribosomal RNA bands were marked under UV light. Blots were hybridized at 42°C in a buffer containing 50% deionized formamide, 1.0 M sodium chloride, 1% sodium dodecylsulfate (SDS), 0.2% polyvinyl pyrrolidone, 0.2% ficoll, 0.2% bovine serum albumin, 50 mM Tris-HCl (pH 7.5), 0.1% sodium pyrophosphate, 10% dextran sulfate and denatured salmon sperm DNA (100 μg/ml). Blots were hybridized with radiolabeled cDNA probes encoding HSP-27, HSP-70 and ubiquitin [36, 40, 41]. cDNA inserts were labeled employing a multiprime labeling system (Amersham Nederland BV, Den Bosch, Holland) to a specific activity of 10^9 cpm/μg DNA using [^{32}P]-dCTP (3000 Ci/mmol, Amersham Nederland BV, Den Bosch, Holland). Filters were washed at room temperature for 5 min in 2 × SSC (1 × SSC = 0.15 M NaCl, 0.015 M sodium citrate) and 0.1% SDS and at 55°C in 0.1 × SSC containing 0.1% SDS for 20 min. Subsequently, filters were wrapped in household plastic wrap and exposed to Kodak X-OMAT AR films (Kodak Nederland B.V., Odijk, Holland) at –80°C for 1–3 days. A glyceraldehyde-3-phosphate dehydrogenase (GAPDH) cDNA probe (1.2 kb PstI fragment of human cDNA, procured from ATCC, USA) was used to rehybridize membranes for reference purposes. Furthermore, filters were re-hybridized with a cDNA insert of 770 bp encoding 28S rRNA in order to calculate the extent of induction of HSP expression after correcting the RNA loading differences. Hybridization signals on autoradiographs were quantified by video scanning in optical density mode using a Bioprofil version 4.6 computer program (Vilber Lourmat, France). Several exposures of the Northern blots were taken to ensure that quantitation of hybridization signals was in linear range. For normalization, optical density (O.D.) of hybridization signal for each gene was divided by O.D. of the corresponding GAPDH or 28S rRNA signal. Induction of each gene was calculated and expressed as % relative mRNA values (mean ± S.E.M.) in TNF-α treated cardiomyocytes as compared to control. Expression was statistically analyzed using student's "*t*"-test and significance was accepted at $p \leq 0.05$.

Polyacrylamide gel electrophoresis and Western blot analysis

After each incubation, medium was removed and cells were washed with PBS, harvested and sonicated in Laemmli buffer (1% SDS, 0.1 M Tris-Cl, pH 6.8, 20% glycerol, 0.001% bromophenol blue, 0.1 M DTT and 5% b-mercaptoethanol). Cell suspension was boiled for 5 min in a boiling water bath, centrifuged, and then cell debris were discarded and the supernatant collected. The protein concentration was measured in the supernatants according to the method of Lowry *et al.* [42] using bovine serum albumin as a standard. Samples of 20 μl each were subjected to one-dimensional sodium dodecyl sulphate-polyacrylamide gel electrophoresis (SDS-PAGE) in a discontinuous system using the method described by Laemmli [39]. The resolving gel concentration was 15% of the total acrylamide:bis-solution (37.5:1). The separated proteins were electrophoretically transferred according to the method of Towbin *et al.* [43] to a 0.45 μm nitrocellulose membrane (Schleicher and Schüll, Dassel, FRG) at room temperature using a transfer cell (Hoefer Scientific, San Francisco, USA) and buffer containing 25 mM Tris/glycine,

0.02% SDS and 10% (v/v) methanol, pH 8.3. After protein transfer, nitrocellulose membrane (controlled by Ponceau S staining) was blocked with 5% skimmed milk in TBS/Tween (10 mM Tris-HCI, pH 8.0, 150 mM NaCl, 0.05% Tween 20), incubated over night with rabbit anti-HSP-25 antibodies [44] in TBS/Tween and after extensive washing the antibodies were detected by means of alkaline phosphatase conjugated secondary antibodies employing a commercially available kit (ProtoBlot for Western Blot AP System, Promega Corporation, Madison, USA). Antibodies used for Western blotting were monoclonal antiubiquitin antibodies (working dilution 1:500) recognizing monomeric ubiquitin and ubiquitin-protein conjugates [31] and polyclonal antibodies against mouse HSP-25 [44] (working concentration I :750).

Results and discussion

Cardiomyocytes in culture

Under normal conditions, fetal cardiomyocytes in culture displayed rhythmic and synchronous contractions for several days. Figure 1 shows a microphotograph of spontaneously contractile fetal cardiomyocytes cultured for a week and stained with anti-actin antibodies. Actin was localized in cross striation in a very specific and regular manner with interspacing confirming one of the typical characteristics of cardiomyocytes [45]. We examined the purity of cardiomyocytes in culture on a regular basis by plating them on glass coverslips and growing them for 72 h prior to staining with anti-actin or anti-titin antibodies [6, 8]. Serum deprivation of cardiomyocytes did not significantly alter the contractility without a sign of cellular damage up to 24 h, the time taken prior to start of TNF-α incubation [6]. However, treatment of cardiomyocytes with TNF-α (25 ng/ml) induced arrhythmias and cessation of spontaneous contractions in a concentration and time dependent manner. 8 h of TNF-α incubation resulted in complete cessation of spontaneous contractions in serum depleted cells. Previously, we have shown that higher doses of TNF-α (10,000 units/ml) caused complete cessation of spontaneous contractions as early as 11 min after starting of treatment [6]. This inhibitory effect of TNF-α on contractility of cardiomyocytes was reversible upon incubation of cells in growth medium containing 10% of FCS.

When cells were treated for a longer period of time with a relatively low dose of TNF-α (25 ng/ml), as in this study, 24 h of incubation resulted in cellular necrosis leaving a number of floating dead cells in the culture medium. From these observations it can be inferred that TNF-α exerts a direct effect on contractility of cultured cardiomyocytes in a time and concentration dependent manner. TNF-α has been shown to have a reversible and directly concentration and time dependent negative inotropic effect on the myocardium [2, 5, 6]. In addition to such direct effects, TNF-α has been reported to be a potent inducer of cell surface adhesion molecules that facilitate adherence of neutrophils to cardiomyocytes and render them more susceptible to neutrophil mediated cellular injury [5, 46, 47]. Therefore, in pathophysiological conditions like myocardial infarction or ischemia, increased local myocardial expression may mimic tissue injury [2, 3]. TNF-α induced intracellular signaling cascade downstream to the ligand receptor binding in cardiomyocytes has been shown to be mediated via $G_{i\alpha}$ and G_{β} proteins and adenylyl cyclase system [17] that may account for the depressed contractile state of cardiomyocytes. However, the intracellular mediators of TNF-α induced cytoprotective events are not yet characterized. Therefore, it is important to understand the intricacy of cellular mechanisms responsible for pleiotropic effects of TNF-α on cardiomyocytes and to determine whether this cytokine exerts its effect(s) directly or whether indirectly by stimulating and releasing other autocrine factors.

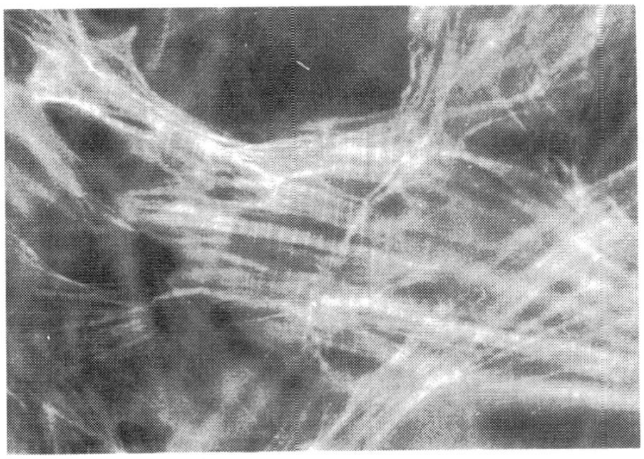

Fig. 1. Fluorescence microphotograph depicting actin microfilaments in cultured cardiomyocytes. Cardiomyocytes were grown on glass coverslips for a week, stained with anti-actin antibodies using rhodamine phalloidin and photographed using a fluorescence microscope equipped with epiellumination as described in Materials and methods. Note the actin-stained microfilaments showing cross striation in a very specific and regular manner with interspacing.

Effect of TNF-α on HSP-27 expression

Apart from its direct role in alteration of contractility of cardiomyocytes, that is whether TNF-α participates in cellular protective mechanisms, we evaluated a number of stress related genes in relation to TNF-α. Figure 2 shows the densitometric analysis of the expression pattern of HSP-27 mRNA in cardiomyocytes in relation to TNF-α. By Northern blot analysis we detected a mRNA species of 0.9 kb encoding HSP-27 in untreated (control, C) as well as TNF-α treated cardiomyocytes [7]. Expression of HSP-27 was drastically induced by TNF-α already as early as within 1 h of

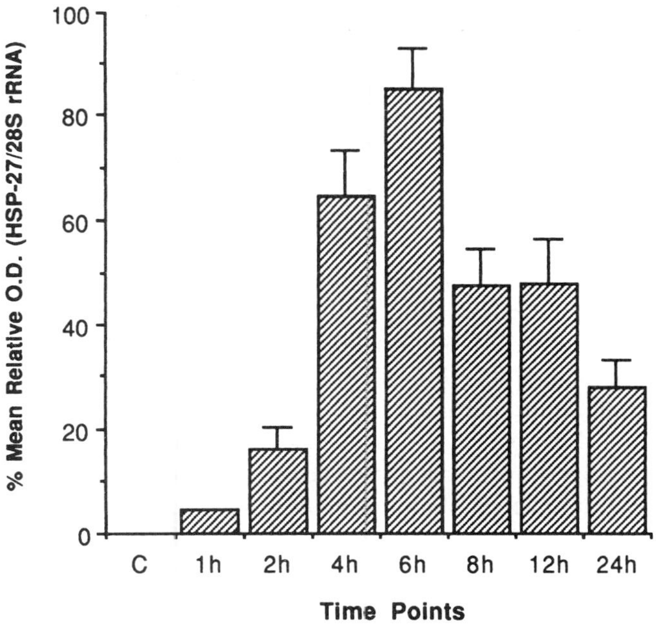

Fig. 2. Effect of TNF-α on HSP-27 gene expression. Bar graph showing quantitative analysis of the TNF-α induced HSP-27 mRNA expression. 15 μg of total RNA extracted from control (untreated) and TNF-α (25 ng/ml) treated cardiomyocytes was hybridized with a rat cDNA insert encoding HSP-27. Filters were rehybridized with a radiolabeled cDNA probe for 28S rRNA for reference purposes. Hybridization signals for mRNA bands of HSP-27 as well as of 28S rRNA were quantitated by video densitometry as described in the text. Data are means of the normalized signal ± S.E.M. (n = 4) and expressed as % mean relative O.D of HSP-27 and 28S rRNA hybridization signals.

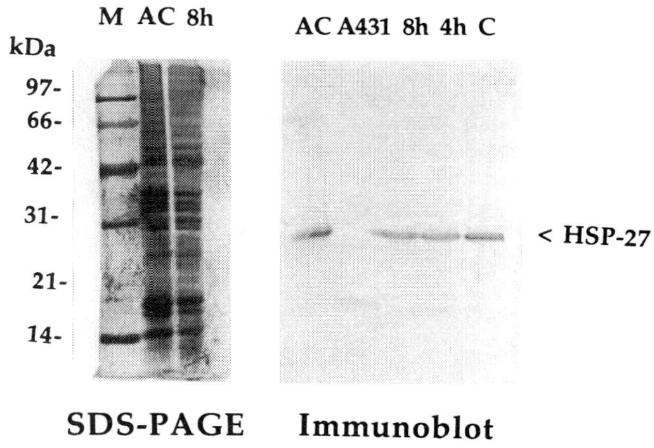

Fig. 3. Immunoblot analysis of HSP-27 accumulation after TNF-α treatment. Proteins extracted from TNF-α treated cardiomyocytes for various time points indicated at the top were fractionated on 15% SDS-PAGE and silver stained (left panel). 5 μg protein from each sample were electrophoresed followed by electrotransfer and Western blot with antiserum recognizing HSP-27. Molecular weight markers (M) are indicated on the left in kilodaltons (kDa). C, control; AC, Ehrlich ascitus carcinoma cell extract, A431, A431 cell extract.

incubation, reaching maximum levels at 6 h. Thereafter the expression remained elevated until 12 h of incubation (Fig. 2). Earlier, TNF-α had been shown to induce increased phosphorylation of HSP-27, but not accumulation of this protein [18–20]. In our experiments we also did not observe any significant accumulation of HSP-27 protein in cardiomyocytes as analyzed by Western blot analysis using anti-HSP-25 antibody [44], not even after 8 h of TNF-α incubation (Fig. 3). As positive and negative controls for the HSP-27 protein we included in our Western blots protein extracts from Ehrlich ascitus carcinoma as well as A431 cells, respectively (Fig. 3).

According to several previous reports, HSP-27 protein levels were unaltered in different cell types treated with TNF-α [18, 19]. In this paper we show that TNF-α induces HSP-27 mRNA transcription but, for unknown reasons, this does not lead to an increase of HSP-27 protein levels. Therefore, increased mRNA levels seem to be a reservoir for quick HSP-27 synthesis when cell needs it. In such a process, TNF-α guided phosphorylation of HSP-27 may play an important role. We do not know yet the role of phosphorylated HSP-27 for cardiomyocytes but several possible explanations can be put forward. HSP-27 is a target of the MAPKAP kinase regulated signal transduction pathway [21]. Phosphorylated HSP-27 seems to differently influence actin polymerization [48], a most likely very important process for heart muscle cells in particular upon cellular injury as for example due to ischemia, heat and TNF-α. Furthermore, HSP-27 has a strong tendency to form aggregates or even high-molecular weight homopolymers [20], which may be intracellularly regulated by various processes (cytoplasmic and/or nuclear localization) which are not yet understood. Mehlen *et al.* [20] have shown that TNF-α induced expression and modifications in phosphorylation, intracellular distribution, and oligomerization of human HSP-27 that can be correlated to a reduced cytotoxicity of this cytokine in L929 cell. Therefore, phosphorylation and intracellular organization of HSP-27 in cardiomyocytes treated with TNF-α, may be related to cytoprotective activity of this protein against the deleterious effects of this cytokine. In addition, a cytoprotective property of the small HSP family of proteins has also been attributed to their role to act as endogenous molecular chaperons giving rise to enhanced tolerance of cells to a stress condition [49].

Effect of TNF-α on HSP-70 expression

TNF-α induced the expression of HSP-70 gene in cardiomyocytes whereas, there was no detectable mRNA expression in control cardiomyocytes (Fig. 4). The mRNA signal of about 2.7 kb hybridizing to the HSP-70 cDNA probe was detectable at 2 h of incubation and the intensity of signal increased dramatically after 2 h of TNF-α incubation, then it remained increased until 12 h, with maximal expression at 8 h [8]. TNF-α incubation of cardiomyocytes for 24 h resulted in decreased mRNA expression of HSP-70 (Fig 4). Several

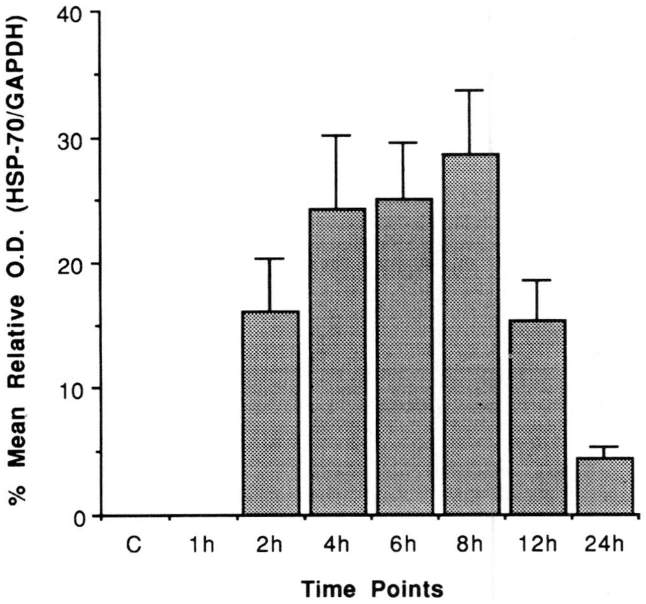

Fig. 4. Densitometric analysis of HSP-70 expression in relation to TNF-α. Bar graph showing quantitative analysis of the TNF-α induced HSP-70 mRNA expression. 15 μg of total RNA extracted from control (untreated) and TNF-α (25 ng/ml) treated cardiomyocytes was hybridized with a cDNA insert encoding human HSP-70. Filters were rehybridized with a radiolabeled cDNA probe for GAPDH for reference purposes. Hybridization signals for mRNA bands of HSP-70 as well as of GAPDH were quantitated by video densitometry as described in the text. Data are means of the normalized signal ± S.E.M. (n = 4) and expressed as % mean relative O.D. of HSP-70 and GAPDH hybridization signals.

groups including our own have shown that myocardial ischemia and reperfusion induce the expression of HSPs that may be involved in the myocardial adaptation and cellular tolerance to ischemic stress [25, 28, 29, 34]. Watanabe *et al.* [51] have found that TNF-α transfected cells expressed an increased amount of HSP-72 supporting a direct regulation of heat shock protein genes by this cytokine.

TNF-α that may participate in cellular injury has been shown to be produced in the postischemic and/or infarcted myocardium and to probably account for the enhanced circulating levels of this cytokine [1–3]. Therefore, it is important to understand the intricacy of cellular mechanisms responsible for the pleiotropic effects of TNF-α on the cardiomyocytes and to determine whether this cytokine exerts its effect(s) directly or whether indirectly by stimulating and releasing other autocrine and/or paracrine factors. It is believed that HSPs play an important role in transient adaptation in cellular activities, thereby coping with the stress period by protecting essential components of the cell and allowing the cell to recover from the stress [23–29]. Thus, elevated expression of HSPs in cardiomyocytes upon TNF-α treatment is an indication for induction of heat shock response as the basis of transient cellular adaptation. TNF-α pretreatment for 24 h in rats has been found to be beneficial for the heart against ischemia-reperfusion injury and this beneficial effect could be mediated via the increased levels of MnSOD [4, 50]. In our cell culture model, the induction of MnSOD, an enzyme involved in detoxification of superoxide anions in mitochondria in response to TNF-α, could also not be ruled out.

Effect of TNF-α on ubiquitin expression

We examined TNF-α treated cultured cardiomyocytes for expression of ubiquitin and ubiquitin-conjugates in order to evaluate the extent of cellular protein damage by this cytokine. Multiple mRNA bands encoding the monomer as well as polymers of ubiquitin were detected in cardiomyocytes [8]. Steady state mRNA levels of ubiquitin were drastically induced (> 10 fold versus untreated cells) by TNF-α being maximal at 6 h of stimulation and thereafter expression of this stress gene declined but remained elevated as compared to control (Fig. 5). Ubiquitin is a highly conserved protein found in all cell types contributing to non-lysosomal and ATP dependent cytosolic proteolysis [25, 30, 31, 36]. In a porcine model of myocardial ischemia and reperfusion, we have earlier reported that ubiquitin mRNAs as well as monomeric and conjugated ubiquitin protein levels were enhanced in the postischemic heart [25]. Myocardial ischemia is known to produce a number of intracellular changes within cardio-

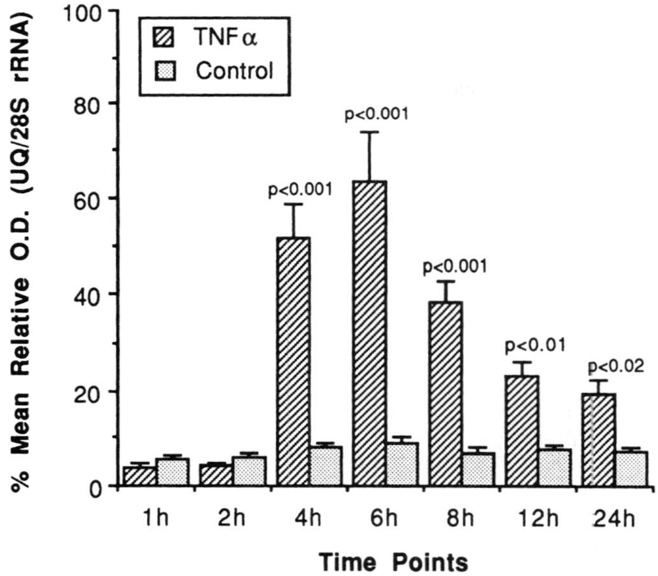

Fig. 5. Densitometric analysis of ubiquitin expression in relation to TNF-α. Bar graphs showing quantitative analysis of the TNF-α induced ubiquitin mRNA expression. Hybridized mRNA bands of ubiquitin as well as of 28S rRNA blots were quantitated by video densitometry as described in the text. Data are means of the normalized signal ± S.E.M. (n = 4) and expressed as % mean relative O.D of ubiquitin and 28S rRNA hybridization signals.

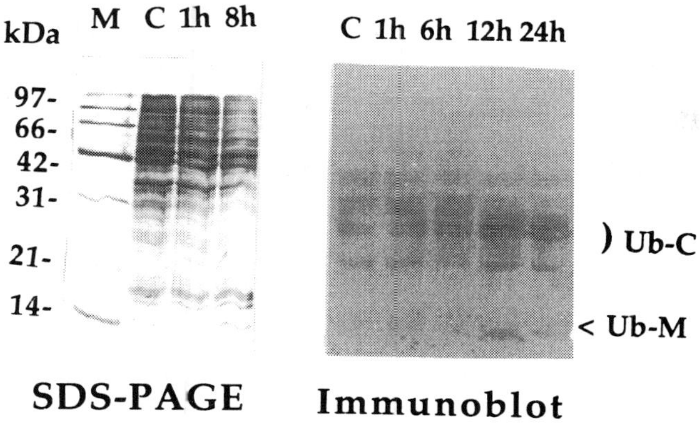

Fig. 6. Immunoblot analysis of ubiquitin and ubiquitin-protein conjugates after TNF-α treatment. Protein staining of SDS-PAGE by Coomassie (left panel). Samples (15 μg protein each) were taken from cells treated with TNF-α for time points given at the top and separated by 15% SDS-PAGE followed by electrotransfer and Western blot with monoclonal anti-ubiquitin antibodies recognizing ubiquitin-protein conjugates (right panel). Molecular weight markers (M) are indicated on the left in kilodaltons (kDa). Ub-M, ubiquitin monomer; UB-C, ubiquitin-protein conjugates; C, control;

myocytes including increased cellular calcium levels, free radical production, decreased intracellular pH, decreased ATP and glucose levels, etc. These cellular alterations lead to metabolic or hypoxic stress and result in cellular injury that may be characterized by denaturation and/or disturbance in three dimensional structure of many proteins. Medina *et al.* [32] have demonstrated that increased protein degradation in muscle atrophy caused by starvation and denervation is due to activation of the ubiquitin system for proteolysis as they found increased levels of polyubiquitin and proteasome mRNAs. Enhanced expression of ubiquitin in cardiomyocytes treated with TNF-α is an indication of cellular protein damage. Figure 6 depicts increased intensity and number of ubiquitin-protein conjugates in TNF-α treated cells as compared to control. Recently, it has been shown that an acute intravenous administration of TNF-α (100 μg/kg body weight) in rats resulted in a time dependent increase in the levels of both free and conjugated ubiquitin in skeletal muscle [52]. Our results on ubiquitin expression indicate that TNF-α induces the non-lysosomal protein degradation pathway for clearance of damaged or denatured proteins in cardiomyocytes .

Conclusion

We have shown that TNF-α is a potent inducer of heat shock protein genes in cardiomyocytes. Induction of genes conferring resistance to the cytotoxic property of TNF-α may provide a means to cardiomyocytes for self-defense under pathophysiological conditions. Hence, the induced expression of cytoprotective molecules such as stress proteins (HSP-27, HSP-70 and ubiquitin) in response to TNF-α may activate protective as well as repair mechanisms in cardiomyocytes for making them more resistant toward a subsequent challenge such as ischemia. It appears that TNF-α on the one hand mimics cellular injury in the heart and on the other hand simultaneously stimulates synthesis of vital proteins like HSPs and other proteins (for instance MnSOD) making TNF-α an interesting and relevant cytokine for the cardiovascular system. Furthermore, the ubiquitin system could play an important role in cytosolic degradation of damaged proteins in TNF-α treated cardiomyocytes where HSPs may counteract the proteolytic events and preserve vital proteins. We conclude that the enhanced transcription of HSP-mRNAs in response to TNF-α in cardiomyocytes may be an important basis for activation of protective and defense mechanisms contributing to cardiac protection, for example against ischemia.

Acknowledgements

The first author is thankful to Prof. Dr. W. Schaper, M.P.I., Bad Nauheim, for supporting the initial stage of this study.

References

1. Maury CP, Tempo AM: Circulating tumor necrosis factor-α (cachectin) in myocardial infarction. Am J Pathol 139: 709–715, 1989
2. Lefer AM, Tsao P, Aoki N, Palladino MJ: Mediation of cardioprotection by transforming growth factor-b. Science 249: 61–64, 1990
3. Herskowitz A, Choi S, Ansari AA, Wesselingh S: Cytokine mRNA expression in postischemic/reperfused myocardium. Am J Pathol 146: 419–428, 1995
4. Eddy LJ, Goeddel DV, Wong GH: Tumor necrosis factor-α pretreatment is protective in a rat model of myocardial ischemia-reperfusion injury. Biochem Biophys Res Comm 184: 1056–1059, 1992
5. Yokoyama T, Vaca L, Rossen RD, Durante, W, Hazarika P, Mann DL: Cellular basis for the negative inotropic effects of tumor necrosis factor-α in the adult mammalian heart. J Clin Invest 92: 2303–2012, 1993
6. Weisensee D, Bereiter HJ, Schoeppe W, Löw-Friedrich I: Effects of cytokines on the contractility of cultured cardiac myocytes. Int J Immunopharmacol 15: 581–587, 1993
7. Löw-Friedrich I, Weisensee D, Mitrou P, Schoeppe W: Cytokines induce stress protein formation in cultured cardiac myocytes. Basic Res Cardiol 87: 12–18, 1992
8. Sharma HS, Weisensee D, Löw-Friedrich I: Tumor necrosis factor-α induced cytoprotective mechanisms in cardiomyocytes: Analysis by mRNA phenotyping. Ann NY Acad Sci, 1996 (in press)
9. Beyaert R, Fiers W: Molecular mechanisms of tumor necrosis factor-induced cytotoxicity. What we do understand and what we do not. [Review]. FEBS Lett 340: 9–16, 1994
10. Tracey KJ, Cerami A: Tumor necrosis factor: A pleiotropic cytokine and therapeutic target. [Review]. Ann Rev Med 45: 491–503, 1994
11. Frater-Schröder M, Risau W, Hallmann R, Gautschi P, Böhlen P: Tumor necrosis factor type α, potent inhibitor of endothelial cell growth

in vitro, is angiogenic *in vivo*. Proc Natl Acad Sci (USA) 84: 5277–5281, 1987

12. Strieter RM, Kunkel SL, Bone RC: Role of tumor necrosis factor-α in disease states and inflammation. [Review] Crit Care Med 21: 447–463, 1993
13. Wiegmann K, Schutze S, Kampen E, Himmler A, Machleidt T, Kronke M: Human 55-kDa receptor for tumor necrosis factor coupled to signal transduction cascades. J Biol Chem 267: 17997–18001, 1992
14. Tartaglia LA, Weber RF, Figari IS, Reynolds C, Palladino MJ, Goeddel DV: The two different receptors for tumor necrosis factor mediate distinct cellular responses. Proc Natl Acad Sci (USA) 88: 9292–9296, 1991
15. Heller RA, Kronke M: Tumor necrosis factor receptor-mediated signaling pathways. [Review] J Cell Biol 126: 5–12, 1994
16. Rothe J, Gehr G, Loetscher H, Lesslauer W: Tumor necrosis factor receptors-structure and function. [Review]. Immunol Res 11: 81–90, 1992
17. Reithmann C, Gierschik P, Werdan K, Jakobs KH: Tumor necrosis factor-α up-regulates Gi alpha and Gi beta proteins and adenylyl cyclase responsiveness in rat cardiomyocytes. Eur J Pharmacol 206: 53–60, 1991
18. Kaur P, Welch WJ, Saklatvala J: Interleukin 1 and tumour necrosis factor increase phosphorylation of the small heat shock protein. Effects in fibroblasts, Hep G2 and U937 cells. FEBS Lett 258: 269–273, 1989
19. Vietor I, Vilcek J: Pathways of heat shock protein 28 phosphorylation by TNF in human fibroblasts. Lymphokine Cytokine Res 13: 315–323, 1994
20. Mehlen P, Mehlen A, Guillet D, Preville X, Arrigo A-P: Tumor necrosis factor-α induces changes in the phosphorylation, cellular localization, and oligomerization of human HSP 27, a stress protein that confers cellular resistance to this cytokine. J Cell Biochem 58: 248–259, 1995
21. Engel K, Ahlers A, Brach MA, Herrmann F, Gaestel M: MAPKAP Kinase 2 is activated by heat shock and TNF-α: *In vivo* phosphorylation of small heat shock protein results from stimulation of the MAP Kinase cascade. J Cell Biochem 57: 321–330, 1995
22. Sharma HS, Weisensee D, Löw-Friedrich I, Schoeppe W, Schaper W: Vascular endothelial growth factor expression in cardiac myocytes *in vitro* and its upregulation by tumor necrosis factor-a. J Cell Biochem 17D: 216, 1993
23. Welch WJ: Mammalian stress response: Cell physiology, structure/function, of stress proteins, and implications for medicine and disease. Physiol Rev 72: 1063–1081, 1992
24. Mestril R, Dillmann WH: Heat shock proteins and protection against myocardial ischemia. [Review]. J Mol Cell Cardiol 27: 45–52, 1995
25. Andres J, Sharma HS, Knöll R, Stahl J, Sassen LMA, Verdouw PD, Schaper W: Expression of heat shock proteins in the normal and stunned porcine myocardium. Cardiovasc Res 27: 1421–1429, 1993
26. Schlesinger M: Heat shock proteins. J Biol Chem 265: 12111–12114, 1990
27. Ciocca DR, Oesterreich S, Chamness GC, McGuire WL, Fuqua SA: Biological and clinical implications of heat shock protein 27,000 (HSP-27): a review. [Review]. J Natl Cancer Inst 85: 1558–1570, 1993
28. Iwaki K, Chi SH, Dillmann WH, Mestril R: Induction of HSP-70 in cultured rat neonatal cardiomyocytes by hypoxia and metabolic stress. Circulation 87: 2023–2032, 1993
29. Sharma HS, Snoeckx LH, Sassen LMA, Knoel R, Andres J, Verdouw PD, Schaper W: Expression and immunohistochemical localization of heat shock protein-70 in preconditioned porcine myocardium. Ann NY Acad Sci 723: 491–494, 1993
30. Jentsch S: Ubiquitin-dependent protein degradation: A cellular perspective. Trends Cell Biol 2: 98–103, 1992
31. Hayashi T, Takada K, and Matsuda M: Post-transient ischemia increase in ubiquitin conjugates in the early reperfusion. Neuroreport 3: 519–520, 1992
32. Medina R, Wing SS, Goldberg AL: Increase in levels of polyubiquitin and proteasome mRNA in skeletal muscle during starvation and denervation atrophy. Biochem J 307: 631–637, 1995
33. Sharma HS, Maulik N, Das DK, Gho BCG, Verdouw PD: Coordinated Expression of heme oxygenase-l and ubiquitin in the porcine heart subjected to ischemia and reperfusion. Mol Cell Biochem, 157: 111–116, 1996
34. Heads RJ, Yellon DM, Latchman DS: Differential cytoprotection against heat stress or hypoxia following expression of specific stress protein genes in myogenic cells. J Mol Cell Cardiol 27: 1669–1678, 1995
35. Knowlton AA: The role of heat shock proteins in the heart. J Mol Cell Cardiol 27: 121–131, 1995
36. Einspanier R., Sharma HS, Scheit KH: An mRNA encoding polyubiquitin in porcine corpus luteum: Identification by cDNA cloning and sequencing. DNA 6: 395–400, 1987
37. Brand T, Sharma HS, Fleischmann KE, Duncker DJ, McFalls EO, Verdouw PD, Schaper, W: Proto-oncogene expression in porcine myocardium subjected to ischemia and reperfusion. Circ Res 71: 1351–1360, 1992
38. Chomczynski P, Sacchi N: Single step method of RNA isolation by acid guanidinium thiocyanate-phenol-chloroform extraction. Anal Biochem 162: 156–159, 1987
39. Laemmli U: Cleavage of structural proteins during the assembly of the head of bacteriophage T4. Nature 227: 680–685, 1970
40. Wu B, Hunt C, and Morimoto R: Structure and expression of the human gene encoding major heat shock protein HSP70. Mol Cell Biol 5: 330–341, 1985
41. Hickey E, Brandon SE, Sadis S, Smale G, Weber LA: Molecular cloning of sequences encoding the human heat-shock proteins and their expression during hyperthermia. Gene 43: 147–154, 1986
42. Lowry O, Rosenbrough N, Farr A, Randall R: Protein measurements with the folin phenol reagent. J Biol Chem 193: 265–275, 1951
43. Towbin H, Staehelin T, Gordon J: Electrophoretic transfer of proteins from polyacrylamide gels to nitrocellulose sheets: Procedure and some applications. Proc Natl Acad Sci (USA) 76: 4350–4354, 1979
44. Gernold M, Knauf U, Gaestel M, Stahl J, Kloetzel PM: Developmental and tissue specific distribution of mouse small heat shock protein HSP25. Dev Gen 14: 103–111, 1993
45. Schaper J, Hein S, Brand T, Schaper W: Contractile proteins and the cytoskeleton in isolated rat myocytes. J Appl Cardiol 4: 423–429, 1989
46. Vaddi K, Nicolini FA, Mehta P, Mehta, IL: Increased secretion of tumor necrosis factor-α and interferon-gamma by mononuclear leukocytes in patients with ischemic heart disease. Relevance in superoxide anion generation. Circulation 90: 694–699, 1994
47. Ikeda U, Ikeda M, Kano S, Shimada K: Neutrophil adherence to rat cardiac myocyte by proinflammatory cytokines. J Cardiovas Pharmacol 23: 647–652, 1994
48. Benndorf R, Hayeß K, Ryazantsev S, Wieske M, Behlke J, Lutsch G: Phosphorylation and supramolecular organization of murine small heat shock protein HSP25 abolish its actin polymerization-inhibiting activity. J Biol Chem 269: 20780–20784, 1994
49. Jacob U, Gaestel M, Engel K, Buchner J: Small heat shock proteins are molecular chaperons. J Biol Chem 268: 1517–1520, 1993
50. Wong GH, Goeddel DV: Induction of manganous superoxide dismutase by tumor necrosis factor: possible protective mechanism. Science 242: 941–944, 1988
51. Watanabe N, Akiyama S, Tsuji N, Sasaki H, Yamauchi N, Okamoto T, Kobayashi D, Niitsu Y: Induction of synthesis of heat shock protein 72 in tumor necrosis factor gene transduced cells. Jpn J Cancer Res 85: 997–999, 1994
52. Garcia MC, Agell N, Llovera M, Lopez SF, and Argiles JM: Tumour necrosis factor-alpha increases the ubiquitinization of rat skeletal muscle proteins. FEBS Lett 323: 211–214, 1993

Molecular and Cellular Biochemistry **160/161**: 225–230, 1996.

Delayed cardioprotection is associated with the sub-cellular relocalisation of ventricular protein kinase Cε, but not p42/44MAPK

Susan Wilson, Wu Song, Kaszala Karoly,[1] Tanya Ravingerova,[2] Agnes Vegh,[1] Juluisby Papp,[1] Shogo Tomisawa,[3] James R. Parratt and Nigel J. Pyne

Department of Physiology and Pharmacology, University of Strathclyde, 204 George Street, Glasgow, G1 1XW, Scotland; [1]Albert Szent-Gyorgi Medical School, Doutef 12, Szeged, Hungary; [2]Slovak Academy of Sciences, Department of Pharmacology, Bratislava, Slovak Republic; [3]Chugai Pharmac Company LTD, 2-1-9 Kyobashi, Chuo-Ku, Tokyo 104, Japan

Abstract

Both noradrenaline administration to rats and rapid cardiac pacing in dogs induces delayed protection of the heart against ischaemia-induced ventricular arrhythmias. In an attempt to establish molecular mechanisms underlying the delayed cardioprotection, we have examined the potential role of two kinases, PKCε and p42/44MAPK. These protein kinases are expressed in the ventricles of the heart and are characterised by their ability to regulate ion-flux and gene transcription.

In the rat p42MAPK is predominantly localised in the high-speed supernatant fraction of the ventricle homogenate, whereas p44MAPK is enriched in the nuclear low speed pellet. A small proportion of the p42MAPK is activated even in hearts from control animals. However, neither kinase is relocalised or activated by noradrenaline administration and this provides preliminary evidence that p42/44MAPK may not play a significant role in delayed protection in this species.

In contrast, noradrenaline does induce the translocation of PKCε to cell membranes, a response that is sustained for up to 4 h. However, PKCε is down-regulated from the cytoplasm after 24 h post noradrenaline treatment. PKCε is also translocated to the membrane in dogs that have been classically pre-conditioned and cardiac paced. In the latter case, translocation of PKCε from the cytoplasm to the cell membrane is evident 24 h after pacing. These results indicate that the release of endogenous mediators may either inhibit down-regulation or elicit an increase in PKCε mRNA expression. Therefore, in dog heart the sub-cellular relocalisation of PKCε persists into the 'second window' and may play a central role in the molecular mechanism governing delayed cardioprotection. It is important in the future to identify either the gene products that are induced or the target protein(s) that are phosphorylated by PKCε. (Mol Cell Biochem **160/161**: 225–230, 1996)

Key words: classical preconditioning, noradrenaline, P42/44MAPK, protein kinase Cε, nuclear transcription, heat shock proteins, immediate early genes

Introduction

Preconditioning is the process in which brief periods of ischaemia can protect the heart against a subsequent prolonged ischaemic episode, if this commences less than 1 h later [1–3]. A second (delayed) form of protection can be induced by brief ischaemic episodes (coronary artery occlusion) by heat stress, bacterial endotoxin, catecholamines and by repeated periods of cardiac pacing. This has been described as the 'second-window' of cardioprotection [4]. This second-window of protection is associated with the induction of immediate early genes, such as c-*myc*, c-*fos* and c-*jun*, leading to the expression of 'so-called' late genes. These include heat shock proteins, e.g. hsp70 (reviewed by Yellon and Baxter [4]). Changes in ion conductance, that lead to calcium overload may also profoundly influence acute myocardial ischaemia.

We have focused our studies upon both PKC and p42/

44MAPK (for review see [5, 6]). Both kinases are known to regulate transcriptional events. We propose that the continued activation of some of these kinases may provides a molecular mechanism that sustains both the expression of cardioprotective genes, such as nitric oxide synthase and the phosphorylation of important cardioprotective proteins. Nitric oxide synthase has already been implicated in the mechanism of delayed cardioprotection, for example against ischaemia-induced ventricular arrhythmias [7].

Activation of the extracellular signal regulated kinase (p42/44MAPK: also known as CRK-1/2) allow communication between transmembrane signals and the nucleus (see [6]). Nuclear localised p42/44MAPK catalyses phosphorylation of transcription factors that activate *cis*-activating regulatory DNA elements and subsequently regulate the induction of c-*fos*. Some of these immediate early and late-induced gene products may play a significant role in delayed cardioprotection. Protein kinase C is comprised of a multiple family of isoforms that can be classified into three sub-groups, namely: (1) calcium-sensitive (α, βI and βII, γ); (2) calcium-insensitive (δ, ε) and (3) non-typical isoforms (ξ,η,ϕ). The activation of some of these kinases involves their rapid translocation from the cytoplasm to cell membranes. There is substantial evidence to indicate that PKC may be involved cardioprotection against infarct [5]. PKC is also involved in the induction of c-*fos* [8]. We have found that the ventricular myocardium of adult rats and dogs contain high levels of PKCε (a calcium-independent PKC isoform) and have therefore assessed its role in delayed cardioprotection. PKCδ and ε are the major isoforms expressed in adult rat ventricular myocytes [9], although PKCδ may be absent in dog heart. Indeed even in rat heart, the levels of PKCε vastly exceeds that of PKCδ.

In this paper, we demonstrate that noradrenaline and cardiac pacing induce delayed cardioprotection in rats and dogs respectively. We provide preliminary evidence to demonstrate that p42/44MAPKs are not activated or redistributed following the administration of noradrenaline and are therefore unlikely to play a significant role in delayed protection in rat heart. However, both noradrenaline and cardiac pacing induce translocation of PKCε to cell membranes. However, whereas PKCε is down-regulated from the cytoplasm after 24 h post noradrenaline treatment, cardiac pacing in dogs induces the sustained translocation of PKCε. This results in significantly larger amounts of PKCε in cell membranes even 24 h after pacing. These results indicate that the release of endogenous cardioprotective mediators may inhibit down-regulation or increase the rate of PKCε mRNA expression. PKCε may therefore play a role in the mechanism of delayed cardioprotection.

Materials and methods

Reagents

All biochemicals were from Sigma Chemical Co. (U.K). Anti-p42/44MAPK and anti-PKCε antibodies were from Affiniti (U.K.). Reporter HRP-anti-mouse and anti-rabbit antibodies were from the Scottish Antibody Production Unit (Carluke, Scotland). ECL kits were from Amersham (U.K.). Male Sprague Dawley rats (100–200 g) were used for isolation of heart tissue.

Delayed cardioprotection by noradrenaline and by cardiac pacing

Male Sprague-Dawley rats (100–200 g) were given noradrenaline (l mg/Kg) by intraperitoneal injection and the animals sacrificed 10, 20, 60 min and 24 h after injection. The hearts were rapidly removed, perfused with buffer A containing 0.25 M sucrose, 10 mM-Tris/HCl, pH 7.4, 1 mM-EDTA, 2 mM-benzamidine and 0.1 mM-PMSF in order to remove as much blood as possible. The ventricular tissue was dissected and homogenised as described below.

Dogs of either sex and weighing 17 kg were anaesthetised using sodium pentobarbitone and a Corat F4 bipolar pacing electrode was introduced into the right ventricle, via the jugular vein. The correct placing of this electrode was confirmed by recording the electrocardiogram. The dogs were allowed to breathe spontaneously and blood pressure was monitored from the carotid artery. The dogs were then paced, at a rate of 220 beats/min for four 5 min periods with a 5 min rest (reperfusion) period between the pacing stimuli. This procedure results in a marked reduction in arrhythmia severity when the left anterior descending coronary artery is occluded 20–24 h later [10, 11] . At this time the dogs were re-anaesthetised, thoractomised and samples of the left ventricular tissue from the area supplied by the left anterior descending coronary artery, were removed and immediately frozen using Wollenberger clamps. The samples were stored in frozen liquid nitrogen and transported to Glasgow in dry ice. Some dogs were also preconditioned by two 5 min periods of coronary artery occlusion with a 20 min reperfusion period between each occlusion. This results in protection against life-threatening ventricular arrhythmias when the same coronary artery is occluded for a longer period of time (30 min) up to 1 h later. Samples were removed and processed as described above. These samples were homogenised as described below.

Preparation of low and high-speed particulate and supernatant fractions

Ventricular tissue samples were weighed and homogenised in ice-cold buffer A, using a Turrax homogeniser. The subsequently obtained homogenate was filtered twice through gauze and centrifuged at 2000 g for 20 min at 4°C in a Beckman Prep 65 ultracentrifuge. The particulate fraction was resuspended in buffer A and is termed here, the low speed nuclear pellet. The low-speed supernatant was recentrifuged at 50000 g for 20 min at 4°C. The high-speed supernatant fraction contains cytoplasmic proteins. The pellet which is resuspended in buffer A and contains a membrane fraction and is termed, the high-speed particulate fraction. The protein concentration of each fraction was assayed using the Bradford protein assay.

Detection of p42/44MAPK and PKCε

After equalising proteins in each sample, low-speed nuclear pellets, high-speed supernatant and high-speed particulate fractions were separately combined with sample buffer containing final concentrations of 0.125 M-Tris/HCl, pH 6.7, 0.5 mM-$Na_4P_2O_7$, 1.25 mM-EDTA, 2.5% (v/v) glycerol, 0.5% (w/v) SDS, 25 mM-DTT, 0.05% (w/v) bromophenol blue). Samples were boiled for 5 min and then subjected to SDS-PAGE. Resolved proteins were transferred to nitrocellulose sheets. These were subsequently blocked in 3% (w/v) bovine serum albumin in 20 mM-Tris buffered saline, pH 7.4 containing 0.03% (v/v) Tween 20 (T-TBS). Blocking was performed at room temperature for 2.5 h. After this time the nitrocellulose sheets were probed with either anti-p42/44MAPK or anti-PKCε antibodies in T-TBS containing 0.2% BSA (w/v). These were incubated at 37°C for 12 h. After this time, the nitrocellulose sheets were washed six times in T-TBS. Detection of MAPK and PKCε reactivity was achieved by incubating nitrocellulose sheets for 2 h at room temperature with reporter HRP-linked anti-mouse antibodies. Antibodies were diluted in T-TBS containing 0.2% BSA (w/v). After washing the blots as described above, to remove excess reporter antibody, immunoreactive bands were detected with the ECL detection kit.

p42/44MAPK are activated in response to both growth factor and G-protein-linked receptor stimulation and is evident by their retarded mobility upon SDS-PAGE: A consequence of phosphorylation by MEK.

Results

Identification of p42/44MAPK in the ventricular low speed nuclear pellet and high-speed supernatant

Immunoblotting isolated ventricular sub-cellular fractions with an anti-MAPK antibody identified both forms of this kinase. In Figs. 1A and 1B, we show that these kinase migrate on SDS-PAGE as 42 and 44 kDa proteins and can be identified in both low speed nuclear and high speed supernatant fractions of the rat heart ventricle. The p42MAPK is predominantly localised in the high-speed supernatant fraction (Fig. 1A), containing cytoplasm, whereas p44MAPK is enriched in the low speed nuclear pellet (Fig. 1B). An additional MAPK-related protein of molecular mass 41 kDa can also be detected in the high-speed supernatant fraction and may represent a proteolysed fragment of p42/44MAPK (Fig. 1A).

Activation and relocalisation of p42/44MAPK

The activation of p42/44MAPK is detected by its retarded mobility on SDS-PAGE. By comparing immunoblots of lysates derived from PDGF-treated smooth muscle cells, which contain both non-activated and activated p42MAPK, we are able to demonstrate that a proportion of the total pool of ventricular p42MAPK is activated in the high-speed supernatant fraction isolated from control animals (Fig. 1A). In contrast, we were unable to detect activated p42MAPK in the low-speed nuclear pellet fraction (Fig. 1B). Neither p42MAPK nor p44MAPKs are activated by the administration of noradrenaline, i.e. we do not detect an increase in the amount of ERK-2 displaying retarded mobility on SDS-

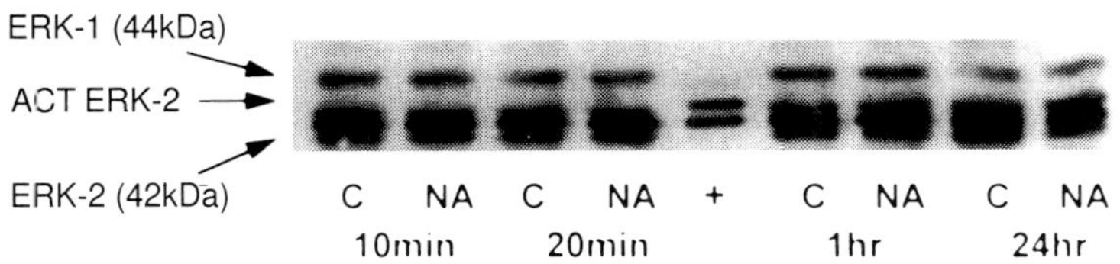

Fig. 1A. The effect of noradrenaline administration on ERK-1/ERK-2 activity. Immunoblot of ERK-1/ERK-2 in cytosolic high-speed supernatant fractions isolated from rat hearts as described in the Materials and methods section. ERK-1, ERK-2 and activated ERK-2 are denoted by arrows. Activated ERK-2 is detected as a 43 kDa immunoreactive band that displays retarded mobility in comparison to ERK-2. C denotes control animals, NA denotes animals administered noradrenaline (1 mg/kg i.p.) 10 min, 20min, 1 h or 24 h prior to removal of the heart. + denotes positive control (PDGF-stimulated smooth muscle cells). This result is representative of an experiment performed three times in hearts isolated from three separate groups of animals.

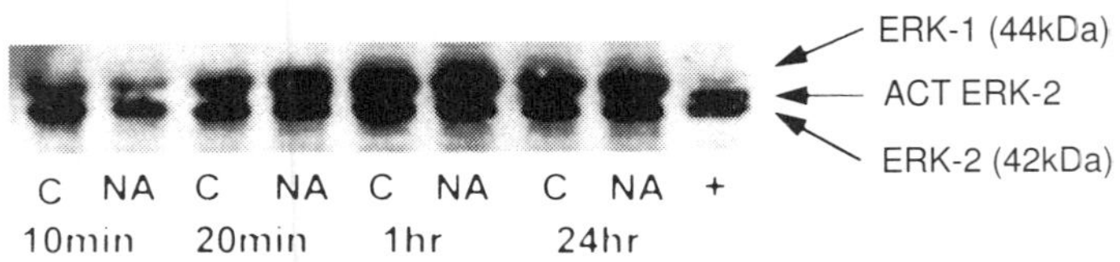

Fig. 1B. The effect of noradrenaline administration on ERK-1/ERK-2 activity. Immunoblot of ERK-1/ERK-2 in low-speed nuclear pellets isolated from rat hearts as described in the Materials and methods section. ERK-1 and ERK-2 are denoted by arrows. No activated ERK-2 is detected. C denotes control animals, NA denotes animals administered noradrenaline (1 mg/kg i.p.) 10 min, 20 min, 1 h or 24 h prior to removal of the heart. + denotes positive control (PDGF-stimulated smooth muscle cells). This result is representative of an experiment performed three times in hearts isolated from three separate groups of animals.

PAGE. Furthermore, there is no major relocalisation of either kinase into the low-speed nuclear pellet fraction during the entire 24 h time-course (Figs. 1A and 1B). This indicates that nuclear translocation of MAPK in intact ventricular myocytes is unlikely to occur. We therefore, conclude from this preliminary evidence, that the p42/44MAPKs is unlikely to play a significant role in the delayed cardioprotection that is induced by noradrenaline.

Translocation of PKCε

In Figs. 2 and 3, we show that immunoblotting isolated subcellular fractions from rat and dog ventricular homogenates with a specific anti-PKCε antibody identified PKCε as a 90 kDa protein. In rat ventricles, PKCε was identified in both high-speed supernatant and particulate fractions (Figs. 2A and 2B). Noradrenaline induces the translocation of PKCε to cell membranes, a response that occurs within 10 min of noradrenaline administration and is sustained for up to 4 h. This is indicated by a reduction of PKCε levels in the high-speed supernatant and a concurrent increase in PKCε levels in high-speed particulate fraction. The early translocation of PKCε correlates with the onset of cardioprotection. However, PKCε is downregulated after 24 h post noradrenaline administration. This is indicated by a reduction in the amount of PKCε in the high-speed supernatant and particulate fractions.

In contrast, we show in Fig. 3 that cardiac pacing in dogs induces a sustained translocation of PKCε to cell membranes, which persists for 24 h after pacing. We also show in Fig. 3 that classical preconditioning of dogs induces the rapid translocation of PKCε to cell membranes (Fig. 3). Both responses are indicated by an increase in the immunoreactivity of PKCε in the high-speed particulate fraction.

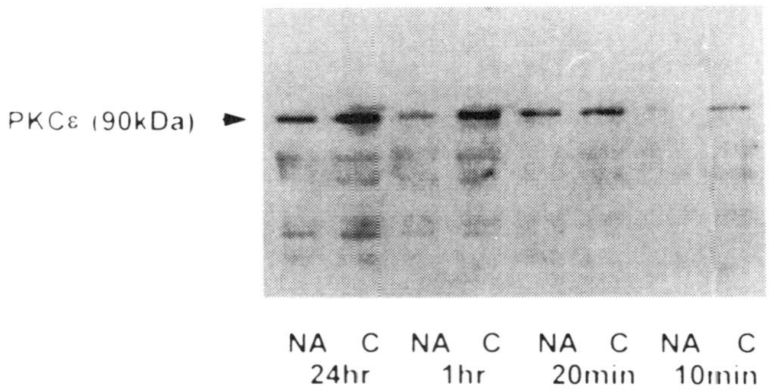

Fig. 2A. Noradrenaline induces the translocation of PKCε from the isolated cytosolic fraction of cardiac tissue. Immunoblot of PKCε in cytosolic high-speed supernatant fractions isolated from rat hearts as described in the Materials and methods section. PKCε is denoted by an arrow. C denotes control animals and NA denotes animals administered noradrenaline (1 mg/kg i.p.) 10 min, 20 min, 1 h or 24 h prior to removal of the heart. This result is representative of an experiment carried out three times in hearts isolated from three separate groups of animals.

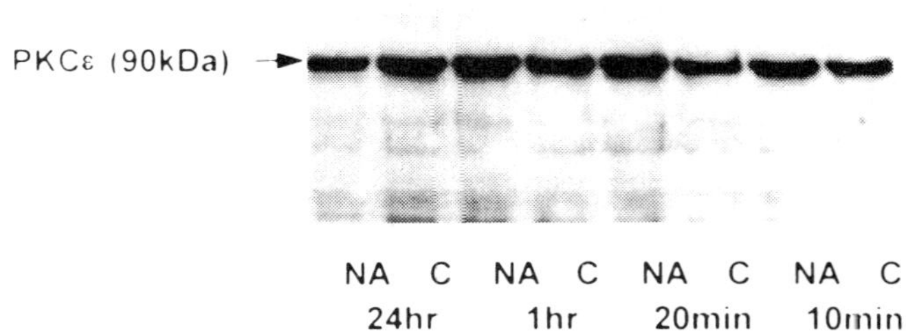

Fig. 2B. Noradrenaline induces the appearance of PKCε in isolated cardiac membranes. Immunoblot of PKCε in membranes isolated from rat hearts as described in the Materials and methods section. PKCε is denoted by an arrow. C denotes control animals and NA denotes animals administered noradrenaline (1 mg/kg i.p.) 10 min, 20 min, 1 h or 24 h prior to removal of the heart. This result is representative of three experiments carried out in hearts isolated from three separate groups of animals.

Discussion

p42/44MAPK

A number of G-protein linked receptor agonists can elicit the activation of cytoplasmic extracellular signal-regulated p42/44MAPK [6]. This involves the GTP loading of $p21^{ras}$ and the recruitment of c-Raf-1 kinase to cell membranes. c-Raf-1 kinase catalyses the phosphorylation of MEK-1, a dual specific kinase that in turn catalyses Thr/Tyr phosphorylation of p42/44MAPK. Activated p42/44MAPKs are crucial intermediates in the initiation of immediate early gene expression and

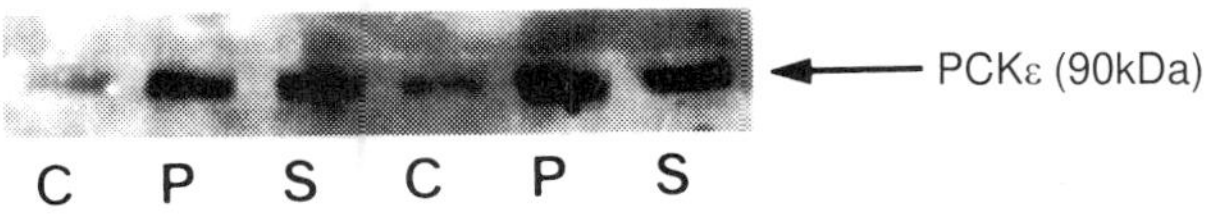

Fig. 3. PKCε is translocated to isolated cardiac membranes after classical preconditioning and during the second window of protection. Immunoblot of PKCε in the membrane fraction isolated from canine hearts as described in the Materials and methods section. PKCε is denoted by an arrow. C denotes control animals, P denotes animals which have been classically preconditioned, as described, immediately prior to removal of the hearts and S denotes hearts removed during the second window of protection, 24 h after rapid cardiac pacing. This result is representative of an experiment carried out four times in hearts isolated from four separate groups of animals.

upon activation, translocate to the nucleus. Nuclear localised p42/44MAPK catalyses phosphorylation of transcription factors that regulate *cis*-activating DNA elements and include $p62^{TCF}$/Elk-l. p42/44MAPK also catalyse the phosphorylation of $p90^{rsk}$, which in turn catalyses the phosphorylation and activation of other transcription factors, including nuclear serum response factor (SRF) and Nur77. Phosphorylated SRF and Elk-1 participate in the subsequent activation of the serum response element (SRE), by eliciting formation of the ternary SRF-SRE-Elk- 1 complex. The activation of this complex can subsequently stimulate c-*fos* expression. However, we demonstrate that in adult rat heart ventricles p42MAPK/MAPKs are not activated and do not undergo relocalisation in response to noradrenaline. This provides preliminary evidence that they do not play a significant role in delayed protection.

Protein kinase C

Protein kinase C is a family of multiple isoforms [12–14]. The activation of some of these kinases involve their rapid translocation from the cytoplasm to cell membranes. This occurs in response to the agonist-induced hydrolysis of membrane phospholipids by specific phospholipases, including PLC and PLD. These phospholipases catalyse the liberation of diacylglycerol, an activator of PKC. Diacylglycerol binds to the cysteine enriched zinc finger binding motif of PKC and elicits a conformational change in the protein, thereby removing a pseudosubstrate domain that lies in close proximity to the catalytic site.

The PKC isoform expression in the heart appears to change with age. Specifically, four PKC isoforms can be detected in whole extracts from neonatal rats (PKCα, δ, ε and ζ). PKCε and δ are the most abundantly expressed isoforms and their expression persists into adulthood, whereas the expression of PKCα and ζ declines markedly [9]. Although low levels of these kinases can be detected in rat heart it is likely that they are expressed in non-myocytes, such as fibroblasts. The nPKC isoforms, include PKCη and ϕ. PKCϕ is predominantly expressed in skeletal muscle and is only minimally expressed in heart [15, 16], whilst PKCη is detected in low levels in both neonatal and adult rats. However, PKCη is also likely to be expressed in non-myocyte cells [17]. PKCε and ζ are expressed in dog heart, whilst the PKCα and δ isoforms appear to be absent [5, 9].

There is substantial evidence to indicate that PKC may be involved in classical preconditioning (reviewed by Cohen *et al.* [18]). For instance, the diacylglycerol analogue, 1, 2-dioctanoyl-*sn*-glycerol can mimic protection against infarction and contractile dysfunction. Chelerythrine, a specific regulatory site inhibitor of PKC can also attenuate these protective effects. Perhaps most importantly, PKC can also regulate long-term expression of late genes via activation of the SRF and may therefore play a role in the 'second-window' of protection. This can be achieved either directly, by translocation to the nucleus or via the p42/44MAPK cascade. In the latter case, membrane localised PKC phosphorylates c-Raf-l kinase, the direct up-stream kinase that catalyses phosphorylation of MEK-1 [19]. PKC isoforms also regulate ion-channel conductances, such as Ca^{2+}. For instance, PMA has been reported to increase inward calcium currents via a dihydropyridine-sensitive L-type channel in neonatal ventricular myocytes [20]. The influx of calcium may reduce intracellular cyclic AMP via inhibition of the type 5 adenylyl cyclase [21]. However, this is not observed in adult ventricular myocytes. PKC has also been demonstrated to catalyse phosphorylation of several important contractile proteins, such as troponin I and C [22]. The phosphorylation of troponin I may reduce myofibrillar actomyosin Mg-ATPase activity and could contribute to a negative inotropic response, i.e. decreased twitch amplitude. Given that PKCε is the predominantly expressed isoform in adult heart, it is likely that it plays a significant role in cardioprotection.

Noradrenaline induces translocation of PKCε to cell membranes in rat ventricles, a response that is sustained for up to 4 h. However, it is down-regulated after 24 h post noradrenaline treatment. Continued membrane translocation of PKCε will eventually elicit its down-regulation. This occurs because diacylglycerol increases its susceptibility to proteolytic degradation and will eventually deplete both membrane and cytosolic pools. This is consistent with the effect of PMA on PKCε levels in adult and neonatal ventricular myocytes [5]. In contrast, cardiac pacing in dogs induces the sustained translocation of PKCε, such that significantly larger amounts of PKCε are found in cell membranes even 24 h after pacing. This may involve the release of endogenous mediators that either inhibit down-regulation or elicit an increase in PKCε mRNA expression. Therefore, in dog heart ventricles the sub-cellular relocalisation of PKCe persist into the 'second window' and may play a central role in the molecular mechanism governing delayed cardioprotection.

Conclusion

We have identified for the first time a sustained translocation of dog ventricular PKCε to cell membranes. This contrasts with the effect of noradrenaline which elicits a transient response. However, we cannot exclude a role for the induction of cardioprotective late genes by PKCε in this case. The p42/44MAPK cascade does not appear to be involved, as evidenced by the failure to detect activated kinase or a significant relocalisation. However, these studies need to be verified by direct measurements of p42/44MAPK activity.

In addition to the established role of PKCε in classical preconditioning, these studies provide the first direct evidence to implicate a role for PKCε in the 'second-window' of cardioprotection. This is supported by a recent study showing that the delayed reduction in infarct size resulting from preconditioning effected by four 5 min occlusions of the left circumflex coronary artery in rabbits is abolished by chelerythine, a general potent PKC inhibitor [23]. Final confirmation of the role of PKCε in delayed cardioprotection will involve either its selective inhibition by pharmacological drugs or the use of genetic strategies, i.e. the inhibition of PKCε mRNA expression.

Acknowledgements

These studies were supported by the European Commission (Network Grant ERB-CLPA CT 92 4.0.0.9), by the British Council, the Hungarian Committee for Technical Development and by Chugai Company, Tokyo, Japan.

References

1. Murry CE, Jennings RB, Reimer KA: Preconditioning with ischemia: A delay of lethal cell injury in ischaemic myocardium. Circulation 74: 1124–1136, 1986
2. Lawson CS, Downey JM: Preconditioning: State of the art in myocardial protection. Cardiovasc Res 27: 542–550, 1993
3. Parratt JR: Protection of the heart by ischaemic preconditioning: Mechanisms and possibilities for pharmacological exploitation. Trends Pharmacol Sci 15: 19–25, 1994
4. Yellon DM, Baxter GF: 'A second window of protection' or delayed preconditioning phenomenon: Future horizons for myocardial protection. J Mol Cell Cardiol 27: 1023–1034, 1995
5. Steinberg SF, Goldberg M, Rybin VO: Protein kinase C isoform diversity in the heart. J Mol Cell Cardiol 27: 141–153, 1995
6. Marshall CJ: Specificity of receptor tyrosine kinase signalling: Transient versus sustained extracellular signal-regulated kinase activation. Cell 80: 179–185, 1995
7. Vegh A, Papp JGy, Parratt JR: Prevention by dexamethasone of the marked antiarrhythmic effects of preconditioning induced 20 h after rapid cardiac pacing. Brit J Pharmacol. 113: 1081–1082, 1994
8. Kariya K, Karns IR, Simpson PC: Expression of a constitutively activated mutant of the isozyme of protein kinase C in cardiac myocytes stimulates the promoter of the myosin heavy chain isogene. J Biol Chem 266: 10023–10026, 1991
9. Rybin VO, Steinberg SF: Protein kinase C isoform expression and regulation in the developing rat heart. Circ Res 74: 299–309, 1994
10. Kaszala K, Vegh A, Parratt JR, Papp J: Time course of pacing induced preconditioning in dogs. J Mol Cell Cardiol 27: A145, 1995
11. Kaszala K, Vegh A, Papp J Gy, Parratt JR: Time-course of the protection against ischaemia and reperfusion-induced ventricular arrhythmias resulting from brief periods of cardiac pacing. J Mol Cell Cardiol 1996, (in press)
12. Nishizuka Y: Intracellular signalling by hydrolysis of phospholipids and activation of protein kinase C. Science 258: 607–614, 1992
13. Pears CJ, Parker PJ: Commentary: Domain interactions in protein kinase C. J Cell Sci 100: 683–686, 1991
14. Hug H, Sarre TF: Review article: Protein kinase C isoenzymes: Divergence in signal transduction? Biochem J 291: 329–343, 1993
15. Chang JD, Xu Y, Raychowdhury MK, Ware JA: Molecular cloning and expression of a cDNA (ηPKC): A new member of the *nn*PKC family expressed in skeletal muscle, megakaryoblastic cells, and platelets. J Biol Chem 268: 14208–14214, 1993
16. Osada S, Mizuno K, Saido TC, Suzuki K, Kuroki T, Ohno S. A new member of the protein kinase C family, ηPKC-0, predominantly expressed in skeletal muscle. Mol Cell Biol 12: 3930–3938, 1992
17. Bogovevitch MA, Parker PJ, Sugden PH: Characterization of protein kinase C isotype expression in adult rat heart: Protein kinase C-ε is a major isotope present, and it is activated by phorbol esters, epinephrine, and endothelin. Circ Res 72: 757–767, 1993
18. Cohen MV, Liu Y, Downey JM: Activation of protein kinase C is critical to the protection of preconditioning. In: CL Wainwright, JR Parratt (eds). Myocardial Preconditioning. Springer, Berlin, 1996, pp 185–206
19. Sozeri O, Vollmer K, Liyanage M, Frith D, Kour G, Mark GE, Stabel S: Activation of the c-Raf protein kinase by PKC phosphorylation. Oncogene. 7: 2259–2262, 1992
20. Dosemeci A, Dhallen RS, Cohen NM, Lederer WJ, Rogers TB. Phorbol ester increases calcium current and simulates the effects of angiotensin II on cultured neonatal rat heart myocytes. Circ Res 62: 347–357, 1988
21. Cooper DMF, Brooker G: Ca^{2+}-inhibited adenylyl cyclase in cardiac tissue. Trends in Pharm Sci 14: 34–37, 1993
22. Venema RC, Kuo JE: Protein kinase C mediated phosphorylation of troponin I and C-protein in isolated myocardial cells is associated with inhibition of myofibrillar actomyosin MgATPase. J Biol Chem 268: 2705–2711, 1993
23. Baxter GF, Goma FM, Yellon DM: Involvement of protein kinase C in the delayed cytoprotection following sub-lethal ischaemia in rabbit myocardium. Brit J Pharmacol 115: 222–224, 1995

Molecular and Cellular Biochemistry **160/161**: 231–239, 1996.

HSP25 in isolated perfused rat hearts: Localization and response to hyperthermia

Brigitte Hoch, Gudrun Lutsch, Wolfgang-Peter Schlegel, Joachim Stahl, Gerd Wallukat, Sabine Bartel, Ernst-Georg Krause, Rainer Benndorf and Peter Karczewski

Max Delbrück Centre for Molecular Medicine, Berlin-Buch, Germany

Abstract

Recent investigations concentrate on the correlation between the myocardial expression of the inducible 70-kDa heat shock protein (HSP70i) by different stress conditions and its possible protective effects. Only few studies have focused on the involvement of small heat shock proteins in this process. We analyzed the location of the small heat shock protein HSP25 in isolated cardiomyocytes as well as its location and induction in isolated perfused hearts of rats. By immunofluorescence microscopy HSP25 was found to colocalize with actin in the I-band of myofibrils in cardiomyocytes of isolated perfused hearts as well as in isolated neonatal and adult cardiomyocytes. Hyperthermic perfusion of isolated hearts for 45 min resulted in modulation of different parameters of heart function and in induction of HSP25 and HSP70i. Temperatures higher than 43°C (44–46°C) were lethal with respect to the contractile function of the hearts. Compared to control hearts perfused at 37°C, significant increases during hyperthermic perfusion at 42°C and 43°C were obtained for heart rate, contraction velocity and relaxation velocity. In response to hyperthermia at 43°C and after subsequent normothermic perfusion for 135 min at 37°C, left ventricular pressure, contraction velocity and relaxation velocity remained significantly elevated. However, heart rate returned to control values immediately after the period of heat treatment. HSP25 is constitutively expressed even in normothermic perfused hearts as shown by Western blotting. Hyperthermia increased the content of HSP25 only in the left ventricular tissue. In contrast, HSP70i was strongly induced in all analyzed parts of the myocardium (left ventricle, right ventricle, septum). Our findings suggest a differential regulation of HSP25 and HSP70i expression in response to hyperthermia in isolated perfused hearts. The constitutively expressed HSP25 seems to be located adjacent to the myofibrils which implies a specific role of this protein even under unstressed conditions for the contractile function of the myocardium. (Mol Cell Biochem **160/161**: 231–239, 1996)

Key words: stress protein induction, HSP25, intracellular location, isolated perfused heart, hyperthermia, contractile function

Introduction

Induction of heat shock proteins (HSPs) has implications for the self-preservation of cells and organisms not only against hyperthermia but also against other stresses like hypoxia, ischemia and reperfusion induced injury [1]. HSPs were postulated to be protective to the cell through stabilizing macromolecular structures and their chaperoning functions (reviewed in [2]). Their induction preconditions the cell or organism to survive a subsequent stress which otherwise would be lethal [3, 4]. Protective effects of stress protein induction against hypoxia/ischemia have been found in various cellular systems [5–8], in whole body heat-stressed animals [9–13], and in heat-treated isolated perfused hearts [14]. In these systems, expression of the inducible 70kDa HSP (HSP70i) correlates with resistance of cellular functions against stress-induced damage or with improved myocardial recovery. Recently, transgenic animal models overexpressing the human HSP70i [15] or the rat HSP70i [16] in mice also show an increased resistance of the transgenic hearts to ischemic injury. Langendorff-perfused hearts from rats isolated from whole body heat-treated animals showed improved

Address for offprints: B. Hoch, Max Delbrück Centre for Molecular Medicine, Dept. Molecular Cardiology, Robert-Rössle-Str. 10, D-13122 Berlin-Buch, Germany recovery from ischemia [17] and cardiac arrest [18]. This kind of heat pretreatment increased both the level of HSP70i and catalase activity. Their decline with time was associated with a loss of protective effects [19]. In isolated perfused hearts from rabbit, a preceding heat stress reduced the size of a subsequently set infarction [20]. Hyperthermic retrograde perfusion of isolated rat hearts induced HSP70i at mRNA and protein level [21] .

Only little is known about the induction and role of the small heat shock protein HSP25 in the myocardium (reviewed in [22]). The small or low molecular weight HSPs (e.g. HSP25 in rat [23]) build one family of the HSPs (reviewed in [24, 25]). HSP25 is rapidly phosphorylated by different stress treatments [26–29], but the biological consequences of this phosphorylation are still a matter of discussion [30]. HSP25 is able to form supramolecular structures [31]. It has been reported to associate with the cytoskeleton and seems to play a role for the organization of actin [32–36]. Cardiac and skeletal muscles belong to those tissues with the highest basic levels of HSP25 [37]. After ischemia in isolated perfused rat heart [38] or after whole body heat shock [39] HSP25 mRNA is elevated few hours after stress treatment in the myocardium. Stunned porcine myocardium showed a significant increase of HSP27 transcripts [40]. However, no changes in HSP27 protein content were observed in this study.

To gain insights into possible interactions between HSP25 and the contractile apparatus in rat heart, we analyzed the location of this protein in isolated neonatal and adult cardiomyocytes as well as in isolated perfused hearts with immunofluorescence microscopy. Furthermore, in perfused hearts changes in the amount of HSP25 compared to the level of HSP70i and alterations of contractile parameters in response to hyperthermia were studied.

Material and methods

Materials

Substances for isolation of rat heart cells and for the perfusion of hearts were obtained from Sigma (Germany), Gibco (U.S.A.) and Biochrom (Germany). Protein molecular weight marker (Rainbow cold marker) was obtained from Amersham (Germany).

Isolation and cultivation of cardiomyocytes

Animal experiments were performed in accordance with the Declaration of Helsinki and internationally accepted principles concerning the care and use of laboratory animals. For isolation of hearts from adults, male albino Wistar rats (250–300 g) kept on standard pellet diet with free access to water were used in all experiments. Cardiomyocytes from adult rats were isolated from ventricles according to [41]. The cells were cultured serum-free on laminin-coated (1 μg/cm^2) cover slides for 2 days at 37°C in a modified medium 199 with Earle's salts, 10^{-7} M insulin, 0.2% BSA, 100 IU/ml penicillin, 100 μg/ml streptomycin, and 10 μM cytosine-β-D-arabinofuranoside. Rat neonatal cardiomyocytes were prepared from ventricles of 1–2 day old Wistar rats by a modified method according to [42]. Cells were cultured on cover slides for 4 days at 37°C in SM20-1 medium supplemented with 10% heat-inactivated calf serum and 2 μM fluorodeoxyuridine [43].

Heat treatment of isolated perfused rat hearts

Hearts from anesthetized (30 mg/kg sodium pentobarbitural, intraperitoneally) and heparinized (500 U/kg body weight) adult rats were excised and cannulated for retrograde aortic perfusion with a modified Krebs-Henseleit solution containing 118 mM NaCl, 4.7 mM KCl, 1.5 mM $CaCl_2$, 1.2 mM $MgSO_4$, 25 mM $NaHCO_3$, 0.05 mM Na_2EDTA, 0.23 mM KH_2PO_4 and 11.1 mM glucose saturated with 95% O_2/5% CO_2 (pH 7.4). Perfusion apparatus and corresponding software were purchased from Fa. Hugo Sachs Electronic (Germany). Measurement of contractile parameters were carried out as described [44]. The left ventricular end-diastolic pressure was set from 12–15 mm Hg. The recorded functional parameters were heart rate, left ventricular pressure (LVP), contraction velocity (dp/dt; measured as left ventricular force development), and relaxation velocity (-dp/dt; measured as decreasing rate of left ventricular force). After a stabilizing period (30 min at 37°C) hearts were heat-treated for 45 min at various temperatures and recovered for 135 min at 37°C. Control hearts were perfused for 210 min at 37°C (Fig. 1). Mean values of contractile parameters obtained during hyperthermic perfusion and at the last 15 min of the recovery period were calculated relative to the respective mean values recorded for 15 min before heat treatment. For control hearts analogous time segments were used for calculation (Fig. 1). Rapid hyperthermic perfusion was achieved by retrograde perfusion with preheated perfusion solution and by parallel increase of the temperature of the water bath controlling the temperature of the water jacket of the Hugo Sachs apparatus. To switch back to normothermic perfusion after the heat treatment, precooled perfusion solution and rapid decrease of the temperature in the water jacket by exchange were used. After perfusion for 210 min, hearts were put into ice-cold perfusion solution. Left ventricle, right ventricle and septum were quickly separated and immediately frozen in liquid nitrogen.

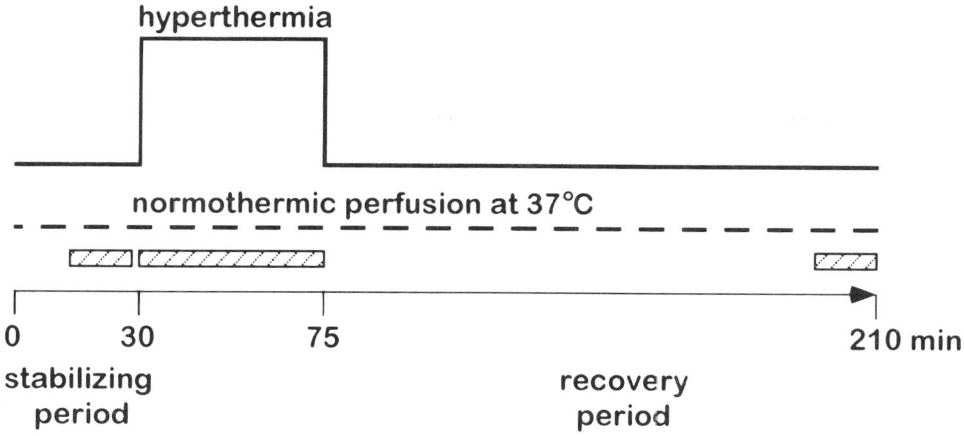

Fig. 1. Protocol for hyperthermic and normothermic (control) perfusion of isolated hearts. After a stabilizing period of 30 min at 37°C, hearts were hyperthermic perfused for 45 min at various temperatures followed by subsequent normothermic perfusion at 37°C for 135 min to recover the hearts (solid line). Control hearts were perfused at 37°C throughout the experimental period (210 min) (dashed line). Hatched bars indicate the time segments of the 15 min period before hyperthermia, the 45 min of hyperthermic perfusion and the 15 min period at the end of the recovery period used for calculation of mean changes in contractile function during hyperthermia and at the end of the experimental period compared to control hearts.

Immunofluorescence microscopy of isolated cardiomyocytes and isolated perfused hearts

Isolated cardiomyocytes were fixed and permeabilized according to a protocol developed for optimal preservation of the cytoskeleton [45] . Cells were treated for 15 min with 4% formaldehyde (freshly prepared from paraformaldehyde) in 0.1 M PIPES, pH 6.8, containing 0.2% Triton X100, 5 mM EGTA and 2 mM $MgCl_2$. Thereafter, 5 min postfixation with ice-cold ethanol was carried out and cells were preincubated for 30 min in 20 mM Tris/HCl, pH 8.0, containing 1% BSA and 0.2% Tween 20 to suppress unspecific antibody binding. As primary antibodies affinity-purified polyclonal anti-HSP25 antibody [46] and the monoclonal anti-actin antibody HHF35 (Ortho Diagnostic Systems, Germany) were used. Primary antibodies were visualized with fluorescein-(DTAF)- and Cy3-labeled species-specific secondary antibodies (Dianova, Germany).

For immunofluorescence microscopy of isolated perfused hearts from adult rats, perfusion was carried out as described above. Hearts were perfused for 210 min at 37°C. During the last 15 min fixation solution (4% formaldehyde in 0.1 M phosphate buffer, pH 7.4) was used instead of Krebs-Henseleit solution. Thereafter, left ventricular free wall was cut into small pieces and fixation was continued for 1 h at room temperature by immersion in the same fixative. After infiltration with 2.3 M sucrose, tissue blocks were frozen in liquid nitrogen and stored until use for cryosectioning. Cryosections of 1 μm thickness were prepared according to the method developed by Tokuyasu [47] with a Leica ultramicrotome Ultracut E equipped with a cryochamber FC4E (Leica, Germany). Double-labeling of cryosections was performed with the same antibodies as described above. Evaluation of immunofluorescence labeling was done with an Axioplan microscope equipped with epifluorescence optics and appropriate filter systems (Zeiss, Germany). Micrographs were taken with Kodak T Max 400 film using an automatic MC100 camera (Zeiss).

Immunoblotting

Total protein of heart tissue was isolated by homogenization in the presence of 50 mM Tris/HCl pH 6.8, 10% glycerol and 6% SDS in a glass potter. After centrifugation (5 min, 14000 g) protein content of the supernatant was determined with the DC Assay kit (BioRad, U.S.A.) according to the manufacturers instructions. After electrophoretic separation on 7.5–18% SDS polyacrylamide gradient gels, proteins were electrotransferred onto nitrocellulose membranes according to [48]. Rainbow cold marker and recombinant HSP25 protein were used as molecular weight markers (not shown). Equal loading of the gels was verified by Ponceau staining (not shown). As primary antibodies a polyclonal anti-HSP25-antibody [46] and a monoclonal anti-HSP70i-antibody (SPA 810, Biomol, Germany) were used. As secondary antibodies horse radish peroxidase-coupled anti-rabbit- and anti-mouse-IgG-antibodies were used (Sigma, Germany). Immunodetection was carried out by a chemoluminescence reaction with the SuperSignal CLHRP Substrate System according to the manufacturers instructions (Pierce, U.S.A.).

Statistical evaluation

Results are given as means ± standard deviations. Mean values were compared using the unpaired *t*-test assuming $p \leq 0.05$ as statistically significant.

Results

Localization of HSP25 in isolated cardiomyocytes and isolated perfused hearts

Cell cultures of neonatal and adult cardiomyocytes from rats were analyzed by immunofluorescence microscopy with respect to the location of HSP25. With antibodies against HSP25 in spread parts of neonatal cardiomyocytes occasionally a cross-striated staining pattern was observed (arrowheads in Fig. 2A) reminiscent to staining of sarcomeres in myofibrils. In comparison, in isolated adult cardiomyocytes a regular staining pattern throughout the whole cell was observed with striations oriented perpendicular to the longitudinal axis of the rod shaped cells (Fig. 2B). By double-labeling with actin antibodies similar staining patterns were observed (not shown) implicating colocalization of HSP25 and actin in myofibrils.

To compare these results with the situation *in situ*, semithin cryosections of isolated perfused hearts after 210 min of normothermic perfusion were double-labeled with HSP25 and actin antibodies and analyzed by fluorescence microscopy. A labeled left ventricular cardiomyocyte is shown in Figs. 2C and D. Comparison of the HSP25 staining pattern (Fig. 2C) with that of actin (Fig. 2D) reveals striking similarity. Because stained regions of actin represent the I-band of sarcomeres and the stained regions are split into two parts by the unlabeled Z-line, it is concluded that HSP25 is located in the I-band of myofibrils (see arrowheads in Figs. 2C and D). In addition, HSP25 staining is found in some cases at the level of the M-line (Fig. 2C).

Functional response of isolated perfused rat hearts to hyperthermia

Mean values for contractile parameters obtained during the last 15 min of the stabilizing period of normothermic perfused hearts (n = 13) were: 255.4 ± 20.5 beats/min for the heart rate, 36.5 ± 11.4 mm Hg for LVP, 819.8 ± 228.4 mm Hg/s for dp/dt and –586.5 ± 148.4 mm Hg/s for -dp/dt. In preliminary experiments, we subjected hearts to hyperthermia ranging from 38–46°C. Temperatures above 43°C led to cessation of contraction either during heat treatment or during the recovery period. Temperatures below 42°C resulted in lower levels of the investigated HSPs (not shown). Therefore heat treatment was performed at 42°C and 43°C.

Heat treatment for 45 min at 42°C and at 43°C altered the contractile parameters of the perfused hearts in different ways. During the hyperthermic period (open bars in Fig. 3), statistical evaluation revealed significant increases of heat-treated hearts (n = 3 for each temperature) compared to control hearts (n = 3) for the heart rate at 42°C (135% ± 6% versus 97.1% ± 6.6%) and at 43°C (137% ± 6% versus 97.1% ± 6.6%) (Fig. 3A), for dp/dt at 42°C (116% ± 9% versus 92.6% ± 10.4%) and at 43°C (126% ± 9% versus 92.6% ± 10.4%) (Fig. 3C) and for -dp/dt at 42°C (142% ± 9% versus 91.5% ± 10.9%) and at 43°C (164% ± 8% versus 91.5% ± 10.9%) (Fig. 3D). No significant difference compared to the controls was detected for the relative mean LVP during hyperthermic perfusion (Fig. 3B).

The analysis of the contractile function of perfused hearts during the last 15 min of the recovery period (hatched bars in Fig. 3) revealed a different pattern. Contractile parameters of hearts heat-treated at 42°C decreased to mean values of control hearts during the recovery period. This suggests a complete reversibility of the contractile function to that of normothermic perfused hearts (Figs. 3A, 3B, 3C and 3D). In contrast, after hyperthermia at 43°C compared to control hearts mean values remained elevated for LVP by 62.2% (Fig. 3B), for dp/dt by 54.5% (Fig. 3C) and for -dp/dt by 55 5% (Fig. 3D) at the end of the recovery period. The heart rate, however, returned rapidly to control values shortly after the hyperthermic perfusion (hatched bars in Fig. 3A).

Stress protein induction in isolated perfused rat hearts in response to hyperthermia

In control hearts, as well as in hyperthermic perfused hearts, a constitutive expression of HSP25 was observed by Western blotting technique (Fig. 4). HSP25 level in normothermic perfused left ventricular tissue was 1.9 fold and 1.3 fold that of right ventricular tissue and septum, respectively (Table 1). After hyperthermia, HSP25 content was found to be elevated significantly ($p \leq 0.05$) in the left ventricle (Table 1, see also Fig. 4), whereas no significant induction could be observed in right ventricular tissue and in the septum.

For the HSP70i level a completely different result was obtained. A high amount of HSP70i was detected after heat treatment in all analyzed tissues, with highest induction values in left ventricular tissue and at 43°C (Table 1, see also Fig. 4). The level of HSP70i was near or below the detection limit in control hearts perfused at 37°C in all analyzed tissues (Table 1, see also Fig. 4).

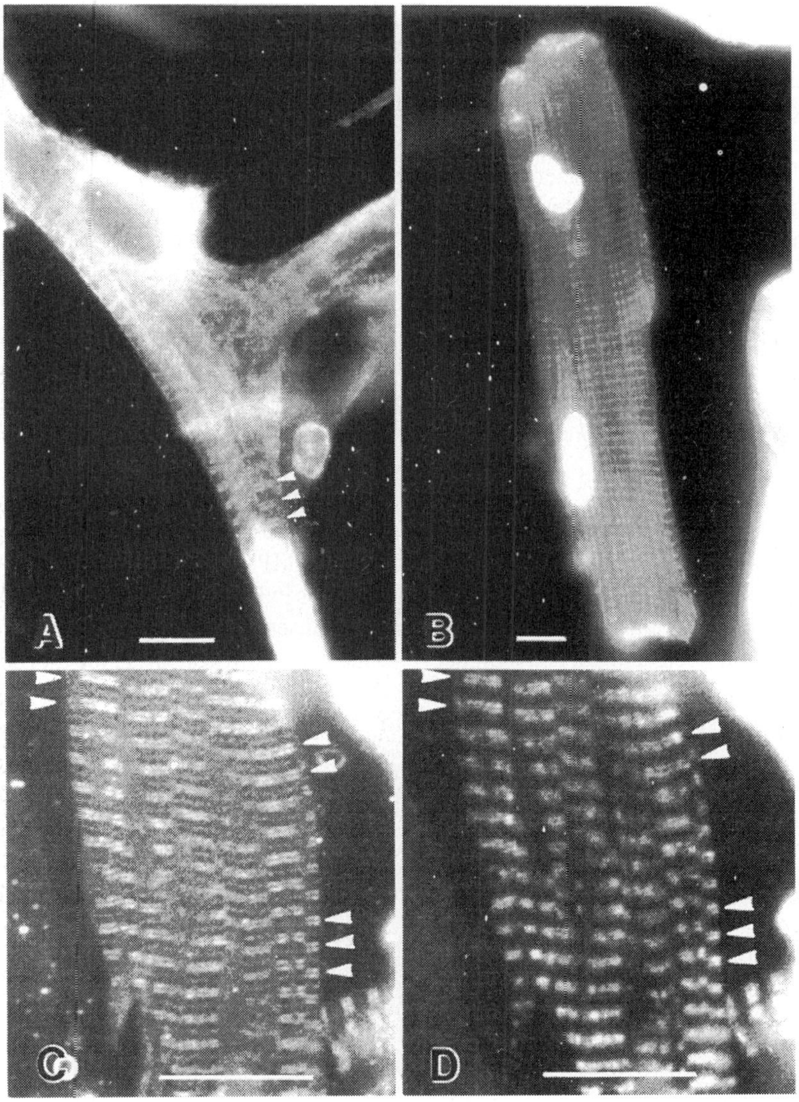

Fig. 2. Immunofluorescence micrographs of isolated cardiomyocytes (A, B) and isolated perfused hearts (C, D). A: Isolated neonatal cardiomyocyte after incubation with HSP25 antibody and DTAF-labeled secondary antibody. Staining of HSP25 in flattened areas of the cell in myofibrillar structures. Arrowheads point to doublets of HSP25 staining. Bar 10 µm. B: Isolated adult rat cardiomyocyte after incubation with HSP25 antibody and DTAF-labeled secondary antibody. HSP25 is located in sarcomeres of myofibrils. Nuclei are stained by DAPI (4′, 6-diamidino-2-phenylindole). Bar 10 µm. C, D: Semithin cryosection of left ventricular cardiomyocyte from isolated perfused rat heart double-labeled with HSP25 (C) and actin antibody (D), visualized by DTAF-(C) and Cy3-labeled secondary antibody (D), respectively. HSP25 and actin staining colocalize in the I-band of myofibrils. Arrowheads point to doublets of HSP25 and doublets of actin staining. Bar 10 µm.

Discussion

Localization of HSP25 in isolated cardiomyocytes and isolated perfused hearts

Isolated neonatal cardiomyocytes differ morphologically from adult cardiomyocytes by their irregular shape and the presence of irregularly arranged myofibrils. In neonatal cardiomyocytes, HSP25 is found only occasionally in myofibrils, whereas in isolated adult cardiomyocytes it occurs mainly in this compartment. There it colocalizes with actin in the I-band of sarcomeres, where it was also found to be located in tissue sections from isolated perfused hearts. This location suggests an involvement of HSP25 in organization of thin myofilaments and myofibrils which gain their final alignment in adult cardiomyocytes. Interaction of HSP25 with actin has been reported previously in several cellular systems. In smooth muscle cells, HSP25 seems to be involved in sustained muscle contraction [49]. HSP25 overexpression in rodent cells results in stabilization of microfilaments under stress conditions [34, 36], and *in vitro* HSP25 inhibits the polymerization of actin [32, 33, 35]. Taken together, these findings support the assumption that this small heat shock protein may be involved in the regulation of actin filament organization.

The distinct location of HSP25 as described here is in accordance with data obtained recently with embryonal, neonatal and adult rat heart tissue [50]. Furthermore, these results show that isolation and cultivation of cardiomyocytes, as well as perfusion of the isolated heart do not substantially influence the location of HSP25.

Functional response of isolated perfused rat hearts to hyperthermia

Hyperthermic perfusion of isolated rat hearts for 45 min resulted in complex changes of functional parameters. During heat treatment, significant increases were obtained for heart rate, dp/dt and -dp/dt compared to control hearts. The response in -dp/dt was more pronounced and reached a higher level of significance. In contrast, LVP did not change significantly. In a similar experimental setup an elevated contractile force was reported only at a distinct time point at the end of brief hyperthermic perfusion [21]. Increase and decrease in heart rate correlated closely with changes of temperature. This immediate and sensitive reaction appears to be characteristic for hyperthermic perfused isolated hearts [21]

After hyperthermia at 42°C, all analyzed functional parameters returned to control values at the end of the recovery period indicating that the applied mode of heat treatment did not cause a persistent modulation of the contractile function. In contrast, after hyperthermia at 43°C only heart rate returned rapidly to control values. LVP, dp/dt and -dp/dt remained significantly elevated when compared to control hearts at the end of the recovery period. It is known that tachycardia induces the liberation of endogenous catecholamines in the myocardium [51]. In isolated perfused rabbit heart noradrenaline release was found to be elevated after a brief period of no-flow ischemia [52]. Therefore, the activation of the cAMP system by liberation of endogenous catecholamines [53] may be involved in the maintenance of ionotropic responses observed after hyperthermia at 43°C (Fig. 3) and may also partly underlay the transient rise in contractile function at 42°C.

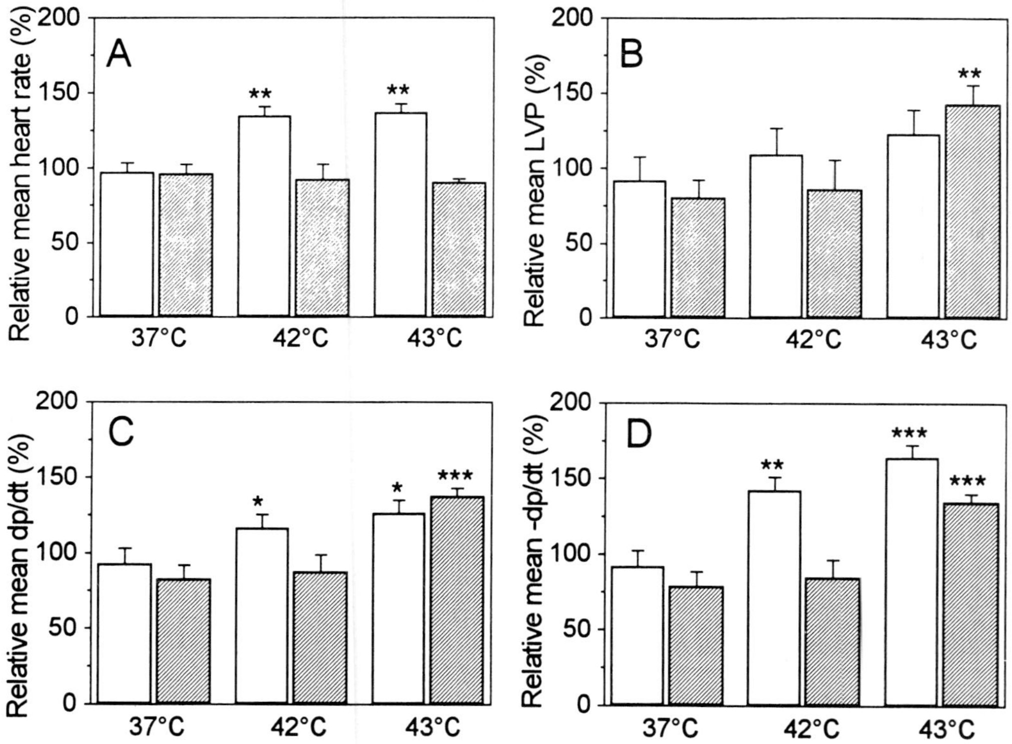

Fig. 3. Comparison of functional parameters of heat-treated isolated perfused hearts. Hearts perfused at 37°C for 30 min were hyperthermic or normothermic (control) perfused for 45 min (open bars) at the indicated temperatures followed by a recovery period of 135 min at 37°C. Heart rate (A), left ventricular pressure (LVP) (B), contraction velocity (dp/dt) (C) and relaxation velocity (-dp/dt) (D) were compared during hyperthermic perfusion and at the end of the recovery period. Mean values of the parameters of contractile function obtained during hyperthermic perfusion and at the last 15 min of the recovery period (hatched bars) were calculated relative to the respective mean values recorded for 15 min before hyperthermia. Mean values were compared using the unpaired *t*-test and significant differences versus control hearts are indicated as follows: *= $p \leq 0.05$; **= $p \leq 0.01$; ***= $p \leq 0.001$.

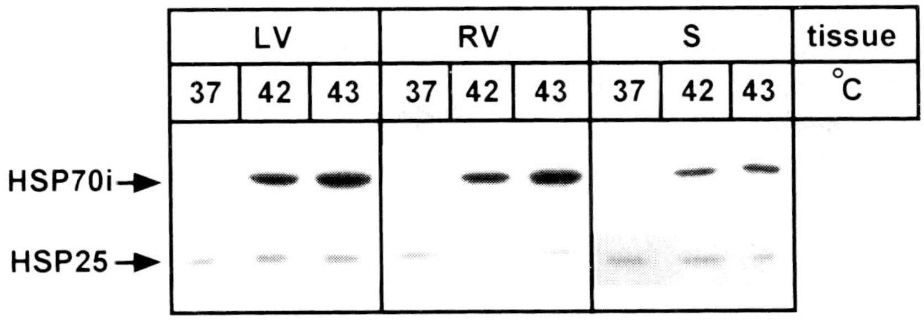

Fig. 4. Detection of HSP25 and HSP70i in normothermic and hyperthermic perfused hearts. At the end of the experimental period total protein of the left ventricle (LV), right ventricle (RV) and septum (S) was isolated, separated on SDS polyacrylamide gradient gels (50 μg total protein per lane) and electrotransferred to nitrocellulose. For detection of HSP25 and of HSP70i an affinity-purified polyclonal HSP25 antibody and a monoclonal anti-HSP70i antibody were used as described in Material and methods. Results of one representative experiment from three experiments are shown.

Table 1. Determination of HSP25 and HSP70i induction in left ventricle, right ventricle and septum of hyperthermic perfused hearts

Heart areal	Condition	HSP25 relative optical density	HSP70i relative optical density
	37°C	0.21 ± 0.03	–
LV	42°C	0.30 ± 0.06	2.39 ± 0.62*
	43°C	0.29 ± 0.02*	3.61 ± 0.59**
	37°C	0.11 ± 0.06	(–)
RV	42°C	0.14 ± 0.09	2.10 ± 0.87
	43°C	0.17 ± 0.06	2.58 ± 0.61*
	37°C	0.17 ± 0.04	–
S	42°C	0.16 ± 0.02	1.48 ± 0.17**
	43°C	0.15 ± 0.06	2.17 ± 0.47*

Temperatures during 45 min of hyperthermic or normothermic (control) perfusion and tissues used for protein isolation are indicated (LV = left ventricle; RV = right ventricle; S = septum). Autoradiograms of immunoblots after chemoluminescence detection (see Fig. 4) were used for densitometric determination of detected protein. Values are means ± standard deviations of relative optical densities of three identically treated hearts (n = 3) and significant differences versus control hearts are indicated as follows: *= $p \leq 0.05$; **= $p \leq 0.01$. A dash indicates that values were below the detection limit. (–) indicates, that one out of three controls showed a weak signal (0.215).

Stress protein induction in isolated perfused rat hearts by hyperthermia

Here we report for the first time data on the increased expression of HSP25 in stressed intact hearts. Both, normothermic and hyperthermic perfused hearts expressed considerable amounts of HSP25. This is in line with earlier investigations according to which HSP25 is constitutively expressed at high level in the myocardium [37]. A significant induction of HSP25 was observed in the left ventricle in response to hyperthermia, whereas in the right ventricle and in the septum there was no change. In contrast, HSP70i showed a strikingly different pattern of expression. It was found to be strongly induced by heat treatment in all heart regions analyzed. Similarly as observed for HSP25, the expression was most pronounced in the left ventricle (Table 1). Interestingly, in unstressed control hearts HSP70i levels were extremely low and in most experiments below the detection limit (Table 1). This is in contrast to McCully *et al.* [21] who reported considerable amounts of HSP70i in normothermic perfused hearts. This may explain in part why these authors observed only a 2.5 fold increase in HSP70i in response to hyperthermia. The differences to our data may also result from the applied heat treatment (15 min at 42°C) which was less severe than that applied by us (45 min at 42°C or 43°C). Furthermore, the negligible expression of HSP70i in controls shows that the mechanical handling of the excised hearts *per se* did not induce expression of heat shock proteins in our experiments.

The expression pattern of HSP25 compared to that of the well characterized, heat-inducible HSP70i indicates a distinct, differential regulation. HSP25 was shown to be located adjacent to myofibrils (this study). But we do not know yet, whether it is loosely associated or a tightly bound component of cardiac myofibrils.

In summary, high constitutive expression of HSP25 suggests that heat stress regulates HSP25 function in the heart not only by induction, but also by other mechanisms, such as phosphorylation/dephosphorylation [27, 29], changes in supramolecular organization [30], and location (reviewed in [24]). Further efforts should be undertaken to elucidate the relevance of these specific properties of HSP25 in the stressed and unstressed myocardium.

Acknowledgements

We thank D. Riege for technical assistance in protein preparation and M. Wegener and H. Schmidt for isolation and cultivation of neonatal rat cells. Recombinant HSP25 protein marker was a kindly gift of Dr. M. Gaestel, MDC, Berlin, Germany. This work was supported by the Deutsche Forschungsgemeinschaft to R.B. and P.K (Bel464/2-1) and to G.L. (Lu499/3-1). Part of this work was supported by a grant of the Deutsche Forschungsgemeinschaft SFB 273, YE5 to R.B.

References

1. Yellon DM, Latchman DS: Stress proteins and myocardial protection. J Mol Cell Cardiol 24: 113–124, 1992
2. Morimoto RI, Tissières A, Georgopoulos C: The Biology of Heat Shock Proteins and Molecular Chaperones. Cold Spring Harbor Laboratory Press, New York, 1994
3. Hahn GM, Li GC: Thermotolerance, thermoresistance and thermosensitization. In: RI Morimoto, A Tissières, C Georgopoulos (eds). Stress Proteins in Biology and Medicine. Cold Spring Harbor Laboratory Press, New York, 1990, pp 79–100
4. Polla BS, Mili N, Kantengwa S: Heat shock and oxidative injury in human cells. In: B Maresca, S Lindquist (eds). Heat Shock. Berlin, Springer-Verlag, 1991, pp 279–290
5. Iwaki K, Chi S-H, Dillmann WH, Mestril R: Induction of HSP70 in cultured rat neonatal cardiomyocytes by hypoxia and metabolic stress. Circulation 87: 2023–2032, 1993
6. Mestril R, Chi S-H, Sayen MR, O'Reilly K, Dillmann WH: Expression of inducible stress protein 70 in rat heart myogenic cells confers protection against simulated ischemia-induced injury. J Clin Invest 93: 759–767, 1994
7. Heads RJ, Latchman DS, Yellon DM: Stable high level expression of a transfected human HSP70 gene protects a heart-derived muscle cell line against thermal stress. J Mol Cell Cardiol 26: 695–699, 1994
8. Heads RJ, Yellon DM, Latchman DS: Differential cytoprotection against heat stress or hypoxia following expression of specific stress protein genes in myogenic cells. J Mol Cell Cardiol 27: 1669–1678, 1995
9. Currie RW, Karmazyn M, Kloc M, Mailer K: Heat shock response is associated with enhanced postischemic ventricular recovery. Circ Res 63: 543–549, 1988
10. Donnelly TJ, Sievers RE, Vissern FLJ, Welch WJ, Wolfe CL: Heat shock protein induction in rat hearts. Circulation 85: 769–778, 1992
11. Currie RW, Tanguay RM, Kingma Jr JG: Heat shock response and limitation of tissue necrosis during occlusion/reperfusion in rabbit hearts. Circulation 87: 963–971, 1993
12. Marber MS, Walker JM, Latchman DS, Yellon DM: Myocardial protection after whole body heat stress in the rabbit is dependent on metabolic substrate and is related to the amount of the inducible 70-kD heat stress protein. J Clin Invest 93: 1087–1094, 1994
13. Hutter MM, Sievers RE, Barbosa V, Wolfe CL: Heat-shock protein induction in rat hearts. Circulation 89: 355–360, 1994
14. Currie RW: Effects of ischemia and perfusion temperature on the synthesis of stress-induced (heat shock) proteins in isolated and perfused rat hearts. J Mol Cell Cardiol 19: 795–808, 1987
15. Plumier J-CL, Ross BM, Currie RW, Angelidis CE, Kazlaris H, Kollias G, Pagoulatos GN: Transgenic mice expressing the human heat shock protein 70 have improved post-ischemic myocardial recovery. J Clin Invest 95: 1854–1860, 1995
16. Marber MS, Mestril R, Chi S-H, Sayen MR, Yellon DM, Dillmann WH: Overexpression of the rat inducible 70-kD heat stress protein in a transgenic mouse increases the resistance of the heart to ischemic injury. J Clin Invest 95: 1446–1456, 1995
17. Currie RW, Ross BM, Davis TA: Induction of the heat shock response in rats modulates heart rate, creatine kinase and protein synthesis after a subsequent hyperthermic treatment. Cardiovasc Res 24: 87–93, 1990
18. Amrani M, Allen NJ, O'Shea J, Corbett J, Dunn MJ, Tadjkarimi S, Theodoropoulos S, Pepper J, Yacoub MH: Role of catalase and heat shock protein on recovery of endothelial and mechanical function. Cardioscience 4: 193–198, 1993
19. Karmazyn M, Mailer K, Currie RW: Acquisition and decay of heat-shock-enhanced postischemic ventricular recovery. Am J Physiol 259: H424–H431, 1990
20. Walker DM, Pasini E, Kucukoglu S, Marber MS, Iliodromitis E, Ferrari R, Yellon DM: Heat stress limits infarct size in the isolated perfused rabbit heart. Cardiovasc Res 27: 962–967, 1993
21. McCully JD, Myrmel T, Lotz MM, Krukenkamp IB, Levitsky S: The rapid expression of myocardial HSP70 mRNA and the heat shock 70 kDa protein can be achieved after only a brief period of retrograde hyperthermic perfusion. J Mol Cell Cardiol 27: 873–882, 1995
22. Das DK, Maulik N, Moraru II: Gene expression in acute myocardial stress. Induction by hypoxia, ischemia, reperfusion, hyperthermia and oxidative stress. J Mol Cell Cardiol 27: 181–193, 1995
23. Uoshima K, Handelman B, Cooper LF: Isolation and characterization of a rat HSP27 gene. Biochem Biophys Res Commun 197: 1388–1395, 1993
24. Knowlton AA: The role of heat shock proteins in the heart. J Mol Cell Cardiol 27: 121–131, 1995
25. Mestril R, Dillmann WH: Heat shock proteins and protection against myocardial ischemia. J Mol Cell Cardiol 27: 45–52, 1995
26. Welch WJ: Phorbol ester, calcium ionophore, or serum added to quiescent rat fibroblast cells all result in the elevated phosphorylation of two 28000 dalton mammalian stress proteins. J Biol Chem 260: 3058–3062, 1985
27. Arrigo A-P: Tumor necrosis factor induces the rapid phosphorylation of the mammalian heat shock protein hsp28. Mol Cell Biol 10: 1276–1280, 1990
28. Gaestel M, Schröder W, Benndorf R, Lippmann C, Buchner K, Hucho F, Erdmann VA, Bielka H: Identification of the phosphorylation sites of the murine small heat shock protein hsp25. J Biol Chem 266: 14721–14724, 1991
29. Saklatvala J, Kaur P, Guesdon F: Phosphorylation of the small heat-shock protein is regulated by interleukin 1, tumour necrosis factor, growth factors, bradykinin and ATP. Biochem J 277: 635–642, 1991
30. Mehlen P, Kretz-Remy C, Briolay J, Fostan P, Mirault M-E, Arrigo A-P: Intracellular reactive oxygen species as apparent modulators of heat-shock protein 27 (hsp27) structural organization and phosphorylation in basal and tumour necrosis factor α-treated T47D human carcinoma cells. Biochem J 312: 367–375, 1995
31. Behlke J, Dube P, van Heel M, Wieske M, Hayeß K, Benndorf R, Lutsch G: Supramolecular structure of the small heat shock protein HSP25. Prog Colloid Polym Sci 99: 87–93, 1995
32. Miron T, Vancompernolle K, Vandekerckhove J, Wilchek M, Geiger B: A 25-kD inhibitor of actin polymerization is a low molecular mass heat shock protein. J Cell Biol 114: 255–261, 1991
33. Rahman DRJ, Bentley N, Tuite MF: The *Saccharomyces cerevisiae* small heat shock protein HSP26 inhibits actin polymerization. Biochem Society Transactions 23: 77S, 1995
34. Lavoie JN, Hickey E, Weber LA, Landry J: Modulation of actin microfilament dynamics and fluid phase pinocytosis by phosphorylation of heat-shock protein 27. J Biol Chem 268: 24210–24214, 1993
35. Benndorf R, Hayeß K, Ryazantsev S, Wieske M, Behlke J, Lutsch G: Phosphorylation and supramolecular organization of murine small heat shock protein HSP25 abolish its actin polymerization-inhibiting activity. J Biol Chem 269: 20780–20784, 1994
36. Lavoie JN, Lambert H, Hickey E, Weber LA, Landry J: Modulation of cellular thermoresistance and actin filament stability accompanies phosphorylation-induced changes in the oligomeric structure of heat shock protein 27. Mol Cell Biol 15: 505–516, 1995
37. Kato K, Shinohara H, Goto S, Inaguma Y, Morishita R, Asano T:

Copurification of small heat shock protein with αB-crystallin from human skeletal muscle. J Biol Chem 267: 7718–7725, 1992

38. Das DK, Engelman RM, Kimura Y: Molecular adaptation of cellular defences following preconditioning of the heart by repeated ischemia. Cardiovasc Res 27: 578–584, 1993
39. Currie RW, Tanguay RM: Analysis of RNA for transcripts for catalase and SP 71 in rat hearts after *in vivo* hyperthermia. Biochem Cell Biol 69: 375–382, 1991
40. Andres J, Sharma HS, Knöll R, Stahl J, Sassen LMA, Verdouw PD, Schaper W: Expression of heat shock proteins in the normal and stunned porcine myocardium. Cardiovasc Res 27: 1421–1429, 1993
41. Volz A, Piper HM, Siegmund B, Schwartz P: Longevity of adult ventricular rat heart muscle cells in serum-free primary culture. J Mol Cell Cardiol 23: 161–173, 1991
42. Halle W, Wollenberger A: Differentiation and behaviour of isolated embryonic and neonatal heart cells in a chemically definite medium. Am J Cardiol 25: 292–299, 1970
43. Wallukat G, Morwinski R, Kuhn H: Modulation of the β adrenergic response of cardiomyocytes by specific lipoxygenase products involves their incorporation into phosphatidylinositol and activation of protein kinase C. J Biol Chem 269: 29055–29060, 1994
44. Schlegel W-P, Bartel S, Karczewski P, Krause E-G: Interaction of distinct signal transduction systems in the heart: Studies on contractility in isolated rat heart. In: Freiburg Focus on Biomeasurement Vol. 9: Pharmacological Evaluation of Cardioprotective Substances. Experimental Induction and Indicators of Myocardial Injury and Myocardial Protection. Publisher: Gesellschaft für Erfahrungstransfer in der Biomesstechnik e.V., Buchenbach, Germany; Biomesstechnik-Verlag, March, Germany, 1995, pp 71–77
45. Vielkind U, Swierenga SH: A simple fixation procedure for immunofluorescent detection of different cytoskeletal components within the same cell. Histochem 91: 81—88, 1989
46. Stahl J, Wobus AM, Ihrig S, Lutsch G, Bielka H: The small heat shock protein HSP25 is accumulated in P19 embryonal carcinoma cells and embryonic stem cells of line BLC6 during differentiation. Differentiation 51: 33–37, 1992
47. Tokuyasu KT: Application of cryoultramicrotomy to immunocytochemistry. J Microscopy 143: 139–149, 1986
48. Benndorf R, Nürnberg P, Bielka H: Growth phase-dependent proteins of the Ehrlich ascites tumor analyzed by one- and two-dimensional electrophoresis. Exp Cell Res 174: 130–138, 1988
49. Bitar KN, Kaminski MS, Hailat N, Cease KB, Strahler JR: HSP27 is a mediator of sustained smooth muscle contraction in response to bombesin. Biochem Biophys Res Commun 181: 1192–1200, 1991
50. Lutsch G, Vetter R, Stahl J, Benndorf R: Localization of HSP25 in the I-band of cardiac myofibrils of adult rat. J Mol Cell Cardiol 27: A111, 1995
51. Yamaguchi N, Champlain J, Nadeau R: Correlation between the response of the heart to sympathetic stimulation and the release of endogenous catecholamine into coronary sinus of the dog. Circ Res 36: 662–668, 1975
52. Wollenberger A, Krause E-G, Shahab L: Endogenous catecholamine mobilization and the shift to anaerobic energy production in the acutely ischemic myocardium. Int Symp Coronary Circulation and Energetics of the Myocardium, Karger, Basel/New York, 1966, pp 200–219
53. Krause E-G, Lehmann I, Mest H-J, Taube C, Förster W: Electrically induced tachyarrhythmia and the effect of propranolol on the release of cyclic AMP and prostaglandin E by the canine left ventricle. Adv Myocardiol 3: 185–191, 1982

Molecular and Cellular Biochemistry **160/161**: 241–247, 1996.

Molecular cloning, sequencing and expression analysis of a fatty acid transport gene in rat heart induced by ischemic preconditioning and oxidative stress

Nilanjana Maulik and Dipak K. Das

Cardiovascular Division, Department of Surgery, University of Connecticut School of Medicine, Farmington, CT 06030, USA

Abstract

In this study, ischemia and oxidative stress-inducible gene expression in heart was examined by subtractive hybridization technique. Total RNA was isolated from ventricular muscle fragments of normal and oxidative stress-induced hearts. Poly $(A)^+$ RNA was purified followed by the construction of a plasmid cDNA library. This was followed by the subtractive screening of oxidative stress-induced cDNA library. The positive colonies were amplified and the plasmid isolated. An aliquot was subjected to restriction cutting with Bam H1 and EcoRl; the fragments corresponding to cDNA insert were separated by electrophoresis, radiolabeled by random-primed DNA synthesis, and used as probes in standard Northern blotting experiments. An aliquot containing the plasmid from the confirmed positives was then subjected to bidirectional partial DNA sequencing (using M13 and T7/T3α primers) by the chain-extension/chain termination method. These sequences were subjected to a computerized search for homologies against all sequences in the updated worldwide Gen Bank and EMBL sequence databases followed by restriction mapping and reading frame identification. Out of 24 putative positive colonies screened, one clone was matched with > 97% homology with FAT gene that has been implicated in binding or transport of long chain fatty acids. cDNA probe synthesized from this clone identified two major transcripts of 4.8 and 2.9 kb. Additional experiments were then performed where isolated perfused rat hearts were subjected to the following treatments: (1) 5 min ischemia; (2) 10 min ischemia; (3) 20 min ischemia; (4) 5 min ischemia followed by 10 min reperfusion (ischemic preconditioning); and (5) 5 min ischemia followed by 10 min reperfusion, repeated four times (4 × preconditioning). RNAs were extracted from these hearts and hybridized with the FAT cDNA probe. The results indicated that FAT gene was induced by oxidative stress, ischemic preconditioning, but not by ischemia. (Mol Cell Biochem **160/161**: 241–247, 1996)

Key words: fatty acid transport protein, gene expression, subtractive hybridization, oxidative stress, ischemia/reperfusion, ischemic preconditioning

Introduction

Adaptation to ischemia is a novel concept for the myocardial preservation during ischemia and reperfusion which functions by stimulating endogenous protective mechanism rather than using the conventional approach of protection by therapeutic interventions using exogenous compounds. The concept was born from the findings that brief, repetitive coronary occlusions would lead to the expression of many stress-inducible genes including the mRNAs of heat shock proteins (HSP), and antioxidants that have been implicated in myocardial protection [1–5]. Induction of the expression of these stress-related genes has been viewed as the third line of defense for the heart [4]. However, it can be speculated that there are many other as yet unknown genes which may also comprise the member of the defense team. In order to exam-

Address for offprints: D.K. Das, Cardiovascular Division, Department of Surgery, University of Connecticut School of Medicine, Farmington, CT 06030-1110, USA ine the identity of the unknown genes, we used the subtractive hybridization technique. DNA sequencing identified one of the positive clones as fatty acid transport (FAT) gene which was first reported to be present in the rat adipose membrane [6]. Subsequently, FAT protein was identified in rat liver [7]. This protein has been implicated in binding or transport of long chain fatty acids [6, 7]. The results of our study demonstrated for the first time the induction of FAT by oxidative stress. Additionally, using the FAT cDNA probe we found several-fold induction of the mRNA of FAT protein in the ischemically preconditioned myocardium.

Materials and methods

Induction of oxidative stress

Male Sprague Dawley rats weighing about 300 gm were injected intraperitoneally with bacterial endotoxin (0.5 mg/kg) (Sigma Chemical Co., St. Louis, MO). The endotoxin was prepared by Sigma Chemical Company from *Salmonella typhimurium* by phenol extraction followed by gel filtration and irradiation. The lipopolysaccharides were composed of a lipid A moiety attached to a polysaccharide chain. The body temperature was monitored after the endotoxin treatment. Rats were sacrificed after 3, 4, and 6 h of endotoxin injection, hearts excised and immediately frozen in liquid nitrogen. Induction of the oxidative stress was confirmed by estimation of malonaldehyde formation, a presumptive marker of oxidative stress and free radical formation, in the heart as described previously [8].

Isolation of total RNA

Total RNA was isolated from experimental as well as control samples by a modified acid-guanidinium-thiocyanate-phenol-chloroform method [9] using Promega's total RNA isolation kit (Promega, Madison, WI). To construct the cDNA library from the experimental tissue, mRNA was purified out of the total RNA as Poly(A)$^+$ RNA using an oligo (dT)-biotin-streptavidin-paramagnetic particles system (Polytract, Promega, Madison ,WI).

Construction of cDNA Library

Six μg poly(A)$^+$ RNA from endotoxin treated hearts (pooled samples) were used from the construction of the cDNA library in a pUC-derived plasmid vector essentially by the method of Okayama and Berg [10] as modified by Gubler and Hoffman [11]. The competent cells were obtained from Gibco BRL (Super Script™ system for plasmid cloning) for transformation. First strand synthesis of cDNA was performed by using a special oligo (dT)-labeled end Not 1 primer. After RNA-primed synthesis of second strand cDNA, Sal 1 adapters were added, followed by digestion with Not 1 to create the compatible terminus at the poly (dT) end of the ds cDNA.

Six ng of large cDNA that was separated by size fractionation on Sepharyl HR-500 columns was used to oriented ligation into a Not l-Sal l-precut pSPORT 1 plasmid and 1/4th of the ligation reaction used to transform 100 μl competent MAX efficiency DH5α *E. Coli* cells.

Transformation efficiency was determined, and the cells were plated at a density of approximately 0.5–1 × 10^3 CFU/85 mm plate for 10–14 h.

Subtractive hybridization

Subtracted cDNA probes were prepared by a modification of the technique described by Rhyner *et al.* [12]. In brief, 6 μg of poly (A)$^+$ RNA extracted from endotoxin-treated hearts were subjected to reverse transcription for 1 h at 42°C using 1,000 units of a RNAse H-M-MLV reverse transcriptase (Superscript™II, Gibco BRL, Grand Island), 1 mg random 9-mer primers (Stratagene, La Jolla, CA), dNTP mixture (0.5 mM concentrated with 1 μCi[α-^{32}P] dCTP tracer), 100 units placental RNAse inhibitor (Promega, Madison, WI) and appropriate buffer in a 125 μl reaction mixture. After removing the RNA strand by alkaline hydrolysis, the cDNA was purified by phenol-chloroform extraction, desalting, and ethanol precipitation. The cDNA was subjected to two rounds of hybridization (to a C_0.t value > 1,000) at 68°C in 30 μl of 2 M phosphate buffer, pH 6.8 with a tenfold excess (w/w) of poly(A)$^+$ RNA from control tissue, and the non-hybridized cDNA separated after each round of hybridization by hydroxyapatite chromatography (BioGel HTP, BioRad, Hercules, CA) on thermojacketed (60°C) columns using 0.05 M phosphate buffer, pH 6.8 as loading buffer and 0.15 M phosphate buffer, pH 6.8 as elution buffer. The cDNA enriched in specifically expressed sequences was desalted, purified and subjected to random-primed second strand synthesis for radiolabeling to high specific activity (> 0.6 × 10^9 cpm/μg). Replicas from cDNA library platings at 0.5–1 × 10^3 CFU/plate (85 mm plates) were made by colony lifts on nylon filters (Bio-Rad, Hercules, CA) and subtractive screening was performed by hybridization with this probe. The positive clones were amplified and the plasmid DNA was isolated. An aliquot was then subjected to restriction cutting with Bam H1 and ECoRl; the fragments separated by electrophoresis (Fig. 1), radiolabeled by random-primed DNA synthesis and used as probes for standard Northern analyses.

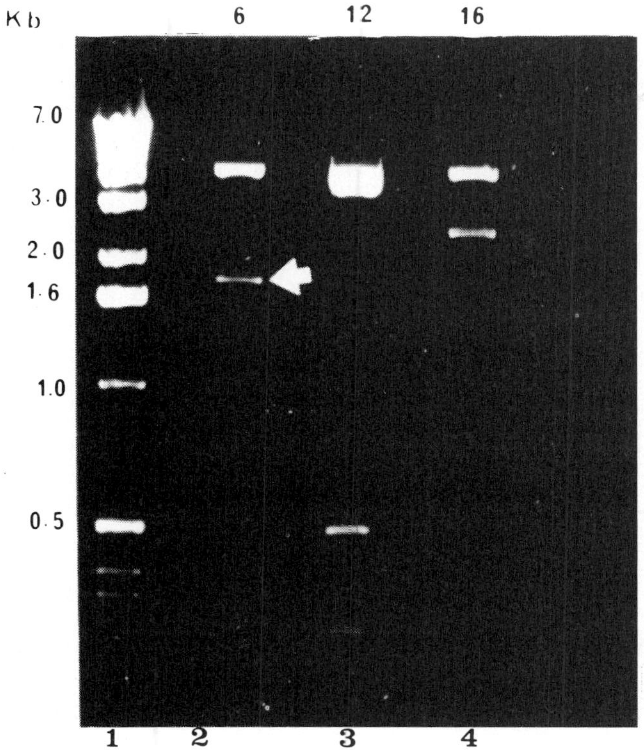

Fig. 1. Identification (sizing) of the cDNA inserts present in the positive clones. Two μl aliquot from clone numbers 6, 12, and 16 were subjected to restriction cutting with Bam H1 and ECoRl for 1 h at 37°C. The bands were separated by 0.7% agarose gel electrophoresis. Lane 1: 1 kb DNA ladder; lane 2: clone 6 which corresponds to FAT gene (1.8 kb)

Identification of genes and DNA sequencing

An aliquot containing the denatured plasmid DNA from the confirmed positive clones were subjected to bidirectional partial DNA sequencing using M13 and 17/T3α primers by Sanger dideoxy-mediated chain extension-chain-termination method [13]. The sequences were subjected to a computerized search for homologies against all sequences in the updated world wide Gen Bank and EMBL sequence databases followed by restriction mapping and reading frame identification using the University of Wisconsin Genetics Computer Group Software (GCG Ver 8.0) on a LAN-attached Vax server.

Northern hybridization

For Northern blot analysis, total RNA was electrophoresed in 1% agarose formaldehyde-formamide gel and transferred to Gene Screen plus hybridization transfer membrane (Biotech Systems, NEN Research Products, Boston, MA) by 18 h standard capillary transfer. The membrane was then baked under vacuum at 80°C for 1 h.

Hybridization

Hybridization was performed in a hybridization oven (Hybaid, Labnet, NJ) using one membrane and 10 ml QuickHyb aqueous exclusion rate-enhancing solution (Stratagene, La Jolla, CA) per roller bottle according to the manufacturer's instructions. QuickHyb hybridization solution has been developed by stratagene for use in Southern and Northern electrophoresis using randomly labeled radioactive and non-radioactive nucleic acid probes. Prior to prehybridization, the membrane was dipped briefly into deionized water to remove excess salt from the membrane. The membrane was covered with a thin film of the QuickHyb solution at all times (33 ml QuickHyb hybridization solution/cm^2 of the blot). The membrane was prehybridized in QuickHyb solution at 68°C for 10–20 min.

The labeled probe and Salmon Sperm DNA mixture was boiled for 2 min and then added to the prehybridization solution. Hybridization was performed for 2 h. The membrane was washed twice for 15 min at room temp with 2 × SSC - 0.1% (w/v) sodium dodecyl sulfate (SDS) buffer (2.5 ml/cm^2). This was followed by high stringency wash for 30 min at 60°C with 0.1 × SSC - 0.1% (w/v) SDS buffer. The membrane was wrapped with a Saran wrap and the wrapped membrane was placed on Kodak X-OMAT™AR film with an intensifying screen at –80°C. The film was developed after 24 h.

Each hybridization was repeated at least four times using different membranes. After each hybridization the residual cDNA was removed and rehybridized with Glyceraldehyde 3 phosphate dehydrogenase (G3PDH) cDNA probe, the results of which served as a loading control.

The autoradiograms were quantitatively evaluated by computerized β scanner. The results of densitometric scanning were normalized relative to the signal obtained for G3PDH cDNA.

Results

Development of oxidative stress after endotoxin treatment

Endotoxin caused the development of oxidative stress to the hearts as evidenced by the increased generation of malonaldehyde. As shown in Fig. 2, malonaldehyde concentration of heart was increased from 158.6 ± 9.12 to 396.3 ± 50 pmol/gm after 1 h of endotoxin treatment. The concentration of malonaldehyde was slightly but not significantly reduced to 295.9 ± 2.25 pmol/gm after 3 h and after 4 h the concentration was 245 ± 44 pmol/gm.

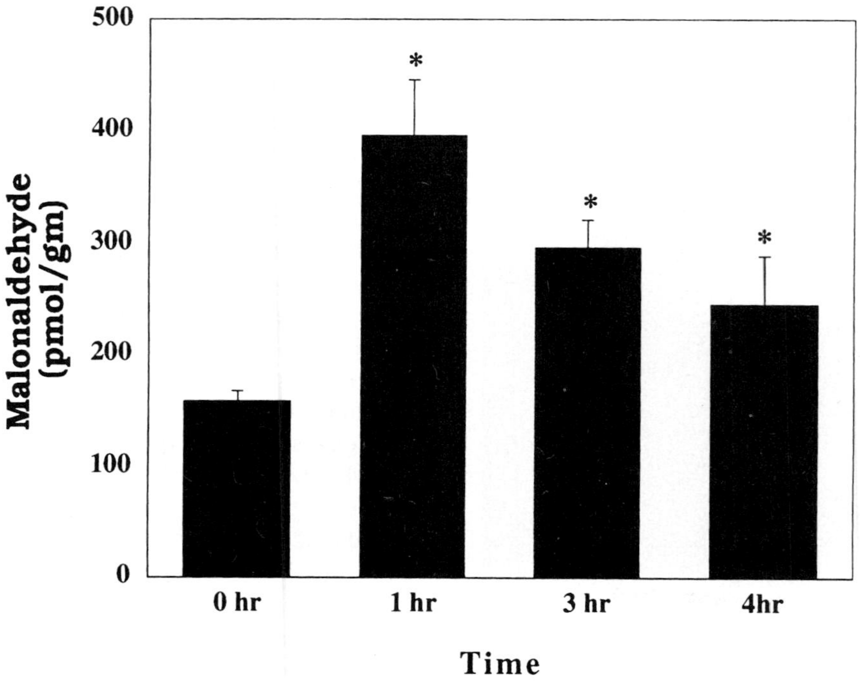

Fig. 2. Effects of endotoxin treatment on malonaldehyde content of heart. Rats were injected with endotoxin (0.5 mg/kg) then they were killed at 0, 1, 3, and 4 h. Hearts were excised, and malonaldehyde was estimated in heart by using high-performance liquid chromatography method. Results are means ± S.E. of 6 different animals per group. Each result was run in duplicate.*p < 0.05 compared with 0 h (baseline).

```
                                              10        20        30
                                     AAACGAAGTCTTAAGACATGCTTGTTTTTA
                                     |||||||| |||||||||||||||||||||
Ratfat TAGCTACATTTTTGGTAAAACCAATCTCCAAAACGAAGACTTAAGACATGCTTGTTTTTA
         1520      1530      1540      1550      1560      1570

             40        50        60        70        80        90
Ox1-Cl TAAAACACACCTATCTGTAGTTGAAGAAACGGTGGTGTGCGCGCTCTCTCTCTTATTGCA
       ||||||||||||||||||||||||||||||||||||||||||||||||||||||||||||
Ratfat TAAAACACACCTATCTGTAGTTGAAGAAACGGTGGTGTGCGCGCTCTCTCTCTTATTGCA
         1580      1590      1600      1610      1620      1630

            100       110       120       130       140       150
Ox1-Cl GATATATATTTATTTATATATTGCAATAAGCCACAGCATATTTCGAAAAGATTAATATGT
       |||||||||| ||| |||||||||||||||||||||||||||| || ||||| |||||||
Ratfat GATATATATTCATTCATATATTGCAATAAGCCACAGCATATTTTGACAAGATCAATATGT
         1640      1650      1660      1670      1680      1690

            160       170       180       190
Ox1-Cl CACTAAGCCTATATTTTTAATAAAATCTTGTATTTTGTTAAGTCC
       |||||||||||||||||||||||||||||||||||||||||||||
Ratfat CACTAAGCCTATATTTTTAATAAAATCTTGTATTTTGTTAAGTCCATCATCTGCAACTGA
         1700      1710      1720      1730      1740      1750

Ratfat GTGGACTTCAATTTCTGCAGAACTAATTATCTTTTTTGGTTCTGATTTACTGATTTTTTT
         1760      1770      1780      1790      1800      1810
```

Fig. 3. Sequence of cDNA insert excised by Bam Hl and ECoRI which showed 97% homology with FAT gene.

Induction of FAT gene as a function of the duration of endotoxin treatment

Out of 24 putative positive clones screened, one clone revealed > 97% homology with the fatty acid transport (FAT) protein (Fig. 3) that had been implicated in binding or transport of long chain fatty acids. cDNA probe systhesized from this clone identified mainly one major transcript which was nearly 4.6 kb. Another transcript of about 2.9 kb was also poorly recognized by this probe, and is not clear in the Figure.

Figure 4 shows the induction of FAT gene by Northern blot analysis of RNA from normal heart (lane 1) and endotoxin-

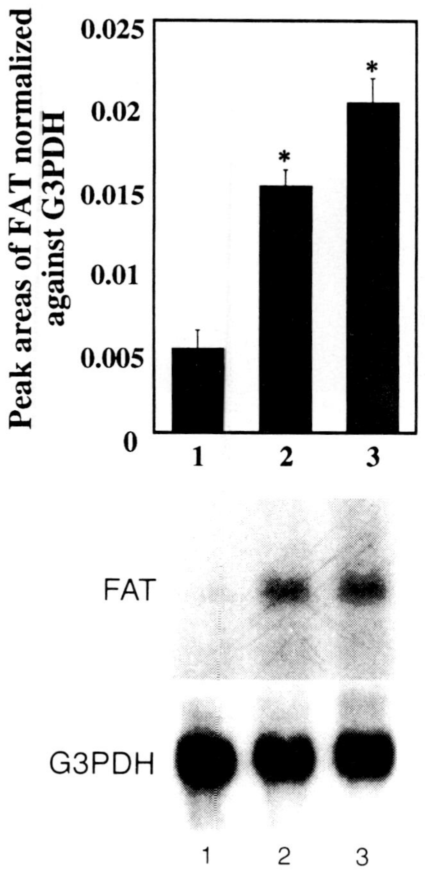

Fig. 4. Quantitative Northern blot analysis demonstrating induction of FAT gene in endotoxin treated hearts. Bottom panel: Experimental group of rats were subjected to exdotoxin treatment while the control group was treated with saline only. Total RNA (10 μg) was isolated from hearts and used for Northern analysis as described in Materials and methods. The blots were first probed with [32P] FAT cDNA (top) and then stripped and probed again with [32P] G3PDH (bottom). The filters were exposed at –80°C with intensifying screen. Lane 1: control; lane 2: 3 h endotoxin treated; lane 3: 4 h endotoxin treated. Top panel: The results of densitometric scanning (mean ± S.E.M.) for 4 different experiments at each time point are shown for each blot. *p < 0.05 compared to control.

treated hearts (lane 2: 3 h; lane 3: 4 h using FAT cDNA probe). The size of the detected mRNA was consistent with that expected from rat transcript which was approximately 4.6 kb. Again, the signal for 2.8 kb was not very strong. Interestingly, the induction for the expression of FAT gene progressively enhanced as a function of the duration of endotoxin treatment. The induction was 2-fold after 2 h (lane 2) and 3-fold after 4 h (lane 3) compared to control (lane 1).

Effect of ischemia and ischemic preconditioning on FAT gene expression

To examine whether the FAT gene could also be induced by ischemia, isolated rat hearts were made ischemic for different periods of time, e.g. 5, 10 and 20 min. Furthermore, a group of isolated hearts were subjected to ischemic preconditioning by inducing 5 min of ischemia followed by 10 min of reperfusion (1 × PC). This process was also repeated 4 times by subjecting the hearts to repeated ischemia and reperfusion (4 × PC). Control experiments were performed by perfusing the hearts with KHB buffer only. As indicated in Fig. 5, the induction of FAT gene was not stimulated by ischemia as compared to control. However, significant enhancement (1.2-fold) in the induction was noticed in the 1 × PC hearts. This induction was further increased (1.8-fold) after 4 × PC, suggesting that it was not ischemia, but oxidative stress mediated by ischemia/reperfusion that was instrumental for the enhanced expression of FAT gene.

Discussion

The results of this study demonstrated the induction of the expression for the FAT gene in rat heart by oxidative stress and ischemic preconditioning. Subtractive hybridization technique identified the induction of FAT gene by oxidative stress. Using the FAT cDNA probe we were also able to demonstrate the induction of mRNA for FAT protein in ischemically preconditioned myocardium.

Only recently FAT protein has been identified in rat adipocyte membrane [6]. Using Northern analysis of RNA obtained from rat adipose tissue and probed with the cDNA identified two major transcripts of 4.8 and 2.9 kb which were abundant in heart, intestine, fat, muscle and testis. The authors also demonstrated strong induction of FAT mRNA during the differentiation process and by treatment of preadipocytes with dexamethasone, conditions that were also associated with an increase of fatty acid transport.

The precise role of FAT protein in heart is not known. It is speculated that this protein binds or transports long-chain fatty acids across the membrane [6, 7]. It is also possible that FAT protein serves as the regulators of fatty acid binding and transport processes. In heart, the majority of ATP is derived from oxidative phosphorylation and the acetyl CoAs necessary for this are derived from the fatty acids. A growing body of evidence supports a role for abnormal fatty acid metabolism in the pathophysiology of myocardial ischemia reperfusion injury [14]. Prolonged ischemia or reperfusion following ischemia have been shown to be associated with the accumulation of long chain nonesterified fatty acids and their thioesters. These amphiphilic substances can induce major changes in membrane function either by inserting free amphiphile molecules into the lipid bilayer or by their detergent-like effects that deplete the membrane lipids [15].

A number of interrelated factors may function in concert to potentiate fatty acid release and their subsequent accumu-

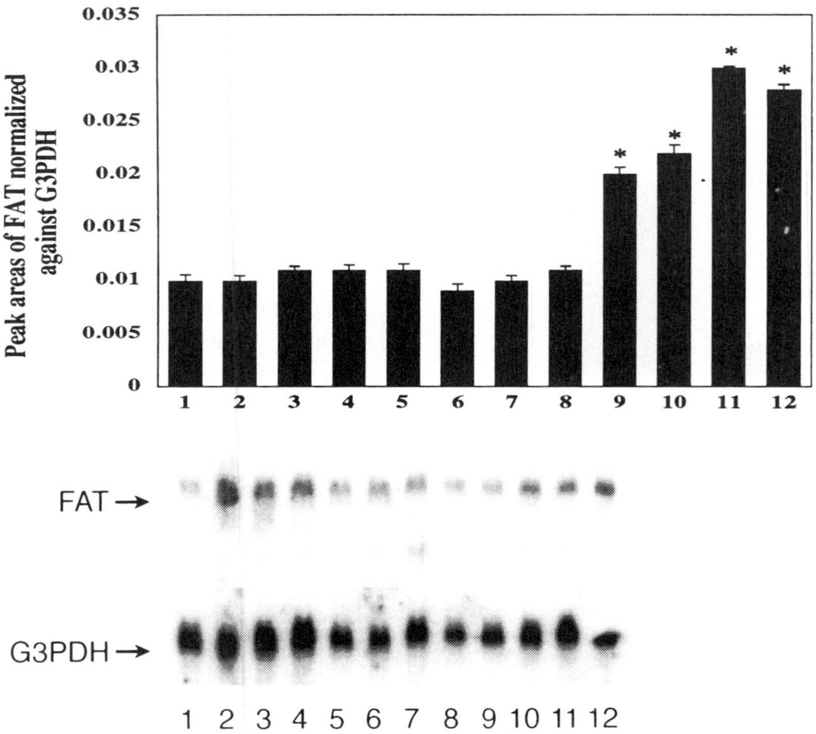

Fig. 5. Quantitative Northern blot analysis demonstrating induction of FAT gene by ischemic preconditioning. Bottom panel: Experimental group of rats were subjected to 5, 10 or 20 min of ischemia as well as preconditioned by repeated ischemia and reperfusion as described in Materials and methods. Control experiments were performed by perfusing the hearts with KHB buffer only. Total RNA (10 μg) was isolated from hearts and used for Northern analysis. The blots were first probed with [^{32}P] FAT cDNA (top) and then stripped and probed again with [^{32}P] G3PDH (bottom). The filters were exposed at –80°C with intensifying screen. Lane 1: control; lane 2 and 3: 5 min ischemia; lane 4 and 5: 10 min ischemia; lane 6 and 7: 20 min ischemia; lane 8: control; lane 9 and 10: 1 × PC; lane 11 and 12: 4 × PC. Top panel: The results of densitometric scanning (means ± S.E.M.) for four different experiments for each time point are shown with each blot. *$p < 0.05$ compared to control.

lation in ischemic-reperfused myocardium. Ischemia-induced inhibition of β-oxidation of fatty acids and loss of sarcolemmal phospholipids may comprise the two most prominant factors that contribute to the accumulation of free fatty acids and their long-chain acyl CoA thioesters [16]. Previous studies have demonstrated that preservation of membrane phospholipids with a corresponding inhibition of fatty acid accumulation provides a means of protecting the heart from ischemia reperfusion injury [16].

The induction for the expression of FAT gene in the ischemic heart strongly suggests a regulatory role of this protein in the ischemic myocardium. The known fact of the generation of oxygen free radicals and development of oxidative stress during the reperfusion of ischemic myocardium [17] and the induction of FAT gene by oxidative stress further implicate a role of this protein in the ischemic reperfused heart. Our results demonstrated strong induction of FAT mRNA in the ischemically preconditioned myocardium. The mechanism of protection afforded by preconditioning remains highly controversial. It is tempting to speculate that the induction of FAT gene may in part be instrumental for the myocardial preservation induced by ischemic preconditioning. Ischemic preconditioning induced by repeated ischemia and reperfusion certainly induces oxidative stress. The results of this study clearly demonstrate that expression of FAT gene is induced by ischemic preconditioning as well as oxidative stress, but not by ischemia.

Acknowledgements

This study was supported by NIH grants HL 33899 and HL 22559 and a grant-in-aid from the American Heart Association.

References

1. Das D K, Engelman R M, Kimura Y: Molecular adaptation of cellular defences following preconditioning of the heart by repeated ischemia. Cardiovasc Res 27: 578–584, 1993
2. Andres J, Sharma HS, Knoll R, Stahl J, Sassen LMA, Verdouw PD, Schaper W: Expression of heat shock proteins in the normal and stunned myocardium. Cardiovasc Res 27: 1421–1429, 1993
3. Sharma HS, Snoeckx LHEH, Sassen LMA, Knoll R, Andres J, Verdouw PD: Immunochemical localization of heat shock protein 72 in the porcine heart: Enhanced transcription during ischemic preconditioning.

Circulation 88: 3045–3058, 1993
4. Das DK, Maulik N, Moraru II: Gene expression in acute myocardial stress. Induction by hypoxic, ischemia, reperfusion, hyperthermia and oxidative stress. J Mol Cell Cardiol 27: 181–193, 1995
5. Das DK, Moraru II, Maulik N, Engelman RM: Gene expression during myocardial adaptation to ischemia and reperfusion. Annal New York Acad Sci 723: 292–307, 1994
6. Abumrad NA, El-Maghrabi MR, Amri EZ, Lopez E, Grimaldi PA: Cloning of a rat adipocyte membrane protein implicated in binding or transport of long-chain fatty acids that is induced during preadipocyte differentiation. J. Biol Chem. 268: 17665–17668, 1993
7. Fitscher BA, Klaaßen-Schluter CM, Stremmel W: Evidence for hepatocyte membrane fatty acid transport protein using rat liver mRNA expression in Xenopus laevis oocytes. Biochem Biophys Acta 1256: 47–51, 1995
8. Maulik N, Watanabe M, Engelman D, Engelman RM, Kagan VE, Kisin E, Tyurin V, Cordis GA, Das DK: Myocardial adaptation to ischemia by oxidative stress induced by endotoxin. Am J. Physiol 269: C907–C916, 1995
9. Chomczynski P, Sacchi N: Singal-step method of RNA isolation by acid guanidinium thiocyanate-phenol -chloroform extraction. Anal Biochem 162: 156–159, 1987
10. Okayama H, Berg P: High-efficiency cloning of full-length cDNA. Mol Cell Biol 2: 161–165, 1982
11. Gubler U, Hoffman BJ: A simple and very efficient method for generating cDNA libraries. Nucleic Acids Res 15: 6304, 1983
12. Rhyner TA, Biguet NF, Berrard S, Borberly AA, Mallet J: An efficient approach for the selective isolation of specific transcripts from complex brain mRNA populations. J Neurosci Res 16: 167–181, 1986
13. Sanger F, Nicklen S, Coulson AR: DNA sequencing with chain - terminating inhibitors. Proc Natl Acad Sci USA 74: 5463–5468, 1977
14. Srimani BN, Engelman RM, Jones R, Das DK: Protective role of intracoronary fatty acid binding protein in ischemic reperfused myocardium. Circ Res 66: 1535–1543, 1990
15. Katz AM: Membrane-derived lipids and the pathogenesis of ischemic myocardial damage. J Mol Cell Cardiol 14: 627–632, 1982
16. Das DK, Engelman RM. Phospholipids and Myocardial Reperfusion injury. In: DK Das (ed.) Pathophysiology of Reperfusion Injury, CRC Press, Florida, 1992, pp. 149–179
17. Das DK, Maulik N. Evaluation of antioxidant effectiveness in ischemia reperfusion tissue injury. Methods Enzymology 233: 601–610, 1994

Molecular and Cellular Biochemistry **160/161**: 249–255, 1996.

Protaglandins and the antiarrhythmic effect of preconditioning in the isolated rat heart

Michael Arad, Tatiana Oxman, Ron Leor and Babeth Rabinowitz

Cardiac Metabolism Laboratory, Heart Institute, Sheba Medical Center, Tel Hashomer and Sackler School of Medicine, Tel-Aviv University, Tel Aviv, Israel

Abstract

Our study evaluated the relationship between the endogenous production of prostacyclin and the antiarrhythmic effect of ischemic preconditioning against ischemic and reperfusion-induced tachyarrhythmia. Langendorff perfused rat hearts underwent 30 min regional ischemia with reperfusion. Preconditioning was induced by a single episode of 5 min ischemia and 15 min reperfusion. Prostaglandin 6-keto $F_{1\alpha}$ (a stable metabolite of prostacyclin) was determined in the coronary effluent.

In the control group the incidence of tachyarrhythmia was 31% during ischemia and 67% during reperfusion. Preconditioning did not affect ischemic arrhythmias but attenuated arrhythmias a reperfusion (8%, $p < 0.01$) and was associated with increased release of prostacyclin prior to reperfusion. Aspirin abolished the antiarrhythmic effect of preconditioning against reperfusion tachyarrhythmias. However, no relationship was found between suppression of prostacyclin production and the occurrence of arrhythmia in individual hearts.

Thus, our findings suggest that metabolites of arachidonic acid via the cyclooxygenase pathway are involved in the protective effect of ischemic preconditioning against reperfusion-induced tachyarrhythmias. (Mol Cell Biochem **160/161**: 249–255, 1996)

Key words: preconditioning, arrhythmia, prostacyclin, aspirin, rat

Introduction

Cardiac synthesis and release of various eicosanoids during and following ischemia of short duration have been documented by several investigators [1–3] and in our previous work [4]. Prostacyclin is the main metabolite of arachidonic acid, produced under these circumstances.

Short term ischemia followed by reperfusion is expected to render the myocardium resistant to damage by a subsequent, longer ischemic period – a phenomenon named ischemic preconditioning [5, 6]. The protective effect of heart preconditioning with ischemia manifests by reduction of infarct size, better recovery of stunned myocardium and attenuation of both ischemic and reperfusion arrhythmias [7]. The antiarrhythmic effect was mostly demonstrated in regionally ischemic and reperfused hearts, either *in vivo* or *in vitro* [8, 9]. Studies by Vegh *et al.* [10, 11] suggested an involvement of products of cyclooxygenase pathway, nitric oxide and kinins in preconditioning-mediated attenuation of ischemic arrhythmias in dog. However, Li and Kloner [12] could not prevent the effect of preconditioning on either ischemic arrhythmia or infarct size by aspirin pretreatment in rat, and thus concluded that these cardioprotective effects were no mediated by prostanoids.

Our study was designed to reassess the question of the involvement of endogenous prostaglandins in the cardioprotective effect of ischemic preconditioning. We chose for this purpose the protective effect of preconditioning against ventricular tachyarrhythmias in the well-established model of regional ischemia and reperfusion in the isolated rat heart, and examined (1) prostacyclin release, as a potential protective agent; (2) the effect of aspirin, as indirect evidence of cause-effect relationship between prostaglandins and preconditioning.

Address for offprints: M. Arad, Heart Institute, Sheba Medical Center, 52621 Tel-Hashomer, Israel

Materials and methods

Preparation

Male Sprague-Dawley rats weighing 250–350 g received 1000 u heparin and 6 mg/100 g Na-pentobarbital intraperitoneally. The hearts were rapidly excised, arrested in ice-cold heparinized saline and mounted on the Langendorff isolated heart apparatus. A modified Krebs-Henseleit solution consisting of NaCl 118, $NaHCO_3$ 25, glucose 11.1, KCl 4.9 $CaCl_2$ 2.7, $MgSO_4$ 1.2, KH_2PO_4 1.2 and NaEDTA 0.5 mM was equilibrated with 95% O_2 and 5% CO_2. The whole system was heated to 37°C by means of water jacketing. The heart was suspended in a sealed chamber containing a silicon funnel for collection and volumetric measurement of the coronary effluent. The perfusate column height was kept at 80 cm H20, producing a coronary perfusion pressure of about 55 mmHg. A stainless steel electrode was inserted into the right ventricular epicardium for bipolar electrocardiogram recording relative to the aortic cannula. Hearts were allowed to beat spontaneously during the entire experiment.

The preparation for experiments with regional ischemia consisted of introduction of 04 silk thread underneath the left coronary artery. The artery was approached close to its origin and usually above any visible bifurcations. Regional ischemia was produced by surgical ligation over a tiny teflon tubing. The artery was compressed without applying traction, and the heart could remain closed in the heart chamber under constant experimental conditions. Reperfusion was initiated by untieing the ligature, thus allowing to repeat the same procedure during the subsequent ischemia.

The coronary perfusion pressure was recorded through Senso Nor 840 transducer at the level of the aortic cannula. The coronary flow was measured volumetrically.

The cardiac rhythm was continuously displayed on a PC monitor. Paper recordings were obtained at the onset of ischemia and reperfusion and elsewhere, as required. Distinction between tachycardia and fibrillation on 'per animal' basis was extremely difficult in this model, because of numerous intermediate grades in the morphology of tachyarrhythmia and a tendency to convert from one type to another (Fig. 1). Therefore we defined ventricular tachyarrhythmia (TA) as tachycardia or fibrillation of at least 5 sec duration. Tachyarrhythmia persisting till the end of the correspondent ischemia or reperfusion was referred to as sustained (STA).

Protocol

Experiments were performed in 4 groups of hearts: Group 1: Control – Sustained Regional Ischemia (RI), n = 13: 45 min stabilization, 30 min RI, 30 min reperfusion (R). Group 2: Preconditioning – Regional ischemia, n = 12: 30 min stabilization, 5 min RI, 15 min reperfusion (R), followed by 30 min sustained RI and 30 min R. Group 3: Control – Aspirin, n = 7 , aspirin 50 μg/ml in Krebs-Henseleit was given for 2 min, 10 min prior to sustained RI, as in group 1. Group 4: Preconditioning – R I + Aspirin, n = 9: 30 min stabilization, 5 min RI, 5 min R, 2 min perfusion with aspirin 50 μg/ml in Krebs-Henseleit, 8 min of normal perfusion (to complete 15 min R), sustained RI and R.

Prostacyclin determination

Samples were collected at minute 30 of baseline perfusion, at minutes 5, 15, 30 of ischemia and minutes 1, 3, 5, 15 of the reperfusion periods, and then kept at -20°C. Release of prostacyclin (PGI_2) was assessed by measurement of its stable metabolite – 6 Keto $PGF_{1\alpha}$. The Du Pont 3[H]-RIA kit was used.

Data analysis

All data are presented as mean ± S.E.M. Coronary flow and release of PGI_2 were expressed per min per g wet heart weight (g). The extent of regional ischemia was defined as percent reduction of coronary flow, compared to baseline. The results were compared with the Fisher Exact test and the analysis of variance. ANOVAs with Duncan intergroup comparisons or repeated measures were used, as required.

Results

Effects of preconditioning

Coronary flow

Was slightly higher in group #2 at baseline, and remained so for the rest of the experiment (Fig. 2). Coronary ligation caused an average reduction of 36 ± 2% in the coronary flow without differences between the groups. Preconditioning had no significant effect on the coronary flow during and after sustained ischemia.

Arrhythmia

No ischemic TA were observed during 5 min regional ischemia. Short TA was recorded at reperfusion in one of the hearts. During sustained RI ventricular arrhythmia developed within 14–17 min from coronary ligation. The incidence of ischemic TA did not differ significantly between the groups (Table 1). Reperfusion arrhythmia developed within 30–90 sec from removal of coronary ligature; ventricular fibrilla-

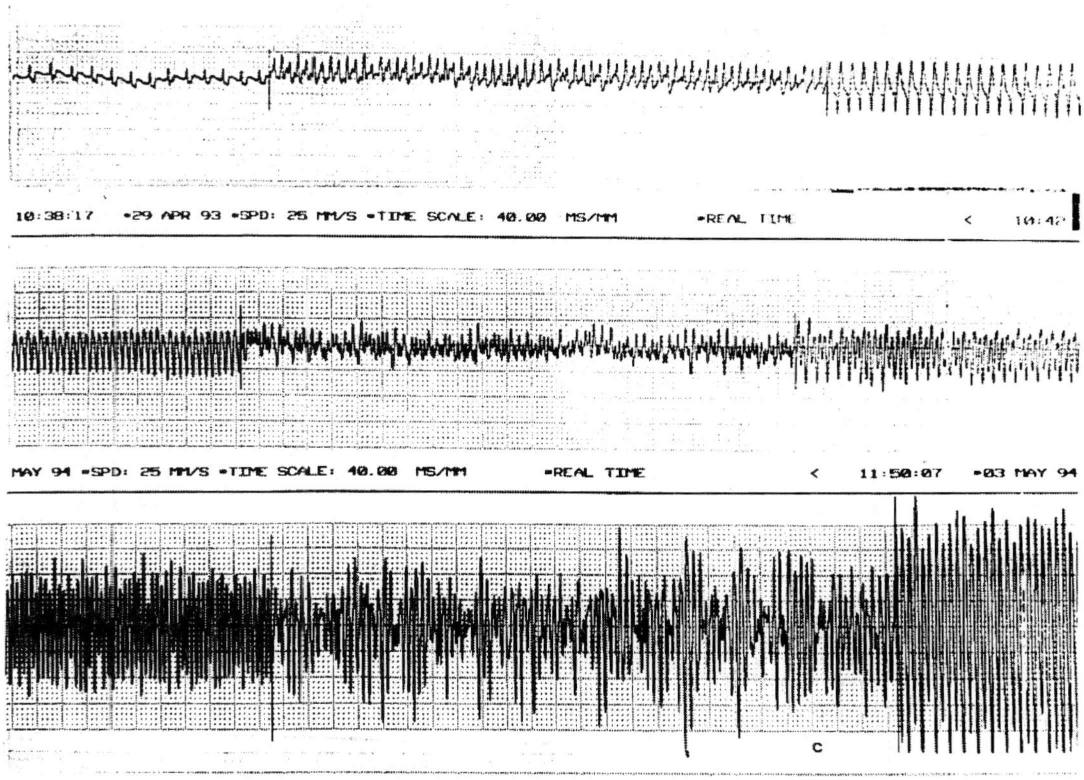

Fig. 1. Examples of reperfusion tachyarrhythmias, recorded in three different isolated rat hearts. *Upper panel*: Sinus rate, polymorphous ventricular tachycardia which later converted to sustained ventricular tachycardia; *Middle panel*: ventricular tachycardia deteriorating into fibrillation and then turning into polymorphous tachycardia; *Lower panel*: another example of spontaneous convertion between different morphologies of ventricular tachyarrhythmias – fibrillation and tachycardia.

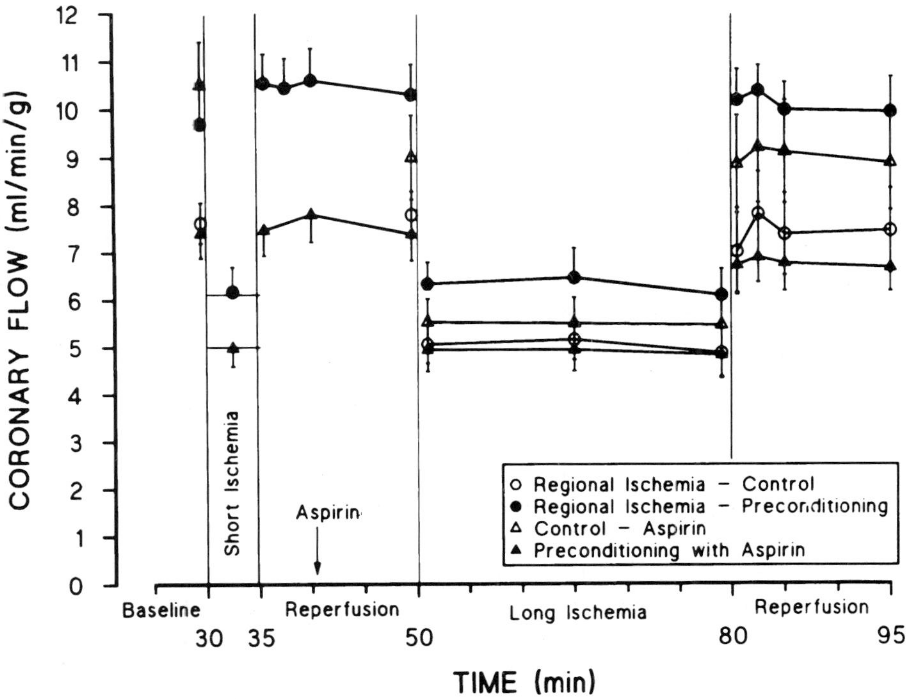

Fig. 2. Mean coronary flow in the course of the experiment in the different groups. There were inter-group differences in the basal coronary flow, but no significant differences in the extent reduction of coronary flow during ischemia or in the reactive hyperemia.

Table 1. Heart rate and arrhythmia incidence

	Heart rate (beats/min)		Arrhythmia (incidence)			
	Basal	Preischemic	Ischemia		Reperfusion	
			TA	STA	TA	STA
#1	276 ± 11	275 ± 11	4/13	1	8/12	4[a]
#2	265 ± 9	271 ± 11	4/12	0	1/12*	0
#3	259 ± 5	261 ± 6	0/7	0	5/7	1
#4	284 ± 7	286 ± 6	0/9**	0	5/9	3

Groups: #1 – Control, #2 – Preconditioning, #3 – Control with Aspirin, #4 – Preconditioning with Aspirin. TA – tachyarrhythmia (total count), STA – sustained TA. [a]In one of the hearts from group #1, sustained ventricular fibrillation started during ischemia and continued throughout reperfusion. This heart was excluded from the analysis of reperfusion TA. *p = 0.012 (group #2 vs all others) **p = 0.094 (groups #3, 4 vs groups #1, 2).

tion or ventricular tachycardia which often deteriorated into fibrillation. This TA was sustained in about half of the cases As may be seen in Fig. 3, preconditioning almost completely abolished reperfusion tachyarrhythmias.

Prostacyclin

Figure 4 illustrates the release pattern of 6-keto $PGF_{1\alpha}$ in regionally ischemic hearts. A trend for decrease in release of 6-keto $PGF_{1\alpha}$ was found during regional ischemia in the control group ($p < 0.01$ in comparison to baseline, starting from minute 15 of ischemia). This decrease did not occur in preconditioned hearts, thus creating a progressive dissociation between prostacyclin release curves of the 2 groups. Increased levels were found at minute 1 of reperfusion, but the levels at that time did not statistically differ between the experimental groups.

Effects of Aspirin

Administration of aspirin prior to sustained ischemia did not significantly affect the spontaneous heart rate or coronary flow (Table 1 and Fig. 2, respectively). No tachyarrhythmias were recorded during sustained regional ischemia. A reperfusion, TA occurred in 5 out of 9 hearts in preconditioned hearts which received aspirin (p = n.s. vs control, Fig. 3). Aspirin caused a pronounced decrease in prostacyclin release during both ischemia and reperfusion (Fig. 5). However, no correlation was found between suppression of prostacyclin release and development of reperfusion TA.

Discussion

Our study evaluated the relationship of production of prostacyclin to the cardioprotective effect of ischemic preconditioning against reperfusion-induced tachyarrhythmia. A single episode of short term regional ischemia prior to prolonged regional ischemia was definitely antiarrhythmic in the isolated rat heart model. This effect was associated with an increased

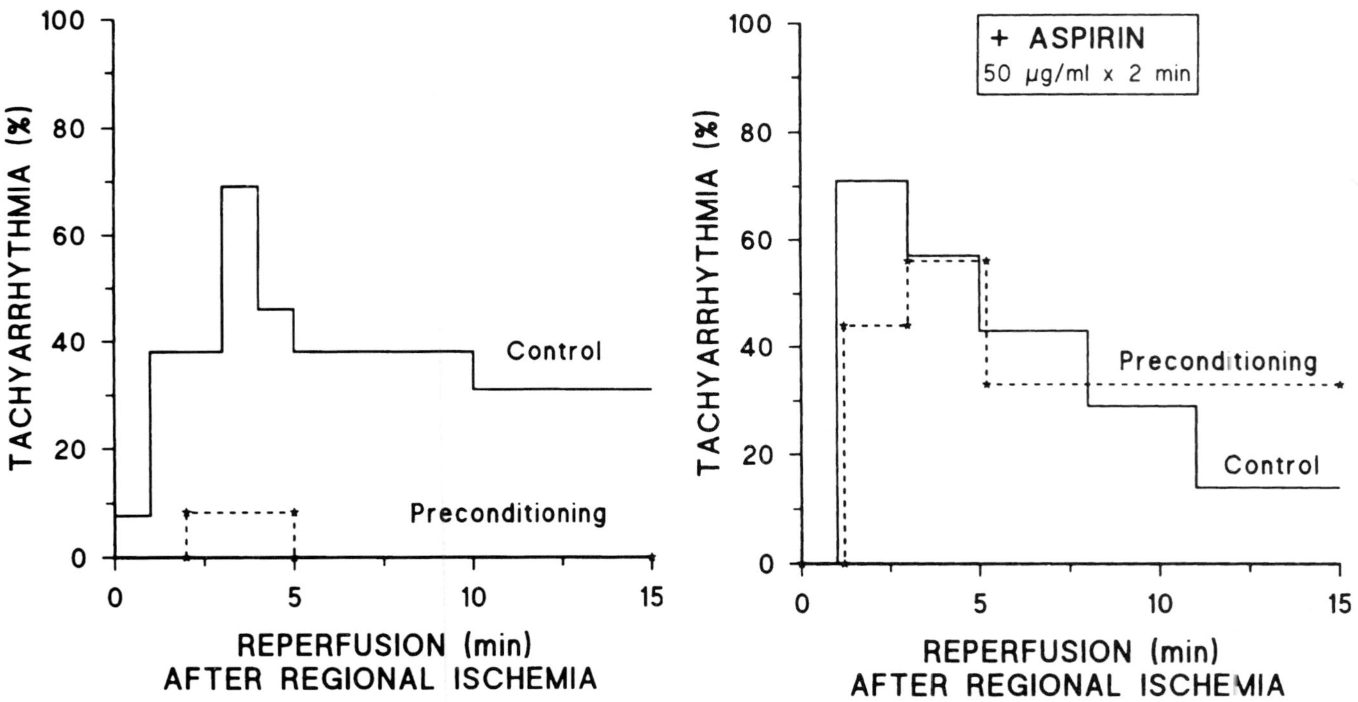

Fig. 3. Tachyarrhythmia incidence as a function of reperfusion duration. Preconditioning provided a clear-cut protection from reperfusion tachyarrhythmias, which was abolished in the presence of aspirin (see Table 1 for statistics). The duration of reperfusion in the curves is limited to 15 min, because no changes in the heart rhythm occurred after that time. Note that one of the control hearts started the reperfusion while in tachyarrhythmia.

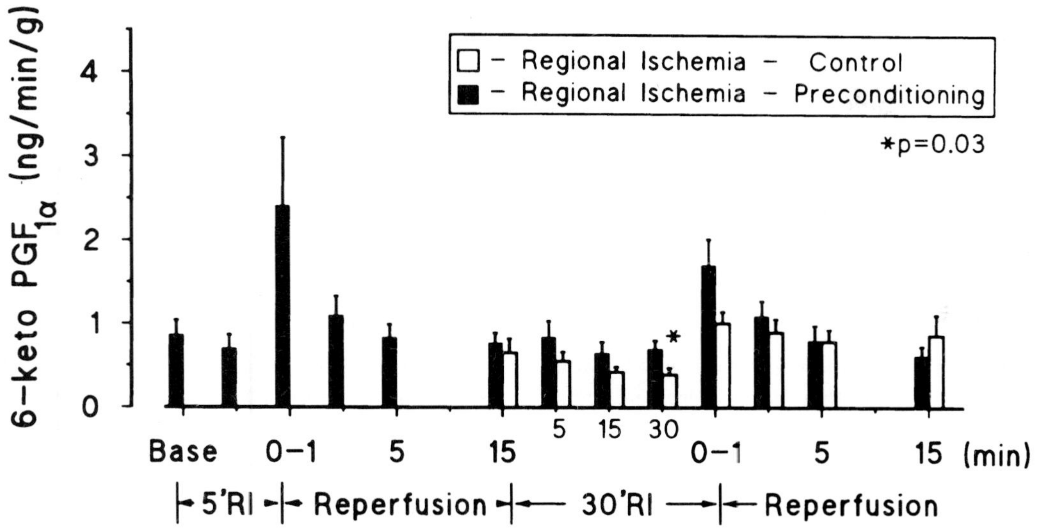

Fig. 4. Release of PGF_{α} (a stable metabolite prostacyclin) in the course of the experiment in groups 1 and 2. Note the difference in the release at minute 30 of ischemia just prior tc onset of reperfusion.

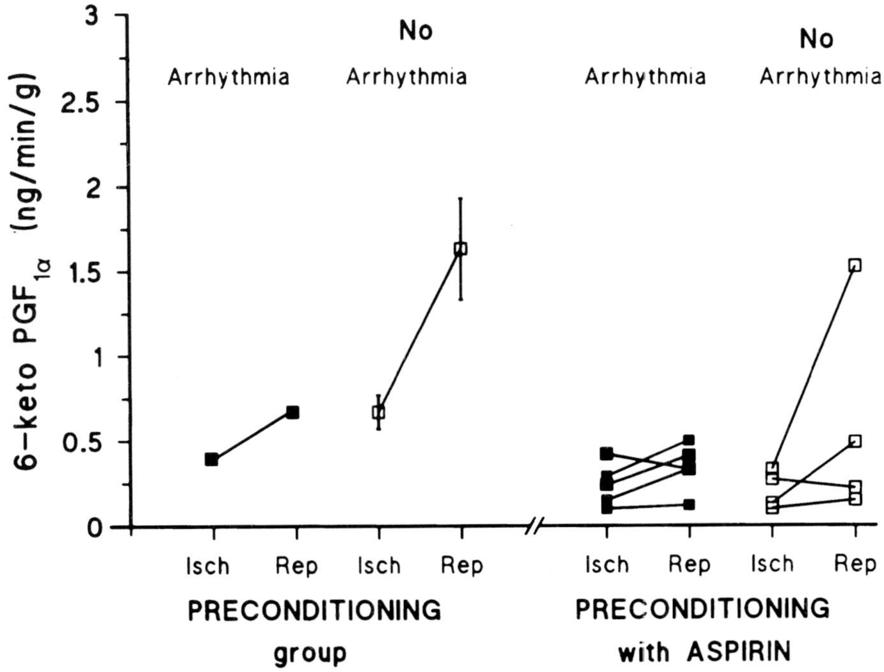

Fig. 5. Relationship between prostacyclin release and reperfusion tachyarrhythmia in preconditioned hearts, with or without aspirin. Within each group, subgroups with or without arrhythmia are separately presented. Increased release was found in the first minute of reperfusion ($p = 0.002$). The only heart in group 2 which developed tachyarrhythmia had the lowest rate of prostacyclin production within its group (the rest of which presented as mean ± S.E.M.). Aspirin pretreatment prior to sustained ischemia reduced the production of prostacyclin ($p < 0.001$), attenuated its increase at reperfusion ($p = 0.007$ for the interaction effect) and increased the incidence of tachyarrhythmia (Table 1). However, no direct relationship could be found between arrhythmia and prostacyclin release in the aspirin-treated hearts.

release of prostacyclin during sustained ischemia, and was abolished by pretreatment with aspirin (Figs. 3 and 4).

Early synthesis of prostaglandins by either the anoxic or the ischemic heart is apparent prior to development of histological or biochemical evidence of damage. It probably constitutes some kind of compensatory response [13, 14]. Various prostaglandins, mainly prostacyclin and thromboxane but also PGE_2 and PGF_{2a} are produced under these circumstances, inducing different, sometimes contradictory effects on platelet activation, the vasculature, oxygen radical formation and electrophysiologic properties [15, 16]. Prostacyclin is produced mainly by the vascular endothelium and possibly by myocytes and cardiac mesenchymal cells [17, 18].

Studies utilizing various blockers of the cyclooxygenase pathway did not lead to decisive conclusions concerning the role of eicosanoids in perpetuation versus protection against the ischemic damage [19]. Pharmacological studies with stable analogs of prostacyclin, performed *in vivo* and *in vitro*, usually reported promising results attributable to coronary dilation, inhibition of platelet aggregation, inhibition of neutrophil activation and free radical formation, reduction of noradrenaline overflow, membrane stabilization, preservation of mitochondrial function and other cytoprotective effects [20–22]. Prostacyclin was found to have protective effects against ischemic and reperfusion arrhythmias [23]. The protective action of prostacyclin, and it antiarrhythmic effect in particular, may be achieved by reduction of the ischemic release of norepinephrine, prevention of ischemia-induced increase in intracellular cyclic AMP [24], and by stabilization of the membrane properties [25]. With the development of the concept ischemic preconditioning it became apparent that endogenous metabolites released during/after brief ischemia, contribute to myocardial protection during the subsequent ischemia [7]. Vegh and coworkers [10] concluded that PGI_2 is involved in the antiarrhythmic effects of preconditioning, based on their observations linking preconditioning-induced prevention of ischemic arrhythmias in the dog heart to products of cyclooxygenase pathway, an effect similar to that of exogenous Iloprost. Other investigators could not confirm a role for prostaglandins in limitation of infarct size or reperfusion arrhythmia in preconditioned hearts: Li and Kloner [12] showed that pretreatment with aspirin prior to repetitive preconditioning did not abolish cardioprotection from arrhythmia or infarct, resulting from 90 min ischemia in rat *in vivo*; Liu *et al.* [26] reported that cyclooxygenase inhibitors don't intervene with the effect of preconditioning on infarct size in rabbit hearts, either *in vivo* or *in vitro*.

As already discussed, manifestations of protection by preconditioning differ in the various experimental models, do not appear concomitantly and seem to be mediated by different mechanisms of cardiac adaptation to repeated ischemia [7, 27]. In the model of preconditioning used in our study, the antiarrhythmic effect was pronounced during reperfusion. The production of prostacyclin was augmented in the preconditioned hearts during regional ischemia, but was not significantly different at reperfusion (Fig. 4). To investigate the possible contribution of prostaglandins to the antiarrhythmic effect of preconditioning, experiments were carried out in preconditioned and control regionally ischemic hearts, pretreated with aspirin. Aspirin, infused between preconditioning and sustained ischemia abolished the protective effect of preconditioning against reperfusion arrhythmia (Fig. 3). However, no relationship between PGI_2 release and the occurrence of tachyarrhythmia was found in aspirin-treated hearts (Fig. 5).

Tachyarrhythmias during ischemia were quite uncommon in our model and appeared to be non-responsive to preconditioning (Table 1). Our findings indicate their possible attenuation by aspirin (in agreement with the results of Fagbemi [28]), but this must be confirmed in further experiments.

We propose that prostacyclin and ischemic preconditioning are not directly interrelated. Increased level of prostacyclin may be considered as just an indicator of increased production of prostaglandins by preconditioned hearts. The mechanism of the antiarrhythmic action of preconditioning in regional ischemia-reperfusion and the exact role played by prostaglandins is unclear, yet. Arrhythmia generation may be influenced by the balance between the different metabolites of arachidonic acid via the cyclooxygenase pathway, and this equilibrium could be modified by either preconditioning or aspirin [16, 23].

Acknowledgement

The biochemical work of this study was performed by Mrs. Ella Elazar and Mrs. Rodica Klein.

References

1. Berger HJ, Zaret BL, Speroff L, Cohen LS, Wolfson S: Regional cardiac prostaglandin release during myocardial ischemia in anesthetized dogs. Circ Res 38: 566–571, 1976
2. Karmazyn M: Contribution of prostaglandins to reperfusion-induced ventricular failure in isolated rat hearts. Am J Physiol 251: H133–H140, 1986
3. Okada T: Hypoxia-induced change in prostaglandin production and coronary flow in isolated rat heart. J Mol Cell Cardiol 23: 930–948, 1991
4. Rabinowitz B, Arad M, Elazar E, Klein R, Har Zahav Y: Epicardial versus endocardial 'in mirror' changes in prostaglandin synthesis after short periods of ischemia and reperfusion. Eicosanoid 5: 163–167, 1992
5. Murry CE, Jennings RB, Reimer KA: Preconditioning with ischemia: A delay of lethal cell injury in ischemic myocardium. Circulation 74: 1124–1136, 1986
6. Murry CE, Richard VJ, Reimer KA, Jennings RB: Ischemic preconditioning slows energy metabolism and delays ultrastructural damage during a sustained ischemic episode. Circ Res 66: 913–931, 1990
7. Lawson CS, Dawney JM: Preconditioning: State of the art of myocardial protection. Cardiovasc Res 27: 542–550, 1993
8. Hagar JM, Hale SL, Kloner RA: Effect of preconditioning ischemia on reperfusion arrhythmias after coronary artery occlusion and reperfusion in the rat. Circ Res 68: 61–68, 1991
9. Shiki K, Hearse D: Preconditioning of ischemic myocardium: reperfusion induced arrhythmias. Am J Physiol 253: H147–H1476, 1987
10. Vegh A, Szekeres L, Parratt JR: Protective effects of preconditioning of the ischemic myocardium involve cyclooxygenase products. Cardiovasc Res 24: 1020–1023, 1990

11. Vegh A, Szekeres L, Parratt J: Preconditioning of the ischemic myocardium – involvement of the L-arginine nitric oxide pathway Br J Pharmacol 107: 648–652, 1992
12. Li Y, Kloner RA: Cardioprotective effects of ischemic preconditioning are not mediated by prostanoids. Cardiovasc Rei 26: 226–231, 1992
13. Alexander RW, Kent KM, Pisano JJ, Keiser HR, Cooper T: Regulation of postocclusive hyperemia by spontaneously synthesized prostaglandin in the dog heart. J Clin Invest 55: 1174–1181, 1995
14. Woditsch I, Schror K: Prostacyclin rather than nitric oxide is a tissue protective factor in myocardial ischemia. Am J Physiol 263: H1390–H1396, 1992
15. Coker SJ, Marshall RJ, Parratt JR, Zeitlin IJ: Does the local myocardial release of prostaglandin E2 or F2 contribute to the early consequences of acute myocardial ischemia? J Mol Cell Cardiol 13: 425–434, 1981
16. Coker SJ, Parratt JR, Ledinghan I McA, Zeitlin IJ: Thromboxan and prostacyclin release from ischemic myocardium in relation to arrhythmias. Nature 291: 323–324, 1981
17. Ahumada G, Sobel BE, Needleman P: Synthesis of prostaglandins by cultured rat heart myocytes and cardiac mesenchymal cells. J Mol Cell Cardiol 12: 685–700, 1980
18. Bolton HS, Chanderbhan R, Bryat RW, Bailey JM, Weglicki WB Vahouny GV: Prostaglandin synthesis by adult cardiac myocytes. Mol Cell Cardiol 12: 1287–1298, 1980
19. Karmazyn M: Synthesis and release of cardiac eicosanoids with particular emphasis on ischemia and reperfusion. Can J Physiol Pharmacol 67: 912–921, 1989
20. Ferrari R, Cargnoni A, Ceconi C, Curello S, Belloli S, Albertini A, Visioli O: Protective effect of prostacyclin-mimetic on the ischemic-reperfused rabbit myocardium. J Mol Cell Cardiol 20: 1095–1106, 1988
21. Ravingerova T, Tribulova N, Ziegelhoffer A, Styk J, Szekeres L: Suppression of reperfusion induced arrhythmias in the isolate rat heart: pretreatment with 7-oxo prostacyclin *in vivo*. Cardiovasc Res 27: 1051–1055, 1993
22. Simpson PJ, Lucchesi BR: Myocardial ischemia: The potential therapeutic role of prostacyclin and its analogues. In R.J. Gryglewski and G. Stock (Eds). Springer-Verlag, Berlin/Heidelberg, 1988, pp 179–194
23. Parratt J: Endogenous myocardial protective (antiarrhythmic) substances. Cardiovasc Res 27: 693–702, 1993
24. Schror K, Addicks K, Darius H, Ohlendorf R, Rosen P: PGI_2 inhibits platelet activation and prevents myocardial damage by inhibition of catecholamine release from adrenergic nerve terminals. Evidence for cAMP as common denominator. Thromb Res 21: 175–180, 1981
25. Szekeres L, Szilvassy, Udvary E, Vegh A: 7-oxo-PGI_2 induced late appearing and long-lasting electrophysiological changes in the heart *in situ* of the rabbit, guinea-pig, dog and cat. J Mol Cell Cardiol 21: 545–554, 1989
26. Liu GS, Stanley AW, Downey J: Cyclooxygenase products are not involved in the protection against myocardial infarction afforded by preconditioning in rabbit. Cyclooxygenase pathway's involvement in preconditioning. Am J Cardiovasc Pathol 4: 157–164, 1992
27. Wainwright CL: Myocardial preconditioning as the heart's self protecting response against the consequences of ischemia. Trends Pharmacol Sci 13: 90–93, 1992
28. Fagbemi SO: The effect of aspirin, indomethacin and sodium meclofenamate on coronary artery ligation arrhythmias in anaesthetized rats. Eur J Pharmacol 97: 283–287, 1984

Molecular and Cellular Biochemistry **160/161**: 257–263, 1996.

Prevention by 7-oxo-prostacyclin of the calcium paradox in rat heart: Role of the sarcolemmal (Na,K)-ATPase

Attila Ziegelhöffer,[1] Tatiana Ravingerová,[1] Andrej Džurba,[1] Narcisa Tribulová,[1] Ján Slezák,[1] Albert Breier[2] and László Szekeres[3]

[1]*Institute for Heart Research, Slovak Academy of Sciences;* [2]*Institute of Molecular Physiology and Genetics, Bratislava, Slovak Republic;* [3]*Institute of Pharmacology, Albert Szent-Györgyi Medical University, Szeged, Hungary*

Abstract

It is demonstrated a fast and significant depression in the sarcolemmal (Na,K)-ATPase activity that occurs as early as 25 sec after the onset of Ca^{2+} depletion, and participates in the development of Ca^{2+}-paradox in the rat heart. Pretreatment of the animals with 7-oxo-prostacyclin (PGI2) 24– 48 h prior to the experiment prevented fairly the Ca^{2+}-depletion-induced depression in (Na,K)ATPase activity and the accompanying structural and functional damage to the heart and sarcolemma during Ca^{2+}-depletion as well as the development of Ca^{2+}-paradox during the subsequent Ca^{2+}-repletion. Pretreatment with PGI_2 was chosen intentionally because previous experiments revealed, that in its late effect the drug is acting via stabilizing the membranes due induction of high activity of (Na,K)-ATPase that has increased affinity to ATP. From results obtained the following may be concluded: If during the phase of Ca^{2+}-deprivation, the capability of heart sarcolemma to maintain sodium extrusion remains preserved, the expected aggravation of Ca^{2+}-overload injury to Ca^{2+}-paradox that would develop during Ca^{2+}-repletion, may be definitely prevented. Sufficiently preserved (Na,K)-ATPase activity, hand in hand with stabilized sarcolemmal structure, may prevent an accumulation of sodium beneath the sarcolemma and consequently also an overexcessive entry of Ca^{2+} into the myocytes. (Mol Cell Biochem **160/161**: 257–263, 1996)

Key words: 7-oxo-prostacyclin, calcium paradox, sarcolemmal (Na,K)-ATPase, subsarcolemmal sodium, subsarcolemmal Na/Ca ratio, calcium overload of the heart

Introduction

Since 1966, when Zimmerman and Hülsman [1] discovered that repletion of calcium to isolated hearts perfused for a brief period with calcium-free medium paradoxically results in collapse of systems controlling calcium entry into myocytes, the phenomenon of calcium paradox became the matter of numerous studies. However, the main question, what may be the prime cause triggering the cascade of events yielding in excess calcium-induced damage of cardiac cells, could still not be answered definitely. The initial conception based on increased entry of extracellular Na^+ during depletion and/or at the very beginning of repletion of Ca^{2+} [2] was weakened by absence of increase in total intracellular Na^+ concentration [3]. Oppositely, the latter finding seemed rather support the role of early ultrastructural alterations observed during Ca2+-depletion on the sarcolemma and intercalated discs [4, 5]. These might indicate that calcium paradox could be caused by rupture of membranous structures that may occur upon reentry of excess Ca^{2+}, i.e. by mechanical damage to the myocyte. Nevertheless, in respect to recently described inhomogeneity in intracellular concentration of sodium [6, 7] as well as to the recognition that precautions taken to prevent the deterioration of Na^+ homeostasis during the depletion of Ca^{2+} may also protect the heart against calcium paradox [8, 9], antagonized again the concept of mechanical damage. Hence, the nature of the true prime impulse triggering the calcium paradox remains still a 'bundle of contradictions' [10].

In the present paper we are attempting to provide some ad-

Address for offprints: A. Ziegelhöffer, Institute for Heart Research, Slovak Academy of Sciences, Dúbravská cesta 9, 842 33 Bratislava, Slovak Republic ditional information about the possible role of sarcolemmal (Na,K)-ATPase and the respective subsarcolemmal compartment of sodium (that may immediately be extruded or participate in the Na/Ca-exchange mechanism) in initiation of calcium paradox. We would like to demonstrate that Ca^{2+}-paradox may be prevented, when during depletion of Ca^{2+} the activity of sarcolemmal (Na,K)-ATPase can be maintained high enough to keep the concentration of Na^+ in the subsarcolemmal compartment on appropriate level. The respective modulation of the sodium pump activity may be achieved by second window preconditioning with 7-oxo-prostacyclin [5, 8, 9].

Materials and methods

Treatment of animals

All studies were performed in accordance with the Guide for the care and use of laboratory animals published by the US National Institutes of Health (NiH publication No 85-23, revised 1985).

Male Wistar rats, specific pathogen free strain (230–280 g body weight) fed on a standard pellet diet and tap water *ad libitum* were used for all studies. The animal were anaesthetized with pentobarbitone sodium (60 mg/kg b.wt.). Ten min after i.p. administration of heparin (500 IU) the animals were sacrificed and the hearts were rapidly excised and arrested in ice-cold buffer. A group of 20 animals was pretreated with a single dose of 5 μg 7-oxo-prostacyclin (PGI_2) in 100 μl per 100 g b.wt. intramuscularly, 48 h before excision of the heart [11]. Ten animals were injected with saline instead of PGI_2 and used as controls to the pretreated group. The excised hearts were than cannulated via the aorta and perfused at 37°C in non-recirculating mode employing the Langendorff technique at a constant perfusion pressure of 75 mmHg. The perfusion medium was Krebs-Henseleit (KH) bicarbonate solution (pH 7.4) gassed with 95% oxygen and 5% carboxide, containing in $mmol.l^{-1}$: NaCl 118.5; $NaHCO_3$ 25.0; $MgSO_4$ 1.2; Na_2HPO_4 1.2; $CaCl_2$ 1.4; KCl 3.0 and glucose 11.5. Precautions were taken to avoid the precipitation of calcium; solution was filtered through a 5 μm porosity filter to remove the contaminants.

In hearts in which the integrity of sarcolemma was investigated in terms of its permeability to lanthanum ions, the perfusion with KH solution was in final stage replaced by perfusion with Hepes solution (pH 7.4) containing in $mmol.l^{-1}$ NaCl 140.0; KCl 4.0; $MgCl_2$ 1.0; $LaCl_3$ 0.05 and Hepes 3.0. This solution was gassed with 100% O_2.

Calcium paradox was induced by 3 min perfusion with calcium free KH solution, followed by 10 min perfusion with normal calcium containing KH medium. Introduction of the calcium paradox was preceeded by 15 min lasting stabilized perfusion of the hearts in oxygenated mode.

An epicardial electrogram (EG) was recorded by means of a pair of stainless steel electrodes attached to the apex of the heart and to the aortic cannula, respectively. Electrical activity of the heart was characterized as the heart rate (HR) calculated from EG. Left ventricular pressure was measured by means of a plastic water-filled catheter placed into the left ventricle and connected to a Statham pressure transducer (P23 Db). Left ventricular developed pressure (systolic minus diastolic pressure), rate of maximal pressure development (dP/dt max), as well as HR were used to assess cardiac function and were monitored continuously on Mingograph ELEMA (Sweden). Coronary flow was measured by a timed collection of the coronary effluent.

Investigation of ATPase activities

Ten per cent (volume/weight) cell free homogenate from freshly excised hearts was prepared by means of a Polytron (PT-20, Switzerland) homogenizer (4 times 15 sec at 0°C, setting 5). The homogenization medium contained in $mmol.l^{-1}$: Tris-HCl 10.0 (pH 7.4); sucrose 10.0; EDTA 0.5 and 1% deoxycholate-Na salt. The homogenate was filtered through two layers of gauze and immediately used for estimation of the ATPase activities.

The specific activities of sarcolemmal (Na,K)-ATPase and Ca^{2+}-ATPase with low affinity to calcium were estimated by measuring the phosphate liberated from splitting of ATP. Reaction was allowed to proceed for 10 min at 37°C in 1 ml of a medium containing in 50 $mmol.l^{-1}$ histidine-HCl buffer (pH 7.0), 45–75 μg of homogenate protein and an optimal composition of stimulatory ions, as follows: 2, 10 and 100 $mmol.l^{-1}$ $MgCl_2$, KCl and NaCl – for the (Na,K)-ATPase and 2.0 $mmol.l^{-1}$ $CaCl_2$ for Ca^{2+}-ATPase. Enzyme reaction was started by the addition of a final concentration of 2 $mmol.l^{-1}$ ATP and it was terminated by 1 ml of ice-cold trichloroacetic acid (0.73 $mmol.l^{-1}$). The high contents of sodium and potassium that were in tissue homogenate already present, does not enable the estimation of (Na,K)-ATPase activity via a direct activation of the enzyme with K^+, Na^+ and Mg^{2+} ions. Therefore, the enzyme activity was estimated as the difference between phosphate splitted from ATP in the presence and in the absence of ouabain (1.0 $mmol.l^{-1}$). All measurements of enzyme activities were performed in three parallel samples. For further details about estimation of the enzyme activities see [11].

Morphological and histochemical investigations

Morphological investigations were preformed in parallel series to all experimental groups. For this purpose, the hearts

(5 from each group) were immediately fixed by perfusing with 2.5% glutaraldehyde in 0.1 mmol.l^{-1} cacodylate buffer (pH 7.3) for 10 min. Than, the fixed tissue was sectioned transmurally, into 1 mm^3 pieces. Samples (n = 5) were taken from the left ventricle of each heart and an additional 3 h fixation was performed. After rinsing in 0.1 mmol.l^{-1} cacodylate buffer and postfixation in 1% buffered osmium tetroxide for 1 h, the tissue pieces were dehydrated in ethanol and propylene oxide, and finally embedded in EPON 812. One µm thick tissue sections (n = 10 per tissue block) were cut and stained with 1% toluidine blue. Sections from randomly chosen blocks from each heart were examined by light microscopy. The representative blocks were thin sectioned and stained with uranyl acetate and lead citrate for examination in electron microscope (Tesla 500). In the average 10 micrographs per section were used for ultrastructural analysis. This procedure is essentially similar to that described in our earlier studies [5, 12].

Histochemical estimation of ATPase was performed in 10 gm thick macro slices prepared in cryostat (Dittes, Duspiva). Freshly obtained slices from the left ventricle were incubated for 30 min at 37°C in 80 mmol.l^{-1} Tris-maleate buffer (pH 7.2), containing in addition (in mmol.l^{-1}) $Pb(NO_3)2$ 2; $MgSO_4$ 2.5; Tetramizol 1; ATP-Na salt 1. After incubation the slices were washed with distilled water and subsequently immersed for 1 min in yellow ammoniumsulfide solution. Finally, the slices were embedded in gelatine and investigated in light microscope.

Statistical procedures

Results were expressed as means ± S.E.M. The one way analysis of variance (Anova) was first applied to test for treatment as a source of variance. Subsequently the data from the pretreated groups were compared with those from the untreated groups using the unpaired Student's *t*-test with $p < 0.05$ considered as significant. The changes in specific activities of enzymes were evaluated by means of the *t*-test only.

Results

Investigation of haemodynamic parameters indicated that in hearts pretreated with PGI_2 did not reveal the alterations typical for calcium paradox (Table 1). A more detailed analysis of single variables indicated: A 14–16% and an 11–27% decrease in HR observed one min after the onset of Ca^{2+}-deprivation and during Ca^{2+}-repletion respectively. Nevertheless, none of these changes was found statistically significant.

In hearts of animals pretreated with PGI_2 significantly high dP/dt values and better recovery of coronary flow (both $p < 0.001$) were registered during the phase of Ca^{2+}-repletion. In the same experimental conditions, i.e. after completed Ca^{2+}-depletion and Ca^{2+}-repletion, all hearts from the group of untreated control animals were failing.

(Na, K)-ATPase and Ca^{2+}-ATPase activities in homogenates from the myocardium were investigated to indicate the degree of Ca^{2+}-caused damage to cardiac sarcolemma (Fig. 1.). Results revealed that after 3 min lasting C^{2+}-deprivation the activity of (Na,K)-ATPase in not pretreated control hearts was depressed to 19.3% ($p < 0.001$) whereas in PGI_2-pretreated hearts it was still exceeding by 37.2% ($p < 0.05$) the activity find in saline-treated controls not subjected to Ca^{2+}-depletion and repletion.

The activity of (Na,K)-ATPase in hearts of control animals pretreated with PGI_2 48 h prior to the experiment was elevated expectedly [11] by 349% ($p < 0.001$). This stimulatory effect on the (Na,K)-ATPase proved to be relatively specific since the activity of the low affinity Ca^{2+}-ATPase estimated parallelly, was inhibited by PGI_2-pretreatment to 88.2% ($p < 0.05$). Moreover, PGI_2 failed to affect the activity of Ca^{2+}-ATPase in Ca^{2+}-depleted hearts ($p < 0.05$).

During Ca^{2+}-deprivation the activity of (Na,K)-ATPase becomes fast depressed. The representative curve (Fig. 2.), obtained by the method of nonlinear curve fitting, indicate that within the first 25 sec after the onset of Ca^{2+} depletion only 33.3% of the enzyme activity remained preserved. In following time intervals the rate of (Na,K)-ATPase depression slowed down and the residual value amounting approxi-

Table 1. Effect of 7 oxo PGI_2 on heart function during Ca-depletion and Ca-repletion in isolated rat hearts

Group (min)	Control value	Ca-depletion (1)	(3)	Ca-repletion (min) (1)	(5)	(10)
Untreated	(n = 8)					
dP/dt	1541.4 ± 59.6	ND	ND	ND	ND	ND
HR	205.2 ± 12.9	172.5 ± 12.8	170.6 ± 13.5	182.5 ± 9.8	129.4 ± 18.7	155.0 ± 19.7
CF	5.8 ± 0.6	7.7 ± 1.0	9.3 ± 1.0	6.3 ± 0.5	2.9 ± 0.4	3.2 ± 0.7
Pretreated	(n = 6)					
dP/dt	1577.3 ± 152.2	ND	ND	1162.7* ± 218.9	858.0* ± 204.1	830.7 ± 184.5
HR	228.3 ± 12.0	198.3 ± 9.8	246.7 ± 16.2	180.0 ± 17.3	175.0 ± 17.0	175.0 ± 15.2
CF	6.1 ± 0.5	6.1 ± 0.04	9.6 ± 0.6	7.0 ± 0.6	6.9* ± 0.4	7.3* ± 0.6

dP/dt - contractility (mm Hy/sec); HR - heart rate (beats/min); CF - coronary flow (ml/min). Data are means ± S.E.M. *$p < 0.001$, pretreated vs untreated groups. ND - not determined.

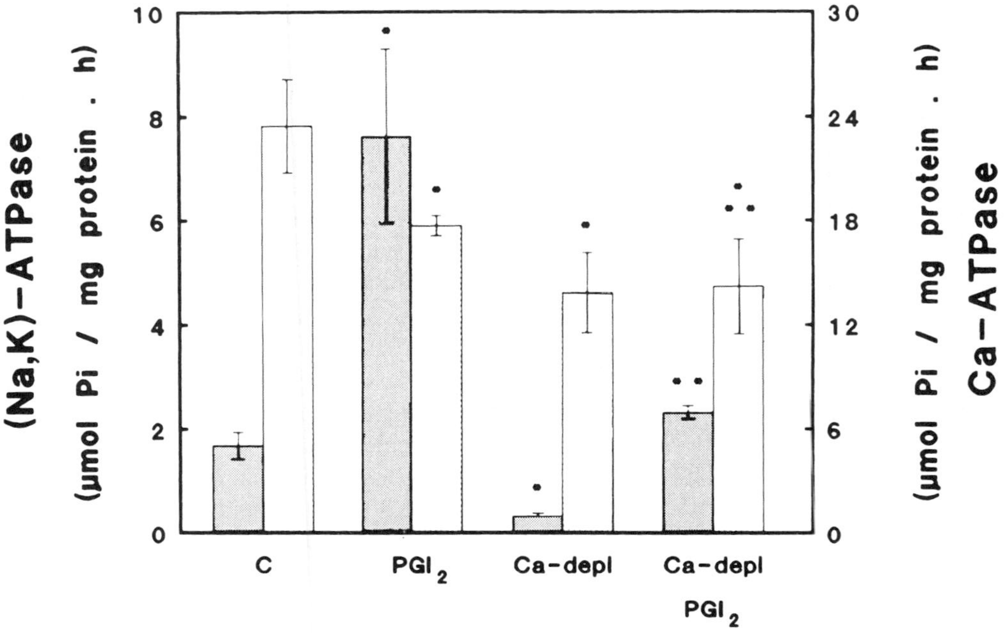

Fig. 1. Changes in activities of (Na,K)-ATPase and Ca^{2+} -ATPase in rat heart homogenates after 3 min lasting depletion of calcium. The effect of 7-oxo-prostacyclin. For experimental conditions see Materials and methods. C - Controls; PGI_2 - Pretreatment with 7-oxo-prostacyclin; Ca- depl - Calcium depletion; Ca - depl PGI_2- Calcium depletion in the 7-oxo-prostacyclin-pretreated group; Bars crosshatched, fine - (Na,K)ATPase; Bars dotted, fine - Ca^{2+}-ATPase; Results are means ± S.E.M. from 10 experiments; *$p < 0.01$ against the respective controls; **$p < 0.01$ against the control as well as the 7-oxo-prostacyclin-treated and the Ca^{2+} -depletion groups; ***-$p > 0.05$ against the 7-oxoprostacyclin-treated and the Ca^{2+}-depletion groups.

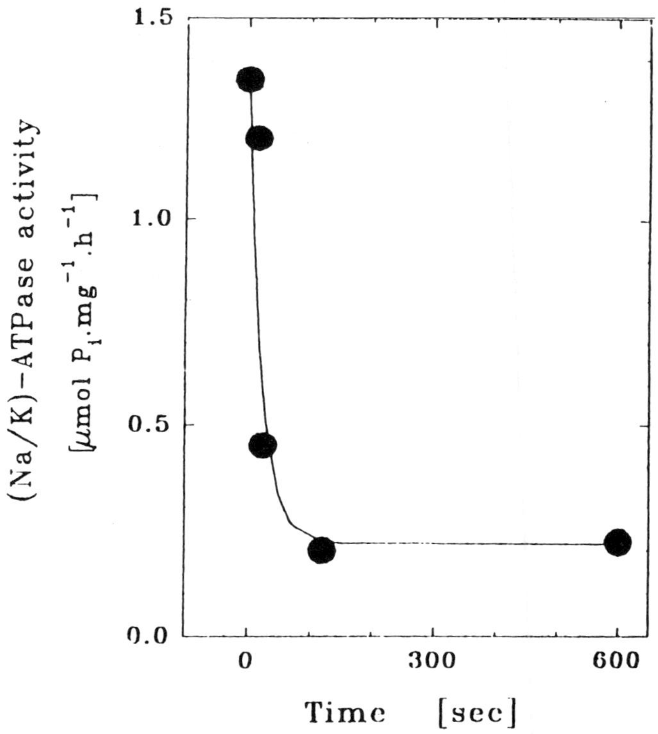

Fig. 2. Time course of the calcium Ca^{2+}-depletion-induced depression in the (Na,K)-ATPase activity of rat heart homogenate. The points represent means with a diameter equal to S. E. M., n = 10. For experimental conditions see Materials and methods.

mately 15% of the original activity was reached after approximately 2 min of Ca^{2+}-depletion. The process of non-linear curve fitting followed the equation $f=[a*exp(-b*t)]+c$, where the sum $a+c$ represents the residual activity in time $t \rightarrow \infty$ and b represent the rate constant.

Morphological investigation of hearts after Ca^{2+}-depletion revealed marked ultrastructural deterioration of the cardiomyocytes. Main characteristics of this deterioration were: swollen mitochondria, partially splitted sarcolemma as well as abundant intracellular deposits of lanthanum (Fig. 4.), in contrast to well preserved ultrastructure and lanthanum-inpermeable sarcolemma (Fig. 3.) in time-matched control hearts. In accordance with the morphological changes, ATPase histochemistry of hearts with Ca^{2+} -paradox exhibited damaged cardiomyocytes with reduced enzyme activity (Fig. 6.). This contrasted strongly with the PGI_2-pretreated hearts (that had elevated (Na,K)-ATPase activity already prior the onset of calcium deprivation, see Fig. 1.) where both, the integrity of cardiomyocytes (Fig. 5.) and the ATPase activities (Fig. 7.) remained fairly good preserved even after replacement of calcium.

Discussion

Calcium free perfusion-induced alterations in cellular homeostasis of cations are initiated by physico-chemical and/

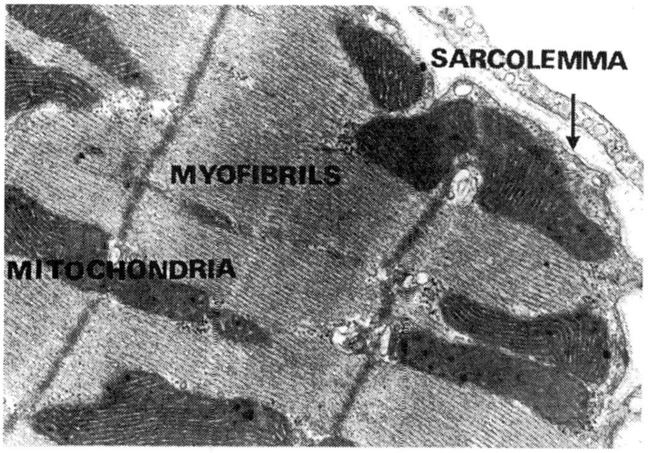

Fig. 3. Electronmicrograph of isolated perfused control heart with well preserved ultrastructure. Length of the normoxic perfusion 35 min. For technical details see the Materials and methods. Magnification × 31,000.

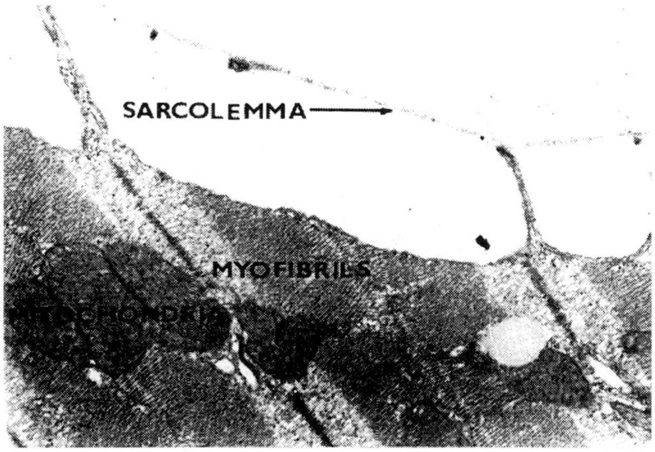

Fig. 4. Electronmicrograph of isolated perfused heart after 15 min of stabilized perfusion followed by 3 min of Ca^{2+}-depletion. Evident structural changes of the sarcolemma with separated glycocalyx and penetration of lanthanum into the cardiomyocyte. For technical details see Materials and methods. Magnification × 23,000.

or chemical forces acting from the extracellular space. Therefore, the structures of sarcolemma that are directly facing the extracellular space may represent the prime target for these forces. Structures, jeopardized this way are also believed to be responsible for disarrangement of influx and internal content of cations that occurs during Ca^{2+} depletion and at the onset of Ca^{2+}-repletion. In these conditions Alto and Dhalla (1979) found minor changes in sarcolemmal ATPase activities only. However, in contrary, using different sarcolemmal preparation Lamers *et al.* (1984) revealed a considerable depression in ATPase activities already at the end of the Ca^{2+}-deprivation period. To bypass the problem which membrane preparation may reflect more genuinely the changes induced by Ca^{2+}-depletion and Ca^{2+}-repletion, in

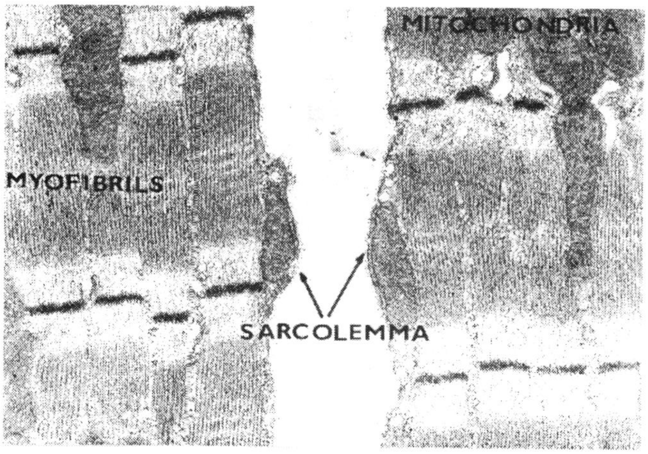

Fig. 5. Electronmicrograph of isolated perfused rat heart with (Na,K)-ATPase activity elevated in consequence of pretreatment of animals with 7-oxo-prostacyclin. The effect of Ca^{2+}-depletion. Experimental conditions were similar to those in Fig. 4. Note the well preserved ultrastructure and sarcolemmal membrane integrity. For technical details see the Materials and a methods. Magnification × 33,500.

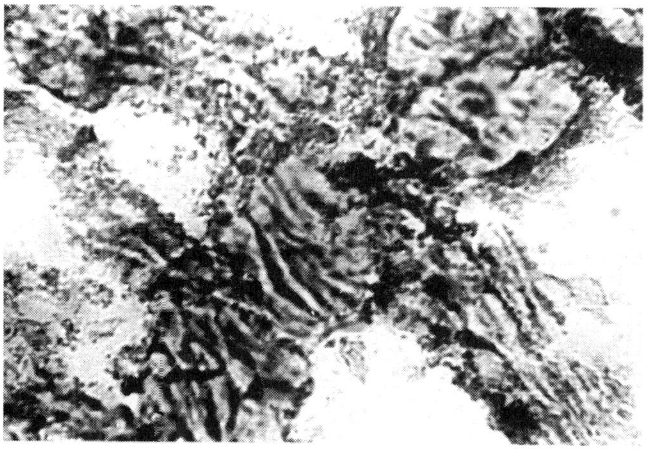

Fig. 6. Microphotography of histochemical detection of sarcolemmal- and myosin-ATPases in isolated perfused rat heart. The effect of calcium paradox. Note, disrupted cardiomyocytes with reduced ATPase activity. For technical data see Materials and methods. Magnification × 120.

present study cell free homogenates of the myocardium were utilized. In corroboration with the results of Lamers *et al.* (1984) we have also found a Ca^{2+}-deprivation-induced depression in cardiac sarcolemmal ATPases and particularly in (Na,K)-ATPase activity (for the preliminary report see [15]). This depression was manifested ($p < 0.05$) as early as 20–25 sec after the onset of Ca^{2+}-depletion (Fig. 2). Moreover, it was further prograding and thus indicating both, structural weakening of the cell membrane [4, 16–20] as well as lowering the capability of sarcolemma to extrude Na^+ ions, respectively. A possible link between depression of the sodium pump activity and function of the intercalated discs remains still to be elucidated.

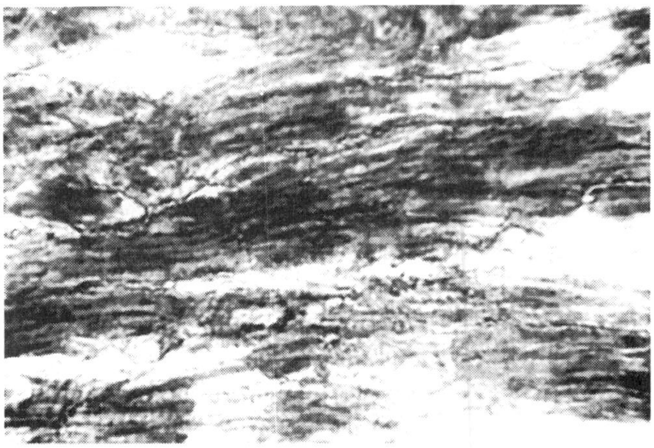

Fig. 7. Microphotography of histochemical detection of sarcolemmal and myosin ATPases in isolated perfused rat heart with (Na,K)-ATPase activity elevated in consequence of pretreatment of animals with 7-oxo-prostacyclin. The effect of experimental conditions applied in induction of complete Ca^{2+}-paradox, i. e. 15 min stabilized normoxic perfusion followed by 3 min of Ca^{2+}-depletion and 10 min of Ca^{2+}-repletion. Note, the integrity of cardiomyocyte and ATPase activity remain fairly preserved. For technical data see Materials and methods. Magnification × 120.

It was established without any doubt [3, 19, 21) that Ca^{2+}-paradox differs from Ca^{2+}-overload by more extensive damage to the cell membrane [4, 5] Figs. 1–4]. This damage involves an extensive separation of the external lamina from the surface coat [16] and it was reported to occur already during Ca^{2+}-depletion. Following repletion of Ca^{2+} the damage is brought to a climax with disruption of the intercalated discs [17, 20] and an overall deterioration of the cardiac cells. From the aspect of time sequence the above events represent three discrete stages [4]: (1) The period of Ca^{2+}-depletion; (2) The early period or the very beginning of Ca^{2+}-repletion; (3) The period following cellular contracture and membrane damage. Although an overall agreement could be achieved about the processes operating in stage three, still much confusion remains in understanding the nature and hierarchy of mechanisms involved in the first two stages. Our finding of depressed (Na,K)-ATPase activity already at the onset of Ca^{2+}-depletion indicated the possibility of an early perturbation in transsarcolemmal extrusion of Na^+ that may represent an essential link in the chain of events yielding in Ca^{2+}-paradox. Consequently we hypothesized that the observed early decrease in (Na,K)-ATPase activity may alter the concentration of Na^+ as well as its ratio to other cations, at least in a narrow subsarcolemmal region [7] and this may have a triggering role in induction of Ca^{2+}-paradox. The above assumption seems to be proved by our further finding that prevention of the Ca^{2+}-depletion-induced depression in (Na,K)-ATPase activity (Fig. 4) may also prevent the development of Ca^{2+} -paradox (Table 1, Figs. 1, 5 and 7). The experimental tactics applied was based on our previous observation that pretreatment of rats with PGI_2 can lead to induction of heart sarcolemmal (Na,K)-ATPase with high activity anc increased affinity to ATP [11, 22, 23].

The exact mechanism by which the high activity of (Na,K)-ATPase may prevent the development of Ca^{2+}-paradox is not yet established definitely. Nevertheless, it may be speculated, that the highly active Na^+-pump is capable to maintain the Na^+ concentration in sub-sarcolemmal region [7] on a level that disables an intensive activation of Na^+/Ca^{2+} exchange system requested to promote a massive and later uncontrolled Ca^{2+}-entry into the myocytes. In this respect, the ratio of Na^+ to other cations present in subsarcolemmal compartment [7] may be even a more important criterion than the actual Na^+ concentration itself. However, an other possibility involving a general membrane stabilizing effect of PGI_2, cannot be excluded either. Nevertheless, the latter possibility seems to be slightly less important because PGI_2 seems to be acting rather selectively, increasing mainly the activity of (Na,K)-ATPase and not the activity of other membrane bound enzymes such as the Ca^{2+} ATPase (Fig. 1) etc. [11].

Whatever may be the reason for the Ca^{2+}-deprivation-induced changes in (Na,K)-ATPase activity, its prevention with PGI_2 indicated [8, 9, 23] that the events subsequently following the depression of the enzyme activity may be similar to those reported after ouabain-induced inhibition of the sodium pump [6, 7]. As concerns the question whether extracellular sodium may contribute in induction of Ca^{2+}-paradox [15, 16, 24] or not [4, 25, 26], the described role of (Na,K)-ATPase does not excludes any possibility. In any case, it is the activity of the latter enzyme that seems to control, whether the local sub-sarcolemmal concentration of Na^+ may reach the threshold level necessary for activation of the Na^+/Ca^{2+} exchanger, i.e. trigger the Ca^{2+}-paradox, or not. It has to be stressed once again, that the induction of Ca^{2+}-paradox may be triggered by a Ca^{2+}-deprivation-caused increase in the Na^+/Ca^{2+} ratio beneath the sarcolemma as well.

From biochemical and histochemical investigation of sarcolemmal ATPase activities as well as investigation of functional parameters and the ultrastructure of the heart in present study the following conclusion may be drawn: If during the phase of Ca^{2+}-deprivation, the capability of heart sarcolemma to maintain sodium extrusion remains preserved (Fig. 1), the aggravation of Ca^{2+}-overload injury to Ca^{2+}-paradox that would occur during Ca^{2+}-repletion may be definitely prevented (Table 1, Figs. 5 and 7). Sufficiently preserved Na-K-ATPase activity, hand in hand with stabilized sarcolemmal structure [4, 5, 19], may prevent an accumulation of sodium beneath the sarcolemma [7] and consequently also an over-excessive entry of Ca^{2+} [2, 24] into the myocytes.

Acknowledgements

The excellent technical assistance of Mrs. Hýbelová, Havránková, Hradecká and Brichtová is gratefully acknowledged.

This work was supported in part by the Slovak Grant Agency for Science (grants No. 2/1258/95, 2/1256/95, 2/1257/95 and 2/1001/95 as well as by the European Communities grant COST CIPA CT 92 4009

References

1. Zimmerman ANE, Hülsmann WC: Paradoxical influence of calcium ions on the permeability of the cell membranes of the isolated rat heart. Nature 211: 646–647, 1966
2. Chapman RA, Tunstall J: The calcium paradox of the heart. Progr Biophys Mol Biol 50: 67–96, 1987
3. Ruigrok TJC: Is an increase of intracellular Na^+ during Ca^{2+} depletion essential for the occurrence of the calcium paradox? J Mol Cell Cardiol 23: 499–503, 1991
4. Altschuld RA, Ganote CE, Nayler WG, Pieper HM: What constitutes the calcium paradox? J Mol Cell Cardiol 23: 765–767, 1991
5. Tribulová N, Slezák J, Ravingerová T: Influencing of calcium paradox by membrane stabilizing agents: The role of sarcolemma and intercalated disc. J Submicrosc Cytol Patol 25: 267–277, 1993
6. Isenberg G, Wendt-Gallitelli MF: X-ray microprobe analysis of sodium concentration reveals large transverse gradients from the sarcolemma to the centre of voltage clamped guinea-pig ventricular myocytes (Abstract). J Physiol (London) 420: 86P, 1990
7. Carmeliet E: A fuzzy subsarcolemmal space for intracellular Na^+ in cardiac cells? Cardiovasc Res 26: 433–442, 1992
8. Ravingerová T, Styk J, Trégerová V, Pancza D, Slezák J, Tribulová N, Ziegelhöffer A, Pissarek M, Szekeres L: Protective effect of 7-oxo-prostacyclin on myocardial function and metabolism during postischemic reperfusion and calcium paradox. Basic Res Cardiol 86: 245–253, 1991
9. Ravingerová T, Tribulová N, Ziegelhöffer A: The late effect of 7-oxo-prostacyclin protects the rat heart against calcium paradox. Life Sci 53: 1309–1316, 1993
10. Chapman RA, Suleiman RS, Rodrigo GC, Tunstall J: The calcium paradox: a role for $[Na]_i$, a cellular or tissue basis, a property unique to the Langendorff perfused heart? A bundle of contradictions! J Mol Cell Cardiol 23: 773–777, 1991
11. Džurba A, Ziegelhöffer A, Breier A, Vrbjar N, Szekeres L: Increased activity of sarcolemmal (Na^+,K^+)-ATPase is involved in the late cardioprotective action of 7-oxo-prostacyclin. Cardioscience 2: 7–9, 1991
12. Slezák J, Tribulová N, Schulze W, Ziegelhöffer A, Ravingerová T, Okoličány J, Trégerová V: Differences in transmural distribution of cardiomyocyte injury. Ca-paradox versus postischemic reperfusion phenomenon. Biomed Biochim Acta 48: 102–107, 1989
13. Alto LE, Dhalla NS: Myocardial cation contents during induction of calcium paradox. Am J Physiol 237: 11713–11719, 1979
14. Lamers JMJ, Stinis JT, Ruigrok TJC: Biochemical properties of membranes isolated from calcium-depleted rabbit hearts. Circ Res 54: 217–226, 1984
15. Ziegelöffer A, Ravingerová T, Džurba A, Tribulová T: Changes in sarcolemmal (SL) ATPase activities during calcium depletion are involved in development of calcium paradox (CaP) in rat heart (Abstract). J Mol Cell Cardiol 24: (Supplement 5) S.98, 1992
16. Baker JE., Bullock GR, Hearse DJ: The temperature dependence of the calcium paradox: Functional and morphological correlates of cellular injury. J Mol Cell Cardiol 15: 393–341, 1983
17. Brusselen P: Effects of sodium on membrane damage induced by calcium paradox in rat heart (Abstract). J Mol Cell Cardiol 22: 507–522, 1991
18. Piper HM, Siegmund B: The calcium paradox and the oxygen-paradox in heart and isolated cardiomyocytes (Abstract). J Mol Cell Cardiol 23: (Supplement 5) S.149, 1991
19. Ruigrok TJC, van Echteld CJA: Pathophysiology of the calcium paradox (Abstract) J Mol Cell Cardiol 24: (Supplement I) S.26, 1992
20. Severs NJ: The structure and function of cardiac intercellular junctions (Abstract). J Mol Cell Cardiol 23: (Supplement V) S.150, 1991
21. Ruigrok TJC, van Echteld CJA, The intracellular Na^+ concentration prior to Ca^{2+} repletion has no bearing on the occurrence of the calcium paradox as originally defined. J Mol Cell Cardiol 23: 769–771, 1991
22. Ziegelhöffer A, Grünermel J, Džurba A, Procházka J, Kolář F, Vrbjar N, Pelouch V, Ošťádal B, Szekeres L: Sarcolemmal cation transport systems in rat hearts acclimatized to high altitude hypoxia. Influence of 7-oxo-prostacyclin. In: B. Ošťádal and N.S. Dhalla (eds). Heart Function in Health and Disease. Kluwer Academic Publishers, Boston, 1993, pp 219–228
23. Ziegelhöffer A, Ravingerová T, Džurba A, Slezák J, Tribulová N, Szekeres L: High Na/K-ATPase activity induced by pretreatment with 7-oxo prostacyclin (PGI_2) is involved in prevention of Ca-paradox (CaP) in the rat heart (Abstract). J Mol Cell Cardiol 23: (Supplement V) S.161, 1991
24. Chapman RA: A rise in intracellular sodium would seem to predispose the heart to the calcium paradox. J. Moll Cell Cardiol 22: 503–507, 1990
25. Slezák J, Tribulová N, Ravingerová T, Ziegelhöffer A, De Nollin S, Ruigrok TJC: The effect of lidocaine on concentration and localization of Ca^{2+} in Ca^{2+}-depleted and Ca^{2+}-repleted hearts (Abstract). J Mol Cell Cardiol 23: (Supplement V) S.161, 1991
26. van Echteld CJA, Kirkels JH, Eijgelshoven HMJ, van der Meer P, Ruigrok TJC: Intracellular sodium during ischemia and calcium free perfusion: A ^{23}Na NMR study. J Mol Cell Cardiol 23: 297–308, 1991

Molecular and Cellular Biochemistry **160/161**: 265–271, 1996.

Donors of nitric oxide mimic effects of ischaemic preconditioning on reperfusion induced arrhythmias in isolated rat heart

Maria Bilińska, Michał Mączewski, Andrzej Beręsewicz
Department of Clinical Physiology, Medical Centre of Postgraduate Education, Warsaw, Poland

Abstract

NO has been implicated in the mechanism of ischaemic preconditioning. To verify this hypothesis further we have attempted to reproduce effects of ischaemic preconditioning by nitric oxide donors administration prior to the ischaemia. The effect of glyceryl trinitrate (GTN) and 3-morpholino-sydnonimine-hydrochloride (SIN- 1), NO donors, on reperfusion induced ventricular tachycardia (VT) and ventricular fibrillation (VF) in Langendorff perfused rat hearts subjected to 10 min regional ischaemia followed by 10 min reperfusion were examined. Results: GTN, 500 μM and SIN-1, 10 μM, administered for 5 min and washed for another 5 min prior to ischaemia (to mimic ischaemic preconditioning), almost completely abolished reperfusion induced VF. GTN and SIN-1, administered at the time of reperfusion, increased the incidence of sustained VF and the duration of VT and VF. When given 5 min before the ischaemia and throughout the ischaemia and the reperfusion, SIN-1 abolished VF. Adenosine, 10 μM, applied according to the above three protocols, did not affect reperfusion arrhythmias, although adenosine induced changes in coronary flow and post-ischaemic reflow were similar to those produced by the NO donors. In conclusions: (1) NO is able to mimic the effect of ischaemic preconditioning on reperfusion arrhythmias in rat heart, supporting the view that NO may be one of the endogenous substances triggering ischaemic preconditioning; (2) In crystalloid-perfused heart, NO may be deleterious when its administration is restricted to the reperfusion period. (Mol Cell Biochem **160/161**: 265–271, 1996)

Key words: nitric oxide, preconditioning, reperfusion arrhythmias, isolated rat heart, SIN-1, glyceryl trinitrate, adenosine, nitric oxide toxicity

Introduction

Ischaemic preconditioning is an endogenous protective mechanism in which brief periods of myocardial ischaemia and reperfusion render a myocardium resistent to a subsequent more sustained ischaemic insult [1]. The mechanism underlying preconditioning is unknown and is currently under intense investigation. There is evidence in support of adenosine being an important mediator of protection in rabbits and dogs, where hearts can be preconditioned by administration of adenosine or adenosine A_1 agonists prior to the sustained ischaemic insult and where adenosine antagonists can block preconditioning-induced protection [2, 3]. However, adenosine mediated mechanism does not appear to play a role in the mechanism of preconditioning in rat heart [4–6], suggesting an important role of other endogenous mediator(s) in this species [2, 3].

Studies in anaesthetized dog have suggested that it may be an increased release of nitric oxide (NO) during the first brief ischaemic period which mediates the protection against reperfusion arrhythmias associated with the subsequent ischaemia/reperfusion [3, 7]. To verify this concept further, in this study we have attempted to mimic the effect of ischaemic preconditioning on reperfusion arrhythmias in rat heart by brief administration of NO donors prior to the ischaemia. In additional experiments, the donors were administered either throughout ischaemia and reperfusion or at the time of reperfusion only. Glyceryl trinitrate (GTN, which must be metabolized to release NO) and 3-morpholino-sydnonimine-hydrochloride (SIN-l, which spontaneously releases NO)

Address for offprints: A. Beręsewicz, Department of Clinical Physiology, Medical Centre of Postgraduate Education, Marymoncka 99, 01-813 Warszawa, Poland were used. Analogous studies were performed with adenosine to differentiate between direct effects of NO and those related to NO-induced changes in coronary flow.

Materials and methods

Isolated heart preparation

Male Wistar rats (250–300 g) fed *ad libitum* were injected with 500 units of heparin sulphate, i.p., 20 min prior to sacrifice. Animals were anaesthetized with sodium pentobarbital, 10–15 mg per rat, i.p. The heart was removed and placed in an ice-cold Krebs-Henseleit buffer. It was then connected to a perfusion apparatus and perfused in the Langendorff mode at a constant perfusion pressure of 75 cm H_2O. The pulmonary artery was cannulated. The heart was perfused in a non-recirculating manner with a prefiltered (5.0 μm Millipore filter) Krebs-Henseleit bicarbonate buffer containing (in mM) 118 NaCl; 23.8 $NaHCO_3$; 4.7 KCl; 1.2 KH_2PO_4; 2.5 $CaCl_2$; 1.2 $MgSO_4$; 11 glucose and 0.01 EDTA. The buffer was equilibrated with 95% O_2 + 5% CO_2 gas mixture giving pH 7.4 and PO_2 550–580 mmHg at 37°C. The heart was enclosed in a small stoppered waterjacketed chamber. The temperature of the perfusate, as well as that of the atmosphere surrounding the heart, was thermostatically controlled. The hearts were not paced.

At the onset of each experiment a ligature was placed around the left main coronary artery. Regional ischaemia and reperfusion were induced by occluding and releasing the snare as described before [8]. The successful induction of ischaemia and the adequacy of reperfusion were confirmed by measuring coronary flow. Coronary flow determinations were also used as an indirect estimate of the size of the ischaemic zone resulting from coronary occlusion. Coronary flow was determined by weighing of one min samples of the effluent. This represented ventricular coronary flow as in rat heart atria receive their blood supply from extracoronary vessels [9]. At the end of each experiment the ventricles were dried to obtain their dry wt.

Experimental protocols

Following 30 min of aerobic perfusion hearts were randomized to one of the following groups (n = 12–18 hearts per group): (1) control untreated group; (2) SIN(PC), GTN(PC) and Ado(PC) groups; in order to mimic ischaemic preconditioning SIN-1 (10 μM), GTN (500 μM) and adenosine (10 μM) were applied for 5 min and washed for another 5 min prior to the regional ischaemia; (3) SIN(A), GTN(A) and Ado(A) groups in which the infusion of the drug was initiated 1 min before the end of 10 min regional ischaemia and it was continued throughout the reperfusion; (4) SIN(B) and Ado(B) groups in which the infusion of the drug was initiated 5 min before the regional ischaemia and it was continued throughout the regional ischaemia and reperfusion; All hearts underwent 10 min regional ischaemia and 10 min of reperfusion during which arrhythmias were quantified.

Evaluation of reperfusion arrhythmias

An epicardial ECG recording was made with an Elema Schönander Mingograph-81 polygraph (Stockholm, Sweden) using silver wire electrodes attached directly to the heart. The electrograms were analyzed for: (i) the mean duration and overall incidence of ventricular tachycardia (VT) and ventricular fibrillation (VF) occurring within the whole reperfusion period, the latter determined as % of those hearts in an experimental group which were in sinus rhythm 30 s before and at the start of the reperfusion; (ii) the incidence of sustained VF (until the end of reperfusion period) – determined as % of those hearts having reperfusion-induced VF; (iii) the incidence of VT and VF within each individual minute of the reperfusion period (continuous presence + new entries) – expressed as % of those hearts in sinus rhythm at the start of the reperfusion. This latter procedure allows to analyze time-course of the reperfusion arrhythmias and gives an indirect estimate of the duration of the arrhythmia [8].The arrhythmias were defined according to the Lambeth Conventions [10].

Exclusion criteria

Excluded from the study were the hearts exhibiting: (i) any rhythm disturbances during the equilibration period; (ii) VT and/or VF during last 30 s of the regional ischaemia and (iii) ischaemic zone resulting from coronary occlusion < 35% of the total ventricular mass. At most one heart was excluded from each experimental group.

Drugs used and their application

Adenosine was obtained from Sigma, glyceryl trinitrate (GTN) from Schwarz-Pharma, Monheim. 3-Morpholino-sydnonimine-hydrochloride (SIN-1) was a gift from Cassella AG, Frankfurt/Main. The concentrated solutions of SIN-1, GTN and adenosine in Krebs-Henseleit buffer were freshly prepared and were administered directly to the aortic cannula by means of the microprocessor controlled microinjection pump. The glassware and tubing containing SIN-1 and GTN were protected from light.

Analysis of data

The parametric data are presented as means (S.E.M.). For coronary flow, heart rate and ischaemic zone size data, comparisons between multiple groups were performed using one-way variance analysis. If a statistically significant difference was established, the basal values were compared to the pre-ischaemic values using paired *t*-test (cf. Table 1). As the duration of VT and VF have no Gaussian distribution, the Kruskal-Wallis test was used to compare duration of VT and VF between control and treatment groups. To compare the incidence of VT and VF between untreated and treated groups an overall C test for $2 \times n$ table was constructed followed by a sequence of 2×2 of Fisher's exact test to compare the individual groups. $p < 0.05$ was considered significant.

Results

Reperfusion arrhythmias and NO donors; comparison with adenosine

In the untreated hearts the overall incidence of VT was 78%, that of VF was 65%, and that of sustained VF was 55% (Fig. 1, inset). In the majority of hearts, VT developed within the first minute of the reperfusion, falling within the next 3–4 min close to zero. The incidence of VF was fairly constant over the whole reperfusion period (Fig. 1). None of the interventions studied affected the incidence and the time-course of reperfusion induced VT (Figs 1 and 2).

SIN-1, 10 μM and GTN, 500 μM applied for 5 min and washed for another 5 min prior to the regional ischaemia greatly reduced the overall incidence and duration of reperfusion induced VF and completely abolished sustained VF (Table 1, Figs 1 and 2).

The severity of reperfusion induced VF was increased, compared to that in the untreated hearts, in the hearts in which SIN-1 (Fig. 1) and GTN (Fig. 2) were administered only at the time of reperfusion. This was evidenced by an increased duration as well as the incidence of VF in those hearts.

When both above protocols were combined in one and the administration of SIN-1 was started 5 min before the regional ischaemia and was continued throughout ischaemia and reperfusion, only an antiarrhythmic effect of the drug was evident (Fig. 1, SIN(B)). Thus, neither the time-course nor the incidence and the mean duration of the reperfusion arrhythmias was significantly different between the experimental groups SIN(PC) and SIN(B).

Irrespective of the protocol of its administration, neither the incidence of reperfusion induced VF (Fig. 3) nor the duration of VT and VF (Table 1) were affected by adenosine, 10 μM.

Hemodynamic data

Applied to mimic preconditioning, SIN-1, GTN and adenosine similarly increased coronary flow (to 149 ± 7.8%, 146 ± 6.7% and 160 ± 8.9% of the basal level, respectively). During the following 5 min washout period, the flow returned to its basal level in SIN-1 and GTN-treated hearts, and it was still increased in adenosine-treated hearts at the start of the regional ischaemia (Table 1). The reflow in the experimental groups of SIN(PC) (Fig. 4), GTN(PC) (not shown) and

Table 1. Haemodynamic variables, occluded zone size and duration of reperfusion induced ventricular tachycardia (VT) and ventricular fibrillation (VF)

Intervention	n	Coronary flow (ml/min/g dry wt) Basal	Coronary flow (ml/min/g dry wt) Pre-ischaemic	Ischaemic reduction of CF (%)	Heart rate (min^{-1}) Basal	Heart rate (min^{-1}) Pre-ischaemic	Mean duration of VT (sec)	Mean duration of VF (sec)
Control	18	59.7 ± 2.8	59.5 ± 3.2	44.5 ± 2.0	283 ± 9.6	283 ± 9.4	14.0 ± 4.4	227.8 ± 62
SIN-1(PC), 10 μM	12	62.3 ± 4.9	61.4 ± 4.0	43.2 ± 2.8	278 ± 9.6	275 ±10.4	10.0 ± 3.3*	11.1 ± 10*
SIN-1(A), 10 μM	12	66.4 ± 4.9	66.4 ± 4.9	45.0 ± 3.3	284 ± 5.4	284 ± 5.4	16.6 ± 3.9	397.8 ± 72*
SIN-1(B), 10 μM	12	63.5 ± 4.1	93.4 ± 5.8*	48.9 ± 3.5	300 ± 7.6	306 ± 7.7	16.5 ± 5.2	17.0 ± 12*
Nitroglycerine(PC), 500 μM	12	53.7 ± 4.5	54.4 ± 4.9	45.2 ± 3.1	288 ± 7.7	284 ± 8.5	6.7 ± 2.0*	11.3 ± 9.2*
Nitroglycerine(A), 500 μM	12	56.4 ± 2.2	57.0 ± 2.2	41.8 ± 2.1	277 ± 10.6	277 ± 10.6	11.3 ± 4.0	407.5 ± 72*
Adenosine(PC), 10 μM	12	54.8 ± 2.8	86.6 ± 4.3*	45.2 ± 1.8	249 ± 8.1	247 ± 7.6	10.0 ± 1.7	185.5 ± 24
Adenosine(A), 10 μM	13	56.8 ± 3.0	56.8 ± 3.0	41.3 ± 1.7	273 ± 7.5	273 ± 5.5	10.7 ± 1.5	192.3 ± 29
Adenosine(B), 10 μM	13	58.1 ± 3.1	93.2 ± 5.2*	45.2 ± 1.2	259 ± 15.6	248 ± 9.2	11.1 ± 4.2	193.3 ± 33

Values are means (S.E.M.) n = number of those hearts which after exclusions were included into the study; Coronary flow and heart rate were measured at 30 min of the stabilization (basal), just before the regional ischaemia (pre-ischaemic), and during the regional ischaemia (ischaemic). SIN-1(PC), Nitroglycerine(PC), Adenosine(PC) = the drugs were administered for 5 min and washed for another 5 min prior to the regional ischaemia; SIN-1(A), Nitroglycerine(A), Adenosine(A) = the drugs were administered 1 min before the end of the regional ischaemia and throughout the reperfusion; SIN-1(B), Adenosine(B) = the drugs were administered 5 min before regional ischaemia and administration continued throughout the rest of the experiment; *$p < 0.05$, compared to the respective basal value (CF and heart rate) or to the respective control (duration of VT and VF)

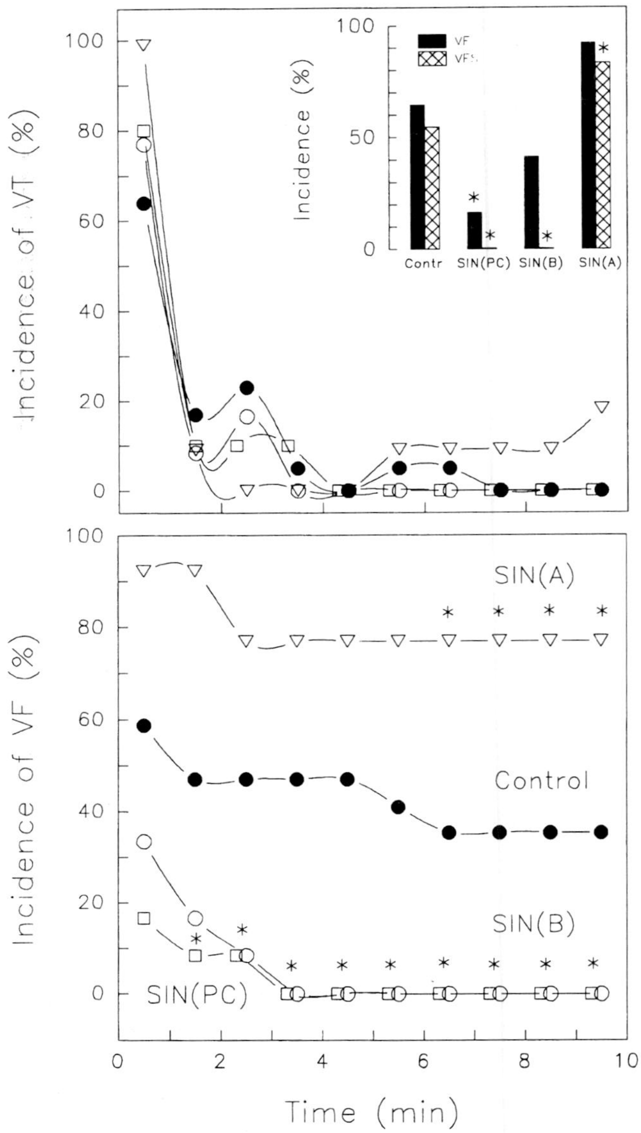

Fig. 1. Effect of SIN-1, 10 μM, on the time course of reperfusion induced ventricular tachycardia (VT) and ventricular fibrillation (VF) in isolated rat hearts. ●—● controls (Contr, n = 18); □—□ SIN-1 applied for 5 min and washed for another 5 min prior to regional ischaemia (SIN(PC), n = 12); Δ—Δ SIN-1 applied 1 min before the end of 10 min regional ischaemia and the administration continued throughout the reperfusion (SIN(A), n = 12); O—O SIN-1 applied 5 min prior to ischaemia and the administration continued throughout the regional ischaemia and reperfusion (SIN(B), n = 12); The overall incidence of VF (dark bar) and sustained VF (VFS, hatched bar) occurring within the whole 10 min reperfusion period is shown in the inset. *p < 0.05, for comparisons vs. controls.

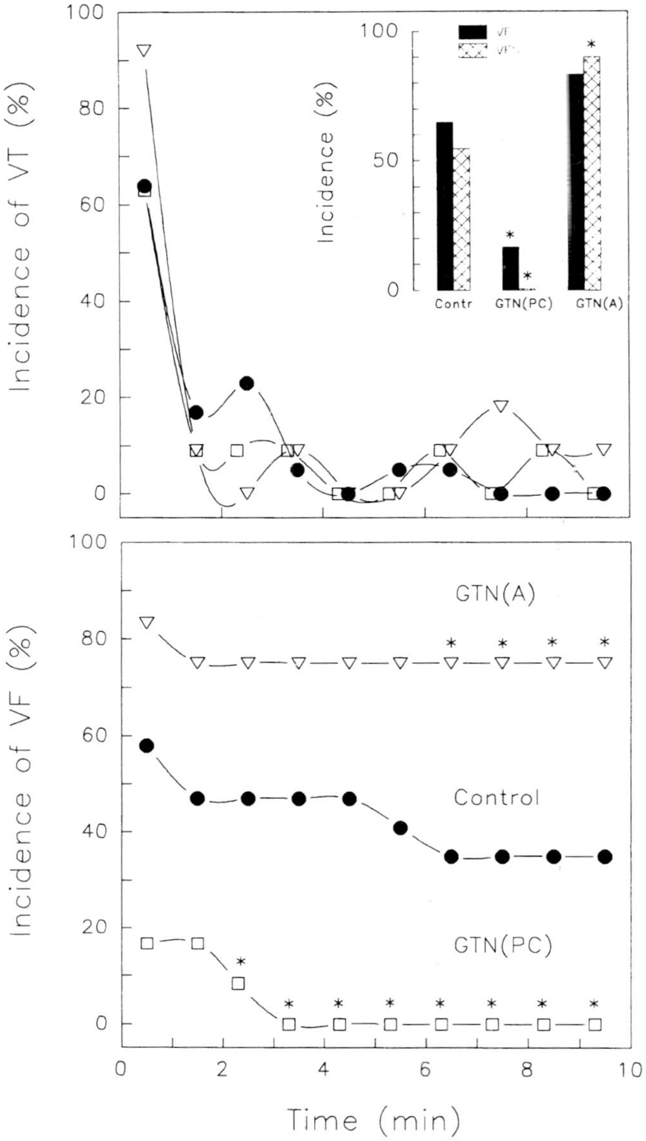

Fig. 2. Effect of glyceryl trinitrate (GTN, 500 μM), on the time course of reperfusion induced ventricular tachycardia (VT) and ventricular fibrillation (VF) in isolated rat hearts. ●—● controls (Contr, n = 18); □—□ GTN applied for 5 min and washed for another 5 min prior to regional ischaemia (GTN(PC), n = 12); Δ—Δ GTN applied 1 min before the end of 10 min regional ischaemia and the administration continued throughout the reperfusion (GTN(A), n = 12); The overall incidence of VF (dark bar) and sustained VF (VFS, hatched bar) occurring within the whole 10 min reperfusion period is shown in the inset. *p < 0.05, for comparisons vs. controls.

Ado(PC) (Fig. 5) was not different from that in the respective control group. Likewise, when administered at the time of reperfusion, SIN-1 (Fig. 4), GTN (not shown), and adenosine (Fig. 5) similarly increased post-ischaemic reflow. The same was true also when SIN-1 and adenosine were applied before the ischaemia.

The coronary ligation caused a similar per cent drop in coronary flow in the control as well as in the treatment groups (Table 1), suggesting the absence of the inter-group differences in the ischaemic zone size.

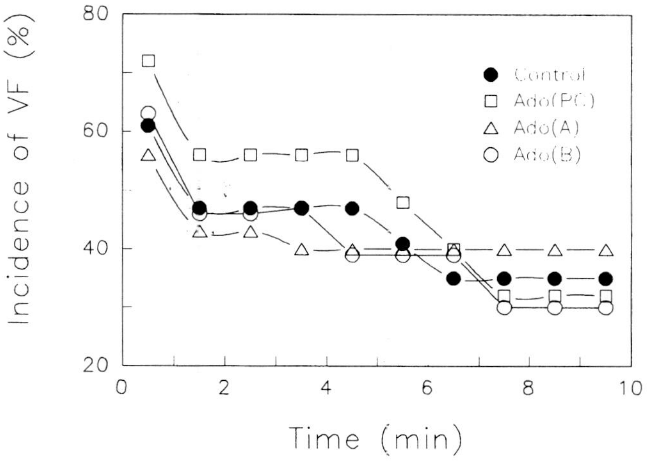

Fig. 3. Effect of adenosine, 10 μM, on the time course of the reperfusion induced ventricular fibrillation (VF) in isolated rat hearts. ●—● controls (n = 18); □—□ adenosine applied for 5 min and washed for another 5 min prior to regional ischaemia (Ado(PC), n = 12); Δ—Δ adenosine applied 1 min before the end of 10 min regional ischaemia and the administration continued throughout the reperfusion (Ado(A), n = 12); O—O adenosine applied 5 min prior to ischaemia and the administration continued throughout the regional ischaemia and reperfusion (Ado(B), n = 12); *p < 0.05, for comparisons vs. controls.

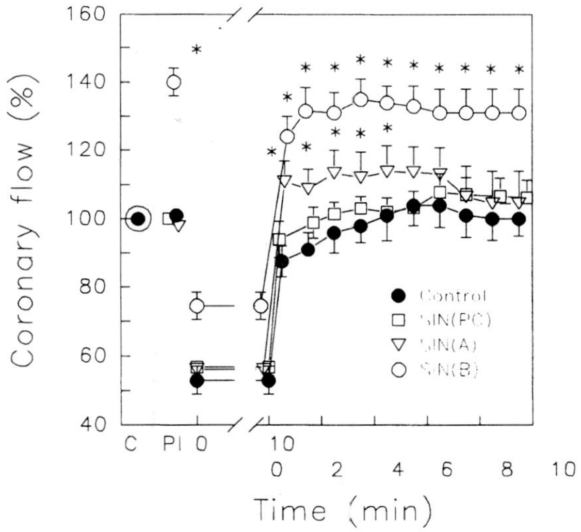

Fig. 4. Effect of SIN-1, 10 μM on coronary reflow following regional ischaemia in isolated rat hearts. SIN(PC) = SIN-1 was administered for 5 min and washed for another 5 min prior to the regional ischaemia; SIN(A) = SIN-1 was administered 1 min before the end of the regional ischaemia and throughout the reperfusion; SIN(B) = SIN-1 was administered 5 min before the regional ischaemia and throughout the ischaemia and reperfusion. Coronary flow was measured at (i) 30 min of the aerobic equilibration (C, expressed as 100%); (ii) just before the regional ischaemia (PI); (iii) during 10 min regional ischaemia; and (iv) during 10 min reperfusion. See Table 1 for the number of hearts in each experimental group. *p < 0.05, for comparisons vs. control

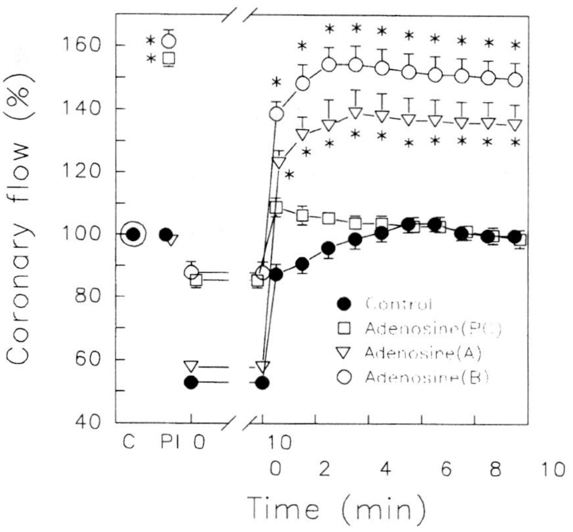

Fig. 5. Effect of adenosine, 10 μM on coronary reflow following regional ischaemia in isolated rat hearts. Ado(PC) = adenosine was administered for 5 min and washed for another 5 min prior to the regional ischaemia; Ado(A) = adenosine was administered 1 min before the end of the regional ischaemia and throughout the reperfusion; Ado(B) = adenosine was administered 5 min before the regional ischaemia and throughout the ischaemia and reperfusion. Coronary flow was measured at (i) 30 min of the aerobic equilibration (C, expressed as 100%); (ii) just before the regional ischaemia (PI); (iii) during 10 min regional ischaemia; and (iv) during 10 min reperfusion. See Table 1 for the number of hearts in each experimental group. *p < 0.05, for comparisons vs. control.

Discussion

We demonstrate that in isolated rat heart: (1) chemically unrelated NO donors, GTN and SIN-1, mimicked the protective effect of ischaemic preconditioning on reperfusion arrhythmias; (2) the donors applied at the time of reperfusion, aggravated reperfusion arrhythmias; (3) adenosine applied according to these protocols, in the concentration producing changes in coronary flow and reflow similar to those produced by the NO donors, did not affect reperfusion arrhythmias. Thus we conclude that in agreement with other reports [11], the severity of reperfusion arrhythmias is independent from coronary flow and reflow, also in the hearts treated with NO donors. Thus the diverse effects of NO donors on reperfusion arrhythmias are likely to result from a direct interaction of NO with intrinsic arrhythmogenic mechanism(s). In crystalloid-perfused heart, an anti platelet [12] and anti-neutrophile actions [13] of NO seem to be of a minor importance.

NO and ischaemic preconditioning

Preconditioning has been repeatedly demonstrated to reduce severity of reperfusion arrhythmias in rat heart [6, 14–16]. Here we demonstrate that GTN and SIN-1, administered for

5 min and washed out for another 5 min prior to ischaemia, can mimic the antiarrhythmic effect of preconditioning. This provides an evidence that, indeed, a brief NO-induced stimulation is able to trigger changes in the heart that confer lasting protection against the subsequent ischaemia/reperfusion. It is not known at this stage what these changes might be.

The following sequence of reactions has been proposed to mediate preconditioning in rabbit heart: Occupation of A_1 adenosine receptor (or any receptor coupled to phospholipase C), G_i protein linked activation of phospholipase C, activation of protein kinase C, and related phosphorylation of proteins [3]. In rat, preconditioning neither appears to be mediated by adenosine [4–6] nor requires functional G_i protein [17, 18], however, protein kinase C seems to be involved [19, 20]. It is conceivable, that in the absence of adenosine mechanism, NO mediated mechanism of preconditioning operates in rat heart. It is not clear if these two mechanisms could share common elements or rather utilize different biochemical pathways.

It has been proposed that NO contributes to the antiarrhythmic effect of preconditioning via elevation of cGMP [7]. Consistent with this hypothesis is report that preconditioning induced by overdrive pacing is accompanied by an increase in cardiac cGMP content [21]. NO can increase myocardial cGMP [22], and cGMP, in turn, can inhibit Ca^{2+} influx [23]. Thus, the inhibition of Ca^{2+} influx by cGMP may be an important protective mechanism to ameliorate disturbances in calcium homeostasis. The reduction of perfusate calcium has been reported to ameliorate reperfusion arrhythmias in isolated hearts [8, 24].

NO may act as an anti-free radical agent [25, 26]. Moreover, NO has been demonstrated to inhibit xanthine oxidase and the enzyme catalyzed production of superoxide [27]. Different anti-free radical interventions have been reported to be protective against reperfusion arrhythmias in rat heart [8, 28, 29]. Thus it is tempting to speculate that the 'preconditioning' effect of NO donors, and possibly a mechanism of the preconditioning as such, involves (directly or via cGMP) some anti-free radical mechanism. Actually, it has been shown that preconditioning, in parallel to its effect on reperfusion arrhythmias, reduced the formation of free radicals in isolated rat heart [16].

Deleterious effect of NO in ischaemic/reperfused heart

In cats and dogs subjected to myocardial ischaemia/reperfusion, the infusion of NO or its precursor, L-arginine, during the early phase of reperfusion, has been consistently demonstrated to reduce infarct size, an effect attributable to the NO-induced reduction in platelet aggregation and neutrophil accumulation in ischaemic cardiac tissue [30, 31]. An intriguing finding of this study is that in crystalloid-perfused hearts, GTN and SIN-1, administered at the time of reperfusion, aggravated reperfusion arrhythmias. However, when SIN-1 was administered both before and after the ischaemia, only antiarrhythmic activity of the drug was apparent. These indicate that in isolated heart: (1) NO is proarrhythmic only when its administration is restricted to the reperfusion period and (2) for the antiarrhythmic activity of NO to develop, its action must be initiated before the ischaemia, suggesting that 'preconditioning' action of NO may overcome the toxic effect of NO.

NO has been implicated in tissue injury, first as a result of its capacity to interact with superoxide to produce toxic intermediate (such as peroxynitrite) [32] and secondly because NO inactivates iron-sulfur centered enzymes which are required for essential metabolic activity [26]. Our results do not support this second mechanism as it does not explain why long lasting exposure to NO was protective while short one was proarrhythmic (SIN(B) vs SIN(A))

The NO donors appeared to aggravate reperfusion arrhythmias when they were given at the time of reperfusion, i.e. when an increased production of free radicals is expected to occur [33, 34]. In crystalloid-perfused heart, neutrophils, platelets as well as other cellular units and metabolites are lacking and a source of oxygen free radicals is limited primarily to that provided by xanthine oxidase. Thus, in crystalloid-perfused heart the protective anti-platelet [12] and anti-neutrophile [13] actions of NO are also lacking. We speculate, therefore, that the proarrhythmic action of NO donors in isolated rat heart exemplifies the toxicity of intermediate(s) of the reaction between superoxide and NO [32], a possibility which needs to be, however, confirmed by measurements of peroxinitrite formation. The toxicity of this origin has been already implicated in myocardial reoxygenation injury in piglets subjected to whole body hypoxia [35].

The fact that NO donors exhibit antiarrhythmic activity only when they are administered before ischaemia, may have two explanations. First, the NO donors applied upon the reperfusion, may not have enough time to prevent rapidly developing reperfusion arrhythmias. Second explanation, which applies also to the 'preconditioning' activity of NO would be that, if NO is to confer protection, some biochemical process must be initiated prior to ischaemia and/or reperfusion.

Conclusion

(1) Although this study does not provide a firm evidence that endogenous NO is involved in the effect of preconditioning in the rat heart, it demonstrates that NO confers lasting protection against reperfusion arrhythmias, supporting the view that NO is one of the endogenous substances triggering ischaemic preconditioning; (2) in special conditions of crys-

talloid-perfused heart, and only when it is given at the time of reperfusion, exogenous NO may be deleterious for ischaemic/reperfused heart.

Acknowledgements

This study was supported by the KBN 6 P2 07 045 04 grant.

The technical assistance of Mr. Marek Wozniak and Mrs. Alicja Protasowicka is gratefully acknowledged.

References

1. Murry CE, Jennings RB, Reimer KA: Preconditioning with ischemia: A delay of lethal cell injury in ischemic myocardium. Circulation 74: 1124–1136, 1986
2. Lawson CS, Downey JM: Preconditioning – state of the art myocardial protection. Cardiovasc Res 27: 542–550, 1993
3. Parratt JR: Protection of the heart by ischaemic preconditioning -mechanisms and possibilities for pharmacological exploitation. Trends Pharmacol Sci 15: 19–25, 1994
4. Li YW, Kloner RA: The cardioprotective effects of ischemic preconditioning are not mediated by adenosine receptors in rat hearts. Circulation 87: 1642–1648, 1993
5. Cave AC, Collis CS, Downey JM, Hearse DJ: Improved functional recovery by ischaemic preconditioning is not mediated by adenosine in the globally ischaemic isolated rat heart. Cardiovasc Res 27: 663–668, 1993
6. Miura T, Ishimoto R, Sakamoto J, Tsuchida A, Suzuki K, Ogawa T, Shimamoto K: Suppression of reperfusion arrhythmia by ischemic preconditioning in the rat: Is it mediated by the adenosine receptor, prostaglandin, or bradykinin receptor? Basic Res Cardiol 90: 240–246, 1995
7. Vegh A, Szekeres L, Parratt JR: Preconditioning of the ischaemic myocardium; involvement of the L-arginine nitric oxide pathway. Br J Pharmacol 107: 648–652, 1998
8. Kopacz M, Karwatowska-Prokopczuk E, Beresewicz A: Reperfusion arrhythmias and purine wash-out in isolated rat and rabbit heart. Effect of allopurinol, dimethylthiourea, and calcium reduction. J Mol Cell Cardiol 25: 859–874, 1993
9. Halpern MH: The dual blood supply of the heart. Am J Anat 101: 1–16, 1957
10. Walker MJA, Curtis MJ, Hearse DJ, Campbell RWF, Janse MJ, Yellon DM, Cobbe SM, Coker SJ, Harness JB, Harron DWG, Higgins A, Julian DJ, Lab MJ, Manning AS, Northover BJ, Parratt JR, Riemersma RA, Riva E, Russel DC, Sheridan DJ, Winslow E, Woodward B: The Lambeth Conventions. Guide lines for the study of arrhythmias in myocardial ischemia, infarction and reperfusion. Cardiovasc Res 22: 447–455, 1988
11. Ibuki C, Hearse DJ, Avkiran M: Rate of reflow and reperfusion induced arrhythmias: Studies with dual coronary perfusion. Cardiovasc Res 26: 316–323, 1998
12. Moncada S, Palmer RMJ, Higgs EA: Nitric oxide: physiology, pathophysiology, and pharmacology. Pharmacol Rev 43: 109–142, 1991
13. Clancy RM, Leszczynska-Piziak J, Abramson SB: Nitric oxide, an endothelial cell relaxation factor, inhibits neutrophil superoxide anion production via a direct action on the NADPH oxidase. J Clin Invest 90: 1116–1121, 1992
14. Shiki K, Hearse DJ: Preconditioning of ischemic myocardium: reperfusion-induced arrhythmias. Am J Physiol 253: H1470–H1476, 1987
15. Hagar JM, Hale SL, Kloner RA: Effects of preconditioning ischemia on reperfusion arrhythmias after coronary artery occlusion and reperfusion in rat. Circ Res 68: 61–68, 1991
16. Tosaki A, Cordis GA, Szerdahelyi P, Engelman RM, Das DK: Effects of preconditioning on reperfusion arrhythmias, myocardial functions, formation of free radicals, and ion shifts in isolated ischemic/reperfused rat hearts. J Cardiovasc Pharmacol 23: 365–373, 1994
17. Liu YG, Downey JM: Preconditioning against infarction in the rat heart does not involve a pertussis toxin sensitive G-protein. Cardiovasc Res 27: 608–611, 1993
18. Lawson CS, Coltart DJ, Hearse DJ: The antiarrhythmic action of ischaemic preconditioning in rat hearts does not involve functional Gi proteins. Cardiovasc Res 27: 681–687, 1993
19. Speechly-Dick ME, Mocanu MM, Yellon DM: Protein kinase C: its role in ischemic preconditioning in the rat. Circ Res 75: 586–590, 1994
20. Li YW, Kloner RA: Does protein kinase C play a role in ischemic preconditioning in rat hearts? Amer J Physiol-Heart Circ Phy 37: H426–H431, 1995
21. Szilvassy Z, Ferdinandy P, Bor P, Jakab I, Lonovics J, Koltai M: Ventricular overdrive pacing-induced anti-ischemic effect: A conscious rabbit model of preconditioning. Am J Physiol 266: H2033–H2041, 1994
22. Shah AM, Lewis MJ: Modulation of myocardial contraction by endocardial and coronary vascular endothelium. Trend Cardiovasc Med 3: 98–103, 1993
23. Mery PF, Pavoine C, Belhassen L, Pecker F, Fischmeister R: Nitric oxide regulates cardiac Ca^{2+} current: Involvement of cGMP-inhibited and cGMP-stimulated phosphodiesterases through guanylyl cyclase activation. J Biol Chem 268: 26286–26295, 1993
24. Hearse DJ, Tosaki A: Free radicals and calcium: Simultaneously interacting triggers as determinants of vulnerability to reperfusion-induced arrhythmias in rat heart. J Mol Cell Cardiol 20: 213–223, 1988
25. Kanner J, Harel S, Granit R: Nitric oxide as an antioxidant. Arch Biochem Biophys 289: 130–136, 1991
26. Henry Y, Lepoivre M, Drapier JC, Ducrocq C, Boucher JL, Guissani A: EPR characterization of molecular targets for NO in mammalian cells and organelles. FASEB J 7: 1124–1134, 1993
27. Fukahori M, Ichimori K, Ishida H, Nakagawa H, Okino H: Nitric oxide reversibly suppresses xanthine oxidase activity. Free Radical Res 21: 203–212, 1994
28. Manning AS, Hearse DJ: Reperfusion-induced arrhythmias: mechanism and prevention. J Mol Cell Cardiol 16: 597–518, 1984
29. Bernier M, Manning AS, Hearse DJ: Reperfusion arrhythmias: Dose-related protection by anti-free radical interventions. Am J Physiol 256: H1344–H1352, 1989
30. Johnson G,III., Tsao PS, Mulloy D, Lefer AM: Cardioprotective effects of acidified sodium nitrite in myocardial ischemia with reperfusion. J Pharmacol Exp Ther 252: 35–41, 1990
31. Nakanishi K, Vinten-Johansen J, Lefer DJ, Zhao ZQ, Fowler WC, Mcgee DS, Johnston WE: Intracoronary l-arginine during reperfusion improves endothelial function and reduces infarct size. Am J Physiol 263: H1650–H1658, 1992
32. Beckman JS, Beckman TW, Chen J, Marshall PA, Freeman BA: Apparent hydroxyl radical production by peroxynitrite: Implications for endothelial injury from nitric oxide and superoxide. Proc Natl Acad Sci USA 87: 1620–1624, 1990
33. Garlick PB, Davis MJ, Hearse DJ, Slater TF: Direct detection of free radicals in the reperfused rat heart using electron spin resonance spectroscopy. Circ Res 61: 757–760, 1987
34. Zweier JL, Flaherty JT, Weisfeldt ML: Direct measurement of free radical generation following reperfusion of ischemic myocardium. Proc Natl Acad Sci USA 84: 1404–1407, 1987
35. Matheis G, Sherman MP, Buckberg GD, Hayborn DM, Young HH, Ignarro LJ: Role of L-arginine-nitric oxide pathway in myocardial reoxygenetion injury. Am J Physiol 262: H616–H620, 1992

Molecular and Cellular Biochemistry **160/161**: 273–282, 1996.

Crucial role of intracellular effectors on glycogenolysis in the isolated rat heart: Potential consequences on the myocardial tolerance to ischemia

N. Lavanchy, S. Grably, A. Garnier and A. Rossi
Laboratoire de Bioénergétique, Université Joseph Fourier, Grenoble, France

Abstract

The role played by *glycogenolysis* in the ischemic heart has been recently put into question because it is suspected that a slowing down of this process could be beneficial for the tolerance of the myocardium to *ischemia*. The role of the *intracellular effectors* that control the rate of glycogenolysis has therefore regained interest. We aimed to understand the role played by those intracellular effectors which are directly related to the *energy balance* of the heart. To this end, we review some of the previously published data on this subject and we present new data obtained from *P-31* and *C-13 NMR* spectroscopic measurement on *isolated rat heart*. Two conditions of ischemia were studied: 15 min global no-flow and 25 min low-flow ischemia. The hearts were isolated either from control animals or from rats pre-treated with isoproterenol (5 mg.kg^{-1} b.w. i.p.) 1 h before the perfusion in order to C-13 label glycogen stores. Our main results are as follows: (1) the biochemically determined glycogenolysis rate during the early phase of ischemia (up to 10–15 min) was larger in no-flow ischemia than in low-flow conditions for both groups, (2) direct measurement of the glycogenolysis rate, as determined by C-13 NMR, after labelling of the glycogen pool in the hearts from isoproterenol-treated rats, confirms the estimations from the biochemical data, (3) glycogenolysis was slower in the hearts from pre-treated animals than in control hearts for both conditions of ischemia, (4) the total *activity of glycogen phosphorylase* (**a** + **b**) increased, by 50%, after 5 min no-flow ischemia, whereas it decreased by 42% after the same time of low-flow ischemia. However, the ratio phosphorylase **a**/**a** + **b** was not altered, whatever the conditions, (5) the concentration of *inorganic phosphate* (Pi) increased sharply during the first minutes of ischemia, to values above 8–10 mM, under all conditions studied. The rate of increase was larger during no-flow ischemia than during low-flow ischemia. The concentration of Pi was thereafter higher in controls than in the hearts from isoproterenol-treated animals.

The calculated cytosolic concentration of free *5'AMP* increased sharply at the onset of ischemia, reaching in a few minutes values above 30 μM in controls and significantly lower values, around 15 μM, in the hearts from isoproterenol-treated rats. (6) The hearts from isoproterenol-treated rats displayed a reduced *intracellular acidosis*, when compared to controls, under both conditions of ischemia.

We conclude that the intracellular effectors, mainly *free AMP*, play an essential role in the control of glycogenolysis *via* allosteric control of phosphorylase **b** activity. The alteration in the concentration of *free Pi*, the substrate of both forms of phosphorylase, can also be considered as determinant in the control of the rate of glycogenolysis.

The attenuation of ischemia-induced intracellular acidosis in the hearts from isoproterenol-treated rats could be a consequence of a reduced glycogenolytic rate and is likely to be related to a better resumption of the mechanical function on reperfusion. (Mol Cell Biochem **160/161**: 273–282, 1996)

Key words: Isolated heart, ischemia, glycogen, catecholamines, intracellular effectors, Pi, AMP

Address for offprints: A. Rossi, Laboratoire de Bioénergétique fondamentale et appliquée, Université Joseph Fourier, BP 53 X, 38 041, Grenoble, France

Introduction

About 30 years ago, Albert Wollenberger and his research group dedicated a series of experimental works to the understanding of the mechanisms involved in the very fast process of glycogenolysis that occurs during ischemia of the myocardium [1–5]. Wollenberger and Krause [4] concluded that: 'the humoral regulation (of glycogenolysis) may not be as essential as the non humoral internal metabolic regulation mechanisms of the myocardial cell.' This assumption has recently regained interest, owing to the possible importance of a reduction of glycogenolysis in the protection of myocardium via ischemic preconditioning [6, 7].

This paper is therefore aimed to bring to light the possible crucial role of intracellular metabolic factors that are directly related to the energy balance of the cell, in the regulation of glycogenolysis. To this end, we will present results obtained with the help of NMR spectroscopy on an experimental model commonly used in the study of myocardial ischemia: the isolated perfused rat heart. The role played by glycogen in the ischemic heart and the mechanisms of glycogenolysis will be discussed initially.

Glycogenolysis was classically considered as a process beneficial for the ischemic cardiomyocyte and several experimental studies, following a paper from Scheuer and Stezoski [8], showed that an increased glycogen content at the start of ischemia exerted a protective effect against ischemic damage [9, 11]. Furthermore, Bricknell *et al.* [12] proposed that glycolytically produced ATP is better able to delay or prevent ischemic contracture than is ATP produced through oxidative phosphorylation. Nevertheless, further studies emphasized the complexity of the role of glycogen. The exact opposite to the conclusions of Scheuer and Stezoski [8] were reached by Lagestrom *et al.* [13]. Furthermore, Neely and Grotyohann [14] showed, in rat heart, that decreasing glycolysis from glycogen exerted a protective effect during a total global ischemia. Indeed, it must be taken into consideration that glycolysis is associated with the production and the intracellular accumulation of end-products such as protons [15, 16] and lactate, which can in turn exert deleterious cellular effects. Thus, as recently discussed in details by Opie [17], the potentially beneficial production of energy from glycolysis can be counterbalanced by the deleterious effect of end-products, so that the final balance largely depends on the importance of the washout resulting from a residual blood flow in the ischemic zone. In addition, it is necessary to bear in mind that the flux of energy (ATP) from exogenous glucose seems more protective than is the flux from glycogenolysis [18]. Finally, recent studies on the *ischemic preconditioning* phenomenon emphasize the complexity of this question, since several observations suggest that partial depletion of glycogen stores, just before a sustained ischemia, can improve the resistance of the myocardial cell to a further sustained ischemic episode [6, 7].

Besides the *still controversial role of the mobilization of glycogen* stores, the exact nature of the mechanisms involved in glycogenolysis during cardiac ischemia is not well known, in spite of the very large amount of data related to this question (for review, see ref. [19–24]). The phosphorolysis of glycogen is under the dependence of both the availability of the other substrate, *inorganic phosphate*, and the activity of *glycogen phosphorylase*. This enzyme can be in an active form (protein structure relaxed) or in an inactive one (so-called tense protein structure) and two basically different mechanisms control its activation level. An allosteric control is exerted by intracellular effectors: 5'-AMP is the main activator, and ATP and glucose-6-phosphate are inhibitors. The second control mechanism involves the phosphorylation of the enzyme from its unphosphorylated **b** form to the **a** form which is spontaneously active (relaxed) and, therefore, cofactor independent. Such a covalent phosphorylase activation is under the hormonal control of catecholamines through the well-known cascade initiated by cyclic AMP production. Although the formation of a hybrid form (phospho-dephosphorylated) with an intermediate sensitivity towards effectors has been detected in skeletal muscle [25–27], the existence of such a form has not been demonstrated in heart muscle. Under physiological conditions the **b** (inactive) form prevails and the effect of the allosteric inhibitors is stronger than that of activators, so that glycogenolysis does not occur. Administration of epinephrine under normoxic conditions has little effect and it seems very doubtful that catecholamine-induced glycogenolysis has any physiological significance for aerobically respiring myocardial cell [28, 29]. On the contrary, the activation of glycogenolysis can be very fast and intense as soon as ischemia or hypoxia occurs, but the trigger for glycogen breakdown under ischemic conditions has not been fully identified. In particular, the relative importance of either hormonal drive or intracellular effectors has not been established. About 30 years ago, Morgan and Parmeggiani [30, 31] observed that the rate of glycogenolysis in the anoxic isolated rat heart did not correlate with the level of phosphorylase **a.** In a careful study of the role of the intracellular factors, these authors proposed that the activation of phosphorylase **b** by elevated levels of 5'AMP and Pi and reduced concentrations of ATP and G6P could play a key role in anoxia-induced glycogenolysis. A few years later, Wollenberger and Krause [4] studying ischemia-induced glycogenolysis in the canine heart supported these findings. However, despite these convergent studies, this question still remains open and deserves attention, owing to the possible importance of glycogenolysis as one of the determinant processes in ischemic preconditioning. In addition, the use of new techniques, such a NMR spectroscopy, allows new approaches to this question through direct evaluation of fluxes (C-13) and measurement of the actual concentrations of the free (mobile and reactive) phosphorylated compounds (P-31). This paper aims to re-evaluate the potential role of the effectors of glycogenolysis.

Materials and methods

Animals and perfusion

Female Wistar rats, weighing 230–250 g and receiving a standard commercial diet (Extralabo, France), were used for all studies. The animals were anaesthetized with sodium pentobarbital (50 $mg.kg^{-1}$ b.w. i.p.). After injection of heparin (1500 i.U. kg^{-1}b.w. i.v.), the hearts were excised and aortically perfused using a non recirculating system. The temperature was fixed at 37°C and the perfusion pressure was maintained at 9.81 kPa. The hearts were allowed to beat spontaneously.

The perfusion medium contained (mM): NaCl 120, KCl 5.6, $CaCl_2$ 2.4, $MgCl_2$ 1.2 and glucose 5.5 mM. The $NaHCO_3$ content was adjusted in order to obtain a pH of 7.4 at the level of the heart after bubbling the medium with carbogen (95% O_2-5% CO_2).

For NMR measurements, the hearts were placed in a 15 mm diameter teflon-stoppered tube and immersed in a small volume of perfusate, the excess being evacuated via a peristaltic pump. The NMR tube was then placed in the bore of the superconducting magnet. Both temperature and gas content of the perfusion fluid were maintained constant by using a thermostatically controlled glass column.

Experimental protocols

Following a 30 min perfusion period, the hearts were subjected to one of the following protocols of ischemia: 15 min global no-flow or 25 min of global low-flow ischemia, the residual coronary flow being fixed at 1.3% of the pre-ischemic coronary flow by means of a peristaltic pump. In experiments where C-13 labelling of glycogen was carried out, the rats were subjected to an injection of isoproterenol (5 $mg.kg^{-1}$s.c.) 1 h before isolation of the heart to deplete glycogen stores. Then, as previously described [32], the hearts were first perfused for 30 min with D [1-C-13] glucose to label glycogen, before being subjected to one of the protocols of ischemia.

P-31 and C-13 NMR measurements

As described in detail in previous papers [32–33], NMR studies were performed using a Bruker WM 250 spectrometer (5.9 Tesla) operating at 101.3 MHz for P-31 and 62.9 MHz for C-13. 2.5 or 3 min P-31 NMR spectra (88 or 130 scans) were recorded throughout the perfusion, using parameters previously indicated [33]. ATP, Pi, and PC contents were determined by integrating the areas under the respective resonances and areas were corrected for partial saturation. The values were first expressed relative to the pre-ischemic ATP area, then converted in µmoles per g using the ATP concentration biochemically measured on separate control hearts after a 25 min normoxic perfusion (4.6 ± 0.3 µmoles/g w.w. mean ± S.E.M.). Intracellular pH was evaluated using the pH-dependent Pi resonance position relative to the pH independent PC resonance position.

Each C-13 NMR spectrum was acquired over a 2.5 or a 5 min period [32]. Bilevel proton decoupling was used (0.5 W during the delay, 3 W during the data acquisition), maintaining however an average temperature of 37°C. Changes in glycogen stores were determined by integrating areas under the C-13 glycogen-glucose peak (101 ppm) and the C-13 glucose peaks of α and β anomers. Corrections for partial saturation and nuclear Overhauser enhancement allowed the quantification of C-13 glycogen through the known concentration of C-13 glucose in the perfusion medium. Results were expressed as µmoles of C-13 glucose/g w.w.

Biochemical measurements

Determination of phosphorylated compounds and creatine contents

It was performed on ventricles freeze-clamped either before ischemia or after 5 min ischemia following perchloric acid extraction as described previously [33]. G6-P, creatine, phosphocreatine were biochemically determined [34, 35]. Adenine nucleotide compound concentrations were evaluated by HPLC [33].

Glycogen content

Glycogen was extracted with 40% KOH in hearts subjected to 5, 10, 15, 20 min of ischemia, then precipitated with alcohol [32]. Following acid hydrolysis, glucose was determined enzymatically [34]. The glycogen content was expressed as mg glucose/g wet weight and converted into µeq.glucose/g w.w.

Determination of glycogen phosphorylase activity

The phosphorylase activity measurement performed on an appropriate tissue extract, was based on the incorporation of the ^{14}C-glycosyl moiety of labelled glucose-l-phosphate into glycogen, according to the technique described by Gilboe *et al.* [36]. The activity of the a form was evaluated on a tissue extract freed from AMP by Dowex 1 × 4 treatment, while the total activity was determined in the presence of added AMP (3 mM).

Expression of results and statistical analyses
The data from biochemical measurements were first calculated as μmol/g. They could be converted to μM (or mM) concentrations considering that the cytosol represents 50 p 100 of the wet weight [37–38]. The concentration of free AMP was calculated from the concentrations of ATP, PC, Cr and protons (through the P-31NMR measurement of the intracellular pH), assuming that the creatine-kinase and the adenylate-kinase reactions work at equilibrium. The Keqs values used for the calculations were taken from Lawson and Veech. [39].

Statistical comparisons were made using an analysis of variance, each parameter being evaluated by one-way or repeated ANOVA for differences among the groups. P = 0.05 was taken as the limit of significance.

Results

Kinetics of ischemia-induced glycogenolysis

C-13 NMR studies
In isolated perfused rat heart, the detection by C-13 NMR of myocardial glycogen with natural abundance C-13, is not possible within the 5 minutes time span convenient for kinetics studies (Fig. 1). Thus, in order to label the glycogen molecule, we devised a protocol in which the glycogen stores were previously depleted by more than 80% under the effect of strong β-stimulation by isoproterenol given *in situ* [40] 1 h prior to the perfusion. The hearts were then perfused for 30 min in the presence of D [1-C-13] glucose (99% C-13 abundance) [32] when glycogen synthase was in a highly active form [40]. The labelled glucose moieties of glycogen gave rise to a readily detectable resonance (Fig. 1). The mobilization of glycogen during ischemia could then be followed, as shown in Fig. 1, while the formation of lactate could also be detected. These experiments allowed the kinetics of labelled glycogen depletion to be monitored under two conditions of ischemia: No-flow and low flow ischemia (Fig. 2 A). From Fig. 2A it may be observed that: (1) in both experimental models, after a few minutes of ischemia, the degradation of labelled glycogen followed linear-like kinetics over about 15 min and (2) the mobilization of labelled glycogen was faster in no-flow than under low-flow ischemia. The NMR measurement of the rate of degradation of C-13 glycogen gives mean values of 1.2 and 0.6 μeq glucose/min/g in controls and hearts from pre-treated rats respectively.

Biochemical determinations
The dynamics of glycogen breakdown during ischemia appeared very different in the 2 groups of hearts: (1) *in control hearts*, a burst of glycogenolysis occured within the first 5 min of ischemia, reaching 3.0 and 2.0 μeq.glucose/min/g under no-flow and low-flow ischemia respectively, whereas (2) in *hearts from isoproterenol pre-treated rats*, mean values of 1.2 and 0.7 μeq glucose/min/g were mobilized under the no-flow and low-flow conditions of ischemia. These values are in good agreement with the direct measurements using C-13 NMR spectroscopy of 1.2 and 0.6 μeq.glucose/min/g respectively.

Glycogen phosphorylase activity

The activities of phosphorylase **a** and **b** were determined under normoxia and at the fifth minute in both protocols of ischemia (Table 1).

Under normoxic conditions, the total phosphorylase activity (form **a** + form **b**) reached in control hearts a mean value of 12.9 μmol/min/g with the **a** form representing 14% of the total activity. These values are in agreement with most of the commonly reported data [41–46] and with our previous determinations in the hearts *in situ* [40]. The determinations in the isolated heart from isoproterenol-treated animals did not reveal any significant differences with controls (14.9 μmoles/min/g and 12%).

However, *after 5 min of ischemia*, alterations in the activity of phosphorylase were different under the two conditions of ischemia. In *no-flow ischemia*, the total activity of phosphorylase increased by 50% in both groups, whereas it decreased by 42% after 5 min of *low-flow ischemia*. The changes in activity of the **a** form paralleled those of the total enzyme, so that the ratio **a/a** + **b** was similar in both conditions. These results demonstrate that, in the isolated rat heart, the phosphorylase is not covalently activated at the 5th min of ischemia, a time when glycogenolysis is highest. This suggests that the catecholamine drive is probably low (or absent) under such experimental conditions.

Alteration in the concentration of cytosolic inorganic phosphate

It is very essential to know the intracellular concentration of free inorganic phosphate (Pi) which is the second substrate of the reaction of glycogen phosphorolysis. The kinetics of the changes in concentration of free Pi could be monitored from P-31 NMR spectra. The concentration of *free cytosolic Pi* increased sharply at the onset of ischemia in both models (Fig. 3): From basal normoxic values of 2.7 and 1.5 mM in controls and pre-treated hearts respectively, concentrations above 9 mM were reached during the first 2–3 min of ischemia. The Pi concentration continued to increase thereafter linearly throughout ischemia under both conditions studied. Higher rates were observed: (1) during no-flow ischemia as compared to low-flow ischemia, and (2) in hearts from control rats than in hearts from isoproterenol-treated animals (Fig. 3).

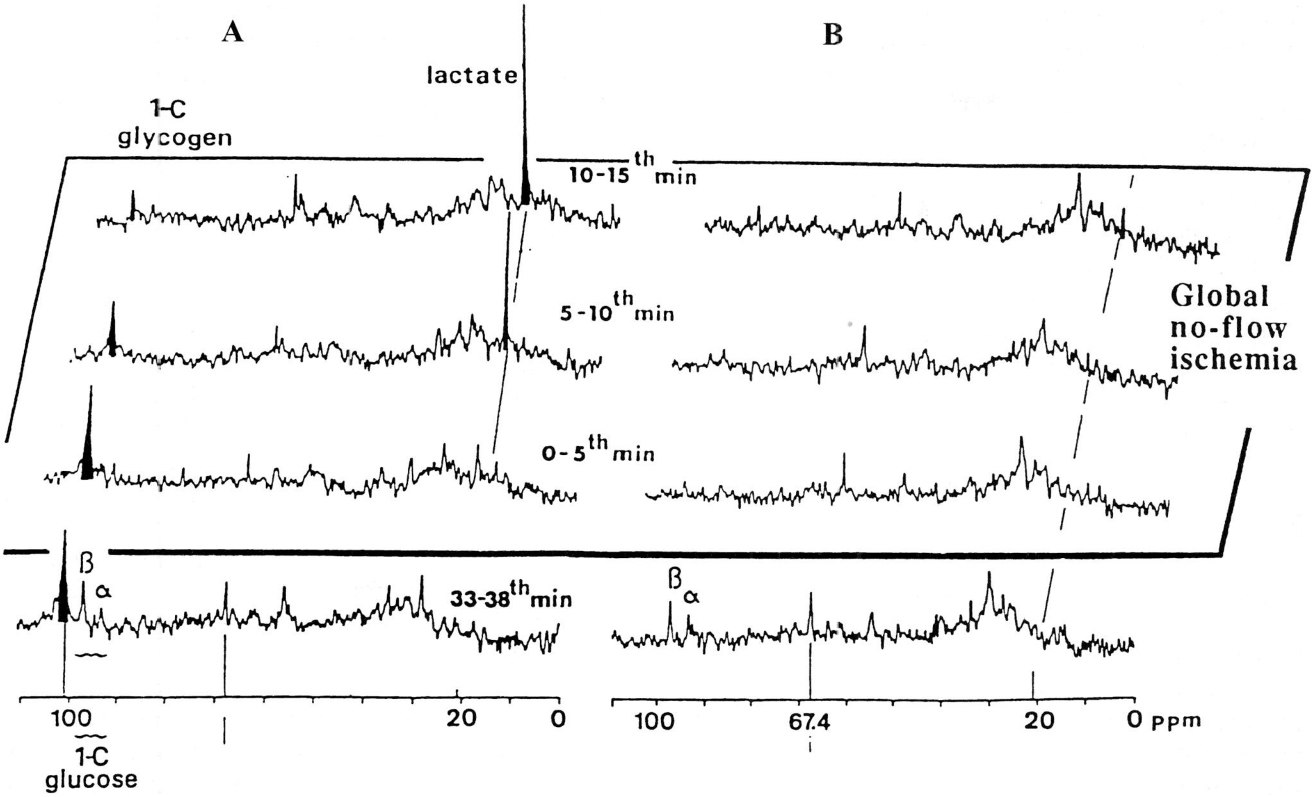

Fig. 1. C-13 NMR detection of glycogen in perfused rat hearts. 5 min bilevel proton decoupled C-13-NMR spectra of isolated rat hearts (62.9 MHz, 730 scans). Dioxanne in D_2O, placed in a capillary tube, provided an external reference (67.4 ppm) for the assignment of resonance frequencies. (A) The heart from the isoproterenol-treated rat was first perfused with 5.5 mM D [1-C-13] glucose (99% C-13 abundance), for 30 min: C-13 labelled glycogen gave a resonance at 101.4 ppm. Then C-13 glucose in the perfusion fluid (α and β anomers) was replaced by natural abundance C-13 glucose for 5 min (33–38th min spectra). (B) In the control heart, C-13 labelling did not occur and natural abundance C-13 glycogen could not be detected.

Alterations in the intracellular concentration of the effectors of glycogen phosphorylase

The effectors of phosphorylase, ATP, AMP, G6P, which are soluble phosphorylated compounds, can be detected by P-31 NMR. However, the concentration of free AMP is far below the threshold of detection by NMR, and it must be determined indirectly through calculations using the two chemical reactions supposed to be at equilibrium in the cell: One catalysed by creatine-kinase and the other by myokinase. The values in Table 2 reveal a rapid and large increase in the concentration of free AMP during ischemia. Total AMP content, measured by HPLC, exhibited roughly the same variations as free AMP concentrations, except under low-flow ischemic conditions where the loss in adenosine is probably higher than under no-flow ischemic conditions.

As the G6P content is low and its resonance not well resolved in NMR spectra, it was determined biochemically on freeze-clamped ventricles. The variation in G6P concentration remained limited, the higher values being observed in no-flow ischemia. The alterations of the ATP and AMP concentrations in the hearts from pre-treated animals were attenuated when compared to control hearts. This is probably a consequence of the fact that the ATP pool was partly depleted before perfusion under the effect of the isoproterenol treatment.

Alterations in intracellular pH

The determination of intracellular pH (pHi) from the resonance frequency of free Pi in P-31 NMR spectra gave the following results: (1) at the end of the periods of ischemia, the pHi reached values of 5.95 ± 0.08 (n = 11) and 5.90 ± 0.07 (n = 10, means ± S.E.M.) in low-flow and no-flow ischemia respectively and (2) the intracellular acidosis was significantly lower in the hearts from isoproterenol-treated rats for low-flow ischemia and for no-flow ischemia (6.30 ± 0.09, n = 5 and 6.25 ± 0.10, n = 6).

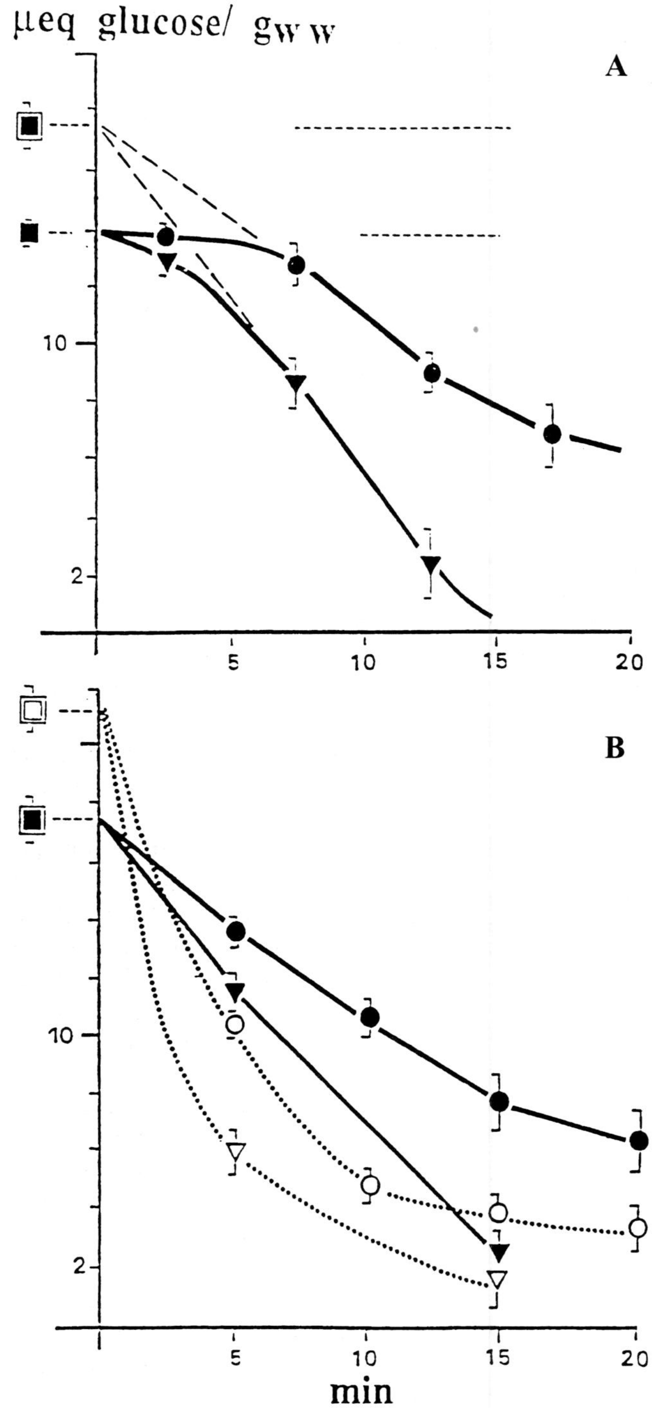

Fig. 2. Kinetic changes in glycogen stores during global ischemia of the perfused rat heart. (A) NMR determinations: C-13 labelled glycogen stores in hearts from isoproterenol-treated rats: Glycogen stores were synthesized from D [1-C-13] glucose during 30 min before ischemia and 80% of the pool was then in the form of 1-C-13 glycosyl moities. Ischemia was started at time zero, on the ordinate: variations of the NMR signal intensity (solid lines), the broken lines indicate the assumed early mobilization of residual abundance glycogen (see text). (B) Biochemical determinations: Glycogen stores in hearts from control (broken lines), and treated hearts (solid lines) ▼ no-flow ischemia, 15 min ● low-flow ischemia, 24 min (1% of pre-ischemic coronary flow), Means ± S.E.M. (n = 7).

Table 1. Glycogen phosphorylase activity

		Normoxia	Ischemia (5th min) no-flow	low-flow
	a	1.85 ± 0.29	3.00 ± 0.25*	0.84 ± 0.09*+
Hearts from		(8)	(9)	(7)
control rats	T	12.87 ± 1.61	19.23 ± 1.76*	7.38 ± 0.54*+
	%	14 ± 1	16 ± 1	11 ± 1
	a	1.86 ± 0.15	2.42 ± 0.20*	0.96 ± 0.09*+
Hearts from		(9)	(7)	(7)
iso-treated rats	T	14.93 ± 1.59	22.36 ± 2.70*	9.00 ± 0.49*+
	%	12 ± 1	11 ± 1	11 ± 1

Means ± S.E.M. (n), $\mu mol.g^{-1}.min^{-1}$, a = phosphorylase a, T= phosphorylase a + b, % = ratio phosphorylase a/total enzyme activity. *$p < 0.05$ vs normoxia; +$p < 0.05$ comparison low-flow vs no-flow.

Discussion

The kinetics of glycogen mobilization determined from biochemical measurements are different in controls and in hearts from isoproterenol-treated rats. The process appears to be linear in the hearts from isoproterenol-treated rats as confirmed by the C-13 NMR technique even though we might suspect that this degradation does not homogeneously affect the glucose moieties situated at different levels in the glycogen particles [47]. Our results suggest that the early breakdown concerns in particular the particles of glycogen exhibiting natural abundance C-13 glucose moities which, being formerly formed, probably remains more closely associated with the various enzymes involved in glycogen metabolism. Both techniques show that the degradation rate is larger in no-flow than in low-flow ischemia. This last observation was also made in control hearts, but, for these hearts, the decrease in glycogen stores is faster at the onset than at the end of ischemia.

The changes in the total activity of phosphorylase are different under the two conditions of ischemia. Indeed, whereas the extracted activity increased (by 50%) after 5 min of no-flow ischemia, it fell (by 42%) following the same duration of low-flow ischemia in control hearts. The same trend occured in the hearts from isoproterenol-treated animals. These changes can probably be explained by the well-known association of glycogen with specific proteins and especially with the enzymes involved in glycogen metabolism [48, 49]. The increase in total activity in no-flow ischemia may be attributed to an increased extractability of the enzyme following its dissociation from glycogen particles [48, 49]. It is likely that in low-flow ischemia some loss in phosphorylase occurs when glycogen particles are degraded [50].

A common feature of the conditions investigated in this study is that the *percentage of **a** form* relative to total activity of phosphorylase *is not altered whatever the conditions.* This means that no covalent alteration of phosphorylase occured under the experimental conditions used in this study.

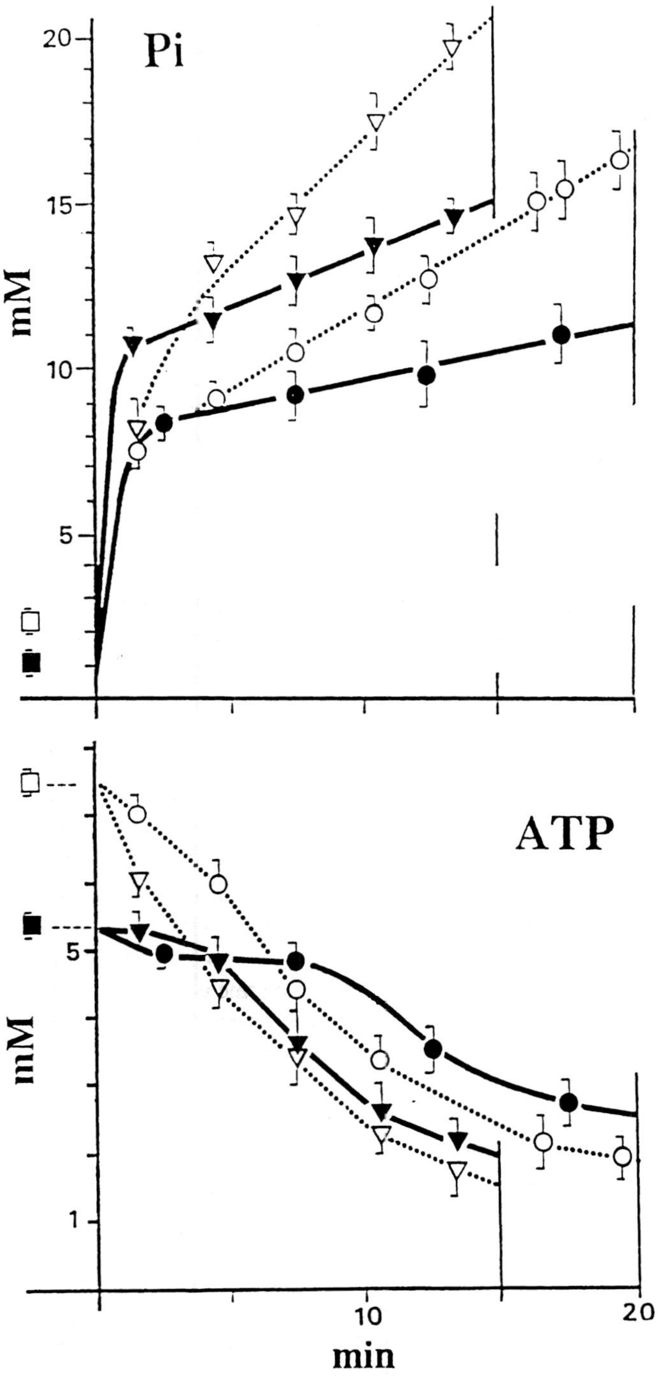

Fig. 3. Time-course of changes in free inorganic phosphate and ATP during ischemia of the isolated rat heart. The Pi and ATP contents, evaluated from the areas of the P-31 NMR resonance peaks, were first expressed as μmol/g using an internal standard (ATP) and then converted into mM (see Materials and methods). ▼▽ no-flow ischemia, 15 min ●○ low-flow ischemia, 24 min. Open symbols: Control hearts; closed symbols: hearts from isoproterenol-treated animals. Means ± SD from 5–10 hearts.

Several authors have described the conversion of phosphorylase **b** into the enzymatically active **a** form on the early stage of ischemia [4, 41, 44, 51, 52]. Others have disputed the role of intracellular cyclic AMP as the trigger of glycogenolysis with the onset of myocardial ischemia [53] and several experiments have suggested that the activity of phosphorylase could simply be increased by the effect of intracellular effectors on the **b** form [4, 30, 51]. Giving a precise description of the different events that could play a role in the induction of glycogenolysis in the course of ischemia, Wollenberger and Krause [4] proposed that: 'The humoral regulation (of glycogenolysis) may not be as essential as the non humoral internal metabolic regulation mechanisms of the myocardial cell.' The present results are clearly in agreement with this conclusion. They suggest that, at least in the model of isolated perfused heart, catecholamines have no significant effect on ischemia-induced glycogenolysis. However, several investigators have proposed that catecholamine release occurs in the myocardium in response to ischemia [1, 5, 28, 44, 52]. Others have reported that there is only a negligible release of noradrenaline in the isolated rat heart during global low-flow ischemia, especially when glucose is used as substrate [54]. In our experiments, we could not detect any increase either in the release of catecholamines or in myocardial cyclic AMP content during ischemia (not shown). Furthermore, our previous experiments demonstrated that the addition of β-adrenergic blockers to the perfusion fluid had no significant effect on the intracellular acidosis that develops during ischemia, unless β-blockers exert ancillary effects (negative inotropic effects, membrane stabilizing activity) [55, 56]. Thus our data suggest that the adrenergic drive seems to be absent in the model of isolated rat heart subjected to global ischemia. This is probably due mainly to the deconnection of the isolated heart from orthosympathic afferences and also to the washout of the small amount of released catecholamines. It is of great interest to bear in mind that cyclic AMP seems not to be involved in the anoxia-induced glycogenolysis in the rat heart in situ [28, 29, 43].

It cannot be excluded that a very early phosphorylation of glycogen phosphorylase could occur transiently and some activation of the kinase, resistant to β blockade [4], could be attributed to a direct catecholamine-independent effect. Calcium ions could play a role by directly activating the phosphorylase **b** kinase [57, 58]. But, even if a transient transformation of phosphorylase **b** to the **a** form does occur during the first few minutes of ischemia, subsequently phosphorylase **b** probably becomes strongly active while phosphorylase **a** is progressively dephosphorylated [4]. We can therefore consider that the rate of glycogenolysis depends chiefly on the concentration of the allosteric effectors of phosphorylase **b** and on the availability of the second substrate, inorganic phosphate. In our experiments, the concentration of the most important positive effector of phosphorylase **b**, namely free AMP, increased from 1 up to 30 μM after 5 min of ischemia. This value is in the range of the Ka for AMP: 50 μM [30, 31]. In parallel to the fast and large

Table 2. Concentration of ATP, AMP, Pi and G6P in perfused hearts subjected to ischemia

	Normoxia		Ischemia (5th.min)			
			no-flow		low-flow	
	C	ISO	C	ISO	C	ISO
AMP						
total μmol.g^{-1}	80 ± 11	50 ± 5*	1120 ± 160	400 ± 117*	210 ± 20	160 ± 16
	(8)	(10)	(8)	(7)	(6)	(6)
free μM	1.0 ± 0.1	0.5 ± 0.1*	30 ± 4	14 ± 4*	44 ± 10	16 ± 4*
	(9)	(10)	(8)	(7)	(6)	(6)
G6P μM	83 ± 12	250 ± 10*	450 ± 26	500 ± 50	367 ± 50	333 ± 83
	(8)	(8)	(8)	(7)	(6)	(6)
Pi mM	2.7 ± 0.2	1.5 ± 0.2*	13.0 ± 0.7	12.0 ± 0.8	9.5 ± 0.3	9.0 ± 0.8
	(14)	(12)	(7)	(6)	(7)	(6)
ATP mM	7.5 ± 0.6	5.5 ± 0.6*	4.5 ± 0.7	4.5 ± 0.7	6.0 ± 1.1	5.0 ± 0.6
	(14)	(10)	(8)	(6)	(7)	(6)

Values for the free soluble forms are converted as μM or mM concentrations assuming that the cytosol pool represents 50 p.100 of the total wet weight. Means ± S.E.M. (n) C = controls, ISO = hearts from isoproterenol treated rats. ATP, G6P and total AMP contents were determined using biochemical methods, free AMP and Pi concentrations were calculated from P 31 NMR spectra (see methods). *$p < 0.05$ vs controls.

increase in AMP concentration, the content of allosteric inhibitors of phosphorylase **b** decreased (ATP) or remained stable (G6P). As a result, a net stimulation of the activity of the **b** form of phosphorylase should rapidly occur.

Furthermore, we have to take into account the alterations in Pi [55]. The Km of phosphorylase **b** for Pi is about 20 mM in normoxia, decreasing to lower values during anoxia [30, 31]. The NMR data demonstrate that the concentration of Pi was far below the Km under normoxic conditions (1.5–2.7 mM), but rose rapidly to values (9 mM) that could be very effective in increasing the rate of glycogen phosphorolysis. The proposition that the Pi concentration controls the rates of glycogenolysis, is supported by the fact that the increase in Pi content during ischemia was lower under conditions where glycogenolysis was also lower (low-flow vs no-flow ischemia and in hearts from isoproterenol-treated rats vs control hearts).

The hypothesis that *intracellular effectors* and *Pi concentration* play a crucial role in the control of the rate of glycogenolysis during ischemia has several potential implications. Owing to the well known relationship between glycolysis and the development of intracellular acidosis [15, 16, 60], any alteration in glycogenolysis can induce a cascade of consequences on cellular homeostasis and function. Therefore the 'protective' effect of glycogenolysis is questionable. In this respect some recent observations on the alterations of glycogenolysis in the ischemic preconditioning phenomenon deserve major attention [6, 7]. Indeed, reduced intracellular acidosis seems to be common to almost all the experimental conditions which lead to a protection against ischemic damage [6, 7, 61, 62]. However, depletion of glycogen before a sustained period of ischemia has generally demonstrated limitation of acidosis in several models of ischemia [33, 60, 62] although improvement of functional recovery has been found to be variable: absent [60], worse [13] or better [63].

Under the experimental conditions of this study, the development of acidosis during ischemia was faster in no-flow than in low-flow ischemia (not shown). Moreover, the intracellular acidosis reached at the end of ischemia, was significantly lower in the hearts from isoproterenol-treated rats, which demonstrated lower glycogenolysis rates. The estimation of an index of mechanical performance (the product Left Ventricular Developed Pressure × Frequency) revealed that, following a 25 min low-flow ischemia, the hearts from isoproterenol treated rats resumed a better functional activity on reperfusion (115 ± 56, n = 6 vs 30 ± 25, n = 9 expressed as a percentage of 25 min normoxic values). Thus, the particular conditions of strong β-stimulation *in situ* brings about a sequence of events during a subsequent ischemia-reperfusion: reduced glycogenolysis, attenuated acidosis and increased mechanical activity on reperfusion, that are similar to those associated with ischemic preconditioning.

In conclusion, we would like to stress the important role that *intracellular effectors* could play during an ischemic episode. The molecules which are directly involved in energy transfer in the cardiomyocyte are obviously the most sensitive sensors of the energy balance inside the cell. Thus, it seems reasonable to hypothesize that these molecules could control the course of the various metabolic events induced by ischemia and, therefore, influence the functional capabilities of the cardiac muscle upon reperfusion.

References

1. Wollenberger A, Shahab L: Anoxia-induced release of noracrenaline from the isolated perfused heart. Nature 207: 88–89, 1965
2. Wollenberger A, Krause EG, Shahab L: Endogenous catecholamine mobilization and the shift to an aerobic energy production in the acutely ischemic myocardium. In: G. Marchetti and B. Taccardi. Coronary

circulation and energetics of the myocardium. Karger, Basel and New York, 1967, p 200

3. Shahab L, Wollenberger A: Freisetzung von Noradrenalin aus demisolierten durchströmten Herzen bei akute Anoxie und nach Gabe von Stoffwech selgiften. Acta Biol Med Germ 19: 939–959, 1967
4. Wollenberger A, Krause EG: Metabolic control characteristics of the acutely ischemic myocardium. Amer J Cardiol 22: 349–359, 1968
5. Wollenberger A, Krause EG, Heier G: Stimulation of 3'-5'-cyclic AMP formation in dog myocardium following arrest of blood flow. Biochem Biophys Res Commun 36: 664–670, 1969
6. Wolfe CL, Sievers RE, Visseren FLJ, Donnelly TJ: Loss of myocardial protection after preconditioning correlates with the time course of glycogen recovery within the preconditioned segment. Circulation 87: 881–892, 1993
7. Schaefer S, Carr LJ, Prussel E, Ramasamy R: Effects of glycogen depletion on ischemic injury in isolated rat hearts: insights into preconditioning. Am J Physiol 268: H935–H944, 1995
8. Scheuer J, Stezoski SW: Protective role of increased myocardial glycogen stores in cardiac anoxia in the rat. Circ Res 27: 835–849, 1970
9. Hewitt RL, Lolley DM, Adrouny GA, Drapanas T: Protective effect of glycogen and glucose on the anoxic arrested heart. Surgery 75: 1–10, 1974
10. Towart R, Schlossmann K, Kazda S: Protection against cardiac anoxia- role and limitations of increased glycogen reserves in the isolated right ventricular strip. Basic Res Cardiol 76: 639–646, 1981
11. Lolley DM, Ray JF, Myers WO, Sautter RD, Tewksbury DA: Importance of preoperative myocardial glycogen levels in human cardiac preservation. J Thorac Cardiovasc Surg 78: 678–687, 1979
12. Bricknell OL, Daries DS, Opie LH: A relationship between adenosine triphosphate, glycolysis and ischaemic contracture in the isolated rat heart. J Mol Cell Cardiol 13: 941–945, 1981
13. Lagerstrom CF, Walker WE, Taegtmeyer H: Failure of glycogen depletion to improve left ventricular function of the rabbit heart after hypothermic ischemic arrest. Circ Res 63: 81–86, 1988
14. Neely JR, Grotyohann LW: Role of glycolytic products in damage to ischemic myocardium. Circ Res 55: 816–824, 1984
15. Gevers W: Generation of protons by metabolic processes in heart cells. J Mol Cell Cardiol 9: 867–874, 1977
16. Dennis SC, Gevers W, Opie LH: Protons in ischemia: where do they come from; where do they go to? J Mol Cell Cardiol 23: 1077–1086, 1991
17. Opie LH: Myocardial ischemia-metabolic pathways and implications of increased glycolysis. Cardiovasc Drugs and Therapy 4: 777–790, 1990
18. Owen P, Dennis S, Opie LH: Glucose flux rate regulates onset of ischemic contracture in globally underperfused rat hearts. Circ Res 66: 344–354, 1990
19. Fischer EH, Heilmeyer LMG, Haschke RH: Phosphorylase and the control of glycogen degradation. Curr Top Cell Regul 4: 211–251, 1971
20. Graves DJ, Wang JH: α-glucan phosphorylases-Chemical and physical basis of catalysis and control. In: P.D. Boyer (Ed.) The Enzymes. Academic 7, New York, 1972, pp 435–482
21. Krebs EG: Phosphorylation and dephosphorylation of glycogen phosphorylase: a prototype for reversible covalent modification. Curr Top Cell Regul 18: 401–419, 1981
22. Madsen NB: Glycogen phosphorylase. In: P.D. Boyer (Ed.) The Enzymes. Academic 17, New York, 1986, pp 365–394
23. Johnson LN: Glycogen phosphorylase: Control by phosphorylation and by allosteric effectors. FASEB J 6: 2274–2282, 1992
24. Fischer EH, Hurd S, Kon P, Seery YL, Teller DC: Phosphorylase: relation of structure to activity. In: W.J. Whelan (Ed.) Control of glycogen metabolism. Academic, London, 1968, pp 19–33
25. Hurd SS, Teller D, Fischer EH: Probable formation of partially phosphorylated intermediates in the interconversion of phosphorylase a and b. Biochem Biophys Res Commun 24: 79-84, 1966
26. Bot G, Kovacs EF, Gergely P: Partial phosphorylation of muscle phosphorylase. I. Formation of a hybrid phosphorylase *in vitro*. Biochem Biophys Acta 370: 70–77, 1974
27. Gergely P, Bot G, Kovacs EF: Partial phosphorylation of muscle phosphorylase. II. Formation of a hybrid phosphorylase *in vivo*. Biochem Biophys Acta 370: 78–84, 1974
28. Williams BJ, Mayer SE: Hormonal effects on glycogen metabolism in the rat heart *in situ*. Mol Pharmacol 2: 454–464, 1966
29. Mayer SE: Effect of catecholamines on cardiac metabolism. Circ Res (suppl to 34–35): 129–135, 1974
30. Morgan HE, Parmeggiani A: Regulation of glycogenolysis in muscle. II. Control of glycogen phosphorylase reaction in isolated perfused heart. J Biol Chem 239: 2434–2439, 1964
31. Morgan HE, Parmeggiani A: Regulation of glycogenolysis in muscle. III. Control of muscle glycogen phosphorylase activity. J Biol Chem 239: 2440–2445, 1964
32. Lavanchy N, Martin J, Rossi A: Glycogen metabolism: A ^{13}C-NMR study on the isolated perfused rat heart. FEBS Letters 178: 34–38, 1984
33. Lavanchy N, Martin J, Rossi A: Graded global ischemia and reperfusion of the isolated perfused rat heart: Characterization by 31P-NMR spectroscopy of the extent of energy metabolism damage. Cardiovasc Res 18: 573–582, 1984
34. Bergmeyer HU, (Ed.) Methods of enzymatic analysis. Academic Press, New York, 1963
35. Eggleton P, Elsden SR, Gough N: Estimation of creatine and diacethyl. Biochem J 37: 526–529, 1943
36. Gilboe DP, Larson KL, Nuttal FQ: Radioactive method for the assay of glycogen phosphorylase. Anal Biochem 47: 20–27, 1972
37. Kauppinen RA, Hiltunen JK, Hassinen IE: Subcellular distribution of phosphogens in isolated perfused rat hearts. FEBS Lett 112: 273–276, 1980
38. Polimeni PI, Buraczewski SI: Expansion of extracellular tracer spaces in the isolated heart perfused with cristalloid solution: Expansion of extracellular space, transsarcolemmal leakage, or both. J Mol Cell Cardiol 20: 15–22, 1988
39. Lawson JWR, Veech RL: Effects of pH and free Mg^{2+} on the Keq of the creatine kinase reaction and other phosphate hydrolyses and phosphate transfer reactions. J Biol Chem 254: 6528–6537, 1979
40. Grably S, Rossi A: Changes in cardiac glycogen synthase and phosphorylase activities following stimulation of beta-adrenergic receptors in rats. Basic Res Cardiol 80: 175–181, 1985
41. Cornblath M, Randle AJ, Parmeggiani A, Morgan HE: Regulation of glycogenolysis in muscle. Effects of glucagon and anoxia on lactate production, glycogen content, and phosphorylase activity in the perfused isolated rat heart. J Biol Chem 238: 1592–1597, 1963
42. Oye I: The role of phosphorylase a and b for the control of glycogenolysis in the isolated working rat heart. Acta Physiol Scand 70: 229–235, 1967
43. Neely JR, Whitfield CF, Morgan HE: Regulation of glycogenolysis in hearts: effects of pressure development, glucose and FFA. Am J Physiol 219 (4): 1083–1088, 1970
44. Dobson JG, Mayer SE: Mechanisms of activation of cardiac glycogen phosphorylase in ischemia and anoxia. Circ Res 33: 412–420, 1973
45. Nuttall FQ, Gannon MC, Bergstrom WJ: Insulin and epinephrine effects on heart glycogen synthase and phosphorylase activity. Am J Physiol 228: 1815–1820, 1975
46. Miller TB, Praderio M, Wolleben C, Bullman J: A hypersensitivity of glycogen phosphorylase activation in hearts of diabetic hearts. J Biol Chem 256: 1748–1753, 1981

47. Laughlin MR, Petit Jr. WA, Barrett J: The time course of myocardial glycogenolysis stimulated by glucagon. J Mol Cell Cardiol 25: 175–183, 1993
48. Wanson JC, Drochmans P: Role of the sarcoplasmic reticulum in glycogen metabolism. J Cell Biol 54: 206–224, 1972
49. Entman ML, Bornet EP, Van Winkle WB, Goldstein MA, Schwartz A: Association of glycogenolysis with cardiac sarcoplasmic reticulum. J Mol Cell Cardiol 9: 515–528, 1977
50. Schulze W, Krause EG, Wollenberger A: On the fate of glycogen phosphorylase in the ischemic and infarcting myocardium. J Mol Cell Cardiol 2: 241–251, 1971
51. Sakai K, Abiko Y: Acute changes of myocardial norepinephrine and glycogen phosphorylase in ischemic areas after coronary ligation in dogs. Jpn Circulation J 45: 1250–1255, 1981
52. Hough FS, Gevers W: Catecholamine release as mediator of intracellular activation in ischemic perfused rat hearts. South Afr Med J 49: 538–543, 1975
53. Podzuweit T, Dalby AJ, Cherry GW, Opie LH: Cyclic AMP levels in ischaemic and non-ischaemic myocardium following coronary artery ligation: relation to ventricular fibrillation. J Mol Cell Cardiol 10: 81–94, 1978
54. Carlsson L, Abrahamsson T, Almgren O: Local release of myocardial norepinephrine during acute ischemia: An experimental study in the isolated perfused rat heart. J Cardiovasc Pharmacol 7: 791–798, 1985
55. Lavanchy N, Martin J, Giacomelli M, Rossi A: Evaluation by ^{31}P NMR of the effects of acebutolol on the ischaemic isolated rat heart. Eur J Pharmacol 125: 341–351, 1986
56. Lavanchy N, Martin J, Rossi A: The role of beta-adrenoceptors in ischemia induced acidosis in the isolated rat heart: a 31-P NMR study in. In N.J. Dhalla, I.R. Innei and R.E. Beamish, (Eds). 'Myocardial ischemia,' Martinus Nijhoff Publishing, Boston, Dordrecht, Lancaster, 1987, pp 199–212
57. Drummond GI, Duncan L: The action of calcium ion on cardiac phosphorylase b kinase. J Biol Chem 241: 3097–3103, 1966
58. Dobson JG, Ross J, Mayer SE: The role of cyclic adenosine 3'-5' monophosphate and calcium in the regulation of contractility and glycogen phosphorylase activity in guinea pig papillary muscle. Circulation Res 39: 388–395, 1976
59. Parmeggiani A, Morgan HE: Effects of adenine nucleotides and inorganic phosphate on muscle phosphorylase activity. Biochem Biophys Res Commun 9: 252–256, 1962
60. Garlick PB, Radda GK, Seeley PJ: Studies of acidosis in the ischaemic heart by phosphorus nuclear magnetic resonance. Biochem J 184: 547–554, 1979
61. Kida M, Fujiwara H, Ishida M, Kawai C, Ohura M, Miura I Yabuuchi Y.: Ischemic preconditioning preserves creatine phosphate and intracellular pH. Circulation 84: 2495–2503, 1991
62. De Albuquerque CP, Gerstenblith G, Weiss RG: Importance of metabolic inhibition and cellular pH in mediating preconditioning contractile and metabolic effects in rat hearts. Circ Res 74: 139–150, 1994
63. Kupriyanov VV, Lakomkin VL, Steinschneider AY, Severina MY, Kapelko VI, Ruuge EK, Saks VA: Relationships between pre-ischemic ATP and glycogen content and post-ischemic recovery of rat heart. J Mol Cell Cardiol 20: 1152–1162, 1988

Molecular and Cellular Biochemistry **160/161**: 283–288, 1996.

Probucol treatment reverses antioxidant and functional deficit in diabetic cardiomyopathy

N. Kaul, N. Siveski-Iliskovic, M. Hill, N. Khaper, C. Seneviratne and P.K. Singal

Division of Cardiovascular Sciences, Department of Physiology, University of Manitoba, St. Boniface General Hospital Research Centre, Winnipeg, Canada

Abstract

Earlier we reported that probucol treatment subsequent to the induction of diabetes can prevent diabetes-associated changes in myocardial antioxidants as well as function at 8 weeks. In this study, we examined the efficacy of probucol in the reversal of diabetes induced myocardial changes. Rats were made diabetic with a single injection of streptozotocin (65 mg/kg, i.v.). After 4 weeks of induction of diabetes, a group of animals was treated on alternate days with probucol (10 mg/kg i.p.), a known lipid lowering agent with antioxidant properties. At 8 weeks, there was a significant drop in the left ventricle (LVSP) and aortic systolic pressures (ASP) in the diabetic group. Hearts from these animals showed an increase in the thiobarbituric acid reacting substances (TBARS), indicating increased lipid peroxidation. This was accompanied by a decrease in the myocardial antioxidant enzymes activities, superoxide dismutase (SOD) and glutathione peroxidase (GSHPx). Myocardial catalase activity in the diabetic group was higher. In the diabetic + probucol group both LVSP and ASP showed significant recovery. This was also accompanied by an improvement in SOD and GSHPx activities and there was further increase in the catalase activity. Levels of the TBARS were decreased in this group. These data provide evidence that diabetic cardiomyopathy is associated with an antioxidant deficit which can be reversed with probucol treatment. Improved cardiac function with probucol may be due to the recovery of antioxidants in the heart. (Mol Cell Biochem **160/161**: 283–288, 1996)

Key words: oxidative stress, superoxide dismutase, glutathione peroxidase, catalase, lipid peroxidation

Introduction

Diabetes mellitus has been shown to be associated with an increased incidence of heart dysfunction and atherosclerosis [1–4]. Although insulin deficiency is the fundamental cause in type I diabetes and treatment with insulin has a significant beneficial effect, occurrence of myocardial dysfunction remains prevalent in the diabetic population [5]. It is therefore necessary to investigate other means of therapy for the treatment and/or prevention of diabetic cardiomyopathy.

There is now ample evidence supporting the involvement of reactive oxygen derived species in the process of diabetic cardiomyopathy [6–7]. Different partially reduced forms of oxygen such as superoxide, hydroxyl radical and hydrogen peroxide are controlled by various cellular defense mechanisms consisting of enzymatic (superoxide dismutase, catalase and glutathione peroxidase) and other non enzymatic processes [8–10]. There is increasing evidence that an enhanced production and/or ineffective scavenging of these reactive oxygen species results in increased oxidative stress which may play a crucial role in the pathogenesis of cardiac abnormalities [10]. Probucol is long known to be a lipid lowering agent having strong antioxidative effects in inhibiting oxidative modification of LDL [11–12]. With a continuous administration, probucol accumulates in tissues and acts as an antioxidant.

In addition to its strong antioxidant properties, probucol has also been shown to enhance endogenous antioxidant enzyme activities [13–14]. Although probucol has been shown to reduce alloxan or streptozotocin (STZ) induced β cell damage [15–17], prevention of diabetic cardiomyopathy with this drug has also been reported [7]. In rats made dia-

Address for offprints: P.K. Singal, Division of Cardiovascular Sciences, St. Boniface General Hospital, Research Centre, 351 Tache Avenue, Winnipeg, MB R2H 2A6, Canada betic with a single injection of STZ, probucol treatment maintained the hemodynamic function [7].

In this study, we recorded hemodynamic function, myocardial antioxidant enzyme activities and lipid peroxidation to examine the reversibility of diabetes-induced myocardial changes by probucol.

Materials and methods

Male Sprague Dawley rats weighing (150 ± 10 gms) were fed *ad libitum* and divided into four groups: controls; control + probucol; diabetic; and diabetic + probucol. Diabetes was induced by single tail vein injection of streptozotocin at a dose of 65 mg/kg (Sigma Chemicals Co. St. Louis, MO). In order to ensure that the animals were diabetic, urine analysis was done after 24 h with Keto Diastix (Ames, Miles Canada). Treatment with probucol (10 mg/kg intraperitoneally) was initiated 4 weeks after the induction of diabetes and was continued for the following 4 weeks (3 injections/week).

Hemodynamic measurements

After 8 weeks, the rats were anaesthetized with sodium pentobarbital (50 mg/kg i.p.) and hemodynamic assessments were made [13]. Left ventricular peak systolic pressure (LVSP), aortic systolic pressure (ASP) and aortic diastolic pressure (ADP) were recorded on a Beckman Dynograph. Measurements were taken 15 min after a steady state had been reached. Blood samples and the hearts were collected for further studies. Blood was immediately centrifuged at 4°C to separate serum for the measurement of glucose levels and insulin.

Assay for insulin and glucose

Insulin levels were assayed in the serum by using standard rat insulin RIA kit supplied by Linco Res. Inc. St. Louis, MO. Serum glucose concentration was determined by using SIGMA kit based on a procedure described earlier [18].

Superoxide dismutase (SOD) assay

Heart homogenate was centrifuged at 20,000 g for 20 min and the supernatant was assayed for SOD activity by following the inhibition of pyrogallol autooxidation [19]. Aliquot of supernatant containing 150 µg protein was added to Tris HCl buffer containing 25 µl pyrogallol (24 mM) and 10 µl catalase (30 µM). The final volume of 3 ml was made up with the same buffer. Changes in absorbance at 420 nm were recorded at 1-min intervals for 5 min.

Glutathione peroxidase (GSHPx) assay

Cytosolic GSHPx was assayed in a 3 ml cuvette containing 2.0 ml of 75 mM phosphate buffer, pH 7.0 [20]. The following solutions were then added: 50 µl of 60 mM glutathione, 100 µl glutathione reductase solution (30 U/ml), 50 µl of 0.12 M NaN_3, 100 µl of 15 mM Na_2 EDTA, 100 µl of 3.0 mM NADPH and 100 µl of cytosolic fraction obtained after centrifugation at 20,000 g for 25 min. Water was added to make a total volume of 2.9 ml. The reaction was started by the addition of 100 µl of 7.5 mM H_2O_2, and the conversion of NADPH to NADP was monitored by a continuous recording of the change of absorbance at 340 nm at 1-min intervals for 5 min.

Catalase assay

Ventricles were homogenized in 0.05 M potassium phosphate buffer (pH 7.4) in a weight to volume ratio of 1:9 and centrifuged at 40,000 g for 30 min. Supernatant, 50 µL, was added to the cuvette containing 2.95 ml of 19 mM H_2O_2 solution prepared in potassium phosphate buffer [21]. The colour development was read at 240 nm on a Zeiss spectrophotometer. Commercially available catalase was used as a standard.

Lipid peroxidation assay

Measurement of lipid peroxidation was done by determining myocardial thiobarbituric acid reactive substances (TBARS) using the method described before [22]. In order to minimize peroxidation during the subsequent assay procedure, 2% butylated hydroxytoluene was added to the TBA reagent mixture [23].

Proteins and statistical analyses

Protein content was determined [24]. Data were expressed as mean ± S.E.M. For a statistical analysis of the data, group means were compared by one-way ANOVA. Values of $p < 0.05$ were considered significant.

Results

General observations

Body weight in diabetic and diabetic + probucol groups was significantly less than in control and control + probucol groups (Table 1). Serum glucose and insulin levels were not significantly different in control and control + probucol groups. In the diabetic group, serum glucose was about 430% and insulin levels were about 30% of the control values (Table 1). Insulin levels in the diabetic + probucol group were higher by about 20% than in the diabetic group, but it was significantly lower than in control and control + probucol groups (Table 1).

Hemodynamics

Left ventricular peak systolic (LVSP) and aortic systolic pressures (ASP) were significantly decreased in the diabetic group (Table 2). However, in the probucol and diabetic + probucol groups, these values were not significantly different from controls (Table 2). All four groups were not different from each other with respect to the heart rate (HR) and aortic diastolic pressure (Table 2).

Antioxidant enzymes and lipid peroxidation

Activities of endogenous antioxidant enzymes superoxide dismutase (SOD), glutathione peroxidase (GSHPx) and catalase (CAT) were examined, and these data are presented in Figs. 1–3 respectively. SOD activity in the diabetic group was about 50% lower than in controls. Hearts from control + probucol treated rats showed significant increase in SOD activity (Fig. 1). Probucol treatment in diabetic animals caused small, but significant increase in SOD activity as compared to diabetic group. This value for the SOD activity in diabetic + probucol group was still less than in the control

Table 1. General characteristics of control and diabetic rats with and without probucol

Parameter Studied	Control	Diabetic	Control + Probucol	Diabetic + Probucol
Body wt (g)	479.4 ± 6.2	297.8 ± 6.1*	471.5 ± 11.5	304.7 ± 4.2*
Serum glucose (mg/dl)	138.4 ± 12	603.7 ± 40*	150.0 ± 5.0	422.6 ± 33*†
Insulin levels (ng/ml)	1.28 ± 0.20	0.38 ± 0.08*	1.09 ± 0.45	0.46 ± 0.06*

Values are mean ± S.E. of 8–10 experiments. Starting weight of animals in all groups was 150 ± 10 g. *Significantly different ($p < 0.01$) from respective control groups. †Significantly different ($p < 0.01$) from the diabetic group.

Table 2. Hemodynamics of control and diabetic rats with and without probucol treatment

Parameter studied	Control	Diabetic	Control + Probucol	Diabetic + Probucol
ASP (mm/Hg)	110.4 ± 10.9	89.5 ± 5.4*	99.5 ± 2.2	95.5 ± 5.0
ADP (mm/Hg)	72.9 ± 7.9	70.7 ± 20.4	79.2 ± 4.4	79.9 ± 25.1
LVSP (mm/Hg)	113.3 ± 9.2	90.6 ± 2.7*	108.5 ± 6.5	102.4 ± 5.2
HR (beats/min)	325.5 ± 12.6	305.4 ± 20.8	328.0 ± 30.4	326.8 ± 18.2

Values are S.E. of 6–6 experiments. ASP – Aortic systolic pressure; ADP – Aortic diastolic pressure; LVSP – Left ventricular peak systolic pressure; HR – Heart rate. *Significantly different ($p < 0.05$) from control.

and control + probucol groups (Fig. 1). Diabetic group showed significant decrease in GSHPx activity compared to control (Fig. 2). GSHPx was significantly improved in diabetic + probucol group and the value was not different from the control levels. Catalase activity was significantly increased in the diabetic animals compared to control and control + probucol treated groups (Fig. 3). Catalase activity in the diabetic + probucol group was significantly higher as compared to all other groups (Fig. 3).

Lipid peroxidation, as indicated by the thiobarbituric acid reactive substances (TBARS), was about 100% higher in diabetic group than in control and control + probucol groups. Diabetic + probucol group had TBARS levels significantly decreased compared to diabetic group, but these values were still higher than in control and control + probucol groups (Fig. 4).

Discussion

Although insulin therapy has revolutionized the management of insulin dependent diabetes, patients sill have to contend with a variety of secondary complications. Studies of the cardiovascular function in the diabetic patients and animals have revealed a significant reduction in the cardiac reserve [25–27]. This depression in contractile function is attributed to different factors such as depressed ATPase activity of the contractile proteins [28], abnormality in the calcium metabolism [29] and other membrane dysfunctions [30–32]. However, the mechanism involved in the development of diabetic cardiomyopathy has not been established in any of the study.

Oxygen free radicals have been implicated not only in diabetes but also in diabetes associated cardiomyopathy. If left unchecked, oxygen free radicals can cause cell and tissue injury. In a normal and healthy situation, a basal level of antioxidant enzyme activity is maintained to protect against deleterious effects of free radicals. However, cardiac structure and function are known to be affected by increased free oxygen radicals or decreased levels of antioxidants [10, 33–35]. In a previous study, we have reported that the treatment with probucol, initiated one day following the induction of diabetes, help maintain cardiac function as well as the anti-

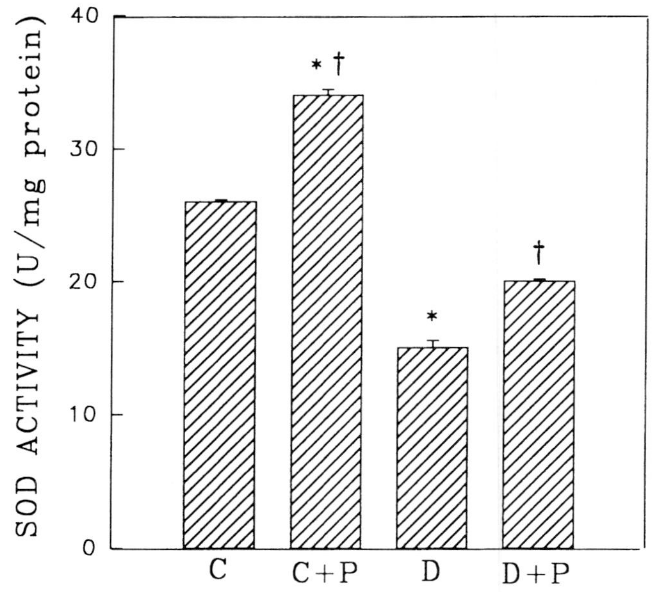

Fig. 1. Effects of probucol on myocardial superoxide dismutase (SOD) activity in control and diabetic rats. Each value represents ± S.E.M. of 8–10 experiments. C, control; C+P, control + probucol; D, diabetic; and D+P, diabetic + probucol. *Significant ($p < 0.05$) from control. †Significant ($p < 0.05$) from the control and diabetic groups.

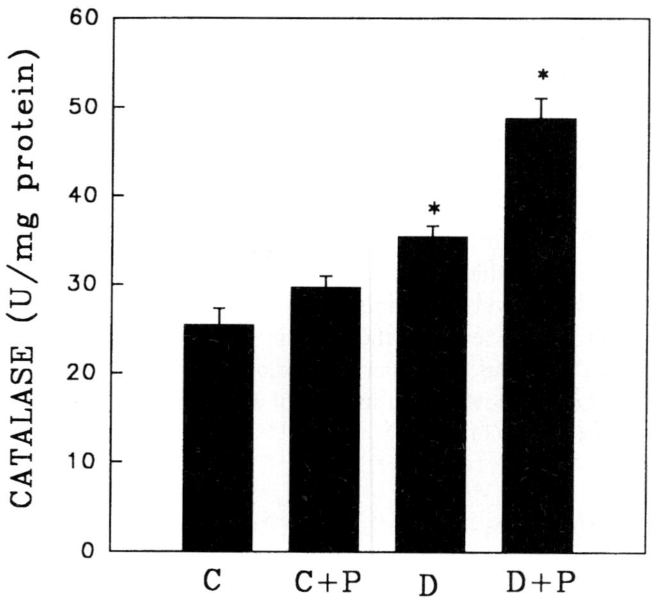

Fig. 3. Effects of probucol on myocardial catalase activity in control and diabetic rats. Each value represents ± S.E.M. of 8–10 experiments. C, control; C+P, control + probucol; D, diabetic; and D+P, diabetic + probucol. *Significant ($p < 0.05$) from all other groups.

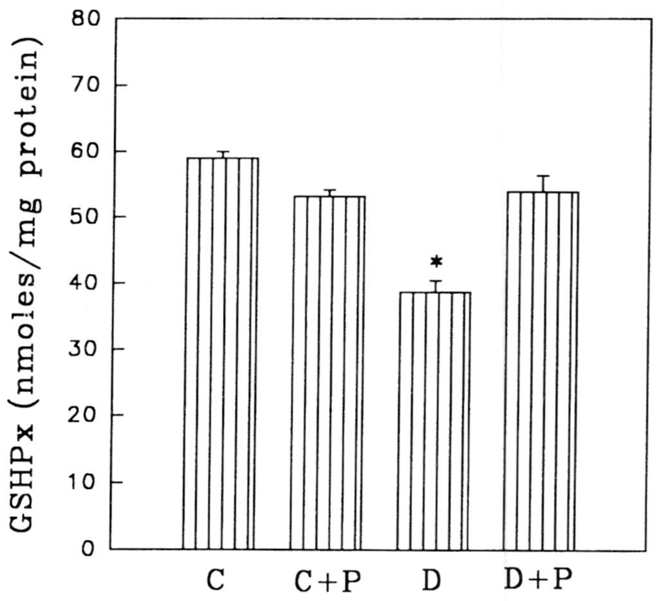

Fig. 2. Effects of probucol on myocardial GSHPx activity in control and diabetic rats. Each value represents ± S.E.M. of 8–10 experiments. C, control; C+P, control + probucol; D, diabetic; and D+P, diabetic + probucol. *Significant ($p < 0.05$) from all other groups.

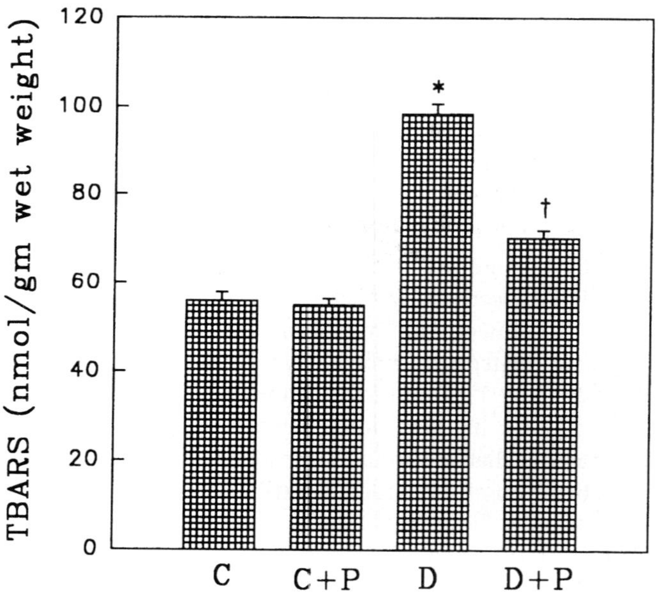

Fig. 4. Effects of probucol on lipid peroxidation in control and diabetic rats. Each value represents ± S.E.M. of 8–10 experiments. C, control; C+P, control + probucol; D, diabetic; and D+P, diabetic + probucol. *Significant ($p < 0.05$) from the Control and Control + Probucol groups. †Significant ($p < 0.05$) from all other groups. TBARS, thiobarbituric acid reactive substances.

oxidant status of the heart [7]. In the same study, it was also reported that probucol increased endogenous antioxidants may have contributed in the observed beneficial effect. Probucol induced enhancement of endogenous antioxidants have been reported in other studies [13, 14]. In this study, we have shown for the first time that probucol treatment initiated 4 weeks after the induction of diabetes can in fact reverse the deterioration of cardiac function as well as myocardial endogenous antioxidants changes caused by diabetes mellitus.

The antioxidant changes in different enzymes showed a

characteristic pattern. CAT has been implicated in the detoxification of high H_2O_2 concentrations whereas GSHPx is effective at lower concentrations of H_2O_2. CAT activity was increased in diabetic as well as in diabetic + probucol group, probably to detoxify H_2O_2. However, hearts from diabetic animals showed decrease in SOD and GSHPx activities and increased lipid peroxidation or oxidative stress. In probucol-treated diabetic animals, not only SOD and GSHPx activities were improved but CAT was also significantly higher. This was accompanied by a decrease in the myocardial lipid peroxidation in the diabetic + probucol treated group suggesting that probucol was effective in decreasing the oxidative stress.

The reasons for the decrease of GSHPx and SOD enzyme levels are not known. However, earlier reports have suggested that reactive oxygen radicals can themselves reduce the activity of these enzymes [36, 37]. Decrease in endogenous antioxidants in diabetic tissues and blood, and reversal of these changes with the addition of antioxidants (SOD, CAT and glutathione) has been reported [38].

Probucol is a lipophilic compound that likely incorporates into the cellular membrane. In this regard, relatively high concentrations of probucol were detected in the hearts of probucol-treated monkeys [39]. Diabetes-prone (BB/W) rats had a significant reduction in the incidence of diabetes when treated with probucol [40]. Probucol also increased the time of onset of diabetes in rats [40]. These beneficial effects of probucol were speculated to be due to its antioxidant activity.

Although beneficial effects of probucol are associated with improved endogenous antioxidants, involvement of other mechanisms cannot be ruled out. In this regard, favourable changes due to probucol treatment in the elasticity of the arterial wall may contribute in the improved hemodynamic function [41]. Diabetic cardiomyopathy is known to be accompanied by increased wall-stiffness of blood vessels. The finding of a decrease in the left ventricular systolic pressure in diabetic animals in this study is in agreement with many previous observations [7, 42].

In addition, probucol has a significant lipid lowering effect [12]. Probucol enters into the core of the lipoprotein particles and may thereby change the nature of the lipoprotein [43]. Studies in experimental diabetes have shown that plasma and lipoproteins contain lipid peroxidation products [6, 16]. In the present study, there was a correlation between the increased lipid peroxide levels and the myocardial dysfunction. Thus lowering of lipids as well as lipid peroxidation may also have a role in the observed beneficial changes. Furthermore, improvement in insulin and glucose levels in probucol treated diabetic rats is also an important observation indicating some reversal of the diabetic condition. This may also play a role in improving heart function. Reduced oxidative stress as well as injury to the pancreas may be one of the mechanisms for these effects of probucol.

Although possibilities of potential benefits of improved diabetic condition and lipid lowering are not ruled out, our data suggest that in diabetic cardiomyopathy the tissue antioxidant status is in deficit which contributes to the deterioration of cardiac function. The reversal of these changes by probucol, an antioxidant, provides evidence of their association with the progression of the disease condition. This study suggests potential usefulness of antioxidant therapy in diabetic cardiomyopathy.

Acknowledgements

Study was supported by the Diabetic Foundation and the Medical Research Council of Canada. Nalini Kaul and Charita Seneviratne received Postdoctoral Fellowship support from Manitoba Health Research Council. Natasha Siveski-Iliskovic and Neelam Khaper received studentship support from the University of Manitoba.

References

1. Kannel WB: Role of diabetes in cardiac disease. In: S. Zoneraich, (ed.) Diabetes and the Heart, Charles C. Thomas, Springfield, Illinois pp 97–120
2. Rubler S, Dlugash J, Yuceoglu UZ, Kumral T, Branwood AW, Grishman A: New type of cardiomyopathy associated with diabetic glomerulosclerosis. Am J Cardiol 30: 595–602, 1972
3. Regan TJ, Lyons MM, Ahmed S, Levinson GE, Oldewurtel HA, Ahman M, Haider B: Evidence for cardiomyopathy in familial diabetes mellitus. J Clin Invest 60: 885– 893, 1977
4. Dhalla NS, Pierce GN, Innes IR, Beamish RE: Pathogenesis of cardiac dysfunction in diabetes mellitus. Can J Cardiol 1: 263–284, 1985
5. Fein FS, Sonnenblick EH: Diabetic cardiomyopathy. Prog Cardiovasc Dis H 27: 255–270, 1985
6. Wohaieb SA, Godin DV: Alterations in free radical tissue defense mechanisms in streptozotocin induced diabetes in the rat: Effects of insulin treatment. Diabetes 36: 1014–1018, 1987
7. Kaul N, Siveski-Iliskovic N, Thomas TP, Hill M, Khaper N, Singal PK: Probucol improves antioxidant activity and modulates development of diabetic cardiomyopathy. Nutrition. In Press, 1995
8. Simmons KJ Defence against free radicals has therapeutic implications. J Amer Med Assoc 251: 2187–2192, 1984
9. Halliwell G, Gutteridge JMC: Lipid peroxidation, oxygen radicals, cell damage and antioxidant therapy. Lancet I: 1396–1397, 1984
10. Kaul N, Siveski-Iliskovic N, Hill M, Slezak J, Singal PK: Free radicals in the heart. J Pharmacol Toxicol Meth 30: 55–63, 1993
11. Parthasarathy S, Young SG, Witztum TL, Pittman RC, Steinberg D: Probucol inhibits oxidative modification of low density lipoproteins. J Clin Invest 77: 641–644,1986
12. Carew TE, Schwenke DC, Steinberg D: Antiatherogenic effect of Probucol unrelated to its hypocholesterolemic effect. Proc Natl Acad Sci, USA 84: 7725–7729, 1984
13. Siveski-Iliskovic N, Kaul N, Singal PK: Probucol promotes endogenous antioxidants and provides protection against adriamycin-induced cardiomyopathy. Circulation. 89: 2829–2835, 1994
14. Siveski-Iliskovic N, Hill M, Chow D, Singal PK: Probucol protects

against adriamycin cardiomyopathy without interfering with its antitumor effect. Circulation 91: 10–15, 1995
15. Chisolm GM, Morel DW: Lipoprotein oxidation and cytotoxicity: Effect of probucol on streptozotocin treated rats. Am J Cardiol 62: 20B–26B, 1988
16. Morel DW, Chisolm GM: Antioxidant treatment of diabetic rats inhibits lipoprotein oxidation and cytotoxicity. J Lipid Res 30: 1827–1834, 1989
17. Matsushita M, Yoshino G, Iwai M: Protective effects of probucol on alloxan diabetes in rats. Diabetes Res Clin Prac 7: 313–316, 1989
18. Bondar RJL, Mead DC: Evaluation of glucose-6-phosphate dehydrogenase from leuconostoc mesenterioles in the hexokinase method for determining glucose in serum. Clin Chem 20: 586–590, 1974
19. Marklund SL: Pyrogallol autoxidation. In: R.A. Greenwald, (ed.) Handbook of Methods for Oxygen Radical Research. Boca Raton, Florida: CRC Press, 1985 pp 243–247
20. Paglia DE, Valentine WN: Studies on the quantitative and qualitative characterization of erythrocyte glutathione peroxidase. J Lab Clin Med 70: 158–247, 1967
21. Clairborne A: Catalase activity. In: R.A. Greenwald (ed.) Handbook of Methods for Oxygen Radical Research. Boca Raton, Florida: CRC Press, 1979 pp 283–284
22. Placer ZA, Cushman LL, Johnson BC: Estimation of product of lipid peroxidation (malondialdehyde) in biochemical systems. Anal Biochem 16: 359–365, 1966
23. Aust SD: Lipid peroxidation. In: R.A. Greenwald (ed.) Handbook of Methods for Oxygen Radical. Boca Raton, Florida: CRC Press, 1985 pp 203–207
24. Lowry OH, Rosebrough NJ, Farr AL, Randall AJ: Protein measurement with the folin phenol regent. J Biol Chem 193: 265–275, 1951
25. Regan TJ, Ettinger PO, Khan ME, Jesuani MU, Lyon MM, Oldewurtel HA, Weber M: Altered myocardial function and metabolism in chronic diabetes mellitus without ischemia in dogs. Circ Res 35: 222–237, 1974
26. Factor SM, Minase T, Sonnenblick E: Clinical and morphological features of human hypertensive diabetic cardiomyopathy. Am Heart J 99: 446–458, 1980
27. Heyliger CE, Pierce GN, Singal PK, Beamish RE, Dhalla NS: Cardiac alpha and beta adrenergic receptor alterations in diabetic cardiomyopathy. Basic Res Cardiol 77: 610–618, 1982
28. Pierce GN, Dhalla NS: Mechanism of defect in cardiac myofibrillar function during diabetes. Am J Physiol 248: E170–E175, 1985
29. Pierce GN, Kutryk MJB, Dhalla NS: Alterations in Ca^{2+} binding by and composition of the cardiac sarcolemmal membrane in chronic diabetes. Proc Natl J Acad Sci, USA 80: 5412–5416, 1983
30. Makino N, Dhalla KS, Elimban V, Dhalla NS: Sarcolemmal calcium transport in streptozotocin-induced diabetic cardiomyopathy in rats. Am J Physiol 253: E202–E207, 1983
31. Penparkgul S, Fein F, Sonnenblick Eh, Scheuer J: Depressed cardiac sarcoplasmic reticular function from diabetic rats. J Mol Cell Cardiol 13: 303–309, 1981
32. Ganguly PK, Pierce GN, Dhalla KS, Dhalla NS: Defective sarcoplasmic reticulum calcium transport in diabetic cardiomyopathy. Am J Physiol 244: E528–E535, 1983
33. Kobayashi A, Yamashete T, Kaneko M, Nishiyama T, Hayashi H, Yamazaki N: Effects of verapamil on experimental cardiomyopathy in the Syrian hamster. JACC 10: 1128–1134, 1987
34. Gupta M, Singal PK: Time course of structure, function, and metabolic changes due to an exogenous source of oxygen metabolites in rat heart. Can J Physiol Pharmacol 67: 1549–1559, 1989
35. Guarnieri C, Flamigni F, Caldarera CM: Role of oxygen in cellular damage induced by reoxygenation of hypoxic heart. J Mol Cell Cardiol 12: 797–808, 1980
36. Hodgson EK, Fridovich I: The interaction of bovine erythrocyte superoxide dismutase with hydrogen peroxide: Chemiluminescence and peroxidation. Biochemistry 14: 5399–5303, 1975
37. Searle AJ, Willson RL: Glutathione peroxide: Effect of superoxide, hydroxyl and bromine free radicals on enzyme activity. Int J Radiat Biol 37: 213–217, 1980
38. Giuliano D, Acampara R, D'Onofrio P: Medical hypothesis: cardiovascular complications of diabetes mellitus-from glucose to insulin and back. Diabetic Metabol 20: 445–453, 1994
39. Marshall RN: Pharmacology and toxicology of probucol. Artery 10: 7–21, 1982
40. Drash AL, Rudert WA, Borquaye S, Wang R, Lieberman I: Effect of probucol on development of diabetes mellitus in BB rats. Am J Cardiol 62: 27B–30B, 1988
41. O'Brien K, Nagano Y, Gown A, Kita T, Chait A: Probucol treatment affects the cellular composition but not antioxidized low density lipoprotein immunoreactivity of plaque from Watanabe heritable hyperlipedimic rabbits. Arteriosclerosis and Thrombosis 11: 751–759, 1991
42. Litwin SE, Raya TE, Anderson PG, Daugherty S, Goldman S: Abnormal cardiac function in the streptozotocin-diabetic rat. J Clin Invest 86: 481–488, 1990
43. Schwartz CJ: Introduction – The Probucol Experience: A review of the past and a look at the future. Am J Cardiol 62: 1B–5B, 1988

Molecular and Cellular Biochemistry **160/161**: 289–295, 1996.

Glycogen phosphorylase isoenzyme BB in diagnosis of myocardial ischaemic injury and infarction

Ernst-Georg Krause, Georg Rabitzsch, Franz Noll,[1] Johannes Mair[2] and Bernd Puschendorf[2]

Department of Molecular Cardiology, Max Delbrück Center for Molecular Medicine, Berlin-Buch, Germany; [1]Institute of Medical Diagnostics, Berlin-Steglitz, Germany; [2]Department of Medical Chemistry and Biochemistry, University Innsbruck, Innsbruck, Austria

Abstract

This review deals with glycogen phosphorylase (GP) and its isoenzyme BB in the diagnosis of ischaemic myocardial injury. Early identification and confirmation of acute myocardial infarction is essential for correct patient care and disposition decision in the emergency department. In this respect, glycogen phosphorylase isoenzyme BB (GPBB) based on its metabolic function is an enzyme for early laboratory detection of ischaemia. In the aerobic heart muscle GPBB together with glycogen is tightly associated with the vesicles of the sarcoplasmic reticulum. Release of GPBB, the main isoform in the human myocardium, essentially depends on the degradation of glycogen, which is catalyzed by GP. Ischaemia is known to favour the conversion of bound GP in the *b* form into GP *a*, thereby accelerating glycogen breakdown, which is the ultimate prerequisite for getting GP into a soluble form being able to move freely in the cytosol. The efflux of GPBB into the extracellular fluid follows if ischaemia-induced structural alterations in the cell membrane become manifest. The clinical application of GPBB as a marker of ischaemic myocardial injury is a very promising tool for extending our knowledge of the severity of myocardial ischaemic events in the various coronary syndromes. The rational roots of this development were originated from Albert Wollenberger's research work on the biochemistry of cardiac ischaemia and the transient acceleration of glycogenolysis mainly brought about by GP activation. (Mol Cell Biochem **160/161**: 289–295, 1996)

Key words: myocardial infarction, glycogen phosphorylase, isoenzyme BB, glycogenolysis, enzyme release, glycogen-glycogenolysis-complex, cardiac ischaemia.

Introduction

During recent years biochemical markers for the detection of acute myocardial injury gained particular interest for the early diagnosis of evolving acute myocardial infarction (AMI), the monitoring of thrombolytic treatment, and for risk stratification in patients with unstable angina. Creatine kinase (CK) MB mass, CK isoforms, myoglobin, cardiac troponin I and troponin T have a comparable early sensitivity for AMI, and all are markedly more sensitive than CK and CKMB activity [1]. However, in the final analysis myoglobin, CKMB mass, CK isoforms, and troponins are not sufficiently sensitive within the first 3–4 hours after the onset of AMI, and the diagnostic performance of the electrocardiography (ECG) is clearly superior to that of biochemical markers during this time interval [1, 2]. However, up to 50% of chest pain patients may have non-diagnostic ECGs at hospital admission [3]. Laboratory parameters are cheaper, easier to perform and to interpret for the non-specialist than other alternative diagnostic methods (e.g. echocardiography, myocardial scintigraphy). Consequently the search is still going on for more rapidly detectable and more sensitive markers which should start to be released in the phase of reversible ischaemic myocardial damage. In this respect, glycogen phosphorylase isoenzyme BB (GPBB) based on its metabolic function and on first clinical results is a very promising enzyme for the early laboratory detection of ischaemic myocardial injury.

Address for offprints: E.-G. Krause, Robert-Rössle-Str. 10, D-13125 Berlin-Buch, Germany

Biochemistry of glycogen phosphorylase

Glycogen phosphorylase (GP) is one of the best studied enzymes in biochemistry. It is a glycolytic enzyme which plays an essential role in the regulation of carbohydrate metabolism by mobilization of glycogen [4]. It catalyses the first step in glycogenolysis in which glycogen is converted to glucose-l-phosphate, utilising orthophosphate. The physiological role of muscle phosphorylase is to provide the fuel for the energy supply required for muscle contraction. Its activity is allosterically regulated by the binding of AMP and phosphorylation. Phosphorylase kinase converts GP *b* into its more active form GP *a*. Phosphorylase exists in the cardiomyocyte in association with glycogen and the sarcoplasmatic reticulum and forms a macromolecular complex (sarcoplasmatic reticulum glycogenolysis complex) [5, 6]. The degree of association of GP with this complex depends essentially on the metabolic state of the myocardium. With the onset of tissue hypoxia, when glycogen is broken down and disappears, glycogen phosphorylase is converted from a particulate into a soluble form, and the enzyme becomes free to move around in the cytoplasma [5–7].

GP exists as a dimer under normal physiological conditions. The dimer is composed of 2 identical subunits. At least 3 GP isoenzymes are found in human tissues that are named after the tissue in which they are preferentially expressed, GPLL (liver), GPMM (muscle), and GPBB (brain) [4]. The three isoenzymes can be distinguished by functional and immunological properties. They are encoded by three distinct genes. The genes of the 3 human GP isoenzymes have been cloned and sequenced [8]. The proteins predicted by the cDNA sequences consist of chains of 846 (LL), 842 (MM) and 862 (BB) amino acid residues. Amino acids 1–830 match and differences are mainly found at the C-terminus, which is the catalytic domain of the protein. In pairwise sequence comparison the brain type protein is 80% identical with the liver type and 83% identical with the muscle type. GPBB has 21 and 16 additional amino acid residues on its C-terminal portion that are not present on the MM and LL isoenzymes, respectively.

Adult human skeletal muscle contains only one isoenzyme, *GPMM*. *GPLL* is the predominant isoenzyme in human liver and all other human tissues except heart, skeletal muscle, and brain. The *isoenzyme BB* is the predominant isoenzyme of human brain. Its molecular weight as a monomer is approximately 94 kD. In the human heart the isoenzymes BB and MM are found, but GPBB is still the predominant isoenzyme in myocardium. By far the highest concentrations of GPBB were found in human brain and heart. The tissue concentrations of GPBB in heart and brain are comparable [9]. Although immunoblot, electrophoresis, and northern blot data are partly conflicting [4, 8–11], there is evidence that GPBB isoenzyme might not be restricted to brain and heart in humans. Much lower GPBB concentrations have been reported, for example, in leukocytes, spleen, kidney, bladder, testis, digestive tract and aorta. However, in all these tissues the isoenzyme LL is by far the predominant GP isoenzyme.

Pathophysiology: Shift to anaerobic energy production in the acutely ischaemic myocardium, activation and release of GP

The energy metabolism of the heart, in particular the biochemistry of cardiac ischaemia as well as the biochemical mechanisms of the neuronal and hormonal regulation of the heart became a main profile of Albert Wollenberger's research work and his group during the sixties [12–17]. Investigating the energy production in the acutely ischaemic myocardium Wollenberger *et al.* [12] demonstrated that in ischaemia the sources of utilizable chemical energy (ATP, phosphocreatine) are rapidly exhausted. They clarified the mechanism and demonstrated the rapidity of cardiac glycogenolysis when the blood circulation was suddenly arrested [13–15]. Of course, the activation of conversion of glycogen to glucose phosphate cannot be faster than the velocity of the first reaction step in this process which is catalyzed by GP. Using the freeze-clamp technique [16] a transient rise in cardiac cAMP levels followed by an activation of GP due to conversion of the nonphosphorylated *b* form into phosphorylase *a* by phosphorylase kinase was for the first time demonstrated during acute ischaemia (Fig. 1) [17]. As β-adrenoceptor blockade diminished the GP conversion a sympathetic activation of adenylyl cyclase by neurotransmitters was postulated at that time [14, 17]. Concomitantly the rate of glycogenolysis was found to be accelerated [13–15]. Furthermore there is evidence that the kinetic properties of GPBB allow a glycogen breakdown catalyzed by the *b* form. Compared with GPMM this isoenzyme is characterized by low values of K_m for the substrate orthophosphate as well as of $K_{0.5}$ for the activator AMP [10]. An ischaemia-induced rise in the levels of intracellular orthophosphate and AMP in myocardium may therefore induce a second, long lasting acceleration of glycogenolysis under these conditions [15, 18]. Indeed cardiac glycogen breakdown was found to continue during post-ischaemic reperfusion when the *a* form of GP declined to pre-ischaemic control levels when the orthophosphate level was still high [19]. Using isolated glycogen particles form canine heart muscle a burst in glycogenolysis could be initiated by either cAMP and calcium ions which was accompanied by a breakdown of glycogen [7]. In experimental studies as well as in patients with acute myocardial infarction the released GPBB was exclusively found in the *b* form [20, 21]. Thus it is suggested that the activity of GPBB (form *b*) catalyses the prolonged degradation of glycogen in the sarcoplasmatic reticulum-glycogenolysis complex in the

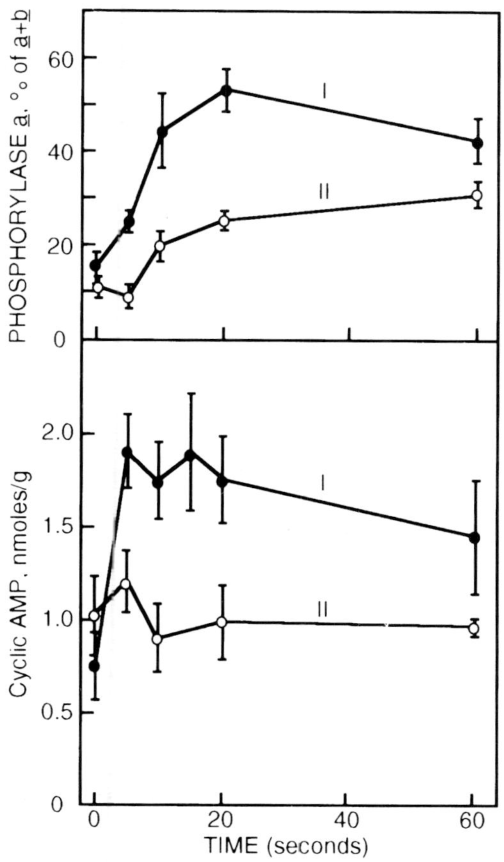

Fig. 1. Cyclic AMP and phosphorylase a levels in the left ventricle of the dog before and after the onset of myocardial ischaemia. I: Dogs not treated with a β-blocker; II: dogs treated with pronethalol. Phosphorylase (a + b) content averaged 14.3 μmole phosphate/g × min in both group I and group II. Adapted from Wollenberger *et al.* [17]

ischaemic area of the myocardium [15, 18] .

In conscious dogs a rapid release of GPBB was measured in the cardiac lymph after a transient ligation of a coronary artery for not longer than 10 min, which did not lead to histological signs of myocardial necrosis [20]. An efflux of GP from the myocardium after hypoxia or substrate depletion has been observed earlier in the isolated perfused rat and rabbit heart (Fig. 2) [22, 23]. The GP release in these experiments correlated with the remaining myocardial glycogen content [23]. In this Langendorff model of the isolated perfused heart the addition of imipramine under aerobic conditions to a cardioplegic perfusion solution only caused a release of CK, but not of GP (Fig. 3) [23]. The myocardial glycogen content remained uneffected as well. Imipramine causes in a certain concentration range a selective increase in the plasma membrane permeability without myocardial hypoxia. On the other hand, the stimulation of glycogenolysis by high doses of adrenaline did not cause a decrease in myocardial GP activity, although the glycogen content of the tissue was greatly diminished by the added adrenaline (Table 1) [22]. These experimental results allow one to conclude that the release of GPBB requires both a burst in glycogenolysis and a concomitantly increased plasma membrane permeability as known for injured cardiomyocytes by ischaemia [24].

Given its molecular mass (94 kD as a monomer) the early release of GPBB raises questions about the mechanisms of its release from ischaemic myocardium. An essential part of an explanation may be its key role in the energy metabolism of ischaemic myocardium. When glycogen is broken down and disappears, GPBB becomes free to move from the peri-SR compartment directly into the extracellular fluid, if cell membrane permeability is simultaneously increased, which is usually the case in ischaemia. A high GPBB concentration gradient, which immediately is formed in the compartment of the sarcoplasmatic reticulum glycogenolytic complex, may be the reason for the high efflux rate of this enzyme. In contrast to other cytosolic proteins, this gradient may at least partly be also realized via *T-tubuli* and may contribute to the efflux of GPBB (Fig. 4).

In summary, the ischaemia-sensitive glycogen degradation, which is regulated by Ca^{2+}, metabolic intermediates and catecholamines, seems to be a crucial prerequisite for the efflux of GPBB. This outlines the specific sensitivity of this enzyme marker to indicate transient imbalances in heart energy metabolism as it is the case during angina pectoris attacks and/or in the infarcting myocardium. Therefore this enzyme is a promising analyte for the detection of ischaemic myocardial injury.

First Clinical Results

Acute myocardial infarction

There have been distinct differences in sensitivities of GPBB in comparison with myoglobin, CKMB mass, CK, and cardiac troponin T within the first 2–3 h after AMI onset [25, 26], and in our patients GPBB was the most sensitive parameter during the first 4 h after AMI onset (Table 2). In the majority of AMI patients GPBB increased between 1–4 h after the onset of chest pain. Therefore, GPBB may be a very important marker for the early diagnosis of AMI. GPBB usually peaks before CK, CKMB or troponin T and returns within the reference interval within 1–2 days after AMI onset (Fig. 5). Basal concentration of GPBB in blood plasma of normal healthy was found to be independent of sex and age; the empirical data distribution of basal GPBB was gaussian [27].

As with soluble markers, such as myoglobin and CKMB, we could demonstrate that time course of GPBB in AMI patients is markedly influenced by whether or not early reperfusion of the infarct-related coronary artery occurs [26]. The well established so called 'wash out' phenomenon after successful thrombolysis leads to a more rapid increase in GPBB

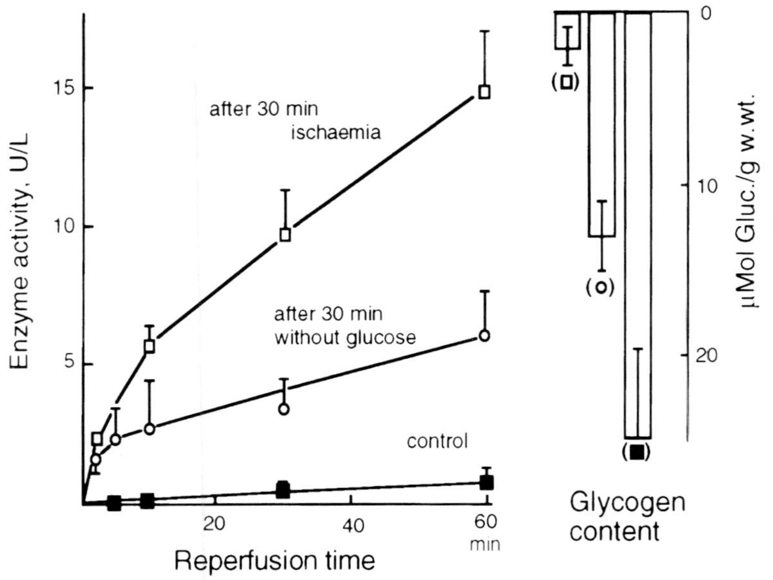

Fig. 2. Release of glycogen phosphorylase and glycogen content of rabbit hearts perfused in the Langendorff mode Enzyme efflux was assayed during reperfusion after global ischaemia (□), after substrate depletion (○), and under aerobic control perfusion conditions (■). Finally the perfused hearts were freeze-clamped and glycogen content was determined. Values are means plus 1 S.E.M. Gluc. = glucosyl residue; W. Wt. = wet weight. Adapted from Krause *et al.* [23].

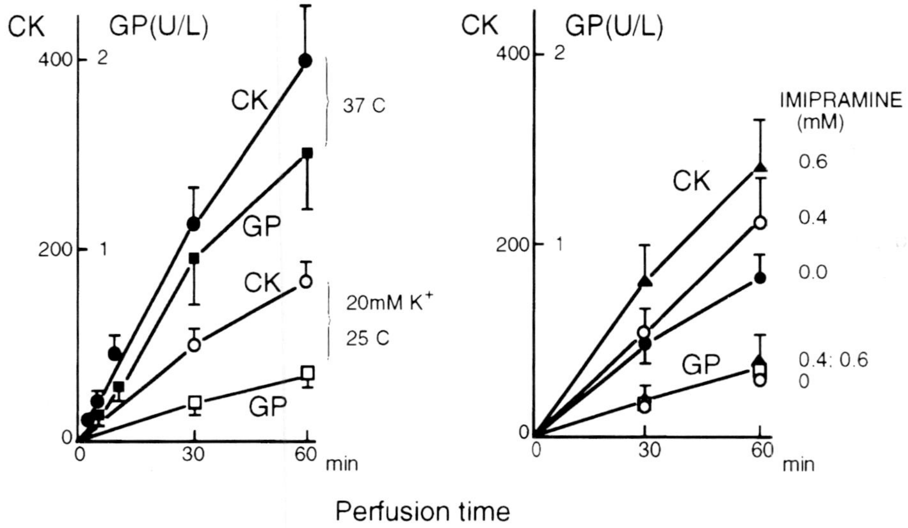

Fig. 3. Release of glycogen phosphorylase (GP) and creatine kinase (CK) from normothermic beating and hypothermic, K^+-arrested, perfused rabbit hearts. The effect of 0.4 and 0.6 mM imipramine was only studied with non-beating hearts at 24°C. Adapted from Krause *et al.* [23].

and results in earlier and higher peak values (Fig. 6). Therefore, GPBB may be useful, alongside with other soluble myocardial proteins, to assess the effectiveness of thrombolytic therapy non-invasively. However, decision limits to detect successful, and what is clinically more important, failed reperfusion, remain to be established in a study controlled with acutely performed coronary angiography.

Unstable angina pectoris

The application of GPBB is not restricted to conventional myocardial infarction. An early release of GPBB was demonstrated in patients with Braunwald class III unstable angina who showed ST-T alterations at rest. Only GPBB was increased above the upper reference limit in the majority of these patients at hospital admission (Fig. 7) [28]. Whether the early GPBB release in these patients was due to minimal necrosis of myocardial tissue or severe reversible ischaemic

Table 1. Activity of glycogen phosphorylase (GP) and glycogen concentration in perfused rat hearts after addition of adrenaline

Treatment	GP a+b (µmol orthophosphate/ min × mg protein)	Glycogen (µmol glucose/ g wet weight)
Controls (n=5)	0.41 ± 0.05	13.4 ± 2.4
Adrenaline (n=5)	0.50 ± 0.09	3.9 ± 0.8

Adapted from Schulze *et al.* [22].

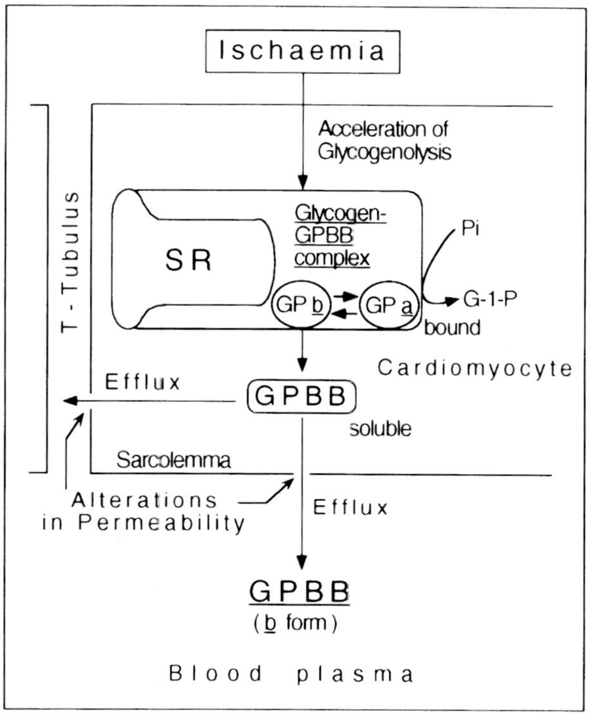

Fig. 4. Scheme of GPBB release from myocardium in ischaemia. Glycogen phosphorylase (GP) together with glycogen is tightly associated with the vesicles of sarcoplasmic reticulum (SR) under normal conditions. A release of GPBB, the main isoform in the myocardium, essentially depends on the degradation of glycogen, which is catalyzed by GP *a* (the phosphorylated, active form of the isoenzyme) and by GP *a* (nonphosphorylated, AMP-dependent form). Ischemia is known to favor the conversion of bound GP *b* into GP *a* thereby accelerating glycogen breakdown, which seems to be the ultimate prerequisite for getting GP into a soluble form. An efflux of GPBB into the extracellular fluid may only follow if ischaemia-induced structural alterations in the cell membrane are manifested. Pi: inorganic phosphate; G-1-P: glucose 1-phosphate. For more details see text. Adapted from Rabitzsch *et al.* [26].

injury is currently not known. As underlined by the receiver-operating characteristic (ROC) curve and ROC area calculations of GPBB and comparison with those of CK, CKMB mass, myoglobin, and troponin T [26], GPBB showed the best diagnostic performance of all markers tested to detect acute ischaemic coronary syndromes (AMI or severe unstable angina at rest with transient ST-T alterations) on admission (Fig. 8). GPBB plasma concentrations in patients with stable angina resembled those of healthy individuals or patients without angina [26].

Table 2. Early sensitivities of GPBB and other biochemical markers before start of thrombolytic treatment in patients with acute myocardial infarction who were admitted within 4 h after the onset of chest pain

	GPBB mass	CK	CKMB mass	Myoglobin	Troponin T
Sensitivity	0.77 (0.55–0.92)	0.20 (0.04–0.48)	0.47 (0.21–0.73)	0.47 (0.21–0.73)	0.40 (0. 16–0.68)

(95% confidence interval)

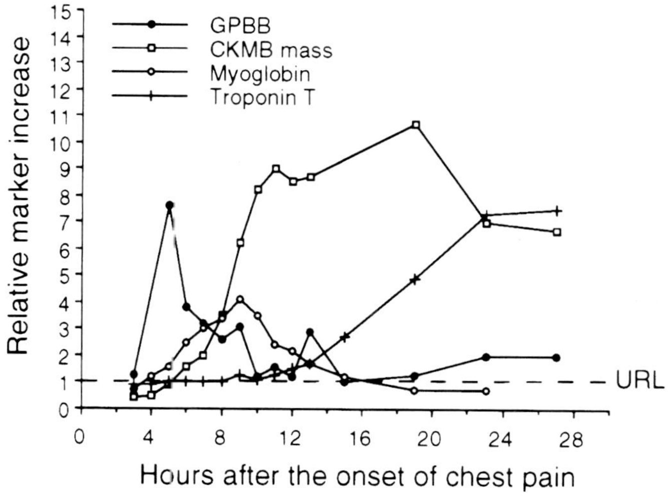

Fig. 5. Glycogen phosphorylase BB, CKMB mass, myoglobin, and cardiac troponin T time courses in a patient with a small non-Q wave myocardial infarction. Data are given as ×-fold increase of the upper reference limit (URL).

Coronary artery bypass grafting (CABG)

GPBB is also a sensitive marker for the detection of perioperative myocardial ischaemia and infarction in patients undergoing CABG [29]. In uncomplicated patients GPBB peaks within 4 h after aortic unclamping and returns to baseline values within 20 h. GPBB release correlates with aortic crossclamping time, which reflects the duration of myocardial hypoxia during cardioplegic cardiac arrest. GPBB time courses of patients with perioperative myocardial infarction (PMI) differ markedly in time to peak values (peaks occur later) and peak concentrations (> 50 µg/L) from uneventful patients. However, also patients with severe episodes of perioperative myocardial ischaemia that do not fulfil standard PMI criteria show markedly elevated GPBB concentrations compared with uncomplicated patients. In patients with emergency CABG GPBB, but not CKMB, correlated with clinical evidence of myocardial ischaemia [29]. In summary, GPBB is a very sensitive marker of perioperative ischaemic myocardial injury in CABG patients.

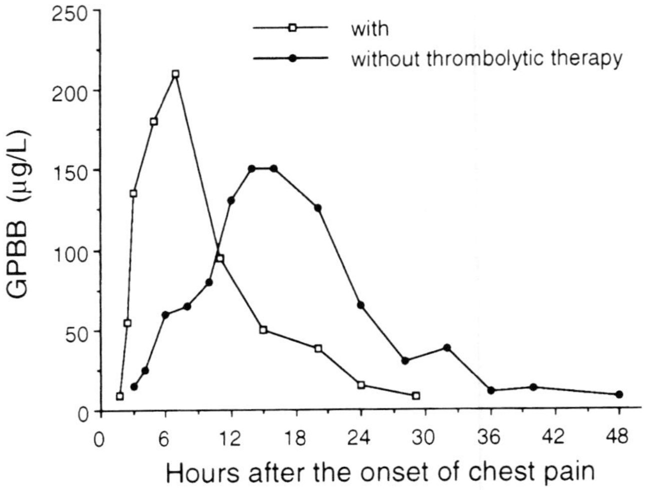

Fig. 6. Influence of early reperfusion of the infarct related coronary artery on glycogen phosphorylase BB time courses GPBB time course in a patient who was successfully reperfused by thrombolytic therapy and in a patient who did not receive fibrinolytic treatment because of contra-indications. The GPBB upper reference limit is 7 μg/L.

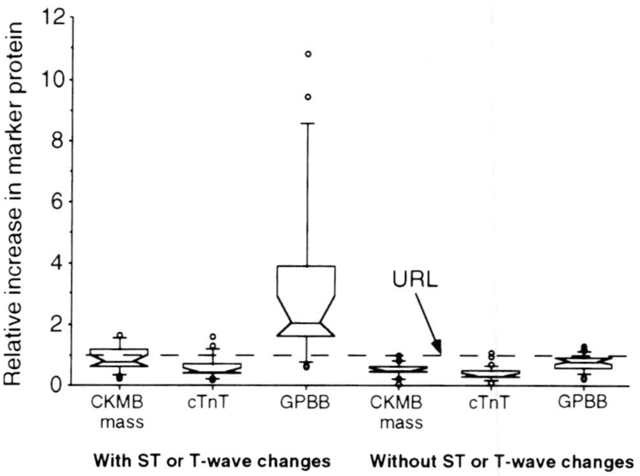

Fig. 7. CKMB mass. cardiac troponin T, and GPBB concentrations on admission to the emergency department in patients with unstable angina. Patients were grouped according to the presence of signs of myocardial ischaemia in the admission ECG recording. Marker concentrations are given as ×-fold increase of the upper reference limit. The data are given as notched box plots. In the box plot the lines represent the 10th, 25th, 50th 75th, and 90th percentile, the notches the 95th confidence bands about the median, values above and below the 10th and 90th percentile are represented as data points. Abbreviations: creatine kinase (CK), cardiac troponin T (cTnT), glycogen phosphorylase (GP). Adapted from Mair *et al.* [28].

Diagnostic specificity

GPBB is not a heart-specific marker and its specificity is limited. However, increases in GPBB are specific for ischaemic myocardial injury when damage to the brain and consequent disturbance of the blood-brain barrier can be excluded. According to experimental studies and clinical observations increases in GPBB do not occur in response to therapeutic

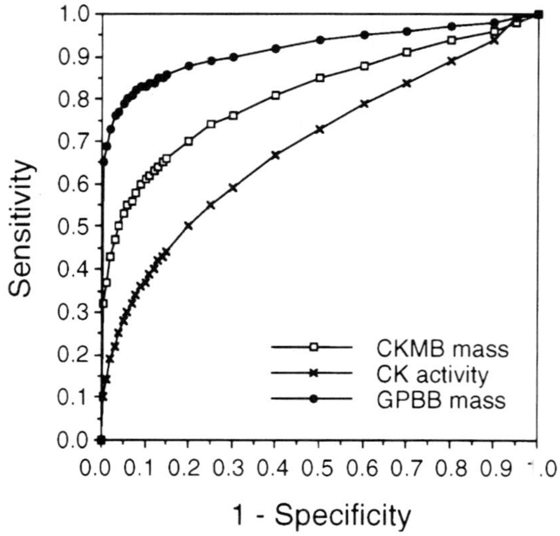

Fig. 8. Receiver operating characteristic (ROC), curves of GPBB. CKMB mass and CK activity for the identification of acute coronary syndromes in non-traumatic chest pain patients at hospital admission. An acute coronary syndrome was defined as either acute myocardial infarction or unstable angina with reversible ST-T alteration in the admission ECG recording. The larger the area under the ROC curve the better is the discriminating power of the parameter. The area under the GPBB ROC curve is significantly greater than those of the other markers. Adapted from Rabitzsch *et al.* [26].

circumstances in which cardiac work is increased and glycogen might be mobilized, such as after administration of catecholamines and glucagon, provided that a concomitant myocardial injury with cell membrane damage does not take place [22, 29]. The diagnostic specificity of GPBB for myocardial injury in non-traumatic chest pain patients was in the range of CKMB [26], which suggests sufficient specificity in clinical practice. Future studies on the diagnostic specificity of GPBB, however, will also have to address the issue in an unselected cohort of patients including severely traumatized patients with and without head injuries, patients with liver damage or renal failure. As long as the diagnostic specificity of GPBB for myocardial damage is not fully delineated, a positive GPBB result should be later confirmed by cardiac troponin I measurement.

Conclusions

The first hints that blood GP increases above its upper reference limit after the onset of AMI earlier than does CK were obtained approximately 2 decades ago, with total GP activity measured by a not very sensitive enzymatic assay [21]. The breakthrough occured later in our laboratory [25] with the development of a sensitive and specific immunoenzymometric assay for the measurement of the isoenzyme GPBB. This newly developed assay was used in all our clinical studies. The challenge still exists to develop a rapid assay that is suitable for bedside or 'stat' use in the routine laboratory. Of course, the first clinical results will have to be confirmed in a larger number of patients but they allow several important conclusions, and there is no doubt that GPBB is a promising marker for the detection of ischaemic myocardial injury. This is probably explained by its function as a key enzyme of glycogenolysis. GPBB has so far been the most sensitive marker for the diagnosis of AMI within 4 h after the onset of chest pain. GPBB was the only marker which was increased in a considerable proportion of AMI patients within 2–3 hours after the onset of chest pain. The application of GPBB is not restricted to conventional AMI. GPBB also increased early in patients with unstable angina and reversible ST-T alterations in the resting ECG at hospital admission. Therefore GPBB could be useful for early risk stratification in these patients. GPBB has also been a sensitive marker for the detection of perioperative myocardial ischaemia and infarction in patients undergoing coronary artery bypass grafting. The diagnostic specificity of GPBB is sufficient for clinical practice, in non-traumatic chest pain patients it was in the range of that of CKMB. Thus, if the first clinical results on GPBB can be confirmed, the future scenario for the laboratory testing for myocardial injury could be the combination of cardiac troponin I or T and GPBB measurement, which combines cardiac specificity with high early sensitivity for ischaemic myocardial damage.

References

1. Mair J, Morandell D, Genser N, Lechleitner P, Dienstl F, Puschendorf B: Equivalent early sensitivities of myoglobin, creatine kinase MB mass, creatine kinase isoform ratios, and cardiac troponin I and T for acute myocardial infarction. Clin Chem 41: 1266–1272, 1995
2. Mair J, Smidt J, Lechleitner P, Dienstl F, Puschendorf B: A decision tree for the early diagnosis of acute myocardial infarction in non-traumatic chest pain patients at hospital admission. Chest 108: 1502–1509, 1995
3. Rozenman Y, Gotsman MS: The earliest diagnosis of acute myocardial infarction. Annu Rev Med 45: 31–44, 1994
4. Newgard CB, Hwang PK, Fletterick RJ: The family of glycogen phosphorylases: structure and function. Crit Rev Biochem Molec Biol 24: 69–99, 1989
5. Meyer F, Heilmeyer LMG Jr, Haschke RH, Fischer EH: Control of phosphorylase activity in a muscle glycogen particle: Isolation and characterization of the protein glycogen complex. J Biol Chem 245: 6642–6648, 1970
6. Entman ML, Kaniike K, Goldstein MA, Nelson TE, Bornet EP, Futch TW, Schwartz A: Association of glycogenolysis with cardiac sarcoplasmic reticulum. J Biol Chem 251: 3140–3146, 1976
7. Entman ML, Bornet EP, van Winkle WB, Goldstein MA, Schwartz A: Association of glycogenolysis with cardiac sarcoplasmic reticulum: II. Effect of glycogen depletion, deoxycholate solubilization and cardiac ischemia: evidence for a phosphorylase kinase membrane complex. J Mol Cell Cardiol 9: 515–528, 1977
8. Newgard CB, Littmann DR, Genderen C, Smith M, Fletterick RJ: Human brain glycogen phosphorylase. Cloning sequence analysis, chromosomal mapping, tissue expression and comparison with the human liver and muscle isozymes. J Biol Chem 263: 3850–3857, 1988
9. Kato A, Shimizu A, Kurobe N, Takashi M, Koshikawa K: Human brain-type glycogen phosphorylase: Quantitative localization in human tissues determined with an immunoassay system. J Neurochem 52: 1425–1432, 1989
10. Will H, Krause E-G, Böhm M, Guski H, Wollenberger A: Kinetische Eigenschaften der Isoenzyme der Glykogenphosphorylase b aus Herz- und Skelettmuskulatur des Menschen. Acta Biol Med Germ 33: 149–160, 1974
11. Proux D, Dreyfus J-C: Phosphorylase isoenzymes in tissues: Prevalence of the liver type in man. Clin Chim Acta 48: 167–172, 1973
12. Wollenberger A, Krause E-G: Activation of α-glucan phosphorylase and related metabolic changes in dog myocardium following arrest of blood flow. Biochim Biophys Acta 67: 337–340, 1963
13. Wollenberger A, Krause E-G, Macho L: Thyroid state and the activity of glycogen phosphorylase in ischaemic myocardium. Nature 201: 789–791, 1964
14. Wollenberger A, Krause E-G, Shahab L: Endogenous catecholamine mobilization and the shift to anaerobic energy production in the acutely ischemic myocardium. In: G. Marchetti, B. Taccardi, (eds). International Symposium on the Coronary Circulation and Energetics of the Myocardium. Karger, Basel 1967, pp 200–219
15. Wollenberger A, Krause E-G: Metabolic control characteristics of the acutely ischemic myocardium. Am J Cardiol 22: 349–359, 1968
16. Wollenberger A, Ristau O, Schoffa G: Eine einfache Technik der extrem schnellen Abkühlung grösserer Gewebsstücke. Pflügers Arch Ges Physiol 270: 399-412, 1960
17. Wollenberger A, Krause E-G, Heier G: Stimulation of 3′, 5′-cyclic AMP formation in dog myocardium following arrest of blood flow. Biochem Biophys Res Commun 36: 664–670, 1969
18. Dobson JG, Mayer SE: Mechanism of activation of cardiac glycogen phosphorylase in ischemia and anoxia. Circ Res 33: 412–420, 1973
19. Kalil-Filho R, Gersdtenblith G, Hansford RG, Chacko VP, Vandegaer K, Weiss RG. Regulation of myocardial glycogenolysis during post-ischemic reperfusion. J Mol Cell Cardiol 23: 1467–1479, 1991
20. Michael LH, Hunt JR, Weilbaecher D, Perryman MB, Roberts R, Lewis RM, Entman ML. Creatine kinase and phosphorylase in cardiac lymph: Coronary occlusion and reperfusion. Am J Physiol 248: H350–359, 1985
21. Krause E-G, Will H, Böhm M, Wollenberger A: The assay of glycogen phosphorylase in human blood serum and its application to the diagnosis of myocardial infarction. Clin Chim Acta 58: 145–154, 1975
22. Schulze W, Krause E-G, Wollenberger A: On the fate of glycogen phosphorylase in the ischemic and infarcting myocardium. J Mol Cell Cardiol 2: 241–251, 1971
23. Krause E-G, Härtwig A, Rabitzsch G: On the release of glycogen phosphorylase from heart muscle: Effect of substrate depletion, ischemia and of imipramine. Biomed Biochim Acta 48: S77–82, 1989
24. Reimer KA, Jennings RB: Myocardial ischemia, hypoxia and infarction. In: HA Fozzard, E. Haber, RB Jennings, AM Katz and HE Morgan (Eds). The Heart and Cardiovascular System. Raven Press, New York, 2nd ed. 1991, pp 1875–1973
25. Rabitzsch G, Mair J, Lechleitner P, Noll F, Hofmann U, Krause E-G, Dienstl F, Puschendorf B: Isoenzyme BB of glycogen phosphorylase b and myocardial infarction. Lancet 341: 1032–1033, 1993
26. Rabitzsch G, Mair J, Lechleitner P, Noll F, Hofmann U, Krause E-G, Dienstl F, Puschendorf B: Immunoenzymometric assay of human glycogen phosphorylase isoenzyme BB in diagnosis of ischemic myocardial injury. Clin Chem 41: 966–978, 1995
27. Rabitzsch G, Noll F, Hofmann U, Krause E-G, Armbruster FP: Basal concentration of the isoenzyme BB of the glycogen phosphorylase in human blood. Clin Chim Acta 214: 109–111, 1993
28. Mair J, Puschendorl B, Smidt J, Lechleitner P, Dienstl F, Noll F, Krause E-G, Rabitzsch G: Early release of glycogen phosphorylase in patients with unstable angina and transient ST-T alterations. Brit Heart J 72: 125–127, 1994
29. Mair P, Mair J, Krause E-G, Balogh D, Puschendorf B, Rabitzsch G: Glycogen phosphorylase isoenzyme BB mass release after coronary artery bypass grafting. Eur J Clin Chem Clin Biochem 32: 543–547, 1994

Molecular and Cellular Biochemistry **160/161**: 297–302, 1996.

Unaltered ryanodine receptor protein levels in ischemic cardiomyopathy

Wolfgang Schillinger, Markus Meyer, Goro Kuwajima,[1] Katsuhiko Mikoshiba,[2] Hanjörg Just and Gerd Hasenfuss

Medizinische Klinik III, Universität Freiburg, Germany; [1]Shionogi Institute for Medical Science, Osaka, [2]Dept. of Molecular Neurobiology, Institute of Medical Science, University of Tokyo, Japan

Abstract

Previous studies on sarcoplasmic reticulum calcium release channel (ryanodine receptor) demonstrated that protein levels are unchanged in myocardium from hearts with end-stage failing dilated cardiomyopathy. In ischemic cardiomyopathy, ryanodine receptor mRNA levels were shown to be decreased but no data on protein levels are available. Accordingly, protein levels of ryanodine receptor, calsequestrin, and sarcoplasmic reticulum calcium-ATPase (SR-Ca^{2+}-ATPase) were measured by Western blot analysis in nonfailing human myocardium (n = 7) and in end-stage failing myocardium due to ischemic cardiomyopathy (n = 14). Protein levels of calsequestrin which is the major sarcoplasmic reticulum calcium storage protein were similar in nonfailing myocardium and in myocardium from end-stage failing hearts with ischemic cardiomyopathy. Ryanodine receptor protein levels, normalized to total protein or calsequestrin were also unchanged in ischemic cardiomyopathy. In contrast, protein levels of SR-Ca^{2+}-ATPase normalized to total protein or calsequestrin were decreased by 31 and 30%, respectively ($p < 0.05$). The data indicate that (1) sarcoplasmic reticulum calcium uptake sites are decreased relative to the release sites in ischemic cardiomyopathy, and (2) alterations of sarcoplasmic proteins are similar in ischemic and dilated cardiomyopathy. (Mol Cell Biochem **160/161**: 297–302, 1996)

Key words: sarcoplasmic reticulum, calcium-ATPase, calsequestrin, human myocardium, heart failure

Introduction

The molecular and functional basis of human myocardial failure has been the subject of many recent publications. The data is still controversial, but evidence accumulates that alterations in intracellular calcium cycling is of significant pathophysiological relevance [1–4]. Many studies indicate that decreased sarcoplasmic reticulum calcium uptake may be a major defect underlying disturbed calcium handling in human heart failure independent of the etiology of failure [5–9]. Whether or not quantitative or qualitative alterations in function or expression of the sarcoplasmic reticulum calcium release site (ryanodine receptor) are involved in disturbed calcium handling of the failing heart is less clear. Regarding the function of the ryanodine receptor, Holmberg and Williams [10], measuring single channel recordings from sarcoplasmic reticulum of failing human hearts with dilated cardiomyopathy, demonstrated normal basal channel properties . In contrast, D'Agnolo *et al.* [11] observed a significant increase in caffein threshold of the ryanodine receptor from hearts with failing dilated cardiomyopathy suggesting altered gating mechanism. Nimer *et al.* [12] observed decreased stimulation of sarcoplasmic reticulum calcium uptake with ryanodine in failing hearts from dilated cardiomyopathy which may indicate altered ryanodine response of the sarcoplasmic reticulum calcium release channel. Regarding expression of the sarcoplasmic reticulum calcium release channel in dilated cardiomyopathy, Brillantes *et al.* [13] observed unchanged mRNA levels, Go *et al.* [14] reported decreased mRNA levels, and Arai *et al.* [9] observed an inverse correlation between ryanodine receptor mRNA levels and atrial natriuretic factor mRNA levels, the latter was used as an indicator for left ventricular hypertrophy and failure. Protein levels of the ryanodine receptor were shown to be unchanged in failing

Address for offprints: G. Hasenfuss, Medizinische Klinik III, Universität Freiburg, Hugstetter Strasse 55, 79106 Freiburg, Germany hearts with dilated cardiomyopathy [7]. In ischemic cardiomyopathy a decrease of ryanodine receptor mRNA levels have been consistently observed [13, 14]. However, no data are available on ryanodine receptor protein levels in ischemic cardiomyopathy.

Accordingly, the present study was performed to compare ryanodine receptor protein levels in myocardium from end-stage failing hearts with ischemic cardiomyopathy and from nonfailing human hearts. In addition, protein levels of the sarcoplasmic reticulum calcium storage protein calsequestrin and the SR-Ca^{2+}-ATPase were measured.

Materials and methods

Patients

Quantification of the ryanodine receptor, SR Ca^{2+}-ATPase, and calsequestrin was performed in myocardium from failing and nonfailing human hearts. Specimens from failing hearts were obtained from 14 patients with end-stage heart failure who were undergoing cardiac transplantation. All patients suffered from ischemic cardiomyopathy (ICM) and were classified stage IV according to the functional classes of heart failure of the New York Heart Association. Clinical characteristics and pretransplant hemodynamics from the time closest to cardiac tranplantation are given in Table 1. Nonfailing human myocardium (NF) was obtained from 7 brain-dead organ donors (2 female, and 5 male) whose hearts could not be used for cardiac transplantation because of technical reasons. None of them had a history of cardiac disease, and their mean age was 39 ± 2 years which is not significantly different from the mean age of the patients with heart failure (Table 1). Immediately after cardiectomy, a piece of the left ventricular wall was excised, washed in a Krebs-Ringer-Solution, and quickly frozen in liquid nitrogen. The samples were stored at –80°C until use.

The present study was reviewed and approved by the Ethical Committee of the University Clinics of Freiburg

Preparation of cardiac tissue homogenates

About 100 mg of myocardium without connective or adipose tissue, endocardium or epicardium were thawed in a ninefold volume of an ice-cold 20 mmol/L Na-HEPES, pH 7.4, 4 mmol/L EGTA, 1 mmol/L DTT. Proteinase inhibitors (0.1 mmol/L Leupeptin and 0.3 mmol/L PMSF) were included to prevent homogenates from proteolytic degradation. Homogenization was performed at 4°C for eight times 15 sec by use of a Polytron-Homogenizer PT-K (Brinkman Instruments) with a PTA-7 unit, followed by 15 strokes of a motorized Potter-Elvehjem. The protein concentrations were determined in triplicate according to Lowry [15] using a standard curve prepared from different concentrations of bovine serum albumin in the above buffer. The yield of protein per gram wet weight was calculated from the protein concentration in the homogenates. It was 136 ± 7 mg/g and 128 ± 2 mg/g in myocardium from nonfailing hearts, and hearts with ischemic cardiomyopathy, respectively (no significant differences between groups). Aliquots of the homogenates were frozen in liquid nitrogen and stored at –80°C until use.

Table 1. Clinical characteristics of transplant recipients

Patient	Age, y	Sex	EF, %	CI, $L*min^{-1}*m^{-2}$	PCW, mm Hg	Previous Medications
1	43	M	23	2.6	24	Digitoxin, captopril, amiodarone, ISMN, phenprocoumone
2	54	M	27	1.6	–	Furosemide, captopril, ISMN, phenprocoumone
3	55	M	27	1.8	32	Digoxin, furosemide, captopril, ISMN
4	38	M	38	2.1	30	Digitoxin, furosemide, captopril, amiodarone, ISMN, phenprocoumone
5	52	M	27	2.0	22	Digitoxin, furosemide, spironolactone, phenprocoumone
6	56	F	26	2.4	22	Digitoxin, furosemide, captopril
7	55	M	36	2.7	10	Digitoxin, furosemide, enalapril, molsidomine, amiodarone, phenprocoumone
8	47	M	–	2.1	21	Digoxin, furosemide, captopril, ISMN
9	57	F	21	2.1	35	Digoxin, furosemide, captopril, ISMN, prazosine, molsidomine, phenprocoumone
10	34	M	16	1.6	38	Digitoxin, furosemide, spironolactone, captopril, metoprolol, ISMN
11	58	M	16	1.8	24	Digoxin, spironolactone, captopril, ISMN, nifedipine
12	52	M	28	2.0	28	Etacryne acid, phenprocoumone
13	50	M	34	2.3	23	Digitoxin, furosemide, captopril, amiodarone, phenprocoumone
14	45	M	30	1.5	27	Digitoxin, furosemide, captopril, phenprocoumone

EF indicates ejection fraction; CI, cardiac index; PCW, mean pulmonary capillary wedge pressure; previous medications, drug therapy at the day before transplantation; and ISMN, isosorbide mononitrate.

Western blot analysis

An aliquot of each sample was solubilized at 37°C in 2% sodium dodecyl sulfate, 5% 2-mercaptoethanol, 10% glycerol, 0.00125% bromphenol blue, and 62.5 mmol/L Tris-Cl, pH 6.8. Equal amounts of protein from all samples were subjected to SDS-polyacrylamide gel electrophoresis according to Laemmli [16] and transferred to nitrocellulose for 2 h at 125 V [17]. The nitrocellulose membranes were blocked in 5% nonfat milk dissolved in TBS (20 mmol/L Tris-Cl, pH 7.4, 150 mmol/L NaCl). The blots were then probed for 2 h with an antibody to ryanodine receptor [18] diluted 1:4000 in TBS, containing 1% bovine serum albumin and 0.1% TWEEN-20, or with antibodies to SR Ca^{2+}-ATPase [19] (1:10000), and calsequestrin [20] (1:2000), respectively. Then, the membranes were incubated for 1 h with a peroxidase-labelled secondary antibody (Amersham Buchler Ltd.) in a dilution of 1:10000. Immunoreactive bands were visualized utilizing an ECL Western Blot Kit (Amersham Buchler Ltd.) and exposure to a Kodak X-OMAT AR x-ray film. The band densities were evaluated by densitometric scanning by use of a 2202 Ultrascan laser densitometer (LKB). Since several blots had to be performed for quantificating each SR protein in all samples, all assay conditions like incubation or exposure times, and reagent concentrations followed a highly standardized procedure. In addition, a normalization procedure was applied in order to be able to account for blot to blot differences that occured in spite of standardization. For that purpose, one nonfailing heart was used as a reference on all blots. Normalization was performed by dividing densitometric units of each sample by the value of the corresponding reference heart from the same blot. For graphical reasons, all normalized values were again multiplied by a constant representing the value of the reference heart shown on the blot in Fig. 1 (first band each). Each individual value represents the mean of two independent determinations. Linearity of the assay was verified at varying exposure times by plotting different amounts of protein subjected to ECL Western blot against corresponding densitometric units (Fig. 2). Comparative analyses in NF and ICM were done within the linear range as shown for each sarcoplasmic reticulum (SR) protein. The amounts of protein used for quantification were 50 µg for SR Ca^{2+}-ATPase, and 80 µg for the ryanodine receptor and calsequestrin, respectively. A more detailed description of the assay procedure has recently been published [7].

Statistical analysis

All data are expressed as mean ± S.E.M. Differences between mean values were tested for significance by an independent *t*-test, accepting $p < 0.05$ as significantly different.

Results

Figure 1 shows Western immunoblots of SR proteins responsible for calcium release into the cytosol (ryanodine receptor), calcium reuptake from the cytosol (SR Ca^{2+}-ATPase), and calcium storage in the lumen of the SR (calsequestrin).

There was no significant difference between nonfailing and

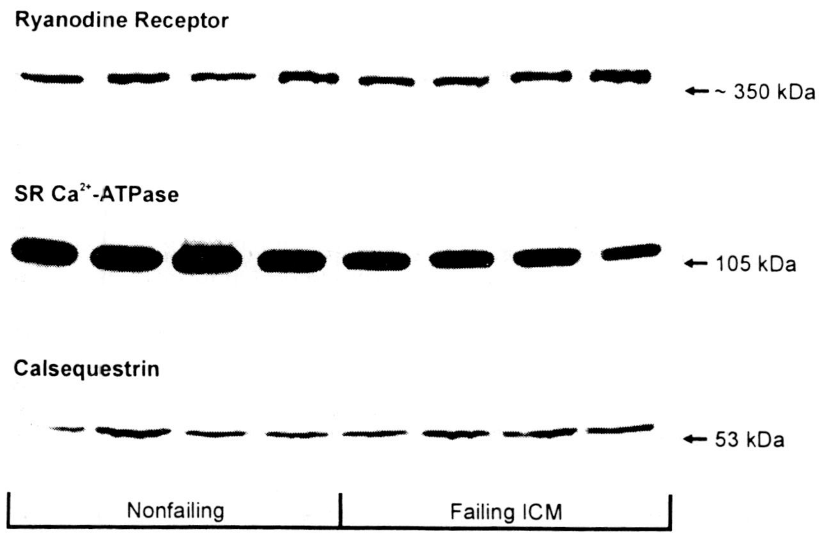

Fig. 1. Western immunoblots of sarcoplasmic reticulum (SR) proteins in human myocardium. Equal amounts of protein from nonfailing myocardium and from myocardium with end-stage heart failure due to ischemic cardiomyopathy (ICM) were subjected to Western blot analyses.

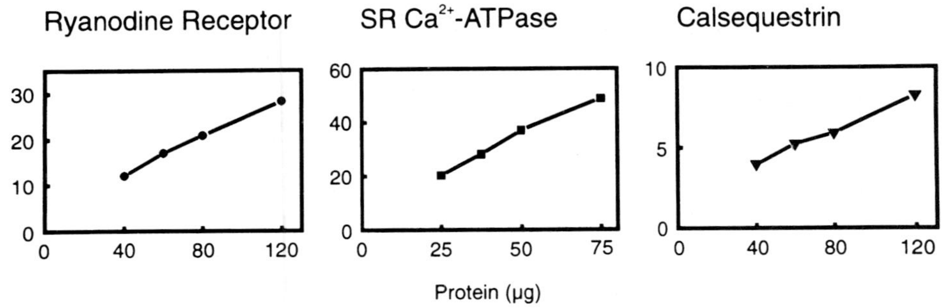

Fig. 2. Linearity of the assay was verified at varying exposure times by plotting different amounts of protein subjected to ECL western blot against corresponding densitometric units as shown for each sarcoplasmic reticulum (SR) protein. Ordinates give relative densitometric units. Comparative analyses in NF and ICM were done within the linear range. The amounts of protein used for quantification were 50 µg for SR Ca^{2+}-ATPase (middle figure), and 80 µg for the ryanodine receptor (left figure) and calsequestrin (right figure), respectively.

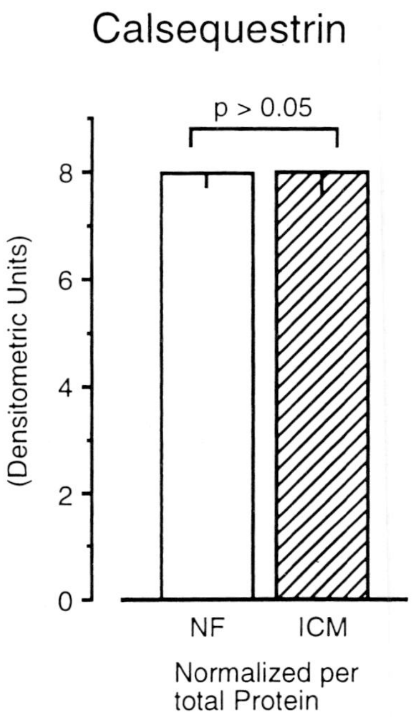

Fig. 3. Bar graphs showing mean protein levels of calsequestrin, the major sarcoplasmic reticulum calcium binding protein, in nonfailing myocardium (NF, n = 7) and in myocardium from end-stage failing hearts due to ischemic cardiomyopathy (ICM, n = 14). All individual values were normalized per total protein recovered per gram wet weight.

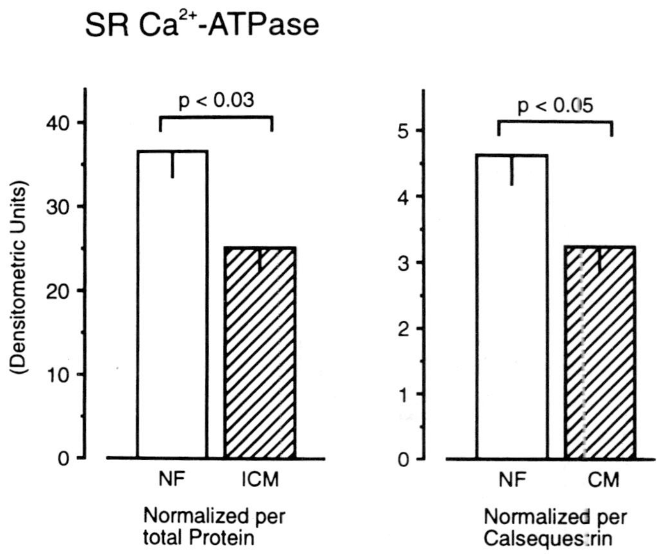

Fig. 4. Bar graphs showing mean protein levels of SR Ca^{2+}-ATPase in nonfailing myocardium (NF; n = 7) and myocardium from end-stage failing hearts with ischemic cardiomyopathy (ICM; n = 14). Left, all individual values were normalized per total protein recovered per gram wet weight, and right values were normalized per calsequestrin protein levels.

failing human myocardium regarding protein levels of the SR calcium storage protein, calsequestrin, in the total protein recovered per gram wet weight (Fig. 3). SR-Ca^{2+}-ATPase protein levels normalized to total protein recovered per gram wet weight or to calsequestrin protein levels were reduced by 31 or 30% ($p < 0.05$), respectively, in end-stage heart failure due to ischemic cardiomyopathy (Fig. 4).

Protein levels of the ryanodine receptor, did not show a significant difference between myocardium from nonfailing hearts and myocardium from failing hearts due to ischemic cardiomyopathy. This finding was independent from normalization of protein levels to total protein recovered per gram wet weight or to calsequestrin protein levels (Fig. 5).

A first series of blots in which proteinase inhibitors were omitted had shown that the ryanodine receptor in ischemic cardiomyopathy is much more susceptible to proteolytic degradation under *in vitro* conditions than nonfailing myocardium or failing myocardium from hearts with dilated cardiomyopathy (Fig. 6).

Discussion

The present study shows that ryanodine receptor protein levels and calsequestrin protein levels are unchanged in ischemic cardiomyopathy and confirms previous findings of reduced SR-Ca^{2+}ATPase protein levels.

The calcium sensitive SR calcium release channel (ryano-

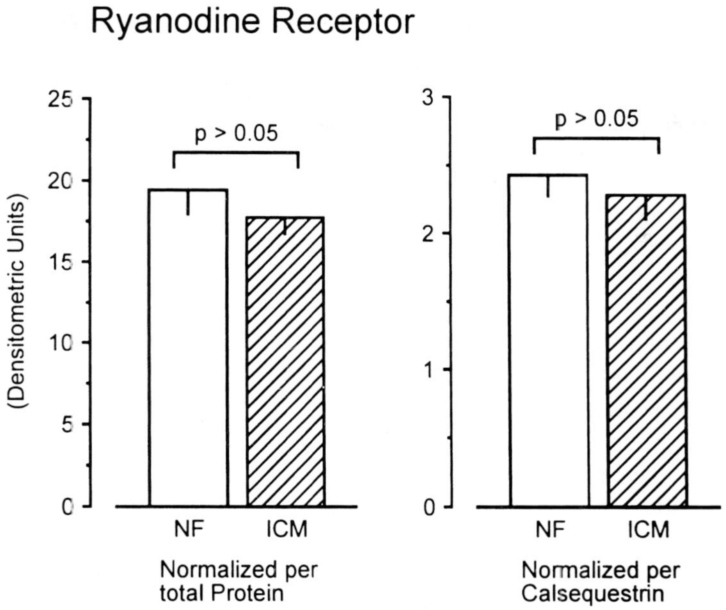

Fig. 5. Bar graphs showing mean protein levels of the ryanodine receptor in nonfailing myocardium (NF; n = 7) and in endstage failing myocardium due to ischemic cardiomyopathy (ICM; n = 14) Normalization was performed per total protein recovered per gram wet (left) weight or per calsequestrin protein levels (right).

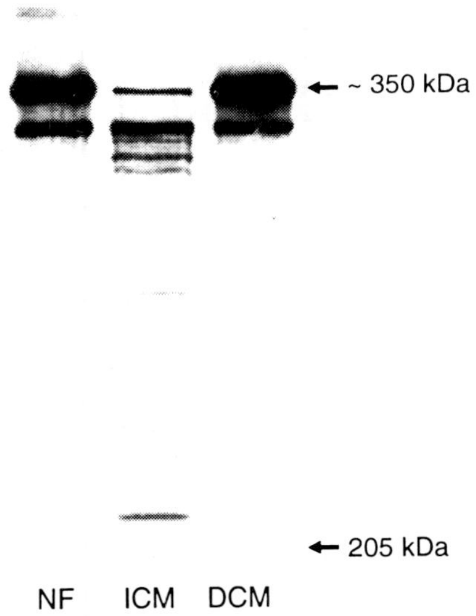

Fig. 6. Western immunoblot demonstrating proteolytic degradation of cardiac ryanodine receptor in ischemic cardiomyopathy (ICM) when myocardial samples were homogenized in the absence of proteinase inhibitors. Homogenates prepared from nonfailing human myocardium and from a failing heart with dilated cardiomyopathy (DCM) did not show markedly higher proteolytic activity under these conditions apart from the appearance of a fainter second band that migrated somewhat faster.

dine receptor) plays a key role in excitation-contraction coupling. The ryanodine receptor is regulated by calcium which enters the cell through voltage gated calcium channels in the sarcolemma. Once activated by calcium influx, the channel opens and releases calcium for activation of contractile proteins [21–23]. This process is termed calcium-induced calcium release [24]. The present data indicate that protein levels of the ryanodine receptor are unchanged in heart failure due to ischemic cardiomyopathy when compared to nonfailing human myocardium. These findings contrasts to previous studies indicating decreased ryanodine receptor mRNA levels [9, 13, 14]. Quantitative differences between mRNA and protein levels in the comparison of failing and nonfailing human myocardium may result from alterations in mRNA or protein synthesis and degradation in the failing heart. Since protein levels are relevant from a functional point of view, determination of mRNA levels may be misleading when function aspects are considered.

Interestingly, proteolytic degradation of the ryanodine receptor under *in vitro* conditions was more pronounced in ischemic compared to dilated cardiomyopathy and compared to nonfailing myocardium. This is consistent with previous data from Holmberg and Williams indicating proteolytic degradation of the ryanodine receptor during ischemia and/or post-ischemic reperfusion [25]. Activation of cellular proteases during ischemia has also been related to the finding of degradation of troponin and cytoskeletal proteins [26, 27]. The pathophysiological role (if any) of increased protease activity in ischemic cardiomyopathy is speculative. In presence of unchanged ryanodine receptor protein levels, increased proteolytic degradation would indicate increased turnover of the ryanodine receptor in myocardium from hearts with ischemic cardiomyopathy. This of course would not be consistent with the finding of decreased mRNA in ischemic cardiomyopathy. Alternatively, the findings of unchanged protein levels but increased protease activity under *in vitro* conditions may indicate that the ryanodine receptor is protected against proteolytic degradation *in vivo*.

The findings of unchanged ryanodine receptor protein levels, unchanged levels of the calcium storage protein calsequestrin but decreased levels of SR-Ca^{2+}-ATPase indicate that a major defect is located in the calcium uptake site of the SR in ischemic cardiomyopathy. Of course, the present data do not exclude the possibility that altered function of the ryanodine receptor may be involved in disturbed excitation-contraction coupling in failing ischemic cardiomyopathy.

The changes in sarcoplasmic reticulum SR protein levels in ischemic cardiomyopathy are similar to those observed in dilated cardiomyopathy [7]. This indicates that altered sarcoplasmic reticulum protein expression occurs independent from the etiology of heart failure. Altered protein levels may, therefore, result from neurohumoral activation or hemodynamic overload rather than from specific disease related mechanisms.

Acknowledgements

This study was supported by DFG grants HA 1233/3-2 and HA 1233/4-1 We are grateful to Dr. K. Campbell for providing the antibody to SR-Ca^{2+}-ATPase and to Dr. C. Heilmann for providing the antibody to calsequestrin.

References

1. Morgan JP: Abnormal intracellular modulation of calcium as a major cause of cardiac contractile dysfunction. N Engl J Med 325: 625–632, 1991
2. Beuckelmann DJ, Näbauer M, Erdmann E: Intracellular calcium handling in isolated ventricular myocytes from patients with terminal heart failure. Circulation 85: 1046–1055, 1992
3. Hasenfuss G, Mulieri LA, Leavitt JB, Allen PD, Haeberle JR, Alpert NR: Alterations of contractile function and excitation-contraction coupling in dilated cardiomyopathy. Circ Res 70: 1225–1232, 1992
4. Pieske B, Kretschmann B, Meyer M, Holubarsch Ch, Weirich J, Minami K, Posival H, Just H, Hasenfuss G: Alterations in intracellular calcium handling associated with the inverse force-frequency relation in human dilated cardiomyopathy. Circulation 92: 1169–1178, 1995
5. Mercadier JJ, Lompre AM, Duc P, Boheler KR, Fraysse JB, Wisnewsky C, Allen PD, Komajda M, Schwartz K: Altered sarcoplasmic reticulum Ca^{2+}-ATPase gene expression in the human ventricle during end-stage heart failure. J Clin Invest 85: 305–309, 1990
6. Hasenfuss G, Reinecke H, Studer R, Meyer M, Pieske B, Holtz J, Holubarsch Ch, Posival H, Just H, Drexler H: Relation between myocardial function and expression of sarcoplasmic reticulum Ca^{2+}-ATPase in failing and nonfailing human myocardium. Circ Res 75: 434–442, 1994
7. Meyer M, Schillinger W, Pieske B, Holubarsch C, Heilmann C, Poslval H, Kuwajima G, Mikoshiba K, Just H, Hasenfuss G: Alterations of sarcoplasmic reticulum proteins in failing human dilated cardiomyopathy. Circulation. 92: 778–784, 1995
8. Schwinger RH, Böhm M, Schmidt U, Karczewski P, Bavendick U, Flesch M, Krause EG, Erdmann E: Unchanged protein levels of SERCA II and phospholamban but reduced Ca^{2+} uptake and Ca^{2+} ATPase activity of cardiac sarcoplasmic reticulum from dilated cardiomyopathy patients compared with patients with non failing hearts. Circulation 92: 3220–3228, 1995
9. Arai M, Alpert NR, MacLennan DH, Barton P, Periasamy M: Alterations in sarcoplasmic reticulum gene expression in human heart failure: A possible mechanism for alterations in systolic and diastolic properties of the failing myocardium. Circ Res 72: 463–469, 1993
10. Holmberg SR, Williams AJ: Single channel recordings from human cardiac sarcoplasmic reticulum. Circ Res 65: 1445–1449, 1989
11. D'Agnolo A, Luciani GB, Mazzucco A, Gallucci V, Salviati G: Contractile properties and Ca^{2+} release activity of the sarcoplasmic reticulum in dilated cardiomyopathy. Circulation 85: 518–525, 1992
12. Nimer LR, Needleman DH, Hamilton SL, Krall J, Movsesian MA: Effect of ryanodine on sarcoplasmic reticulum Ca^{2+} accumulation in nonfailing and failing human myocardium. Circulation 92: 2504–2510, 1995
13. Brillantes AM, Allen P, Takahashi T, Izumo S, Marks AR: Differences in cardiac calcium release channel (ryanodine receptor) expression in myocardium from patients with end stage heart failure caused by ischemic versus dilated cardiomyopathy. Circ Res 71: 18–26, 1992
14. Go LO, Moschella MC, Watras J, Handa KK, Fyfe BS, Marks AR: Differential regulation of two types of intracellular calcium release channels during end-stage heart failure. J Clin Invest 95: 888–894, 1995
15. Lowry OH, Roseborough NJ, Farr AL, Randall RJ: Protein measurements with the Folin phenol reagent. J Biol Chem 193: 265–275, 1951
16. Laemmli UK. Cleavage of structural proteins during the assembly of the head of bacteriophage T4. Nature 227: 680–685, 1970
17. Towbin H, Staehelin T, Gordon J: Electrophoretic transfer of proteins from polyacrylamide gels to nitrocellulose sheets: procedure and some applications. Proc Natl Acad Sci USA. 76: 4350–4354, 1979
18. Kuwajima G, Futatsugi A, Niimobe M, Nakanishi S, Mikoshiba K: Two types of ryanodine receptors in mouse brain: Skeletal muscle type exclusively in Purkinje cells and cardiac muscle type in various neurons. Neuron 9: 1133–1142, 1992
19. Jorgensen AO, Arnold W, Pepper DR, Kahl SD, Mandel F, Campbell KP: A monoclonal antibody to the Ca^{2+}-ATPase of cardiac sarcoplasmic reticulum crossreacts with slow type I but not with fast type II canine skeletal muscle fibers: An immuno-cytochemical and immunochemical study. Cell Motil Cytoskeleton. 9: 164–174, 1988
20. Kreisel W, Hildebrandt H, Schiltz E, Köhler G, Spamer C, Dietz C, Mossner W, Heilmann C: Immuno-gold electron microscopical detection of heat shock protein 60 (hsp60) in mitochondria of rat hepatocytes and myocardiocytes. Acta Histochem 96: 51–62, 1994
21. Barry WH, Bridge JHB: Intracellular calcium homeostasis in cardiac myocytes. Circulation 87: 1806–1815, 1993
22. Endo M: Calcium release from the sarcoplasmic reticulum. Physiol Rev 57: 71–108, 1977
23. Wagenknecht T, Grassucci R, Frank J, Saito A, Inui M, Fleischer S: Three dimensional architecture of the calcium channel/foot structure of sarcoplasmic reticulum. Nature 338: 167–170, 1989
24. Fabiato A: Calcium-induced release of calcium from the cardiac sarcoplasmic reticulum. Am J Physiol. 245: C1–C14, 1983
25. Holmberg SR, Williams AJ: The calcium-release channel from cardiac sarcoplasmic reticulum: Function in the failing and acutely ischemic heart. Bas Res Cardiol 87 (Suppl 1): 255–268, 1992
26. Toyo-Oka T, Ross J Jr: Ca^{2+} sensitivity change and troponin loss in cardiac natural actomyosin after coronary occlusion. Am J Physiol 240: H704–H708, 1981
27. Jennings R. Reimer KA, Steenbergen C: Myocardial ischemia revisited. The osmolar load, membrane damage, and reperfusion. J Mol Cell Cardiol 18: 769–780, 1986

Molecular and Cellular Biochemistry **160/161**: 303–306, 1996.

Myocardial infarction and nitric oxide

Richard J. Bing and Hiroshi Suzuki

Department of Experimental Cardiology, Huntington Medical Research Institutes, Pasadena, CA 91101, USA

Abstract

The report deals with the induction of the inducible form of nitric oxide synthase (iNOS) in infarcted heart muscle of rabbit and man. In the rabbit, nitric oxide synthase was significantly increased in the infarcted area beginning on the third day following ligation of a coronary artery. iNOS induction occured primarily in macrophages. In man, iNOS immunoreactivity was also primarily localized in macrophages on the seventh day following death from myocardial infarction. Of the specific inhibitors of iNOS in infarcted heart muscle, S-methylisothiourea (SMT) was the most potent. Its greatest effect occured in the normal non-affected area of the heart. Dexamethasone and cyclosporin A failed to inhibit iNOS. Apoptosis of macrophages commenced two days following ligation of a coronary artery. Mol Cell Biochem **160/161**: 303–306, 1996)

Key words: infarcted heart, myocardial infarction, nitric oxide synthase, apoptosis

Introduction

Numerous reviews on the biophysics in molecular biology of nitric oxide have been published [1, 2]. We know that it is an inorganic free radical gas with an unpaired electron and has remarkable physiological and pathological functions [3] It is formed in macrophages from L-arginine by the inducible form of nitric oxide synthase resulting in the formation of L-citrulline and nitric oxide (NO) [2]. Only activated macrophages synthesize NO. In contrast to the constitutional NO synthase (cNOS), which is active in vascular endothelial cells, iNOS requires *de novo* synthesis for its expression and is calcium independent. The L-arginine: NO pathway in macrophages is a primary defense mechanism against tumor cells, invading microorganisms and other pathogens. In addition, inhibition of iNOS is dependent on various cytokines such as tumor necrosis factor and some interleukins [2].

The role of the constitutive form of nitric oxide (cNOS) in the regulation of physiological conditions has been well studied. The role of inducible form of NO (iNOS) is more difficult to define, since it is activated primarily under pathological conditions, for example during an immunological response. Whenever macrophages are activated, iNOS is induced, creating its effect by producing nitric oxide against invading bacterial and cancer cells and even against the body's own cells. This is the case in myocardial infarction, where the presence of activated macrophages leads to NO production with important molecular, biological and pathological consequences.

This paper describes some of the results obtained in this laboratory on iNOS and nitric oxide in infarcted heart muscle in experimental myocardial infarction in the rabbit and man.

Nitric Oxide Synthase (NO) in myocardial infarction

The presence of macrophages in infarcted heart suggested to us that these cells express the inducible form of NOS (iNOS). Following ligation of the first anterolateral branch of the circumflex coronary artery in rabbits and beginning with postoperative day 1, iNOS activity was elevated in the infarcted area of the left ventricular wall as determined with the method of Bredt [4, 5]. On the second post-operative day the differences between the infarcted and non-infarcted areas became significant and iNOS activity reached its highest value on the third day. Twenty-one days after surgery, the activity of iNOS had disappeared. In the area of risk, no significant changes in iNOS activity were noted. It was also noted that the NO inhibitors L-NAME and L-NMMA diminished the production of NO by 43%.

Address for offprints: R.J. Bing, Huntington Medical Research Institutes, Department of Experimental Cardiology, 99 N. El Molino Avenue, Pasadena, CA 91101, USA

Macrophages and induction of iNOS in infarcted heart muscle

It was apparent from the studies of Hibbs that activated macrophages participate in the induction of iNOS [6]. Infarcted heart tissue is heterogeneous, consisting of necrotic and intact cardiomyocytes, endothelial cells, inflammatory immune cells, including macrophages and neutrophils; all of these are potential sources of iNOS activity. The induction of iNOS in cardiomyocytes has been demonstrated [7]. However, using monoclonal antibodies derived from mice, Wildhirt *et al.* reported that macrophages are the major source of increased iNOS activity in infarcted rabbit and human myocardium [8]. The selectivity of the antibodies to macrophage iNOS was confirmed by Western Blot and determined on both activated and inactivated macrophages. iNOS was colocalized in macrophages in identical sections. The height of enzymatically determined iNOS activity on post-operative day 3 coincided with the number of infiltrating macrophages.

In patients who had died of myocardial infarction, large numbers of macrophages were present twenty-five days after infarction when neutrophils had already disappeared [9]. Immunohistochemical localization of both the constitutive and inducible forms (cNOS and iNOS) and identification of macrophages were made by an indirect immune peroxydase method and with monoclonal antibodies to iNOS and cNOS. In normal myocardium, the constitutive form of NO synthase (cNOS) was seen in endothelial cells of intramyocardial blood vessels and endocardium but it was absent in infarcted tissue. On day 7 following death from myocardial infarction, iNOS immunreactivity was localized in infiltrating macrophages and non-necrotic cardiomyocytes within or close to the infarcted region. cNOS was visualized primarily in preserved vascular endothelial cells [9].

The question arises as to the possible impact of induction of iNOS in infarcted myocardial tissue on cardiac function. Since NO activates guanylate cyclase, resulting in production of cyclic GMP, which possesses negative inotropic action, a decline in contractility of the heart can be expected. Mery *et al.* have shown that in cardiomyocytes cyclic GMP formation is partially responsible for the inhibition of cardiac L-TYPE Ca^{2+} current via activation of a cGMP-dependent protein kinase [10]. Additional studies by Balligand *et al.* [11] and by Schulz *et al.* [12] demonstrated that induction of the Ca^{2+} independent iNOS leads to depressed contractility of rat cardiomyocytes and isolated rat heart preparations via cGMP dependent mechanism. De Belder and Moncada furthermore ascribed the negative inotropism, myocardial dilatation and *in situ* toxicity in inflammatory heart disease to increased generation of NO by immunological induction of a high output NO pathway [13]. Induction of NO synthase within the inflammatory cells infiltrating the heart of murine Coxsackie B-3 induced myocarditis, has now been shown [14].

Effects of iNOS inhibitors

Several competitive inhibitors of NOS have been identified. The most widely used are the L-arginine analogues which inhibit both constitutive and inducible forms of the enzyme [15]. Recently, two selective inhibitors of iNOS have been introduced, isothiourea and aminoguanidine [16].

S-methylisothiourea (SMT) was found to be 10- to 30- fold more potent as an inhibitor of inducible NOS (iNOS) than the L-arginine analogues [17]. Aminoguanidine which incorporates the guanido group of L-arginine linked to hydrazine also inhibits iNOS but to a lesser degree than SMT [18].

Occlusion of a coronary artery elicits different responses of the myocardium to these inhibitors dependent on the localization within the infarcted heart. In the infarcted area both circulation and respiration have ceased leading to cell necrosis. Adjoining to this region is the area of risk, in which respiration as measured with tetrazolium, still takes place, although the region is now ischemic. In the non-infarcted, normal area, both circulation and cellular respiration remain intact. The inhibitory effect of SMT and guanidine, both specific iNOS inhibitor and of L-NNA, which inhibits both isoforms of NO synthase, differ with the affected area of the myocardium [18a]. SMT is the strongest inhibitor with the lowest EC50; its greatest degree of inhibition takes place in the normal portion of the myocardium. IC50 of L-NNA is identical in all three myocardial regions. Aminoguanidine is the least effective inhibitor. Here, too, the inhibition is most pronounced in the normal portion of the heart. Possibly, the lesser degree of inhibition in the ischemic portion of the myocardium is due to defective binding of the inhibitor with the substrate (arginine).

Failure of dexamethasone and cyclosporin A to inhibit iNOS activity is not surprising. Both are non-competitive, but interfere with protein synthesis and hence with the production of the inducible form of nitric oxide. Cai *et al.* also showed that dexamethasone and cyclosporin A reduced iNOS activity and cardiac allografts and attenuate iNOS mRNA induction by interleukin-1 beta and interfere in -gamma in cultures of activated rats ventricular myocytes [19]. The induction of iNOS RNA in cardiomyocytes was also abolished by dexamethasone in lipopolysaccharide-injected rats [20].

Dexamethasone also prevented induction of iNOS in heart muscle when administered three hours after coronary ligation [21].

Apoptosis in infarcted heart tissue

The term apoptosis was coined in 1972 to define a particular form of cell death which is distinct from necrosis [22]. A number of agents, amongst them, nitric oxide, have been shown to induce apoptosis [23–25]. In myocardial ischemia of 30 min followed by reperfusion of 4 h, Gottlieb *et al.* found apoptosis in cardiomyocytes after 30 min of ischemia followed by 4 h of reperfusion [26]. After 5 h of ischemia without reperfusion, no apoptosis was noted.

Using DNA electrophoresis (DNA fragmentation) and TUNEL staining (terminal deoxynucleotidyl transferase TdT-mediated dUTPBiotin nick end labeling), together with immunohistochemical identification of macrophages, Suzuki *et al.* noted apoptosis 2 or 3 days following ligation of a coronary artery, at a time when the activity of iNOS was at its height [27]. Apoptosis was primarily confined to infiltrating macrophages, suggesting a relationship between NO induction and the appearance of apoptosis. The relationship between nitric oxide and apoptosis has also been suggested [28, 29]. However, definite proof of the relationship between NO formation in apoptosis in myocardial infarction is lacking.

The appearance of apoptosis in macrophages of infarcted myocardium is an example of an existing mechanism trying to establish homeostasis under pathological circumstances. In contrast to necrosis, apoptosis is a programmed process to eliminate cells which pose a threat to the organ. Macrophages located in infarcted heart tissue may be harmful to cardiomyocytes by prolonged production of NO, resulting in diminished contractility [7]. Apoptosis can be teleologically explained as a homeostatic mechanism that affects the selective deletion of cells whose survival could prejudice the welfare of the organ [30]. This certainly applies to the role of macrophages in infarcted heart muscle.

Most of the work on nitric oxide within recent years has defined its mode of origin, its molecular biological and physical properties. The results presented here emphasize the importance of nitric oxide in myocardial infarction. The future will undoubtedly reveal more of its effect on this disorder and will suggest possible approaches in its treatment.

Acknowledgement

Supported by grants from the Hoover Foundation and the Council for Tobacco Research.

References

1. Kuo PC, Schroeder RA: The emerging multifaceted roles of nitric oxide. Annals of Surgery 221: 220–235, 1995
2. Knowles RG, Moncada S: Nitric oxide synthase in mammals. Biochem J298: 249–258, 1994
3. Lowenstein CJ, Dinerman JL, Snyder SH: Nitric Oxide: A physiological messenger. Ann Int Med 20: 227–237, 1994
4. Dudek RR, Wildhirt SM, Conforto A, Pinto V, Suzuki H, Winder S, Bing RJ: Inducible nitric oxide synthase activity in myocardium after myocardial infarction in rabbit. Biochem Biophys Res Commun 205: 1671–1680, 1994
5. Bredt DS, Hwang PM, Snyder SH: Localization of nitric oxide synthase indicating a neural role for nitric oxide. Nature 347: 768–770, 1990
6. Hibbs JB Jr., Varinz, Taintor RR: L-arginine is required for expression of the activated macrophage effector mechanisms causing selective metabolic inhibition in target cells. J Immunol 138: 550–565, 1987
7. Yang X, Chowdhury N, Cai B, Brett J, Marboe C, Sciacca RR, Michler RE, Cannon P: Induction of myocardial nitric oxide synthase by cardiac allograft rejection. J Clin Invest 94: 714–721, 1994
8. Wildhirt SM, Dudek RR, Suzuki H, Pinto V, Narayan KS, Bing RJ: Immunohistochemistry in the identification of nitric oxide synthase isoenzymes in myocardial infarction. Cardiovasc Res 29: 526–531, 1995
9. Wildhirt SM, Dudek RR, Suzuki H, Narayan KS, Winder S, Choe J, Bing RJ: Expression of nitric oxide synthase isoforms after myocardial infarction in humans. Endothelium In press
10. Mery PF, Lohrnan SF, Walter U, Fischmeister R: Ca^{2+} current is regulated by cyclic GMP-dependent protein kinase in mammalian cardiac myocytes. Proc Natl Acad Sci USA 88: 1197–1201, 1991
11. Balligand JL, Ungureanu P, Kelly RA: Abnormal contractile function due to induction of nitric oxide synthesis in rat cardiac myocytes follows exposure to activated macrophage-conditioned medium. J Clin Invest 9[illegible]: 2314–2319, 1993
12. Schulz R, Panas DL, Catena R: The role of nitric oxide in cardiac depression induced by interleukin-1β and tumor necrosis factor alpha. Br J Pharmacol 114: 27–34, 1995
13. DeBelder A, Moncada S: Cardiomyopathy: Arolefornitricoxide. Int J Cardiol 50: 263–267, 1995
14. Lowenstein CJ, Allen G, Walker A, Rose N, Snyder S, Heroskowitz A: Nitric oxide inhibits viral replication. Circulation 88: I-G-H, 1993
15. Knowles RG, Moncada S: Nitric oxide sythases in mammals. Biochem J 298: 249–258, 1994
16. Furfine ES, Harmon MF, Paith JE, Knowles RG, Salter M, Kiff RJ, Duffy C, Hazelwood R, Oplinger J, Garvey EP: Potent and selective inhibition of human nitric oxide synthases. J Biol Chem 269: 26677–26683, 1994
17. Szabo C, Southan GJ, Thiemerman C: Beneficial effects and improved survival in rodent models of septic shock with S-methylisothiourea sulfate, a potent and selective inhibitor of inducible nitric oxide synthase. Pro Natl Acad Sci USA. 91: 12472–12476, 1994
18. Misko TP, Moore WM, Kasten TP, Nickols GA, Corbett JA, Tilton RG, McDaniel ML, Williamson JR, Currie MG: Selective inhibition of the inducible nitric oxide synthase by aminoguanidine. Eur J Pharmacol 233: 119–125, 1993

18a. Suzuki H, Wolf W-P, Akiyama K, Horstman D, Grant P, Cannavino C, Bing RJ: Effect of inhibitors of inducible form of nitric oxide synthase in infarcted heart muscle: Proc Assoc Am Phys 108: 173–178, 1996

19. Cai B, Roy DK, Sciacca R, Michler RE, Cannon PJ: Effects of immunosuppressive therapy on expression of inducible nitric oxide synthase (iNOS) during cardiac allograft rejection. Int J Cardiol 50: 243–261,
20. Balligand JL, Ungureanu-Longrois D, Simmons WW, Pimental D, Malinski TA, Kapturezak M, Taha Z, Lowenstein CJ, Davidoff AJ, Kelly RA, Smith TW, Michel T: Cytokine-inducible nitric oxide synthase (iNOS) expression in cardiac myocytes. J Biol Chem 269: 27580–27588
21. Dudek RR, Wildhirt S, Pinto V, Giesler G, Bing RJ: Dexamethasone inhibits the expression of an inducible nitric oxide synthase in infarcted rabbit myocardium. Biochem Biophys Res Commun 202: 1120–1126, 1994
22. Kerr JFR, Wyllie AH, Currie AR: Apoptosis: A basic biological phe-

nomenon with wide-ranging implications in tissue kinetics. Br J Cancer 26: 1239–1257, 1972
23. Gerschenson LE, Rotello RJ: Apoptosis: A different type of cell death. FASEB J 6: 2450–2455, 1992
24. Sarih M, Souvannavong V, Adam A: Nitric oxide synthase induces macrophage death by apoptosis. Biochem Biophys Res Commun 191: 503–508, 1993
25. Albina JE, Cui S, Mateo RB, Reichner J S: Nitric oxide- mediated apoptosis in murine peritoneal macrophages. J Immunol 150: 5080–5085, 1994
26. Gottlieb RA, Burleson KO, Kloner RA, Babior BM Engler RL: Reperfusion injury induces apoptosis in rabbit cardiomyocytes. J Clin Invest 96: 1621–1628, 1994
27. Suzuki H, Wildhirt SM, Dudek RR, Narayan KS, Bing RJ: Induction of apoptosis in myocardial infarction and its possible relationship to nitric oxide synthase in macrophages. Tissue Cell 28: 89–97, 1996
28. Cui S, Reichner JS, Mateo RB, Albina JE: Activated murine macrophages induce apoptosis in tumor cells through nitric oxide-dependent or-independent mechanisms. Cancer Res 54: 2462–2467, 1994
29. Pinskey DJ, Yang Y, Szabolcs M, Lia OH, Scicza RR, Cannon P: Nitric oxide induces apoptosis of adult cardiac myocytes. Abstract 2703 Annual Meeting American Heart Association, 1995
30. Kerr JFR: Programmed cell death. In: The cellular and molecular biology of apoptosis. M. Lavin, D. Watters, (eds). Harwood Academic Publisher, 1992, pp 1–15

Molecular and Cellular Biochemistry **160/161**: 307–313, 1996.

Estrogen effects in the heart

T. Pelzer, A. Shamim and L. Neyses

Department of Medicine, University of Würzburg, Würzburg, Germany

Abstract

Gender specific differences in cardiovascular disease are largely mediated by sex hormones. The use of estrogens significantly reduces the overall incidence of heart disease in postmenopausal women. Beneficial effects of estrogens on plasma lipoprotein levels are clearly established. However, these do not explain the magnitude of risk reduction seen in clinical studies. Thus, additional and currently unknown functions of estrogens must be operative. Elucidation of the exact estrogen action in the heart will have important implications in the treatment of cardiovascular disease. It will probably enhance the therapeutic repertoire in treating heart disease, the most common cause of death in industrialized countries. We will review the current understanding of the function of estrogens in the heart and discuss potential strategies on how to apply these data to clinical practice. (Mol Cell Biochem **160/161**: 307–313, 1996)

Key words: myocardium, hypertension, gene expression, estrogens, cardiac hypertrophy, signal transduction, genetic program

Introduction

Gender specific differences in the incidence of cardiovascular disease were first recognized more than 50 years ago [1]. Whereas the risk of heart disease in men increases constantly with age, pre-menopausal women have a significantly lower risk that increases rapidly to levels comparable with their male counterparts after menopause if they don't receive estrogens. In fact, the cardiovascular mortality in postmenopausal women receiving estrogen substitution is 30–50% less than in their untreated counterparts [2]. Thus, estrogens are among the major determinants of cardiovascular disease. From this viewpoint it seems tempting to employ estrogens in the prevention and therapy of heart disease. However, at present the therapeutic application of estrogens in heart disease is limited by the fact that the underlying mechanisms are only partially understood. The beneficial effect of estrogens on lipid metabolism [3] and the formation of atherosclerotic plaques in arterial vessels [4] has long been described, but the magnitude of these effects does not fully explain the protective action [5]. Thus, additional and presently unknown mechanisms must be operative. In our current understanding, the effects of estrogens are either genomic (i.e. mediated by changes in gene expression) or non-genomic (i.e. independent of changes in gene expression) and will be discussed in this order. The genomic action of estrogens in the heart also affects directly the gene expression of cardiac myocytes; these effects are currently a major interest of our own laboratory.

Genomic effects of estrogens in the heart

The genomic effects of estrogens in the heart can be grouped somewhat arbitrarily into effects on the coronary vasculature and the myocardium. We will first discuss the general principles of estrogen mediated gene expression, which are common to both systems and then focus separately on vascular and myocardial effects. The role of estrogens in the cardiovascular system was first discovered by pharmacological studies in isolated preparations of blood vessels. Thus, the actions of estrogens on the vascular system are currently better defined than the ones on the myocardium.

Basic mechanisms of estrogen receptor mediated gene expression

Estrogens are members of the steroid hormone family. Biochemical studies in the 1970s and 1980s demonstrated that steroids reversibly bind to specific intracellular receptor proteins. Ligand binding induces a conformational change of the

Address for offprints: L. Neyses, Department of Medicine, University of Würzburg, Joseph-Schneider Strasse 2, D-97080 Würzburg, Germany receptor. This facilitates binding of the hormone-receptor complex to highly conserved DNA sequences, termed hormone response elements (HRE), in the promoter region of specific target genes. In the case of the estrogen receptor (ER), hormone binding is followed by binding to a short cis-acting sequence (consensus: 5′ AGGTCA 3′, [6]) termed 'estrogen response element' (ERE). In a 'classical' ERE both sequences are arranged as palindromes, separated by three random base pairs. ER bound to the ERE transactivates or transrepresses transcription of the target gene by interaction with the basic transcription machinery or additional transcription factors as illustrated in Fig. 1. Screening of cDNA expression libraries in *E. coli* with ER specific antibodies facilitated cloning and further characterization of the estrogen receptor protein at the molecular level [7]. Deletion mutants revealed that the ER, like other steroid hormone receptors, contains different domains which confer different properties on the protein. The hormone binding domain (HBD, [8]) also contains a dimerization surface responsible for homo-

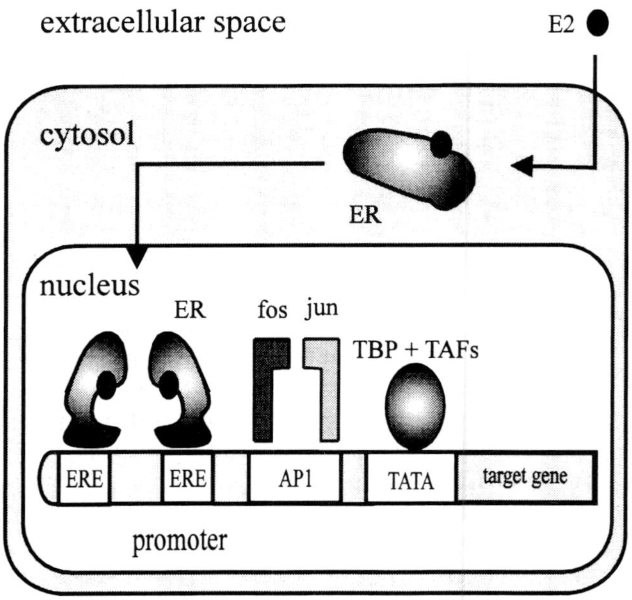

Fig. 1. Model of the general estrogen receptor function. Extracellular estrogen (E2) passes the lipid bi-layer of the cell membrane easily due to its lipophilic properties. Once in the cytosol, E2 binds to the hormone-binding domain of the estrogen receptor (ER), which is otherwise complexed with factors such as HSP 27 (not shown). The receptor ligand complex translocates into the nucleus, where the ER binds to specific DNA consensus motivs, termed estrogen response elements (ERE), in the promoter/enhancer region of specific target genes. The 'classical ERE' consists of two half-sites oriented head to tail as a palindrome, to which the ER binds as a homo-dimer. As discussed in the text, there are additional high-affinity DNA binding sites for the ER. DNA binding occurs also in coordination with other transcription factors, which are represented by a *fos-jun* complex at the AP-1 site. The ER complex then interacts with the basal transcription machinery (**T**ATA-box **B**inding **P**rotein = TBP and **TA**TA-box **A**ssociated **F**actors = TAFs) to regulate the transcriptional activity of the target gene.

dimerization of the ER. The HBD is connected to the DNA binding domain (DBD) by a flexible hinge region. The DBD contains two zinc-binding domains, one that makes contact with the major groove of DNA and a second that is probably responsible for the exclusive homo-dimerization of the ER [9]. Transcriptional activation of target genes is mediated by two separate transactivation domains, termed region AF-1 and AF-2. Transactivation from the AF-2 region is ligand-dependent, whereas the AF-1 region operates ligand-independent. The mechanism by which the ER regulates expression of target genes has already been studied in detail. Phosphorylation of specific amino acid residues is a widespread mechanism in the regulation of signal transduction pathways like in the ras/raf-MAP-kinase system. Likewise, the transcriptional activity of the ER is also regulated by phosphorylation. Migliaccio *et al.* were able to show that the affinity of ER to the ERE is increased by tyrosine phosphorylation [10]. In addition, the ER can also be phosphorylated by MAP-kinase, which increases binding to the ERE and thus transactivates gene expression [11]. The transcriptional activation of target genes by the ER is thought to be mediated by direct or indirect physical interaction of ER and additional factors with the basal transcription machinery [12]. As discussed above, transcriptional activation by the ER is also dependent on the presence of specific DNA binding motifs in the regulatory region of the target gene. The DNA sequence requirements for ER binding to EREs have been determined in detail by transfection experiments of estrogen responsive promoter elements [6]. Binding of the ER to 'classical' (i.e. palindromic) EREs as shown in Fig. 2A is the best established mechanism for ER mediated gene expression. Recent data suggest that the ER may also bind promiscuously to direct repeats of ERE half-sites as illustrated in Fig. 2B [13]. Currently there are no studies indicating that this mechanism does regulate gene expression *in vivo*. In addition estrogen response elements may not be strictly mandatory for transcriptional regulation by the ER. The ovalbumin promoter contains an ERE half-site with striking similarity to an AP-1 site, which binds transcription factors of the *jun-fos* gene family. Transcriptional activation of this particular promoter is enhanced by co-transfection with wild type ER. However, a mutated ER that lacks the DBD is equally efficient to the wild type ER in transactivating this particular promoter [14]. These findings could be explained by protein-protein interactions between the ER and *fos-jun* complexes. Transcriptional transactivation may either result from stabilisation of the *fos-jun* complex or directly from the ERs transactivation domains. Similar studies on the related vitamin-D3 receptor also indicate interaction with *fos-jun*. In this case, the interleukin-2 promoter activity in Jurkat cells was downregulated by binding of the vitamin-D3 receptor to its response element [15]. Also promoter sequences that do not contain EREs have been reported to be transcriptionally regulated by the ER like

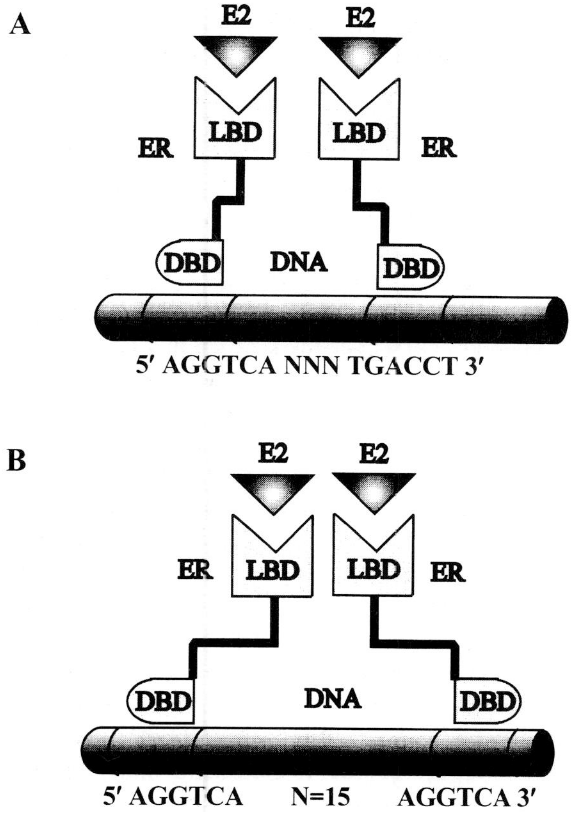

Fig. 2. Model of estrogen receptor binding to estrogen response elements. Panel A illustrates DNA binding of the estrogen receptor (ER) to a 'classical' estrogen response element. As discussed in the text, the ER contains a ligand binding domain (LBD), which binds estrogens like 17β-estradiol (E2) with high affinity. Ligand binding increases the affinity of the receptor-ligand complex for DNA by a conformational change of the receptor. DNA binding occurs at the consensus motiv 5′ AGGTCA 3′ by the ERs DNA binding domain (DBD). The LBD and DBD are connected by the flexible hinge region represented by the black line. DNA binding of the ER occurs in this model as a homo-dimer to a palindrome of two ERE half sites oriented head to tail. As stated in the text, dimerization is mediated by two dimerization interfaces, one located in the LBD and a second is located in the DBD. Panel B illustrates DNA binding of the ER to a non-canonical ERE. In this special case, an ER homo-dimer is shown bound to two ERE half-sites spaced by 15 random base pairs. Please note, that the half-sites are now oriented head to tail. Dimerization of the ER occurs in this case probably only by the LBD and is facilitated by the flexible hinge region between the LBD and DBD [14].

the brain creatine kinase promoter. This particular promoter is regulated by estrogens in the absence of palindromic or nonpalindromic EREs by interaction of the ER with an AT-rich sequence that binds the ubiquitous factor TARP [16]. Meanwhile, null-mutations for the estrogen receptor have been generated in the mouse system [17]. Mice of both sexes are viable but infertile. Infertility in males was unexpected, but the testes of receptor deficient males are small and the morphology of seminiferous tubules is altered. Although the animals are not described to have a prominent cardiovascular phenotype, this does not argue against the relevance of estrogens in the heart, which may be important under circumstances not encountered in the transgenic animals (e.g. hypertension, atherosclerosis etc.). Hemodynamic studies and analysis of gene expression in knockout animals will be a prerequisite in dissecting genomic and non-genomic effects of estrogens in mice.

Estrogen action on the vasculature

Estrogens have been known as vasoactive substances since 1884, when MacKenzie and coworkers described hyperemia of mucosal membranes in association with the menstrual cycle and pregnancy [18]. A number of hemodynamic and pharmacological studies have since proven these results. Estrogens acutely dilate arterial vessels, including coronary arteries *in vitro* and *in vivo*, within 5–10 min [19, 20]. The rapid onset of vasorelaxation strongly suggests a nongenomic mechanism, which does not require *de novo* protein synthesis. Since the basal vascular tone is regulated by the release of nitric oxide and prostaglandins, estrogens might induce a rapid release of NO and prostacyclin from smooth muscle cells and endothelial cells. The first evidence that estrogens modulate transcriptional regulation of gene expression in the cardiovascular system derives from studies that showed specific and saturable binding of radioactively labeled 17β-estradiol to aortic tissue [21]. The same authors also demonstrated up-regulation of the progesterone receptor which is well known to be transcriptionally upregulated by the ER in non-vascular tissues. Recent results have demonstrated that human vascular smooth muscle cells contain functional ER, which is able to regulate the expression of a reporter gene in an estrogen and ERE dependent manner [22]. Furthermore, Hishikawa and co-workers presented evidence that endothelial nitric oxide synthase (eNOS) is transcriptionally up-regulated by the ER in cultured human endothelial cells [23]. Earlier studies stating a reduced mitotic activity of vascular cells in animals or humans treated with estrogens could not differentiate between effects on the serum lipid profile and truly genomic effects [24]. More recently, several investigators convincingly showed the anti-proliferative action of physiological E2 levels in vascular smooth muscle cells *in vitro* [25]. The exact mechanism by which estrogens either promote proliferation, i.e. in breast cancer cells, or repress cell growth are still to be defined. As will be discussed in the next section of this review, there are several candidate genes for this important action of estrogens in the heart.

Target genes in the vasculature

Currently there are no data on genes in the vascular system except perhaps eNOS that have been proven to be regulated by E2 and the ER. However, there is no shortage of potential target genes in view of the hemodynamic and biochemical data available. Table 1 gives a brief overview of these candidate genes. It is a major challenge to think about strategies to identify the genes whose function might correlate with the observed hemodynamic and biochemical data.

Estrogen action on the myocardium and cardiac connective tissue

Until recently, the myocardium was thought to lack estrogen receptors. Surprisingly, studies in our own lab have shown that cardiac myocytes express ER mRNA and protein as determined by Northern and Western analysis [26]. Furthermore, addition of E2 to cultured neonatal cardiac myocytes resulted in nuclear translocation of the ER. These estrogen receptors are transcriptionally competent, since they induce transcription of a luciferase reporter gene whose promotor contains a trimer of EREs. Also the endogenous progesterone gene, which is well known to be E2 responsive in non-cardiac tissues, is up-regulated by the ER in cardiac myocytes. These data indicate that estrogens are able to regulate gene expression in the myocardium. Thus, transcriptional regulation of several genes relevant to the myocardium may occur via estrogen receptors.

Table 1. Potential target genes for estrogen regulation

Target gene	Reference #
angiotensin-I converting enzyme	43, 44
nitric oxide synthase isoform 1	42
nitric oxide synthase isoform 2	55, 56
nitric oxide synthase isoform 3	23, 42
prostacyclin synthase	57
protaglandin cyclooxygenase	57
c-fos	33, 34
c-myc	32
c-jun	34
egr-1	34
estrogen receptor	58
progesterone receptor	58
elastin	59
collagen	59
connexin 43	39, 40
transforming growth factor α	60
α-myosin heavy chain	27, 28
β-myosin heavy chain	27, 28
interleukin 6	61
insulin-like growth-factor	62
heat shock protein 27	63, 64

The table gives an overview of genes relevant to the cardiovascular system which are likely but not proven to be transcriptionally regulated by estrogens. These genes have either been shown to be regulated by E2 in non-cardiac tissues or they contain estrogen response elements in their regulatory regions. Numbering refers to references.

Target genes in the myocardium

Evidence that E2 mediates the expression of important genes in the myocardium comes from studies that show estrogen dependent expression of the myosin heavy chain genes (MHC) in the rat heart. Cardiac hypertrophy in spontaneously hypertensive rats (SHR) leads to a shift in the expression of isomyosins in the heart from the V1 (α/α) to the V3 (β/β) isoform. Myofibrillar ATPase activity and contractility of cardiac myocytes expressing the V3 isoform is significantly reduced. Morano *et al.* were able to show that female SHR rats are able to maintain at least equal ratios of V1 versus V3 isomyosin, whereas in male SHR rats the V3 isoform is clearly predominant [27]. Furthermore, substitution of ovarectomized female rats with physiological levels of 17β-estradiol prevented the decline of V1 isomyosin seen in unsubstituted animals [28]. The exact mechanism by which estrogens and also testosterone [29] regulate MHC expression is at present not well understood. Besides changes in the expression of contractile proteins, cardiac hypertrophy leads to increases in overall protein synthesis and induces specific changes in myocardial gene expression. In a first approximation, these may be summarized as the hypertrophic gene program (review in [30]). In order to get a better understanding on the role of estrogens in cardiac hypertrophy we are currently investigating the effects of E2 on overall protein synthesis of the rat heart. Changes in myocardial gene expression during hypertrophy are thought to be mediated in part by proto-oncogenes. The proto-oncogenes *c-fos*, *c-jun*, *egr-l* and *c-myc* are induced rapidly and transiently after exposure of cardiac myocytes to external hypertrophic stimuli [31]. These transcription factors bind to defined DNA sequences located in the promoter/enhancer region of specific downstream target genes where they regulate transcriptional activity of the gene. Studies in non-cardiac tissues indicate that the expression of several proto-oncogenes is regulated by estrogens [32, 33, 34]. We have shown in a previous study that *egr-1*, which is transcriptionally regulated by E2 in non-cardiac tissues, is required for the transduction of hypertrophic stimuli in cardiac myocytes [35]. Thus, transcriptional control of cardiac gene expression during hypertrophy could be regulated by interference of the ER with proto-oncogene expression. An alternative or additional point of convergence between ER and proto-oncogene mediated gene expression pathways may be the target promoter of downstream genes. This hypothesis is supported by reports on direct ER interactions with *c-fos* and *c-jun* proteins as discussed before. Transcriptional control of cardiac gene expression by estrogens potentially includes a variety of additional genes important for cardiac function. Table 1 presents an overview of these potential target genes. These genes are either known to be regulated by estrogens in non-myocardial tissues or they contain EREs in their promoter region. None of these has actually been shown to be regulated by estrogens in the myocardium. At first, the estrogen receptor gene itself is upregulated in a feed back loop, a mechanism already described for other genes e.g. the myogenic basic-helix-loop-helix transcription factors of the myoD gene family [36, 58]. Differences in gender dependent gene expression may thus be explained by the fact that males, which express ERs at lower levels, first have to upregulate their estrogen receptors until maximal transcriptional activation of ER responsive target genes occurs. Cardiac hypertrophy is associated with an increase in life threatening arrythmias. Propagation of the electrical impulse in the heart is facilitated by highly specialized intercellular channels, which are formed by proteins of the connexin (CX) gene family [37, 38]. The expression of the predominant connexin in the heart (CX43) may be regulated by estrogens, since the promoter of the rat CX43 gene contains several ERE half-sites [39]. Furthermore, promoter constructs of CX43 are E2 responsive in non-myocyte cell lines [40]. Previous studies also indicate altered connexin expression in the hypertrophied rat myocardium [41]. Further genes which might be transcriptionally regulated by the ER in the myocardium are the calcium-dependent and calcium- independent isoforms of nitric oxide synthase as well as the angiotensin-I converting enzyme (ACE). The activities of these enzymes are either known to be regulated by E2 in vascular cells or they contain EREs in their promoter regions [42–44].

Non-genomic effects of estrogens in the cardiovascular system

Basic mechanisms

Non-genomic effects of steroid hormones are independent of receptor-mediated gene expression. This mechanism does not require *de novo* protein synthesis and is thus characterized by its rapid onset of action within minutes after application of the hormone. In contrast, the genomic action of E2 may require several hours for target gene expression before significant effects can be observed. Acute and chronic administration of E2 results in increased cardiac output, lowered systolic and diastolic blood pressure, decreased total peripheral resistance and increased myocardial perfusion in animal models. It also acutely potentiates the acetylcholine induced vasorelaxation in postmenopausal women at physiological concentrations [45] and increases cerebral blood flow in migrane attacks [46].

Non-genomic estrogen effects in vascular cells

Evidence that E2 can rapidly alter coronary vascular tone in an ER independent fashion comes from studies in ovarectomized hypercholesteraemic monkeys. Upon administration of acetylcholine, coronary arteries of estrogen treated animals showed vasorelaxation whereas animals not substituted with E2 reacted with a marked coronary vasoconstriction [47]. Coronary arteries from male and female rabbits denuded of the endothelium also reacted with vasorelaxation after treatment with E2 [48]. Other investigators report that vasorelaxation in uterine arteries can be antagonized by inhibitors of nitric-oxide synthase (NOS [49]). These data might actually reflect two independent non-genomic mechanisms: Increased NOS activity by E2 and interaction of E2 with ion channels of vascular cells. E2 at supraphysiological levels (10–30 µM) has calcium- antagonistic properties since it relaxes strips of coronary arteries by reducing intracellular free calcium concentrations through an inhibition of Ca^{2+} influx [50, 51]. Since the endothelial nitric oxide synthase gene is transcriptionally upregulated by E2, these data also demonstrate a potential overlap between genomic- and non-genomic effects of E2 on NO production. Recently, White *et al.* presented data indicating that the E2 mediated and endothelium independent relaxation of coronary arteries results from an increased outward current through Ca^{2+} and voltage-activated K^+ channels. These data were obtained at physiological levels of 17β-estradiol, identical levels of 17α-estradiol did not result in measurable effects [52].

Non-genomic estrogen effects in the myocardium

At present our knowledge about non-genomic effects of estrogens in the myocardium itself is very limited. Jiang *et al.* report on a calcium antagonistic effect of E2 in guinea-pig cardiac myocytes [50]. However, the concentrations of E2 in these experiments were in the micromolar range which exceeds the physiological level by three orders of magnitude. E2 at physiological levels might for instance also regulate the electrophysiological properties of cardiac myocytes as it does in vascular smooth muscle cells. The techniques to answer these questions are at hand and initial studies may open an additional and promising field of research.

Perspectives

Starting from mainly clinical and epidemiological observations on gender specific differences in cardiovascular disease our knowledge on the role of sex hormones in heart disease has increased tremendously in recent years. Significant progress has been made especially in the molecular analysis of nuclear steroid hormone receptors. We now have a good understanding of how this family of transcription factors regulates gene expression (reviews in [53, 54]). A considerable number of potential target genes for estrogen action in the heart have been cloned. Thus, the previously somewhat vague function of estrogens in the heart can now be precisely defined. Investigation of estrogen dependent signalling pathways that regulate cardiac gene expression will likely result in the identification of specific targets for future therapeutic intervention. Estrogens are already established drugs in the prevention or treatment of osteoporosis and gynaecological disorders. But the therapeutic application of these compounds in heart disease is at present limited mainly for two reasons. First, the question if estrogens, even in combination with progesterones, increase the incidence of breast cancer has not been finally answered. Second, the application of female sex hormones in male patients causes a variety of side effects reaching from cosmetical problems and infertility to an increased risk of testicular cancer. Further research should thus be directed to the elucidation of the regulatory pathway of estrogen mediated gene expression in the heart. The identification of downstream and eventually heart specific target genes of estrogens and synthesis of novel estrogen-analogs should circumvent the unfavourable side effects of estrogens in men and women.

Acknowledgements

The authors are grateful to S. Oberdorf-Maass for excellent technical assistance and to the members of the laboratory for conceptual discussions.

Part of this work has been made possible by a grant from the 'Deutsche Forschungsgemeinschaft.'

References

1. Glendy RE, Levine SA, White PD: Coronary disease in youth: Comparison of 100 patients under 40 with 300 persons past 80. JAMA 109: 1775–1781, 1937
2. Stampfer MJ, Colditz GA, Willett WC, Manson JE, Rosner B, Speizer FE, Hennekens CH: Postmenopausal estrogen therapy and cardiovascular disease. N Engl J Med 325: 756–762, 1991
3. Bush TL, Barrett-Connor E, Cowan LD, Criqui MH, Wallace RB, Suchindran CM, Tyroler HA, Rifkind BM: Cardiovascular mortality and noncontraceptive use of estrogen in women: results from the Lipid Research Clinics Program follow-up study. Circulation 75: 1102–1109, 1987
4. Hong MK, Romm PA, Reagan K, Green CE, Rackley CE: Effects of estrogen replacement therapy on serum lipid values and angiographically defined coronary artery disease in postmenopausal women. Am J Cardiol 69: 176–178, 1992
5. Adams MR, Clarkson TB, Kaplan JR, Koritnik DR: Ovarian secretions and artheriosclerosis. In: JN Gutmann, AH DeCherney, PM Sarrel, (eds). Ovarian secretions and cardiovascular and neurological function. Raven Press, New York, 1990
6. Klein-Hitpass L, Schorpp M, Wagner W, Ryffel GU: An estrogen responsive element derived from the 5′ flanking region of the xenopus vitellogenin A12 gene functions in transfected human cells. Cell 46: 1053–1061, 1986
7. Walter P, Green S, Krust A, Bornert JM, Jeltsch JM, Staub A, Jensen E, Scrace G, Waterfield M, Chambon P: Cloning of the human estrogen receptor cDNA. Proc Natl Acad Sci USA 82: 7889–7893, 1985
8. Kumar V, Green S, Staub A, Chambon P: Localisation of the estradiol-binding and putative DNA-binding domains of the human estrogen receptor. EMBO J 5: 2231–2236, 1986
9. Waterman ML, Stuart A, Nelson C, Greene GL, Evans RM, Rosenfeld MG: A single domain of the estrogen receptor confers deoxyribonucleic acid binding and transcriptional activation of the rat prolactin gene. Mol Endocrinol 2: 14–21, 1988
10. Migliaccio A, Di Domenico M, Green S, de Falco A, Kajtaniac EL, Blasi F, Chambon P, Auricchio F: Phosphorylation on tyrosine of in vitro synthesized human estrogen receptor activates its hormone binding. Mol Endocrinol 3: 1061–1069, 1988
11. Kato S, Endoh H, Masuhiro Y, Kitamoto T, Uchiyama S, Sasaki H, Masushige S, Gotoh Y, Nishida E, Kawashima H, Metzger D, Chambon P: Activation of the estrogen receptor through phosphorylation by mitogen-activated protein kinase. Science 270: 1491–1494, 1995
12. Jacq X, Brou C, Lutz Y, Davidson I, Chambon P, Tora L: Human TAFII30 is present in a distinct TFIID complex and is required for transcriptional activation by the estrogen receptor. Cell 79: 107–117, 1994
13. Kato S, Sasaki H, Suzawa M, Masushige S, Tora L, Chambon P, Gronemeyer H: Widely spaced, directly repeated PuGGTCA elements act as promiscuous enhancers for different classes of nuclear receptors. Mol Cell Biol 15: 5858–5867, 1995
14. Gaub MP, Bellard M, Scheuer I, Chambon P, Sassone-Corsi P: Activation of the ovalbumin gene by the estrogen receptor involves the *fos-jun* complex. Cell 63: 1267–1276, 1990
15. Alroy I, Towers TL, Freedman LP: Transcriptional repression of the interleukin-II gene by vitamin D3: Direct inhibition of NFATp/AP-1 complex formation by a nuclear hormone receptor. Mol Cell Biol 15: 5789–5799, 1995
16. Sukovich DA, Mukherjee R, Benfield PA: A novel, cell-type-specific mechanism for estrogen receptor mediated gene activation in the absence of an estrogen-responsive element. Mol Cell Biol 14: 7134–7143, 1994
17. Lubahn DB, Moyer JS, Golding TS, Couse JF, Korach KS, Smithies O: Alteration of reproductive function but not prenatal sexual development after insertional disruption of the mouse estrogen receptor gene. Proc Natl Acad Sci USA 90: 11162–11166, 1993
18. MacKenzie J: Irritation of the sexual apparatus. Am J Med Sci 87: 360, 1884
19. Ueland K, Parer JT: Effects of estrogens on the cardiovascular system of the ewe. Am J Obstet Gynecol 96: 400, 1966
20. Silva de Sa MF, Meirelles RS: Vasodilatation effect of estrogen on the human umbilical artery. Gynecol Invest 8: 307, 1977
21. Horwitz KB, Horwitz LD: Canine vascular tissues are targets for androgens, estrogens, progestins and glucocorticoids. J Clin Invest 69: 750, 1982
22. Karas RH, Patterson BL, Mendelsohn ME: Human vascular smooth muscle cells contain functional estrogen receptor. Circulation 98: 1943–1950, 1994
23. Hishikawa K, Nakaki K, Marumo T, Suzuki H, Kato R, Saruta T: Up-regulation of nitric oxide synthase by estradiol in human aortic endothelial cells. FEBS Lett 360: 291–293, 1995
24. Rhee CY, Spaet TH, Stemerman MB, Lajam F, Shiang HH: Estrogen suppression of surgically induced vascular intima hyperplasia in rabbits. J Lab Clin Med 90: 77, 1977
25. Vargas R, Wroblewska B, Rego A, Hatch J, Ramwell PW: Estradiol inhibits smooth muscle cell proliferation of pig coronary artery. Br J Pharmacol 109: 612–617, 1993
26. Grohé C, Briesemeister G, Stimpel L, Karas RH, Vetter H, Neyses L: Functional estrogen receptors in myocardial and myogenic cells. Circulation 90: I-538, 1994

27. Morano I, Gagelmann M, Arner A, Ganten U, Ruegg JC: Myosin isoenzymes of vascular smooth and cardiac muscle in the spontaneously hypertensive and normotensive male and female rat: A comparative study. Circ Res 59: 456–462, 1986
28. Malhotra A, Buttrick P, Scheuer J: Effects of sex hormones on development of physiological and pathological cardiac hypertrophy in male and female rats. Am J Physiol 259: H866–H871, 1990
29. Morano I, Gerstner J, Rüegg JC, Ganten U: Regulation of myosin heavy chain expression in the hearts of hypertensive rats by testosterone. Circ Res 66: 1585–1590, 1990
30. Neyses L, Pelzer T: The biological cascade leading to hypertrophy. Eur Heart J 16 (suppl. N): 8–11, 1996
31. Neyses, Nouskas J, Luycken J, Fronhoffs S, Oberdorf S, Williams RS, Sukhatme VP, Vetter H: Induction of immediate-early genes by angiotensin-II and endothelin- 1 in adult rat cardiomyocytes. J Hypertens 11: 927–934, 1993
32. Dubik D, Dembinski TC, Shiu RPC: Stimulation of the *c-myc* oncogene expression associated with estrogen-induced proliferation of human breast cancer cells. Cancer Res 47: 6517–6521, 1987
33. Weisz A, Rosales R: Identification of an estrogen response element upstream of the human *c-fos* gene that binds the estrogen receptor and the AP-1 transcription factor. Nucleic Acids Res 18: 5097–5106, 1990
34. Cicatiello L, Sica V, Bresciani F, Weisz A: Identification of a specific pattern of 'immediate-early' gene activation induced by estrogen during mitogenic stimulation of rat uterine cells. Receptor 3: 17–30, 1993
35. Neyses L, Nouskas J, Vetter H: Inhibition of endothelin- 1 induced myocardial protein synthesis by an antisense oligonucleotide against the early growth response gene- 1. Biochem Biophys Res Commun 181: 22–27, 1991
36. Thayer MJ, Tapscott SJ, Davis RL, Wright WE, Lassar AB, Weintraub H: Positive autoregulation of the myogenic determination gene myoD1. Cell 58: 241–248, 1989
37. Beyer EC, Paul DL, Goodenough DA: The connexin family of gap junction proteins. J Membrane Biol 116: 187–194, 1990
38. Willecke K, Hennemann H, Dahl E, Jungbluth S, Heynkes R: The diversity of connexin genes encoding gap junctional proteins. Eur J Cell Biol 56: 1–7, 1991
39. De Leon JR, Buttrick PM, Fishman GL: Functional analysis of the connexin-43 gene promoter *in vivo* and *in vitro*. J Mol Cell Cardiol 26: 379–389, 1994
40. Yu W, Dahl G, Werner R: The connexin-43 gene is responsive to estrogen. Proc R Soc Lond 255: 125–132, 1994
41. Bastide B, Neyses L, Ganten D, Paul M, Willecke K, Traub O: The gap junction protein connexin-40 is preferentially expressed in vascular endothelium as well as conductive bundles of rat myocardium and is increased under hypertensive conditions. Circ Res 73: 1138–1149, 1993
42. Weiner C, Lizasoain I, Baylis SA, Knowles RG, Charles IG, Moncada S: Induction of calcium-dependent nitric oxide synthases by sex hormones. Proc Natl Acad Sci 91: 5212–5216, 1994
43. Proudler AJ, Ahmed A1 H, Crook D, Fogelman I, Rymer JM, Stevenson JC: Hormone replacement therapy and serum angiotensin-converting-enzyme activity in postmenopausal women. Lancet 346: 89–90, 1995
44. Testut P, Soubrier F, Corvol P, Hubert C: Functional analysis of the human somatic angiotensin-I converting enzyme gene promoter. Biochem J 293: 843–848, 1993
45. Gilligan DM, Badar DM, Panza JA, Quyyumi AA, Cannon RO: Acute vascular effects of estrogen in postmenopausal women. Circulation 90: 786–791, 1994
46. Brass LM, Kisiel D, Sarrel PM: A correlation between estrogen and middle cerebral artery blood velocity at different times of the menstrual cycle in women with catamenial migraines. J Cardiovasc Technol 9: 68, 1990
47. Williams JK, Adams MR, Klopfenstein HS: Estrogen modulates responses of atherosclerotic coronary arteries. Circ Res 81: 1680–1687, 1990
48. Jiang C, Sarrel PM, Lindsay DC, Poole-Wilson PA, Collins P: Endothelium-independent relaxation of rabbit coronary artery by 17β-estradiol *in vitro*. Br J Pharmacol 104: 1033–1037, 1991
49. Van Buren G, Yang D, Clark KE: Estrogen-induced uterine vasodilatation is antagonized by L-nitroarginine methyl ester, an inhibitor of nitric oxide synthesis. Am J Obstet Gynecol 16: 828–833, 1992
50. Jiang C, Poole-Wilson PA, Sarrel PM, Mochizuki S, Collins P, McLeod KT: Effect of 17β-estradiol on concentration, Ca^{2+} current and intracellular free Ca^{2+} in guinea-pig isolated cardiac myocytes. Br J Pharmacol 106: 739–745, 1992
51. Han SZ, Karaki H, Ouchi Y, Akishita M, Orimo H: 17β-estradiol inhibits Ca^{2+} influx and Ca^{2+} release induced by thromboxane in porcine coronary artery. Circulation 91: 2619–2626, 1995
52. White RE, Darkow DJ, Falvo Lang JL: Estrogen relaxes coronary arteries by opening BkCa channels through a cGMP-dependent mechanism. Circ Res 77: 936–942, 1995
53. Evans RM: The steroid and thyroid hormone receptor superfamily. Science 240: 889–895, 1988
54. Beato M, Herrlich P, Schütz G: Steroid hormone receptors: Many actors in search for a plot. Cell 83: 851–857, 1995
55. Xie QW, Cho H, Calaycay J, Murnford R, Swiderek KM, Lee TB, Ding A, Troso T, Nathan C: Cloning and characterization of inducible nitric oxide synthase from mouse macrophage. Science 256: 225–228, 1992
56. Xie QW, Whisnant R, Nathan C: Promoter of the mouse gene encoding calcium-independent nitric oxide synthase confers inducibility by interferon-gamms and bacterial lipopolysaccharides. J Exp Med 177: 1779–1784, 1993
57. Chang WC, Nakao J, Orimo H, Murota SI: Stimulation of prostaglandin cyclooxygenase and prostacyclin synthase activities by estradiol in rat aortic smooth muscle cells. Biochem Biophys Acta 620: 472, 1980
58. Katzenellenbogen BS, Nardulli AM, Read LD: Estrogen regulation of proliferation and hormonal modulation of estrogen and progesterone receptor biosynthesis and degradation in target cells. Prog Clin Biol Res 322: 201–211, 1990
59. Cembran J, Lillo M, Val J, Mardones J: Influence of sex difference and hormones on elastin and collagen in aorta of chickens. Circ Res 8: 527, 1960
60. Kenney NJ, Saeki T, Gottardis M, Kim N, Garcia-Morales P, Martin MB, Normanno M, Ciardiello F, Day A, Butler ML, Salomon DS: Expression of transforming growth factor α antisense mRNA inhibits the estrogen-induced production of TGFα and estrogen-induced proliferation of estrogen-responsive human breast cancer cells. J Cell Physiol 156: 497–514, 1993
61. Pottratz ST, Bellido T, Mocharia H, Crabb D, Manolagas SC: 17β-estradiol inhibits expression of human interleukin-6 promoter-reporter constructs by a receptor-dependent mechanism J Clin Invest 93: 944–950, 1994
62. Michels KM, Lee WH, Seltzer A, Saavedra JM, Bondy CA: Up-regulation of pituitary [125] insulin-like growth factor-I (IGF-I) binding and IGF-I binding protein-2 and IGF-I gene expression by estrogen. Endocrinology 132: 23–29, 1993
63. Fuqua SAW, Blum-Saringos M, McGuire WL: Induction of the estrogen-regulated '24K' protein by heat shock. Cancer Res 49: 4126, 1984
64. Mendelsohn ME, Zhu Y, O'Neill S: The 29-kDa proteins phosphorylated in thrombin-activated platelets are forms of the estrogen receptor-related 27 kDa heat shock protein. Proc Natl Acad Sci USA 88: 11212–11216, 1991

Molecular and Cellular Biochemistry **160/161:** 315–324, 1996.

Xenoreactive natural antibodies and induced antibodies – their effects on beating cardiomyocytes as a model of a xenograft

Ursula Müller-Werdan, Bernd Koidl,[1] Arnd Autenrieth,[2] Dieter Klein,[2] Karl Werdan[3] and Claus Hammer[2]

Department of Internal Medicine III, University of Halle-Wittenberg, Klinikum Kröllwitz, Ernst-Grube-Straße 40, D-06097 Halle, Germany; [1]Institute for Medical Physics and Biophysics, University of Graz, Harrachgasse 21, A-8010 Graz, Austria; [2]Department of Experimental Surgery, University of Munich, Klinikum Großhadern, Marchioninistraße 15, D-81377 München, Germany; [3]Department of Cardiac Intensive Care Medicine, University of Halle-Wittenberg, Klinikum Kröllwitz, Ernst-Grube-Straße 40, D-06097 Halle, Germany

Abstract

Xenotransplantation has been complicated by hyperacute rejection reactions, which are supposedly triggered by preformed natural antibodies (PNAb) of the recipient organism, whereas the role of antibodies specifically induced by previous antigen contact (IAb) is less clear.

Primary cultures of spontaneously beating neonatal rat cardiomyocytes were used as a model of the heart to elaborate the effects of both PNAb and IAb from xenogeneic species and to investigate into their mechanisms of action. An experimental setup allowing for rapid medium exchange under continuous observation was employed.

Sera containing PNAb reproducibly bring about a stereotype pattern of altered contractility including an initial increase in beating frequency followed by a temporary cessation of beating within the first minutes after administration. After recovery of spontaneous contractions, the cells within the monolayer exhibited a dissociation of the synchronicity of the beating persisting for several hours. The temporary pause in beating was prevented by a very high extracellular calcium concentration, but not by extracellular electrical stimulation sufficient to trigger contractions in control cells. Electrophysiological measurements carried out in adult ventricular guinea pig heart muscle cells under the same experimental conditions revealed an increase of the excitation threshold of the cells after application of sera containing PNAb due to an enhanced input resistance. These results indicate that the effect of PNAb is the consequence rather of a generally reduced excitability of the cell than of the inhibition of a singular ionic conductance. After specific absorption of PNAb directed against rat antigens beating of neonatal rat cardiomyocytes ensued without interruption. Sera specimens devoid of complement produced similar effects on contractility, although the duration of the standstill period was significantly shorter. The increase in input resistance visualized in guinea pig myocytes was absent after removing PNAb against guinea pig antigens but not after absorbing PNAb directed against rat epitopes.

Signs of a permanent cytotoxicity after the administration of PNAb were lacking in all experiments.

IAb against rat heart tissue raised in rabbits stopped the contractions of neonatal rat cardiomyocytes within 30 min after administration irreversibly and lead concentration-dependently to a destruction of the cells. (Mol Cell Biochem **160/161:** 315–324, 1996)

Key words: antibodies, neonatal rat cardiomyocytes, hyperacute rejection reaction, electrophysiology, xenotransplantation, contractility, complement

Address for offprints: U. Müller-Werdan, Department of Internal Medicine III, University of Halle-Wittenberg, Klinikum Kröllwitz, Ernst-Grube-Straße 40, D-06097 Halle, Germany

Introduction

The lack of human donor organs renders xenotransplantation a promising attempt for the future [1–4]. However, organs transplanted between widely divergent species are jeopardized by humoral hyperacute rejection reactions, which are far more difficult to overcome than the cellular allograft rejections. Preformed natural antibodies (PNAb) against surface antigens of divergent species are present in the serum of every mammalian organism shortly after birth – without prior specific antigen contact – and are thought to trigger the cascade of reactions prompting the destruction of xenogeneic organ transplants [5, 6]. By binding to and activating endothelial and parenchymal cells of the donor organ, PNAb may initiate both immediate complement-dependent blood coagulation and a complement-independent release of mediator substances [7, 8] and functional disturbances not necessarily complement-mediated [9, 10].

The complement-dependent part of rejection cascades may be overcome by engineering animals transgenic for human complement regulatory proteins [11]. However, our studies with isolated cardiomyocytes reveal that complement-independent effects of PNAb may still pose an obstacle in the clinical setting. Neonatal rat cardiomyocytes in culture are a well-established model. Halle and Wollenberger were among the first scientists elaborating cell culture techniques for neonatal rat cardiomyocytes [12]. They make a suitable subject for investigating aspects of cell function, as the cells retain the capacity for spontaneous contractions, and as cultures are homogenous, containing less than 20% fibroblasts. The impact of diluted, dialysed, electrolyte- and pH-adjusted sera of various xenogeneic species on these cells was put to test. To investigate into the mechanism of action of PNAb electrophysiological measurements were carried out in adult guinea pig ventricular cardiomyocytes under the same experimental conditions.

Using cultured rat cardiomyocytes, experiments were performed to elaborate the different mode of action of naturally occurring PNAb, which rather non-specifically recognize several antigens, in comparison to the antibodies induced by antigen contact (IAb) which very selectively bind to the epitope they recognize. Understanding of the different reactivities of IAb and PNAb may be important beyond the issue of xenotransplantation, as PNAb seem to be involved in the pathogenesis of autoimmune diseases [13].

Materials and methods

Materials

Fetal calf serum, CMRL-1415-ATM medium, horse serum and collagenase (Worthington, CLS II) were purchased from Biochrom (Berlin, F.R.G.), bovine insulin (24–25 I.U./mg) and dexamethasone from Sigma (Taufkirchen, F.R.G.). Bovine serum albumin and trypsin (1:250) were obtained from Serva (Heidelberg, F.R.G.), gentamycin from Byk-Essex (München, F.R.G.). All other chemicals used in the experiments were of analytical grade and were purchased from Merck (Darmstadt, F.R.G.) or Boehringer (Mannheim, F.R.G.). The culture plastic flasks (175 cm^2 and 25 cm^2) and multidish-6-well plates came from Nunclon (Roskilde, Denmark) and Falcon (Becton Dickinson Labware, New Jersey, USA), respectively.

Methods

Preparation of sera specimens

The sera were obtained from man and pig. The titers of PNAb were determined as hemagglutinines of rat and pig erythrocytes in a serial dilution assay from 1:2 to 1:4096 to quantitate the amount of PNAb present in the sera.

To eliminate electrolyte and pH imbalances, the sera were dialysed against the serum-free culture medium (see below) (dialysis membrane Nr. 94571-25, pore size 10000 kD, Reichelt Chemietechnik, Heidelberg, F.R.G.) and diluted 1:2 with it prior to the experiments.

Concentrations of sodium, potassium and calcium of the experimental specimens were determined spectrometrically.

The pH of the serum samples was controlled before and after adding sera to the myocytes. All experiments were performed in a thermo-equilibrated chamber at 37°C.

Aliquots of sera were inactivated prior to the experiments by incubating them at 56°C for 30 min ('inactivation').

Selected sera were exhaustively absorbed with rat heart homogenates prior to the experiments ('absorption'): one serum specimen at a time was coincubated with a cell sample at 37°C for 30 min, the mixture being continuously shaken. Thereafter, the suspension was centrifuged and the supernatant used for the experiments. The hemagglutination titer against rat erythrocytes of these specimen was 0.

Preparation and cultivation of neonatal rat cardiomyocytes

Preparation and cultivation of monolayer cultures of spontaneously contracting NRCM were carried out as described in detail [14]. Briefly, hearts of 1–3-day-old Sprague-Dawley rats (Savo, Kissleg, F.R.G.) were removed after sacrificing the animals and placed in phosphate-buffered saline (PBS) under sterile conditions. The hearts were cut into small pieces and the cells disaggregated by repeated incubations in trypsin (0.12%)-collagenase (0.03%)-PBS (Ca^{2+} and Mg^{2+} – free) at 37°C. Muscle cells were separated from nonmuscle cells by the differential attachment technique [15]. The suspension of muscle cells in CMRL-1415-ATM medium containing 10%

fetal calf serum, 10% horse serum, and 0.02 mg/ml gentamycin was alloted into multidish-6-well plates or 25 cm^2-flasks (1.0–1.5 × 10^5 cells/cm^2). After 24 h of growth, in which the cells formed a monolayer culture of spontaneously contracting cells, the medium supplemented with fetal calf serum and horse serum was replaced by serum-free CMRL-1415-ATM medium of pH 7.40 containing 10^{-7} M dexamethasone, 5 × 10^6 M insulin, 4 × 10^{-7} M iron-saturated transferrin, 4 × 10^{-7} M bovine serum albumin and 0.02 mg/ml gentamycin ('serum-free culture medium'). The cells were cultured for 3 days, the medium being replaced by fresh medium every 24 h (Fig. 1).

For the experiments testing the impact of various sera on contractile function, the culture medium had to be exchanged rapidly and thoroughly by sucking it off with a glass pipette.

Measurement of contractility in neonatal rat cardiomyocytes [13]

Figure 2 illustrates the experimental set-up used to measure contractility in neonatal rat cardiomyocytes [14] as first described by Kaumann *et al.* in 1977 [16]: The beating cells in the multidish-6-well plate or 25 cm^2-flask were observed through an inverted phase-contrast microscope (Diavert, Leitz, Wetzlar, F.R.G.) in a temperature controlled chamber (37°C) at a 300-fold magnification. It was recorded by a CCD video camera (Sony XC 77CE, Sony Corporation, Japan). The image was displayed on the screen of a monitor (Grundig BG 23T, F.R.G.) and simultaneously transduced to a VHS video recorder (Bauer Video Cassette Recorder VRP 30). Contraction of the cells during beating causes changes in light intensity around the cell wall, the dark cell body contrasting with the light surroundings. These light intensity changes were recorded by a photocell (BPY 61 Siemens, F.R.G.) attached to the monitor screen. The output signal by the photocell was amplified, electronically filtered and monitored on a conventional recorder giving the amplitude of pulsation, the first derivative yielding the contraction and relaxation velocities. For medium exchange, the culture medium was sucked off completely under surveillance by phase contrast microscopy and replaced by fresh medium within 30 sec under continuous observation.

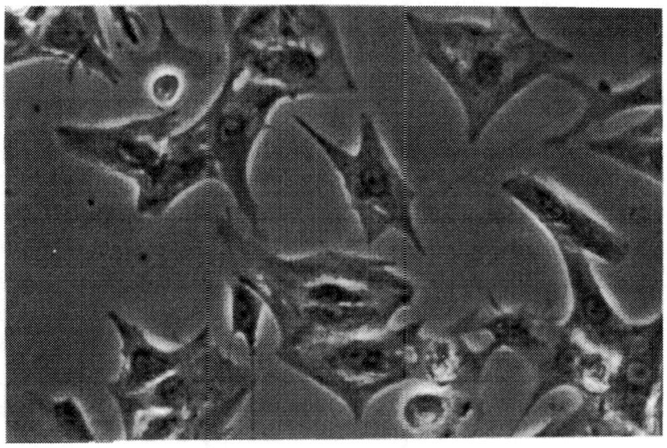

Fig. 1. Isolated neonatal rat cardiomyocytes (NRCM) in culture visualized by phase-contrast microscopy at a 300-fold magnification.

Ventricular adult guinea pig cardiomyocytes' assay: electrophysiological measurements

Single cardiomyocytes of adult guinea pigs were prepared as described by Piper *et al.* [17]. The experimental set-up has been described in detail [18]. The experiments were performed 24 h after preparation of the cells on an inverted microscope (Axiovert 10, Zeiss, Oberkochen, FRG). During the experiments, the cells were kept in an experimental bathing chamber mounted on the stage of the microscope. The serum aliquots were dialysed against the experimental Tyrode solution (composition in mM: NaCl 137, KCl 5.4, $CaCl_2$ 1.8, $NaHCO_3$ 2.2, NaH_2PO_4 0.4, $MgCl_2$ 1.1, Hepes/Na 10, glucose 5.6) and diluted with it for the electrophysiological analyses. A microperfusion system was used (Micro-Metric Instruments, Tampa, FL, USA) for superfusing the cells.

For registration of transmembrane potentials, patch electrodes (Hamill *et al.* [19]) were used. Whole cell recordings were made using an axopatch 200 patch clamp amplifier (Axon Instruments, Foster City, USA). Voltage clamp pulses were generated via an IBM compatible computer connected to a D/A and A/D converter (Digidata 1200, Axon Instruments). Data acquisition and analyses were performed using the pCLAMP software (Axon Instruments). The patch pipette filling solution contained (in mM): KCl 110, EGTA 11, $MgCl_2$ 2, $CaCl_2$ 1, HEPES/K^+ 10, ATP/K^+ 4.3, pH 7.4.

Signals representing voltage and current were A/D-converted by a PCM audioprocessor (Sony, Tokyo, Japan) and stored on a video recorder (Blaupunkt, Hildesheim, FRG). For data evaluation the recordings were replayed to a digital storage oscilloscope (W+W electronic, Basel, Switzerland) and from there transferred to a HP 1000 A-900 minicomputer (Hewlett Packard, Sunnyvale, CA, USA) and finally plotted on a HP 7475 A plotter.

Preparation of IAb

Rabbits were immunized by subcutaneous administration of rat-heart-homogenates centrifuged and resuspended in salt solution on days 1, 8 and 23. On day 31 serum was obtained. Prior to the administration to cardiomyocyte cultures, the sera were dialysed against culture medium and diluted 1:2 with the medium.

Determination of cytotoxicity

After the experiments, the cells were washed and dissolved in detergent (0,2 M NaOH). As parameters of cytotoxicity,

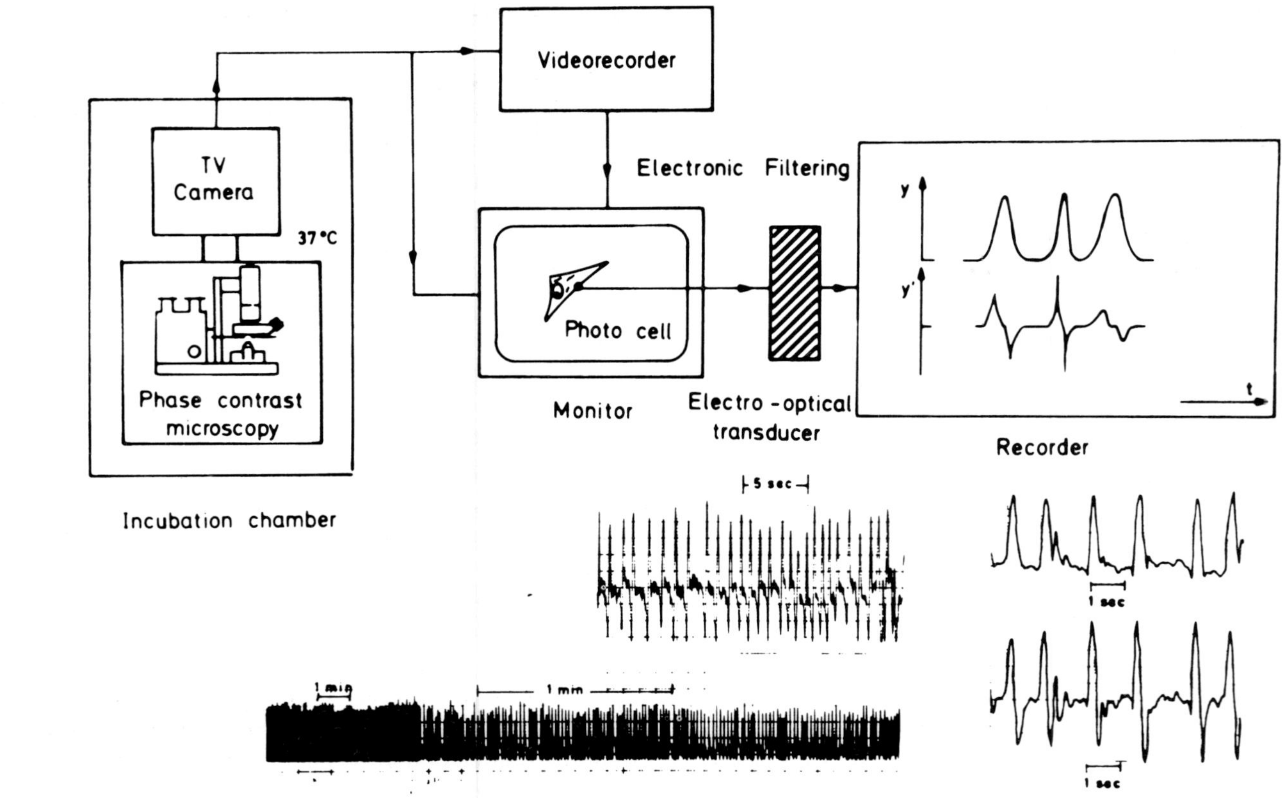

Fig. 2. Monitoring system for measurement of beating of NRCM in culture, for further explanation see 'Materials and methods'.

potassium content of the alkali solution was determined by a spectrometrical assay. Protein in the corresponding samples was measured according to Lowry [20].

Statistical evaluation

Values in Table 1 are given as mean ± S.D. All experiments were carried out at least in triplicate. The normality of distributions was checked with David's test. The unpaired two-tailed Student's *t*-test was used for groups with equal variances. In case of unequal variances based on Bartlett's test, the Welch's approximate t-test was used. Statistical significance was assumed at $p < 0.05$.

For the analysis of synchronicity of beating analysis of variances according to Hartley and the Mann-Whitney test were employed.

Results

Acute effects of sera containing PNAb on contractile parameter of neonatal rat cardiomyocytes

To study the effect of sera containing PNAb on neonatal rat cardiomyocytes, spontaneously beating and electrically driven cells were incubated with dialysed and diluted sera specimen. These cells were kept under continuous surveillance by phase contrast microscopy. As a control, the effect of a rapid medium exchange was monitored. As variables of contractility beating frequency, amplitude of contraction, contraction velocity and relaxation velocity were monitored. Contractile activity was not persistently impaired by rapid medium exchange and contractions continue, although some temporary disturbances may occur.

Table 1. Analysis of experiments with human sera given to spontaneously beating NRCM

	Untreated (n = 22)	Inactivated (n = 5)	Absorbed (n = 9)
Occurrence of standstill after addition of sample [sec]	62 ± 32	69 ± 23**	0*
Duration of standstill [sec]	145 ± 88	93 ± 33*	0*

*of statistical significance
**not of statistical significance
(x– ± S.D.)
For further experimental detail see 'Materials and methods'.
[Ca2+] in the incubation medium was 1.3 mM.

Impact of sera containing PNAb

The application of human sera diluted 1:2 with the serum-free culture medium (Fig. 3A) leads within a few seconds to a reproducible temporary increment in beating frequency. However, after – on average – 62 sec (range: from 15–110 sec) the contractions stop completely and remain stopped for

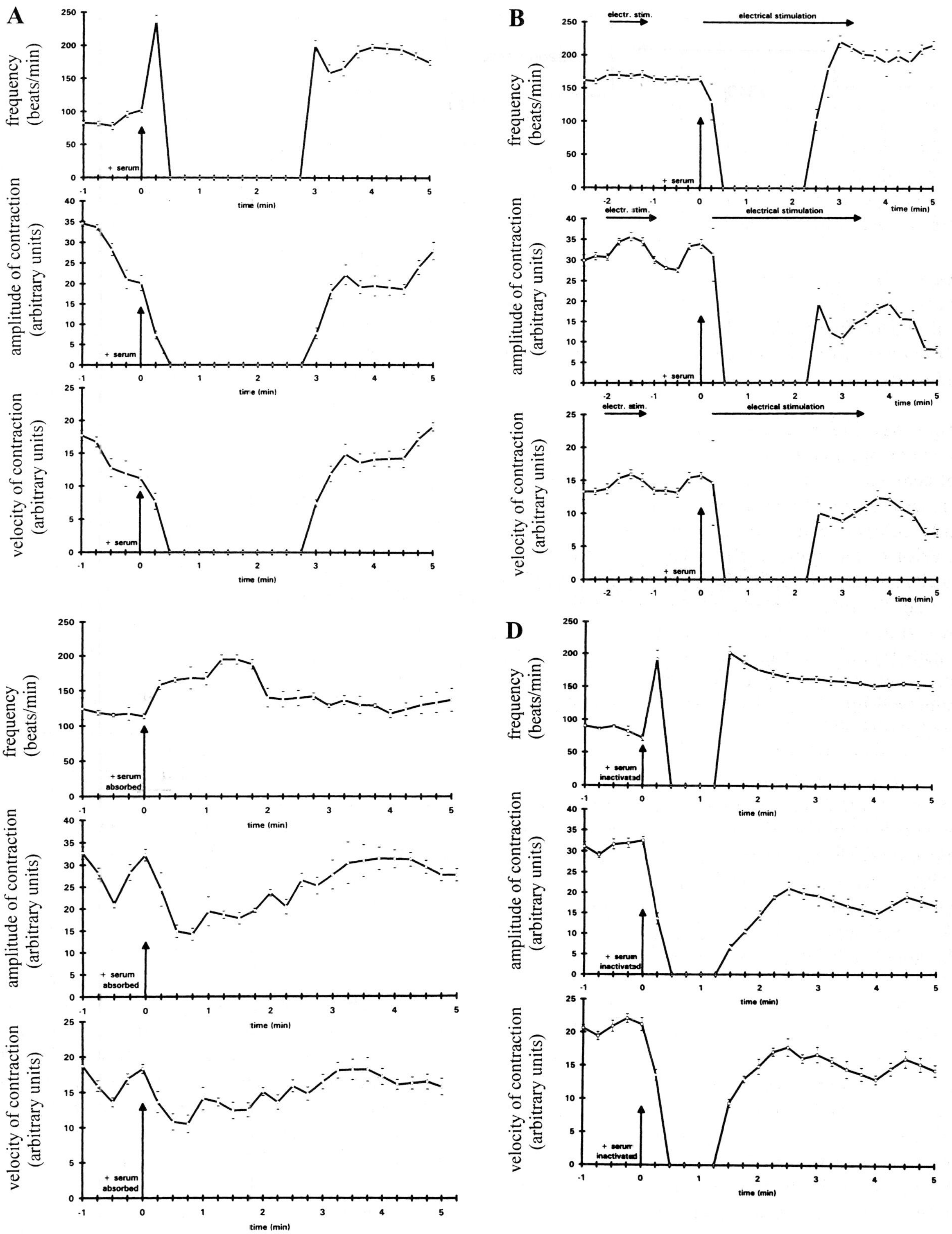

Fig. 3. Effect of the application of xenogeneic serum samples (indicated by vertical arrows) on neonatal rat cardiomyocytes: (a) administration of dialysed and diluted human serum on a rat cardiomyocyte; (b) effect of extracellular electrical stimulation before the addition of serum and during the standstill period, the horizontal arrows on top indicating periods of electrical stimulation; (c) application of adjusted human serum, PNAb against rat antigens having been absorbed prior to the experiment; (d) effect of inactivated human serum specimen. From top to bottom in each graph beating frequency (beats/min), amplitude of pulsation (arbitrary units) and contraction velocity (arbitrary units) of a rat cardiomyocyte are given, the abscissa indicating time (min). For each 15 sec period, values were meaned and standard deviations were obtained as a measure of beating irregularity.

145 sec (mean value) (range: 30–315 sec) (confer Table 1). Thereafter, contractility slowly recovers, the recovery period taking about 2 min (representative experiment given in Fig. 1A). The decrease in beating frequency – negative chronotropy – goes along with a reduction in contraction and relaxation velocity, two parameters associated with inotropy.

The same qualitative phenomenon can be observed when sera of other species (baboon, cynomolgus, pig, dog, rabbit, hamster) are used (data not shown).

Influence of high calcium concentration

In 6 experiments, the experimental solution containing PNAb was adjusted to a free extracellular calcium concentration of 3,5 mM: in these experiments, cellular contractions continued uninterrupted.

Effect of extracellular electrical stimulation

A further attempt was taken to surmount the PNAb-induced stop of contractions by extracellular electrical stimulation: Cells superfused with serum-free medium (calcium concentration 1.3 mM) at a constant rate, exhibited regular contractions. Prior to the application of serum, the cells were electrically stimulated for 50–70 sec at a frequency slightly higher than the rate of spontaneous contractions and responded without disturbance. The administration of human serum halted contractions. Electrical stimulation re-started at the same voltage did not restore contractility, the cessation in contractions lasting for the expected period of time (confer Fig. 3B).

Effect of absorption of PNAb

In an attempt to find the cause for the temporary cessation of contractions, sera were treated prior to being administered to isolated rat cardiomyocytes: Human serum or pig serum that had been absorbed to rat tissue homogenates beforehand did not stop spontaneous contractions, although an initial increase in the beating frequency occurred as it did occur with the addition of non-treated human sera (for an example see Fig. 3C).

Effect of inactivation

To identify a possible role of complement for the reversible halt of contractions, some sera specimen were inactivated (see Materials and methods). Although devoid of the heat-labile complement factors, human serum samples produced similar effects on contractility as the non-treated serum (Fig. 3D) (Table 1), although the duration of the standstill period was significantly shorter in the case of inactivated sera.

Chronic effects of sera containing PNAb on contractility of neonatal rat cardiomyocytes

Although the cessation of beating after administration of diluted human serum to rat cardiomyocytes was completely reversible in all wells, analysis of beating frequency of cell monolayers after the standstill period (after 2, 4, 6, 8, 10, 12 h; at least 5 visual fields per well) revealed a statistically significant increase in variances of beating frequency as opposed to controls lasting for at least 12 h. This finding is in favour of a 'dissociation' of the contractile activity of the cells making up the monolayer, which is evident by microscopic observation, too.

However, morphological signs of cytotoxicity were absent in all experiments and no significant fall in potassium content was determined in the presence of PNAb (data not shown).

Effect of sera containing PNAb on adult guinea pig ventricular cardiomyocytes – action potential recordings and voltage clamp experiments

The membrane potential of a guinea pig ventricular cell was measured by a patch electrode in the whole cell configuration. The resting potential amounted to –70 mV. The amplitude elicited by slightly superthreshold stimulation at an interval of 2 sec was 125 mV.

The main effect of superfusing the cells with pig serum diluted 1:4 was a strong reduction of the excitability of the cell (see Fig. 4): From the moment when the serum reached the cell, the stimulation was no longer sufficient to elicit action potentials. This is evident between stimuli #3 and #4 in Fig. 4. This failure of the stimulation to trigger action potentials could be overcome by a strong increase of the duration and strength of stimulation. In the two action potentials preceding the loss of excitability a stimulus artefact which was not visible in the control (action potential #1) can be seen. This increases beat by beat and finally reaches a maximum. A closer inspection of the whole recording in addition shows that the noise overlaying the pen recording increased during this time. Both effects indicate an increase of the input resistance of the cell. This increase could be verified by measuring the steady state current voltage relationship of the cell in voltage clamp experiments before and during addition of serum (Fig. 5). The reduction of the slope of this relationship by the serum indicated also an increase in the input resistance.

After absorption of antibodies directed against guinea pig epitopes, these phenomena were no longer seen, whereas after removing antibodies against rat antigens, they same electrophysiological effects were found.

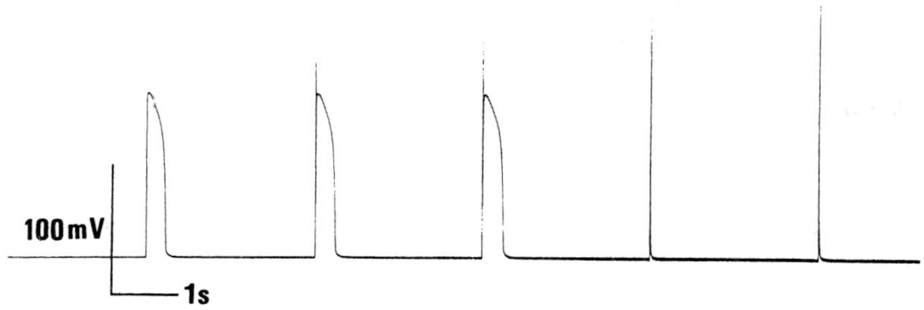

Fig. 4. Action potentials of a guinea pig ventricular cell under the influence of dialysed pig serum diluted 1:4. For explanation see 'Results'.

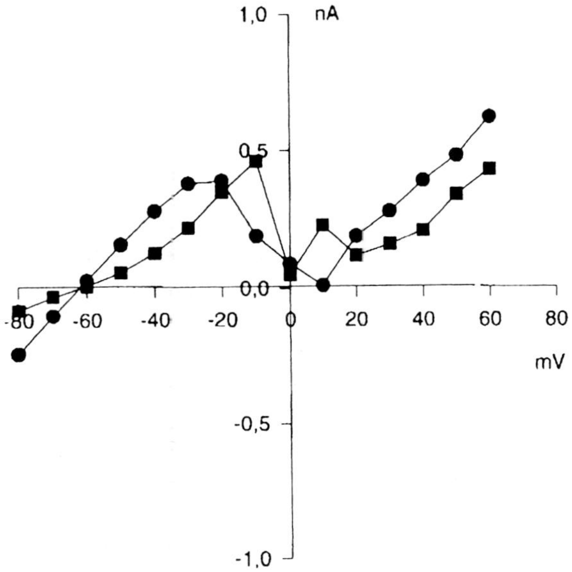

Fig. 5. Current-voltage-relationship of adult guinea pig cardiomyocytes in the absence (round symbols) or presence (square symbols) of diluted pig serum containing PNAb.

Effects of IAb on neonatal rat cardiomyocytes

Sera containing IAb stopped the contractions within less than 30 min and lead to a concentration-dependent efflux of potassium from the rat cardiomyocytes (Fig. 6) indicating severe cell damage. Cytotoxicity was in parallel documented by electron microscopy (not shown).

Discussion

The role of PNAb and IAb in hyperacute rejection – lessons from isolated cardiomyocytes

Wollenberger was one of the first to use cardiomyocytes in culture. As a measure of inotropy, contraction and relaxation velocities can be used [11]. Numerous studies have been performed on this model to investigate the effect of catecholamines, ouabain, toxins, mediators and activated mediator cells on contractile function [21, 22]. The pattern of a disturbed contractility induced by xenogeneic sera containing PNAb (initial increase in beating frequency, temporary stop in beating, spontaneous recovery of contractile activity) does not resemble those seen with the action of catecholamines, ouabain, calcium antagonists or substances inducing calcium overload such as pore forming toxins (e.g. streptolysin O, staphylotoxin [22]), complement [23] or cardiodepressant factor [24].

Wallukat *et al.* reported the effects of (auto-) antibodies on contractile function of cardiomyocytes [25, 26].

In neonatal rat cardiomyocytes, anti-β-adrenoceptor autoantibodies from patients with myocarditis and dilated cardiomyopathy increase the frequency of spontaneous beating, but do not cause contractions to stop [25]. Furthermore, anti-β_2-receptor-autoantibodies from patients with allergic asthma bronchiale inhibit the β_2-adrenoceptor-mediated positive chronotropic action in NRCM [26]. These data prove that antibodies may indeed exert specific effects on neonatal rat cardiomyocytes, although the pattern of effects is clearly distinct from the one seen in the present study.

Absorption of PNAb from the xenogeneic sera eliminates the effects on contractility of rat cardiomyocytes. As can be deduced from the electrophysiological measurements the main effect of addition of xenogeneic serum is an increase of the excitation threshold in the cell. Preliminary voltage clamp experiments and the increase of the input resistance seen in the voltage recordings indicate that this effect is the result rather of a general reduction of excitability of the cell than of the inhibition of a specific ionic conductance. In an attempt to interpret these data, one might tentatively assume that large molecules, most likely PNAb, stick to the cell membrane rather non-specifically and thus lead to an increased excitation threshold. Thereafter, the membrane property may be restored by e.g. pinocytotic uptake of those membrane-bound proteins or by some other unknown mechanism.

In keeping with this hypothesis, in the presence of very high free extracellular calcium concentrations, the transsarcolemmal ion flux, which in neonatal rat cardiomyocytes is the major trigger of beating activity [27], could be sufficient to elicit contractions.

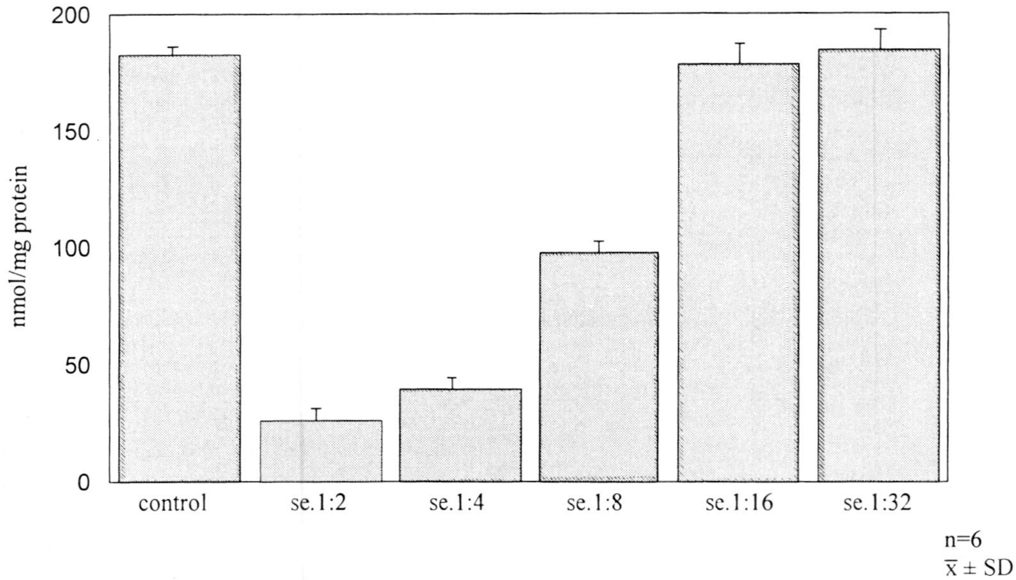

Fig. 6. Concentration-dependent fall in potassium content (given in nmol/mg protein on the ordinate) of neonatal rat cardiomyocytes in the presence of serum containing IAb. Serial dilutions of dialysed rabbit serum with culture medium (dilutions of serum (se.) 1:2, 1:4, 1:8, 1:16, 1:32) were added to neonatal rat cardiomyocytes and potassium content was determined after an incubation period of 90 min in comparison to cells kept in pure culture medium (control).

A more subtle impairment of contractility of the cell monolayers after the recovery from standstill persists for at least 12 h: as opposed to the rather synchronized beating of control cells, a higher variance of beating frequency is conspicuous for serum-treated cell monolayers. The cause of this dissociation of contractile activity is not known for the moment, but could well be linked with the altered electrophysiological properties of the cell membranes under the influence of PNAb-containing serum.

Of note is that complement has a direct action on neonatal rat cardiomyocytes, enhancing contraction velocity via increased calcium influx, which is time-dependently corrected by compensatory mechanisms [23]. However, the PNAb effects presented in this paper do not seem to be complement-dependent, as inactivated sera produce similar effects.

IAb differ from PNAb in that they highly specifically recognize the antigens they are directed against. In this cell culture model IAb in dependance on their concentration exhibit a strong cytotoxic action leading to cell death. These result are in keeping with the fundamental differences in the biological properties of the two antibody species [28].

Presumed consequences for the clinical situation

According to the experiments shown here, PNAb are unlikely to contribute to chronic rejection. However, cessation of beating of a cardiac xenograft immediately after perfusion with the recipient's blood, which cannot be overcome by pacemaker action, might create some problems in a clinical setting. Moreover, the functional dissociation of the syncytial heart tissue may give rise to arrhythmias for some hours after implanting a xenograft. These problems deserve further attention.

However, the presence of IAb in the recipient's blood either by previous attempts to transplant an organ from the same species or by other contact to animals of the species (e.g. food, agriculture) – should according to these experiments be detrimental in xenotransplantation.

Acknowledgement

We thank Prof. Dr. Martin Schreiber, University of Munich, for his valuable help with the statistical evaluation of the long-term observation experiments of neonatal rat cardiomyocytes. This study was supported by Deutsche Forschungsgemeinschaft (Mu 1010/1-4) and Austrian Science Research Fund (P 9045). Part of the results were published in Transplantation [1] and in Transplantation Proceedings [2, 3].

References

1. Müller-Werdan U, Klein D, Zander M, Werdan K, Hammer C: Beating neonatal rat cardiomyocytes as a model to study the role of xenoreactive natural antibodies in xenotransplantation. Transplantation 58: 1403–1409, 1994
2. Müller-Werdan U, Koidl B, Autenrieth A, Klein D, Zander M, Werdan K, Hammer C: Preformed natural antibodies: their impact on contractile activity of cardiomyocytes. Trans Proc 27: 2109–2111, 1995

3. Autenrieth AE, Müller-Werdan U, Koidl B, Werdan K, Hammer C: Different action of xenoreactive natural (PNAB) and immune antibodies (IAB) on beating cardiomyocytes (BCM). Trans Proc, in press
4. Hammer C: Xenotransplantation – is it just around the corner? Nephrol Dial Transplant 9: 1724–1729, 1994
5. Hammer C, Müller U, Werdan K: Beating cardiomyocytes for detection of xenogeneic rejection. Eur Surg Res 24: 316–317, 1992
6. Müller U, Klein D, Zander M, Werdan K, Hammer C: Preformed natural antibodies: their impact on beating neonatal rat cardiomyocytes. Transplant Proc 26: 1000–1002, 1994
7. Hammer C: Preformed natural antibodies (PNAB) and possibilities of modulation of hyperacute xenogeneic rejection (HXAR). Transplant Proc 21: 522–523, 1989
8. Platt JL: A perspective on xenograft rejection and accomodation. Immunol Rev 141: 127–149, 1994
9. Hammer C, Csapo C, Höbel G, Müller U, Werdan K: Preformed natural antibodies: Action on endothelial cells. Transplant Proc 24: 590–591, 1992
10. Hammer C, Höbel G, Csapo C. Fixation of mammalian natural antibodies to porcine endothelial cells. Transplant Proc 24: 588–589, 1992
11. McCurry KR, Kooyman DL, Alvarado CG, Cotterell AH, Martin MJ, Logan JS, Platt JL: Human complement regulatory proteins protect swine-to-primate cardiac xenografts from humoral injury. Nature Medicine 1: 423–427, 1995
12. Halle W, Wollenberger A: Differentiation and behavior of isolated embryonic and neonatal heart cells in a chemically defined medium. Am J Cardiol 25: 292–299, 1970
13. Varela F, Andersson Å, Dietrich G, Sundblad A, Holmberg D, Kazatchkine M, Coutinho A: Population dynamics of natural antibodies in normal and autoimmune individuals. Proc Natl Acad Sci USA 88: 5917–5921, 1991
14. Werdan K, Erdmann E: Preparation and culture of embryonic and neonatal heart muscle cells: modification of transport activity. Methods in Enzymology 173: 634–662, 1989
15. Blondel B, Roijen I, Cheneval JB: Heart cells in culture: a simple method for increasing the proportion of myoblasts. Experientia 27, 356–358, 1971
16. Kaumann AJ, Wittmann R, Birnbaumer L, Hoppe BH: Activation of myocardial β-adrenoceptors by the nitrogen-free low affinity ligand 3,4-dihydroxy-α-methylpropiophenone (U0521). Naunyn-Schmiedeberg's Arch Pharmacol 296: 217–228, 1977
17. Piper HM, Probst I, Schwarz P, Hütter FJ, Spieckermann PG: Culturing of calcium stable adult cardiac myocytes. J Mol Cell Cardiol 14: 397–412, 1982
18. Koidl B, Flaschberger P, Schaffer P, Pelzmann B, Bernhart E, Mächler H, Rigler B: Effects of the class III antiarrhythmic drug ambasilide on outward currents in human atrial myocytes. Naunyn-Schmiedeberg's Arch Pharmacol 353: 226–232, 1996
19. Hamill OP, Marty A, Neher E, Sakmann B, Sigworth FJ: Improved patch-clamp techniques for high resolution current recordings from cells and cell-free membrane patches. Pflugers Arch 391: 85–100, 1981
20. Lowry OH, Rosebrough NJ, Farr AL, Randall RJ: Protein measurement with the folin phenol reagent. J Biol Chem 193: 265–275, 1951
21. Werdan K, Müller U, Reithmann C: Negative inotropic cascades in cardiomyocytes triggered by substances relevant to sepsis. In: G. Schlag G, H. Redl (eds). Pathophysiology of shock, sepsis and organ failure. Springer-Verlag, Heidelberg, 1993, pp. 787–834
22. Müller-Werdan U, Reithmann C, Werdan K: Cytokines and the heart – Molecular mechanisms of septic cardiomyopathy. Landes Company, Austin, 1996
23. Berger H-J, Taratuska A, Smith TW, Halperin JA: Activated complement directly modifies the performance of isolated heart muscle cells from guinea-pig and rat. Am J Physiol 265: H267–H272, 1993
24. Hallström S, Koidl B, Müller U, Werdan K, Schlag G: A cardiodepressant factor isolated from blood blocks Ca^{2+} current in cardiomyocytes. Am J Physiol 260: H869–H876, 1991
25. Wallukat G, Morwinski R, Kowal K, Förster A, Boewer V, Wollenberger A: Autoantikörper gegen myokardiale β-Adrenoceptoren im Serum von Patienten mit Myokarditis (MC) und dilativer Kardiomyopathie (DCM): β-adrenerger Agonismus ohne Desensibilisierung. Z Kardiol 80 [Suppl 3]: 141, 1991
26. Wallukat G, Wollenberger A: Circulating autoantibodies from patients with allergic asthma interfering with β_2-adrenoceptor stimulation in cultured neonatal rat heart myocytes. Naunyn-Schmiedeberg's Arch Pharmacol 344: R 99, 1991
27. Chin TK, Friedman WF, Klitzner TS: Developmental changes in cardiac myocyte calcium regulation. Circ Res 67: 574–579, 1990
28. Casali P, Notkins AL: $CD5^+$ B lymphocytes, polyreactive antibodies and the human B-cell repertoire. Immunol Today 10: 364–368, 1989

Index to Volumes 160 & 161